German Development of the Swept Wing

1935–1945

German Development of the Swept Wing

1935–1945

Edited by
Hans-Ulrich Meier

With contributions from
Burghard Ciesla
Hans Försching
Hans Galleithner
Werner Heinzerling
Bernd Krag
Helmut Schubert

Translated by
Egon Stanewsky

LIBRARY
OF FLIGHT

Ned Allen, Editor-in-Chief
Lockheed Martin Corporation
Bethesda, Maryland

Published by the American Institute of Aeronautics and Astronautics, Inc. (AIAA) in cooperation with the Deutsche Gesellschaft für Luft- und Raumfahrt - Lilienthal-Oberth e.V. (DGLR).

Published by the American Institute of Aeronautics and Astronautics, Inc., 1801 Alexander Bell Drive, Reston, Virginia, U.S., 20191-4344 in cooperation with the Deutsche Gesellschaft für Luft- und Raumfahrt - Lilienthal-Oberth e.V. (DGLR).

1   2   3   4   5

Cover design by Virginia S. Kozlowski.
The front cover image shows the original results of the worldwide first measurements of a swept wing at high speeds (copyright GOAR [10146-P8065-30]). These were proposed by *Albert Betz* and carried out by *Hubert Ludwig* in November 1939 in the 11-cm × 11-cm high-speed wind tunnel of the AVA Göttingen. Essentially this result was also a basic milestone for the development of the Boeing 777.

Library of Congress Cataloging-in-Publication Data

Pfeilflügelentwicklung in Deutschland bis 1945. English
   German development of the swept wing: 1935–1945 / edited by Hans-Ulrich Meier; with contributions from Burghard Ciesla ... [et al.]; translated by Egon Stanewsky.
      p. cm. -- (Library of flight)
   Includes bibliographical references and index.
   ISBN 978-1-60086-714-9 (alk. paper)
   1.   Airplanes--Wings, Swept-back--Research--Germany--History--20th century.   I. Meier, H. U. (Hans Ulrich) II. Ciesla, Burghard. III. Title.
   TL673.S9P44 2010
   629.134'32--dc22   2010020064

DEDICATION

To my friend and colleague for more than 40 years,
Tuncer Cebeci of The Boeing Company,
who, from the start of the German edition,
never tired in his support for this history of aerodynamics.

H.-U. M.

FOREWORD

German Development of the Swept Wing: 1935–1945 is a remarkable book for anyone who wants to understand how many dedicated individuals used science, mathematics, experimental and theoretical research, and technology over a relatively short period of time to produce an engineering achievement— high-speed flight, the basis of success for all modern aircraft. This example and history lesson is particularly valuable today when engineers are assigned a goal without all of the necessary information and resources to produce a desirable outcome. Because of the subject matter, the swept wing and area rule, this book is especially valuable for the education of current and future aerospace engineers. It is not written with a parochial view, but acknowledges and integrates international contributions to this subject.

As described in the preface to the German-language edition, *Die Pfeilflügelentwicklung in Deutschland bis 1945*, the editor, Prof. Dr.-Ing. Hans-Ulrich Meier, led a group of distinguished colleagues (see their biographies at the back of the book) in six different disciplines to produce for the first time a work that showed "the way from the actual idea to the product, that is, the application of the swept wing to the realization of high-speed flight." Information was gathered from many original-source documents and reports not to be found in other books. The outstanding photos and graphics that were selected show key individuals and results. "The aim of this book is, therefore, to inform the scientifically interested reader in detailed and intelligible terms about research results that are difficult to access."

The English-language version is also a testimony to the vision and contributions of an international effort, as described in the preface to the English-language edition. When Professor Roger L. Simpson was AIAA President (2005–2007) and Professor Joachim Szodruch was President of the Deutsche Gesellschaft für Luft- and Rahmfahrt (DGLR), both societies agreed to have more intensive cooperation and joint ventures. When Professor Simpson received a copy of the German edition as a gift and read it, he knew that this work should be translated into English to be available to many more people. Soon it was determined that AIAA would jointly publish the English-language edition with DGLR and the Deutsches Zentrum für Luft- and Rahmfahrt

(DLR) would provide funding for the translation into English. Fortunately, Dr.-Ing. Egon Stanewsky, who is very knowledgeable about the subject and had spent some years in the United States, agreed to perform the translation. As mentioned in the preface to the English-language edition, both AIAA and DGLR made contributions to the final production of this edition. All of the outstanding efforts and financial and staff support are gratefully acknowledged and appreciated.

Roger L. Simpson, Ph.D., P.E.
The Jack E. Cowling Professor
of Aerospace and Ocean Engineering
Virginia Polytechnic Institute
and State University
President, AIAA 2005–2007

Prof. Dr.-Ing. Joachim Szodruch
Member of the German Aerospace
Center (DLR) Executive Board
President, DGLR 2001–2009
CEAS 2009

A NOTE FROM THE SERIES EDITOR

This collection of essays exploring the history of swept-wing technology and related developments belongs in the *Library of Flight* series because of the great impact the sweep concept has had on modern aircraft design—no transonic or faster aircraft looks right to us today without sweep in the wing. Hollywood puts sweep into its fanciful spaceships, even though sweep has little utility in space. Sweep arose from our enchantment with ever higher speeds as a commercial and military advantage. In aerospace history, sweep was the enabling invention for breaking through the second great speed barrier into the realm of near-sonic (transonic) speeds. The first great barrier broken, of course, was by escape from thick-coming obstacles along the ground, escape by means of altitude to flight itself. Sweep then got us to the sound barrier, and the axial flow turbine engine got us past it. One cannot doubt that we will continue onto higher speeds in transonics and even perhaps to relativistic speeds.

Perhaps more significant than its impact on planform fashions and platform performance is that the underlying theory marked the emergence of a richer and more practical role for aerodynamics—richer because sweep theory required a multidimensional view of airflows, twice the simple two-dimensional view prevailing at that time. Not only did swept-wing airflows gain utility from a third spatial dimension, but consideration of other parameters like viscosity, compressibility, and intrinsic energy of the flow entered the theory much more prominently than before and were applied directly to practical designs. While German design sought to harvest the insights of the new multidimensional aerodynamics with largely passive innovations like sweep, it set the stage for much work on active control of multidimension flows, increasingly active today with work on flow control by active devices, including synthetic jets, dielectric barrier discharge devices, and others. This anthology recounts this story thoughtfully and carefully, so it is a fitting addition to this series.

The *Library of Flight* is part of the growing portfolio of information services from the American Institute of Aeronautics and Astronautics (AIAA). It augments the Institute's other two book series, the *Progress* in aerospace series of technical monographs and the *Education* series of textbooks, with the best of a growing variety of other topics in aerospace from aviation policy, to case studies, to studies of aerospace law, management, and beyond.

Ned Allen
Editor-in-Chief
Library of Flight

BRIEF TABLE OF CONTENTS

xi

TABLE OF CONTENTS

CHAPTER 6   EXPERIENCE GAINED DURING DEVELOPMENT AND TESTING OF
THE FIRST SWEPT-WING JET AIRCRAFT 443
BERND KRAG

PREFACE TO THE ENGLISH-LANGUAGE EDITION

On the basis of partly unknown reports and other sources, we have tried to analyze and summarize the status of swept-wing technology in Germany at the end of WWII. Our research concentrated on the phenomenon of drag divergence at high subsonic *Mach* numbers. As a result, swept wing, super-critical airfoil, and area rule have been found as solutions to shift drag to higher *Mach* numbers. The output can be seen in the fundamental shapes of existing military and civil airplanes.

In researching this topic, we learned a lot about the fundamentals of high-speed aerodynamics performed in the time frame considered. We are convinced that young scientists can still learn from the results achieved at that time. History shows that a lot of problems we face are not new and some have already been solved.

We have tried to achieve consistency of terminology in the translation. In order not to lose the original character of the book, we decided to include the references without translation. A list of often-used translations of German institutions and locations is given in the appendices. For readers who want to gain access to original or translated reports, we included an appendix listing appropriate institutions/archives as well.

The reviews of the German edition were excellent; however, it was empha-sized that the understanding of these important breakthroughs in aircraft development would be limited to German-language speakers. It was *Roger Simpson* who contacted us to propose translation and publication through AIAA in order to promote distribution worldwide. With the financial support of the DLR, organized by board members *Klaus Hamacher* and *Joachim Szodruch*, the realization of the whole project was facilitated considerably. *Rodger Williams* (Director of Publications, AIAA) agreed to the proposal of *Joachim Szodruch* (President, DGLR) and *Peter Brandt* (Secretary General, DGLR) to a joint AIAA–DGLR publication of the translated book.

With financial support from the DLR, it was possible to engage *Egon Stanewsky* as our competent translator. He was Scientist Associate at the Lockheed-Georgia Company, in Marietta, Georgia, and until his retirement in 2000 he continued his research activities at transonic speeds in leading positions at the DLR. In a first step, *Britta Rath* converted the German printed version of the manuscript from QuarkXPress into MSWord. *Richard Sanderson*, an aerodynamic specialist, who has not lost his English accent when speaking German even after working 40 years in Germany, was ready to improve our German-English, acting as our proofreader. With the professional and experienced help of *Pat DuMoulin* (AIAA, Senior Editor) we finished the translation of the manuscript at the end of 2009. It was a pleasure working with her. The copyeditor *Meredith Perkins* carefully revised the final manuscript. All original figures of the German edition have been included, and the compositor *Techset* added English-language labels. Some of the archival photographs are smaller than in the German edition; readers are encouraged to view the original documents listed in the figure captions or the archives listed in Appendix C. *Janice Saylor* (Marketing Strategist, AIAA) did an excellent promotion for the book. The authors would like to express their special appreciation to all of those mentioned herein. They helped us considerably with their contributions to the development of this English-language edition.

In the name of the authors,
Hans-Ulrich Meier
September 2010

PREFACE TO THE GERMAN EDITION

In 2003, the world commemorated the 100th anniversary of the first successful powered flight of the Wright brothers. This marked an important milestone in the development of flight technology—after the glider flight of *Otto Lilienthal*, which marked the beginning of mankind's flight. The utilization and further development of the first flight vehicles took place, however, not in the United States but primarily in Europe where the military quickly recognized the strategic importance of aircraft development for warfare. The first World War (WWI) led with the setup of the Air Corps in Germany and similar organizations within the Allies, to enormous technology thrusts in aerodynamics, airframe construction, and propulsion. After WWI, knowledge gained during the series production of military aircraft entered civil aviation. Conditions of the Versailles Treaty did not allow Germany the development of large aircraft and high-powered flight engines; thus, it became for the German aviation industry increasingly more difficult to keep abreast with the foreign competition. To bypass the restrictions imposed by the Versailles Treaty, some companies founded subsidiaries in neighboring foreign countries. Less restricted by the conditions of the Versailles Treaty was aeronautical research. It was, most of all, thanks to *Ludwig Prandtl* and his collaborators at the Aerodynamische Versuchsanstalt Göttingen (AVA) (Aerodynamic Test Establishment Göttingen) that, despite all restrictions, revolutionary research results in the area of fluid mechanics were obtained that received worldwide recognition. Together with the Institut für Aerodynamik of the TH-Aachen (Institute of Aerodynamics of the Institute of Technology Aix-la-Chapelle), which was headed by the *Prandtl*-pupil *Theodore von Kármán*, there were two centers of international importance for aeronautical research, which attracted students and scientists from all industrialized nations.

The stepwise relaxation of the restrictions related to the construction of aircraft and the economic recovery that commenced in the second half of the 1920s allowed the research institutions a modest extension of their research facilities and an increase in personnel. Because of the tense financial situation of the Weimar Republic, caused by the imposed reparations, the construction of modern test facilities initially had to be deferred. After overcoming the world economic crisis, the modernization and the construction of new facilities commenced, especially at the Deutsche Versuchsanstalt für Luftfahrt (DVL) (German Aeronautical Test Establishment) in Berlin. After the National Socialists assumed power at the beginning of 1933, the extension and construction of new facilities were continued at an increased rate. In 1935, the Luftfahrtforschungsanstalt Herman Göring (LFA) (Aeronautical Research Establishment) in Braunschweig was founded. The focus of the German aeronautical research, especially in aerodynamics, was, at that time, the high-speed flight at higher altitudes (with the paradigm "faster, higher, farther"). Extending and modernizing test facilities was an important prerequisite in order to catch up on the lead held by the western industrialized countries in aircraft development and to gain a leading position in some areas. Here, high-speed aerodynamics played a dominant role. The introduction of the swept-wing concept at the 1935 Volta Congress in Rome, Italy by *Adolf Busemann* represented an important milestone. Only in Germany was the value of this ingenious idea to the development of high-speed aircraft recognized. With the jet engine and the swept wing, one had two key technologies for the realization of high-speed flight in hand. Thanks to the strong interest shown by the Reichsluftfahrtministerium (RLM) (State Ministry of Aeronautics/German Air Ministry) to high-speed flight, this technology was generously sponsored and led to unique research activities. On the basis of these research results, several high-speed aircraft projects were developed in Germany. After the war these were applied to the aircraft development in many countries and even today have not lost any of their importance.

The history of the development of the swept wing has often been described within the last 50 years by several German and non-German authors. The foremost topic was always the basic theoretical swept-wing concept of *Adolf Busemann* and the experimental confirmation of the 1939 theory by *Hubert Ludwieg* at the AVA. In "Luftfahrtforschung in Deutschland," Vol. 30 of the series "Deutsche Luftfahrt" by Bernard & Graefe Publisher of Bonn, Germany (see also trans. E. A. Hirschel, H. Prem, G. Madelung: *Aeronautical Research in Germany—from Lilienthal until Today*, Springer-Verlag, Berlin/Heidelberg, 2004), fundamental ideas concerning the development of the swept wing as the basis of high-speed aircraft have already been described. But up to now, there has been no detailed documentation showing the way from the actual idea to the product, that is, the application of the swept wing to the realization of high-speed flight. The aim of this book is, therefore, to inform the scientifically interested reader in detailed and intelligible terms about research results that are difficult to access. On this basis and with the aid of typical examples, the transfer of German know-how after WWII in the development of high-speed aircraft can be understood objectively.

Together with colleagues of six different disciplines, research was conducted in domestic and foreign libraries and archives to analyze scientific publications, documents, and original reports, previously not available or used.

The study of these documents allowed an assessment of the research results gained and the technical progress achieved with the utilization of the swept wing on high-speed aircraft, missiles, and follow-on projects during 1935–1945 in Germany.

Within a historic review, the merits of *Ludwig Prandtl* and his collaborators at Göttingen in developing the fundamentals for the realization of high-speed aircraft will be reviewed first. The meeting of the worldwide, leading scientists involved in fluid mechanics at the 1935 Volta Congress in Rome was used by the 33-year-old *Prandtl* student *Adolf Busemann* to introduce his ideas concerning the reduction of drag in supersonic flow of the swept wing. As director of the new Institute für Gasdynamik (Institute of Gas Dynamics) of the Luftfahrtforschungsanstalt Hermann Göring, Braunschweig (LFA) (Aeronautical Research Establishment, Braunschweig), *Adolf Busemann*, as well as his colleagues at the AVA and the DVL, had an opportunity to verify his swept-wing concept by wind-tunnel investigations. In a critical analysis, it will be shown which of the new facilities and measuring techniques for the investigation of high-speed projects were suitable and employed for reliable measurements.

Not generally known in expert circles, new design criteria and computational methods in 1940 led to the successful development of high-speed airfoils whose basic concept was very similar to that of today's "supercritical" (transonic) airfoils. These new airfoils were first employed on German jet aircraft. Principal investigations on different swept wings, carried out in 1939 by *H. Ludwieg*, had shown a distinct drag reduction in compressible flows but also a decrease in lift and negative effects on the flight mechanical stability. Further problems were discovered in swept-wing investigations within the low-speed range at high lift. The corresponding task definition initiated extensive investigations regarding the wing optimization within the entire speed range. As with the airfoil development, new computational and design methods for swept wings supported efficient procedures regarding experimental investigations in the wind tunnel.

Within the framework of an RLM-sponsorship, industry built its own facilities including speed wind tunnels to investigate new configurations. Here, independent of research establishments, important new knowledge in design aerodynamics, which has, up to now, not lost its importance to the development of aircraft, was gained. Among the latter the transonic area rule was discovered, formulated, patented, and applied in practice by *Otto Frenzl* 1943 to 1945 at the Junkers Company. Besides the swept wing and the super-critical airfoil, the area rule belongs to the few very basic concepts of design aerodynamics. *Werner Heizerling* will consider in detail this development in his documentation, which until today is applied to many high-speed aircraft to postpone the steep drag rise occurring when approaching the speed of sound to higher velocities thus enabling higher flight speeds to be achieved.

In the international literature, wing sweep is still correctly presented as a German development during the time from 1935 to 1945; however, the discovery, patent, and first application of the area rule is not ascribed to *Otto Frenzl* (1944) but to *Richard T. Whitcomb* (1952). Also of great practical importance were the investigations carried out at Junkers and Heinkel concerning the low-drag attachment of engine nacelles on the wings and the optimization of swept wings for fighter aircraft at Messerschmitt.

In 1935, it was stated at the Volta Congress that a crucial prerequisite for the realization of high-speed flight would be the development of new propulsion systems. In his contribution about the development of German turbojet engines, *Helmut Schubert* will only deal with important data because, with his cooperation, the problem was already extensively documented in "Luftfahrtforschung in Deutschland" (vol. 30 of "Deutsche Luftfahrt"). In his historic review of the state of aircraft engine construction in Germany during the 1930s, he shows the limits of propeller propulsion for high-speed flight. Thereafter, *Schubert* addresses the development of the axial compressor for jet engines, which had already reached production readiness in Germany in 1944. Important research activities concerning engine integration and the treatment of problems related to engine intakes will be cited. A detailed analysis will be omitted here because the most important results in this area were already compiled by *Dietrich Küchemann*, among others, on more than 1000 printed pages in the Göttingen monographs that were written after the war under the supervision of *Albert Betz* at the AVA.

More serious problems in the aeroelastic behavior of high-speed aircraft occurred in near-sonic and transonic flow. As a result of the effect of compressibility and following the introduction of the swept-wing concept, the coupling of bending and torsion due to the new geometry of the wing was given special emphasis. *Hans W. Foersching* documents the revolutionary and guiding research and development work concerning the aeroelastic behavior in high-speed flight in Germany up to the end of WWII at the threshold of a new age in aeronautical technology. This relates especially to the basic contributions to the aerodynamic theory of the unsteady oscillating wing in compressible subsonic and supersonic flow, the theory of wing flutter, and the measurement of the natural vibration parameters in ground vibration tests. Without this work, the achievement of high-speed aircraft with swept wings would not have been possible at that time. Many of these pioneering investigations attracted great attention among the victorious powers after the end of WWII and were immediately utilized in aircraft development. They formed the basis for the further development of aeroelastics, a fundamental field of knowledge in modern aircraft design.

During initial aerial combat in WWII, crashes occurred during attack and escape maneuvers without recognizable enemy encounter. To find the cause, extensive flight tests were carried out by the test centers of the Luftwaffe

(German Air Force). It was found that flow separation due to suddenly appearing shock waves and correspondingly changing pressure distributions on wings, fuselage, and stabilizers led to a loss in stability and control that could not be handled by regular pilots. *Hans Galleithner*, an experienced flight-test engineer, has analyzed and summarized the documents about the flight tests of high-speed aircraft during the time considered. Of interest are the results of flight testing German aircraft after the war at test centers in the United States and United Kingdom. The comparison between German and the first generation of jet aircraft in the United States and United Kingdom has never been shown in such a comprehensive way.

As is already known, up to the end of the war only two high-speed aircraft with a genuine swept wing were built. These were the Junkers Ju 287 and the experimental aircraft Messerschmitt P 1101. The supersonic experimental aircraft DFS 346 was in an advanced stage of design when the Siebel company was occupied by the Americans. Evidence that the low-aspect-ratio wing was of interest to high-speed flight is shown by the designs of *A. Lippisch* and the application to antiaircraft rockets. Only a glider version of the delta-wing aircraft Lippisch P13 was built but not tested in Germany. As *Bernd Krag* shows in a comprehensive analysis, the P13 became the starting point for quite a number of successful delta-wing aircraft and was certainly the "godfather" of the development of the first supersonic civil passenger aircraft, the "Concorde." It will be shown step by step how, starting with the German example and using their ideas and experience, an independent new design was developed. The development and flight testing of this "pioneer generation" of jet aircraft revealed which problems German engineers would have had to deal with if one of the planned projects with swept wings had actually been achieved.

In Chapter 7, *Bernd Krag* investigates the application of the swept-wing technology to the development of rockets. Antiaircraft rockets had to operate at far higher speeds than the target. Therefore, swept wings were intended for rockets from the beginning. The first "real" swept wing was indeed realized on the experimental rocket F 25 "Feuerlilie (Fire Lily)" of the LFA with *Alfred Busemann* having been the "godfather" of the development. The demand for natural stability, high maneuver loads, good controllability as well as a simple design could also be met by low-aspect-ratio wings. For the supersonic rocket "Wasserfall (Waterfall)" of the Heeresversuchsanstalt in Peenemünde (HVA) (Test Establishment of the Army Peenemünde), wings with an aspect ratio smaller than one proved to be advantageous. On the basis of this experience, all modern antiaircraft rockets were equipped with low-aspect-ratio wings.

Burghard Ciesla and *Bernd Krag* consider in a concluding chapter on the basis of examples, the know-how transfer in the area of high-speed aerodynamics from Germany to the victorious allied nations. Despite the fact that until the spring of 1945 flight tests were only carried out with one prototype of the jet bomber Ju 287 with swept wings, almost all of the basic knowledge was available. The takeover of German and Austrian scientists after the WWII and the resulting extensive utilization of German scientific–technical research data are unique in the annals of history. The victors appropriated, more or less systematically, scientific, technical, commercial, and economically useful intellectual property and took over scientists and technicians—voluntarily or by force—into their service. The professional competence counted more than the political past. The principle "legitimated by usefulness" held. Those responsible and active in the science transfer acted according to this pragmatic pattern worldwide. Nobody in the world and no power wanted to do without the technical expertise of the Germans.

The authors of this book made it their business to give a presentation and analysis of important scientific results of German high-speed research from 1935 to 1945 in an as objective a way as possible. In preparing this book, the evaluation of documents from domestic and foreign archives played an important role allowing a more in-depth analysis of high-speed research results than possible in Volume 30 of this book series.

The documentation presented by us is aimed at people interested in aircraft development but is also a store of scientific sources for the young researcher. Here we refer to, for instance, the high-quality test results from the 3-meter high-speed wind tunnel of the DVL that might still today constitute an important basis for the validation of up-to-date computational methods. The complete literature survey concerning the research results can be found in the summarizing reports of the Göttingen monographs issued by *Albert Betz* in 1946. Unfortunately, we could only find and evaluate a part of these reports in our investigation. In the case of further investigations, interested parties should contact the DLR libraries.

A critical analysis of further research results of this time would certainly be beneficial to the education of our younger generation of scientists, and even today, still contribute to the solution of present-day problems in aircraft design.

Burghard Ciesla
Hans Försching
Hans Galleithner
Werner Heinzerling
Bernd Krag
Hans-Ulrich Meier
Helmut Schubert
November 2005

ACKNOWLEDGMENTS

It was *Volker von Tein*, at that time a member of the Board of Directors of DLR in charge of the department of aeronautics and energy, who in April 2001 suggested that the development of the swept wing in Germany should be documented in a book. Chapters 5–8 were conceived and edited by *Bernd Krag* who has contributed considerably to the success of our work. During the research into this book, we visited domestic and foreign archives and libraries. We must stress the unselfish support of *Heinz Fuetterer*, long-time administrator of the DLR archives in Göttingen. Besides his advice, he provided many documents. My daughter, *Marlies Frank*, has supported me with her profound knowledge in data processing, initially by generating a computer concept as the basis for our work, and by continuously building up our database.

We enjoyed great cooperation, an essential prerequisite for the success of this research, from the employees of the following institutions:

DLR-Libraries/Archives:
Braunschweig: Dipl.-Math. *Ulrich Heidemann, Gabriele Thieme*; Cologne-Porz: *Astrid Bölt, Dr. phil. Katharina Hein-Weingarten*, now Science & Engineering Application Datentechnik (S.E.A.)

DLR-Institutes:
Göttingen: Prof. *Dr. Andreas Dillmann, Dr.-Ing. Fritz Kiessling, Christine Meyer*, Dipl.-Librarian *Daniela Rieke*; Braunschweig: Prof. *Dr.-Ing. Cord-Christian Rossow, Dr.-Ing. Ralf Heinrich, Dr. Boris Kolesnikow, Ute Maassen, Dr.-Ing. Andreas Bergmann*; Stuttgart: Telecom Service Dipl.-Ing. *Michael Scharfy*
State Archive Berlin: *Dr. Klaus Dettmer*
Federal Archive Berlin: *C. Lorenz, Dr. W. Lenz*
Archive of the History of the Max-Planck-Society, Berlin: *Dr. Marion Kazemi, Susanne Uebele*
German Museum München: Dipl.-Ing. *Hans Holzer, Gudrun Bauer-Seuml, Christiane Hennet, Margrit Prussat, Wolfgang Schinhan*
State and University Library Dresden (SLUB), Branch Library Transportation Sciences: Department Speaker *Michael Kern*
Institute of Fluid Mechanics, TU (Institute of Technology) Brunswick: Prof. *Dr.-Ing. Dietrich Hummel, Dr.-Ing. Horst Saathoff*
Institute of Aerodynamics, RWTH Aachen (Institute of Technology): *Dr.-Ing. Matthias Meinke*
Federal Archive Koblenz (WTS): *Renate Resmini*

Archive of the Museum Peenemuende: *Dirk Zache*, M.A., Dipl.-Ing. *Manfred Kanetzki*, Dipl.-Ing. *Peter Profe*

Caltech-Galcit: Wind Tunnel Chief *Gerald Landry*

Boeing Historical Archives Long Beach: Prof. *Dr. Tuncer Cebeci* (Distinguished Fellow Boeing Company), Historian *Patricia M. McGinnis*, *Morgan M. (Mac) Blair* (formerly of North American Rockwell), *Lowell F. Ford*, *Jim Turner*

Seattle: Manager/Historian *Michael J. Lombardi*

Archive Accademia Nazionale dei Lincei: (Documents of the Volta-Congress) Rome: Direttore Generale *Dott. Antonio Luigi Cocuzzi*

Manchester Museum of Science: Curator Air & Space *Nick Forder*

Especially important and helpful were conversations, discussions, and detailed correspondence with the following experts:

Prof. em. *Dr.-Ing. Dr. h.c. mult. Werner Albring*, *Larry Davis* (formerly NAA), *Sven Frank*, *Dr.-Ing. Kurt Graichen*, Dipl.-Ing. *Karl Koessler*, Prof. *Dr.-Ing. Peter Hamel*, Prof. *Dr.-Ing. Ernst-Heinrich Hirschel*, Dipl.-Ing. *Horst Prem*, *Stephen Ransom*

We are grateful to the estates of the following comtempory witnesses for trustfully lending us valuable material:

Prof. *Dr.-Ing. Siegfried Erdmann* (formerly Peenemuende): by Prof. *Dr. W.J. Bannink*, TU Delft;

Dr.-Ing. Werner Krueger (formerly AVA Göttingen): by the Vice President of the Central Administrative Tribunal (Oberverwaltungsgerricht), *Henning Krueger*, Frankfurt (Oder)

Only with the aid of the many professional comments and corrections of our editors Dipl.-Ing *Klaus Peters* and *Walter Amann*, always patient and open to innovations, with the latter also having been responsible for the production and layout of this work, was the successful completion of this book possible.

Hans-Ulrich Meier

Chapter 1

HISTORIC REVIEW OF THE DEVELOPMENT OF HIGH-SPEED AERODYNAMICS

HANS-ULRICH MEIER

1.1 DEVELOPMENT OF THE FUNDAMENTALS OF HIGH-SPEED AERODYNAMICS UP TO LUDWIG PRANDTL

Very early, namely with the introduction of artillery, a connection between the propagation of sound and velocity was recognized, because when firing a shell it was not possible to directly correlate the flash at ignition with the bang of the explosion. Today, for instance, hikers, sailors, and golfers know that the distance of a thunderstorm front can be estimated by counting the seconds between the observation of lightning and hearing the thunder. At a speed of sound of about 330 m/s, the front is, after counting to "three," still 1 km away. The first person who carried out an estimation of the speed of sound was in 1687, the famous natural scientist *Isaac Newton* (1643–1727). He based his estimates on the "elasticity" of air; however, he made the incorrect assumption of an isothermal change of state within the sound wave. It took almost another 100 years before the famous French mathematician *Pierre Simon Marquis de Laplace* developed an exact relation for the speed of sound (c) assuming an adiabatic change of state. He proved that in the case of an ideal gas, the speed of sound was only dependent on the temperature (T) of the gas:

$$c = (\kappa RT)^{1/2} = (\kappa p/\rho)^{1/2}$$

(κ = adiabatic exponent, R = gas constant). Within a flowfield, the pressure (p), the density (ρ), and the temperature (T) generally change from location to location.

Hence, the speed of sound (c) also varies from location to location. The nondimensional relation of the local flow velocity (U) to the local speed of sound is termed "local *Mach* number":

$$M = U/c$$

If $M < 1$, the flow is called subsonic flow; if $M > 1$, it is called supersonic flow. The Austrian universal scholar *Ernst Mach* was most likely the first

1

person to understand and experimentally investigate the physical fundamentals of supersonic flow in a systematic way. *Ernst Mach* (1838–1916) had in 1867, after studies in mathematics, physics, and philosophy in Vienna, accepted a professorship in experimental physics at the German University in Prague. *Ernst Mach* [1, 2] became, among other things, known for his guiding work as an experimental physicist in the area of gas dynamics. Here, he visualized and analyzed with new investigative methods (stroboscopic methods, Schlieren, and shadowgraph pictures of density distributions) supersonic flows on shells. In his honor, the famous Swiss gasdynamicist *Josef Ackeret* introduced the term "*Mach* number," and, in remembrance of these pioneering investigations, the terms "*Mach* cone" and "*Mach* angle" are used today in supersonic flow in gasdynamics.

The first investigations concerning the treatment of flows of compressible media, that is, gasdynamics, were all conducted in the 19th century. In the beginning, there were *St. Venant* and *L. Wantzel* [3] with their work regarding the outflow of gases in the presence of large pressure differences. In 1859, *Georg Friedrich Bernhard Riemann* (1826–1866) [4] first presented at the Koenigliche Gesellschaft der Wissenschaften zu Göttingen (Royal Society of Sciences at Göttingen) his important work "Concerning the Propagation of Plane Air Waves of Finite Amplitude." However, his attempt to compute a shock wave in a generally valid form failed, as he did not take the change of entropy of the gas across the shock wave into account. It was twelve years later that *William John Macquorn Rankine* (1820–1872) [5], two years prior to his death, reported on this problem at the Royal Society. In his contribution, "On the Thermodynamic Theory and Waves of Finite Longitudinal Disturbances," which was published in the *Philosophical Transactions of the Royal Society*, he formulated the equations for a normal shock wave with a nonisentropic change of state across the shock.

During our investigations with regard to other important scientific results, we also found that these equations for the shock relations were "discovered" for a second time by the French ballistic researcher *Pierre Henry Hugoniot* [6]. In the literature it is, therefore, common today to call all solutions to this flow problem Rankine–Hugoniot equations. At the same time the work of *P. Molenbroek* [7], who treated the motion of gases based on potential theory assuming the existence of a velocity potential, originated.

To generate a continuous supersonic flow, a nozzle that accelerates the flow in the convergent part up to sonic velocity is required while, thereafter, the divergent part allows the formation of the supersonic flow. It was the Swedish scientist *Carl G. P. de Laval* (1845–1912), who, as one of the pioneers in the development of steam engines, utilized this possibility of flow acceleration. In 1890, he built a 30-cm-diam turbine stage with 200 blades! Steam was led via special nozzles to the blades such that the turbine accelerated to very high speeds. By means of these Laval nozzles, an essential improvement in the energy conversion from the kinetic energy of the steam to the turbine blades was reached. Despite the fact that it was up to *Aurel Boleslav Stodola* (1858–1942) to work out the scientific fundamentals of a Laval-nozzle, the name is inseparably connected with the researcher and entrepreneur *de Laval*. The publishing of a further important treatise on gaseous jets in compressible flow by *C. A. Tschapligin* [8] in 1904 falls into the 20th century. In the next chapter we report in detail about the German activities—especially about those performed under the guidance of *Ludwig Prandtl* in Göttingen.

1.2   LUDWIG PRANDTL'S COLLABORATORS IN GÖTTINGEN

At the start of his scientific career, *Ludwig Prandtl* carried out various important investigations in the area of elasticity and strength. In his doctoral thesis, he addressed an interesting problem from the field of the stability of the elastic equilibrium, namely the tilting (buckling) weight of tall girders. After taking up his activity in Göttingen in 1904, *Prandtl* worked primarily on problems related to fluid mechanics. *J. C. Rotta* [9] drew up a detailed and fascinating documentation of the work of this most important aerodynamicist from 1904 to 1925. Therefore, we deal only with the most important studies that were a basis for the development of the design aerodynamics of high-speed aircraft.

During the first years, three doctoral theses, among others, were written under the guidance of *Prandtl*, which were all based on his famous work "Concerning the Fluid Motion with Very Little Friction," published in 1904 [10] and today belong to the classic areas of the application of boundary-layer theory. In their theses, *Heinrich Blasius* treated the laminar boundary layer on a flat plate (1907) [11], *Ernst Boltze* transferred the computations to bodies of revolution (1908) [12], and *Karl Hiemenz* computed the laminar boundary layer on a cylinder in crossflow (1910) [13]. These studies on boundary-layer theory are mentioned here as examples of scientific fundamentals because they, and studies based on them, were—and still are—of essential importance to the development of the entire aircraft industry. We will return later to some special problems, such as boundary layer separation on swept wings.

During these years, *Prandtl* also turned to flows with considerable density changes, that is, to gasdynamics. Here, he treated initially a task related to the construction of turbines where, at high flow velocities, shock waves occur (1906) [14]. The solution, known as Prandtl–Meyer expansion, describes the turning of a sonic or supersonic flow around a corner. It is based on a theory by *Ludwig Prandtl* [15], which was addressed by *Theodor Meyer* (1907) [16] within the framework of a doctoral thesis. First consider the supersonic flow on a concave wall, that is, with a pressure increase. Considering Fig. 1.1, it becomes obvious that the flow inside the second field is not only deflected

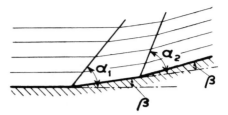

Fig. 1.1 Supersonic flow over a concave wall.

with respect to the first by an angle β but also, as a consequence, the velocity behind the first compression wave is smaller and consequently the *Mach* angle α_2 larger than α_1. For an expansion wave on the contrary, the velocity behind the first wave is larger, and, therefore, the *Mach* angle is smaller. The wave fronts converge or diverge according to the amount $(\beta + \alpha_2 - \alpha_1)$. For the radius of curvature zero, one recognizes, when going to the limit (see Fig. 1.2), that in case of a wall that forms a convex corner of angle $-\beta$, the flow remains unaffected by the corner up to a plane forming the angle α_1 with respect to the freestream direction and then expands by an angle of $(\beta + \alpha_1 - \alpha_2)$. Here, velocity and pressure remain constant along each ray, and the flow turns into the direction of $-\beta$ again into a parallel flow with constant velocity.

Besides the fundamental contributions of *Ludwig Prandtl* to the treatment of viscous flow on the basis of the boundary-layer theory of 1904, he worked on a further unresolved problem, namely, the determination of the aerodynamic quantities of a wing of finite span. Contrary to the wing of infinite span, the flow here is three-dimensional. The pressure differences between the lower and upper surfaces of the wing will try to equalize thereby forming two discrete vortices of opposite sense of rotation and whose axes nearly coincide with the direction of the freestream. These two vortices are of the circulation strength Γ. With these so-called "free" vortices and the "bound" vortex on the wing, a "horseshoe vortex" develops [17] (Fig. 1.3). This physical model of the three-dimensional flow about a wing of finite span and its analytical formulation was given for the first time by *Prandtl* [18] after *F. W. Lanchester* [19] had earlier carried out some related qualitative investigations. This theory, generally known as "lifting-line theory," was extended by *Prandtl* by the "lifting-wing theory." The task of the lifting-wing theory,

Fig. 1.2 Supersonic flow over a convex wall.

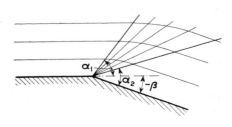

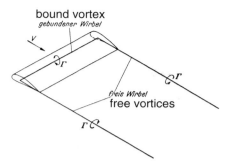

Fig. 1.3   Vortex system of a wing of finite span according to *L. Prandtl.*

i.e., the computation of the lift distribution and the drag of a wing of a given shape, turned out to be most difficult. With the aid of his famous "Integro-Integral equation" for the circulation distribution, *Prandtl* succeeded in providing theories for both mathematical formulations for the model of the "horseshoe vortex," whose solutions remained for many years the basis for the computation of the rectangular (unswept) as well as the swept wing. In his article concerning the 50th birthday of *Ludwig Prandtl, Theodore von Kármán* [20] wrote in 1925:

> Prandtl's main achievement for the flight technique is, however, without doubt, the wing theory, namely the discovery and determination of the so-called "induced drag," which at once opened the way for a rational design method for aircraft.

In a seminar on applied mechanics during the winter semester of 1922–1923 at the University of Göttingen, *Jacob Ackeret* lectured for the first time about aerodynamic forces on wings in supersonic flow. For simplification, it was assumed that the only slightly cambered wing was infinitely thin and had a very large aspect ratio. This study was initiated and supervised by *Prandtl*. In the publication of this contribution by *Ackeret* [21] in 1925, one may read "in a still unpublished lecture, Prof. Prandtl communicated approximate formulas for the subsonic region which allowed, in a simple form, an estimate of the effect of compressibility."

Ludwig Prandtl had presented his thoughts about the "correction of compressibility effects" frequently in seminars but, in fact, not before *H. Glauert* [22] (1928). Even so, *Glauert* had learned the basics of lift theory from *Prandtl* visiting Göttingen in the early 1920s, but his simple method to correct the lift at *Mach* numbers smaller than 0.7 was the first publication within this important field. In the literature, these compressibility corrections are therefore called the Prandtl–Glauert rule. The rule allows, for instance, the transfer of results of measurements in incompressible flow to compressible subsonic flow.

In 1927, *Carl Wieselsberger*, who, after receiving his doctorate at Technische Hochschule (TH) (Institute of Technology Munich), became the assistant of *Prandtl*, visited the Aerodynamische Versuchsanstalt (AVA) Göttingen (Aerodynamic Test Establishment, Göttingen) on his way from Japan. This circumstance occasioned a photograph that shows *Prandtl* with his scientific collaborators (Fig. 1.4). *Carl Wieselsberger* had followed in 1922, upon an invitation by the Japanese government to advise on the construction of new wind tunnels. In this advisory function, he participated decisively in the set up of the Aeronautical Research Institute of Tokyo University.

In 1928, *Adolf Busemann* [23, 24] published suggestions for the graphical determination of two-dimensional supersonic flow. These deliberations formed the basis for the Prandtl–Busemann Method of Characteristics, which, beginning in the early 1930s, allowed the computation of supersonic flow around airfoils and in Laval nozzles with the aid of this numerical-graphical difference method. Thus an effective tool, for instance, for the design of supersonic wind tunnels became available for the first time. Figure 1.5 shows the velocity map (hodograph) developed by *Busemann*. The principle is based

Fig. 1.4 *Ludwig Prandtl* and collaborators on the occasion of the visit of *Carl Wieselsberger* from Japan, sitting (from left): *O. Tietjens, J. Ackeret, A. Betz, C. Wieselsberger, L. Prandtl, W. Tollmien, R. Langer*; standing: *O. Flachsbart, W. Buehl, J. Nikoradse, A. Busemann, R. Seiferth, H. Muttray, H. Peters, D. Schrenk.*

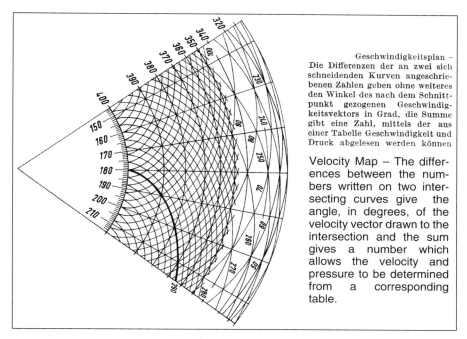

Geschwindigkeitsplan –
Die Differenzen der an zwei sich
schneidenden Kurven angeschrie-
benen Zahlen geben ohne weiteres
den Winkel des nach dem Schnitt-
punkt gezogenen Geschwindig-
keitsvektors in Grad, die Summe
gibt eine Zahl, mittels der aus
einer Tabelle Geschwindigkeit und
Druck abgelesen werden können

Velocity Map – The differences between the num-
bers written on two inter-
secting curves give the
angle, in degrees, of the
velocity vector drawn to the
intersection and the sum
gives a number which
allows the velocity and
pressure to be determined
from a corresponding
table.

**Fig. 1.5   Prandtl–Busemann Method of Characteristics, a numerical–graphical differ-
ence method for the computation of supersonic flow about airfoils and through Laval
nozzles (*Prandtl*, Stroemungslehre = flow theory).**

on the idea that the steady change in the velocity direction in supersonic flow
may be represented, as in Figs. 1.1 and 1.2, by a series of sudden individual
deflections. Here, the step-like line thus generated will, as in computational
approximations, be replaced by a polygon. If the angle β of the deflections
will be chosen to be always equal, such as 2 deg, then only velocity directions
will occur, which differ in a fixed direction by a whole multiple from the
angle β. If one draws the velocity vectors all from the origin zero, a regular
net of circular symmetry is obtained. The nodes are the permitted velocities.
Every line between two neighboring nodes represents one of the transitions
that occur in a 2-deg wave. All waves considered here originate at a wall; all
corresponding conditions are, therefore, located in this "velocity diagram" on
the bold curve. At that time, all two-dimensional supersonic flows could be
discussed by the method just described. In that respect, *Busemann* [25]
addresses the flow around an inclined flat plate in supersonic flow (Fig. 1.6)
and the symmetrical flow about a biconvex airfoil (Fig. 1.7). The numbers
given here represent the ratios of the total pressure weakened by the (shock)
losses, p_0', to the freestream total pressure, p_0. Figure 1.8 shows a schlieren
picture of the real flow about such an airfoil. (This method of flow visualization
was already developed at the end of the 18th century by the German scientist

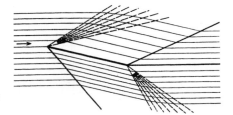

Fig. 1.6 Flat plate at an angle of attack in supersonic flow according to *A. Busemann*.

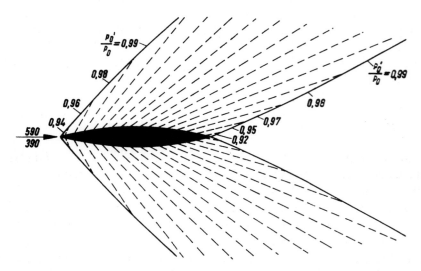

Fig. 1.7 Supersonic symmetrical flow about a biconvex airfoil; pressure p_0' is the total pressure reduced by the loss due to the shock wave.

Fig. 1.8 Schlieren picture of the flow about a biconvex airfoil in supersonic flow.

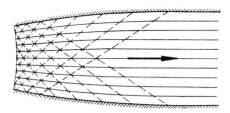

Fig. 1.9 Graphic determination of a nozzle contour to generate a parallel supersonic flow according to *A. Busemann*.

Toepler and not, as is often assumed in the English literature, by Mr. Schlieren!) It should be noted here that, contrary to the usual schlieren pictures, lighter areas generally denote density increases and darker areas denote density decreases. An especially important application of *Busemann's* method was the design of supersonic nozzles (Fig. 1.9), where, in the divergent part, parallel flow was to be established. Here, the incoming waves were not to be reflected at the wall so that the jet was not disturbed by waves deflecting it. The nozzle flows were checked with the aid of schlieren photographs.

1.3 VOLTA CONGRESS 1935 IN ROME, A BREAKTHROUGH IN HIGH-SPEED AERODYNAMICS

To honor the great Italian physicist *Alessandro Volta*, the "Fondazione Alessandro Volta" was founded in 1930 at the Reale Accademia d'Italia. In 1800, *Alessandro Volta* (1745–1827) had reported in a letter to the president of the academy an invention that concerned the generation of a constant direct current (DC). *Volta* used the findings of *Luigi Galvani* (1737–1798), who had learned as professor of anatomy that one could observe a twitching when removing a nerve from a frog's leg while the frog was suspended from a copper hook and touched by an iron rod. It became clear to *Volta* that *Galvani* had established the effect of electricity that was produced by the two metals (copper and iron) touching. After recognizing that these were charged after the separation of the two metals, he amplified the effect by arranging several zinc and copper plates alternately one on top of the other with an additional material, for instance, wet paper, between each pair of plates. *Volta* had thus invented the battery without a clear idea of the functioning of galvanic elements.

The "Fondazione Alessandro Volta" was furnished by the Societá Generale Edison di Elettricità with the generous donation of 8M Lira, today corresponding to 0.5M Euro, in order to arrange, according to the statutes, a yearly conference in the area of physical, mathematical, historical, and ethical/philosophical sciences. Here, leading scientists from all over the world were to be invited for creative discussions and an exchange of knowledge in a particularly agreeable atmosphere. In 1931, the Academy invited a first congress on nuclear physics. Up until 1934 this was followed by three further congresses on mechanical and ethical topics, respectively. By the end of 1934, the Academy sent out the invitation to the 5th Volta Congress [26], which was to deal exclusively with problems of high speed in flight technology.

1.3.1 PREPARATION FOR THE CONGRESS

The choice of the topic of the congress was certainly influenced by the speed records that were successfully established at the time in Italy with seaplanes. Although the attainment of sonic speeds seemed, in 1935, still beyond reality, an international speed record of 709 km/h over 3 km was established by *Francesco A. Agello* on October 23, 1934 with the seaplane Macchi MC 72 ("Macchi-Castoldi").

The MC 72 was originally planned to participate in the "Schneider Trophy" competition that took place for the last time in 1931. The plane flew, for the first time in this year, but could not participate in the competition. Due to the high moment of torque of the propeller, it tended to run in circles during take off from the water. Unfortunately a safe takeoff was hardly possible, and several accidents occurred. Because the Fiat AS6 motor with a power of 2280 kW (3057.5 hp) consisted of two 12-cylinder engines of the type Fiat AS5 in tandem arrangement (Fig. 1.10), the two "individual engines" were redesigned to drive counter-rotating metal propellers. After three years of further developmental effort, the problem with the moment of torque was solved. As demonstrated in Fig. 1.11, the plane was in fact designed around the engine and the pilot with the obvious objective to establish a new speed record for seaplanes. Hence, the solution was to overcome the large aerodynamic drag by sheer power. The pilots also reported the occurrence of strong buffeting when reaching high speeds. Based on today's state of knowledge, this may definitely be explained by massive flow separations partly caused by compressibility effects in the air. This then demonstrated the

Fig. 1.10   Seaplane Macchi-Castoldi (today Aermacchi) MC 72. In 1934, *Francesco Agello* established the speed record for seaplanes of 709 km/h, which still holds today.

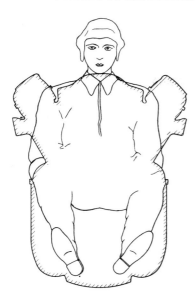

Fig. 1.11   Cross-sectional drawing of the Fiat engine of the record plane MC 72 compared to the body size of the pilot, fuselage diameter, and hence its drag optimization was obtained this way.

limits in the flow about complete aircraft and the performance of propeller propulsion. The motivation and the objective for this Volta Congress initiated by the Italians were to find solutions to these problems. The speed record for seaplanes of 709 km/h achieved in 1934 still stands today.

A world record for landplanes with piston engines was achieved by *Fritz Wendel* on April 26, 1939, with a Messerschmitt Me209 VI at a speed of 755 km/h (filed with the Federation Aeronautic International, FAI, as Bf 109R). Only 30 years later, on August 16, 1969, *Darryl Greenamyer* was able to reach a new speed record of 777 km/h with a Grumman F-8F-2 Bearcat considerably modified for the record attempt.

Interesting in this context are the visionary thoughts and remarks of the congress chairman, *G. Arturo Crocco*, in his welcoming speech. He believed in mankind's thirst for research and that man would, one of these days, fly with a flight vehicle within the stratosphere with a speed of more than 3000 km/h. What he could not know was that only 15 years later his vision would, based on scientific knowledge gained at the congress, be far surpassed.

Participation in this event was by personal invitation only; the lecturers were carefully and skillfully selected by *Crocco*, who was charged by the academy with the entire organization. In early December 1934 *Ludwig Prandtl* received an unofficial invitation by *Crocco* on behalf of the president of the Reale Accademia d'Italia. On this occasion, he asked *Prandtl* to contribute to the meeting of experts with the topic "A Homogeneous Theory of Lift in Compressible Fluids" (Fig. 1.12). *Crocco* added a preliminary list of topics to be presented for an evaluation. With *Crocco*, who had obviously chosen *Prandtl* as his advisory expert, the latter conducted a lively discussion

a)

REALE ACCADEMIA D'ITALIA

CONVEGNO VOLTA 1935

Rom den 4 Dezember 1934.XIII

Herrn Professor L. PRANDTL
Universitaet
          GOETTINGEN

Sehr verehrter Herr Kollege!

Ich habe die Ehre Ihnen im Vertrauen mitzuteilen dass Sie einen Brief des Praesidenten der Kgl. Akademie von Italien erhalten werden mit der Einladung an der Volta Zusammenkunft des naechsten Jahres (30.September- 6.Oktober,1935) teilnehmen zu wollen, und einen Bericht ueber das folgende Thema vorzulegen:

"Einheitliche Theorie des Auftriebs in zusammendruekbaren Fluessigkeiten

          Da diese Einladung,die auf diplomatischen Wege erfolgt, ziemlich lange Zeit in Anspruch nehmen wird, so sende ich Ihnen diesen Brief einzig und allein zu dem Zweck um Ihnen mehr Zeit fuer die Vorbereitung Ihres Berichtes zu ueberlassen und Ihnen einige Erklaerungen zu dem Thema zu geben.

          Bevor ich jedoch dieses Argument naeher behandle, moechte ich meine persoenliche Hochschaetzung Ihres Werkes zum Ausdruck bringen und zugleich die hertzliche Bitte an Sie richten dieser Zusammenkunft beiwohnen zu wollen.

          Was den Gegenstand des Themas betrifft, so beehre ich mich Sie zu benachrichtigen, dass die Praesidentschaft beim Vorschlag dieses Themas gedacht hat Sie um die Beweisfuehrung zu bitten wie und bis zu welchem Punkt und mittels welchen Verfahren heute oder in der naechsten Zukunft eine Theorie des Auftriebs zu bauen moeglich ist, die Zusammendrueckbarkeit des Mediums in einer allgemeinen Weise beruecksichtigt, um das Verhalten tragender Flaechen zu den verschiedenen Werten als besondere Zufaelle davon ableiten zu koennen.

          In einer so beschaffenen Theorie waeren als besondere Zufaelle inbegriffen die gewoehnliche aerodynamische Theorie des Auftriebs fuer den unendlich langen und den endlichen Tragfluegel, einschliesslich der Theorie der Induktion, Ihre angenaehrte Theorien und die des verstorbenen Professors Glauert und anderer fuer die Geschwindigkeiten nahe der Shallgeschwindigkeit, und endlich die von Professor Ackeret und anderen Forschern fuer Unterschallgeschwindigkeiten in einfachen Faellen.

Fig. 1.12   a) Letter of invitation by *G. Arturo Crocco* of the Reale Accademia d'Italia to *Ludwig Prandtl* concerning the 5th Volta Congress "High Speed in Aeronautics" from September 30 to October 6, 1935, in Rome; b) translation.

b)

Rome on the 4th of December 1934
Professor L. PRANDTL
University GÖTTINGEN

Dear Colleague,

I have the honor to <u>confidentially</u> tell you that you will receive a letter from the president of the Royal Academy of Italy with an invitation to participate in the VoltaMeeting of the next year (30 September – 6 October, 1935) and to give a report on the following subject:

<u>Homogeneous Theory of Lift in Compressible Fluids</u>

Since this invitation, following diplomatic procedures, will take a long time to arrive, I am sending you this letter solely for the purpose of giving you more time for the preparation of your report and to give you some explanations on the topic.

However, prior to the closer treatment of this matter, I would like to express my personal deep respect for your work and, at the same time, sincerely beg you to attend this meeting.

Concerning the matter of the subject, I have the honor to inform you that the presidency, when proposing this topic, thought to request from you the presentation of evidence of how and to what point and by what method, today or in the near future, a lifting theory could possibly be build-up that considered the compressibility of the medium in a general form in order to be able to derive from it the behavior of lifting surfaces also for special cases.

To be included in such a theory were, as special cases, the ordinary aerodynamic lifting theory for the infinite- and finite-span wing, including the theory of induction, your approximate theories and the ones of the late Professor *Glauert* and others for speeds close to the speed of sound, and, finally, the ones of Professor *Ackeret* and other researchers for subsonic flow in simple cases.

I beg you, at the same time, to kindly let me know whether this subject, whose difficulty equals your capabilities, is in your judgment suitable for a useful treatment, and to propose eventual changes also with regard to the title.

In order for you to see the relation of the topic of your lecture to the topics of other presentations, allow me to attach a list of temporary titles of all aerodynamic topics.

In expecting your kind answer, allow me dear colleague to send you my sincerest greetings.

Your obedient servant,

[signature]

President of the Volta Meeting 1935

Two Attachments

1 – Excerpt from the provisional arrangement of the Volta Meeting

2 – List of the temporary titles of the aerodynamic lectures

Attachment 2

REALE ACCADEMIA D'ITALIA

<u>LIST OF THE TEMPORARY TITLES OF THE AERODYNAMIC LECTURES</u>

1 – Homogeneous Lift Theory in Compressible Fluids

2 – Dynamic Lift at Supersonic Speeds

3 – Dynamic Lift at Speeds Close to the Speed of Sound

4 – The Problem of Drag in Compressible Fluids

5 – Questions of Testing Techniques at High Speeds

6 – Wind Tunnels for High Speeds

7 – Final Results of High-Speed Tests

8 – Test Results on Propellers at High Speeds

Fig. 1.12 **(Continued)**

a) (Continued)

Ich bitte Sie gleichzeitig mir guetigst mitteilen zu wollen
ob dieser Gegenstand, dessen Schwierigkeit Ihre Faehigkeiten
gleichkommt, geeignet ist fuer eine nutzbringende Behandlung und mir
Ihr Urteil darueber und Ihre eventuellen Aenderungsvorschlaege, auch
bezueglich des Titels, freundlich mitteilen zu wollen.

Damit Sie sehen koennen in welcher Beziehung das Thema Ihres
Berichtes zu den Themen anderer Berichte steht erlaube ich mir Ihnen
die Liste der provisorischen Titel aller aerodynamischen Themen
beizuschliessen.

In Erwartung einer guetigen Antowort erlaube ich mir Ihnen,
sehr verehter Herr Kollege, meine hochachtungsvollsten Gruesse zu
uebersenden.

                    Ihr ergebenster

                                    Gen. G. Arturo Crocco

                         Praesident der Volta Zusammenkunft 1935
Zwei Beilagen:
1 - Auszug der provisorischen Anordnung der Volta Zusammenkunft
2 - Liste der provisorischen Titel der aerodynamischen Berichte.

REALE ACCADEMIA D'ITALIA

CONVEGNO VOLTA 1935

LISTE DER PROVISORISCHEN TITEL DER AERODYNAMISCHEN BERICHT

1 - Einheitliche Theorie des Auftriebs in zusammendrückbaren
Flussigkeiten.

2 - Dynamischer Auftrieb bei Überschallgeschwindigkeiten.

3 - Dynamischer Auftrieb bei Geschwindigkeiten nahe der
Schallgenschwindigkeit.

4 - Das problem des Winderstandes in zusammendrückbaren Flüssigkeite

5 - Fragen der Versuchstechnik bei hohen Geschwindigkeiten.

6 - Windkanäle für hohe Geschwindigkeiten.

7 - Endgültige Ergebnisse von Hochgeschwindigkeits Versuchen.

8 - Versuchsergebnisse an Propellern bei hohen Geschwindigkeiten.

Fig. 1.12 (Continued)

about the list of proposed presentations [27] prior to the congress. Concerning his own topic "Homogeneous Theory of Lift in Compressible Fluids," *Prandtl* remarked, among other things, on December 12, 1934:

> There is, however, only one theory for the case where the maximum flow velocity on the wing is below the speed of sound and one completely different theory in case the lowest velocity on the wing is already above the speed of sound. For the interim regime, there is no theory. This theory may, therefore, not be called homogeneous and there is also no way in sight how such a theory could be established...

His topic was therefore fixed as "General Considerations about the Flow of Compressible Fluids." *Prandtl's* suggestion to also include the drag of projectiles, because especially good flow visualizations were available for them, was not accepted by *Crocco* in order not to tone down the clear reference of the congress to aeronautics. On December 27, 1934, *Ludwig Prandtl* received his official letter of invitation signed by the president of the Royal Italian Academy, *Guglielmo Marconi*, and the president of the 5th Volta Congress, *G. A. Crocco*. *Prandtl* agreed to participate, after he received the official letter of invitation via the Reichsluftfahrtministerium (RLM) (State Ministry of Aeronautics) in mid-February 1935 and the RLM asked him to take on a lecture (Fig. 1.13). It is certainly not generally known that *Prandtl* proposed to the chairman of the congress that his former collaborator, *Adolf Busemann*, at that time professor of fluid mechanics at the Institute of Technology Dresden, should be invited to give a lecture on the topic of "Aerodynamic Lift at Supersonic Speeds." *Crocco* responded very positively to the suggestion of *Prandtl* and invited *Busemann*. Thus, *Busemann*, as a young scientist, was given the opportunity to present and discuss his ideas pertaining to the theoretical treatment of supersonic flow with the "fathers" of modern aerodynamics, *Theodore von Kármán, Ludwig Prandtl*, and *Geoffrey Ingram Taylor*.

Starting in March 1935, a lively correspondence developed between *Prandtl* and the lecturers of the second main session "Aerodynamics," *G. I. Taylor* and *A. Busemann*, about professional questions and the contents of their contributions. *Taylor* wrote in a handwritten letter dated March 14, 1935 to *Prandtl* that he especially missed the titles and names of further lectures. To avoid overlapping or repetition by the lecturers, he informed *Taylor* on March 26, 1935 briefly about the contents of his presentation and mailed him unofficially a list of the lecturers and their topics. While *G. I. Taylor* sent his completed manuscript on April 14 to *Prandtl*, *Busemann* wrote on April 20 that he "still had no idea of the contents of his Volta presentation" (Fig. 1.14). In May, *E. Pistolesi* requested publications concerning his topic "Aerodynamic Lift at High Subsonic Speeds." *Prandtl* answered him immediately and

a)

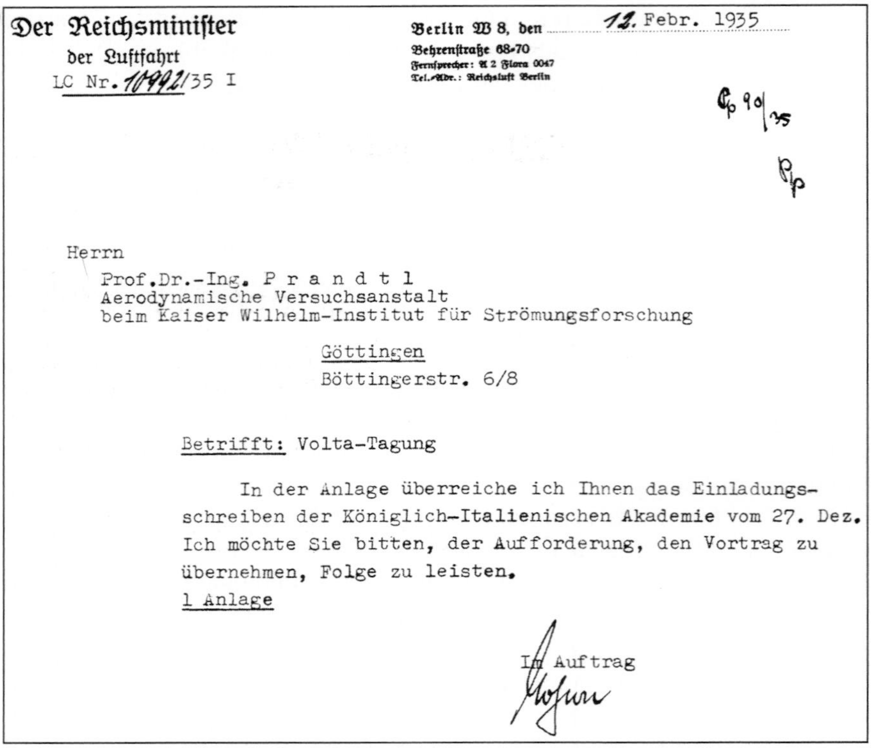

𝔇er Reidjsminifter
    ber Luftfahrt
   LC Nr. *10992*/35 I

Berlin W 8, ben .... *12.* Febr. 1935
Behrenstraße 68-70
Fernsprecher: A 2 Flora 0047
Tel.-Adr.: Reichsluft Berlin

Herrn
   Prof.Dr.-Ing. P r a n d t l
   Aerodynamische Versuchsanstalt
   beim Kaiser Wilhelm-Institut für Strömungsforschung

               Göttingen
               Böttingerstr. 6/8

      Betrifft: Volta-Tagung

        In der Anlage überreiche ich Ihnen das Einladungs-
schreiben der Königlich-Italienischen Akademie vom 27. Dez.
Ich möchte Sie bitten, der Aufforderung, den Vortrag zu
übernehmen, Folge zu leisten.
    1 Anlage

                        Im Auftrag

b)

The State Minister of
Aeronautics

Berlin on 12 February, 1935

             Prof. Dr.-Ing. Prandtl
             Aerodynamic Test Establishment
             at the Emperor Wilhelm Institute of Flow Research
             Goettingen
             Boettingerstr. 6/8

Subject: Volta Congress

    I am sending to you in the attachment the letter of invitation of the Royal
Italian Academy, dated 27th of December. I would like to ask you to accept the
request and take on this presentation.
    1 Attachment

                           On behalf of the Minister

Fig. 1.13 **a) Letter of the State Ministry of Aeronautics (RLM) to** *Ludwig Prandtl* **with
the request to participate in the 5th Volta Congress; b) translation.**

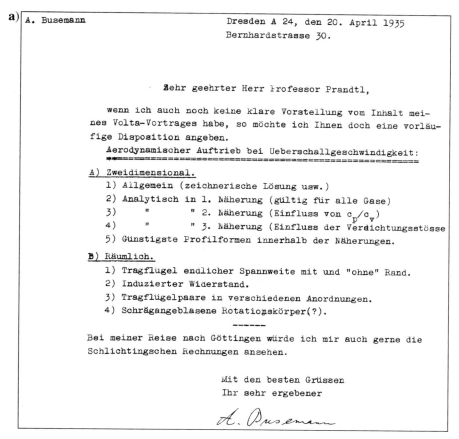

Fig. 1.14   a) Letter outlining the table of contents of *A. Busemann's* lecture "Aerodynamic Lift at Supersonic Speeds" to *L. Prandtl*; b) translation.

informed him about research in this area. After the submission of his Volta lecture on June 8, *L. Prandtl* wrote to *A. Busemann*:

> As attachment I am sending you a copy of my Volta lecture which I have just finished requesting you to review it and, most of all whether in your opinion in the last pages I have taken too many cherries away from you. After answering my questions, I would appreciate it if you could give my copy back to me, since I also want to pass it on to Taylor. By the way, how far along are you with your lecture?

On June 12, *J. Ackeret* requested photographs and drawings of the Göttingen supersonic wind tunnel, which *Prandtl* had then sent to him. To better understand the political situation in 1935, the forwarding of the information about the Volta Congress by the State Ministry of Aeronautics (RLM) to the Zentrale für Wissenschaftliches Berichtswesen (ZWB) (Center of Scientific-Publications) is worth mentioning; the ZWB hereupon ordered *Prandtl* and

b)

A. Busemann   Dresden A 24, April 20, 1935
                                               Bernhardstrasse 30.

Dear Professor Prandtl,
   Although not having a clear idea about the contents of my Volta lecture, I would like to give you a preliminary disposition.

<u>Aerodynamic Lift at Supersonic Speeds</u>

A) Two-dimensional
   1)   General (graphical solution etc.)
   2)   Analytical to a 1st approximation (valid for all gases)
   3)   Analytical to a 2nd approximation (effect of c_p/c_v)
   4)   Analytical to a 3rd approximation (effect of shock waves)
   5)   Most favorable airfoil shapes within the approximations

B) Three-dimensional
   1)   Wing of finite span with and "without" tip
   2)   Induced drag
   3)   Wing pairs in various arrangements
   4)   Bodies of revolution in oblique flow

On my trip to Goettingen, I would also like to look at Schlichting's computations.

                                               With best regards
                                               Very sincerely yours

                                               A. Busemann

Fig. 1.14   (Continued)

Busemann to publish their manuscripts in a special issue "High-Speed Flight" of the Deutsche Luftfahrtforschung (LuFo) (German Aeronautical Research) prior to the publication of the Volta proceedings. *Prandtl* had considerable problems with this procedure but no chance of a direct intervention as both manuscripts had to be cleared by the RLM via the ZWB. At least he obtained a postponement of the German publication until the professional meeting of the Deutsche Versuchsanstalt für Luftfahrt (DVL) (German Test Establishment of Aeronautics) on October 11 in Berlin was to take place, that is, five days after the end of the Volta Congress.

1.3.2 AT THE CONFERENCE

In the only group photograph in the proceedings of the Volta Congress [26] we can, unfortunately, only identify a few important persons (Fig. 1.15). The attachment to the Volta proceedings contains all active participants with a photograph, a short description of their professional career, and a listing of

Fig. 1.15 Group photograph of the participants to the 5th Volta Congress at the Palazzo della Farnesina, located at the Tiber, built in the style of the Italian Renaissance and adorned with the famous frescos "Psyche" of Raphael.

their most important publications, however, mainly in Italian. From the latter, abridged versions from the scientists of especial importance to the present investigation are here given in the original version [26]. In the case of *Adolf Busemann,* this concerns the early periods of his scientific career, and so his biography will be supplemented later.

Under the general conditions offered by the Accademia d'Italia, *G. A. Crocco* (Fig. 1.16) had created a unique condition for the success of the congress with the choice of the congress location and the selection of the scientists to be invited. After his academic education at the Engineering School in Rome and his military career, he had taken over all important positions within the Italian aeronautical organizations. As general and professor, he was able to employ all possibilities that state-owned institutions offered.

1.3.2.1 CONFERENCE SESSION TOPICS. The program of the meeting was divided into three main sessions. In the first session, problems primarily related to the achievement of speed records with single-engine aircraft were described with the speed record achieved in 1934 by the Italian *Francesco Agello* with the Macchi MC 72 taking center stage:

H. E. Wimperis (London)	British Technical Preparation for the Schneider Trophy Contest 1931
M. Casoldi (Milan)	Italian High-Speed Seaplane

C ROCCO G. ARTURO (Italia). Accademico d'Italia. Nato a Napoli, è oggi Professore d'Aeronautica generale nella R. Scuola d' Ingegneria Aeronautica di Roma. Già ufficiale del Genio Militare e poi tenente generale del Genio Aeronautico, si occupò sin dal 1903 del progresso aeronautico italiano, svolgendo teorie matematiche sulla propulsione e sulla stabilità delle macchine aeree e progettando e costruendo dirigibili, aeroplani, idroplani, impianti aerodinamici, armi e metodi di tiro aereo. Fu direttore generale al Ministero dell'Economia Nazionale e poi al Ministero dell'Aeronautica. Presiede il Comitato Tecnico del Registro Italiano Navale ed Aeronautico, il Reparto aeronautico del Comitato per l'Ingegneria del Consiglio Nazionale delle Ricerche e l'Associazione Italiana di Aerotecnica. Medaglia d'oro della Académie des Sciences di Parigi, medaglia d'oro dell'Istituto Lombardo di Scienze e Lettere, premio Santoro della R. Accademia dei Lincei, Membro ordinario della R. Accademia dei Lincei. Ha numerose pubblicazioni, da ufficiale e da professore, che trovansi elencate nel vol. IV dell'« Annuario » della R. Accademia d'Italia.

Indirizzo abituale: Roma, Via Alessandro Torlonia 23.

Fig. 1.16   *G. Arturo Crocco*, president of the 5th Volta Congress.

C. F. Bona (Turin)	Italian Engines for High-Speed Seaplanes
G. H. Stainforth (London)	British Methods of High-Speed Flying and Training of Pilots
M. Bernasconi (Milan)	Italian Training Methods for Pilots in High-Speed Flight

The second session concerned aerodynamics with emphasis on compressible flow. The following list of lecturers again clearly demonstrates the outstanding personalities in this field that were gathered, with *Adolf Busemann* and *Eastman N. Jacobs*, ages 34 and 35, belonging to the younger generation. Three of the lecturers had passed through the school of *Prandtl*, *Adolf Busemann* having developed an especially warm personal contact with him.

L. Prandtl (Göttingen)	General Considerations About the Flow of Compressible Fluids
G. I. Taylor (London)	Well-Established Problems in High-Speed Flow
Th. von Kármán (Pasadena)	Problem of Resistance in Compressible Flows
E. Pistolesi (Pisa)	Lift at High-Subsonic Speeds
A. Busemann (Dresden)	Aerodynamic Lift at Supersonic Speeds
M. Panetti (Turin)	Problems at High Speeds

| G. P. Douglas (London) | Research on Model Airscrews at High Speeds |
| J. Ackeret (Zurich) | Wind Tunnels for High Speeds |

Supplementing high-speed aerodynamics, topics of thermodynamics were treated in the third main session:

G. Constanzi (Rome)	Flying in the Stratosphere
H. R. Ricardo (London)	Propulsion for High Altitudes: Thermo- and Gasdynamics
A. Anastasi (Rome)	Propulsion for High Altitudes: Mechanics and Cooling
M. Roy (Paris)	Delivery of External Air in the Case of Jet Engines
N. A. Rinin (Leningrad)	Rocket Engines without Injection of External Air

It is obvious from the structure of the program that *G. A. Crocco* succeeded, especially in the area of aero- and gasdynamics, to have the actual state of knowledge presented by the world's most renowned experts.

1.3.2.2 LINEARIZED THEORY FOR FLOWS AT SUPERSONIC SPEEDS. *Ludwig Prandtl*, (Fig. 1.17) dealt in his survey lecture with the linearized theory for flows at supersonic speeds. He presented classical examples: the Prandtl–Meyer supersonic corner flow and the approximate graphical solutions of *Adolf Busemann* for general two-dimensional supersonic flow. Because *G. I. Taylor* lectured after him, he mentioned only his contribution [28] concerning the computation of the rotationally symmetric flow about a cone apex of finite cone angle in supersonic flow. *Prandtl* pointed out that solutions for the general case of a body of revolution in supersonic flow, based on the linearized approximate method given by *Th. von Kármán* and *N. B. Moore* [29], were mainly applicable to very slender bodies.

Important to the congress was *L. Prandtl's* new approach to treat compressible flows on the basis of the acceleration potential. How new this idea was follows from the correspondence with *G. A. Crocco* about the change of the manuscript. *Prandtl* further addressed the ideas presented at the congress after his return to Göttingen and published them one year later [30]. Schlieren pictures obtained from a test series of transonic flow over a slightly wavy wall, unpublished at the time of the congress, were presented by *L. Prandtl* in a supplement.

Later, the wavy part was replaced by a flat circular arc section. To make the flow conditions clearly recognizable, small grooves were scratched into the cylindrical surface with a file. Incidentally, that was a method already employed in the calibration and optimization of the nozzles of the 6×7 cm Supersonic Wind Tunnel of the AVA. The examples show in Fig. 1.18 that

RANDTL prof. LUDWIG (Germania). Nato nel 1875 a Freising nella Baviera Superiore. Laureatosi ingegnere meccanico nel Politecnico di München, divenne assistente di A. Föppl; nel 1900 conseguì il dottorato in Filosofia all' Università di Monaco. Nel 1901 fu nominato professore di Meccanica al Politecnico di Hannover e nel 1904 professore di Fisica tecnica e più tardi di Meccanica applicata all'Università di Göttingen. Nel 1907–1908 costruì la prima galleria aerodinamica, nel 1915–1917 una seconda più grande galleria, l'attuale Istituto per esperienze aerodinamiche, che nel 1925 fu ingrandito e trasformato, divenendo il «Kaiser Wilhelm Institut für Strömungsforschung». Ha molti lavori nel campo della elasticità e della resistenza dei materiali, dell'idrodinamica, dell'aerodinamica e della termodinamica, pubblicati in varie Riviste. I risultati delle esperienze e delle ricerche effettuate nel Laboratorio di Göttingen sono: «Ergebnisse der Aerodynamischen Versuchsanstalt zu Göttingen», I–IV Lieferung, München 1921–1932. Ricordiamo pure: «Vier Abhandlungen zur Hydrodynamik und Aerodynamik» in collaborazione A. Betz, Göttingen 1929; «Abriss der Strömungslehre», Braunschweig 1931; «Theory of viscous fluids». Div. G dell'opera: «Aerodynamic Theory», diretta da W. Durand, Berlin 1935.

Del resto le ricerche del Prandtl sono note a tutti i cultori dell'aerodinamica moderna, che riconoscono unanimemente in lui uno dei più insigni maestri di questa disciplina.

Indirizzo abituale: Göttingen, Calsowstr. 15.

Fig. 1.17 *Ludwig Prandtl*, head of the AVA Göttingen and the Emperor Wilhelm Institute of Flow Research (KWI).

only a small supersonic region has formed while in Fig. 1.19 this region has expanded further and is terminated by a distinct shock wave. *L. Prandtl* supplemented these investigations with investigations of the flow in a curved channel. Because of the centrifugal forces, a pressure increase occurs between the convex and concave surfaces (from the inside to the outside) and, therefore,

Fig. 1.18 Shock configurations on a circular-arc airfoil, $p_0 = 1.60$ atm (23.51 psi).

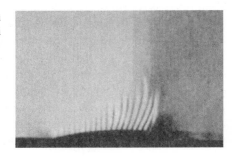

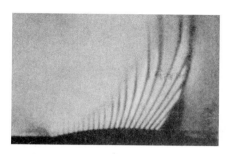

Fig. 1.19 Shock configurations on a circular-arc airfoil, $p_0 = 1.88$ atm (27.63 psi).

at the convex side a higher velocity than on the concave one. With these investigations *Prandtl* indicated which flow phenomena had his primary interest and that he saw the largest problems of fluid mechanics in the transonic flow regime.

To show the up-to-date schlieren pictures, L. *Prandtl* had to first submit them to the RLM for publication, and in doing so, he emphasized in the letter of September 14 that "the whole thing had no military character!"

1.3.2.3 PROBLEMS OF HIGH-SPEED AERODYNAMICS. As the next lecturer, *G. I. Taylor* (Fig. 1.20) discussed known and clearly defined problems of high-speed aerodynamics. In doing so, he first explained the analogy of flow and electrical fields that he developed. As an application, he showed the flow around a cylinder. In supersonic flow, *G. I. Taylor* was primarily interested in the conical flowfield around a nose cone in axisymmetric flow. In this region, computations agreed quite well with experimentally determined angles (schlieren photographs, s.f. [26]). The good agreement resulted from the fact that, in this case, viscosity played a very minor role and that in this comparison of computation and experiment nearly inviscid compressible flow is considered.

1.3.2.4 DRAG IN SUPERSONIC FLOW. A highly interesting report concerning the determination of drag in compressible flows was presented by *Th. von Kármán* (Fig. 1.21). He investigated the effect of compressibility on viscous drag. Under the assumption that in turbulent compressible boundary layers the exchange coefficients for momentum and heat are equal and, therefore, the turbulent *Prandtl* number is $Pr_t = 1$, *von Kármán* was able to give a corresponding equation for the relation between heat transfer and friction coefficient. The practical importance of this relation is that the application was not restricted to the flow about a flat plate but also proved to be a useful approximation to the flow about slender bodies, that is, in flows with moderate pressure gradients. This way, *Nusselt's* heat-transfer number for compressible flows, already independently given by L. *Prandtl* and G. I. *Taylor* in 1910,

TAYLOR prof. GEOFFREY INGRAM F. R. S. (Inghilterra). Nato nel 1886. Effettuò i suoi studi al Trinity College di Cambridge. Fu in aviazione durante la guerra, conseguendo nel 1915 il brevetto di pilota. Attualmente socio (fellow) della Royal Society e professore per le ricerche alla Royal Society a Yarrow. Membro della Bristish Aeronautical Research Committee. Autore di numerose pubblicazioni in riviste scientifiche, di matematica, idrodinamica, fisica e ingegneria, fra le quali citiamo le seguenti, come aventi attinenza col Convegno Volta: « A Mechanical Method for Solving Problems of Flow in Compressible Fluids », " Proc. Roy. Soc. A. ", 121, 1928; « Progress during 1924–8 ins Calculation of Flow of Compressible Fluid », " R. and M. Aeronautical Research Committee", 1928; « Recent Work on Flow of Compressible Fluids », " Journal Lond. Math. Soc. ", vol. V, 1930; « Applications to Aeronautics of Ackeret's Theory of Aerofoils Moving at Speeds greater than that of Sound », " R. and M. Aeronautical Research Committee ", 1932; « Air Pressure on a Cone Moving at High Speeds » in collaborazione con T. W. MacColl, " Proc. Roy. Soc. A. ", vol. 139, 1933; « Strömung um einem Körper in einer kompressiblen Flüssigkeit », " Z. A. M. M. ", Band 10, 1930.

Indirizzo abituale: Cambridge, Farmfield, Huntingdon Road.

Fig. 1.20 *Geoffrey Ingram (G. I.) Taylor*, **Cambridge University, England.**

could be extended, and it could, in addition, be shown that the *Reynold's* analogy remained valid independent of the *Mach* number. In the practical application, however, the extremely strong effect of shock boundary-layer interaction on drag, known at the time but not amenable to computation, could not be taken into account.

To complete his argumentations about the determination of drag in supersonic flow, *Th. von Kármán* addressed wave drag in the second part of his lecture. Together with *N. B. Moore*, he had in a 1932 treatise, already quoted by *L. Prandtl*, given a solution for the drag in axisymmetric flow based on the theory of small disturbances (Small-Disturbance theory) and a source-sink distribution on the body axis. As late as 1971, *A. Busemann* [31] reported in his survey article, "Compressible Flow in the Thirties," the enthusiastic reception given by the "Volta attendants" to *von Kármán's* demonstration of the optimization of a low-drag shell of given caliber and nose extent.

1.3.2.5 LIFT AT SUPERSONIC SPEEDS. In his presentation, "Lift at Supersonic Speeds," *E. Pistolesi* considered a Prandtl–Glauert relation that allowed an approximate determination of the angle of attack and the airfoil thickness by a transformation from incompressible to compressible flow conditions, an

Karman (DE) dott. TEODORO (Stati Uniti). Nato a Budapest. Laureato ingegnere meccanico a Budapest, in filosofia a Göttingen, dottore *ad honorem* in ingegneria a Berlino. Membro dell'Accademia delle Scienze di Göttingen e della R. Accademia delle Scienze di Torino. Lettore di meccanica applicata e aerodinamica all'Università di Göttingen negli anni 1909–12; nominato nel 1913 Capo del nuovo Istituto Aeronautico dell'Università di Aachen. Dal 1930 è Direttore del Laboratorio Aeronautico Guggenheim nel California Institute of Technology a Pasadena, e Direttore delle Ricerche dell' Istituto Guggenheim per i dirigibili di Akron (Ohio).

Le sue pubblicazioni sono numerosissime e molto importanti. Ne ricordiamo qui solo alcune, riguardanti fondamentali argomenti di aerodinamica, come la teoria dei vortici alterni, la teoria della turbolenza, la teoria delle correnti ultrasonore: « Ueber den Mechanismus des Widerstandes, den ein bewegter Körper in einer Flüssigkeit erfährt », " Ges. d. Wiss. zu Göttingen" 1912; «Ueber laminare und turbulente Reibung », " Z. A. M. M. ", 1921, Heft 4, Bd. 1; « Mechanische Aehnlichkeit und Turbulenz », " Nachr. d. Ges. d. Wiss. ", Göttingen, 1930; « Resistance of Slender Bodies Moving with Supersonic Velocities » in collaborazione con N. B. Moore. Trans. of A. S. M. E., Applied Mechanics Section, 1932–33; « General Aerodynamic Theory » in collaborazione con J. M. Burgers, vol. II of " Aerodynamic Theory " edited by W. F. Durand, 1935. Egli ha anche importanti pubblicazioni su problemi di elasticità e resistenza dei materiali e delle strutture.

Indirizzo abituale: Pasadena (Stati Uniti). Guggenheim Aeronautical Laboratories, California Institute of Technology.

Fig. 1.21 *Theodore von Kármán*, director of the Guggenheim Aeronautical Laboratory of the California Institute of Technology (Caltech), Pasadena, California, United States.

interesting consideration that did not further enter the literature, because flow separation at high angles of attack for given airfoils of finite thickness still had to be determined in the wind tunnel.

1.3.2.6 Cylindrical Flowfields. As *Th. von Kármán* [32] wrote in his memoirs, after this followed the most important presentation of the conference—that of a young man, *A. Busemann* (Fig. 1.22) from Germany, a pupil of *Ludwig Prandtl*. Like *Busemann*, another young participant in the congress, *Carlo Ferrari* [33], at the time age 32 and already a professor of aerodynamics at the University of Turin, called *Busemann* in his report about the Volta Congress, written in 1996 at the age of 93, a "rising star of modern aerodynamics." *Ferrari* described *von Kármán* as sharp-witted; *Prandtl* as thorough, going into detail; and *G. I. Taylor* as especially able to analyze experimental results and to formulate, based on these results, simplified theoretical approaches. *Adolf Busemann* was, as far as he remembered, very reserved but precise and concise in his statements especially when he reported

Busemann

B USEMANN prof. ADOLF (Germania). Nato a Lu-
becca nel 1901. Effettuati gli studi di costru-
zione di macchine al Politecnico di Braunschweig, fu assistente di O. Föppl nel Laboratorio
resistenza materiali dal 1922 al 1925 ed effettuò
lavori sui perfezionamenti delle macchine di prova e
sull'isteresi elastica. Nel 1925 passò a Göttingen sotto
la guida di Prandtl ed effettuò dapprima ricerche
sulla corrente nelle giranti delle turbo–macchine,
quindi sulle correnti ad alta velocità, elaborando,
insieme a Prandtl, il calcolo grafico delle correnti iper-
acustiche. I suoi studi di gasdinamica gli ottennero la
libera docenza nel 1930. Nel 1931 passò a Dresda,
assistente del professore R. Mollier, pubblicando lavori
di termodinamica e meccanica dei fluidi. Ritiratosi Mollier per età dall'insegnamento,
egli ha tenuto nello scorso anno per incarico la Cattedra illustrata dal 1873 dai
nomi di Gustavo Zeuner e di Riccardo Mollier. Oltre i lavori citati, il Busemann
ha trattato l'Elastomeccanica in « Handbuch der Physik » e la Gasdinamica in
« Handbuch der Experimentalphysik ».

Indirizzo abituale: Dresda A. 24, Bernhardstr. 30.

Fig. 1.22   *Adolf Busemann*, **professor at the Institute of Technology (TH) Dresden.**

about his own research results. This might have been the reason why only a few participants understood his work in its depth, its logical consequences, and its practical applications. Maybe also because, compared to his honorable colleagues, the 34-year-old *Adolf Busemann* was a "new boy" in the international world of science.

Born in 1901 in Luebeck, he studied at the Institute of Technology (TH) Braunschweig and obtained his doctorate in engineering (Dr.-Ing.) at age 24, only one year after he finished his studies with an engineering diploma (Dipl.-Ing.). His scientific career commenced in 1925 on entering the Emperor-Wilhelm Institute (KWI) in Göttingen, which was headed by *Ludwig Prandtl*. The latter sponsored *Busemann's* professional development by giving him the opportunity to qualify in 1930 to give lectures at the University of Göttingen. *Prandtl* supported his appointment as assistant professor at the TH Dresden, although he greatly appreciated *Adolf Busemann* as a colleague and considered his departure from the KWI a loss for the research activities in the field of compressible flow.

Now, on the occasion of the Volta Congress, *Adolf Busemann* had the opportunity to present to his mentor the solutions he saw for the problems in supersonic aerodynamics. In his deliberations on the topic "Aerodynamic Lift at Supersonic Speeds," he first asked the question whether, after the large increase in drag and the reduction in lift occurring when the flight speed

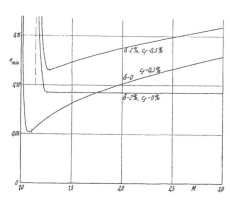

Fig. 1.23 **Most favorable glide ratio of a biconvex airfoil with 2% thickness and an average skin-friction coefficient of** $c_f = 0.3\%$ **dependent on the** *Mach* **number** M.

approached the speed of sound, levels of aerodynamic coefficients could be reached that allowed flight at supersonic speeds. For this, *Busemann* concentrated on pure supersonic flow with no subsonic regions being present. In his elegant estimation of the characteristics of a biconvex airfoil, he came to results about the change in the glide ratio with *Mach* number as shown in Fig. 1.23. (To obtain a better copy, the figure was taken from the German "Volta" publication of *Busemann* [34]!) If one selects a skin-friction coefficient of $c_f = 0.003$, which corresponds to the value of a flat plate at the *Reynolds* numbers considered, one obtains at a *Mach* number of $M = 1.08$ the minimum glide ratio of $\varepsilon = 0.05$. Particularly clear is the influence of the thickness of the airfoil because at a thickness-to-chord ratio of $\delta = 2\%$ for a noninclined airfoil pure supersonic flow is actually only possible at *Mach* numbers above $M = 1.14$. *Busemann* pointed out that a tremendous technical effort was required to reach this thickness ratio and cope with these angles of attack. For instance, the thickness ratio of a razor blade is about 0.3% without, however, forming a circular arc. To decide when the airfoil thickness and when friction causes the large glide ratios, the limits for disappearing skin friction and disappearing airfoil thickness, respectively, are presented in the figure. After these ingenious simple boundary-value considerations, *Adolf Busemann* developed with the aid of a potential approach, a method for the determination of the aerodynamic forces at larger deflections based on a common investigation with *O. Walchner* [35], and then proceeded to the second part of his lecture that was to change the world of high-speed aerodynamics.

In his contribution "Cylindrical Flow Fields," *Adolf Busemann* [34] again made it clear that he could transfer his pronounced analytical thinking to the solution of complex, physical problems. The basic consideration of why the cylindrical flowfield about a wing in an obliquely approaching flow can be converted into a plane two-dimensional flow if the pressure forces on the pressure side of the wing are to be computed is presented in

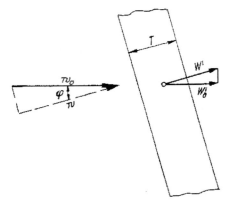

Fig. 1.24   Wing in oblique flow.

detail in the second chapter of the manuscript of his lecture. Because the original version of this important contribution to aerodynamics is not easily accessible in the literature, it is added as Exhibit 1.1 (see translation in the chapter appendix). The stagnation pressure q_0 of the incoming flow (freestream) is composed of the density of the gas ρ_0 and the flow velocity w_0 as follows:

$$q_0 = 1/2 \; (\rho_0 w_0^2)$$

In the case of a wing obliquely approached under an angle φ (swept wing), see Fig. 1.24, one has to distinguish between the actual stagnation pressure q_0 of the freestream and the effective stagnation pressure q that does not include the axial component of the approaching flow (freestream). In a similar way, one distinguishes between the *Mach* number of the freestream

$$M_0 = w_0/c$$

and the effective *Mach* number

$$M = w/c$$

with the speed of sound of the undisturbed freestream being

$$c = (\kappa R \, T)^{1/2}$$

and κ being the adiabatic exponent, R the gas constant, and T the static temperature. For the swept wing, the following relation can, according to the sketch in Fig. 1.25, be derived as

$$M = M_0 \cos \varphi$$

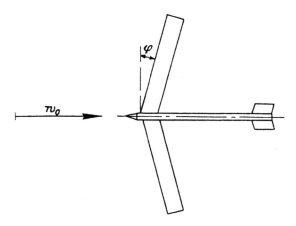

Fig. 1.25   Swept wing.

because the density ρ and the speed of sound c are independent of the obliqueness of the wing relative to the freestream. With this ingeniously simple consideration, *Adolf Busemann* could show that with increasing sweep angle φ, at a given *Mach* number, the component of the velocity normal to the leading edge of the wing decreased. The wing "sees," so to speak, only the normal component of the freestream velocity, and the commencement of the compressibility effects is shifted to a higher *Mach* number. In the transonic speed range (high subsonic flow) the first shock waves occur only at higher freestream *Mach* numbers. In transonic flight, the so-called critical *Mach* number is, therefore, in case of the swept wing increased, and the drag rise is shifted to higher freestream *Mach* numbers. For supersonic flight, one obtains correspondingly a shift in the occurrence of wave drag to higher *Mach* numbers and hence a reduction in total drag. *Adolf Busemann* considered in his further deliberations the computation of the optimal glide angle ε for a lifting system as sketched in Fig. 1.25. For that purpose, he determined the most favorable sweep angle φ retaining the known considerations of the plane two-dimensional flow about the wing as far as possible. The airfoil chord was assumed to be constant, and the airfoil section and the angle of attack were unchanged and taken normal to the wing axis. A relatively simple relation for the glide angle ε is given in Eq. (32a) (see Exhibit 1.1, translated in the chapter appendix). Two comments of *Busemann* again confirmed his strongly developed sense for interpreting the analytically obtained results with regard to their technical realization:

- At higher *Mach* numbers M_0, one does not necessarily drop below the most favorable value of the glide ratio of the flow at the *Mach* number M. The form effect will, indeed, become smaller, but the skin-friction drag will be somewhat higher. However, as long as the form is decisive, a glide ratio will certainly be achieved below the one that would have been obtained in a flow with the *Mach* number *M*.

- The consideration of the glide ratios could lead to the idea of an artificial increase of the *Mach* number as being desirable in order to then reduce it by the sweep. Such a swept autogiro does, however, not mean an improvement as it is not the drag caused by the desired lift that is of importance but the performance (power) required to generate this lift. If one multiplies, therefore, the glide ratios with the velocities, the apparent advantage disappears instantly.

But this remark is even more generally valid: If one obtains at supersonic speeds again glide ratios compatible with the ones at lower speeds, it does not mean that one can use them technically. They require, corresponding to the higher speed, increased power and, in turn, increased weight, etc. The technical tasks commence, therefore, only after the magnitude of the glide ratio is known.

It should be mentioned here that already in 1924 *Max Munk* [36] pointed out the effect of the normal velocity component in case of the flight at yaw of a rectangular wing at subsonic speeds. This was obviously unknown to *Adolf Busemann* because *Munk* had, when sweeping the wing and giving it a V-position, only considered the effect on stability but not the effects of the *Mach* number. His pupil, *R. T. Jones* [37, 38], could have remembered *Munk's* work when he discovered in 1945 the advantages of wing sweep for high-speed planes in the United States for the second time. This was 10 years after *Adolf Busemann's* presentation at the 1935 Volta Congress! Obviously *Busemann's* swept-wing concept was known worldwide since 1935, and the records were available to everybody, but the prerequisites for the application and utilization for transonic or supersonic flight were for many foreign scientists and engineers not yet given. Even *Th. von Kármán* wrote in his 1967 book *The Vortex Street* (p. 263):

> With regard to future aircraft developments the most important lecture at the conference was the one given by a young man by the name of *Dr. Adolf Busemann* from Germany, a pupil of *Prandtl*, who is now working in the United States for NASA. *Busemann* analyzed the nature of lift at supersonic speeds and assisted in establishing the fundamentals for the construction of the first jet aircraft during the Second World War by showing, for the first time, how the characteristics of the swept wing could solve many problems in the near-sonic speed range. I have to admit that I, at the time, did not pay much attention to this suggestion.

Subsequent to the contributions on the theoretical treatment of compressible flows, three lectures on wind-tunnel investigations followed. *E. N. Jacobs* reported about the last results from the 11- and 24-in. high-speed wind tunnels. Airfoil measurements that were later to become an important basis for comparative measurements, most of all in the newly constructed high-speed wind tunnel of the DVL in Berlin, were reported. We will consider this in detail later.

The analyses of the force measurements were supplemented by Tschlieren photographs. The investigations clearly revealed problems when reaching the critical *Mach* number. Better results had been hoped for from the "24-in." high-speed wind tunnel just completed as wall interference should, at the same model size, have been smaller. *Panetti* of the Institute of Technology (TH) Turin among others introduced a model support for rotating models and an interferometer for free-flight models. These investigations were supplemented by the report on test results that *G. P. Douglas* of the National Physical Laboratory in Teddington, England, had obtained on propellers at high subsonic speeds. Pointing the way forward, he analyzed the problems at the blade tips of the investigated propellers when approaching the speed of sound. The noise generated, besides being perceptible, also led to heavy loads on the pilot and the airframe, and there was a marked reduction in performance due to shocks developing at the blade tips. All results showed an acceptable agreement with the theory according to Prandtl–Glauert as long as the speed of sound was not reached locally, and, hence, shock-induced separation did not occur. A change in the form of the blade tips to lower the critical *Mach* number by applying *Busemann's* idea sweeping the blades was first used in Germany in the 1940s. The large interest in improving the performance of propellers was, without doubt, caused by the objective to reach still higher speeds than in the Italian record flights of 1934.

1.3.2.7 HIGH-SPEED WIND TUNNELS. During the last lecture of the second session, *Jakob Ackeret* (Fig. 1.26) described the state of knowledge related to the design and operation of high-speed wind tunnels. *Ackeret* studied at the Institute of Technology (TH) Zurich under *Boleslav Stodola*. From 1921 to 1926, he worked with *Ludwig Prandtl* at the Aerodynamic Test Establishment Göttingen (AVA) predominantly on problems of compressible flow, before he took up a job, at age 28, as a lecturer and in 1931 as a full professor of aerodynamics at the Eidgenössische Technische Hochschule Zürich (ETH) (Swiss Institute of Technology Zurich). In Italy, he had already made a name for himself as a design engineer for high-speed and supersonic wind tunnels. On the occasion of an excursion to the Italian Research Center in Guidonia near Rome, the congress participants could personally convince themselves of the high technical standard of *Ackeret's* wind-tunnel designs. It was at the supersonic wind tunnel of Guidonia where, in 1938, the famous aerodynamicist *Antonio Ferri* had conducted his near-sonic experiments and thereby conceived the first transonic test section. Because of his education by *Stodola* in the area of fluid machinery,

ACKERET prof. JACOB (Svizzera), nato nel 1898 a Zurigo. Laureato ingegnere meccanico nel Politecnico Federale di Zurigo. Nel 1920–21 assistente del prof. Stodola; nel 1921–26, capo reparto nel Laboratorio Aerodinamico di Göttingen. Dal 1927–31 libero docente, dal 1931–34 straordinario e dal 1934 ordinario al Politecnico di Zurigo e Direttore di quell'Istituto di Aerodinamica.

Pubblicazioni principali: « Luftkräfte auf Flügel die mit grösserer als Schallgeschwindigkeit bewegt werden », " Z. für Flugtechnik ", 1925, S. 72–74; « Artikel Gasdynamik » im " Handbuch der Physik ", Verl. Springer Berlin 1927, S. 289–345. Bd. VII; « Ueber Luftkräfte bei sehr hohen Geschwindigkeiten insbesondere bei ebenen Strömungen », " Helv. physica acta ", 1928, S. 301–322; « Der Luftwiderstand bei sehr grossen Geschwindigkeiten », « Schweiz. Bauzeitung », 1929, S. 179–183. Bd. 94. « Experimentelle und theoretische Untersuchungen über Hohlraumbildung (Kavitation) » im " Wasser. Technische Mechanik und Thermodynamik ", 1930, Bd. 1. S. 1–22 und 63–72.

Indirizzo abituale: Zurigo, Institut für Aerodynamik E. T. H.

Fig. 1.26   *Jacob Ackeret*, professor at the ETH Zurich.

Ackeret was an expert concerning turbines and compressors and, with the knowledge in compressible flows acquired at the AVA, predestined for wind-tunnel construction. Because of the increasing power requirements with increasing *Mach* number, he concerned himself extensively with the optimization of diffusers because flow separation led to considerable energy losses (low total pressure recovery) and a bad flow quality in the test section. This is a problem, up to today, solved with the aid of subscale model wind tunnels when building new large facilities. In the case of supersonic wind tunnels, *Ackeret* addressed the "storage wind tunnel," designed at the AVA Göttingen, whose principle structure is shown in Fig. 1.27. To avoid

Fig. 1.27   Basic outline of a supersonic wind tunnel with vacuum-storage (suck-down) operation as developed at the AVA Göttingen. Dried air was taken from a gasometer.

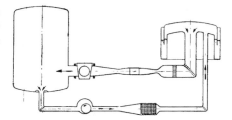

C. Wieselsberger

WIESELSBERGER prof. CARL (Germania). Nato nel 1887 ad Ebersthal (Baviera Inferiore). Direttore dell'Istituto Aerodinamico della Technische Hochschule di Aachen. Effettuò i suoi studi di ingegneria al Politecnico di Monaco e conseguì il dottorato nel 1912 con una ricerca scientifica nel campo dell'aeronautica. Dal 1912–1922 fu collaboratore del prof. Prandtl all'Istituto Aerodinamico di Göttingen e durante la guerra prestò servizio in aviazione. Nel 1922 conseguì la laurea in filosofia all'Università di Göttingen e da quell'anno al 1930 fu consulente scientifico presso gli Istituti aerodinamici del Governo giapponese, specie nell'Istituto delle Ricerche dell'Università Imperiale di Tokyo, nei " Reports " del quale pubblicò vari lavori. Nel 1930 fu nominato professore e direttore dell'Istituto aerodinamico al Politecnico di Aachen. Molte pubblicazioni nel campo dell'aerodinamica egli ha effettuate nella rivista " Zeitschrift für Flugtechnik und Motorluftschiffahrt ", nella " Zeitschrift für angewandte Mathematik und Mechanik ", nei " Rendiconti " degli Istituti aerodinamici di Göttingen, di Aachen, di Tokyo (già citati) e su altre riviste scientifiche.

Indirizzo abituale: Aquisgrana, Aerodynamisches Institut der Techn. Hochschule.

Fig. 1.28 *Carl Wieselsberger*, **professor at the Institute of Technology (TH) Aachen.**

icing of the model, a facility had been created that allowed dried air to be extracted from a bell-shaped air vessel. It was *Carl Wieselsberger* (Fig. 1.28) of the Institute of Technology (TH) Aachen, who already during the discussion of the schlieren pictures presented by *Prandtl* at the Volta Congress, first pointed out the possible occurrence of condensation shocks due to a too high humidity of the air [39]. This important conclusion was obviously either not understood or was ignored by many participants of the meeting and also did not enter the written documentation of the contributions to the discussion.

Returning to Aachen, *Wieselsberger* asked his doctoral student *Rudolph Hermann* [40] to deal with this problem. His research results later entered the concept of the Peenemünde supersonic wind tunnel and contributed decisively to the quality of this research facility. As an example of a continuous test facility, *Ackeret* described the supersonic wind tunnel at Zurich (Fig. 1.29). By means of a skillful optimization of the resistance of the cooler and a 13-stage axial compressor with an adiabatic efficiency of 80%, the drive power for a test section cross section of 40 by 40 cm and a velocity in the test section of twice the speed of sound was successfully limited to 700 kW. The

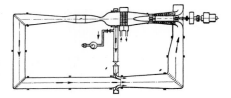

Fig. 1.29 Continuously operating super-sonic wind tunnel; velocity increase by means of an ejector, according to *Ackeret.*

wind-tunnel design for the Italian Aerodynamic Research Center Guidonia, conceived by *Ackeret*, was based on almost identical characteristics.

1.3.2.8 REALIZATION OF HIGH-SPEED FLIGHT. The third main session treated problems of "thermodynamics and combustion," whose solutions were recognized as important prerequisites for the realization of high-speed flight. Further details of the lectures will not be dealt with here, as all five contributions showed, most of all, that for high-speed flight at high altitudes the development of new propulsion systems was first needed. Thus, *H. E. Wimperis* remarked during the final discussion:

> In order to fly fast at high altitudes, we must still await the development of new means of propulsion, such as, for instance, jet engines. However, we are still far removed from the application of these new engines. But one of these days we will—thanks to the outstanding work of this congress—know how to do it. The remaining task will then be to convince the passengers to fly that fast at high altitudes.

That the development and realization of high-speed aircraft would not be determined by the passengers of commercial airlines, but purely by military applications, was certainly already suspected by many participants of the congress. Thus the start of the lecture of *Adolf Busemann* was postponed because the congress participants had to listen first to the declaration of war by the "Duce" *Benito Mussolini* on the occasion of the Italian invasion of Abyssinia, today's Ethiopia. A total of about 300,000 Italian soldiers participated in the invasion. The course of the war was essentially determined by the ultramodern Italian Air Force at that time. For the scientists from France and England, the situation was becoming especially precarious as their governments pursued their own interests in this part of Africa. While the English congress participants were obliged by their governments to attend the lectures only, the French were not under any restrictions.

1.3.3 AFTER THE CONGRESS

As a summarizing evaluation of the success of the Volta Congress, two comments of young participants at the time are cited here. *Adolf Busemann* completed his 1971 article about compressible flows in the 1930s with the words:

> The researchers in the area of compressible flows also remained after the war a large international family where members respected and accepted each other. They made every effort to improve international relations between scientists and avoided giving each other the blame not having first made peace before improving air transportation worldwide.

At the age of 93, *Carlo Ferrari*, the last surviving witness to the times, wrote in the conclusion of his report on the Volta Congress, published in 1996, the conciliatory words:

> I believe that the 5th Volta Congress will remain in everlasting remembrance for all students of fluid mechanics. For all who had the luck to participate, their strong impressions will not be forgotten as long as they live. All participants were connected by a deep friendship. Although the signs of WWII already became clear at the meeting, our friendship was not destroyed by the tragic events of the following years.

After WWII it was *Theodore von Kármán* who took the initiative and founded in Brussels a post-university institution for young scientists under the auspices of the Advisory Group for Aerospace Research and Development (AGARD). In the meantime, the von Kármán Institute for Fluid Dynamics (VKI), named after him, has educated several generations of fluid-mechanics engineers, many of whom have started working in distinguished positions internationally. After 1986, contacts were taken up with the new NATO member candidates from the former Eastern block countries so that today, for instance, students from Hungary, the native country of *von Kármán*, study at the VKI.

Because many of the invited participants to the 5th Volta Congress were to play an important part during or after WWII in the further development of high-speed aerodynamics, the original of the list of participants is reproduced in Exhibit 1.2 (see translation in the chapter appendix).

1.4   EXPERIMENTAL VERIFICATION OF BUSEMANN'S SWEPT-WING THEORY BY HUBERT LUDWIEG AT THE AVA GÖTTINGEN

The lecture of *Busemann* at the Volta Congress and the publication of his contribution by the German Academy of Aeronautical Research immediately led thereafter to his unusual popularity among his German professional colleagues. Also the State Ministry of Aeronautics (RLM) suddenly became aware of him. Shortly after his return from the Volta Congress, *A. Busemann* received invitations from the Test Establishment of the Army in Peenemünde and the Deutsche Forschungsanstalt für Luftfahrt Braunschweig (DFL) (German Research Establishment of Aeronautics) to present his ideas concerning problems in high-speed aerodynamics. The management of the Institute of Gas Dynamics of the DFL in Braunschweig was assigned to him in 1936. In this position, he could later directly influence the swept-wing development. *Busemann's* decision to give up his chair at the TH Dresden after such a short time was surely also influenced by the fact that two high-performance test facilities for investigations at high subsonic and supersonic speeds were promised to be put into operation in 1937. The implementation of his ideas in Germany was, at first, slow moving. In this regard, two lectures that *Ludwig Prandtl* and *Adolf Busemann* gave on the occasion of the second scientific meeting of the Full Members of the German Academy of Aeronautical Research on November 12, 1937, concerning questions of high-speed aerodynamics are particularly remarkable. *Prandtl* repeated and summarized his Volta lecture in a generally intelligible form and primarily addressed the physical fundamentals. In his own way, he described complex relations in gasdynamics—without the aid of equations—and dealt here again with the example of the biconvex airfoil in symmetrical supersonic flow. However, *Busemann*, as a follow-on speaker, surely did not fully approve of this conclusion.

Prandtl [41] argued, among other things:

> One can, of course, also give such a body an angle of attack and carry out for this configuration the design and computations. Besides drag they also give lift. The lift coefficient is here, at speeds slightly above the speed of sound, higher than for very large speeds and, therefore, if one takes into account the unavoidable drag due to surface friction, the ratio of lift to drag for a given angle of attack is better at lower supersonic speed than at high supersonic speed. Unfortunately, the small airfoil glide ratios which the aeronautical engineer is used to at speeds below the speed of sound are here not reached anywhere so that, therefore, not only when exceeding the speed of sound, where quite a large drag can be expected, but also flying at speeds above the speed of sound will have very little prospect. So flying at supersonic speeds will probably remain a dream. It is a very tempting dream if one thinks of a military application since the aircraft noise can here only propagate within the Mach cone; so the aircraft would already have dropped its bombs before anyone would get to hear it at all.

In his report on the objectives of the high-speed technology, *Busemann* [42] addressed specific problems of the high-speed aerodynamics and emphasized in his opening remarks:

> As I will show later, we are with today's attainable speeds at the threshold of an era where the experiences for aerodynamics in low-speed flight

must be substantially supplemented. To penetrate this area, which will surely put higher demands on the performance of the engine; it is of primary important to achieve all possible performance savings by a favorable geometry. This is the objective of the extended aerodynamics, the so-called gasdynamics.

Busemann made it clear that, although the Italians were with a flight speed of over 700 km/h owners of the world speed record for seaplanes, shells, on the other hand, achieved speeds of more than 3600 km/h. In his opinion the achievement of supersonic flight depended primarily on sufficiently powerful propulsion and the geometric design of the aircraft/missile. He described test results of airfoil measurements in the near-sonic range or at low supersonic speeds, available at the time, which had been carried out within the context of possible performance increases of propellers for high-speed aircraft in the United States and at the AVA in Göttingen. Because reliable measurements in the transonic speed range were not available in 1937, he interpolated for a biconvex airfoil the American measurements and results from measurements at Göttingen at supersonic speeds (Fig. 1.30). The

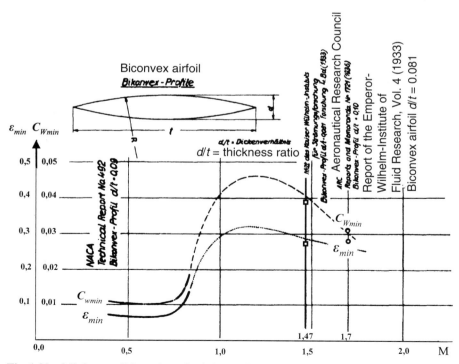

Fig. 1.30   Minimum glide ratio and minimum drag coefficient at various *Mach* numbers according to *A. Busemann* [42].

dashed respectively dotted connecting curves were based on measured drag respectively glide ratios of shells. It should, in essence, be shown here that the reduction in drag for the biconvex airfoil when exceeding *Mach* number 1 could be derived from the results of measurements on long bodies of revolution available at that time. However, he pointed out simultaneously that, on the one hand, the performance of the propulsion system was given by the product of drag and speed, and, on the other hand, the drag was obtained by multiplying the drag coefficient by the stagnation pressure and the wing area. The drag coefficient may, therefore, decrease while the necessary drive power increases continuously with speed. If one imagines the drag coefficients to be multiplied by the third power of the speed, the decrease of the drag coefficient would hardly be recognized. *Busemann*, therefore, saw a realization of supersonic flight only at large altitudes where the aircraft at the higher speed could just carry its weight. For this he demanded a jet engine that had to provide the performance at an almost constant degree of efficiency and, therefore, permitted the performance to grow constantly with the speed.

During the last part of his lecture, *Busemann* dealt with the existing wind tunnels at home and abroad. He talked in the case of supersonic configurations for the first time about manned and unmanned aircraft and pointed out the necessity to determine basic information about aerodynamics and flight mechanics in wind-tunnel tests. Here, he described various test facilities in Switzerland and Italy (built by *J. Ackeret*), at the NPL in England, in the Soviet Union, and in the United States. He introduced the German level of wind-tunnel technology only on a very restricted scale. He mentioned two new, almost operational storage facilities for high-speed investigations of the DFL in Braunschweig, but did not inform the Academic Forum about new enormous facilities in Braunschweig and Berlin that were being projected and partly already under construction. Here, it becomes clear that all high-speed and supersonic projects were already labeled top secret in 1937. We will still address the facilities for the development of the projects at high subsonic speeds (aircraft) and supersonic speeds (missiles) in a separate chapter.

At the AVA Göttingen, a wind tunnel for subsonic and supersonic speeds had been projected in 1937 and went into operation in 1939. It was equipped with an open test section with a nominal subsonic cross section of 11 by 11 cm. The air was sucked via a silica gel dryer from the atmosphere into a 150 m³ vacuum tank. The *Mach* number was controlled by an adjustable diffuser. A test chamber separated the freejet from the environment.

After the commissioning of the wind tunnel, the verification of the swept-wing concept of *Adolf Busemann* was carried out in compressible subsonic flow, on the suggestion of *Albert Betz*, director of the AVA, as one of the

Fig. 1.31   *A. Betz*, **director of the AVA [archive MPG (Max Planck Society for the Promotion of Science) Berlin].**

first research projects (Fig. 1.31). According to *Busemann's* principle, a swept wing with a sweep angle φ (Fig. 1.32, left), behaves like an infinite-span sheared wing (Fig. 1.32, right). However, the center section and the wing tips of the swept wing must be omitted, which leads inevitably to reductions compared to the infinite-span sheared wing. If one, at first, neglects the skin friction, the span-wise component of the freestream velocity, $v \sin \varphi$, is ineffective, and the wing behaves as if it were approached only by the normal component, $v \cos \varphi$. Here, one has to take into account that the effective angle of attack α and the effective ratio of airfoil thickness to chord, d/t are larger than the corresponding ones measured in the flight direction. It follows immediately that the wing behaves with respect to the compressibility effect as if it were approached by the effective *Mach* number $M_{\text{eff}} = M \cos \varphi$. The critical *Mach* number M_{crit}, where shocks first appear on the wing and the drag strongly rises, also increases. (Please also compare the contribution of *H. Ludwieg* [43] concerning the history of the

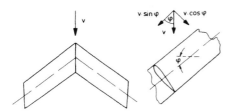

Fig. 1.32   Explanation of the swept-wing principle by a comparison with the sheared wing [43].

swept wing.) Here three different applications of flight technology can be distinguished:

1) *Subsonic flight (M < 1):* If an unswept wing shows typical compressibility effects, like a drag rise, one can reduce these effects again by sweeping the wing, that is, reducing the effective *Mach* number.

2) *Supersonic flight with a subsonic leading edge:* When flying at not too high supersonic *Mach* numbers, one can reduce the effective *Mach* number to a value below 1 by a sufficiently strong sweep so that the supersonic wave drag and also the additional drag associated with high subsonic speeds disappear. As the normal component of the freestream *Mach* number is less than 1, one talks about flight conditions with a subsonic leading edge.

3) *Supersonic flight with a supersonic leading edge:* If flight at higher supersonic *Mach* numbers is required, the effective *Mach* number can also be reduced to a value somewhat above 1. Hereby, the drag at a given lift can, at least for very thin wings, be reduced.

An interesting correspondence between *A. Betz* and *A. Busemann*, which *H. Fütterer* found in the Göttingen archives, shows that the swept-wing concept of *A. Busemann* after his lecture in 1935 at the Volta Congress finally found its "users." In the letter of November 10, 1939 to *Busemann*, *Betz* informed him (Fig. 1.33) that he was only made aware by *Lorenz* from the RLM of his Volta lecture and the publication in "German Aeronautical Research (LuFo Vol. 12)" after the preparations for tests at the AVA in Göttingen were already completed (Fig. 1.33). *Betz*, furthermore, indicated his patent application for the swept wing to secure this development for research. Because he accepted *Busemann* without restrictions as the original inventor of the swept wing, he asked him to point out a possibility to adhere to the patent right. In addition, he reported about first contacts with the Messerschmitt company, which had shown a certain interest in the "matter." On November 14, 1939, *Busemann* (Fig. 1.34) answered and also informed about planned tests with swept-wing models at the LFA in Braunschweig and contacts with the Messerschmitt and Heinkel companies. It is documented here for the first time that at the latest in 1939—but presumably much earlier—the application of his swept-wing concept in compressible subsonic flow was clear to *Adolf Busemann* and that he had offered to the Messerschmitt and Heinkel companies specific measurements in the LFA wind tunnels. Therefore, he only asked the rather inconsequential question in his correspondence with *A. Betz*:

> It would only be interesting if the same Messerschmitt company would provide means for the same investigations at two different locations once through *Dr. Winter* and once through *A. Lippisch*.

a)

Herrn

Prof.Dr.A. B u s e m a n n
Luftfahrtforschungsanstalt
  Hermann Göring

B r a u n s c h w e i g

                                    B.871/39.   10.11.1939.

                        Lieber Herr Busemann!

                In Ihrem Artikel Lufo 12 "Aerodynamischer Auftrieb
        bei Ueberschallgeschwindigkeit" haben Sie,wie ich jetzt fest-
        stellte,auf die Wirkung der Pfeilform hingewiesen. Mir war
        diese Stelle nicht mehr in Erinnerung,ich wurde erst aus
        einer Unterhaltung mit Herrn Dr. L o r e n z darauf aufmerk-
        sam. Vor einiger Zeit habe ich nun ohne Kenntnis dieser
        Stelle Messerschmitt und Heinkel die Anwendung der Pfeilform
        zur Umgehung der Schwierigkeiten bei großen Geschwindigkeiten
        empfohlen,und,um uns zu sichern,die Sache schnell zum Patent
        angemeldet. Die Patentanmeldung dürfte auf Grund Ihrer Vor-
        veröffentlichung aber nicht aufrecht zu erhalten sein. Ande-
        rerseits wäre es auch dem Ministerium erwünscht,wenn der
        Industrie gegenüber ein gewisser Schutz vorhanden wäre. Bevor
        ich deshalb die Anmeldung zurückziehe,wollte ich sie Ihnen
        zeigen,ob Sie irgendeine Möglichkeit sehen,einen Schutz auf-
        rechtzuerhalten. Vielleicht haben Sie aber auch selbst sich

                                                            schon

Fig. 1.33 a) *A. Betz* to *A. Busmann*: **Request for a joint patent application concerning the swept wing, November 10, 1939 (GOAR); b) translation.**

1.4.1 FIRST WORLDWIDE INVESTIGATION ON SWEPT WINGS IN COMPRESSIBLE SUBSONIC AND SUPERSONIC FLOW

Because *A. Busemann* did not have any objections to the swept-wing measurements at the AVA, *A. Betz* put the physicist *Hubert Ludwieg* in charge of carrying out the measurements. *Ludwieg* had in the High-Speed Aerodynamics Branch, headed by *O. Walchner*, already gained extensive experience in this relatively new experimental field (Fig. 1.35). Together with the experienced experimentalist *H. Strassl*, they conceived two models (Fig. 1.36) of relatively

b)

Prof. Dr. A. Busemann
Aeronautical Research Establishment Hermann Göring
Braunschweig

                          B, 871/39   10.11.39

Dear Mr. Busemann,

In your article Lufo 12 "Aerodynamic Lift at Supersonic Speeds" you have, as I now realized, pointed out the effect of sweep. I did not remember this passage but became aware of it in a conversation with Dr. Lorenz. I have, without knowing this passage, recommended to Messerschmitt and Heinkel the application of the swept form to evade the difficulties at high speeds and, to be on the safe side, quickly filed a patent application. The patent application may, due to your publication, not to be adhered to. It would, on the other hand, be desirable to the Ministry if there would be a certain protection towards industry. Before I, therefore, withdraw the application, I wanted to show it to you and ask whether you see some possibility to keep up the protection. But, may be, you yourself have already done something concerning the protection in that matter.

Fig. 1.33   (Continued)

simply design. Both wings possessed the same airfoil, Goe 623, taken normal to the wing centerline. With a thickness ratio of 12% and a maximum-thickness location of 30% of the airfoil chord, this was not a really favorable airfoil for high-speed flight. On the other hand, this airfoil selection made sure that the swept-wing effect would be particularly pronounced in these tests. The swept wing resulted from the trapezoidal wing simply by rotating the two wing halves by 45 deg about the wing center. The model dimensions were, with 80-mm span for the trapezoidal wing and 56.5-mm span for the swept-wing model, extremely small, but at a test section size of 110 by 110 mm the models were relatively large with not-to-be-neglected blockage effects. This fact was clearly of importance when interpreting the test results even if it was, at first, solely a test of the principle.

The test results [44] of these first worldwide swept-wing investigations (Fig. 1.37) show immediately the strong drag reduction in case of the swept wing, as compared to the trapezoidal wing, at the *Mach* number of $M = 0.9$ and, in the area of high angles of attack, also already at $M = 0.7$. Thus, the superiority of the swept-wing, as compared to the straight wing, was confirmed for the first time. The results were immediately reported by *Betz* to the German aircraft industry and aroused their great interest. *Ludwieg* remarked

a)   (Continued)

B.871/39.   10.11.39.   2.

schon etwas in dieser Richtung schützen lassen.

Da Messerschmitt immerhin ein gewisses Interesse
für die Sache gezeigt hat,habe ich einen Versuch vorbereitet,
der die grundsätzliche Richtigkeit der Ueberlegung bestäti-
gen soll(2 sonst gleiche Flügel,der eine mit,der andere ohne
Pfeilform im Hochgeschwindigkeitskanal von Walchner). Wie
mir Herr Walchner sagt,könnten die Versuchsergebnisse in
etwa 4 Wochen vorliegen. Nun sagte mir Herr Dr.Jenissen,daß
Sie in dieser Richtung auch etwas in Vorbereitung haben. Ich
wollte Sie deshalb fragen,ob es noch Sinn hat,daß wir diese
Versuche durchführen. Ich bin der Ansicht,daß es noch Sinn
hätte,wenn wir dadurch früher ein Ergebnis bekommen als es
bei Ihnen zu erwarten ist,auch wenn die Genauigkeit infolge
der kleinen Abmessungen zu wünschen übrig läßt.

Herr Walchner sagte mir,daß er im Auftrag von
Messerschmitt auch den Widerstand von kreisrunden Drähten
und von Profildrähten gemessen habe. Der Bericht wird gerade
angefertigt. Herr Jenissen sagte,daß diese Dinge auch zu
Ihrem Arbeitsgebiet gehören. Es wäre daher gut,wenn wir da
auch unsere Erfahrungen austauschen würden. Im wesentlichen
war bei den Walchnerschen Versuchen herausgekommen,daß die
runden Drähte bei hohen Geschwindigkeiten nicht viel
schlechter wurden,wohl aber die Profildrähte.

Es wäre mir lieb,wenn Sie zu den obigen Fragen
bald Stellung nehmen würden,insbesondere zu der Frage,ob
wir die Versuche mit dem Pfeilflügel,die eben in Vorbereitung sind,noch durchführen sollen.

Mit besten Grüßen und Heil Hitler!
Ihr
A. Betz

1 Anlage.

Fig. 1.33   (Continued)

b) (Continued)

Since Messerschmitt showed, after all, a certain interest in the matter, I have prepared a test that should confirm the correctness of the considerations (two otherwise identical wings, one with, the other without sweep in the high-speed wind tunnel of Walchner). As told to me by Walchner, the test results could be available in about four weeks. Now Dr. Jenissen told me that you were also preparing something in that direction. Therefore, I wanted to ask you whether it still made sense for us to conduct these tests. I am of the opinion that it would make sense if we thereby would obtain results earlier than can be expected from you even if, due to the small dimensions, the accuracy leaves something to be desired.

Mr. Walchner told me that, under contract with Messerschmitt he also measured the drag of circular and of profiled wires. The report is just now being prepared. Mr. Jenissen said that these matters also belong to your area of work. It would, therefore, be good if we also here could exchange our experience. Walchner's tests essentially showed that the circular wires did not get worse at high speeds; the profiled wires did.

I would appreciate your earliest comment on the above questions, especially on whether we should still carry out the tests with the swept wing which are presently being prepared.

With best regards and "Heil Hitler"

Yours

A. U. Betz

Fig. 1.33 **(Continued)**

with respect to the measurements that, besides the wind-tunnel corrections that had not been carried out, further effects had to be taken into account in the interpretation of the results. Measured in the flow direction, the swept wing had, by the factor of $\sqrt{2}$, a thinner airfoil and by the same factor higher *Reynolds* number *Re* than the basic straight wing. The *Reynolds* numbers of the wind-tunnel tests, based on the model chords, were at $Re = 3 \times 10^5$ and $Re = 4.2 \times 10^5$ at least 100 times smaller than for a possible full-scale design. In addition, no practically verifiable experience concerning shock boundary-layer interaction effects was available, which would certainly have triggered the lift drop and the increase in the drag in case of the trapezoidal wing at angles of attack of $\alpha > 3$ deg and the freestream *Mach* number of 0.7.

It is apparent from the December 1939 cover letter of *Betz* that he had immediately sent the first test results to *Busemann* (LFA-Braunschweig) after the evaluation of *Ludwieg's* swept-wing measurements. At the same time, *Betz* informed his colleague about planned measurements in the low

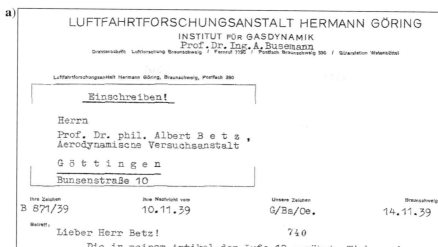

LUFTFAHRTFORSCHUNGSANSTALT HERMANN GÖRING
INSTITUT FÜR GASDYNAMIK
Prof. Dr. Ing. A. Busemann
Drahtanschrift Luftforschung Braunschweig / Fernruf 7790 / Postfach Braunschweig 390 / Güterstation Watenbüttel

Luftfahrtforschungsanstalt Hermann Göring, Braunschweig, Postfach 390

Einschreiben!

Herrn

Prof. Dr. phil. Albert B e t z ,
Aerodynamische Versuchsanstalt

G ö t t i n g e n

Bunsenstraße 10

Ihre Zeichen	Ihre Nachricht vom	Unsere Zeichen	Braunschweig
B 871/39	10.11.39	G/Bs/Oe.	14.11.39

Betreff:

Lieber Herr Betz!   740

Die in meinem Artikel der Lufo 12 erwähnte Wirkung der
Pfeilform habe ich auch den Firmen Messerschmitt und Heinkel
empfohlen und mit Messerschmitt verabredet, darüber eine Messung im Windkanal vorzunehmen. Das Modell wird bereits angefertigt, doch kann ich wegen der Überlastung des Kanals kein Ergebnis innerhalb 4 Wochen versprechen. Ich glaube daher, daß
die Parallelmessung in Ihrem Kanal nicht zu spät kommt. Interes-
sant wäre es nur, wenn dieselbe Firma Messerschmitt einmal durch
Dr. Winter und einmal durch Lippisch Mittel für dieselbe Untersuchung an zwei verschiedenen Stellen bereitstellt.

Zur Untersuchung günstigster Aufhängungen sind in Dresden
Widerstände an runden Drähten senkrecht und geneigt zur Strahl-
achse gemessen worden. Diese Ergebnisse könnte ich Ihnen zur
Verfügung stellen. Profildrähte haben wir nicht untersucht,
weil sie nicht in so geringen Abmessungen zu haben sind, wie
dies für die Messungen im Kanal erforderlich gewesen wäre. Für
die späteren größeren Kanäle kämen dagegen auch profilierte
Drähte in Frage.

Die Frage, ob man die Pfeilform noch schützen könnte, halte
ich ebenfalls für sehr schwierig. Auf jeden Fall wäre wohl durch
meine Vorveröffentlichung der Fall versperrt, daß die Komponente
der Mach'schen Zahl senkrecht zum Flügel größer als 1 ist. Während in meiner Veröffentlichung versucht wurde, die Komponente
der Mach'schen Zahl von oben her der 1 anzunähern, hätte man
bei geringeren Geschwindigkeiten das Bestreben, mit der Normal-

LUFTFAHRTFORSCHUNGSANSTALT
HERMANN GÖRING
BRAUNSCHWEIG

Fig. 1.34   a) Reply of *A. Busemann* to *A. Betz* (GOAR); b) translation.

- 2 -

Herrn Professor Albert Betz, Göttingen

komponente der Mach'schen Zahl möglichst weit von 1 wegzu-
kommen. Im gewissen Sinne ist das ein Unterschied, doch muß
man eine Nichtigkeitsklage von Seiten der Industrie be-
fürchten. Ich selber habe keinerlei Schutzrechte auf die
Pfeilform angemeldet.

Mit den besten Grüßen
Ihr

A. Busemann

b)

AERONAUTICAL RESEARCH ESTABLISHMENT HERMANN GOERING
INSTITUTE OF GASDYNAMICS
Prof. Dr.-A. Busemann

Brunswick 14.11.39

Prof. Dr. phil. Albert Betz
Aerodynamic Test Establishment
Göttingen

Dear Mr. Betz,

I have also recommended to the Messerschmitt and Heinkel companies the effect of the swept form that I mentioned in my article in Lufo 12, and I have made an appointment with Messerschmitt to carry out related measurements in the wind tunnel. The model is already being built, however, due to an excess load on the wind tunnel, I cannot promise a result within four weeks. Therefore, I believe that the parallel measurements in your wind tunnel will not be too late. It would only be of interest if the same Messerschmitt Company would provide means for the same investigation at two different locations once through Dr. Winter and once through A. Lippisch.

To investigate the most favorable (model) suspensions, the drag of circular wires, normal and inclined to the jet axis, were measured at Dresden. I could make these results available to you. We did not investigate profiled wires as they are not available in such small dimensions as needed for the measurements in the wind tunnel. For the future larger wind tunnels, profiled wires were also a possibility.

I consider the question, whether the swept form could still be protected, also to be a very difficult one. Due to my publication, the case would definitely be blocked where the component of the *Mach* number normal to the wing is larger than 1. While in my publication it was attempted to approximate the component of the *Mach* number, coming from above, to 1, at lower velocities the desire is to have the normal component of the *Mach* number as far away from 1 as possible. In a certain sense this is a difference; however one must fear an invalidity suit by industry. I myself have not applied for any patent rights on the swept form.

Yours with best regards,

A. Busemann

Fig. 1.34   (Continued)

Fig. 1.35   *Hubert Ludwieg*, High-Speed Aerodynamics Branch, AVA Göttingen.

supersonic regime and sent him his new wording for a common patent application that he had formulated as follows:

> Aircraft with swept wings, characterized by the fact, that it is used at flight speeds close to the speed of the sound and that the sweep is so strong that the free-stream velocity component normal to the wing axis remains considerably below the speed of sound.

It can be assumed that *Betz* had not only informed *Busemann* but also simultaneously *Willy Messerschmitt*, Messerschmitt Corporation, about these

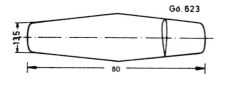

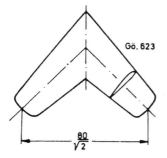

Fig. 1.36   First swept-wing model and comparative trapezoidal wing in the High-Speed Wind Tunnel Göttingen (*H. Ludwieg*) [43].

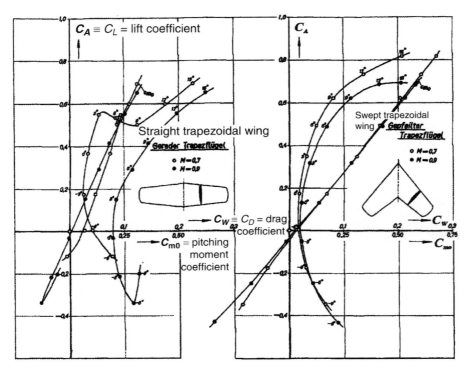

Fig. 1.37   First swept-wing measurements at $M = 0.7$ and $M = 0.9$ according to H. Strassl and H. Ludwieg [44].

important test results. In the letter of *Messerschmitt* of January 17, 1940 (Fig. 1.38), the farsighted interest of an aircraft manufacturer in the development of the swept wing was expressed for the first time. Because sensational results, naturally, made many skeptics appear on the scene, the test program was, on the suggestion of *Ludwieg* [45], considerably extended to qualitatively backup the results and to refute possible objections. In addition to Wings 1 and 2, already tested, four additional wings were included into the model series (Fig. 1.39). The swept wings were generated by turning the trapeze centerline by angles of $\varphi = 15$, 30, 45, and −45 deg. The measurements were again carried out in the high-speed wind tunnel of the AVA; the open-jet nozzle had for the supersonic measurements dimensions of 110×130 mm. These measurements were also financed by AVA's own means and carried out within the branch of O. *Walchner* by H. *Ludwieg*; the test program was, however, already coordinated with the Messerschmitt Company. It was, by the way, *Ludwig Boelkow*, who had, as a young engineer from the Institute of Technology (TH) Berlin-Charlottenburg, started work at the Department of Aerodynamics and Flight Mechanics of the Messerschmitt

a)

WILLY MESSERSCHMITT
PROF.DR.ING. E H., DIREKTOR DER MESSERSCHMITT A.G.

AUGSBURG , den 17.I.4o

Herrn
Professor Dr.Albert B e t z
Aerodynamische Versuchsanstalt

G ö t t i n g e n

Bunsenstr.1o

Sehr geehrter Herr Professor B e t z !

Ich danke für die Übersendung Ihrer Messergebnisse
an Pfeilflügeln. Über den Gewinnst durch Verwendung
grosser Pfeilung, bin ich ausserordentlich erstaunt.
Aus diesem Grunde sehe ich es daher als dringend notwendig an, die so günstigen Versuchsergebnisse in
einem grossen Windkanal zu wiederholen.

Die von Ihnen vorgeschlagene veränderliche Pfeilung
durch Krümmung des Flügels bzw. durch Veränderung des
Pfeilwinkels während des Fluges, dürfte konstruktiv
eine Reihe Schwierigkeiten machen. Ich bin aber der
Überzeugung, wenn sie zwingend notwendig werden, so
dürften wir auch diese Schwierigkeiten überwinden.

Ich danke nochmals für die Übersendung Ihrer Messungen
und Ihrer Anregung und verbleibe mit den besten Grüssen
und

Ihr

Fig. 1.38   a) Letter of thanks by *Willy Messerschmitt* **to** *Albert Betz* **for transmitting the results of the first swept-wing measurements (excerpt from the letter of November 1, 1940, GOAR); b) translation.**

b) WILLY MESSERSCHMITT   Augsburg, 17.I.40
Prof. Dr.-Ing hon., Director of the
Messerschmitt Corporation

                         Prof. Dr. Albert Betz
                    Aerodynamic Test Establishment
                            <u>Göttingen</u>

Dear Professor Betz!

I thank you for sending your test results on swept wings. I am extraordinarily surprised about the gain by utilizing large sweep. Therefore, I consider it as urgently necessary to repeat such favorable test results in a large wind tunnel.

Variable sweep by bending the wing or changing the sweep angle during flight suggested by you would make a number of structural difficulties. But I am convinced that we will overcome these difficulties if it should become urgently necessary.

I thank you again for sending your test results and for your suggestion and remain with the best wishes and...

                                                          Yours

Fig. 1.38   (Continued)

Corporation and had interceded with his boss, *Riclef Schomerus*, for further swept-wing investigations.

As important test results, the lift-vs-drag curves of all six wings are shown in Fig. 1.40 at *Mach* numbers of $M = 0.8$ and 1.21; they were interpreted by

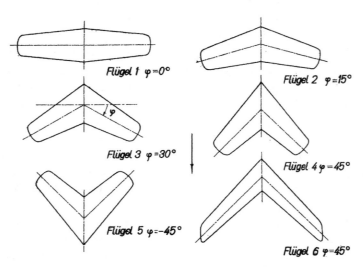

Fig. 1.39   Trapezoidal wing model and five swept-wing models for the second test phase in the high-speed wind tunnel of the AVA (Wings 1–6) [44] (legend: Flügel = Wing).

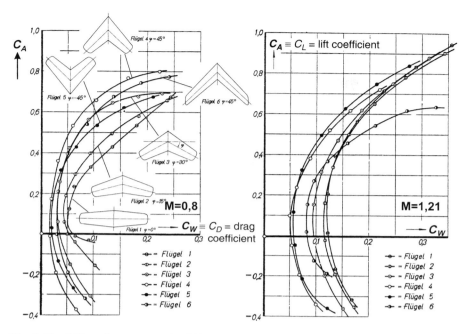

Fig. 1.40   Lift vs drag curves for Wings 1–6 at *Mach* numbers *M* = 0.80 and *M* = 1.21 (*H. Ludwieg*) [45] (legend: Flügel = wing).

Ludwieg as follows: One observes at $M = 0.8$ a strong reduction in the drag coefficient C_W (C_D) with increasing sweep angle φ. This also applies to higher C_A (C_L) values despite the decreasing aspect ratios with an increasing sweep angle, which causes a higher induced drag. One recognizes, furthermore, that the forward-swept Wing 5 ($\varphi = -45$ deg) has, as expected, with regard to drag almost the same aerodynamic quality as the swept Wing 4 ($\varphi = 45$ deg). Also Wing 6, which has in the freestream direction the same airfoil section as the unswept trapezoidal Wing 1, is shown to be strongly superior to the unswept Wing 1. From these results one may obviously conclude that the reduction in drag was reached by the sweep itself and less by the reduction of the thickness ratio of the airfoil section measured in the flight direction. The lift-vs-drag curves of these wings show that also at supersonic speeds a strong improvement in the wing aerodynamics can be achieved with increasing sweep angle φ.

The drag coefficients C_W (C_D) are plotted in Fig. 1.41 as function of the sweep angle φ with the *Mach* number M as parameter for the small C_A (C_L) values of interest in high-speed flight. The C_W (C_D) values strongly decrease at the higher *Mach* numbers with φ while at the lower *Mach* numbers 0.5 and 0.7 first an increase and only at higher sweep angles a decrease can be

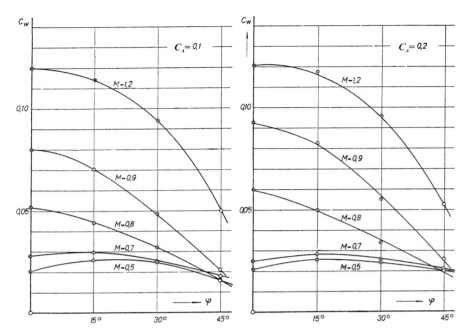

Fig. 1.41   Drag coefficient C_W (C_D) for Wings 1 to 4 dependent on the sweep angle φ at lift coefficients C_A (C_L) = 0.10 and C_A (C_L) = 0.20 for various *Mach* numbers (*H. Ludwieg*) [45].

observed. However, this, at first surprising increase in the drag coefficients, is only due to the unfavorable form of the wing tips of the Models 2 to 4, whose blunt wing areas are inclined under the angle φ with respect to the flow direction. This results in wake (suction) drag, which increases approximately proportional to sin φ while the effective velocity is proportional to cos φ and, therefore, drops only slightly at low φ-values. At low *Mach* numbers where the sweep effect is still small, the negative influence of the wing tips, therefore, outweighs at small φ-values the favorable influence of sweep.

Of course the question of how sweep affected the moments and the location of the pressure point was of special interest to the project aerodynamicists and, up to this investigation, completely unanswered. The lift coefficients C_A (C_L) are plotted in Fig. 1.42 vs the pitching moment coefficient C_m referred to the wing tip and to the largest airfoil chord; the parameter is the *Mach* number. The left graph applies to the trapezoidal Wing 1 ($\varphi = 0$ deg), the right one to the Wing 4 ($\varphi = 45$ deg). One recognizes that the moments for the trapezoidal wing at $M = 0.8$, and even more pronounced at $M = 0.9$, strongly deviate from the ones measured at $M = 0.5$ and $M = 0.7$. For the swept Wing 4, on the other hand, the curves for $M = 0.5, 0.7, 0.8,$ and 0.9 almost completely coincide, at least at small lift coefficients C_A (C_L).

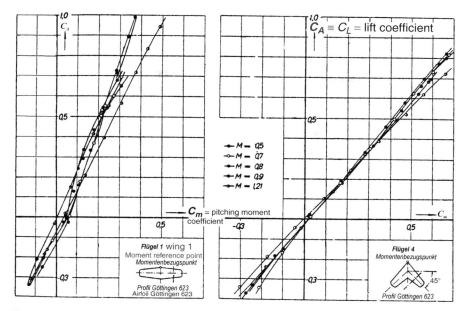

Fig. 1.42   Pitching moment curves for Wings 1 and 4 (H. Ludwieg) [45].

Concerning the development of the gradient $\delta C_A/\delta C_m$ ($\delta C_L/\delta C_m$), the following may be said: In incompressible flow, $\delta C_A/\delta C_m$ ($\delta C_L/\delta C_m$) for the infinite-span wing must decrease proportional to $\cos \varphi$, because the effective freestream velocity, and hence the stagnation pressure, decrease proportional to $\cos \varphi$ while the effective angle of attack increases with $1/\cos \varphi$.

In Fig. 1.43, the lift coefficients, converted to approximately the same aspect ratio, are plotted dependent on the angle of attack for the two *Mach* numbers $M = 0.5$ and $M = 0.8$. From the data at $M = 0.5$, which can be regarded as nearly incompressible, one clearly recognizes that $\delta C_A/\delta \alpha$ ($\delta C_L/\delta \alpha$) does not decrease with $\cos \varphi$ but remains nearly constant. Of course, this result can only apply to certain aspect ratios because for large aspect ratios one approaches more and more the conditions for the infinite-span sheared wing whose values for $\delta C_A/\delta \alpha$ ($\delta C_L/\delta \alpha$) must decrease proportional to $\cos \varphi$. At the *Mach* number $M = 0.8$, the gradient $\delta C_A/\delta \alpha$ ($\delta C_L/\delta \alpha$) for Wings 1 and 2 is, due to the compressibility effect, noticeably reduced. On the other hand, the strongly swept wings show still roughly the same values for $\delta C_A/\delta \alpha$ ($\delta C_L/\delta \alpha$) as in the case of incompressible flow.

1.4.2   *RESULTS AND THE AFTERMATH*

The results obtained during these measurements represent a breakthrough in high-speed aerodynamics. To use the new knowledge in the development of high-speed aircraft, special efforts were required to combine all research

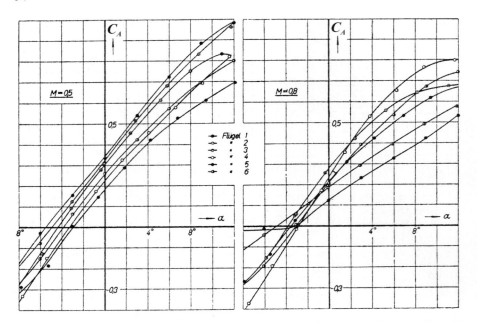

Fig. 1.43   Lift coefficient C_A (C_L) for Wings 1–6 dependent on the angle of attack α at *Mach* numbers of $M = 0.5$ and $M = 0.8$ (*H. Ludwieg*) [45] (legend: Flügel = wing).

activities at universities and research institutions and within the industry. This task of coordination was taken on by the RLM. In compliance with the most severe security regulations, the information exchange and the coordination of the work succeeded excellently, at least at the beginning of WWII, because all funding was also carried out via this institution. For example, strictly secret reports were specifically conveyed to the specialists by the Center of Scientific Publications (ZWB) of the General Air Material Master (Generalluftzeugmeister) in Berlin-Adlershof. Scientific coordination committees contributed considerably to the information transfer and the avoidance of duplicating efforts.

The RLM provided sufficient financial resources for the various research tasks. On the other hand, getting qualified personnel to staff the development departments of research and industry turned out to be more difficult as the war progressed. There were, in addition, increasing problems in procuring material and restrictions in providing energy, for example, for the large wind tunnels.

REFERENCES

[1] *E. Mach, P. Salcher*, "Photographische Fixierung der durch Projektile in der Luft eingeleiteten Vorgänge," Akademie der Wissenschaften, Wien, (IIa) XCV, 1887, pp. 764–780.

[2] E. Mach, L. Mach, "Weitere ballistische-photographische Versuche," Akademie der Wissenschaften, Wien, (IIa) IIC, 1889, pp. 1310–1326.

[3] B. St. Venant, L. Wantzel, "Mémoire et expérience sur l'écoulement déterminé par les différences de pression considérables," Journal de l'École Polytechnique, Cahier 27, 1839, pp. 85–122.

[4] B. Riemann, "Über die Fortpflanzung ebener Luftwellen von endlicher Schwingungsweite," Gesammelte Werke, 2. Auflage, 1892, pp. 156–181.

[5] W. J. M. Rankine, "On the Thermodynamic Theory of Waves of Finite Longitudinal Disturbances," Philosophical Transactions of the Royal Society, Vol. 160, London, 1870, pp. 277–288.

[6] H. Hugoniot, "Mémoire sur la propagation du mouvement dans les corps et spécialement dans les gases parfait," Journal de École Polytechnique, Cahier 57, 1887, pp. 1–97.

[7] P. Molenbroek, "Über die Bewegung eines Gases unter Annahme eines Geschwindig-keitspotentials," Arch. Math. Phys., Gruner Hoppe, Vol. (2) IX, 1890, pp. 157–195.

[8] C. A. Tschapligin, "Über Gasstrahlen," Wiss. Ann. Univ. Moskau Math. Phys., Vol. XXI, 1904, pp. 1–121.

[9] J. C. Rotta, "Die Aerodynamische Versuchsanstalt in Göettingen, ein Werk Ludwig Prandtls. Ihre Geschichte von den Anfängen bis 1925," Vandenhoek & Ruprecht, Göettingen, 1990, pp. 1–332.

[10] L. Prandtl, "Über die Flüssigkeitsbewegung bei sehr kleiner Reibung," Verhandlgn. d. III Intern. Math. Kongr. Heidelberg, Aug. 8–13, 1904, B.G. Teubner Verlag, Leipzig, 1905, pp. 485–491.

[11] H. Blasius, "Grenzschichten in Flüssigkeiten mit kleiner Reibung," Zeitschr. f. Math. u. Phys. 56, 1908, pp. 1–37.

[12] E. Boltze, "Grenzschichten an Rotationskörpern in Flüssigkeiten bei kleiner Reibung," Diss., Göettingen, 1908.

[13] K. Hiemenz, "Die Grenzschicht an einem in den gleichförmigen Flüssigkeitsstrom eingetauchten geraden Zylinder," Dinglers Polytechn. Journal 326, 1911, pp. 321–324.

[14] L. Prandtl, "Zur Theorie des Verdichtungsstoßes," Zeitschr. Ges. Turbinenwes. 3, 1906, pp. 241–245.

[15] L. Prandtl, "Neue Untersuchungen über die strömende Bewegung der Gase und Dämpfe," Phys. Zeitschr. 8, 1907, pp. 23–30.

[16] Th. Meyer, "Über zweidimensionale Bewegungsvorgänge in einem Gas, das mit Überschallgeschwindigkeit strömt," Mittlng. Forsch. Ing.-Wes. Heft 62, 1908.

[17] H. Schlichting, E. Truckenbrodt, "Aerodynamik des Flugzeuges," Bd. II, 2. Auflage, SpringerVerlag, Berlin, Heidelberg, New York, U.S., 1969.

[18] L. Prandtl, "Tragflügeltheorie, I. und II. Mitteilung," Nachrichten der Königlichen Gesellschaft der Wissenschaften Göettingen, Math. Phys. Klasse, 1918, pp. 451–477 and 1919, pp. 107–137.

[19] F. W. Lanchester, "Aerodynamics," London, 1907, "Aerodonetics," London, 1908, Deutsche Übersetzung: Bd. I and II, Berlin/Leipzig, 1909 and 1911.

[20] Th. von Kármán, L. Prandtl, "Zeitschr. f. Flugtechnik und Motorluftschifffahrt (ZFM)," 16, Heft 3, 1925, pp. 37–38.

[21] J. Ackeret, "Luftkräfte auf Flügel, die mit größerer als Schallgeschwindigkeit bewegt werden," Zeitschrift für Flugtechnik und Motorluftschifffahrt, 3. Heft, 16. Jahrgang, 1925, pp. 72–74.

[22] H. Glauert, "The Effect of Compressibility on the Lift of an Airfoil," Proc. Roy. Soc. Vol. CXVIII, London, 1928, pp. 113–119.

[23] A. Busemann, "Zeichnerische Ermittlung von ebenen Strömungen mit Überschallge-schwindigkeit," ZAMM, Bd. 8, 1928, pp. 423–42.

[24] A. Busemann, "Verdichtungsstöße in ebenen Gasströmungen," ed. by A. Gilles, L. Hopf, Th. v. Kármán, "Vorträge aus dem Gebiet der Aerodynamik und verwandte Gebiete," Verlag J. Springer, Berlin, 1929, pp. 162–169.

[25] L. Prandtl, A. Busemann, "Näherungsverfahren zur zeichnerischen Ermittlung von ebenen Strömungen mit Überschallgeschwindigkeit," Festschrift zum 70, Geburtstag von Prof. Dr. A. Stodola, Verlag Füßli, Zürich, 1929, pp. 499–509.

[26] G. A. Crocco (ed.), "Convengno di Scienze Fisiche, Matematiche e Naturali," Tema: Le Alta Velocita in Aviazione, Reale Accademia d'Italia, Fondazione Alessandro Volta, Sept. 30–Oct. 6, 1935, Rome, Italy, Verlag: Reale Accademia d'Italia, 1936, pp. 1–695.

[27] Schriftwechsel aus dem Nachlass von W. Tollmien, "Archiv zur Geschichte der Max-Planck-Gesellschaft," Berlin.

[28] G. I. Taylor, J. W. Macoll, "The Air Pressure on a Cone at High Speeds," Proceedings of the Royal Society A., Vol. 139, London, 1933, pp. 278–298.

[29] Th. von Kármán, N. B. Moore, "Resistance of Slender Bodies Moving with Supersonic Velocities with Special Reference to Projectiles," Tran. ASME, New York, U.S., 1932.

[30] L. Prandtl, "Theorie des Tragflügels im zusammendrückbaren Medium," Luftfahrtforschung 13, 1936, p. 313.

[31] A. Busemann, "Compressible Flow in the Thirties," Annual Review of Fluid Mechanics, Vol. 3, Jan. 1971, pp. 1–12.

[32] Th. von Kármán, "Die Wirbelstraße, mein Leben für die Luftfahrt," Hoffmann und Kampe Verlag, 1968, p. 263.

[33] Carlo Ferrari, "Recalling the 5th VOLTA Congress: High Speeds in Aviation," Annual Review Fluid Mechanics, Vol. 28, 1996, pp. 1–9.

[34] A. Busemann, "Aerodynamischer Auftrieb bei Überschallgeschwindigkeit," Vortrag auf der 5, Volta-Tagung in Rom, 1935, Luftfahrtforschung Bd. 12, Nr. 6, 1935, pp. 210–220.

[35] A. Busemann, O. Walchner, "Profileigenschaften bei Überschall," Forschung auf dem Gebiet des Ingenieurwesens, VDI Verlag GmbH, Berlin NW7, Bd. 4, Heft 2, 1933, pp. 87–92.

[36] Max Munk, "Note on the Relative Effect of Dihedral and the Sweep Back of Airplane Wings," NACA TN 177, 1924.

[37] R. T. Jones, "Properties of Low-Aspect Pointed Wings at Speeds Below and Above the Speed of Sound," NACA Report No. 835, 1945.

[38] R. T. Jones, "Wing Plan Forms for High-Speed Flight," NACA Report No. 863, 1945.

[39] Peter W. Wegener, "The Peenemünde Wind Tunnels: A Memoir," Yale University Press, New Haven, CT, U.S., 1996, p. 25.

[40] R. Hermann, "Der Kondensationsstoß in Überschall-Windkanaldüsen," Luftfahrtforschung, Band 19, Lieferung 6, 1942, pp. 201–209.

[41] L. Prandtl, "Die Rolle der Zusammendrückbarkeit bei der strömenden Bewegung der Luft," Deutsche Akademie der Luftfahrtforschung, Heft 30, 1937, pp. 1–16.

[42] A. Busemann, "Aufgaben der Hochgeschwindigkeitstechnik," Deutsche Akademie der Luftfahrtforschung, Heft 30, 1937, pp. 17–36.

[43] H. Ludwieg, "Geschichte des Pfeilflügels," DGLR-Jahrestagung 1981, Vortrag-Nr. 81–70, 1981, pp. 1–14.

[44] H. Strassl, H. Ludwieg, "Verringerung des Widerstandes von Tragflügeln bei hohen Geschwindigkeiten durch Pfeilform," AVA-Bericht 39/H/18, 1939, pp. 1–5.

[45] H. Ludwieg, "Pfeilflügel bei hohen Geschwindigkeiten," AVA, LGL-Bericht 127, 1940, pp. 44–52.

APPENDIX

Fig. A1.1 Lufo Volume 12, No 6/1935, pp. 210, 214, and 215 [34]; a) original document and b) translation.

a) ORIGINAL DOCUMENT

210 Luftfahrtforschung

Aerodynamischer Auftrieb bei Überschallgeschwindigkeit.

Von A. Busemann, Dresden.

Vorgetragen auf der 5. Volta-Tagung in Rom (30. 9. bis 6. 10. 1935).

Die guten Gleitzahlen der Tragflügel bei geringen Geschwindigkeiten verschlechtern sich mit der Annäherung der Fluggeschwindigkeit an die Schallgeschwindigkeit. Daher entsteht die Frage, ob man bei Überschallgeschwindigkeit wieder zu großen Auftrieben bei geringen Widerständen gelangt und wie ein solcher Auftrieb zu erzielen ist[1]).

Eine grundsätzliche Beschränkung wird den folgenden Überlegungen dadurch auferlegt, daß es sich stets um reine Überschallströmungen handeln soll, in denen keine Gebiete mit Unterschallgeschwindigkeit vorkommen. Bedingt ist dies dadurch, daß für die Strömungen mit gemischten Über- und Unterschallgeschwindigkeit nur ganz vereinzelte Lösungen bekannt geworden sind. Anderseits treten in einer Überschallströmung nur dann Gebiete mit Unterschallgeschwindigkeit auf, wenn man Körper mit stumpfen Vorderenden oder stark angestellte Körper verwendet. In der Nähe des Staupunktes an der Vorderseite dieser Körper erhält man Unterschallgeschwindigkeit und zugleich hohe Drücke, denen man an der Rückseite keine gleich hohen entgegensetzen kann. Es ist daher kaum anzunehmen, daß gerade in diesen gemischten Strömungen besondere Vorteile herauszuholen sind. Immerhin bleibt dies eine unbewiesene Behauptung, weil sich ein Beweis mit den der reinen Überschallströmung angepaßten Überlegungen naturgemäß nicht bringen läßt.

Gliederungen.

I. Ebene Überschallströmungen.
1. Zeichnerische Lösungen.
2. Strömung um die Ecke.
3. Auftrieb und Widerstand.
4. Oberflächenreibung.
5. Günstigste Gleitzahlen.
6. Einfluß der Profildicke.
7. Einfluß der Reibung.
8. Potenzreihe für die Druckdifferenz.
9. Aerodynamische Kräfte bei größeren Ablenkungen.
10. Die Bedeutung der höheren Glieder.
II. Zylindrische Strömungsfeder.
1. Schräg angeblasener Tragflügel.
2. Pfeilförmige Tragwerke.
III. Kegelige Strömungsfeder.
1. Besonderheiten der Überschallströmung.
2. Tragflügelenden.
3. Randwiderstand (induzierter Widerstand).
4. Differentialgleichung für kegelige Strömungsfeder.
5. Achsial angeblasene Kegelspitze.
6. Störung der Potentialströmung durch Verdichtungsstöße.
7. Kegelige Strömungsfeder an Tragflügelrändern.
IV. Zusammenfassung.
V. Schrifttum.

I. Ebene Überschallströmungen.

1. Zeichnerische Lösungen.

Die Behandlung ebener Überschallströmungen ist am weitesten entwickelt. Dies erklärt sich nicht allein durch die Erleichterungen bei Erniedrigung der Dimensionszahl. Vielmehr beruht der Fortschritt hier wesentlich auf der Tatsache, daß sich die Differentialgleichung der Gasströmung bei Beschränkung auf die Ebene linear schreiben läßt, wenn man eine Berührungstransformation auf die Ebene der Geschwindigkeitskomponenten anwendet (vgl. auch III, 4). Dies ist auch die Grundlage des zeichnerischen Verfahrens zur Verfolgung ebener Überschallströmungen, über das L. Prandtl auf der Tagung im Rahmen seines Vortrages besonders berichtete [1, 2]. Gerade die Bewegung von Körpern durch ruhende Luft liefert für dies Verfahren besonders angenehme Verhältnisse, weil die relativ zum Körper gleichmäßig zuströmende Luft saubere Anfangsbedingungen schafft.

---

[1]) Für die 5. Volta-Tagung, die »Hohe Geschwindigkeiten in der Luftfahrt« zum Gegenstand hat, wurde der Verfasser durch deren Präsident General Crocco aufgefordert, diese Frage zu beleuchten. Um unerwünschte Überschneidungen mit anderen Vorträgen dieser Tagung zu vermeiden, die zum Teil Versuchsanlagen und Versuchsergebnisse betrafen, wurden allein die aus der Theorie hervorgehenden Möglichkeiten behandelt.

Die eigentlichen zeichnerischen Lösungen würden erst notwendig, wenn eine gegenseitige Beeinflussung von mehreren tragenden Teilen zu beachten wäre. Tatsächlich gibt es auch gewisse Besonderheiten, die das Vorhandensein von mehreren Körpern voraussetzen. Man kann z. B. keinen zweidimensionalen Einzelkörper angeben, der bei endlicher Tiefe einen endlichen Querschnitt hat, ohne in der reibungslosen Überschallströmung einen Widerstand zu erzeugen. Für zwei derartige Körper kann man eine Strömung ohne Widerstand herstellen. Die einander zugewendeten Seiten der Körper müssen eben und parallel zur Anblasung sein. Die einander abgewendeten Seiten können dagegen durch besondere Formgebung so eingerichtet werden, daß jeder Körper die vom anderen erzeugten Störungen aufhebt. Indem die beiden Körper gegenseitig ihr Wellenfeld glätten, kommt in diesem Falle kein Wellenwiderstand zustande. Für das hier allein zu behandelnde Auftriebsproblem sind mir ähnliche Besonderheiten allerdings nicht bekannt. Ich möchte mich daher auf Einzelflügel beschränken.

2. Strömung um die Ecke.

Der einzelne Tragflügel schneidet mit seiner Vorderkante die ebene Strömung in zwei getrennte Bereiche, die sich erst an der Hinterkante des Tragflügels wieder vereinigen. Hier können sie aber keine Rückwirkung auf den Tragflügel ausüben, weil jede Wirkung in der reinen Überschallströmung auf den sich stromabwärts erstreckenden Machschen Kegel beschränkt bleibt. Das Wellenfeld im oberen Strömungsbereiche ist daher allein von der Gestalt der Oberseite des Tragflügels abhängig, ebenso das Wellenfeld im unteren Bereich von der Gestalt der Tragflügelunterseite. Dieser Fall entspricht aber genau den Voraussetzungen für die Anwendung der Prandtlschen Strömung »um die Ecke«. Bei dieser Strömung ist der statische Druck allein abhängig von der Richtung der Geschwindigkeit und daher der Druck an der Tragflügelfläche nur abhängig von der Neigung der einzelnen Flächenelemente. Mit Hilfe der von Th. Meyer berechneten Beziehungen zwischen Geschwindigkeit und Druck für die Strömung um die Ecke hat J. Ackeret zum erstenmal die Kräfte an Tragflügeln ermittelt [3, 4].

Für die nachfolgenden allgemeinen Untersuchungen ist es zweckmäßiger, statt der punktweise berechneten Beziehungen zwischen Druck und Neigungswinkel einen linearisierten Ausdruck zu verwenden, der bei Benutzung bestimmter Bezugsgrößen sogar von der Art des verwendeten Gases unabhängig wird. Die Drücke sollen mit dem Staudruck q der Strömung verglichen werden, der sich aus der Dichte ϱ des Gases, und der Strömungsgeschwindigkeit w in folgender Weise zusammensetzt:

$$q = \frac{1}{2} \varrho w^2 \quad \ldots \ldots \ldots \quad (1)$$

Bei geringen Geschwindigkeiten gibt der Staudruck den Überdruck am Staupunkt an. Bei hohen Geschwindigkeiten hat er eine endliche Bedeutung nicht mehr, er ist nur eine durch Definition festgelegte Bezugsgröße. Um die Höhe der Geschwindigkeit zu kennzeichnen, vergleicht man sie mit der Schallgeschwindigkeit c des gerade im ungestörten Strömung. Dies Verhältnis wird als Machsche Zahl M bezeichnet und ist bei Überschallgeschwindigkeit maßgebend für den Machschen Winkel α:

$$M = \frac{w}{c} = \frac{1}{\sin \alpha} \quad \ldots \ldots \ldots \quad (2)$$

Die Strömung um die Ecke liefert folgende Differentialbeziehung zwischen Druck p und Stromlinienwinkel β:

$$d\,p = \frac{\varrho\,w^2}{\sqrt{\left(\dfrac{w}{c}\right)^2 - 1}} \cdot d\,\beta \quad \ldots \quad (3)$$

$$c'_a = \left[(B_{1u}+B_{1o})C_1+(B_{2u}-B_{2o})C_2+(B_{3u}+B_{3o})\left(C_3-\frac{1}{2}C_1\right)\right]+\bar{\beta}\left[(B_{0u}+B_{0o})C_1+2(B_{1u}-B_{1o})C_2+3(B_{2u}+B_{2o})\left(C_3-\frac{1}{2}C_1\right)\right]$$

$$+\bar{\beta}^2\left[3(B_{1u}+B_{1o})\left(C_3-\frac{1}{2}C_1\right)\right]+\bar{\beta}^3\cdot2\left(C_3-\frac{1}{2}C_1\right)-\frac{1}{2}D\left[(\bar{\beta}+\beta_{vu}')^3+|\bar{\beta}+\beta_{vu}'|^3+(\bar{\beta}+\beta_{vo}')^3-|\bar{\beta}+\beta_{vo}'|^3\right]\ .\ (21\,\text{a})$$

$$c'_{i\!e} = \left[(B_{2u}+B_{2o})C_1+(B_{3u}-B_{3o})C_2+(B_{4u}+B_{4o})\left(C_3-\frac{1}{6}C_1\right)\right]+\bar{\beta}\left[2(B_{1u}+B_{1o})C_1+3(B_{2u}-B_{2o})C_2+4(B_{3u}+B_{3o})\left(C_3-\frac{1}{6}C_1\right)\right]$$

$$+\bar{\beta}^2\left[(B_{0u}+B_{0o})C_1+3(B_{1u}-B_{1o})C_2+6(B_{2u}+B_{2o})\left(C_3-\frac{1}{6}C_1\right)\right]+\bar{\beta}^3\left[4(B_{1u}+B_{1o})\left(C_3-\frac{1}{6}C_1\right)\right]+\bar{\beta}^4\cdot2\left(C_3-\frac{1}{6}C_1\right)$$

$$-\frac{1}{2}D\cdot\bar{\beta}\left[(\bar{\beta}+\beta_{vu}')^3+|\bar{\beta}+\beta_{vu}'|^3+(\bar{\beta}+\beta_{vo}')^3-|\bar{\beta}+\beta_{vo}'|^3\right]\ .\ .\ .\ .\ .\ .\ .\ .\ .\ (21\,\text{b})$$

Hierin bedeuten die Abkürzungen B_{0u} bis B_{0o} die auf die Profiltiefe bezogenen Integrale der durch den ersten Index angegebenen Potenz der Neigungen β_x' bzw. β_o' nach folgender Anweisung:

$$B_{nx}=\frac{1}{T}\int_0^{S_x}(\beta_x')^n\cdot ds_x\ .\ .\ .\ .\ .\ .\ (22)$$

mit $n=0, 1, 2, 3, 4$ und $x=u, o$.

Um auch an dieser Stelle wieder anschaulichere Werte anzugeben, sei das sichelförmige Kreisbogenprofil mit den Pfeilhöhen f_o und f_u herangezogen (vgl. Abb. 3). Die relativen Pfeilhöhen $\delta_o=\frac{f_o}{T}$ und $\delta_u=\frac{f_u}{T}$ ergeben in der Differenz das Dickenverhältnis $\delta=\delta_o-\delta_u$. (Für das früher angegebene symmetrische bikonvexe Profil gilt dann $\delta_u=-\delta_o$ und $\delta=2\,\delta_o$). Für dieses Profil und alle in bezug auf die Hochachse symmetrischen Profile verschwinden die Integrale der ungeraden Potenzen. Für die geraden Potenzen findet man:

$$\left.\begin{aligned}B_{0u}&=1+\frac{8}{3}(\delta_u)^2, & B_{0o}&=1+\frac{8}{3}(\delta_o)^2\\[4pt]B_{2u}&=\frac{16}{3}(\delta_u)^2 & B_{2o}&=\frac{16}{3}(\delta_o)^2\\[4pt]B_{4u}&=\frac{64}{5}(\delta_u)^4 & B_{4o}&=\frac{64}{5}(\delta_o)^4\\[4pt]\beta_{vu}'&=-4\,\delta_u & \beta_{vo}'&=-4\,\delta_o\end{aligned}\right\}\ \ldots\ (23)$$

Bei den geringen Auftriebsbeiwerten der Überschallprofile könnte es auch in Frage kommen, die Schubspannungen genauer in Auftrieb und Widerstand zu unterteilen. Man erhält dann

$$c_a''=-2\,c_f\cdot\bar{\beta}\quad\text{und}\quad c_w''=2\,c_f,$$

wenn man nicht noch mehr Glieder berücksichtigen muß.

10. Die Bedeutung der höheren Glieder.

Die Genauigkeit, die man mit den höheren Gliedern erreicht, rechtfertigt häufig ihre Anwendung nicht, weil in den Voraussetzungen schon größere Vernachlässigungen vorhanden sind (Beeinflussung der Grenzschicht durch die Druckdifferenzen usw.). Wenn hier doch noch die beiden nächsten Potenzen angegeben sind, so liegt der Wert darin, daß man an ihnen sieht, wann sich die besonderen Eigenschaften der verschiedenen Gase bemerkbar machen können und wann die Verdichtungsstöße in die Betrachtungen hineinzuziehen sind. Trotzdem gehört zu einer anständigen Verwendung jeder Näherungsrechnung, daß man die Grenzen ihrer Gültigkeit abschätzen kann. Für die im Anfang gebrauchte Näherung ist dies mit dem quadratischen Glied schon möglich. Geht man an die Berechnungen von Extremwerten, so sucht man die Stelle, an der sich die Einflüsse aller berücksichtigten Glieder für kleine Änderungen aufheben. An solchen Stellen möchte man gern wissen, ob man nun besser nach oben oder unten abweichen darf. Dies zeigen die höheren Glieder. Auch sonst kann der Einfluß der Glieder niedrigerer Ordnung verschwinden, während die höheren einen Beitrag liefern. Das quadratische Glied diente seinerzeit [5] zur Aufklärung des gemessenen Abtriebes am Kreisabschnittprofil für den Anstellwinkel Null. Der Verdichtungsstoß bleibt z. B. allein übrig beim nicht angestellten unendlich dünnen Kreisbogenprofil.

II. Zylindrische Strömungsfelder.

1. Schräg angeblasener Tragflügel.

In der eigentlichen ebenen Strömung sollen die Stromlinien in Ebenen verlaufen und alle diese Strömungsebenen durch Verschiebung normal zu diesen Ebenen ineinander übergehen. Verzichtet man auf die Bedingung, daß die Stromlinien in Ebenen verlaufen, so gibt es auch dann noch Strömungen, deren Zustände und Geschwindigkeitsvektoren bei der Verschiebung solcher Ebenen in der Richtung ihrer Normalen erhalten bleiben. Diese Strömungen verlangen zylindrische Wände als Grenzbedingungen wie die ebene Strömung. Man unterscheidet gewöhnlich nicht zwischen den ebenen und den zylindrischen Strömungsfeldern, weil die Potentialströmungen sich im zylindrischen Fall nur um eine konstante Geschwindigkeitskomponente senkrecht zur Ebene von den ebenen Strömungen unterscheiden. Man erhält demnach aus einer zylindrischen Strömung eine ebene Strömung, wenn man den Beobachter mit einer bestimmten Geschwindigkeit achsial bewegt. Alle Erscheinungen, die von der achsialen Geschwindigkeit der zylindrischen Grenzen unabhängig sind, bleiben dieselben wie bei der ebenen Strömung. Die Reibung in der Grenzschicht dagegen erfährt eine Veränderung.

Das zylindrische Strömungsfeld um den schräg angeblasenen Tragflügel (Abb. 4) kann man nach diesen Überlegungen soweit in eine ebene Strömung verwandeln, als es sich um die Berechnung der Druckkräfte auf den Tragflügel handelt. Die achsiale Geschwindigkeitskomponente fällt für die Erzeugung von Drücken völlig fort. Sie ändert jedoch die Bezugsgrößen der Strömung. Man muß bei einer Schräganblasung um den Winkel φ unterscheiden zwischen dem wirklichen Staudruck q_0 der Strömung und dem wirksamen Staudruck q, die die achsiale Komponente der Anblasegeschwindigkeit nicht einhält. Zwischen beiden besteht die Beziehung:

$$q=q_0\cdot\cos^2\varphi\ .\ .\ .\ .\ .\ .\ .\ .\ (24)$$

Genau in gleicher Weise gibt es eine wirkliche Machsche Zahl $M_0=\frac{w_0}{c}$ und daneben eine wirksame Machsche Zahl $M=\frac{w}{c}$ mit der Beziehung:

$$M=M_0\cos\varphi\ .\ .\ .\ .\ .\ .\ .\ (25)$$

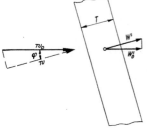

Abb. 4. Schräg angeblasener Tragflügel.

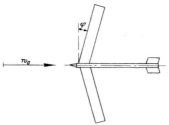

Abb. 5.   Pfeilförmiges Tragwerk.

Denn die Dichte ϱ und die Schallgeschwindigkeit c werden von der Schrägstellung des Tragflügels gegenüber der Windrichtung nicht betroffen.

2. Pfeilförmige Tragwerke.

Bei den ebenen Strömungen ergab sich, daß die besten Gleitzahlen bei bestimmten Machschen Zahlen erreicht werden, die wenig über der Schallgeschwindigkeit liegen. Es wäre bedauerlich, wenn damit das letzte Wort über die günstigsten Gleitzahlen überhaupt gesprochen wäre. Nun zeigt die Gleichung (25), daß sich die wirksamen Machschen Zahlen durch Schrägstellung der Tragflügel erniedrigen lassen. Es müßte daher lohnen, allgemein die pfeilförmigen Tragwerke (Abb. 5), auf ihre Gleitzahl bei Überschallgeschwindigkeit hin zu untersuchen.

Die Pfeilform der Tragwerke ist schon dadurch günstig, daß die Druckwirkungen in der Richtung des Auftriebes voll zur Geltung kommen, während sie in Richtung des Widerstandes nur mit einer Komponente in die Flugrichtung fallen (vgl. Abb. 4). Wenn man nun durch die Verringerung der wirksamen Machschen Zahl größere Flächenbelastungen bei gleichem Anstellwinkel oder gleiche Flächenbelastungen mit geringeren Anstellwinkeln erreicht, wird der Einfluß der Schubspannungen der Reibungsschicht relativ geringer. Absolut kann man natürlich die Reibung durch die Pfeilform nicht beeinflussen.

3. Berechnung des günstigsten Pfeilwinkels.

Um alle Überlegungen der ebenen Strömung um den Tragflügel, soweit es geht, verwenden zu können, soll das Profil und der Anstellwinkel unverändert senkrecht zur Tragflügelachse gemessen werden. Der ganze Tragflügel ist nur um den Winkel φ in die Pfeilstellung hineingeschwenkt (vgl. Abb. 5). Der Winkel φ ist damit der Winkel der Schräganblasung für den einzelnen Tragflügel. Änderungen gegen früher treten dadurch ein, daß die Auftriebs- und Widerstandsbeiwerte c_{w0} und c_{a0} auf den wirklichen Staudruck q_0 zu beziehen sind. Außerdem ist die Widerstandsrichtung neu zu orientieren. Da der Auftrieb allein von den Druckkräften herrührt, gilt für ihn:

$$c_{a_0} = c_a \cdot \cos^2 \varphi = 2\,C_1 \cdot \bar{\beta} \cdot \cos^2 \varphi \quad . \quad . \quad . \quad (26)$$

Denn c_a bezog sich auf den wirksamen Staudruck q. Ebenso ist C_1 auf die wirksame Machsche Zahl bezogen.

Die Änderung des Widerstandsbeiwertes ist leichter bei seiner Unterteilung auf Druck- und Reibungsanteil aufzustellen. Zunächst kann man den Reibungsanteil ganz unverändert hinschreiben:

$$c''_{w_0} = 2\,c_f \quad . \quad . \quad . \quad . \quad . \quad . \quad . \quad (27\,a)$$

Denn hier ist Betrag und Richtung von der Pfeilform unabhängig. Beim Druckwiderstand $W_0{}'$ ergibt die veränderte Orientierung $W_0{}' = W' \cdot \cos \varphi$, weil nur ein Teil des Druckwiderstandes in die Flugrichtung fällt (vgl. Abb. 4). Für c_{w0} kommt dann ebenso wie bei c_a die Unterscheidung von wirklichem und wirksamem Staudruck hinzu.

Daher erhält man nach Gleichung (6c):

$$c'_{w_0} = c'_{w} \cdot \cos^3 \varphi = C_1 \cdot \cos^3 \varphi\,(B_{2u} + B_{2o} + 2\,\bar{\beta}^2) \quad (27\,b)$$

Die Vereinigung der Anteile nach Gleichung (27a) und (27b) liefert:

$$c_{w_0} = 2\,c_f + C_1 \cdot \cos^3 \varphi\,(B_{2u} + B_{2o} + 2\,\bar{\beta}^2) \quad . \quad . \quad (28)$$

Der Faktor C_1 in den Gleichungen (26) bis (28) bedeutet:

$$C_1 = \frac{2}{\sqrt{M^2 - 1}} = \frac{2}{\sqrt{M_0{}^2 \cdot \cos^2 \varphi - 1}} \quad . \quad . \quad . \quad (29)$$

Nach diesen Abänderungen der Werte c_{a0} und c_{w0} zur Erfassung der Pfeilform kann man auch die Gleitzahl des pfeilförmigen Tragwerkes berechnen:

$$\varepsilon = \frac{c_{w_0}}{c_{a_0}} = \left[\frac{1}{2}\,c_f\,\frac{\sqrt{M_0{}^2 \cdot \cos^2 \varphi - 1}}{\cos^2 \varphi} \right.$$
$$\left. + \frac{1}{2}\,(B_{2u} + B_{2o}) \cos \varphi \right] \frac{1}{\bar{\beta}} + \bar{\beta} \cdot \cos \varphi \quad . \quad . \quad (30)$$

Bei Vernachlässigung der Reibung ergibt sich sofort eine Verbesserung, wie zu vermuten war. Die Verbesserung auch bei Berücksichtigung der Reibung kommt indirekt heraus, so daß man zunächst besser den günstigsten Anstellwinkel für die Pfeilform sucht. Sicher geht man nur dann in die Pfeilform, wenn man die günstigste Machsche Zahl des gerade angeblasenen Tragflügels überschritten hat, dann kann man aber den Anstellwinkel $\bar{\beta}_{\mathrm{opt}}$ erreichen, ohne die Grenze $\bar{\beta}_{\max}$ nach Gleichung (13) zu überschreiten. Für die Pfeilform gilt dabei:

$$\bar{\beta}_{\mathrm{opt}} = \sqrt{ \frac{1}{2}\,c_f\,\frac{\sqrt{M_0{}^2 \cdot \cos^2 \varphi - 1}}{\cos^3 \varphi} + \frac{1}{2}\,(B_{2u} + B_{2o}) } \quad (31)$$

Bei diesem Anstellwinkel erhält man die niedrigste Gleitzahl für gegebene Pfeilform:

$$\varepsilon_{\min} = \sqrt{ 2\,c_f\,\sqrt{M_0{}^2 - \frac{1}{\cos^2 \varphi}} + 2 \cos^2 \varphi\,(B_{2u} + B_{2o}) } \quad (32)$$

Solange der günstigste Anstellwinkel nicht durch die Bedingung Gleichung (13) unerreichbar ist, bekommt man auf jeden Fall bedeutende Verbesserungen der Gleitzahl. Nicht nur das von der Form abhängige Glied, sondern auch das Reibungsglied verringert sich beträchtlich. Schreibt man nun statt $\cos \varphi$ das Verhältnis $\dfrac{M}{M_0}$, so ergibt sich:

$$\varepsilon_{\min} = \sqrt{ 2\,c_f\,\frac{M_0}{M}\,\sqrt{M^2 - 1} + 2\,\frac{M^2}{M_0{}^2}\,(B_{2u} + B_{2o}) } \quad (32\,a)$$

Bei den höheren Machschen Zahlen M_0 kommt man nicht unbedingt unter den günstigsten Wert der Gleitzahl für die Strömung mit der Machschen Zahl M. Der Formeinfluß wird zwar geringer, aber der Reibungseinfluß wird etwas erhöht. Solange die Form jedoch entscheidend ist, kommt man sicher unter die Gleitzahl, die man bei der Strömung mit der Machschen Zahl M selbst erreichen würde.

Diese Betrachtung der Gleitzahlen könnte dazu verleiten, eine künstliche Erhöhung der Machschen Zahl als erwünscht anzusehen, um sie dann durch Pfeilform wieder erniedrigen zu können. So ein pfeilförmiger Autogiro bedeutet aber trotzdem keine Verbesserung. Denn es kommt ja nicht auf den Widerstand an, den der gewünschte Auftrieb verursacht, sondern es kommt auf die Leistung an, die erforderlich ist, um den Auftrieb erzeugen zu können. Multipliziert man die Gleitzahlen daher mit den Geschwindigkeiten, so fällt der scheinbare Vorteil sofort wieder weg. Diese Bemerkung gilt aber noch allgemeiner: Wenn man bei Überschallgeschwindigkeit wieder Gleitzahlen erhält, die mit denen bei geringen Geschwindigkeiten vergleichbar sind, so ist damit noch nicht gesagt, daß man sie technisch verwerten kann. Die kosten entsprechend den höheren Geschwindigkeit gesteigerte Leistungen, diese wieder größere Gewichte usw. Die technischen Aufgaben beginnen daher erst, nachdem jetzt die Größe der Gleitzahlen bekannt ist.

b) TRANSLATION

Aeronautical Research
Aerodynamic Lift at Supersonic Speeds
By A. Busemann, Dresden
Presented at the 5th Volta Congress in Rome (30.9. to 6.10.1935)

The good glide ratios of wings at low speeds deteriorate when the flight speed approaches the speed of sound. Therefore, the question arises whether one can attain at supersonic speeds again high lift at low drag and how such lift can be achieved.

A basic restriction will be imposed on the following considerations, namely, that we will always concern ourselves with pure supersonic flow without any embedded subsonic velocities. This is due to the fact that for flows with mixed supersonic and subsonic regions only very isolated solutions have become known. In supersonic flow, on the other hand, areas of subsonic flow only arise if bodies with blunt leading edges or bodies at high angles of attack are employed. In the vicinity of the stagnation point at the face of these bodies, one obtains subsonic velocities and simultaneously high pressures which are not compensated at the rear side with equally high pressures. It can, therefore, hardly be assumed that these mixed flows would bear any special advantages. Still, this remains an unproven claim because proof can naturally not be given by considerations pertaining to pure supersonic flow.

Contents

I. Two-dimensional Supersonic Flow
1. Graphical Solutions

The treatment of two-dimensional supersonic flow is developed further. This is not only explained by the relief due to the reduction to two dimensions. Rather, progress is due to the fact that when restricting oneself to two-dimensional flow, the differential equation of gaseous flow (gasdynamics) may be written in linear form if a transformation to the plane of the velocity components is applied (also compare III.4). The latter is also the basis of the graphical method for the determination of two-dimensional supersonic flow about which *L. Prandtl* reported separately in his lecture at this meeting. The movement of bodies through quiescent air especially provides agreeable conditions for this method because the uniformly approaching air, relative to the body, constitutes clean initial conditions.

The actual graphical solutions were only necessary if one had to account for the mutual interference of several load-carrying components. Indeed certain special features exist that require the presence of several bodies. For instance, one cannot define a two-dimensional single body that possesses a finite depth with a finite cross section without generating drag in inviscid supersonic flow. For two such bodies, one can generate a flow without drag. The sides of the body which do not face each other must be flat and parallel to the incoming flow. The sides facing each other may, on the other hand, be formed by special shaping in a way that each body cancels the disturbances generated by the other. By both bodies smoothing each other's wave fields, no wave drag will in this case be generated. For the lift problem, to be solely treated here, similar special features are, however, not known to me. Therefore, I want to restrict myself to the singular wing.

2. Corner Flow

A single wing divides with its leading edge the plane two-dimensional flow into two separate regions which join again at the trailing edge of the wing. Here, they cannot exert an effect on the wing since any influence in a pure supersonic flow remains restricted to the *Mach* cone extending downstream.

The wave field in the upper flow region is, therefore, only dependent on the shape of the upper surface of the wing. Similarly the wave field in the lower area is dependent on the shape of the lower surface of the wing. But this corresponds exactly to the prerequisites for the application of *Prandtl's* flow "about a corner." For this flow, the static pressure is only dependent on the direction of the velocity (vector) and the surface pressure is, therefore, only dependent on the slope of the individual surface elements. With the aid of the relation between the direction and the pressure of the flow around a corner, given by *Th. Meyer*, *J. Ackeret* determined for the first time the forces on wings.

For the following general investigations, it is expedient to employ, instead of the relation between the pressure and the slope computed point-by-point, a linearized expression which is, when using certain reference quantities, even independent of the kind of gas considered. The pressure should be referenced to the stagnation pressure q of the flow, which is composed of the density ρ of the gas and the flow velocity w as follows:

$$q = 1/2 \, \rho w^2 \tag{1}$$

At low speeds, the stagnation pressure gives the overpressure at the stagnation point. At high speeds, it has no longer this meaning but is only a defined reference quantity. In order to show the magnitude of the speed, one compares it to the speed of sound c of the gas at freestream conditions. The ratio is termed *Mach* number, M, and essentially determines, at supersonic speeds, the *Mach* angle α:

$$M = w/c = 1/ \sin \alpha \tag{2}$$

The flow about a corner yields the following differential relation between the pressure p and the streamline angle β:

$$dp = \frac{\rho w^2}{\sqrt{\left(\dfrac{w}{c}\right)^2 - 1}} \, d\beta \tag{3}$$

The final equations are given in the original report.

II. Cylindrical Flowfields
1. Obliquely Approached Wing

In an actual plane two-dimensional flow, streamlines should run in planes and all these flow planes brought to coincide by shifting them normal to each other. If the condition that streamlines run in planes is neglected, there still will be flows whose conditions and velocity vectors will be maintained when shifting such flowfields in the direction of their normal. These flows require cylindrical walls as boundary conditions similar to two-dimensional flow. One generally does not distinguish between plane two-dimensional and cylindrical flowfields, since the potential flow in case of the latter only distinguishes itself from a two-dimensional flow by a constant velocity component normal to the plane considered. One obtains, therefore, out of a cylindrical flow, a plane two-dimensional flow if the observer is moving with a certain velocity in an axial direction. All phenomena that are independent of the axial velocity of the cylindrical boundaries remain the same as in two-dimensional flow. The friction in the boundary layer will, however, experience a change.

The cylindrical flowfield around a wing in oblique flow (Fig. 4) can, according to these considerations, be converted into a two-dimensional flow as far as the computation of the pressure forces on the wing are concerned. The axial velocity component is totally unimportant for the generation of pressures. It changes, however, the reference quantities of the flow. In oblique flow under an angle φ, one must distinguish between the real stagnation pressure q_0 of the flow and the effective stagnation pressure q that does not contain the axial component of the freestream speed. The following relationship exists between the two:

$$q = q_0 \cos^2 \varphi \tag{24}$$

In exactly the same way, there is a real *Mach* number $M_0 = w_0/c$ and in addition an effective *Mach* number $M = w/c$ with the relation:

$$M = M_0 \cos \varphi \tag{25}$$

The density ρ and the speed of sound c are not affected by the obliqueness of the wing with respect to the flow direction.

2. Swept Wings

It was shown that in two-dimensional flow the best glide ratios were reached at certain *Mach* numbers slightly above the speed of sound. It would have been regrettable if here the last word about the most favorable glide ratios would have actually been spoken. Now, Equation (25) shows that the effective *Mach* number can be lowered by sweeping the wing. Therefore, it

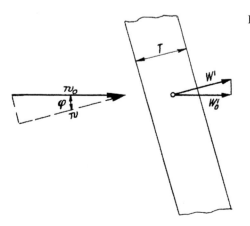

Fig. 4   Wing in oblique flow.

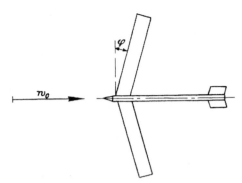

Fig. 5 Swept wing.

should be generally beneficial to investigate swept wings (Fig. 5) with respect to their glide ratios at supersonic speeds.

The swept form of the wings is already favorable due to the fact that the pressure fully acts in the direction of lift while in the direction of drag only the components in the flight direction are effective (compare Fig. 4). If now by reducing the effective *Mach* number, higher wing loading (surface loads) at the same angles of attack or the same wing (surface) loading at lower angles of attack is obtained; the influence of the shear stresses in the boundary layer becomes relatively smaller. In absolute terms, friction can, naturally, not be influenced by the form of the sweep.

3. Computation of the Most Favorable Sweep Angle

In order to be able to utilize, as far as possible, all considerations related to the two-dimensional wing, the airfoil section and the angle of attack shall still be measured normal to the wing axis. The entire wing is only turned by the angle φ into the swept position (compare Fig. 5). The angle φ is thus the angle of obliqueness of the incoming flow with respect to the single wing. Changes as compared to former considerations occur due to the fact that the lift and drag coefficients c_{w0} and c_{a0} have to be referred to the actual stagnation pressure q_0. Since the lift only arises from the pressure forces, the following relation holds:

$$c_{a0} = c_a \cos^2 \varphi = 2\, C_1\, \bar{\beta} \cos^2 \varphi, \qquad (26)$$

since c_1 was referred to the effective stagnation pressure q. Similarly is C_1 referred to the effective *Mach* number.

The change in drag coefficient is easier to be established when split into its pressure and friction parts. First, one may write down the friction part completely unchanged:

$$c_{w0}'' = 2\, c_f, \qquad (27a)$$

since here, amount and direction are independent of the form of the sweep.

For the pressure drag W_0' (D_0') gives the change in orientation in $D_0' = D' \cos \varphi$, since only a part of the pressure drag falls into the flight direction (compare Fig. 4). For c_{d0}' one has, similar to c_1', to account for the difference between the actual and the effective stagnation pressure. Therefore, one obtains [according to Eq. (6c); equation not shown]:

$$c_{d0}' = c_d' \cos^3 \varphi = C_1 \cos^3 \varphi \, (B_{2u} + B_{2o} + 2\bar{\beta}^2) \tag{27b}$$

(with the indices u and o referring to the lower and upper surfaces of the wing).

Joining the parts according to Equations (27a) and (27b) provides:

$$c_{d0} = 2c_f + C_1 \cos^3 \varphi \, (B_{2u} + B_{2o} + 2\bar{\beta}^2) \tag{28}$$

The factor C_1 in the Eq. (26) to (28) stands for:

$$C_1 = 2/\sqrt{M^2 - 1} = 2/\sqrt{M_0^2 \cos^2 \varphi - 1} \tag{29}$$

After these changes of the coefficients c_{10} and c_{d0} to account for sweep, one may also compute the glide ratio of the swept wing:

$$\varepsilon = c_{d0}/c_{10} = \left[\frac{1}{2} c_f \frac{\sqrt{M_0^2 \cos^2 \varphi - 1}}{\cos^2 \varphi} + \frac{1}{2}(B_{2u} + B_{2o}) \cos \varphi \right] \frac{1}{\bar{\beta}} + \bar{\beta} \cos \varphi \tag{30}$$

Neglecting friction, one immediately obtains an improvement, as assumed. The improvement is also indirectly obtained when considering friction so that the most favorable angle of attack for the form of sweep is better first sought. One will certainly select then the swept form only if the most favorable *Mach* number for the wing in two-dimensional flow has been exceeded; but then one may reach the angle of attack $\bar{\beta}_{opt}$ without exceeding the boundary $\bar{\beta}_{max}$ according to Equation (13) (equation is given in the original report [32, p. 211]). For a given form of sweep this results in:

$$\bar{\beta}_{opt} = \sqrt{\frac{1}{2} c_f \frac{\sqrt{M_0^2 \cos^2 \varphi - 1}}{\cos^2 \varphi} + \frac{1}{2}(B_{2u} + B_{2o})} \tag{31}$$

At this angle of attack, one obtains the lowest glide ratio for a given sweep:

$$\varepsilon_{min} = \sqrt{2c_f \sqrt{M_0^2 - \frac{1}{\cos^2 \varphi}} + 2\cos^2 \varphi (B_{2u} + B_{2o})} \tag{32}$$

As long as the most favorable angle of attack is unobtainable under the conditions of Eq. (13), important improvements in the glide ratio are achieved in any case. Not only the part dependent on the form but also the part related to friction is considerably reduced. If one now writes instead of cos φ the ratio M/M_0, one obtains:

$$\varepsilon_{min} = \sqrt{2c_f \frac{M_0}{M} \sqrt{M^2 - 1} + 2\frac{M^2}{M_0^2}(B_{2u} + B_{2o})} \tag{32a}$$

At the higher *Mach* numbers M_0, one does not necessarily fall below the most favorable value of the glide ratio for the flow with the *Mach* number M. Indeed, the influence of the form becomes smaller but the effect of friction is somewhat increased. However, as long as the form is decisive, one reaches glide ratios below the one that would be reached in a flow with the *Mach* number M itself.

This consideration of the glide ratios could lead people to regard an artificial increase in *Mach* number as desirable in order to then be able to reduce it by sweep. Such a sweeplike autogiro does still not mean an improvement. It is not the drag which is generated by the desired lift that is important but the power necessary to be able to generate this lift. If one, therefore, multiplies the glide ratios with the velocities, the apparent advantage immediately disappears. But this remark is even more generally valid—achieving again at supersonic speeds glide ratios comparable with the ones at low speeds does not mean that they are technically realizable. Corresponding to the higher velocities, they require increased power and hence increased weight, etc. The technical tasks, therefore, only commence after the magnitude of the glide ratio is now known.

Fig. A1.2   Invited participants of the 5th Volta Congress, September 30–October 6, 1935 in Rome (* lecturers); a) original document and b) translation.

a) ORGINAL DOCUMENT

PRESIDENZA DEL CONVEGNO

Presidente della Reale Accademia d'Italia

S. E. sen. march. GUGLIELMO MARCONI

Presidente della Classe delle Scienze Fisiche, Matematiche e Naturali

S. E. prof. GIANCARLO VALLAURI

Presidente del Convegno

S. E. ten. gen. prof. G. ARTURO CROCCO

PARTECIPANTI

Furono invitati a partecipare al Convegno gli stranieri:

ACKERET JAKOB (*Zurigo*).*
BÉNARD HENRI (*Parigi*).
BURGERS J. M. (*Delft*).
BUSEMANN ADOLF (*Dresda*).*
DÉVILLERS RENÉ (*Parigi*).
DOUGLAS G. P. (*South Farnbo-rough–Hants*).*
DUPONT PAUL (*Parigi*).
JACOBS EASTMAN N. (*Hampton Va.*).*
DE KÁRMÁN THEODOR (*Pasadena California*).*
MARGOULIS WLADIMIR (*Parigi*).
PRANDTL LUDWIG (*Göttingen*).*
PYE D. R. (*Londra*).

RICARDO HARRY R. (*Londra*).*
RININ N. A. (*Leningrado*).*
ROY MAURICE (*Reichshoffen–Bas Rhin*).*
STAINFORTH GEORGE HEDLEY (*Londra*).*
TAYLOR GEOFFREY INGRAM (*Cambridge*).*
TOUSSAINT ALBERT (*Saint–Cyr–l'École*).
VILLAT HENRI (*Parigi*).
WIESELSBERGER CARL (*Aquisgrana*).
WIMPERIS HARRY EGERTON (*Londra*).*
WITOSZYŃSKI CZESLAW (*Varsavia*).

b) TRANSLATION

President of the Congress
General Professor G. ARTURO CROCCO

PARTICIPANTS

ACKERET JAKOB (*Zurich*)*
BENARD HENRI (*Paris*)
BURGERS J. M. (*Delft*)

BUSEMANN ADOLF
  (*Dresden*)*
DEVILLERS RENE (*Paris*)

DOUGLAS G. P.
  (*South Farnborough-Hants*)
DUPONT PAUL (*Paris*)
JACOBS EASTMAN N.
  (*Hampton Va.*)*
DE KÁRMÁN THEODOR
  (*Pasadena Ca.*)
MARGOULIS WLADIMIR
  (*Paris*)
PRANDTL LUDWIG
  (*Göttingen*)*
PYE D. R. (*London*)

RICARDO HARRY R. (*London*)*
RININ N. A. (*Leningrad*)*
ROY MAURICE (*Reichshoffen Bas
  Rhine*)*
STAINFORTH GEORGE HEDLEY
  (*London*)
TAYLOR GEOFFREY INGRAM
  (*Cambridge*)*
TOUSSAINT ALBERT
  (*Saint-Cyr-l'Ecole*)
VILLAT HENRI (*Paris*)
WIESELSBERGER CARL
  (*Aix-la-Chappell*)
WIMPERIS HARRY EGERTON
  (*London*)*
WITOSZYNSKI CZESLAW
  (*Warsaw*)

Chapter 2

HIGH-SPEED FLIGHT AND ITS AERODYNAMIC AND GASDYNAMIC CHALLENGES

HANS-ULRICH MEIER

2.1 INTRODUCTION

Two important results concerning the realization of high-speed aircraft have already been treated in detail:

1935   *Adolf Busemann's* theory: Drag reduction in supersonic flow by sweeping the wing.

1939   At the suggestion of *Albert Betz, Hubert Ludwieg* conducted the first investigations on swept wings at high subsonic and supersonic speeds at the Aerodynamische Versuchsanstalt Göttingen (AVA) (Aerodynamic Test Establishment). The theory of *A. Busemann* was thus experimentally confirmed.

Prior to the introduction of swept wings, important development and research results were, however, still necessary for the design, construction, and testing of high-speed aircraft so that design requirements could be established, applied, and integrated into a network:

- Construction and calibration of new wind tunnels, as needed for the optimization of airfoils and wing designs, generation of data concerning load assumptions, and verification of designs and projects.
- Development and application of new test and measuring techniques for wind tunnel and flight tests.
- Improvement of the design methods for high-speed aircraft with swept wings. Here, new knowledge was required in aero- and gasdynamics as well as in flight mechanics on the basis of new wind-tunnel data.
- The area rule according to *Otto Frenzl.*
- Aeroelastic behavior of the swept wing in compressible subsonic and transonic flow to avoid flutter.
- Development of new turbojet engines up to production readiness.
- Development of new propulsion concepts.

In this chapter, the extent to which these requirements could be achieved up until May 1945 will be investigated. In our search for documents, it emerged that many original documents in German and foreign archives were no longer traceable or their access was very difficult. Therefore, important results of wind-tunnel tests, flight tests, and calculations are reproduced in this book to provide the reader with direct access to the research and development results. Also as many references as possible are being quoted to ease further, more detailed investigations.

2.2   TEST FACILITIES FOR THE EXPERIMENTAL INVESTIGATION OF HIGH-SPEED AIRCRAFT

2.2.1   REMARKS CONCERNING THE POLITICAL CONSTRAINTS

To be able to understand and judge the importance of the development and construction of new test facilities, a brief review of the historical development of German aeronautics is useful. Interestingly enough, no less a person than *Adolf Bäumker* [1] had prepared an extensive documentation that he presented to *Ludwig Prandtl* on the occasion of his 70th birthday on February 4, 1945. His report covered the complete time period from *Otto Lilienthal's* glider flights in Germany and the first powered flight of the Wright brothers in the United States in December 1903 up to the current situation during WWII in Europe.

Since 1927, *Adolf Bäumker* had been responsible for aeronautical research and development at the State Ministry of Transportation. After the takeover by the National Socialists in 1933, he became head of aeronautical research at the State Commissioner's Department of Aeronautics and subsequently branch head for aeronautical research within the Technological Office of the Reichsluftfahrtministerium (RLM) (State Ministry of Aeronautics). *Bäumker* can, therefore, be described as the first manager of the aeronautical sciences in Germany. A detailed biography of *Adolf Bäumker* has been written by *Katharina Hein* [2] in 1995 where the activities of *Bäumker* under the different political conditions, including the postwar period in the United States and the German Federal Republic, are described.

It is interesting that in May 1944 he had drawn up his report on the history of German aeronautics, classified as "secret," and, therefore, one may be under the impression that a statement of accounts for the postwar period should be presented. After the end of the war, he was able to use his management abilities in the United States in different military departments, an almost "sliding" transition from the National Socialist Regime to democracy! To be able to understand the intellectual atmosphere between 1933 and the end of WWII, some passages of the *Bäumker* report are quoted. Concerning the outcome of WWI he wrote:

> The German aeronautical technology possessed during all of WWI in many of the most important areas a creative superiority over its

opponents. The air superiority of the Allies was primarily based on the number of enemy aircraft. While Germany had built until the end of WWI 47,637 aircraft, the Allies possessed approximately 180,000 planes.

The years 1919–1932 can be considered the second stage of the development of aeronautical research. According to the Versailles Treaty, the development, construction, and procurement of powered aircraft were prohibited in Germany starting in 1918. The realization of important research results—particularly in the field of the aerodynamics—could, therefore, initially only be carried out with unpowered planes, which triggered, in 1920, on the "Wasserkuppe," a hilltop in the Rhön Mountains, an unprecedented triumphant success of the glider movement. In 1922, Germany was, internationally, in a leading position in this field. The first turning point in all of German aeronautics occurred in 1922 with the termination of the ban on civil aircraft construction previously laid down in the Versailles Treaty. *Bäumker* described the situation as follows:

> The impossibility of "choking" German energies was the cause for this slowly emerging development, the faith of all the people in the inexhaustible sources of her strength in the ultimate source of being. For the "People of the Middle"—that is, the middle of Europe—the geopolitical forces also acted with the elemental force of a natural phenomenon. All collapses created an increased will for reconstruction; all sufferings led to increased strength.

Important to the development of German aeronautical research was undoubtedly the fact that in Germany the self-government of science, as in the Kaiser-Wilhelm-Gesellschaft zur Förduring der Wissenschafter (KWG) (Emperor-Wilhelm Society for the Promotion of Sciences) had also in aeronautical research been chosen from the start as the operating principle. The head of the Deutsche Versuchsanstalt für Luftfahrt (DVL) (German Test Establishment of Aeronautics) in Berlin-Adlershof, *Friedrich Bendemann*, in 1912, succeeded in obtaining for the new research center the nonofficial character of a "Registered Association" to secure scientific liberty. From 1919, the AVA Göttingen also became the "Institute of Fluid Research" in the KWG, which also held AVA's more application-oriented part in trust. This was a management principle for research that *Bäumker* defended with surprising success up to the end of the war in 1945.

Bäumker verified in his documentation that the provision of state funds to aeronautical research was still very restricted despite the partial lifting of the restrictions of the Versailles Treaty, especially in comparison with the United States and European countries like France and England. Almost 50% of the direct research funds of approximately 20 million Imperial Marks were

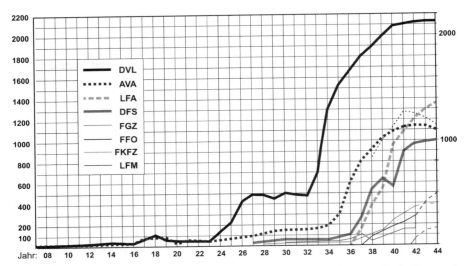

Fig. 2.1 Development of personnel within the various Aeronautical Research Establishments and Institutes between 1907 and 1942 [1] (source: *Bäumker* 1944).
Legend: DVL   German Test Establishment of Aeronautics, Berlin-Adlershof
          AVA   Aerodynamic Test Establishment, Göttingen
          DFS   German Research Institute for Gliding, Darmstadt
          LFA   Aeronautical Research Establishment Hermann Göring, Braunschweig
          FGZ   Research Establishment Count Zeppelin, Ruit near Stuttgart
          FFO   Air Radio Research Institute, Oberpfaffenhofen
          DVG   Radio-telegraphic and Air-Electricity Test Station, Gräfelfing
          LFM   Aeronautical Research Establishment, München

received by the DVL in Berlin from 1926 to 1932. However, these funds served primarily to maintain the staff and not to construct new facilities, as is indicated in Fig. 2.1, which shows the development in German aeronautical research. At this time, the United States had already started with the construction of new test facilities with the focal points of construction being Langley Field (Virginia) and Wright Field (Ohio). In England, new aeronautical research centers had been built at Farnborough and Teddington. France with Chalais-Meudon, Italy with Guidonia, and Russia with ZAGI followed. After the takeover of power by the National Socialists, the government could, nevertheless, fall back on a profound knowledge in the field of aerodynamics and gas dynamics, which had primarily been acquired at the AVA and DVL and the Institutes of Technology Aix-la-Chappell, Braunschweig, and Stuttgart. Here, the book of *J. C. Rotta* [3] should be noted, which comprehensively describes AVA's history from its beginnings in 1907 up to 1925.

The aerospace industry made increasing use of scientific methods of operation also. *Bäumker* wrote:

Men like *Junkers, Dornier, Rohrbach, Heinkel, Focke, Blume, Messerschmitt, Tank* and others drew the fundamentals of their creative work from general technical experience enabled by completed university studies and other professional activities. *Junkers* and *Focke* combined a lot of research-oriented ambition with a practical and realistic way of thinking. *Dornier* and *Rohrbach* were originally experienced in bridge construction and shipbuilding. *Heinkel, Focke, Blume* and *Tank*, originating from the first generation of wartime pilots, were from the very beginning devoted to aircraft construction. *Messerschmitt* was the first designer who started with gliding before coming to powered flight. The time of empiricism, which followed the times of the hobbyist before WWI, was over. Thus science, as supporter and forerunner, stepped into its own rights everywhere.

The development of aeronautical research from 1933 to 1939 can be characterized by two development cycles:

1) The first stage covered the time from February 1933 up to the declaration of the German air sovereignty and the establishment of the German Air Force in March 1935 with the execution of a crash program as a transitory measure.
2) The second stage from March 1935 up to the onset of the war in September 1939 and the declaration of the German air and military sovereignty was characterized "by the generous set up of a new branch of the sciences in a specific systematic procedure." In today's language, this means "System Technology."

Already in March 1933, *Bäumker* established a Zentrale für Wissenschaftliches Berichtswesen (ZWB) (Center of Scientific Publication) and founded in 1935 the Lilienthal-Gesellschaft für Luftfahrtforschung (LGL) (Lilienthal Society of Aeronautical Research) after the breakup of the Wissenschaftliche Gesellsehaft für Luftfahrt (WGL) (Scientific Society for Aeronautics). The scientific part of the Rhoen-Rositten Institute was integrated into the field of aeronautical research as Deutsches Forschungsinstitut für Segelflug (DFS) (German Research Institute of Gliding) and settled in Darmstadt-Griesheim.

In March 1935, the government issued the succinct order:

> German aeronautical research has to reach by, at the latest 1938, the state in its performance capabilities comparable to the leading foreign nations and, thereafter, take over the leadership in important areas .... Since Germany could at this late stage in no way come up with the enormous funds for aerodynamic research spent by foreign countries, she should strive to create a compensation through the construction of modern wind tunnels of a performance corresponding to the future development needs. The current increase of the flight speed shows that for the construction of new wind tunnels, importance should be attached to high flow velocities. The construction program for the new aeronautical research establishments in Germany should be set up so wide-ranging and in-depth that

advantages in the scientific knowledge in comparison to foreign countries would be secured for the future.

For the realization of this demanding requirement, *Bäumker* had almost unlimited financial means for investments into aeronautical research at his disposal. Initially, he attached priority to investments for the construction of new test facilities related to aerodynamics and gasdynamics. The requirements that arose in this area of aeronautics due to orders from industry soon grew such that the construction of their own wind tunnels of medium performance and medium or small nozzle diameters was urgently recommended to all large aircraft companies to solve their day-to-day questions. *Bäumker* expected to achieve a relief for research that way because the number of people employed in the aircraft industry had grown since the accession of power from 4000 to 300,000 in 1938.

It soon became very obvious to *Bäumker* that the realization of the enormous investment projects would not fail due to financing but because of a lack of sufficient professional personnel. With regard to his objectives, approximately 6000 people were needed for the set up of the research organization. So the idea arose to reproduce certain facilities of the DFL and AVA, even if in a changed form, at a different location. *Bäumker* (Fig. 2.2) selected Braunschweig as a location because the relation between research and teaching offered particularly favorable conditions. *Bäumker* had, in addition, the notion that a particularly close cooperation between the AVA and the newly founded Deutsche Forschungsanstalt für Luftfahrt (DFL) (German Research Establishment of Aeronautics), in 1938 renamed as Luftfahrt Forschungsanstalt Hermann Göring (LFA) (Aeronautical Research Establishment Hermann Göring), would be possible. The overall installation of the DFL/LFA as a new research center would distinguish itself by a generous setup. Here, a site, difficult to be seen from the outside, was selected

in order to be protected against observations by unauthorized persons from the ground and from the air. Yes, one wanted to be able to organize

Fig. 2.2 Foundation of the German Aeronautical Research Establishment (DFL), July 1935 (from left to right: *Wimmer*, *Bäumker*, *Hoffmeister*, DLR Archive Porz Wahn).

professional conferences without people attending seeing the total of the installations. This was especially of a certain importance with regard to foreigners visiting the research establishment.

The existence of the LFA in its entirety was actually not known to the Allies during the entire war.

At the suggestion of *Bäumker*, an organizational form for the new research establishment as Registered Association (Club) was selected, similar to the DVL. A Board of Directors was installed as the supervisory authority in accordance with the statutes. Chairmen of the board were *A. Bäumker* from 1936 to 1938, *Kurt Tank* from 1938 to 1942, and *Walter Georgii* from 1942 until the end of the war. Members of the board were, at different times, among others, *Willy Messerschmitt, E. O. Müller* (DVL), *F. Seewald* (DVL), *Carl Wieselsberger, E. Zindel* (Junkers).

Five institutes were set up within the DFL/LFA. The organizational structure is presented in detail in the commemorative publication *25 Years German Research Establishment of Aeronautics, e.V. (Registered Association), DFL, Braunschweig, 1936–1961* [4]; it is partly presented here:

Management (Director): DFL/LFA	*Prof. Dr. phil. Hermann Blenk* *Prof. Dr.-Ing. E. Schmidt* (Deputy)
Institute of Aerodynamics	*Prof. Dr.-phil. Hermann Blenk*
Institute of Gasdynamics	*Prof. Dr.-Ing. Adolf Busemann* Branch Trauen (Rockets and Jet Propulsion): *Dr.-Ing. E. Saenger*
Institute of Strength	*Prof. Dr.-Ing. B. Dirksen*
Institute of Engine Research	*Prof. Dr.-Ing. E. Schmidt*
Institute of Weapons Research	*Dr.-Ing. W. Thomé* (1936–1938) *Dr.-Ing. P. Hackeman* (1938–1942) *Prof. Dr.-Ing. Th. Rossman*
Management and Workshops	*Graduate Engineer (Dipl.-Ing.) G. Loew* (1936–1938)

The group photograph of 1937 (see Fig. 2.3) was taken by *Adolf Bäumker* on the occasion of visiting the LFA, and it shows all former heads of institutes of the LFA in Braunschweig. The commemorative publication of 1961 on the occasion of the 25-year existence of the DFL/LFA includes a list of the scientific collaborators. Their activities within the DFL/LFA up to the end of the war are briefly described, and information about their professional careers after 1945 is given. From this unique document follows that more than 70% of the executive personnel of the Institutes of Aerodynamics and Gasdynamics moved abroad after the war—predominantly to the United States. Because these were all well-known scientists, who were predominantly involved in the development of high-speed aircraft, they are listed in Table 2.1.

Fig. 2.3   Heads of institutes and guests of the DFL/LFA 1937; from left to right: Government Construction Councilor *Paul Horn*, *Hermann Blenk*, *Bernhard Dirksen*, *Ernst Schmidt*, *Adolf Busemann*, *Wilhelm Thomé*, *Hubert Schwenke*, *Gottfried Loew* (photograph by *A. Bäumker*, DLR Archive Porz Wahn).

TABLE 2.1 ACTIVITY AFTER THE END OF THE WAR, MAY 1945

Department	Scientist	Company
Institute of Aerodynamics	Head: *Prof. Dr.-phil.* *H. Blenk*	Ministry of Supply, Station Völkenrode
Department of Flight Mechanics	Head: *Dr.-phil. Dipl-Ing.* *G. Braun*	Wright Air Development Center, Dayton
Propulsion Aerodynamics	Head: Graduate Engineer *Eike*	Institute of Technology (TH) Berlin
Physics Laboratory	Head: *Dr.-phil. W. Kerris*	Wright Air Development Center, Dayton
Subsonic Aerodynamics	Head: *Dr.-Ing. Th. Zobel*	Wright Air Development Center, Dayton
Institute of Gasdynamics	Head: *Prof. Dr.-Ing.* *A. Busemann*	NPL Teddington, UK, 1947 NACA, USA
Department of Model Tests	Head: *Dr.-Ing. W. Knackstedt*	Wright Air Development Center, Dayton
Theoretical Gasdynamics	Head: *Dr.-Ing. G. Guderley*	Wright Air Development Center, Dayton
Branch Trauen	Head: *Dr.-Ing. E. Saenger*	Ministere de l' Air, Paris
Jet Propulsion	Head: *Engineer J. Winkler*	deceased

A comprehensive documentation of the personnel development of the research establishments AVA Göttingen and DVL in Berlin-Adlershof is, unfortunately, not available. It would, however, certainly provide a similar picture. To avoid shortfalls in the set up of the LFA, the provision of professionally qualified personnel was necessary. It was, therefore, laid down at the foundation of the LFA that all other research establishments within Germany had to commit themselves to prepare and train the first part of the scientific and technical personnel in their institutes and workshops for the new tasks. The rapid increase in personnel at the research establishments, AVA and DVL, dramatically accelerated between 1936 and the end of the war (see Fig. 2.1). A similar increase could be constituted for the DFS. *Hermann Blenk*, as former head of an institute, told the Commission of Inquiry of the British Intelligence Objectives Subcommittee (BIOS) in 1945: "concerning the approximately 1500 employees of the LFA, about 200 persons were detached for military training. Of the 150 scientific collaborators (scientist associates), who had a university education, only 50 were suitable for the execution of demanding research tasks." Because a part of these particularly qualified scientists were withdrawn from the AVA and the DVL and thus were lacking there, the insufficient staff at all research establishments becomes particularly obvious. In addition, highly qualified scientists were needed and employed for the formidable development of the test facilities. Important research tasks had, therefore, to be postponed or delayed.

Bäumker had tried to present the leading persons of the German aeronautical research establishments in a table (Fig. 2.4). Here, it must be remarked that this information is not complete [*Bäumker* had, for instance on the instruction of *E. Milch*, taken on the management of the Luftfahrtforschungsanstalt München (LFM) (Aeronautical Research Establishment München) himself]. *Bäumker* emphasized in his concluding considerations the major role which scientists like *Ludwig Prandtl, Wilhelm Hoff, Carl Wieselsberger*, and *Arthur Pröll* had played during the reconstruction of aeronautical research after WWI. He further explained:

> The year 1935 signaled a highlight in the intellectual concentration of the leading men of Aeronautical Research towards the new task. The year 1937 then brought the farthest-reaching organizational and work-related progress towards the newly set objectives. During the years of 1938 and 1939, the "large building" of the new research was completed in its essential parts, intellectually as well as materially; work within the extended frame commenced. WWII, starting in the autumn of 1939, interrupted these happy times of free and in its objectives independent research, aiming the intellectual work—to a slow but quite relentlessly increasing degree—at its own objectives and laws.

However, as pointed out herein, the completion of the "large building" was in a material sense for *Bäumker* in 1939 still a long way off. He initiated in

a)

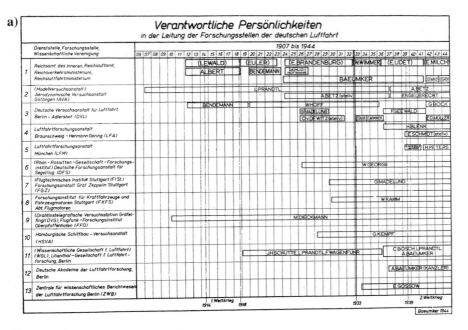

b)

Responsible Personalities
as heads of the Research Locations of German Aeronautics

Office, Research Location,
Scientific Association

1   State Office of the Interior, State Air Office, State Ministry of Transportation,
    State Ministry of Aeronautics
2   (Model-test Establishment)
    Aerodynamic Test Establishment, Göttingen (AVA)
3   German Test Establishment of Aeronautics, Berlin-Adlershof (DVL)
4   Aeronautical Research Establishment Hermann Göring, Brunswick (LFA)
5   Aeronautical Research Establishment Munich (LFM)
6   (Rhön-Rossitten Society Research Institute)
    German Research Institute for Gliding (DFS)
7   (Flight Technical Institute Stuttgart (FISt))
    Research Establishment Count Zeppelin (FGZ)
8   Research Institute of Motor Vehicles and Motor-vehicle Engines Stuttgart
    (FKFS), Aircraft Engines Branch
9   Radio-telegraphic and Air-Electricity Test Station (DVG), Air Radio Research
    Institute (FFO)
10 Test Establishment for Ship Building Hamburg (HSVA)
11 (Scientific Society for Aeronautics (WGL)), Lilienthal-Society for Aeronautical
    Research, Berlin
12 German Academy of Aeronautical Research, Berlin
13 Center of Scientific Publications concerning Aeronautical Research Berlin (ZWB)

**Fig. 2.4   a) Heads of the Research Institutes of German Aeronautics from 1907–1944
according to** *A. Bäumker* **[1] (see also Fig. 2.1); b) translation.**

1940 the LFM and planned wind-tunnel projects that were to outshine everything available up to that point.

The major test facilities that were important or should be used in the development of high-speed aircraft are introduced herein.

2.2.2 LOW-SPEED WIND TUNNELS

2.2.2.1 LARGE VARIABLE-PRESSURE WIND TUNNEL (TUNNEL K VI) OF THE AVA GÖTTINGEN.

A detailed description of the technical concept of the Wind Tunnel K VI of the AVA shortly after it was put into operation is given by H. Winter [5]. W. Krüger [6] documented the technical condition after the operative years 1936–1945. Thereafter, this test facility was dismantled. This tunnel consisted of an open-jet test section with a simple air return leg and two large interchangeable elliptical nozzles (see Fig. 2.5). The dimensions of the axes of the ellipses were 4×5.4 m and 4.7×7.0 m, respectively, with corresponding contraction ratios of 1:4.5 and 1:3. The tunnel could be operated at an overpressure of up to 3 bars primarily to investigate *Reynolds* number effects on airfoil flow. The vacuum operation had primarily been conceived for propeller investigations because, corresponding to the reduced air density, the power requirements of the propeller to be investigated was to be considerably reduced and, hence, the attainable *Mach* number increased. For this generation of the airflow, a single-stage fan had a 7.5-m diameter with 13 adjustable blades. The main drive motor allowed a continuous performance of 2200 kW, which still allowed a speed of approximately 60 m/s to be reached in the large test section. Special attention was paid to the flow quality, because, related to the design, the contraction ratio in case of a pressure tunnel was relatively small. Up to this day, stagnation pressure fluctuations present a problem in open-jet test sections, which has to be solved individually for each test facility. R. Seiferth [7], at that time head of the Institute of Wind Tunnels of the AVA, recognized the triggering cause as the formation of unsteady circular vortices at the nozzle exit. Because there is no useful photograph of Tunnel K VI at the AVA, the solution of the problem is shown with the aid of the installation of a nozzle ring in the LAF 8-m wind tunnel (Fig. 2.6), whose mode of operation R. Seiferth explains as follows:

> At lateral distances of approximately 15 cm, the rear half of the metal strips were cut to generate strips situated in flow direction. The sections thus generated are alternately bent towards the inside and the outside. One can imagine that the individual metal strips push the edge of the jet partly to the outside and partly to the inside. The vortices emanating from these little wings provide a strong mixing in the outer regions of the jet. The vortex rings emanating from the nozzle exit are more or less dissolved impinging, at least not as a whole, upon the edge of the collector.

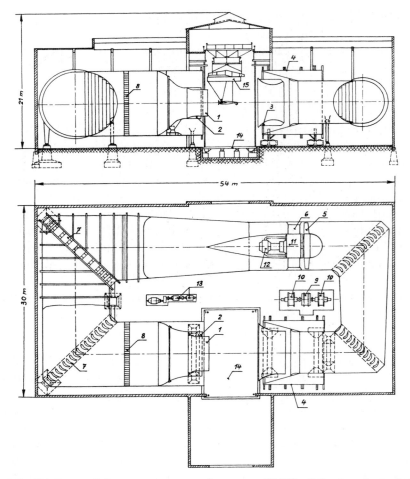

Fig. 2.5 Wind tunnel for overpressure and vacuum (K VI) of the Aerodynamic Test Establishment Göttingen (AVA) [6].

Legend: **1 Small nozzle 4.0 m × 5.4 m, elliptical**   **9 Synchronous motor (2900 kW)**
                2 Small nozzle 4.7 m × 7.0 m, elliptical   **10 Control generators**
                3 Collector (bell mouth)   **11 DC motor**
                4 Movable tube to change the test   **12 Ventilator for fan drive**
                      section   **13 Rotational compressor to charge**
                5 Fan   **and evacuate**
                6 Guide vanes (stator)   **14 Lifting platform**
                7 Water-cooled turning vanes   **15 Six-component balance**
                8 Flow straightener

The stagnation pressure fluctuations q are plotted vs the velocity for the small nozzle in Tunnel K VI (Fig. 2.7) without and with nozzle ring with the metal strips, as in Fig. 2.6, bent 40 deg outwards and 25 deg inwards. Due to the use of the nozzle ring, optimized for this wind tunnel, the stagnation pressure fluctuations could be considerably reduced.

Fig. 2.6 "Seiferth wings" to reduce pressure fluctuations; LFA 8-m wind tunnel. (There is no usable picture of tunnel K VI.)

Seiferth's most important investigation concerning the design of the nozzle ring for Tunnel K VI took place in early 1942, and the results were immediately thereafter realized in the large 8-m wind tunnel of the LFA in Braunschweig. Even if the largest part of the test results of the LFA was lost at the end of the war, many of the world's open-jet test sections have, in the meantime, been improved in their flow quality by so-called "Seiferth wings." Also related are *Seiferth's* computations of resonance cases on the basis of Helmholtz's vortex theorems where he compares the ring vortices periodically emanating from the nozzle wall with the vibration phenomenon of an open organ pipe.

A further important contribution to the improvement of the jet quality was made by *H. Bäuerle* [8] by optimizing the dimensional proportions between nozzle and collector (bell mouth). The testing technique for the investigation of the wind-tunnel turbulence with the aid of hot wires was considerably improved by *H. Schuh* [9]. Interesting also is the remark of *H. Schuh* that not

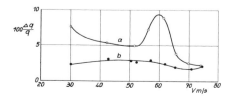

Fig. 2.7 Stagnation pressure fluctuations $q/\Delta q$ vs velocity in the cross section of the open jet of the small nozzle of the closed AVA-tunnel K VI [7]: line (a) without nozzle ring, line (b) with nozzle ring.

only the turbulence intensity (u'/U), but also the turbulence structure, as for instance the integral length scale as measure of the size of the eddies, has an important influence on laminar-turbulent boundary-layer transition. The correlation of drag measurements on spheres, carried out by *H. Seiferth* [10] in various wind tunnels, had, therefore, led to unsatisfactory results. This is a problem which has, up to now, neither been solved theoretically nor experimentally for wind-tunnel flows with low turbulence intensities.

2.2.2.2 5- × 7-m WIND TUNNEL OF THE DVL, BERLIN. In 1932 the DVL had started with the construction of the Göttingen-type 5- × 7-m open-jet wind tunnel and put it into operation for the first time at the end of 1935. The construction had already been initiated and approved by *Adolf Bäumker* prior to the assumption of power by the National Socialists in 1933. The dimensions of the tunnel arose from the set objective to investigate engines with air-cooled motors. In addition, wing investigations were to be realized at *Reynolds* numbers as high as possible. For these reasons, an elliptical cross section of the jet of 5 × 7 m was chosen, which could be enlarged, if required, to 6 × 8 m (see Fig. 2.8). The test facility was described in detail by *M. Kramer* [11] after it commenced operating; here, two features of this wind tunnel are emphasized:

- Due to the thorough design of the screens and the flow straightener in the settling chamber as well as the large contraction ratio, an excellent flow quality was achieved.
- The tunnel was equipped with an automatic six-component balance and test equipment for propellers and engines with a performance of up to 480 kW.

The six-component balance was accommodated in a room above the jet with a load-carrying structure taking up the forces of the model suspended by wires. All parts of the suspension system, including the wires, exposed to the jet were profiled in order to minimize drag. The drag of the suspension system, without the pretightening wires, was thus only about one-third of the minimum drag of the wing models. The six-component balance was, in principle, an electrically controlled moving-weight balance with considerably increased accuracy. For this balance, a remote transmission was developed that allowed a numerical registration of all test results.

2.2.2.3 8-m WIND TUNNEL A3 OF THE LFA, BRAUNSCHWEIG A summarizing report concerning the technical state of the Wind Tunnel A3 is also contained in the documentation of the Göttingen Monographs of *F. Ehlers* [12]. This test facility was built during 1937–1940 and put into operation at the end of 1940. The Göttingen-type open-jet wind tunnel had a circular nozzle with a diameter of 8 m and reached a maximum velocity in the test section of approximately 90 m/s at a drive performance of 8500 kW. Special constructive features are presented in the plan form (Fig. 2.9). This facility was primarily conceived

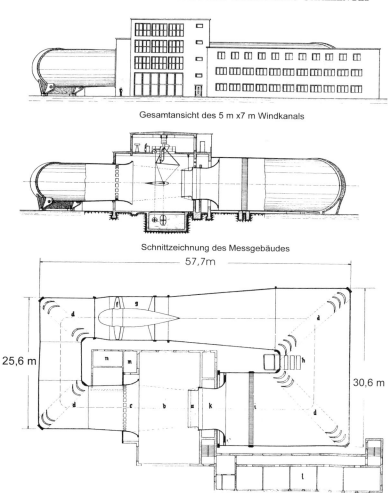

Gesamtansicht des 5 m x7 m Windkanals

Schnittzeichnung des Messgebäudes

a Vorsatzdüse   f Gebläseschaufeln   k Hauptdüse
b Messstrecke   g Leitschaufeln   l Bürogebäude
c Auffangtrichter   h Entlüftungsklappen   m Pumpenraum
d Umlenkschaufeln   i Gleichrichter   n Schalterraum
e Antriebsmotor

Fig. 2.8   5- × 7-m wind tunnel of the DVL, Berlin-Adlershof [11], general view, sectional view, and plan form.

Legend: General view of the 5- × 7-m wind tunnel; sectional view of the test building:

a. Ancillary nozzle   h. Ventilation flaps
b. Test section   i. Flow straightener
c. Collector (bell mouth)   k. Main nozzle
d. Turning vanes   l. Administration building
e. Drive motor   m. Pumps
f. Fan blades   n. Switches
g. Guide vanes

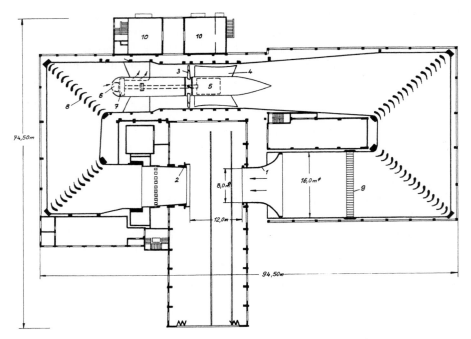

Fig. 2.9   8-m wind tunnel A3 of the LFA in Braunschweig [12], plan form.

for full-scale engine installations as well as for propeller and cooler tests. The size of the facility is clearly demonstrated in Fig. 2.10. A six-component balance was put into operation in the spring of 1942. It corresponded in essence to the suspension balance of the 5- × 7-m subsonic wind tunnel of the DVL in Berlin. Components of the Test Facility A3 were disassembled after WWII within the course of the operation surgeon (OS) and brought to the central spare parts store at Staughton Airfield, England.

Fig. 2.10   Full-scale Messerschmitt Me 109 in the 8-m wind tunnel A3 of the LFA (DLR Archive Porz Wahn).

2.2.2.4   SMALL SUBSONIC WIND TUNNELS OF THE AVA (TUNNELS I AND IV), THE LFA (TUNNEL A1), AND THE DVL (2.5-m TUNNEL).   These wind tunnels all had a similar basic structural concept because all three facilities were based on the so-called "Göttingen-type" design with open-jet test sections as had been introduced by *Ludwig Prandtl* with Tunnel I in 1917. The maximum attainable velocities depended primarily only on the installed drive power and amounted to *Mach* numbers of $M \approx 0.2$. For all three research establishments, they presented an important prerequisite for the execution of basic tests and preliminary tests in the case of complex industrial orders. The LFA Tunnel A1 was, in addition, intensively employed by the Institute of Fluid Mechanics of the Technische Hochschule (TH) (Institute of Technology) Braunschweig (Director *Hermann Schlichting*) for boundary-layer investigations and parametric studies on wing-fuselage combinations.

The smaller test facilities, which have played an important role in the development of the swept wing and the utilization of the large test facilities, are listed in Table 2.2. Measurement techniques, such as especially accurate drag balances, methods to determine laminar-turbulent boundary-layer transition, and methods for unsteady pressure and force measurements, were developed in these wind tunnels.

2.2.3   HIGH-SPEED WIND TUNNELS

2.2.3.1   11- × 13-cm HIGH-SPEED WIND TUNNEL OF THE AERODYNAMIC TEST ESTABLISHMENT (AVA), GÖTTINGEN.   All high-speed wind tunnels of Göttingen operated according to a similar intermittent test principle, as already suggested by *Prandtl* in 1912, where air was sucked from the atmosphere into a vacuum vessel (sphere). Concerning the high-speed wind tunnels according to *Prandtl's* principle, a detailed documentation was written by *Heinz Fütterer* [13] of the Göttingen archives. The main advantage of *Prandtl's* test arrangement is a nearly constant condition of the flow approaching the nozzle. A disadvantage compared to discharge from a pressure vessel is the relatively

TABLE 2.2   SMALL SUBSONIC WIND TUNNELS OF THE RESEARCH ESTABLISHMENTS
AVA, LFA, AND DVL

Research Center	Designation	Test Section (TS) Diam.	Velocity	Remarks
AVA	Tunnel I	open TS, d = 2.25 m	U = 55 m/s	1. WT of Type Göttingen
AVA	Tunnel IV	open TS, d = 1.25 m	U = 70 m/s	Propeller investigations
LFA	Tunnel A1	open TS, d = 2.50 m	U = 58 m/s	Basic tests
DVL	2.5 m Tunnel	closed TS, d = 2.5 m	U = 55 m/s	Basic tests

Fig. 2.11   KWI 6- × 7.3-cm wind tunnel without dryer. In the background is *O. Walchner* (photograph dated 1934, GOAR PS34-4).

small *Reynolds* numbers. Initially, the difficulties in the computation of the supersonic nozzles had to be overcome. Twenty years later, *Adolf Busemann* [15] succeeded together with *Prandtl*, in developing an approximate method for the graphical determination of two-dimensional supersonic flows based on a thesis of *Adolf Steichen* [14], which had been supervised by *Prandtl*. This was the basis for a pragmatic breakthrough in the computation of supersonic nozzles by the so-called method of characteristics.

Otto Walchner, who had taken on the activities of *Busemann* after the latter's change in 1931 to the TH Dresden, successfully tested in 1932–1933 a supersonic flow with a *Mach* number of $M = 1.47$ in a test section of 6×7.3 cm (Fig. 2.11). In 1935, *Walchner* changed from the Kaiser-Wilhelm-Institut (KWI) to the AVA. The KWI-tunnel served primarily as a pilot tunnel for a new, enlarged facility of the AVA. In 1939, an 11×13 cm subsonic/supersonic wind tunnel was put into operation at the AVA (see Figs. 2.12 and 2.13).

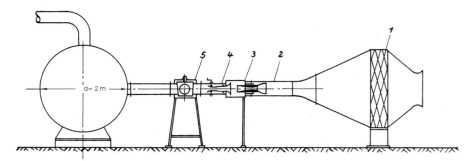

Fig. 2.12   Elevation sketch of the subsonic/supersonic wind tunnel of the AVA, cross-sectional area of the test section 11×13 cm (supersonic), 11×11 cm ($M < 1$) (DLR Archive Göttingen).

Legend: **1 Silica gel filter (Dryer)**

2 Nozzle box

3 Test chamber

4 Variable diffuser

5 Quick-opening valve

Fig. 2.13   11- × 13-cm high-speed wind tunnel with dryer (right) and quick-opening valve (left) (photograph dated 1939, GOAR F/41/114).

After opening the fast-acting valve (5 in Fig. 2.12), air was drawn from the atmosphere and first dried in a silica gel dryer (1 in Fig. 2.12). The air then streamed through a nozzle box (2 in Fig. 2.12), which was set up to accommodate various Laval nozzles ($M = 1.2$ to 3.2), or a nozzle for subsonic flow ($M = 0.5$ to $M < 1.0$), into the open-jet test section (3 in Fig. 2.12). A three-component balance was attached to the side of the associated test chamber. The air then entered the variable diffuser (4 in Fig. 2.12), which allowed, at supersonic speeds, the pressure in the test chamber to be adjusted such that the jet exhausted interference-free from the nozzle. In subsonic flow, the velocity in the test section was directly controlled by the diffuser by regulating the mass-flow rate in the minimum cross section (second throat) where sonic velocity occured. The air then discharged into a vacuum tank with a volume of approximately 150 m³. Dependent on the *Mach* number and the vacuum storage tank, the run time varied between 10 and 20 s. This short run time placed, of course, particularly high demands on the measuring technique. Therefore, a three-component balance was developed at the AVA that only needed a time of approximately 3 to 5 s for the measurement of one component of force. The model was attached to an angular lever and could, alternatively, be rotated about one of three axes (see Fig. 2.14). The rotation of the model caused by the resultant aerodynamic force could be undone by tightening a pair of springs. The forces exerted by the springs allowed the moment of the resultant aerodynamic force about three different axes to be measured which, in turn, allowed the determination of lift, drag, and pitching

Fig. 2.14   11- × 13-cm high-speed wind tunnel, test section with model (photograph dated 1940, GOAR F/41/112).

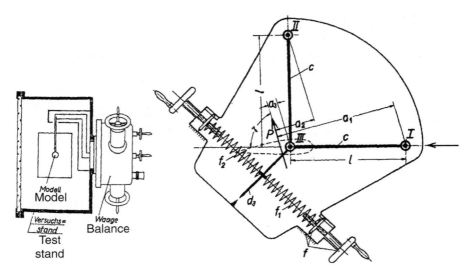

Fig. 2.15 **Schematic representation of the AVA three-component spring balance for high-speed measurements according to** *W. Froessel* **and** *E. Kunze***.**
Legend: c-Balance crossbars
I, II, III-Moment axes of rotation (Pivots)
f, f_1, f_2-Measuring equipment
d_3-Zero index

moment (see Fig. 2.15). A congenially simple but, for the time, accurate solution to a measuring problem.

- The first swept-wing measurements in this high-speed wind tunnel at high subsonic and supersonic speeds were carried out in the autumn of 1939 by *Hubert Ludwieg*.
- This high-speed wind tunnel was handed over in 1946–1947, together with vacuum pumps, controls, and balance, to *S. Goldstein* of the University of Manchester in England. Also included were all design drawings (129!), wiring diagrams, and parts lists. One set of the documents remained, at first, with the AVA in Göttingen to be able to respond to questions.
- *H. Fütterer* has found documents in the Göttingen Archives concerning the so-called OS which showed that, besides the knowledge (Göttingen Monographs), all important transportable facilities, instruments, drive systems, pumps, compressors, etc. had been brought to England in connection with this operation. Just for preparing the design documents, the drawing office had in August 1946, 34 employees: 8 graduate engineers (Dipl.-Ing.), 12 engineers, and 14 technical draftsmen. To be able to check back, initially 1430 blueprints in 37 folders were taken on order of the OS management on April 25, 1947, to the former LFA in Voelkenrode. The further destiny of these documents is not known!

2.2.3.2 21.5-cm DIAMETER WIND TUNNEL FOR HIGH-SUBSONIC FLOW OF THE AVA, GÖTTINGEN. After convincing experimental confirmations of *Busemann's* swept-wing theory, a second high-speed wind tunnel became necessary. Because, in addition to further wing measurements, most all of the contract measurements for the Messerschmitt Company on swept-wing configurations and for other companies investigations on projectiles, missiles, and bombs had to be carried out. Therefore, a second test section with a circular open jet with a diameter of 21.5 cm was connected to the existing vacuum tank in 1941. The setup follows from the elevation sketch shown in Fig. 2.16. For this facility, a new control of the diffuser in the form of a body of revolution was utilized. In this wind tunnel, *Mach* numbers of $M = 0.5$ to 0.95 were attained. This high-speed wind tunnel was, together with 169 design drawings, transferred to the National Physical Laboratory (NPL) in Teddington, United Kingdom.

2.2.3.3 25- × 25-cm HIGH-SPEED WIND TUNNEL OF THE AVA, GÖTTINGEN. In 1943, the construction of a further intermittent high-speed wind tunnel commenced (see Fig. 2.17). To achieve acceptable run times, the volume of the vacuum tank had to be increased to 400 m³, and the performance of the pumps increased to 600 kW. Corresponding to the higher mass-flow rate, a new dryer with two horizontally placed silica gel filters was also designed. As a further innovation for this tunnel, "honeycomb nozzles" for the control of the super-sonic flow at *Mach* numbers of $M > 2$ had been developed. They consisted of

Fig. 2.16 **21.5- × 21-cm wind tunnel of the AVA for investigations at high subsonic speeds, photograph and elevation sketch (DLR Archive Göttingen).**

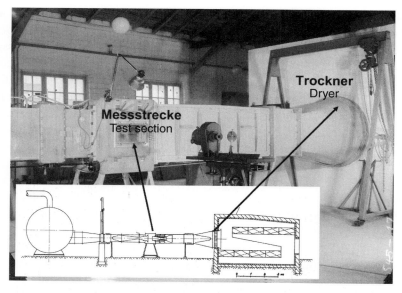

Fig. 2.17 25- × 25-cm high-speed wind tunnel of the AVA (photograph dated 1945, GOAR PS 45-17).

a nozzle box with the clear dimensions of the open jet. The box contained a large number of about 15-mm long Laval nozzles with an outer diameter of 5-mm. Using the honeycomb nozzles, a considerably better parallelism of the flow was obtained without any further corrective measures, however, in exchange for a much higher wind-tunnel turbulence level.

This wind tunnel was nearly completed but not put into operation. According to our enquiries, it was dismantled and within the scope of the OS set up again at the Royal Aeronautical Establishment (RAE) in Farnborough, United Kingdom.

2.2.3.4 STEAM-JET-DRIVEN HIGH-SPEED WIND TUNNEL OF THE AVA (0.765-m DIAMETER). The facility, conceived as ejector wind tunnel, worked according to the principle of the jet pump. Decisive for the choice of steam-jet propulsion was the demand for a high-speed wind tunnel that could be realized within a short time period, despite the delivery difficulties of the turbine industry existing at that time. Because the boiler plant of a potash mine in Reyershausen near Göttingen had been shut down due to flooding and was available for these purposes, the new wind tunnel was set up there (see Fig. 2.18) [16]. The air passed from the intake (1) through a flow straightener (2), a screen (3), and a nozzle (5) with a contraction ratio of 6:1 into the closed-wall test section (6) of 76.5-cm diameter. Thereafter, the air entered the diffuser with an adjustable drop-shaped body to control the flow velocity. Within the diffuser, the steam was injected through an annular nozzle (15)

Fig. 2.18   Steam-ejector driven high-speed wind tunnel of the AVA in Reyershausen [16].

driving the air through the diffuser into the open. Upstream of the nozzle, the steam pressure was reduced to about 2.5 to 4 bar. A pressure-proof chamber (7) contained the three-component balance.

Preparations for the test activities commenced in the autumn of 1943. At a test-section diameter of 0.765 m, *Mach* numbers of $M = 0.5$ to 0.9 at a maximum run time of approximately 3 min were reached. As a rule, only run times of 20 s were needed. Thus, 12 to 15 measurements per hour could be carried out in a permanent operation. Because the air sucked in could not be dried, disturbing condensation effects frequently occurred when testing in the higher *Mach* number range; these effects were eliminated by increasing the temperature of the incoming air to 40° C. This improvement was, however, not implemented any more. The short time to complete the facility during the war and its successful operation impressively document the technical improvisations used to find goal-oriented solutions to the investigation of high-speed aircraft.

Here, we point out the high-speed wind tunnel of the Junkers Company that was also conceived with a steam-jet drive system according to the ejector principle. The available test section cross-sectional area amounted to approximately 900 cm² and permitted a run time of approximately 3 min. Even if this facility did not belong to one of the research centers, important research results related to the development of high-speed aircraft have, nevertheless, been achieved; for example, by *Otto Frenzl* the so-called area rule that will be addressed by *W. Heinzerling* in Sec. 2.3.4.

2.2.3.5 HIGH-SPEED WIND TUNNEL (2.7-m DIAMETER) OF THE DVL, BERLIN. For the first time, the design of the high-speed wind tunnel of the DVL in Berlin-Adlershof was introduced in detail to all interested designers and scientists by *H. Matt* [17] on the occasion of the "High Speed" conference hosted by the Lilienthal Society at Göttingen in the autumn of 1940. In 1946, *H. Matt* [18] documented the technical condition of this facility at the end of WWII in the Göttingen Monographs.

The DVL high-speed wind tunnel was developed as a so-called Göttingen-type closed-return tunnel with the air flowing through a horizontally placed pipe system of 140 m in length (Fig. 2.19). The pipe system was in two areas bridged by buildings, including the test building and the machine room. To avoid dust development and to minimize skin friction along the walls of the pipes, all areas exposed to the flow were covered with a smooth hard glaze. Located downstream of the fan was a nonadjustable annular slit formed as a diffuser that allowed cooling of the tunnel air in conjunction with a mixing region. Here, part of the warm tunnel air was blown off and replaced in the mixing region by cold air. The fan was built as a two-stage, counter-rotating axial fan. Two dc motors served as drive of the fan, each one of them acting on one drive shaft. Their maximum performance amounted to 6500 kW per

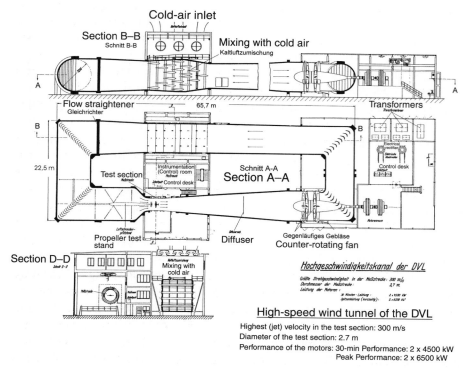

Fig. 2.19   High-speed wind tunnel (2.7-m diameter) of the DVL, Berlin-Adlershof [18].

motor. The armature voltage, employed to control the rotational speed of the motors, was provided by mercury-vapor rectifiers that were, in turn, fed by transformers. Three-phase current of 30 kV was supplied to the transformers with the voltage being kept constant within 1/10% of the set value. Both drives provided the same torque at each operational condition assuring that the spin of both fan stages was cancelled. This was a speed control of a high-performance facility exemplary at the time. By the way, the available power of 13,000 kW was never needed, not even for the largest of the installed models. The quality index of the wind tunnel η_p (the ratio of drive performance to flow energy) was for this DVL facility an average of $\eta_p = 0.1$, and thus substantially more favorable at that time than for all other wind tunnels.

Prior to the start of the operation of the wind tunnel, *B. Göthert* had already recognized that the reliability of the test results depended decisively on the quality of the corrections for blockage effects of the model due to the solid wind-tunnel walls. In this regard, an extensive calibration of the wind tunnel, similar to the determination of the *Mach* number distribution in the entire test section, was needed. To establish a comparison of the turbulence level in the test section with that of other DVL subsonic wind tunnels, tests with spheres were carried out determining the *Reynolds* numbers, Re_{krit}, for which the pressure of the undisturbed flow ($c_p \approx 0$) was still just reached at the back of the spheres. The critical *Reynolds* numbers thus determined were, without exception, still more favorable than those for the other subsonic wind tunnels of the DVL known for their low turbulence levels. This result was later confirmed by many exceptionally sensitive measurements on laminar airfoils and bodies of revolution.

Besides excellent basic investigations concerning the swept-wing development at the DVL, this facility was intensively used by the aeronautical industry because proven novel equipment, to be routinely used, was at their disposal. The measurements of the forces on wings and complete aircraft were carried out with three-component moving-weight balances with the wind-tunnel models mounted on two very slender forward-swept supports that introduced the lift and drag forces via levers into the balance (see Fig. 2.20). For the experimental determination of the considerably lower drag on symmetric bodies of revolution, a special test setup was developed where the disturbance of the flow about the model was minimized by an adapted model support (see Fig. 2.21). A similar installation was designed for drag measurements on wings, especially for a comparative series of wings having different sweep angles whose small drag differences put particularly high demands on the accuracy of the measurements. Independent of the direct force measurements, all drag measurements could be checked by momentum-loss measurements. The pressures, measured by a wake rake behind the model, were displayed by a Göttingen-type mercury-filled multiple-tube

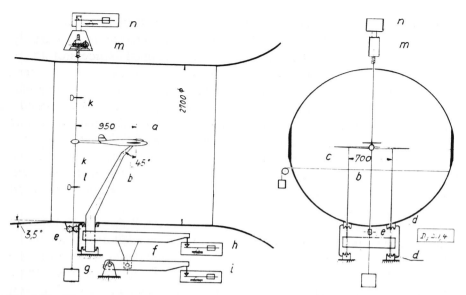

Fig. 2.20   Three-component force measurements in the DVL high-speed wind tunnel (2.7-m diameter) [18]; a = model, b = supports, c = 2-mm wire, d = membrane, e = guide reels, f = balance rods, g = spring joint, h = drag balance, i = lift balance, k = moment transmitting wire, l = oscillation damper, m = angle of attack, and n = pitching moment balance.

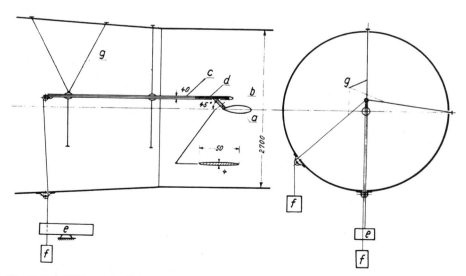

Fig. 2.21   High-resolution drag measurements on symmetric bodies of revolution in the DVL high-speed wind tunnel (2.7-m diameter) [18]; a = model, b = strut, c = support tube, d = sled, e = moving-weight balance, f = pre-load, and g = pre-stress.

pressure gage, and evaluated by means of a photographic-visual method, (also see *B. Göthert* [19]). To simplify this measurement procedure, the DVL developed a momentum-loss balance that proved itself successful in test operations (see Fig. 2.22). Here, each measured total pressure p_{t1} of each wake probe was led to a "diving bell" whose diameter d_f corresponded to the area covered by the associated measuring station d_s. As a reference pressure, the pressure p_t of the undisturbed flow was chosen. All diving bells were suspended from a moving-weight balance whose register display was a direct measure of the drag. With the aid of these directly indicating balances, it was possible to determine even the smallest drag changes that were, for instance, caused by a local upstream shift of the boundary-layer transition region due to surface disturbances.

A high-quality schlieren system [20] was also part of the equipment of the wind tunnel. The utilization of 600-mm diameter field mirrors, and thus a wide field of vision, and a far-reaching automation of the technique to always take four differently exposed photographs with a horizontal and a vertical aperture made this schlieren system, delivered by the Carl Zeiss Company located in Jena, Germany, an important aid in the identification of shock phenomena. The wind tunnel was thus excellently equipped for both research tasks as well as the efficient execution of industry orders.

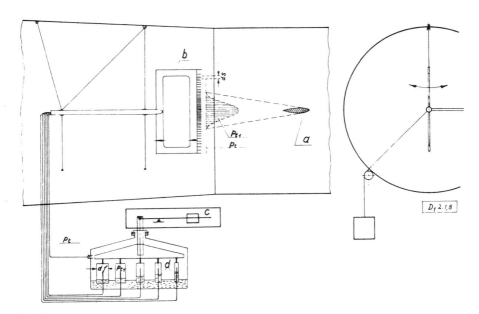

Fig. 2.22 Momentum-loss balance for wake measurements in the DVL high-speed wind tunnel (2.7-m diameter) [18]; a = wing model, b = momentum rake (wake rake), c = moving-weight balance, and d = diving bell, p_{t1} disturbed total pressure, p_t (undisturbed) total pressure.

2.2.3.6 HIGH-SPEED WIND TUNNEL A2 (2.8-m DIAMETER) OF THE LFA, BRAUNSCHWEIG. Detailed documentation is available by *Th. Zobel* [21] in 1943 and by *H. Matt* [22] in the Göttingen Monographs (1946). While the 2.7-m diameter high-speed wind tunnel of the DVL was essentially intended for airfoil and swept-wing measurements, due to its excellent flow quality, especially well suited for these tasks, the high-speed wind tunnel A2 of the LFA Braunschweig was, in addition, to allow the investigation, for instance, of jet engines and their installation on aircraft. Here, considerable amounts of fresh air were needed for the combustion process. In addition, there was such a strong heat and smoke development during the measurements that a closed-return operation of the facility was impossible. Therefore, the possibility of an "Eiffel-type operation" besides the closed-circuit operation, could also be used. For this mode of operation, the selected arrangement of the air ducting within a vertical plane was particularly suitable because, when drawing outside air, contamination could be kept at acceptable limits due to the large ground clearance. By moving back the upper corners and the roof, indicated in Fig 2.23 by the dashed line, the full cross section of the vertical leg could be used to blow off or draw in fresh air. The photograph in Fig. 2.24 shows the test section with the "schlieren windows." The test section was constructed as a closed steel cylinder (circular cross section) of 2.8 m in diameter and a length of 4 m. The contraction ratio of the nozzle was 1:6.8 and thus similar to the one of the DVL 2.7-m high-speed wind tunnel.

The drive of the fan was set up similar to that of the DVL facility employing two dc motors each with 6000-kW peak performance. Analogous to the trial of the DVL 2.7-m high-speed wind tunnel, the wind-tunnel quality index η_p, i.e., the ratio of the drive performance/flow energy, was also determined for the LFA A2 tunnel. The bad wind-tunnel quality indices were striking, for example at a *Mach* number of $M = 0.85$ with $\eta_p = 0.16$ almost twice as high as those for the DVL facility (see Fig. 2.25). Hence, almost double the drive power was needed to reach the same *Mach* number

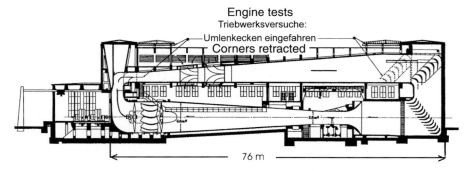

Engine tests
Triebwerksversuche:
Umlenkecken eingefahren
Corners retracted

76 m

Fig. 2.23 High-speed wind tunnel A2 (2.8-m diameter) of the LFA [22, 23], Braunschweig.

Fig. 2.24   View of the test section of the LFA high-speed wind tunnel A2 (DLR Archive Porz Wahn).

in the LFA test section without models. It was found, as an essential reason for this, that the flow in the diffuser of the LFA A2 facility was separated over its entire length. Considerable losses occurred, in addition, due to the extremely unfavorable flow ducting and a separation of the flow immediately downstream of the fan. This caused considerable relatively low-frequency pressure fluctuations, which had already led during the start of the operation of the facility to great difficulties (blade fracture on the fan). Until these technical difficulties were solved (end of 1941), the A2 tunnel was, therefore, only to be operated at smaller *Mach* numbers. The still relatively poor flow quality disqualified this facility to a large degree for basic tests concerning the swept-wing development. Further reasons for the latter were based on the bad *Mach* number distributions in the test section. As indicated by the plot of the pressure distributions along the tunnel wall shown in Fig. 2.26, the relatively large differences in the wall pressures at the individual measuring stations could be explained by the waviness of the wall of the cylindrical test section that showed nearly the same development as the pressure distributions. On the other hand, the utilization of the high-speed wind tunnel A2 for engine investigations, including engine-integration measurements, proved extremely successful because for these

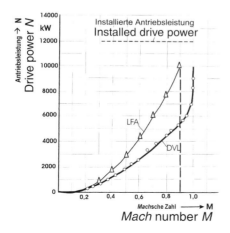

Fig. 2.25   Drive power required for the empty test section up to the speed of sound ($M = 1$); DVL high-speed wind tunnel (2.7-m diameter) [11], LFA high-speed wind tunnel A2 (2.8-m diameter) [15].

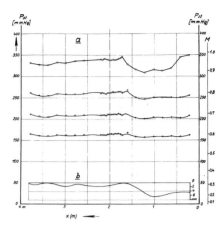

Fig. 2.26 Pressure distributions along the tunnel wall (p_{st} = static pressure) [22]; a = pressure distribution, b = waviness of the wall.

investigations the turbulence intensity was of minor importance. Thus, important investigations concerning the development of the Cruise Missile Fi 103 (V1) were, for instance, carried out in the "Eiffel-type configuration" of this facility on original engines for the Argus and Fieseler Companies. In addition, the aerodynamic investigations for the two antiaircraft rockets "Fire Lily" and "Pike" were performed in this facility. It must, nevertheless, be realized that the high-speed wind tunnel A2, developed at large financial and technical expense, had some conceptual faults, which could have been removed only with considerable effort. Regarding the situation, *Matt* [23] has summarized his suggestions for improvements in 1946 in a report to the Ministry of Aircraft Production. That proposal, however, cannot be found today.

Concerning the instrumentation of tunnel A2, besides the three-component standard balance, which was of similar design and setup as the one of the DVL, especially the interference refractor—in today's language, "interferometer"—developed by *Zobel* [24], should be pointed out. A picture from an interferometer essentially shows lines of equal pressure within the flowfield that can, with the aid of some reference pressures directly measured within the flowfield, also be quantitatively evaluated. His objective to replace pressure distribution measurements by this visual method, thus considerably reducing tunnel occupancy time by employing simpler models for airfoil and wing measurements, could, however, not be realized due to the considerable vibrations of the tunnel and tunnel dimensions being too large.

Still to be pointed out is the further development of piezoelectric methods to measure large and unsteady forces. Through the development and construction of suitable quartz elements with relief devices to prevent charge fading, very high pressure peaks and high-frequency oscillations could be measured, for example as in propeller investigations.

High-speed wind tunnel A2 was—as was tunnel A3—dismantled after WWII. Components such as the drive system were shipped within the framework of the OS to the central spare-parts store at Staughton Airfield, England.

2.2.3.7 HIGH-SPEED WIND TUNNELS A6 AND A7 OF THE LFA, BRAUNSCHWEIG.

High-speed wind tunnels A6 and A7 were conceived and built by *Adolf Busemann* after his appointment as director of the Institute of Gasdynamics of the LFA in 1936. *W. Knackstedt* [25] has described these two facilities in detail in 1946 on orders of the British Military Government in the Göttingen Monographs. The two facilities were arranged such that the entire machine plant and a vacuum tank (spherical vessel with a capacity of 1000 m^3) could be used jointly.

Wind tunnel A6 (Fig. 2.27) operated according to the Prandt principle and drew atmospheric air into a vacuum tank (d) achieving a maximum *Mach* number in the test section of $M = 0.94$. A radial fan evacuated the air in the spherical vessel (d) within four minutes to about 0.10 bar. Connected to this vessel via a long conical pipe (q) and a quick-opening revolving gate valve (t) was the test section (r). The latter was connected via a 70-cm diameter pipe to a funnel-shaped intake (s) of about 2.5×3.5 m, which provided the connection to the free atmosphere. The test section had a width of 40 cm and a height of 36 cm. For the investigation of thick airfoils at high angles of attack, a nozzle of 30-cm width and 60-cm height could be used. While the side walls were arranged parallel, the height of the test section widened by about 5 deg to compensate the displacement effect of the boundary layer on the nozzle walls. The velocity was controlled by a variable diffuser located in the flow direction upstream of the quick-opening valve (t), such that the sonic velocity occurred just within the minimum cross section as long as the difference in pressure between the atmosphere and the vacuum tank remained above the critical one. Corresponding to the cross section of the variable diffuser, a steady flow of constant velocity was established in the test section as long as the velocity in the diffuser remained sonic. A disadvantage in the operation of the tunnel was the fact that the air drawn in could not be dried as in the case of the Göttingen high-speed wind tunnel. The influences on the precision of the test results due to condensation effects could thus not be taken into account.

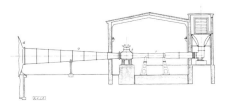

Fig. 2.27 LFA high-speed wind tunnel A6 ($M < 1$) [25].

Fig. 2.28 Model of an Arado Company high-speed aircraft in the LFA A6 tunnel (DLR Archive Porz Wahn).

Lift and moment measurements in the *Mach* number range $M = 0.5$ to 0.8 were carried out with a complete model of the aircraft AR 240 in 1940 under contract to the Arado Aircraft Manufacturer GmbH (limited liability company). The model shown in the test section of high-speed wind tunnel A6 (Fig. 2.28) was a first for the new central workshop of the LFA in Braunschweig. For these and further contract measurements, the LFA delivered uncorrected test results to the customer. The latter then had the wind-tunnel corrections performed at the DVL in Berlin. This procedure clearly indicates that the cooperation between the LFA and the DVL was not optimal despite all efforts of coordination by the RLM.

Supersonic wind tunnel A7 with a test section of 25×25 cm could be operated continuously, and the contour of the supersonic diffuser could be continuously adjusted. The mode of operation was generally the same as for tunnel A6 (Fig. 2.29). The test section was closed with solid plane side walls and top and bottom walls comprised of nozzle blocks mounted in between to achieve supersonic flow and designed by the method of characteristics of Prandtl–Busemann. Nozzle blocks for the *Mach* numbers $M = 1.4$, 1.8, 2.3, 2.5, and 3.0 were available. The quality of the nozzles was checked with the aid of schlieren pictures of a circular cone of 40-deg cone angle, which was moved along the tunnel axis. The angle of the shock waves thus determined allowed the control and correction of the velocity distribution along the tunnel axis.

For the optical observation of the flow in the test section, an interference refractor according to *Mach-Zehnder* (*Mach-Zehnder* interferometer) was available, which together with the Zeiss-Jena Company, *Zobel* had further developed to a high technical perfection. The tunnel A7 possessed a three-component balance of Göttingen design (see Fig. 2.15).

The supersonic wind tunnel was dismantled by the Allies after WWII and continued operating at the University of Bristol. After the shutdown of the facility, the test section of the tunnel (see Fig. 2.30) could be acquired by the

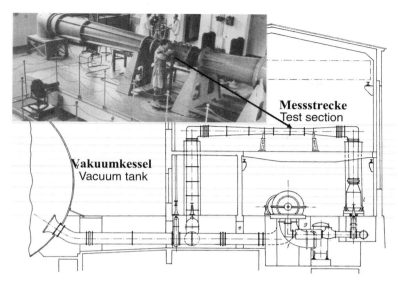

Fig. 2.29   LFA high-speed wind tunnel A7 ($M > 1$) [25].

German Museum in München on the initiative of *Heinzerling*. It may be visited there as a particularly interesting exhibit. The Zeiss "interference refractor" of *Zobel* was brought back to Germany in 1986 as a gift of the Surrey University to the German Museum.

2.2.3.8 HIGH-SPEED WIND TUNNELS A9a AND A9b OF THE LFA BRAUNSCHWEIG. Both wind tunnels, A9a and A9b, were designed and constructed at the Institute of Gasdynamics headed by *Adolf Busemann*. Here, the experiences gained with the facilities A6 and A7 provided an important basis. The tunnels were to supplement the existing high-speed wind tunnels in Germany. This also has been the subject of a detailed documentation by *Knackstedt* [26] in 1946 for the Göttingen Monographs. Both facilities represented at the time, in construction and design, an absolute novelty.

The facility A9a was conceived as a continuous open-jet high-speed wind tunnel (see Fig. 2.31). For special investigations, such as on jet engines, the A9a tunnel could also be altered to allow an Eiffel-type operation. For this mode of operation, the diffuser (r) could be rotated about its vertical axis by

Fig. 2.30   Nozzle and test section of the LFA high-speed wind tunnel A7 with interferometer at the German Museum, München.

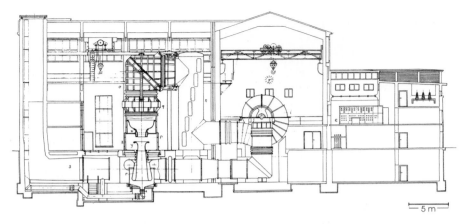

⊏⎯5 m⎯⊐

Fig. 2.31   LFA high-speed wind tunnel A9a with open-jet test section.

180 deg and connected to the exhaust shaft (s). The open jet had a diameter
of 1 m. With the open-jet test section, the concept of *Prandtl* who wanted to
minimize wind-tunnel corrections this way was adopted.

The contraction ratio down to the nozzle exit was approximately 12:1 so
that good flow characteristics with regard to wind-tunnel turbulence could be
expected. The stagnation temperature of the tunnel air could be controlled up
to a level where condensation effects could be prevented. The adjustment of
the velocity of the jet, hence the *Mach* number, was accomplished by changing the number of revolutions per minute (rpm) of the drive motors. The rpm
change was initially carried out by the operator. An automatic speed control
was under development. The pressure level in the chamber surrounding the
free jet (plenum chamber) could be varied in a range between 0.3 bar and
1.7 bar; the *Reynolds* number could hence be changed independently of the
Mach number.

Experiences gained in conjunction with the smaller facility A6 were thus
taken into account in the case of the A9a facility:

*   Continuous operation in the case of airfoil and wing investigations
*   Large open jet with a diameter of 1 m
*   No wind-tunnel wall interferences (but open-jet corrections for larger
    models)
*   Low wind-tunnel turbulence due to a large contraction ratio
*   Avoidance of condensation effects by high stagnation temperatures
*   Freely selectable *Mach* numbers up to $M \approx 1$ by rpm control of the drive
    motor
*   Variable pressure in the test section (change of the *Reynolds* number)

The measuring technique applied was initially confined to conventional
solutions such as pressure measurements with the aid of multitube gauges

and the application of momentum-loss measurements with a pitot rake (wake rake). For schlieren pictures to determine shock locations, the equipment of supersonic tunnel A7 was employed and thus available immediately after the start of the operation of the facility. Until the completion of a six-component balance, a three-component balance was used for force measurements on wings with end plates passing through the jet crosswise. Drag measurements on bodies of revolution, projectiles, and jet engines—here also thrust measurements—were carried out with the aid of an annular-spring force gauge developed at the Institute of Gasdynamics.

Problematic were, at first, the large noise development and resonance vibrations caused by the open-jet test section at certain *Mach* numbers. Such phenomena had already been observed by *Seiferth* in subsonic wind tunnels, and practical solutions had been developed. For instance, serrated rings, corresponding to the *Seiferth* wings, were attached at the nozzle exit, which caused the vortices to leave the nozzle not as closed vortex rings but as individual vortices. As a second measure to reduce the resonance vibrations, the length of the open jet was varied with great success, which simultaneously confirmed *Seiferth's* "organ-pipe theory." The most important wind-tunnel calibrations were thus completed by the end of the war. The tunnel operation was, however, still strongly restricted in its efficiency due to the manual control and the partially improvised instrumentation.

Supersonic tunnel A9b was a closed wind tunnel. The basic setup of the tunnel was similar to that one of tunnel A9a. The test section had a cross section of 94×94 cm. It consisted of a 5-m-long rigid box with two opposite walls of its four walls, consisting of 2.5-mm sheet steel, being adjustable. The adjustment of the nozzles was accomplished by 25 spindles. The nozzle contours for four *Mach* numbers within the range of $M = 1.1$ to 1.8 were determined with the aid of the *Prandtl–Busemann* method of characteristics. Required spindle positions for further *Mach* numbers could arbitrarily be interpolated. The required width of the following diffuser contraction had previously been determined in model tests such that within the diffuser throat a normal shock could still be established. Because the adjustment of the nozzle contour had to be correlated with the respective diffuser position, it was planned to combine certain spindles in groups and to drive them by a motor. This way a quick change in the *Mach* number setting at an optimum pressure recovery in the diffuser could be accomplished. This technology for the attainment of an optimum pressure recovery was, for the first time, utilized in this facility for an industrial application. The pressure level could be varied in the A9b supersonic wind tunnel, as in the A9a facility, between 0.3 bar and 1.7 bar.

High-speed wind tunnels A9a and A9b were connected to a common machine plant and accommodated in a generously designed building. Each of the two wind tunnels had the fan required for its operation and range of operation, each consisting of a one-stage, double-flow radial fan with a cast-iron spiral casing as well as impeller wheels and blades made of steel. The maximum mass-flow rate of the A9a fan was 712,000 m³/h at 1610 rpm and a pressure ratio of 1.24. (The corresponding technical values for the A9b fan were 692,000 m³/h; 1810 rpm; pressure ratio 1.4.) The drive motors of the fans were dc motors each of 4000 kW continuous and 6000 kW peak performance with the number of revolutions being controllable between zero up to 600 rpm. The two compressors with their drives could be coupled to attain the high performances required for a continuous supersonic operation. The facility A9b was only put into operation in 1944. A complete calibration could, up to the end of the war, not be carried out any more.

The overall concept of high-speed wind tunnels A9a and A9b was an example of operating systems still applied worldwide today. This was surely also the reason why the British completely dismantled the entire facility. This included the Allgemeine Elektrizitäts Gesellschaft (AEG) (General Electric Co.) drive systems, the Gute Hoffnung Hütte (GHH) (Mettallurgic Plant) compressors, the entire wind-tunnel facility with the machine hall including the cooling and drying systems and the overhead crane. Especially important for a renewed operation was, of course, the use of the instrumentation developed at the LFA, and particularly the schlieren system and the interference refractor (interferometer), which was further developed by *Zobel*. These optical systems, produced by the Zeiss–Jena Company, were based on two spherical mirrors of 80-cm diameter and a focal length of 4 m. Wind tunnels A9a and A9b possessed a three-component balance. The reconstruction of the entire facility was carried out in England after the war in the newly founded wind tunnel center, the RAE in Bedford, under the name "3 ft × 3 ft Supersonic Wind Tunnel," where it was again put into operation in 1950 as the first large facility. At the same time, a newly built test section for the transonic speed range, based on the experiences with the A9a wind tunnel, went into operation. This was the first transonic wind tunnel in England that could be used for industrial investigations! For operating the supersonic test section, the two compressors were, as planned by the LFA, coupled. The "Eiffel-type operation" employed at the LFA was not used any more.

During the reconstruction of high-speed facilities A9a and A9b, one could fall back on the individual components of wind tunnels A2 and A3 gathered at the Central Store Staughton Airfield. *Peter Hamel* gives more details in [27].

2.2.3.9 PEENEMÜNDE 40- × 40-cm SUPERSONIC WIND TUNNEL. Numerous publications have come out concerning the Test Establishment of the Army HVA Peenemünde, which are almost completely included in a catalog of the Museum Peenemünde issued by *Johannes Erichsen* and *Bernhard M. Hoppe* [28].

The history of the Peenemünde 40- × 40-cm supersonic tunnel starts, interestingly enough, in Göttingen where already in 1918 *Ludwig Prandtl* designed a small intermittently working wind tunnel for supersonic flows with the following characteristics:

- *Low drive power*: Air was sucked in from the atmosphere through an appropriate test section into a vacuum tank. The tank was evacuated by a pump of relatively low drive power.
- *Open-jet test section*: Wind-tunnel corrections for a given angle of attack are in the case of an open-jet test section only about one third of the ones for a closed-wall test section.

Wieselsberger, a pupil of *Prandtl*, had built in the mid-thirties together with his assistant *Hermann* at the TH Aachen (Aix-la-Chapelle) a 10- × 10-cm wind tunnel where the first measurements on the Rocket A-3 were carried out. The construction was based on the *Prandtl* design. Because of the aerodynamic development of the considerably larger Rocket A-4—later known as "V2"—a larger wind tunnel was needed. Consequently, in 1936 *Walter Dornberger* demanded a larger wind tunnel for the newly founded research laboratory in Peenemünde. Entrusted with this task was *Hermann*, who reported directly to the Head of Development, *H. H. Kurzweg*.

A chronological development of the Peenemünde supersonic wind tunnels has, among others, been provided by *F. Zwicky* [29]:

1937   Foundation of the Institute of Aerodynamics of the HVA Peenemünde with 16 employees, Head: *Dr. R. Hermann*. Design and construction of a 40- × 40-cm supersonic tunnel.

1939   "Start of the operation of the supersonic tunnel" (see Fig. 2.32). A detailed documentation concerning this facility was published in 1951 in the INTERAVIA by *Dr. Werner Kraus* [30], formerly Head of the "thermodynamics" branch of the Institute of Aerodynamics of the WVA in Kochel. It represents, according to the enquiry of the author, up to today, the most comprehensive description of the facility by a contemporary witness. *Peter P. Wegener* [31] describes many interesting experiences of this time in Peenemünde and Kochel. However, he delivers few technical details of the wind tunnels. The report of *Hermann Kurzweg* [32], the deputy director of the WVA in Kochel, at the first AGARD conference "Guided Missiles" in Munich in 1956, is much more detailed.

Almost at the same time, the Göttingen 11- × 13-cm high-speed wind tunnel (Fig. 2.12), was built, which was in design almost identical to the Peenemünde facility (Fig. 2.32). The most important technical innovations and knowledge were realized in both facilities:

- Installation of a variable diffuser: As a result, the efficiency and the flow quality were considerably increased. (Variable diffusers were, as already

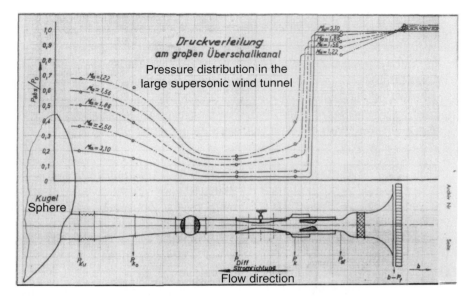

Fig. 2.32 Plan form of the Peenemünde 40- × 40-cm supersonic wind tunnel; wall pressure distributions for the *Mach* numbers $M = 1.2$ to 3.1 (German Museum München, Figure No. 29127).

mentioned, also utilized in conjunction with the *Busemann* high-speed wind tunnels of the LFA in Braunschweig.)

- Drawing in air via a dryer: Avoidance of condensation shocks. These problems had already been discussed by *Wieselsberger* with *Prandtl* and *Busemann* on the occasion of the Volta Congress. After his return to Aachen (Aix-la-Chapelle), *Wieselsberger* requested his collaborator *Hermann* to find a solution to this problem, certainly an important reason to commission him with the construction of the Peenemünde high-speed wind tunnel.

- The Peenemünde wind tunnel was equipped with the most sensitive schlieren system (mirror diameter 50 cm) at the time, which was initially used to calibrate, as had been done in Göttingen, the nozzle blocks designed according to the *Prandtl–Busemann* method. Carl Zeiss Jena developed and built a *Mach–Zehnder* interferometer also utilizing the experiences of *Zobel* of the LFA in Braunschweig (Fig. 2.33).

Fig. 2.33 Test section of the 40- × 40-cm Peenemünde Supersonic Tunnel with the schlieren system of the Zeiss-Jena Company [31].

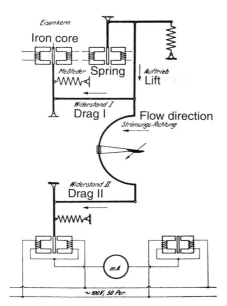

Fig. 2.34   Electromechanical balance with inductive force transducers according to Wieselsberger [33].

- With the successful development of inductive force gages (Fig. 2.34), new technical ground was entered with the design of the balance technology utilizing the developments at the TH Aachen [33] (Aix-la-Chapelle).

Because of the increasing number of contracts, the number of employees was increased to about 200. Although the complete test facility was not damaged during the first great air raid of the Royal Air Force in August 1943, the Institute of Aerodynamics with 180 employees (among them 30 scientists) was transferred to Kochel/Bavaria to the facility only known by the cover name Wasserbau Versuchsanstalt, Kochel (WVA) (Hydraulic Engineering Test Organization). Prior to the shipment of the high-speed wind tunnel, *Siegfried F. Erdmann* [34] was able to convince, among others, *Wernher von Braun* of the necessity to still realize a reconstruction to achieve an $M = 9$ supersonic flow. At $M = 9.5$, he succeeded in taking the first schlieren pictures of such a hypersonic flow. The knowledge gained was to support the project of a hypersonic wind tunnel (1×1 m, $M = 7$ to 10). Kochel had been selected as the location, because here a sufficiently high energy supply for the planned hypersonic wind tunnel (drive power > 50 MW) was available.

1943   The 40×40-cm Supersonic Wind Tunnel was put back into operation in October of 1943, increasing the maximum attainable *Mach* number to $M = 4.4$. At Kochel, the WVA was managed as follows:

Technical management: *Rudolph Hermann* and *Hermann Kurzweg* with the following branch heads:
Testing: *R. Lehnert*

Basics: *P. Wegener* with Dip.-Ing. (Graduate engineer) *H. U. Eckert*
Applied Mathematics: *W. Heybey*
Thermodynamics: *W. Kraus*
Optical Test Methods: *E. Winkler*
Head of Operations: *Herbert Graf*

1945   The Peenemünde 40- × 40-cm Supersonic Wind Tunnel was first
        visited by the Combined Intelligence Objectives Subcommittee
        (CIOS) Group 183.

Zwicky, representative of the USSTAF (U.S. Strategic Air Force in Europe
during WWII), director of Development of the Aerojet Engineering Corporation,
professor of Astrophysics, California Institute of Technology (Caltech), describes
the WVA in Kochel after his visit from May to July 1945 as follows:

> The value of the tunnel is such that we proposed to move it to the United
> States. We also suggested that about 50 German experts be taken along.
> Their potential knowledge concerning the operation of the wind tunnel,
> the construction of more powerful tunnels, and the aerodynamics of vari-
> ous missiles at subsonic, transonic, and supersonic speeds is such that the
> author estimated that several years of work could be saved if the service
> of German personnel could be exploited. This can be done without divulg-
> ing to them any essential military secrets, a fact which we demonstrated
> by assigning to them problems, the solution of which is of interest to us.

The Peenemünde 40- × 40-cm supersonic tunnel, including the vacuum
pumps with drives, controls, the complete instrumentation (schlieren system,
interferometer), and the filters of the air dryers, was completely disassembled
and shipped to Silver Spring, Maryland, near Washington, D.C. Here, a completely new research center for the development of missiles, the Naval
Ordnance Laboratory (NOL), was set up with the aid of former employees of
the WVA under the management of the Navy. In the NOL Wind-Tunnel
Brochure [35] this facility is listed as Aeroballistic Tunnel No. 1. In its construction, it is identical to the Peenemünde 40- × 40-cm supersonic tunnel as
described by *Kraus* [30]. *Hermann Kurzweg*, head of the Aerodynamic
Department of the HVA Peenemünde and the WVA Kochel became, by the
way, associate technical director for Ballistics Research and chief of the
Aeroballistic Research Department of the NOL.

For the interested reader, the so-called "CIOS reports" are certainly of
importance [36, 37]. *R. D. Hiscocks* and *J. L. Orr* of the National Research
Council, Canada, as well as *J. J. Green* of the Air Transport Board, Canada,
were in Northern Germany from June 17 to 29, 1945 to gather first impressions concerning progress in the area of high-speed aerodynamics. They
visited the LFA in Braunschweig and the AVA as well as the Emperor-Wilhelm
Institute (KWI) in Göttingen. Because the DVL was located within the sphere

of influence of the USSR, information concerning the DVL was only indirectly available via scientists, who escaped after the war to Western Germany (such as *Göthert*, who had fled by foot from Berlin to Göttingen). In late autumn of 1945, *R. D. Hiscocks* et al. again inspected the research institutes in Braunschweig and Göttingen. Afterwards, they visited during their nine-week information trip almost all German institutions and industrial companies related to aeronautics, like the Walter works in Kiel; the Focke-Wulf plant in Detmold; the Blohm & Voss plants in Hamburg; and the Bavarian Motor Company plants (BMW) in München. A summarizing experience report [38] about this informative journey appeared in 1948 or 1949. This documentation contains interesting illustrations, but, unfortunately, is no longer available in Germany in its entirety. The authors present, moreover, important research and development results in a generally comprehensible form. Even if the conclusions and comments must, due to the enormous amount of information, be considered critically, a good survey of the 1945 state of the German aeronautical development as of 1945 is provided.

A critical report concerning German research in the area of high-speed aerodynamics was given by *R. Smelt* [39] in October of 1946 at the 696th Lecture of the Royal Aeronautical Society (R.Ae.S.) in London. *Smelt* was an Associated Fellow of the R.Ae.S and had, as a recognized aerodynamicist, shortly after the war, the opportunity to get to know the Research Center LFA, up to this time undiscovered, and to become familiar with relevant work in high-speed aerodynamics in other centers in Germany. In his lecture, he explained:

> In his opinion the research centers known from the pre-war time, i.e., the DVL, Berlin, and AVA, KWI, Göttingen, and the LFA, Braunschweig, were the supporting columns of high-speed aerodynamics.
>
> Important experimental research contributions from Braunschweig were only available at the end of 1942 since the new test facilities had still to be put into operation and calibrated. He particularly pointed out the experimental investigations at Peenemünde. Although these investigations were carried out for the Army on rockets and missiles, the results however provided—in his opinion—important input to the development of high-speed aircraft.
>
> The new facilities for model investigations in the high-subsonic speed range provided, in fact, important development results, whose reliability, however, suffered from the too low *Reynolds* numbers and uncertainties in the wind-tunnel corrections. In comparison to the work in the United Kingdom, adequate flight tests to check the wind-tunnel measurements were missing in Germany.

Smelt specifically referred to the "pragmatic" wind-tunnel corrections of *Göthert* at the DVL in Berlin. It is interesting that in the case of the supersonic wind tunnels he only mentioned the 40- × 40-cm facility at Peenemünde considering as one of the most important advances in German wind-tunnel technology the development of a second throat within the supersonic diffuser downstream of the test section. With this development he associated considerable energy savings related to the fan drive and a better flow quality particularly in the case of open test sections. He did not say a word about the high-speed wind tunnels A7, A9a, and A9b of the LFA, which were based on a similar design principle as the Peenemünde wind tunnel. Perhaps at that time, he did not want to mention the fact that these facilities in Braunschweig were to be dismantled and reconstructed at the RAE in Bedford, United Kingdom.

2.2.4 EXAMPLES OF PROJECTED TEST FACILITIES THAT DID NOT GO INTO OPERATION

In the development of both high-speed aircraft as well as rockets and missiles, it soon became very clear that, in addition to the simulation of the correct *Mach* number, investigations at higher *Reynolds* numbers, which occur in flight, were necessary. For this, various large facilities were projected, which had partly reached in 1945 an astonishing level of completeness despite the war situation. As a simulation problem, it was recognized that the location of the laminar-turbulent transition in the wind tunnel was directly influenced by the wind-tunnel turbulence level.

2.2.4.1 LARGE LOW-TURBULENCE TUNNEL OF THE AVA GÖTTINGEN.
For the validation of computational methods to determine the position of the laminar-turbulent boundary-layer transition and for the development of laminar airfoils, a low-turbulence, low-speed wind tunnel was built for the Institute of Theoretical Aerodynamics of the AVA Göttingen in Reyershausen. It was executed as an Eiffel-type wind tunnel with a closed-wall test section; see Fig. 2.35. The air was led via a square inlet with a cross section of 11 × 11 m and a flow straightener (1) and wire screens (2) into a circular nozzle (3) with a diameter of 3 m. Downstream of the nozzle was the 8-m long test section (5). The test-section cross section was realized by inserting two vertical walls (7) at a distance of 1.5 m as side walls into the circular cross section of 3-m diameter. That way, the test section obtained an almost rectangular cross section of about 4.5 m^2. This resulted in a contraction ratio of larger than 27:1 for the wind tunnel. The facility had gone into the trial phase in 1943. First information about the wind-tunnel turbulence level, obtained with the aid of measurements on spheres, suggested an extremely low-turbulence level. Measures to remedy low-frequency stagnation pressure fluctuations due to flow separation in the diffuser were under preparation.

2.2.4.2 MAGNETIC SUSPENSION GUIDE RAILS FOR FLIGHT TRAJECTORY MEASUREMENTS
An alternative to the investigation of the aerodynamic

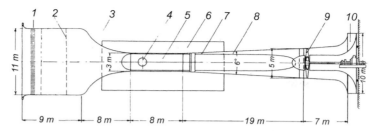

Fig. 2.35   Low-turbulence wind tunnel for incompressible flow of the AVA.
Legend:

1 Flow straightener

2 Fine-meshed screens

3 Nozzle transition from a quadratic
to a circular cross section
(Contraction ratio 25:1)

4 Turntable synchronized with the
angle-of-attack adjustment

5 Test section of 4.5-m² cross section

6 Eiffel chamber (Plenum chamber)

7 Vertical inserts (walls)

8 Diffuser

9 Fan rotor

10 Air outlet (exit)

characteristics of aircraft in compressible subsonic flow, employing large models, was suggested in 1942 by *Otto Mühlhäuser* [40]. His idea was to build a special rail facility where the models to be examined were to be towed through the free atmosphere. Here, he wanted to use the "magnetic-suspension guidance system," developed by *H. Kemper*, for his development. *Kemper* worked at the AVA Göttingen since 1938. *Mühlhäuser* was at the time head of the Institute of Equipment Development of the AVA and had presented a detailed feasibility study concerning his suggestion. After the war more than 30 years later, *Kemper's* ideas were again employed in the development of the magnetic suspension train.

The test track, Fig. 2.36, was to have a track width (gauge) of 2.5 m and was to be based on 1.8-m high supports. To accelerate the vehicle to the test

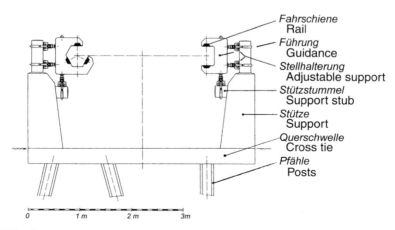

Fig. 2.36   Cross section of the test track with magnetic suspension [41].

Fig. 2.37   Test vehicle with test track; project study of *Mühlhäuser*, 1942, AVA [40].

speed, a catapult in conjunction with the "reaction-type" engine of the test vehicle was to be employed. A sketch of the test vehicle on the test tracks, Fig. 2.37, specifies *Kemper's* ideas. With the aid of a model facility, he experimentally checked some of his design data and came to the following conclusions:

At a total length of the test track of 20 km, a model span of 3 m, and a speed of 320 m/s, a test-section diameter of a "transonic wind tunnel" of 8 m would have been required. He estimated the drive power needed here to be 60,000 kW, according to *Mühlhäuser* 15 to 20 times the amount needed for the flight-test track. Also, the construction costs of the wind tunnel were, with 70 million Reichsmark, almost twice as high as for the flight-test track. This alternative test method is included in our documentation to make two things clear:

- For fundamental research and the pursuit of new ideas, sufficient means and personnel were also still available at the research institutes in 1942.
- As the description of the next projects shows, a degree of megalomania was apparently beneficial when conceiving and realizing large-scale projects.

2.2.4.3   HIGH-SPEED WIND TUNNEL (3-m DIAMETER) OF THE LUFTFAHRTFORSCH-UNGSANSTALT MÜNCHEN (LFM) (AERONAUTICAL RESEARCH ESTABLISHMENT MUNICH). In 1939 just at the outbreak of the WWII, *Ludwig Prandtl* had undertaken an informative journey to Cambridge. Here, he realized that the goal of superior German aeronautical research was out of the question. Consequently, the LFA was conceived and already in 1939 approved by the State Secretary *Erhard Milch*. The research establishment was to be set up in Southern Germany to support the successful aircraft manufacturers Messerschmitt and Dornier in the development of their high-speed aircraft. After the successful trial of the swept-wing technology in the wind tunnel, new test facilities were to be employed also for the investigation of individual engines and engine integration. Because the change in *Reynolds* number in the high-speed wind tunnels of the DVL and the LFA was only possible by varying the model dimensions, this was to be accomplished in the LFM wind

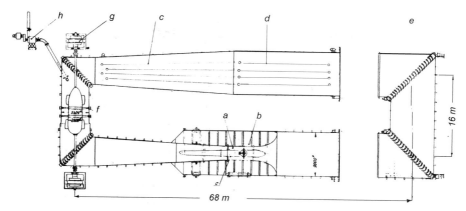

Fig. 2.38   Schematic of the LFM high-speed wind tunnel [41].
Legend:

a Test section	e Removable part for Eiffel-type operation
b Vacuum (Plenum) chamber	f Fan
c Adjustable wall	g Motor
d Cooling walls	h Turbocharger

tunnel by changing the pressure level in the test section. In a schematic drawing (Fig. 2.38), which *Matt* [41] shows in his documentation, some of the new design features are indicated. The horizontally placed air-line was designed in its entire length as a welded cylindrical steel tube. The entire air-line was firmly anchored at the position of the fan in order to be able to compensate or avoid heat expansions and stresses within the overall system. The test section had a circular cross section with a nominal diameter of 3 m but widening conically with a cone angle of 0.5 deg over 5 m. The effect of the wall boundary layers was, in addition, to be compensated by adjustable upper and lower walls. The two-stage counter-rotating fan required a total drive power of 13,000 kW with one motor for each stage. With the aid of a multistage turbocharger, the pressure level could be adjusted between 0.3 and 1.8 bar. In addition to the closed-loop operation, the facility could also be operated—as in the case of the LFA 2.8-m tunnel—in the Eiffel-type mode. For that purpose one side of the air-line, together with the associated corners, was designed to be moved mechanically out of the tunnel circuit by 8 m, (see Fig. 2.38). Many of the new design features (such as the cooling of the tunnel air, the adjustability of the test-section wall, and the fan arrangement) could not be tested and their applicability proven due to the closure of the tunnel after the end of the war. Unfortunately, further information, beyond the one given by *Matt* [41], and details concerning the renewed use of this facility after its dismantling after the end of the war, are not available to the author.

The story of this test facility would not be complete if the role of *Bäumker* was not further described. Under *Bäumker's* chairmanship, the LFM was founded in December of 1940. On September 3, 1940, *Bäumker* [42] wrote to *Albert Betz*, director of the AVA in Göttingen, and made it clear that no further investments in Göttingen were scheduled. The not yet functioning vacuum operation of the large Wind Tunnel VI was downgraded in its priority by *Bäumker* and the completion of the large cold-air tunnel was recommended. Concerning the situation of the LFA, he commented:

> The development of the Aerodynamic Institute of the Aeronautical Research Establishment LFA has reached a certain completion. The medium-size tunnel with a test section diameter of 3 m operates in a permanent mode. The high-speed facility with a 2.8-m test section diameter is about to be put into operation. A certain preparatory time will, however, still be necessary before definitely putting this tunnel into operation. The large tunnel with 8-m test section diameter and a medium flow velocity is presently performing its first trial runs. Within some months, one may also count on the commencement of the operation of this tunnel. The smaller supersonic facilities A6 and A7 at Braunschweig can be considered operational.

This statement did not—as we know today—correspond to the technical status. Not all the new facilities were yet operational, because, for instance, the instrumentation and sufficient calibrations were missing. As will be shown later, the 2.8-m high-speed wind tunnel had enormous technical difficulties. However today, we do not know whether the executives of the LFA deliberately misinformed *Bäumker* about the real situation. The fact is that *Bäumker* initiated a new large-scale construction site at a time when the lack of qualified personnel nearly stopped the research demanded by him. He conceived a new facility center in Southern Germany, which was gigantic in its magnitude and would have been unique worldwide. (See the following description of the 8-m high-speed wind tunnel at Ötztal.)

The start of the construction of the test facilities at the LFM was postponed due to the extension of WWII to the years 1941–1942. *Bäumker* himself took on the management of the LFM on the order of *Milch*. *Milch* expected a good cooperation of *Bäumker* with the acting temporary head, *Heinrich Peters*, in pressing ahead with the realization of the planned facilities. *Bäumker* had obviously fully identified himself with the task of again setting up a new research center. *Helmuth Trischler* [43] thought that particularly the progress in the engine technology—and certainly also the war situation at the time— had put him into an "ecstasy of planning." In his imagination, the plans for the test facilities of the LFM included for them to become an unsurpassed worldwide center of excellence. The increasing defeat of the German Armed Forces led, of course, with time to personnel and material shortfalls which caused considerable delays in the completion of the LFM 3-m high-speed wind tunnel. A letter must be seen in this context, which *Bäumker* wrote in early December to the Fl. Stabsing. (Flight Staff Engineer) *Helmut Schelp* of the Development Office of the General Air Equipment Master (Generalluftzeugmeister) in Berlin [44], referring to negotiations in December 1942 with representatives of the Construction Offices and the Armament Command in the presence of *Schwaiger*. It is remarkable that *Bäumker* resisted the obviously realistic opinion of the research management of the RLM, namely, that the facilities in Ottobrunn (near München) would only gain their importance in the postwar period. *Bäumker* explained, among other things:

> From the negotiations it clearly followed that the Armament Command intended to help us in carrying out the construction. The provision of workers from concentration camps was discussed. Until December 20 the possibilities of a deployment should be examined and then new negotiations take place.

The following description of the construction of the 8-m high-speed wind tunnel in the Ötztal again shows the unrealistic planning situation.

2.2.4.4   HIGH-SPEED WIND TUNNEL (8-m DIAMETER) OF THE LFM IN THE ÖTZTAL.   The project of the large high-speed wind tunnel, the so-called "Construction Project 101, Messerschmitt München" of the LFM, has been described by *Ernstfried Thiel* [45] in detail. The LFM was founded in 1940 with the support of the RLM and planned a large high-speed wind tunnel with a maximum *Mach* number of $M \approx 1$ in a test section of 8-m diameter. The largest wind tunnel in the world at that time would have required a drive power of 76 MW. To be able to provide this power, the propulsion was directly provided by water power via two water (Pelton) turbines and two axial fans of 15-m diameter each. For that reason, the Ötztal in Austria was chosen as location where the corresponding water quantities and fall heights (530 m) could be provided. The facility was primarily planned for the investigation of full-scale engines under realistic operating conditions. However, the simulation of the highest possible *Reynolds* numbers had, in addition, been recognized as important for the development of high-speed aircraft. In the large high-speed wind tunnels of the DVL in Berlin and the LFA in Braunschweig attainable *Reynolds* numbers were lower by a factor of 10 compared to full-scale *Reynolds* numbers. The most important technical specifications, see Fig. 2.39, were defined as follows:

Test-section diameter = 8 m
Test-section cross section = 50.3 m²
Contraction ratio = 9:1
Maximum velocity = 300 m/s

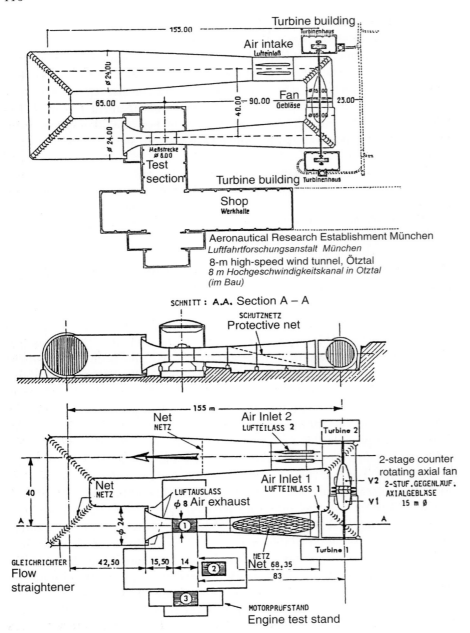

Fig. 2.39 Total view of the 8-m high-speed wind tunnel [45]; top: LFM-project Ötztal, Austria; bottom: arrangement in Modane-Avrieux (Tunnel SI MA), France.

Maximum *Reynolds* number = 12×10^6/m
Maximum shaft drive performance = 76 MW

The LFM was in 1940 directly appointed by *Bäumker* of the Research Department of the RLM with *Fritz Schwaiger*, as head of the RLM Group "Research Facilities," taking on the responsibility for the administrative execution of the wind-tunnel project. The technical supervision was in the hands of an executive board member of the LFM, *Heinrich Peters.*

Information about the enormous wind-tunnel project had, up to the end of the war, not been made accessible to the Allies. The gigantic wind-tunnel components, already partially erected, and the associated buildings and foundations under construction, were first discovered by American troops in May of 1945. By the end of May, a French group of the Securité Militaire Française (SMF) discovered this construction site.

In early June 1945, *F. L. Wattendorf* [46] suggested in a memorandum, after his visit with *H. L. Dryden* to Southern Germany, to transfer the facilities from Ötztal (LFM) and Kochel (Peenemünde) and the jet-engine test stand of BMW in München, as a nucleus of a new research center of the Air Force, to the United States. *Marcel Pierre* [47] reports that there were, at first, discussions concerning the competences in the further use of the facility, which was 25 percent completed in May 1945. Many individual components of the wind tunnel, still stored by the individual suppliers, were ready for collection. While the parts stored in the American Occupational Zone were confiscated by the Americans and partly shipped to the United States, the French started to systematically dismantle the Ötztal facilities. These actions were approved by the French military government and the military commander of Innsbruck. The Dingler Company with its factory located in the French occupational zone in Zweibrücken had been charged during the war with the construction of the entire facility in the Ötztal. Now, the Dingler Company was made responsible by the French officials for the transport of all material already assembled or stored at the Ötztal to Modane. The dismantling started in October of 1945, and the first of 13 freight trains left the Ötztal on December 12, 1945, and the last one with working equipment, crane facilities, hoisting gear, etc. left on June 23, 1946, in the direction of Modane. After this "fait accompli" by the French side, the Americans officially agreed to the reconstruction of the wind tunnel in Modane in October of 1946, after a year of controversies and tough negotiations. The contemporary witness *Ernstfried Thiel* [45] who worked at the Dingler Company Zweibrücken between 1945 and 1957 and later at Dornier has reported on these events and further details.

As *Frank L. Wattendorf* later reported, this bilateral agreement was an example of a first pragmatic solution to a problem that led to international cooperation in the field of aeronautical sciences. Summarizing, it can be stated that the new German test facilities were the initiators of four new research centers:

Organization and Location	Facility	Post-WWII Transfer Location
Heeresversuchsanstalt, HVA, Peenemünde since 1944 WVA, Kochel	40- × 40-cm supersonic wind tunnel	Naval Ordnance Laboratory (NOL) Silver Spring, MD, United States
BMW, München	Engine test facility	AEDC/Tullahoma, TN, United States
Luftfahrtforschungsanstalt LFA, Braunschweig	High-speed wind tunnels A9a, A9b	RAE/Bedford, United Kingdom
Luftfahrtforschungsanstalt München, Ötztal	8-m high-speed wind tunnel	ONERA/Modane–Avrieux, France

2.2.5   WIND-TUNNEL CORRECTIONS FOR THREE-DIMENSIONAL MODELS

2.2.5.1   EFFECT OF WIND-TUNNEL TURBULENCE ON RESULTS MEASURED IN LOW-SPEED WIND TUNNELS.

For the interpretation of wind-tunnel measurements in both compressible as well as incompressible flow, the influence of the wind-tunnel turbulence on the boundary layers on the wind-tunnel models is of great importance today. In that regard, one is reminded of the research concerning the drag laws of spheres, where the test results of *Eiffel* in Paris and *Prandtl* [48] in Göttingen lead to contradictory observations. Particularly confusing was that the drag measurements agreed well for small spheres. Here, *Prandtl* soon found a physically well-founded explanation on the basis of his boundary-layer theory. It was already known that in flow through pipes and on flat plates, after exceeding certain critical *Reynolds* numbers, R, a change in the flow condition from a laminar to a turbulent boundary-layer flow very suddenly occurred, which led to a change in the drag laws (laws of resistance). Of course, this change in the state of the boundary layer also occurs in the flow about a sphere. *Prandtl* explained the sudden drop in drag with laminar boundary-layer separation occurring earlier and already upstream of the "Equator" thus generating a large region of separated flow. The turbulent boundary layer, on the other hand, only separates downstream of the equator forming a vortex area with a smaller diameter behind the sphere. By chance all contradicting sphere measurements had been carried out at *Reynolds* numbers where—according to today's knowledge—the transition from a laminar to a turbulent boundary-layer flow occurred. In the investigations of *Eiffel* compared to *Prandtl's* measurements, the sudden drop in drag occurred at substantially lower *Reynolds* numbers.

Prandtl suspected the cause to be the higher turbulence level of the wind-tunnel flow in case of the so-called Eiffel-type tunnel causing the transition from a laminar to a turbulent boundary-layer flow to occur already at lower *Reynolds* numbers. To prove his theory, *Prandtl* [49] requested *Wieselsberger*

[50] to conduct an extensive test program with spheres of different diameters and furnished the final proof by his famous visualization of the flow about the spheres with the aid of smoke injections. For this, he attached a thin wire upstream of the laminar separation region, which caused the laminar boundary layer to turn into a turbulent one. For all velocities investigated, this lowered the drag coefficient to a value that had without the wire ring only been achieved at a *Reynolds* number of 2.7×10^5. With this knowledge, a method to judge and compare low-speed wind tunnels with the aid of the so-called critical *Reynolds* number, R_k, for which a sudden decrease in drag is measured, had been introduced. This method gained international acceptance and was also practiced in Germany in conjunction with the new low-speed wind tunnels. As an example, one may consider corresponding investigations in the DVL low-speed wind tunnels in Berlin.

Before the "large" wind tunnel of the DVL, already completed in 1933 with a nozzle cross section of 5×7 m and 6×8 m and a jet velocity of 65 m/s, could be used for airfoil investigations, the flow quality of the wind tunnel and the reliability of the instrumentation first had to be demonstrated by a comparison with an internationally accepted test facility. For this, *Doetsch* [51, 52] carried out pressure distribution measurements on the airfoils NACA 2412 and NACA 2418 and compared the results with results of measurements in the variable density tunnel at Langley [53] (Fig. 2.40). This wind tunnel had been developed in 1921 by *Max Munk*, a former collaborator of *Ludwig Prandtl* in Göttingen, and suggested to the NACA Board. At a maximum total pressure of 20 bar, the tunnel dimensions had to stay small in comparison to a wind tunnel operating at atmospheric pressure, which necessarily resulted in relatively high turbulence levels of the wind-tunnel flow. The NACA carried out extensive systematic airfoil measurements in this wind tunnel, which were published in 1933 in the now classic Report NACA TR 460.

Doetsch established as a result of his comparative measurements that not only the maximum lift and the drag but also the lift-curve slope, $dc_a/d\alpha \, (dc_l/d\alpha)$,

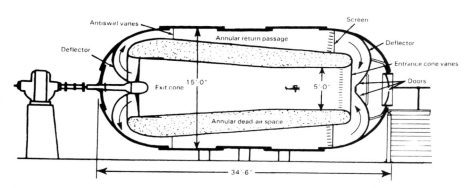

Fig. 2.40   Schematic of the plan form of the NACA 5-ft pressure tunnel for low speeds [53] (*Max Munk*, 1923).

Fig. 2.41 **Lift curve slope** $dc_a/d\alpha = dC_A/$
$d\alpha = f(R_k)$ **for the airfoil NACA 2412 depen-**
dent on the wind-tunnel turbulence [53]
(wing aspect ratio $\Lambda \to \infty$**). The critical**
parameter R_k **characterizes the** *Reynolds*
number R (*Re*) **of the flow about a sphere**
where a sudden drop in drag due to the
change from a laminar to a turbulent sepa-
ration can be noticed. The parameter R_k
increases with decreasing turbulence inten-
sity and approaches a constant value.

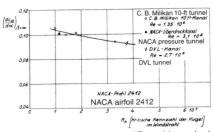

R_k [Critical parameter (*Reynolds* number)
for the sphere in an air stream]

and the pitching moment (neutral-point position) were to a considerable
degree dependent on the turbulence level of the open jet. Figure 2.41 shows,
as an example, the change in the gradient $dC_A/d\alpha$ ($dC_L/d\alpha$) with the critical
Reynolds number of the sphere, R_k, for the airfoil NACA 2412 at an infinite
aspect ratio Λ. (At the critical *Reynolds* number R_k of the sphere—dependent
on the turbulence intensity of the wind-tunnel flow—a sudden drop in the
aerodynamic drag occurs.) The results confirmed that for the test results to be
transferable to flight not only the condition of equal *Reynolds* numbers, but
also the condition of the turbulence level of the test-section flow to be the
same as in the atmosphere must be met. The condition of the turbulence being
equal to the turbulence in the atmosphere ($R_k \approx 4 \times 10^5$) was essentially met
by the DVL tunnel ($R_k = 3.7 \times 10^5$), while the NACA pressure tunnel
($R_k = 1.2 \times 10^5$) pointed to a substantially higher turbulence intensity. The
test results from the DVL wind tunnel were, under the condition of equal
(*Reynolds*) numbers, transferable to flight. In addition, proof was furnished
that, at the low turbulence level of the flow, the DVL tunnel was especially
suitable for the investigation of laminar airfoils. Further related investigations
were carried out by *H.-U. Meier* et al. [54].

2.2.5.2 **WIND-TUNNEL WALL CORRECTIONS IN COMPRESSIBLE FLOW.** For the successful use of the new high-speed wind tunnels, the review and development
of wind-tunnel corrections were of special importance because compared to
incompressible flow, corrections were larger. They increased with increasing
Mach number with $1/\beta^3$ and reached considerable magnitudes when approaching the speed of sound. $\beta = (1 - M^2)^{1/2}$ for $M = 0.9 \to 1/\beta^3 = 12.07$! At the
LGL Committee Meeting "High Speed," *Wieselsberger* presented, upon the
request of the Committee Chairman *Blenk*, a summarizing report [55] on the
contributions concerning wind-tunnel corrections presented on September 3,
1940, at the LFA in Braunschweig, see Appendix D.

   Wieselsberger concentrated his remarks essentially on corrections of

- Freestream flow velocity
- Freestream flow direction

Concerning the corrections of the freestream flow velocity, *Wieselsberger* first summarized the corrections for incompressible freestream velocities, which were required due to the displacement effect of the model in a closed test section. To determine the velocity correction at the location of the model in a closed tunnel (test section) with a circular cross section, the normal velocities at the tunnel wall due to the so-called image doublet were determined; see Fig. 2.42. These normal velocities had to be cancelled with the aid of an auxiliary flow because otherwise the wall would have to be penetrated by the flow. The correction becomes simpler in the case of a tunnel of a rectangular cross section because here the conditions at the boundary of the air stream are satisfied by mirror imaging. Concerning the principle of mirror imaging, reference is made to *H. Glauert* [56], who described mirror imaging in its application to the determination of the wind-tunnel interferences in the following apt sentences:

> The concept of images, used in aerodynamic problems, can be appreciated by considering a few simple examples. If two airplanes are flying side by side, there will be no flow across the vertical plane of symmetry between the airplanes and this plane could be replaced by a rigid wall without altering the flow in any way. Thus, the problem of an airplane flying parallel to a vertical wall can be solved by introducing an image airplane on the other side of the wall and by considering the new problem of two airplanes flying side by side. Similarly the interference experienced by introducing an airplane flying close to the ground can be solved by introducing an inverted image airplane below the ground. This method of introducing the appropriate image or set of images to represent the constraint of the boundary of a stream is capable of a very wide application, and is the method used for analyzing most problems of wind-tunnel interference.

The auxiliary flow was produced by suitable source and sink distributions along the tunnel wall, whose normal velocities just cancelled the normal velocities generated by the doublet. In the case of an open jet, a reduction of

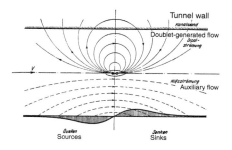

Fig. 2.42   Concerning the illustration of the method of tunnel correction computations according to *Wieselsberger* [55].

the freestream velocity occurs, opposed to the closed test section, because the free jet mixes with the quiescent air surrounding it. The ambient pressure will be imposed on the test section because there cannot be any pressure jump across an open-jet boundary. This problem was also solved with a corresponding source-sink distribution. It should still be mentioned here that the correction is positive in case of solid wind-tunnel walls. However, corrections for free-jet boundaries are negative and approximately only one-third as large, this being one reason for *Ludwig Prandtl* to prefer in the case of compressible flows wind tunnels with open-jet test sections.

The corrections turned out to be even more complex for wind-tunnel models, for example, models of blunt bodies, where an essential share of the drag is generated by the momentum loss in the wake. Inevitably the blockage effect of the model is considerably stronger than in the case of a pure potential flow because due to the reduced flow velocity in the wake, the velocity in the freestream must be increased. The larger blockage was also represented by appropriate sources at the location of the model with the strength of the "additional source" having been determined according to the ratio of the frontal (drag) area of the model to the cross-sectional area of the test section. An accurate description of the drag correction in high-speed wind tunnels was given by *Hubert Ludwieg* [57] in 1944. He considered in the wind-tunnel corrections, and wind-tunnel interferences, the effects resulting from

- Blockage due to the volume of the model
- Blockage due to the wake (represented by a source)
- Gradient of the flow in the wind tunnel due to the blockage

Even today these correction effects must be determined, and in doing so, it still depends essentially on the intuition of the wind-tunnel engineer concerning the model representation that has to precede the determination of the wall interferences. A model representation is, however, no longer necessary if the walls of the test section are, for instance, equipped with orifices to measure the wall pressures. Today, wall pressure methods are available for the computation of the entire field of interference velocities; see *H. Holst* [58].

All correction methods for incompressible freestream velocities and slender models were ascribed to compressible flow with the aid of the *Prandtl* rule such that a model in a tunnel in incompressible flow is assigned to an altered model in a tunnel of a different diameter in compressible flow. The *Prandtl* rule reads in this case as follows:

> At corresponding points of the compressible and the incompressible flow the same potential and the same velocity exist in the freestream direction (x-direction) but a velocity reduced by $(1 - M^2)^{1/2}$ ($M = Mach$ number)

in the directions normal to the freestream direction. So the following relations hold:

$$x_{incompr.} = x_{compr.}$$

$$y_{incompr.} = y_{compr.} \, (1 - M^2)^{1/2}$$

$$z_{incompr.} = z_{compr.} \, (1 - M^2)^{1/2}$$

On the left side of Fig. 2.43, an airfoil with a large aspect ratio s/c (s = span, c = chord) is shown in a closed tunnel in incompressible flow together with the corresponding potential lines. Here, it should particularly be noted that due to the boundary conditions the potential lines must be normal to both the surface of the airfoil and the tunnel wall so that no normal velocity components occur.

If the disturbance velocities generated by the model are no longer small relative to the freestream velocity, the *Prandtl* rule just discussed is no longer applicable. (The rule designated in the literature of the time as *Prandtl* rule is today known as the *Prandtl–Glauert* rule; see Chapter 1.) The conversion of the model cross sections is then no longer similar, but the model must initially be deformed in an unknown way. Here, *Göthert* [59] suggested at the 127th LGL meeting in Braunschweig concerning the correction that one defines an area of high disturbance velocities in the vicinity of the model. Outside this area, the disturbances were assumed to be small. Therefore, it was possible to apply to this part of the flow, and especially to the flow along the tunnel wall, the *Prandtl* rule. For the computation of the large disturbance velocities,

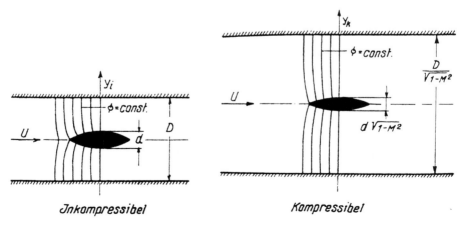

Fig. 2.43 **The effect of compressibility according to *Wieselsberger* [55]; ϕ = velocity potential.**

Göthert introduced a test parameter as a measure for the effective model blockage. For closed wind tunnels, he took the wall pressure measured vertically above the model and arranged the validation of this new correction method in measurements on different-size wings of the same airfoil in the DVL high-speed wind tunnel.

After the start of the operation of the DVL high-speed wind tunnel, the effect of the model size on the blockage of the test section was, at first, to be determined purely empirically. It was, in addition, important to obtain reliable data to check existing correction methods. At a diameter of the test section of 2.7 m, the model dimensions could be varied systematically. Here, *G. Richter* [60] carried out investigations on four model wings of different chord but the same airfoil section NACA 0015-64 in order to determine by a comparison of the pressure distributions the largest model size for which the measurement still delivered reliable results. Simultaneously the correction method of *Göthert* [61] for the stagnation pressure and the *Mach* number was to be verified by the experimental data. The measurements were carried out on four rectangular wings mounted between elliptical endplates. The span of all wings was 1370 mm (about $\frac{1}{2}$ of the test-section diameter), the wing chords were 350, 500, 700, and 1000 mm, respectively. The pressure distributions were measured in the center sections at *Mach* numbers between 0.2 and 0.86 at angles of attack of −1 to +5 deg. In all tests, the static pressures at the test-section walls were measured over the entire test-section length above and below the model. Only such measurements were employed for further evaluation where the speed of sound was not reached anywhere in the test. As an example, Figs. 2.44 and 2.45 show pressure distributions of the wings with chords of 500 and 700 mm at the same *Mach* numbers and similar angles of attack, which clearly indicate the different development of the pressure distributions for both wings. As a result of these investigations, it was concluded that, within the angle-of-attack range investigated and up to $M = 0.86$, the effect of the tunnel wall on the measured aerodynamic airfoil characteristics can, up to a model chord of 500 mm, be determined with sufficient accuracy by a correction of the stagnation pressure and the *Mach* number according to *Göthert*. The limiting value of the *Mach* number for this airfoil of $M = 0.86$, for which the speed of sound was not reached at the tunnel wall, corresponded to about 60% of the tunnel blockage (due to the model and the suspension system) that would have resulted in one-dimensional flow in the same limiting *Mach* number of $M = 0.86$. After these investigations in the DVL high-speed wind tunnel, this reference value was used as the allowable model size for all further airfoil and wing investigations and also introduced at the 2.8-m LFA high-speed wind tunnel at Braunschweig. How careful the flow qualities of the 2.7-m high-speed wind tunnel of the DVL in Berlin were examined is shown by another investigation of *Göthert* and *Richter* [62]. At low *Mach* numbers ($M = 0.15$ and 0.18), they investigated the high-speed models in the

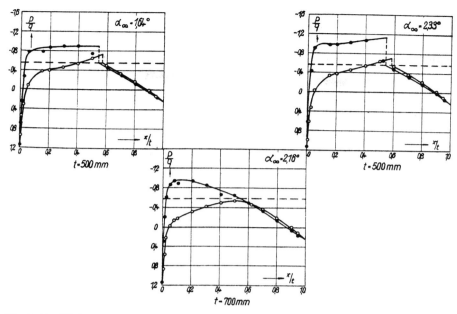

Fig. 2.44   High-speed measurements on the airfoil NACA 0015-64 with different model chords (t = 500 mm, 700 mm) at a *Mach* number of *M* = 0.76 [60].

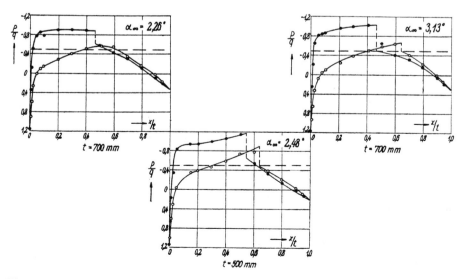

Fig. 2.45   High-speed measurements on the airfoil NACA 0015-64 with different model chords (t = 500 mm, 700 mm) at a *Mach* number of *M* = 0.78 [60].

open-jet test section of 2.15- × 3-m low-speed wind tunnel of the DVL. At low angles of attack, the lift coefficients C_A (C_L) obtained in the high-speed wind tunnel agreed quite well with the ones determined in the low-speed wind tunnel.

For the correction of the angle of attack $\Delta\alpha$, *Wieselsberger* [55] presented in his survey at the 127th LGL meeting a relation that was based on experience in incompressible flow. He showed that the correction of the angle of attack, hence the correction of the induced drag, in compressible flow was of the same order as the incompressible flow. He set up the angle-of-attack correction in the following form:

Closed wind tunnel $\Delta\alpha = (C_A F/(8F_k))\,(1 + K_G)$ $C_A = C_L$

Open-jet $\Delta\alpha = (C_A F/(8F_k))\,(1 + K_F)$

Here, K_G and K_F are functions of the location, the *Mach* number M and the tunnel radius R; F/F_k is the ratio of the wing area to the tunnel cross-sectional area and C_A (C_L) the lift coefficient. The coefficients K_G and K_F are plotted for various distances downstream of the wing (model) in Fig. 2.46 (downwash correction for $\Delta\alpha$). It must be noted that for the tunnel with solid closed walls the angle-of-attack correction $\Delta\alpha$ is positive and for the open-jet test section it is negative. The absolute corrections are of a similar order of magnitude.

As a contribution to the discussion during the LGL meeting, *Göthert* [63] again commented on the question of an open or a closed test sections in high-speed wind tunnels. Here, he referred to airfoil measurements in compressible subsonic flow, which were being conducted by *A. Naumann* [64] in the high-speed wind tunnel Aachen on the airfoil NACA 23012-64 in the closed tunnel and in the open-jet test section. *Göthert* was able to explain and nearly eliminate the discrepancies observed in the drag measurements by a correction of the freestream velocity (*Mach* number).

Because in the case of solid wind-tunnel walls a higher and in a free jet a lower lift occurs, the change in lift as well as the position of the center of

Fig. 2.46 Coefficients K_G and K_F for the correction of the angle of attack at various positions downstream of the model (downwash corrections) [55].

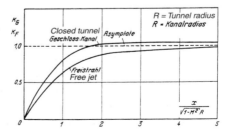

pressure was not only dependent on the *Mach* number, but also to a large degree on the ratio of the wing chord to the tunnel height (*t/h*). Also in that regard, *Wieselsberger* indicated correction possibilities because in practical measurements in high-speed wind tunnels the airfoil chord was no longer small relative to the tunnel height and the error arising here had to be taken into account.

Summarizing, it can be concluded that the corrections for the new high-speed wind tunnels were at a high level and a meaningful interpretation of the test results was, therefore, possible. Also the discussions concerning the principle investigations of *Ludwieg* and *Strassl* were thus placed on an impartial basis. The importance of the national cooperation in this research area becomes already clear by the fact that *Wieselsberger* of the TH Aachen (Aix-la-Chapelle) moved to the AVA in Göttingen, his former sphere of activity under *Prandtl*, and was able to coordinate the work of scientists like *Göthert*, *Richter* of the DVL, *von Baranoff* and *Wendt* of the LFA, and *Pabst* of the Focke–Wulf Company in Bremen.

2.2.6 CONDENSATION EFFECTS IN HIGH-SPEED WIND TUNNELS

If humid air or water vapor flows through a Laval nozzle, slightly downstream of the minimum nozzle cross section the formation of fog due to the temperature decrease is observed. In this case, downstream of the minimum nozzle cross section a compression shock may be observed spreading X-wise. This phenomenon—the connection between the atmospheric humidity and the so-called X-shocks—was first discovered in the 10×10 cm supersonic tunnel of the TH Aachen (Aix-la-Chapelle) by *Wieselsberger*. He discussed this problem on the occasion of a lecture by *Prandtl* at the Volta Congress in 1935. *Wieselsberger* thus explained the shock formation in schlieren pictures shown by *Prandtl*, which were obviously due to condensation shocks. After the lecture of *J. Ackeret* [65] at the Volta Congress, *Wieselsberger* argued during the following discussion:

> In the case of Laval nozzles, which are utilized in wind tunnels to generate supersonic flow, a shock wave occurs, as Prof. Prandtl already reported during his presentation, downstream of the minimum cross section area for which one has, up to now, not yet found an explanation. This shock wave is here particularly undesirable since it means a severe disturbance of the flow. We have thoroughly investigated this shock wave with the aid of the schlieren method without having, up to now, found a means to avoid it. One result of these investigations may, nevertheless, be noteworthy. It turns out that this shock is not always located in the same position but shifts with time parallel to the tunnel axis and is, indeed, dependent on the humidity of the air streaming through the nozzle. At the same humidity, we have also obtained the same shock location. Perhaps

this observation will contribute to learning something more about the
nature of this unexplained shock wave.

Thus, an important problem of supersonic wind tunnels was recognized
and formulated; the X-shock due to condensation was associated with throttling losses, which caused a reduction in the *Mach* number and a pressure rise
at the nozzle exit. After *Oswatitsch* [66] had theoretically treated in 1941
condensation phenomena in supersonic flows at the KWI, Göttingen, the
primary objective now was to quantify the effect of the atmospheric humidity
on the accuracy of the test results obtained in supersonic tunnels.

R. Hermann [67] presented in 1942 results that contained pragmatic solutions to the condensation problems in the new supersonic tunnel of the HVA,
Peenemünde. To quantify the problem, he first analyzed schlieren pictures that
were taken in 1935 at different *Mach* numbers and degrees of relative humidity in the supersonic wind tunnel at Aachen (Aix-la-Chapelle). Here, the
humidity of the air had been changed by spraying water into the air sucked in.
In Fig. 2.47, soft (weak) X-shocks are clearly seen—as defined by *Hermann*—at
low humidity and strong X-shocks at high humidity. It may be assumed from
the photographs that the X-shocks move upstream with increasing humidity.
To check the parallelism of the flow within the supersonic nozzles investigated,
the *Mach* angle μ was determined upstream and downstream of the nozzle exit
from schlieren pictures. The distribution of the *Mach* angle μ along the nozzle
axis in areas where, according to the theory, parallel flow should prevail, is
plotted for weak and strong shocks, respectively, in Fig. 2.48. It can be recognized that with increasing strength of the X-shocks the *Mach* angles μ generally increase, that is the *Mach* numbers decrease. Similar qualitative results
were obtained for the measured wall pressures. *Hermann* still carried out theoretical considerations to clarify the experimental results with the aid of the
"normal condensation shock," but could not—as *Oswatitsch*—suggest any
correction as solution to the condensation problem. However, because the
strong throttling effects due to the X-shock led to changes in the static pressure at the nozzle wall, and hence to a strong impairment of the precision of
the measurements in the HVA Peenemünde supersonic tunnel, a pragmatic
solution had to be found. Experimentally it was found that with dried air
(absolute humidity < 0.2 g/m^3) at a *Mach* number of $M = 1.9$, an X-shock
could no longer be observed. A silica-gel type air drier was, therefore, developed for the HVA supersonic tunnel, which guaranteed a sufficient drying of
the relatively humid Baltic Sea air (Fig. 2.49). Thereafter, it could definitely be
proved that with dried air an isentropic expansion occurred in the nozzle.
Hence, such good parallel flows were now achieved in the wind-tunnel nozzles
designed by the *Prandtl–Busemann* Method of Characteristics that only a
small displacement correction had to be applied to account for the increasing
wall boundary-layer thickness.

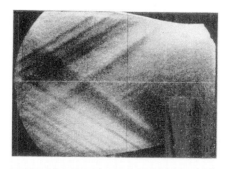

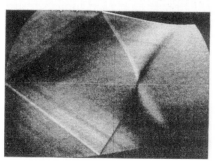

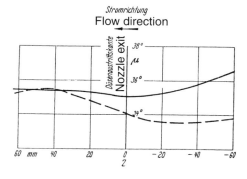

Fig. 2.47   Weak, medium, and strong X-condensation shocks (from the top). Measurements: 1935 in the 10 cm × 10 cm supersonic wind tunnel of the TH Aachen (Aix-la-Chapelle) at $M = 1.7$.

Stromrichtung
Flow direction

Düsenaustrittskante
Nozzle exit

$38°$

μ

$36°$

$34°$

60 mm   40   20   0   -20   -40   -60
z

Fig. 2.48   Distribution of the *Mach* angle μ along the nozzle axis upstream and downstream of the nozzle exit at $M = 1.7$ (dashed curve: weak condensation shock; solid line: strong condensation shock) [67].

Fig. 2.49   Air dryer of the 40 × 40 cm super-sonic wind tunnel of the HVA Peenemünde [30].

2.2.7 CONCLUSIONS

- The program concerned with the construction of new wind tunnels, initiated in 1933 by *Adolf Bäumker*, provided the essential prerequisites for basic experimental investigations on swept wings, new propulsion systems (jet engines), engine integration and high-speed aircraft.
- On the basis of the existing test results, it may be stated that, following the basic investigations of *Hubert Ludwieg* in 1939 in the 11- × 11-cm high-speed wind tunnel in Göttingen, the mainstay of the facilities utilized in the further development of swept wings was the 2.7-m high-speed wind tunnel of the DVL in Berlin. Here, *Bernhard Göthert*, a careful experimentalist, who interpreted the test results with a congenial intuition and who had developed first pragmatic wind-tunnel corrections, played a decisive role. *Göthert* succeeded in always staying in contact with the theoreticians to improve the computational methods related to airfoils and wings in compressible flows.
- The 2.8-m high-speed wind tunnel of the LFA did not play a decisive role in the development of swept wings because the flow quality of this facility did not meet the requirements of the tasks at hand. However, the 2.8-m tunnel could take on important tasks in the investigation of new propulsion systems.
- The high lift problems of the swept wing were mainly addressed at the AVA Göttingen using tunnel VI.
- Further smaller subsonic wind tunnels were available at the universities and in industry for the treatment of basic problems and configuration

studies. From the beginning of his investment program *Bäumker* had supported the construction of smaller facilities in order to relieve the research institutes in the execution of industrial tests.

2.3 PROGRESS IN DESIGN AERODYNAMICS

2.3.1 ORGANIZATION OF RESEARCH AND STATE OF KNOWLEDGE AT THE BEGINNING OF THE WAR IN 1939

The sensational test results of experiments with different swept-wing configurations in 1939 in the high-speed wind tunnel of the AVA Göttingen caused the RLM in Berlin to provide at short notice both material and personnel resources for the development of high-speed aircraft. The RLM had already realized at the start of the war that Germany could only compensate the high productivity of the Allies, and hence the output of aircraft, in the long run, a technical superiority. To achieve this objective, the financial means available for the development of high-speed aircraft were nearly unlimited. On the other hand, with the entry into the war, the recruitment of well-trained personnel quickly became more difficult, which led to delays in the execution of research projects.

The Göttingen swept-wing measurements were, at first, primarily considered as pure basic investigations, because, although they had been carried out at the correct *Mach* number, the *Reynolds* numbers were much too low. The set objectives of the research in the field of aero- and gasdynamics very quickly became clear and were formulated in close cooperation with the aeronautical industry:

- Development of high-speed airfoils (drag minimization at optimum lift)
- Optimization of swept-wing configurations
- High-lift augmentation aids for the swept wing
- Stability investigations and check of the controllability of high-speed aircraft with swept wings
- Integration of jet engines

The actual state of knowledge in mid-1940 in the field of aerodynamics/gasdynamics in Germany was impressively documented for many branches on the occasion of the 12th "High Speeds" Meeting of the "General Flow Research" Committee of the LGL on September 3, 1940, in Braunschweig and on September 26–27 in Göttingen (Fig. 2.50). It was a meeting of German experts in the field of high-speed aerodynamics at an unusually high scientific level. Altogether 67 experts of all important research and test establishments, universities, as well as the aerospace industry participated. The individual special contributions and the contributions to the discussions were published in the LGL report 127 "High Speeds," which was of high quality in both

a)

Ausschuß
Allgemeine Strömungsforschung

Bericht über die Sitzung

Hochgeschwindigkeit

am 3. September 1940 in Braunschweig (s. Anhang)
und am 26. und 27. September 1940 in Göttingen

Bericht 127

Lilienthal-Gesellschaft für Luftfahrtforschung

Berlin SW11, Prinz-Albrecht-Straße 5 (Haus der Flieger) · Fernruf: 12 00 47. App. 1702

Fig. 2.50   a) Cover sheet of the report on the LGL 127 Meeting (Archive MPG Berlin), b) translation.

b)

Secret

Committee
General Flow Research
Report on the Meeting
High Speeds

on September 3, 1940, at Braunschweig (see Appendix)
and on September 26 and 27, 1940, at Göttingen
Report 127

Lilienthal Society for Aeronautical Research
Berlin SW11, Prinz-Albrecht Street 5 (House of Pilots)

Fig. 2.50   (Continued)

contents and presentation; it was, however, accessible only to a select group of experts as it was declared "Secret Command Object." This meeting of the "General Flow Research" Committee was chaired by the then director of the LFA in Braunschweig, *Blenk* (Fig. 2.51). Here only the test programs for the high-speed wind tunnels of the LFA and DVL were initially discussed, because the realization of these programs could be assessed in the report on the next LGL meeting (156) "Questions Concerning High Speeds" on October 29 and 30, 1942, in Berlin.

Fig. 2.51   *Blenk* and *Zobel* on a visit of the wind tunnel facilities at Chalais-Meudon, Paris (DLR Archive Porz-Wahn).

In that regard, *Zobel* [68] of the LFA in Braunschweig emphasized in his comments that, at the time, only two large high-speed wind tunnels were available in Germany, but that the number of open questions concerning compressible flow would still increase. He suggested, therefore, coordinating the research projects especially with regard to the existing or soon to be available test facilities. Here, *Zobel* stated:

> The LFA has for the execution of investigations in the high-speed wind tunnel a three-component (standing) balance as well as an optical facility based on the interference and Schlieren methods at their disposal. Both can be used simultaneously without mutual interference. Applying the optical methods, all pressure orifices can be omitted in two-dimensional pressure distribution measurements. Thus, an essential simplification and time savings in the model construction and the running of the tests can be achieved. Due to these instrumentation-related requirements, mainly such research projects which are not amenable to the standard testing techniques but are particularly suitable for optical methods should be carried out in the high-speed wind tunnel of the LFA.

Beginning in October 1940, the program for the high-speed wind tunnel of the LFA was arranged, according to the following tasks organized into main groups:

- Pre-tests concerning the trial of the balances and the optical methods and related to this; force and pressure distribution measurements on the airfoils Goe 622, 623, 624, NACA 4412, and a 12%-thick symmetrical Joukowsky airfoil
- Pure research tasks
  - Airfoil systematic on 16 symmetrical NACA airfoils
  - High-speed airfoils NACA 1412-64.23012-64
  - Boundary-layer investigations on a flat plate and an elliptical cylinder
  - Boundary layers at different degrees of roughness
- Applied research
  - Capped wing
  - Swing wing
  - (Interrupter) Lateral control
  - Cooler investigations
  - Industrial measurements

Zobel further remarked that, of course, important industrial measurements had to be scheduled into this program dependent on their urgency and the state of the preparations. In addition, he pointed to a fan defect of the wind tunnel that, due to the related partial-load operation, led to a cancellation of all high-speed measurements until early 1941.

Göthert [69] of the DVL in Berlin-Adlershof reported, on the other hand, a successful startup of the operation of the 2.7-m high-speed wind tunnel in 1939. Contrary to the LFA in Braunschweig, the DVL first concentrated on the implementation of proven, conventional test techniques such as pressure distribution and momentum-loss measurements. The drag and moment balance, used up to September 1940, was to be supplemented by a lift balance up to the end of the year. With the aid of a schlieren system, valuable information, for instance, about the location of shock waves in airfoil investigations, had already been gained.

The DVL emphasized in the wind-tunnel planning the development of high-speed airfoils:

- *Purely subsonic speeds*: These airfoils were to have local supersonic regions only at high subsonic *Mach* numbers (referred to today as transonic airfoils).
- *Local supersonic speeds*: These airfoils were already to operate at local supersonic speeds, but still keep, however, their acceptable aerodynamic qualities, planning to control the location of the shock wave by boundary-layer control.

Concerning the airfoil development in compressible subsonic flow, *Göthert* remarked:

> For the preparation of test series on airfoils for purely subsonic flow, K. H. Kawalki [70] carried out computations for airfoil shapes, systematically being varied, at various angles of attack, with the airfoils being characterized as suitable for a favorable high-speed behavior due to their small excess velocities in compressible flow. For the specification of the airfoil shapes, the NACA system was retained, so that the main airfoil parameters, such as, for instance, the thickness ratio, the location of the maximum thickness, the nose radius, etc. were varied within this system.

Further main tasks of the DVL were:

- *Flap investigations*: To determine the influence of the *Mach* number on the rudder effectiveness, pressure distribution measurements on a rectangular wing with flap, rudder-moment measurements, as well as momentum-loss measurements to determine drag, were planned.
- Fuselage investigations
- *Effect of the wing planform and the span-wise thickness distribution*: Continuation of the work of the AVA Göttingen on swept wings. Here, extensive force measurements concerning the effect of the form of sweep and the plan form were planned.
- Stability investigations

- *Oscillatory investigations*: It was planned in conjunction with the Junkers Company to first gain experience concerning the effect of the *Mach* number on the oscillatory behavior of aircraft wings by an oscillation test with a rectangular wing. At that time, all flutter computations were still based on the aerodynamic characteristics of wings in incompressible flow. (See the aeroelastic problems of swept wings in Chapter 4.)
- Propeller tests

Prof. *Blenk* emphasized in his closing remarks that lately visible progress had been achieved in both theory and experiments. However, the wealth of questions was still so large that a uniform opinion on how to proceed could not be expected. The permanent exchange of ideas between the officials in charge in the industry and at the research institutes would, in any case, be of the greatest importance for further research work.

During 1942, more than two years later, the next large meeting of the LGL took place in Berlin on October 29 and 30. As seen from the meeting report LGL 156, in comparison to LGL 127, many of the planned research and development programs had actually been carried out at the DVL Berlin and the LFA Braunschweig. From the list of participants in this LGL meeting, it becomes obvious that in 1942—despite the heavy air raids of the Allied Air Forces—almost twice as many scientists and design engineers, 139, participated than two years earlier at the "High Speeds" meeting, 67. Conspicuous was also the strong showing of expert officials, referees, of the RLM and of over 50 experts from industry. Similar to 1940 at Braunschweig and Göttingen, the main share of the lectures, 13 out of 21, was presented by the DVL with *Göthert* again delivering five lectures single-handedly. Because original test reports of the DVL and also of the LFA were only available on a very limited scale, the following description of the aerodynamic computational and experimental results up to 1942, especially for the transonic speed range, is primarily based on the LGL 127 and LGL 156 reports. Judging the actual situation in 1942, it must be stated that at the AVA Göttingen, despite the difficult personnel situation in research, work still could be carried out, for example, concerning the development of high-lift systems and flutter safety, on an astonishingly large scale.

2.3.2 *High-Speed Airfoils*

The results of the experimental airfoil investigations at high subsonic speeds were summarized in 1945–1946 by *Hubert Ludwieg* [71] in the Göttingen Monographs. Basically the same methods as in incompressible flow were available for the determination of the airfoil characteristics at high speeds. Three-component force measurements, pressure distribution measurements, and wake measurements were carried out, the latter to indirectly determine the airfoil drag.

In wind-tunnel flows at high subsonic speeds, however, the following fundamental difficulties occurred in executing airfoil measurements:

1) The *Reynolds* number increased proportional to the speed and the model size while the required performance (power) increases with the third power of the speed, but only with the second power of the model size. It was, hence, more economical to achieve large *Reynolds* numbers by large wind-tunnel dimensions. The ratio of the model size to the test-section cross section had, in addition, at high speeds to be reduced so that the influence of the jet boundary on the measurements would not become too large.

2) The wind-tunnel wall corrections became substantially more uncertain, especially when the freestream velocity approached the speed of sound.

3) The interference between the suspension system (support) and the model increased considerably because, due to the aerodynamic forces increasing with *Mach* number, the suspension system had to be reinforced.

Different methods were employed for the investigation of airfoils:

1) Investigations on low-aspect-ratio wings ($\Lambda \leq 5$): The effect of the finite aspect ratio was, however, substantially larger in incompressible flow, and the conversion to infinitely large or any other aspect ratio became appreciably more inaccurate.

2) Investigations on low-aspect-ratio wings with endplates to simulate quasi-two-dimensional flow condition: This method had the disadvantage that the relatively large endplates contributed at high speeds considerably to the total drag, and the flow at the wing tips (ends) was disturbed.

3) Pressure distribution measurements to determine lift and pitching moments: These required large models to be able to equip the model with a sufficiently large number of pressure orifices. Here, an area with approximately two-dimensional flow conditions in the center of the wing could be investigated. One could then select large-chord low-aspect-ratio wings with endplates to obtain high *Reynolds* numbers. One had to determine the corresponding airfoil drag by momentum-loss measurements. Although the measurement of the total and static pressures did, in principle, not suffice for the drag determination in compressible flow, *Göthert* [72] had been able to prove already in 1940 that the measurements could, under the assumption of a constant stagnation temperature, be reliably evaluated.

4) Due to density gradients present in compressible flowfields, optical methods, such as the schlieren and interference methods, could be used to determine the airfoil characteristics: Contrary to the pressure distribution measurements these two optical methods provided information about the

entire flowfield. As already mentioned in the description of the new facilities for high-speed investigations in Sec. 2.3.1, especially that the interference method had been further developed by *Zobel* at the LFA to application maturity in large wind tunnels, this interference method was sufficiently accurate in the determination of the density gradients to compute from these reliable pressure distributions. Disadvantageous was that just when exceeding the speed of sound locally with areas of shock waves and flow separations occurring, the determination of pressures and forces from the actually accurate density determinations became uncertain. Because for these flowfields the simple adiabatic relation between pressure and density was no longer valid, additional assumptions would have become necessary. Therefore, one had at first to be content with qualitative results.

2.3.2.1 **EFFECT OF THE THICKNESS RATIO.** *Adolf Busemann* [73] lectured at the 127th LGL meeting in Göttingen once again about the compressibility effect on thin, low-camber airfoils at subsonic speeds. He had already introduced his form of the *Prandtl* rule, later also called "Streamline Analogy," at the first session of the 2nd Science meeting of the full members of the German Academy of Aeronautical Research on April 26, 1940. It allowed now airfoils with d/t (d/c) [ratio of airfoil thickness d to chord t (c)] of 15% to be computed (designed), which were, at the time, of industrial interest regarding realistic wings. After the Academy lecture, *Günter Bock*, director "Flight Vehicles (Flugwerk)" of the DVL, Berlin, pointed out wind-tunnel investigations on a symmetrical airfoil of 15% thickness and a chord of 500 mm in the 2.7-m high-speed wind tunnel (Fig. 2.52). The measurements had been carried out for three different angles of attack. The lift coefficient according to the *Prandtl* rule is shown for comparison. *Bock* pointed out the good agreement between the experimental results and the data according to the *Prandtl* rule up to the velocities where shock waves first appeared because, according to *Busemann's* extension of the *Prandtl* rule, the effect of the compressibility on the lift became even larger. *Bock* in his comparison between theory and computation did not elaborate any more on *Busemann's* theory especially because its application was, at the time, associated with a very large computational effort. This may also have been the reason why the value of the *Busemann* extension of the *Prandtl* rule was only recognized at the end of the war.

Also interesting in this regard was the discussion of *Ludwig Prandtl* with *Göthert* and *Busemann* concerning their extensions of the *Prandtl* rule documented in the proceedings of the 127th LGL meeting, 1940, on page 56:

> *Prandtl* asked *Göthert*, why he had in the comparison between compressible and incompressible flow assumed a thicker airfoil in the case of

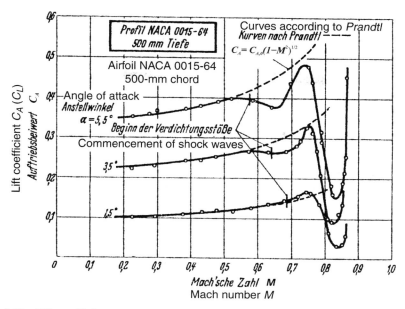

Fig. 2.52 **Lift coefficients** C_A (C_L) **for three different angles of attack** α **dependent on the** *Mach* **number** *M*. **Discussion** *Busemann/Bock* **[73], test: DVL 2.7-m high-speed wind tunnel.**

compressible than in incompressible flow while, according to the *Prandtl* rule, the airfoil had to be made thinner for a comparable pressure distribution in a compressible environment.

Göthert: "If one carries out a suitable affinity distortion with the streamline rather than with the potential, the incompressible flow about an airfoil represents the compressible flow about a thicker airfoil. This can be proved by computations."

Busemann comments: "One can see this without one line of computation."

A lively discussion ensued between *L. Prandtl* and *B. Göthert*. At the printing of the LGL Report 127, *L. Prandtl* added to the documentation of the discussion the following comment:

I absolutely agree with *Göthert*'s explanations at the High-Speed Meeting after knowing these preceding explanations. At the meeting, I still did not know his method so that I was not able to immediately cope with it. *Busemann's* method, where all lateral dimensions remain unchanged and only the dimensions in the free stream direction have to be enlarged by $1/(1 - M^2)^{1/2}$ when going from the compressible flow to the flow with constant volume, does exactly the same as *Göthert's* method, since the latter results from the former by a similar reduction of all dimensions proportional to $(1 - M^2)^{1/2}$.

After *Bernhard Göthert* had already in various reports [74, 75] presented the test results from the DVL high-speed wind tunnel, he added a detailed "discussion" of the test results in a further publication and provided a comparison with tests in other wind tunnels. On the basis of tests on the series of symmetrical airfoils NACA 0 00 10 t/l -1.1 30 at the DVL, he reported on investigations concerning the effect of compressibility on the characteristics of standard symmetrical airfoils. [At the DVL, the airfoil nomenclature of the NACA was largely adopted. Here, it concerns an airfoil with a relative thickness of 10%, a camber of $f/t = 0\%$, a maximum-camber location of $x_f/t = 0\%$, a nose radius $r = (\rho/t)/(d/t)^2 = 1.1$ and a maximum-thickness location of $x_d/t = 30\%$.] All airfoils had a chord of t (c) = 500 mm. To determine lift and pitching moment, pressure distribution measurements were carried out. The drag was determined by momentum-loss measurements in the wake.

The lift-curve slopes $\partial C_A/\partial\alpha$ as function of the *Mach* number M are plotted in Fig. 2.53. The lift development up to the critical *Mach* number M^*, for which the speed of sound is just reached locally on the airfoil, agrees for the thin airfoils exceptionally well with the slope calculated by the *Prandtl–Glauert* method according to $1/(1 - M^2)^{1/2}$ as long as the local velocity has not reached or exceeded the speed of sound anywhere on the airfoil. However, for the thicker airfoils, the measured lift increases stayed below the computed ones. Already for the 15%-thick airfoil a somewhat reduced increase compared to the one computed by *Prandtl–Glauert* was noticeable. For the 18%-thick airfoil, the theory had to fail completely, which becomes immediately obvious from the measured pressure distributions in Fig. 2.54. Plotted here is the measured pressure difference $p = p_0 - p_\infty$ (the local static pressure minus the static pressure of the freestream) related to the stagnation pressure of the freestream $q_\infty = \rho_\infty U^2_\infty/2$. While for the 9%-thick airfoil a shock wave only occurred at 60% chord on the suction side, already at x/t $(x/c) = 0.35$ the 18%-thick airfoil had a massive shock boundary-layer interaction on both the suction side as well as the pressure side, which surely led to a flow separation and hence to a loss in lift. Also for *Göthert* these investigations were primarily of scientific interest because it had already definitely been proven that for high-speed flight only thin airfoils could be used. The neutral-point location $\partial C_{m,t/4}/\partial C_A$ $(C_A = C_L)$, important to the stability, which for the two symmetrical airfoils investigated here was equivalent to the center-of-pressure location, was investigated by *Göthert* in the vicinity of the zero-lift condition (see Fig. 2.55). In plotting the neutral-point location $\partial C_{m,t/4}/\partial C_A$ $(C_A = C_L)$ [the moment coefficient is based on the quarter-chord location, $t/4$ $(c/4)$] at different *Mach* numbers M and the zero-lift coefficient C_A $(C_L) = 0$, it was to be observed that the neutral-point location thus defined only made sense in the range where the pitching-moment curve $C_{m,t/4} = f(C_A)$ was tangent to a straight line through the origin. At high angles of attack, deviations from this "straight line"—especially at high *Mach* numbers—occurred that often had, as a

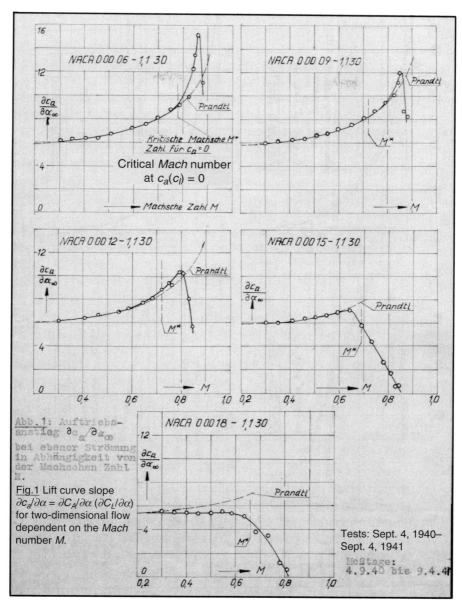

Fig. 2.53 Lift-curve slopes $\partial c_a/\partial\alpha = \partial C_A/\partial\alpha(\partial C_L/\partial\alpha)$ for symmetrical airfoils of different relative thickness ratios of 6% to 18% dependent on the *Mach* number [76], tests: DVL 2.7-m high-speed wind tunnel; axes: $\partial c_a/\partial\alpha_\infty$ $(\partial C_l/\partial\alpha_\infty)$, *Mach* number *M*.

consequence, variations of the neutral point with angle of attack so that the definition of the neutral point lost its meaning in this range. It can be observed in Fig. 2.55 that the neutral point moves at subcritical *Mach* numbers for all airfoils investigated steadily in the direction of the wing leading edge, that is,

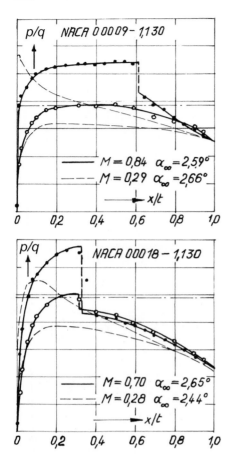

Fig. 2.54 Pressure distributions on two airfoils with 9% and 18% relative thickness [76] (dashed line = incompressible flow; angle of attack $\alpha \approx 2.5$ deg).

in an unstable direction. This neutral-point movement increases with increasing airfoil thickness. Although detailed knowledge concerning shock boundary-layer interaction was still not available, *Göthert* interpreted the neutral-point movement correctly as a boundary-layer effect. Physically reasonable information was provided by the detailed pressure distribution measurements, which allowed for instance reliable statements concerning the location of the shock waves. The resulting changes in the pressure distributions allowed clear conclusions concerning the movements of the neutral-point position at *Mach* numbers above the critical one (at supercritical *Mach* numbers). One recognizes in Fig. 2.55 that for the airfoils with thickness ratios of 6%, 9%, and 12%, the neutral point moves very quickly towards the airfoil trailing edge when exceeding the critical *Mach* number. The behavior of the very thick airfoils (15%, 18%) was principally in the opposite way. *Göthert* concluded that when using thick airfoils at high subsonic speeds, one had to expect besides strong lift losses also control (trimming) problems, which could no longer be balanced by control surface deflections.

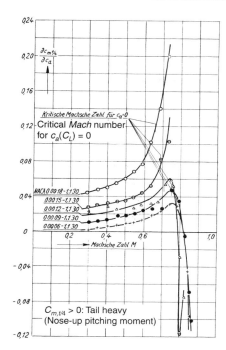

Fig. 2.55   Neutral-point location $\partial C_{m,t/4}/\partial C_A$ (∂C_L) of symmetrical airfoils of different relative thickness ratios of 9%–18% dependent on the *Mach* number [tests: DVL 2.7-m high-speed wind tunnel; axes: $\partial c_{m,t/4}/\partial c_a$ ($\partial C_{m,t/4}/\partial C_A$) ($\partial C_L$); *Mach* number M] [76].

Göthert investigated the effect of compressibility on drag, the second important aerodynamic parameter in high-speed flight. The drag of the rectangular wings in the airfoil investigations was determined by momentum-loss measurements, which were considered in detail at the LGL meeting in Göttingen [77]. The flow in the wake was surveyed at a distance of two wing chords at 11 different spanwise positions. The arithmetic average of the 11 readings was then established for further evaluation. For all airfoils investigated the drag coefficient C_{Wp} at zero angle of attack was plotted as function of the *Mach* number M in Fig. 2.56. The drag coefficients were below the critical *Mach* numbers M* constant over the entire range and only increased systematically with the airfoil thickness. After exceeding the critical *Mach* number M*, the drag curves showed the sharp rise already observed at other DVL measurements only at a *Mach* number M, which was by a certain amount (ΔM up to 0.04) higher than M*. This phenomenon was explained by *Göthert* on the basis of the measured momentum-loss data and pressure distribution measurements as follows:

> That, at first, despite exceeding the critical *Mach* number, a pronounced drag rise does not occur is probably due to the fact that shock waves are, at first, only very weak and that the boundary layer can still negotiate the rise in pressure generated by the shock. The strong drag rise commencing at higher *Mach* numbers indicates a separated boundary layer—mostly

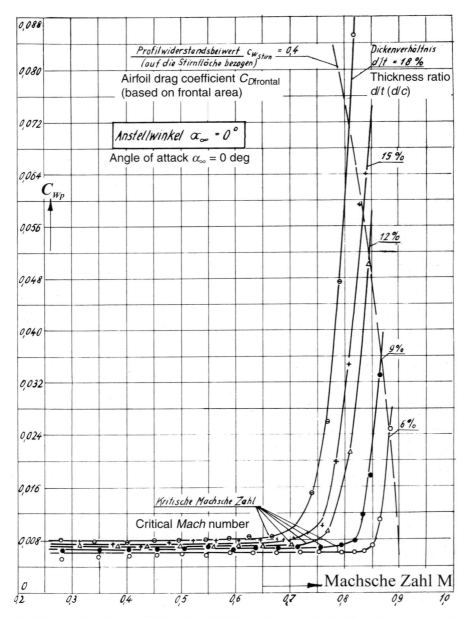

Fig. 2.56 Airfoil drag coefficient C_{Wp} (C_{Dp}) dependent on the *Mach* number at an angle of attack of $\alpha = 0$ deg (tests: DVL 2.7-m high-speed wind tunnel; axes: C_{Wp} (C_{Dp}), *Mach* number *M*) [76].

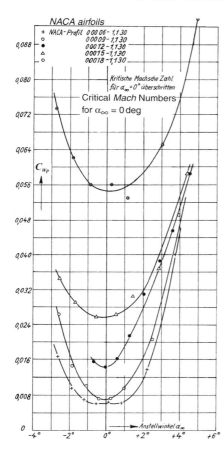

Fig. 2.57 Increase of the airfoil drag coefficient C_{Wp} (C_{Dp}) when exceeding the critical *Mach* number M^* at an angle of attack of $\alpha = 0$ deg [tests: DVL 2.7-m high-speed wind tunnel; x-axis: angle of attack α_∞; y-axis: C_{Wp} (C_{Dp})] [76].

as a result of an oblique shock or a shock that has moved far downstream towards the wing trailing edge.

The plots of the drag coefficients C_{Wp} (C_{Dp}) in Fig. 2.56 are valid for an axisymmetrical freestream at an angle of attack of $\alpha = 0$ deg. A cross plot of these results at a *Mach* number of $M = 0.8$, where all airfoils had reached their critical *Mach* number M^*, showed, of course, an enormous drag rise from $C_{Wp} = 0.008$ for the 9% airfoil to $C_{Wp} = 0.056$ for the 18% airfoil. In complementing these results by drag coefficients of airfoils of different thickness ratios at small angles of attack in the range $\alpha = -3$ to +5 deg (Fig. 2.57), it became clear that for the thin airfoils the drag rise with increasing angle of attack was steeper than for the thick airfoils.

2.3.2.2 COMPUTATION OF HIGH-SPEED AIRFOIL SECTIONS ACCORDING TO K. H. KAWALKI. After *Göthert's* investigations, it became clear that for the design of favorable high-speed airfoils one had to try to reach sonic velocity at freestream Mach numbers as high as possible, that is, to attain a shift in the critical Mach number M*. It seemed, therefore, reasonable to use airfoils with very low excess velocities because then the disturbances on lift, pitching moment, and drag due to the compressibility of the air only occurred very late. In that regard, *K. H. Kawalki* of the DVL had in November of 1940 already introduced corresponding airfoil developments at the LGL 127 "High Speeds" meeting. His computations, which were based on the potential method of *T. H. Theodorsen* and *J. E. Garrick* [79] led him to elliptical airfoils. His computational method allowed him to investigate the most important airfoil parameters, such as the nose radius, the thickness ratio, and the maximum-thickness location, independent of each other. First he could show that "thickened airfoil noses" ($\rho/\rho_{Ell} > 1$) caused a slight reduction of the velocity in the center region. Here, he avoided a velocity peak at the nose and aimed at a velocity increase as monotonous as possible up to the airfoil mid-chord at a lower maximum velocity—compared to the ellipse (Fig. 2.58). An important result was the following: the thicker the airfoil, the larger the nose thickening had to be. Another important result was provided by *Kawalki's* investigation of the effect of the maximum-thickness location on the airfoil velocity distribution. The velocity distributions in Fig. 2.59 show for an airfoil of 9% thickness ($d/t = d/c = 0.09$) the largest reduction of the maximum velocity for the airfoil with a maximum-thickness location of x/c = 0.48 and a thickened airfoil nose of $\rho/\rho_{Ell} = 1.5$. At the location of the maximum thickness, the downstream

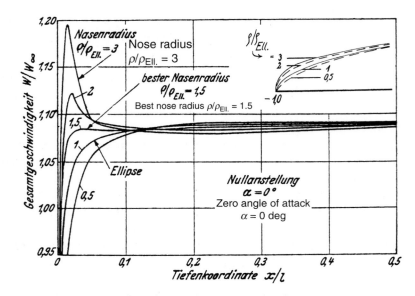

Fig. 2.58 Velocity distribution on bisymmetrical airfoils with different nose radii [thickness ratio d/l (d/c) = 0.09] according to *K. H. Kawalki* [78] [axes: total velocity W/W_∞ (U/U_∞), chord x/l (x/c)].

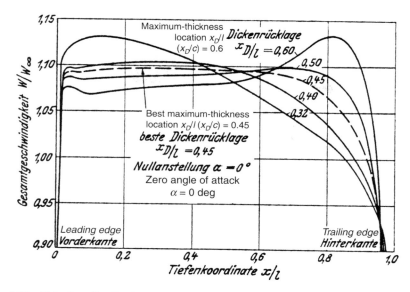

Fig. 2.59   Velocity distribution on bisymmetrical airfoils with different maximum-thickness locations [thickness ratio d/l (d/c) = 0.09] according to $K.$ $H.$ $Kawalki$ [78] [axes: total velocity W/W_∞ (U/U_∞), chord x/l (x/c)].

part of the airfoil was formed by a second- and fourth-order parabola in a way that for smaller thickness ratios [d/t (d/c) < 0.12] the maximum curvature was only located downstream of the trailing edge and that the curvature along all of the downstream part of the airfoil contour monotonously increased. Further investigations showed that the effect, which an inclination of a symmetrical airfoil exerted on the velocity distribution, mainly depended on the form of the airfoil nose. The symmetrical airfoils were, therefore, superposed by camber. By the superposition of camber, it was achieved that, at a given lift coefficient C_A (C_L) which was required for high-speed flight at a low angle of attack, even smaller excess velocities occurred. For the selection of the so-called skeleton lines (mean lines) to realize the camber effect, $Kawalki$ could fall back on a 1934 report of $F.$ $Weinig$ [80]. This way, high-speed airfoils were developed, which showed a further reduction of the excess velocities compared to the symmetrical airfoils developed.

Through a systematic variation of the airfoil parameters, $Kawalki$ was able to develop both optimum symmetrical and cambered high-speed airfoils. This investigation of $Kawalki$ was to play an important role regarding the objections of the German aircraft industry against a patent application of NASA under the name "Wing Profiles" at the German Patent Office in München!

2.3.2.3   OBJECTION (1984) AGAINST THE U.S. PATENT DOCUMENT NASA No. 22 54 888 "SUPERCRITICAL AIRFOILS," BASED ON THE COMPUTATIONAL METHODS OF K. H. KAWALKI (1940).   On November 9, 1972, the National Aeronautics and Space

Administration (NASA), Washington, D.C., filed the U.S. Patent No. 22 54 888 concerning the airfoil development for the flight at high Mach numbers in subsonic flow. The publication of the NASA patent took place on April 12, 1984. The patent application coincided, interestingly enough, with the development of the Airbus A310, the first civil aircraft equipped with a swept wing based on supercritical airfoils.

According to *John D. Anderson* [81], the development of the so-called "supercritical airfoils" had been carried out by *Richard Whitcomb*, at that time head of the 8-ft Transonic Wind Tunnel at NASA Langley. Because a patent application would have led other important countries in the world to considerable financial burdens for Airbus Industrie, the European aeronautical industry filed an appeal at the German Patent Office, München, against granting the patent. It follows from the argument of July 5, 1984, that essential features of the patent claims of NASA were already known and published. Of the altogether 18 literature quotations of the appeal, the oldest and most essential ones were the research reports [82, 83] of *Kawalki*. As is well known, a summarizing presentation of these research results had been given by *Kawalki* in 1940 at the LGL "High Speeds" meeting in Göttingen. His reports as well as the LGL report 127 were available to all eastern and western Allies after the war. According to the objection, the characteristic claims of the NASA patent concerning a supercritical airfoil, were among others:

An airfoil with a freestream *Mach* number within the range of $M = 0.7 - 1.0$ with an upper surface shaped such that, among other things:

- Location of minimum curvature exists
- Minimum velocity increase of the air flowing with sub- and supersonic velocity exists
- Curvature exists that keeps the *Mach* number (supersonic flow) almost constant and reduces the supersonic flow at the trailing edge to almost sonic velocity

Characteristic of Claim 1 of this patent was, among other things, the alleged invention of a minimum curvature in the direction of the airfoil trailing edge downstream of the point of maximum thickness of the airfoil and upstream of the 50% location of the chord. Starting at this position, the curvature was to continuously increase upstream and downstream of this point. In countering the NACA patent claim, it was, therefore, also referred to the work of *Kawalki*.

> With regard to the specifications of Claim 1 of this patent, we refer to the counterarguments [82, 83] that deal in detail with the assessment of transonic airfoils. These research reports had been prepared in 1940 at the German Aeronautical Test Establishment in Berlin-Adlershof and had been classified "For Official Use only" and "Secret" during the war. Both reports were published after the war in the United States of America

as well as in Great Britain. Furthermore, numerous copies of these research reports were presented to the aircraft companies located in the area of the Federal Republic of Germany, which later provided these reports to the German Museum in München. These reports were accessible in the library of the German Museum to everyone before the priority day (Prioritätstag) of the patent.

In addition to the systematic investigations of *Kawalki* [70, 82, 83] concerning the effect of the airfoil shape on the velocity distribution, *Kawalki* presented at the meeting "Questions of High Speeds" of the LGL in 1942 in Berlin further results of his optimization of "high-speed airfoils" [84]. He described airfoils of lowest excess velocities at given lift coefficients, an essential task of design aerodynamics. At a given thickness distribution, the camber was determined, which caused the lowest excess velocity at a desired lift coefficient. Furthermore the effect of the thickness distribution on the excess velocity at the most favorable camber was described. This way, the airfoil most suitable for high-speed flight at a given lift coefficient and minimum excess velocity was determined. Of course the computational effort for this design task related to the airfoil development was, at the time, really enormous, but the results constituted an excellent basis for the experimental investigation in the wind tunnel because the computations were based on potential methods without consideration of the influence of the boundary layer.

2.3.2.4 FURTHER EXAMPLES OF AIRFOIL INVESTIGATIONS PRESENTED AT THE LGL CONFERENCE 1940 IN GÖTTINGEN. *Richter* [85] reported on measurements on the airfoil NACA 00 12-63 in the DVL 2.7-m high-speed wind tunnel. These investigations were carried out within the context of systematic airfoil investigations at the DVL Institute of Aerodynamics under the Branch Head *Bernhard Göthert*. According to the designs of *Kawalki*, an airfoil with a maximum-thickness location had been selected, which was certainly far ahead of the current state of knowledge concerning high-speed airfoils. It was the objective of these measurements to show the effect of the airfoil contour on the change of the airfoil coefficients with *Mach* number M at high freestream speeds by a comparison with test results for the thicker airfoil NACA 00 15-64. A setup for pressure distribution and momentum-loss measurements was chosen, which allowed a correction of the angle of attack and the freestream speed according to *Göthert*. The principle test setup in the 2.7-m high-speed wind tunnel, which *Göthert* kept for all further airfoil measurements, is shown in Fig. 2.60. In this figure, the dimensions of the airfoil and the endplates were indicated, and in Fig. 2.61, the flow conditions for the measurements were shown. From the development of the *Reynolds* number R with the *Mach* number M, it becomes clear that the increase in the *Reynolds* number at high *Mach* numbers is reduced due to the increase in the stagnation temperature. The dependence of the lift coefficient on *Mach* number for

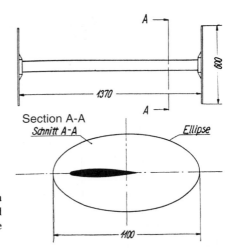

Fig. 2.60 Dimensions and test setup of a rectangular wing with endplates for airfoil measurements [chord t (c) = 500 mm] in the DVL 2.7-m high-speed wind tunnel [85].

various angles of attack is plotted in Fig. 2.62. For each angle of attack, the theoretical lift dependence according to the *Prandtl–Glauert* equation was presented. The initial values of the lift coefficients at the *Mach* number $M \to 0$ were selected such that the curve fitted the measured data as well as possible. There is truly quite good agreement of the measured data with the *Prandtl–Glauert* curves up to the critical *Mach* number that is the *Mach* number where locally the speed of sound was reached on the airfoil. As could be expected from the computations of *Kawalki*, a comparison with the thicker airfoil NACA 00 15-64 (see also Fig. 2.52) showed a lower lift in the area of the critical *Mach* number. An overlapping of the pressure distributions on the pressure and suction sides occurred for the airfoil NACA 00 15-64 at lower freestream *Mach* numbers, which caused heavy lift losses and a forward shift of the pressure point to a position upstream of the wing leading edge.

Further test results of investigations on high-speed airfoils were obtained and introduced by *O. Knappe* [86] of the Ernst Heinkel Company in Rostock. Already in 1937, the company had started with the construction of its own high-speed wind tunnel. The latter was a continuous Eiffel-type wind tunnel

Fig. 2.61 High-speed measurements on the airfoil NACA 0012-63: *Reynolds* number and stagnation temperature (average of all test days) dependent on *Mach* number in the DVL 2.7-m high-speed wind tunnel [85] (axes: *Re*; $t_{Stag.}$, *Mach* number M).

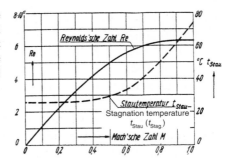

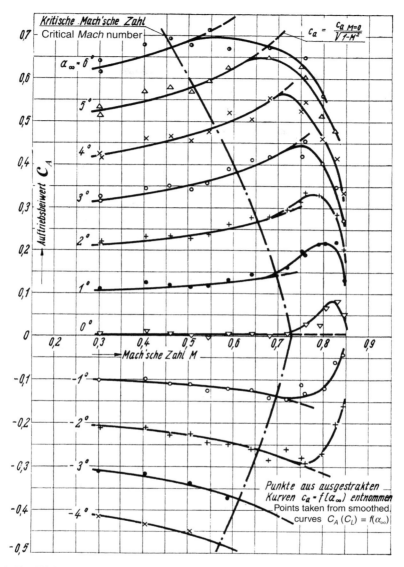

Fig. 2.62   High-speed measurements on the airfoil NACA 0012-63: lift coefficients C_A (C_L) as function of the *Mach* number at different angles of attack in the DVL 2.7-m high-speed wind tunnel [85] (at the critical *Mach* number M^* the speed of sound is reached locally); axes: lift coefficient C_A (C_L), *Mach* number M.

with a test section of 0.13-m² cross section (0.36 × 0.36 m); see Fig. 2.63. The ratio of the inlet area to the test-section cross section was 16:1. Because of this favorable contraction ratio and the installation of a flow straightener, the turbulence level was extremely low. The drive of the two counter-rotating fan wheels consisted of two cage-rotor motors of 185 kW each. The

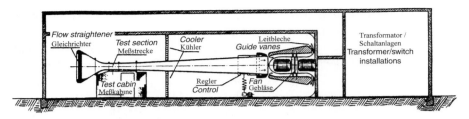

Fig. 2.63   High-speed wind tunnel of the Ernst Heinkel Aircraft Company in Rostock; test-section cross section 0.13 m², O. Knappe [86].

enormously high noise level due to the counter-rotating axial compressor wheels was disadvantageous. In addition, condensation effects at the chosen design principle could not be avoided. The corrections of the angle of attack and the *Mach* number M were performed according to the method of *Göthert*. The first model investigated was a symmetrical airfoil of a thickness of d/t $(d/c) = 0.09$ and a maximum-thickness location of x_D/t $(x_D/c) = 0.42$. Because of the small chord, the tests were carried out in a *Reynolds* numbers range of $R = 0.4$ to 1.3×10^6. The first tests showed, correspondingly, that a laminar boundary-layer separation developed in the regime of shock-free flow, which became turbulent with increasing speed. This caused unexpectedly high angle-of-attack sensitivities with increasing speed, which resulted in considerable "slope changes" in the pitching-moment development. It was then tried to separate these boundary-layer effects from the pure compressibility effects with the aid of a tripping wire. The development of the drag with increasing angle of attack is plotted in Fig. 2.64 for natural and artificial boundary-layer transition. The sketch in the insert shows the arrangement of the tripping wire. Here, a problem was addressed, which is still being controversially discussed today. The results of *Knappe* showed definitely a larger airfoil drag in the case of artificial boundary-layer transition by means of a tripping wire. Here however, it has to be taken into account that, according to the current state of knowledge, the turbulent boundary layer experiences a momentum loss due to the tripping wire, which led to an increase in the

Fig. 2.64   Drag coefficient C_W (C_D) dependent on the angle of attack α at a medium-high (moderate) *Mach* number [86].

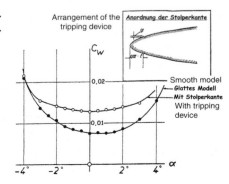

boundary-layer thickness. This boundary-layer thickening was strongly dependent on the airfoil pressure field and considerably affected the shock location. Remarkable was, nevertheless, that *Knappe* had already dealt with these boundary-layer problems in 1940. The discussion concerning the wind-tunnel simulation at *Reynolds* numbers of the full-scale aircraft has also led to the construction of the European Transonic Wind Tunnel (ETW).

Hubert of Messerschmitt in Augsburg reported on an interesting investigation on another symmetrical airfoil of 9% relative thickness with a cambered flap. This concerned the airfoil NACA 000 9-63 with a standard cambered flap at 25% chord. The tests were carried out in the high-speed wind tunnel A6 of the Institute of Gasdynamics of the LFA in Braunschweig. The test section of 40×40 cm permitted, according to the *"Göthert* Criteria," a maximum wing chord of 80 mm. With an airfoil thickness of 7.2 mm, the tunnel blockage was 1.8%. The model had to be designed such that despite the small dimensions, the largest possible number of pressure orifices could be accommodated on the model surface. Here, different model construction methods were tested by the Uher Company in München, and then a model composed of two half-shells was manufactured. Special attention was attached to the disturbance-free construction of the pressure orifices. Considering the intermittent test operation, a diameter of 0.3 mm was selected for the pressure orifices. The long settling time required multiple wind-tunnel runs at a high vacuum. Test results were reported for a *Mach* number range of $M = 0.48$ to $M = 0.86$ at angles of attack of $\alpha = -4$ to $\alpha = +4$ deg and flap angles of 0, 2, and 4 deg. Even if a comparison with other test results was difficult, because no wind-tunnel corrections were applied, a certain agreement with the trends of the measurements published up to that day was established. Up to reaching the local speed of sound, an increase in the aerodynamic forces and moments according to the *Prandtl–Glauert* approximation was noticed. Strong changes occurred at even higher *Mach* numbers in the pressure distributions, which caused a shift in the pressure point location and the neutral-axis location in the direction of the wing trailing edge (downstream). As reported elsewhere, the Messerschmitt Company principally sent all test results of the LFA Braunschweig to the DVL Berlin–Adlershof for the computation of the wind-tunnel corrections. An exchange of experience between the LFA and the Messerschmitt Company did, therefore, not take place any more. Also obvious was that the knowledge of *Hubert* that due to the small model size and hence the limited number of pressure orifices on the flap, rudder forces and moments could not be determined, played a decisive role here.

2.3.2.5 COMPARISON OF HIGH-SPEED MEASUREMENTS OF *GÖTHERT*, DVL BERLIN-ADLERSHOF. Following several publications on symmetrical airfoils investigated, *Göthert* had in a research report [87] carried out a critical analysis of the results and a comparison with measurements in other wind tunnels. As already shown during the discussion of corrections for wind-tunnel wall interferences, *Göthert's* approach as an experimenter was characterized by a special pragmatism. Because it was clear to him that wind-tunnel tests in the near-sonic range only offered the possibility to provide approximate results due to reducing the model size relative to the test-section size, he carried out a systematic comparison of the test results from different wind tunnels and obtained with different model sizes. Here, it had to be considered that for particularly small models the interference forces and moments became larger due to the relatively large supports. The wall interferences were, in addition, being determined for the wind tunnels in a different manner. At the DVL, the pressure distributions along the tunnel wall of the closed test section were also measured in a simple and reliable way so that the attainment of sonic velocity could be checked. Because *Göthert* could not vary the *Reynolds* number in the DVL high-speed wind tunnel independent of the *Mach* number, he tried to gain insight into the physical relations, and especially into boundary-layer effects, by a comparison with results on considerably smaller models from other wind tunnels. Figure 2.65 clearly shows by a comparison of drag coefficients measured on similar airfoils in the AVA Subsonic Wind Tunnel with an open-jet test section and the closed 2.7-m high-speed wind tunnel of the DVL that the drag of the AVA-wing was more than twice as high as the one of the DVL-wing. The rise in the drag coefficients due to compressibility effects occurred at similar *Mach* numbers. The same phenomenon was observed by *Göthert* on an 18%-thick airfoil when he carried out measurements on models of 60- and 500-mm chords, with the airfoil drag of the small model having been determined by force measurements (as in the AVA measurements) and of the large model with the aid of momentum-loss measurements. Striking with regard to these measurements was that the drag coefficients of the small wing were, at low speeds, more than twice as high as the ones of the large wing. Here, *Göthert* tried to find an explanation:

> This effect cannot be explained solely by the difference in the *Reynolds* numbers. (At a *Mach* number of $M = 0.5$ the *Reynolds* number for the 500-mm chord wing is $Re = 4.8 \times 10^6$ and for the 60-mm chord wing $Re = 0.58 \times 10^6$). Despite careful processing of the model surface, it is apparently not possible to achieve on the small model the same surface quality as in the case of the large wing model. I exclude blaming the wall effects of the wind tunnel for the differences in the drag coefficients!

This interpretation of the test results was certainly not self-evident because no information concerning a flow separation on the 60-mm chord model was available. A rough estimate of the higher drag coefficients can be carried out on the basis of the skin-friction law for the flat plate, which indicates that the skin-friction coefficient in the *Reynolds* numbers range $R = 10^5 - 10^7$ decreases with $1/R^{1/5}$. The experimental results were confirmed by computations of

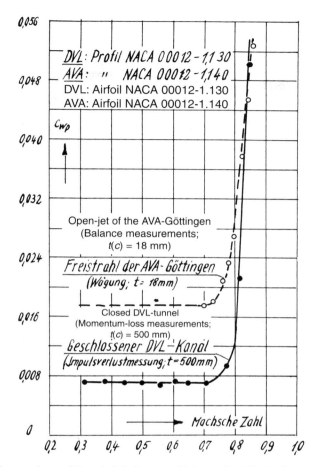

Fig. 2.65 Comparison of the airfoil drag coefficients C_{Wp} (C_{Dp}) of two airfoils with a thickness ratio of 12% [76]; AVA: Measurement on a trapezoidal wing in the 11- × 11-cm high-speed wind tunnel [three-component force measurements with a spring-loaded balance, chord $t(c)$ = 18 mm]; DLR: rectangular wing with endplates [momentum-loss measurements, chord $t(c)$ = 500 mm]; axes: C_{Wp} (C_{Dp}), *Mach* number.

Ralf Heinrich [88] of the DLR Institute of Aerodynamics and Flow Technology in Braunschweig (Fig. 2.66). These computations were carried out with the DLR block-structured FLOWer code [89]. The entire boundary-layer flow was assumed to be fully turbulent, and the *Baldwin–Lomax* turbulence model was used. It will be shown later that the boundary layer in the tests with the large model [t (c) = 500 mm] was still laminar over large parts of the chord. Also the example of the comparison of a measured and a computed pressure distribution shows an astonishingly good agreement (Fig. 2.67), despite the still unknown parameters. This result allows the conclusion that, under the management of *Göthert*, extremely reliable airfoil data were determined in

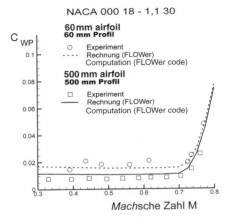

Fig. 2.66 **Comparison of the airfoil drag coefficients** C_{Wp} (C_{Dp}), $(\alpha = 0 \text{ deg})$, **of a normal-size wing** $[t\ (c) = 500\ \text{mm}]$ **and a particularly small wing** $[t\ (c) = 60\ \text{mm}]$ **measured in the DVL 2.7-m high-speed wind tunnel (1941) [76] with computations of** *Ralf Heinrich*, **DLR, Braunschweig, 2005; axes:** C_{Wp} (C_{Dp}), *Mach* **number.**

the DVL 2.7-m high-speed wind tunnel that today, after more than 60 years, represents an interesting database for the validation of CFD codes.

In this connection the comparisons of NACA and DVL investigations on symmetrical airfoils of the same classification [maximum-thickness location x_d/t $(x_d/c) = 30\%$] but different airfoil thickness ratios d/t $(d/c) = 6\%$, 9%, 12% are interesting. The drag measured at the NACA (see Fig. 2.68), at the same *Mach* numbers but smaller *Reynolds* numbers, was higher than the DVL values. However, it was also clearly noticeable that the drag decreased strongly with the relative thickness. The results were a further confirmation of the necessity of measurements in high-speed wind tunnels, which allow a simulation of large *Reynolds* numbers.

2.3.2.6 **DEVELOPMENT OF SPECIAL AIRFOILS FOR THE ATTAINMENT OF HIGH CRITICAL *MACH* NUMBERS AT THE DVL.** As had already been predicted by *Kawalki*, airfoils for high-speed flight were to be developed with excess velocities as small as possible, because then the critical *Mach* numbers were very high and the compressibility effect on lift, pitching moment, and drag commenced

Fig. 2.67 **Comparison of pressure distributions measured at** $\alpha = 0$ **deg on a normal-sized wing** $[t(c) = 500\ \text{mm}\ (1941)]$ **[75] with computations of** *Ralf Heinrich*, **DLR, Braunschweig, 2005, for a symmetrical freestream at** $M = 0.7$; **axes:** c_p, **airfoil chord** x.

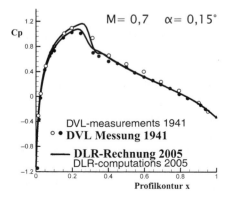

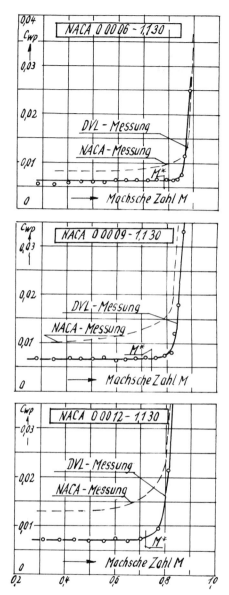

Fig. 2.68   Comparison of drag coefficients C_{Wp} (C_{Dp}) **of airfoil measurements in the DVL 2.7-m high-speed wind tunnel and the NACA 0.28-m wind tunnel [87]; axes:** C_{Wp} (C_{Dp})**, Mach number; "Messung" means measurements.**

only very late. With the aid of computations based on potential theory (flows without friction), it was found that the most favorable maximum-thickness location was in the range between 0.4 and 0.5 chords downstream of the leading edge and the most favorable form of the leading edge was the ellipse. If these airfoils were to have a nonzero lift coefficient C_A (C_L), one had to superpose the symmetrical airfoil by camber such that, at the desired lift, the flow was shock-free. For this purpose, *Göthert* [90] selected together with

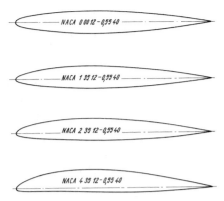

Fig. 2.69 Survey of the airfoil shapes with camber investigated in the DVL 2.7-m high-speed wind tunnel [90].

Kawalki the airfoil series NACA 100 *f/l* 35 12-0.55 40 with the camber ratios of 100 *f/l* (*f/c*) = 0, 1, 2, and 4 for wind-tunnel measurements (Fig. 2.69). Here, 100 *f/l* (*f/c*) designated the camber, 35 the location of maximum camber, 12 the thickness ratio, and 40 the location of the maximum thickness in per cent of the wing chord. The value 0.55 designates the nose radius ρ according to equation $(\rho/l)/(t/l)^2 = (\rho/c)/(t/c)^2 = 0.55$. (Note: *Kawalki* used ρ as nose radius, which normally designates the density.) The airfoils of a chord of $l(c) = 0.35$ m and a span of $b = 1.34$ m (with endplates) were manufactured of metal and equipped with a number of pressure orifices in the center section to determine lift and pitching-moment coefficients. The drag was determined by momentum-loss measurements. In these measurements, the drag coefficients C_W (C_D) were determined for various lift coefficients and *Mach* numbers. As expected, the airfoil with nearly shock-free flow at each lift coefficient had the highest critical *Mach* number. This was at C_A (C_L) = 0 the symmetrical, at C_A (C_L) = 0.2 the 1%-cambered, at C_A (C_L) = 0.4 the 2%-cambered, and at C_A (C_L) = 0.6 the 4%-cambered airfoil. *Ludwieg* explained the particularly low airfoil drag coefficients of $C_{Wp} = C_{Dp} = 0.005$ as follows:

> Within the range of a shock-free entry, a largely laminar boundary layer will be maintained. This laminar influence will not be affected by the compressibility until the critical *Mach* number is reached. After exceeding the critical *Mach* number, the steep drag rise occurs within the range of shock-free entry only after a very small delay.

Surprising was that also at high lift coefficients C_A (C_L), the most favorable behavior regarding the steep drag rise was exhibited by the symmetrical airfoil (Fig. 2.70). At C_A (C_L) = 0.6, the critical *Mach* number for the symmetrical airfoil was already reached at $M = 0.44$, but, nevertheless, the drag rise commenced only at $M = 0.72$. Here, an example was provided where the critical *Mach* number was no longer an indicator for the beginning of the drag rise. The measured pressure distribution clearly showed that a shock appeared

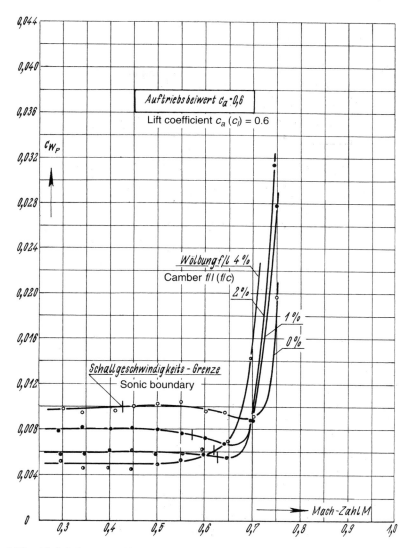

Fig. 2.70   Airfoil drag coefficients C_{Wp} (C_{Dp}) at a lift coefficient C_A (C_L) = 0.6 dependent on the *Mach* number M for differently cambered airfoils [90]; axes: C_{Wp} (C_{Dp}), *Mach* number M.

shortly after exceeding the critical *Mach* number. This phenomenon was explained by *Ludwieg* with the fact that, despite the shock wave, boundary-layer separation did not occur. The relatively small shock losses were compensated, and partly even overcompensated, by a reduction of the skin-friction drag. The reduction in skin-friction drag was explained by a shift of the transition point location towards the airfoil trailing edge. To better under-stand the effect of the boundary-layer development on the airfoils investi-gated, the positions of the boundary-layer transition (transition point locations)

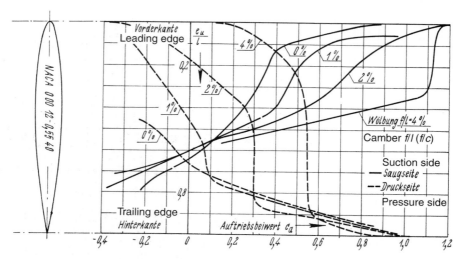

Fig. 2.71   Boundary-layer transition location at a *Mach* number of $M = 0.5$ dependent on the lift coefficient C_A (C_L) according to *Göthert* [90]; axes: l_u/l (c_t/c) = transition location, lift coefficient c_a (c_l).

were observed by dust deposits on the model surface. This method is similar to the so-called sublimation technique, which is still being employed today. That way, the transition point locations at a *Mach* number of $M = 0.5$ for the four wing models at different lift coefficients were determined (Fig. 2.71). One sees that the transition points at negative lift coefficients were located on the upper surface of all wings considerably downstream of the maximum-thickness location. With increasing lift coefficient, the transition points moved steadily towards the airfoil leading edge corresponding to the shift in the pressure minimum. With the aid of the transition point locations, lift ranges could be identified where the transition point on the suction and pressure sides of the airfoil were, on average, closest to the trailing edge of the airfoil and, therefore, airfoil drag lowest. The position of the transition point changes at the *Reynolds* numbers of the tests considered ($M = 0.3 \rightarrow Re = 2.3 \times 10^6$; $M = 0.8 \rightarrow Re = 4.5 \times 10^6$) only a little with the *Mach* number as long as the critical *Mach* number was not exceeded. As shown in Fig. 2.72 for the zero-camber wing, the transition point shifted with increasing *Mach* number in purely subsonic flow steadily in the direction of the leading edge due to the increase in the pressure gradient (according to the *Prandtl–Glauert* rule). As soon as the local supersonic regions developed on the airfoil, the character of the pressure distribution changed such that the pressure minimum shifted towards the wing trailing edge. According to *Göthert's* interpretation of the test results, these pressure distribution changes shifted the transition points towards the wing trailing edge. To show the steadiness of the lift dependence at high *Mach* numbers, the angles of attack corresponding to a lift coefficient

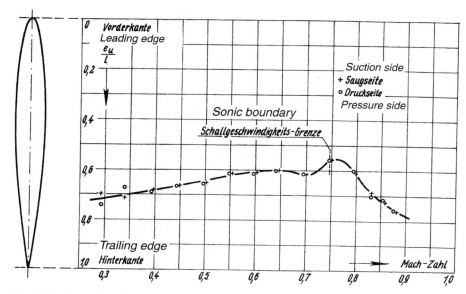

Fig. 2.72 **Location of the transition region dependent on *Mach* number at a constant angle of attack** $\alpha = 0.025$ **deg according to *Göthert* [90]; axes:** l_u/l (c_t/c), ***Mach* number.**

of C_A (C_L) = 0.6 were plotted in Fig. 2.73 as function of the *Mach* number. For the lift coefficients of C_A (C_L) = 0, 0.2, 0.4, and 0.6, a similar trend could be observed. According to the *Prandtl–Glauert* rule, the angle of attack was expected to decrease steadily with increasing *Mach* number. All measured curves showed a good agreement with the *Prandtl–Glauert* rule below the critical Mach numbers. However at *Mach* numbers of $M = 0.7$ to $M = 0.8$, the angle of attack changed rapidly towards higher values. Here, a distinctly stronger angle-of-attack change with increasing camber could be observed. It could again be noticed that, without doubt, the critical *Mach* number was not a unique indicator for the way the angle of attack changed. A plot of the pitching moments C_m referred to the 1/4-chord line; Fig. 2.74 shows as a result that, again, the symmetrical airfoil proved to be the most favorable one independent of the respective lift coefficient.

After the strong effect of the maximum-thickness location had been determined, the present investigations on the symmetrical airfoils NACA 000 12-1.1 30 and NACA 0 00 12-1.1 40 were supplemented by investigations on the airfoil NACA 0 00 12-0.55 50. This test was conducted in the summer of 1943 and allowed *Göthert* [91] a critical comparison of these three airfoils. As expected, the *Mach* number where the steep rise in the drag coefficient C_W (C_D) commenced (drag-rise *Mach* number) for the standard airfoil with the 30% maximum-thickness location was strongly reduced with increasing lift coefficient C_A (C_L). For the airfoil with the 40% maximum-thickness location, this decrease was already noticeably reduced whereas for the airfoil

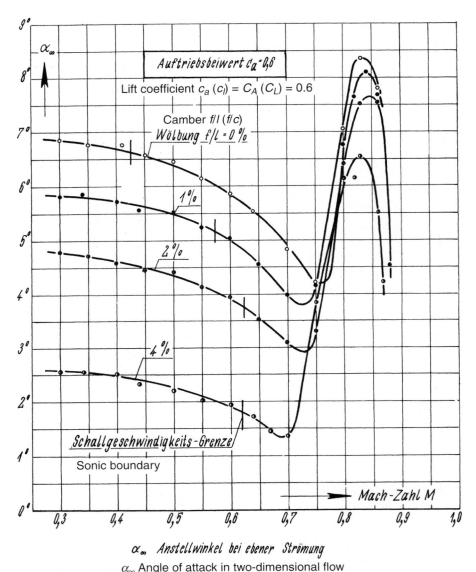

α_∞ *Anstellwinkel bei ebener Strömung*
α_∞ Angle of attack in two-dimensional flow

Fig. 2.73 Angle of attack α **(two-dimensional flow) for a lift coefficient of C_A $(C_L) = 0.6$ dependent on *Mach* number for differently cambered airfoils according to *Göthert* [90].**

with the 50% maximum-thickness location the drag-rise *Mach* number practically did not change any more with lift. For this airfoil, the shocks did, as already explained, obviously not lead to a separation. This aero- and gasdynamically favorable airfoil with a 50% maximum-thickness location thus showed a way to further improve airfoil qualities at supercritical speeds. One

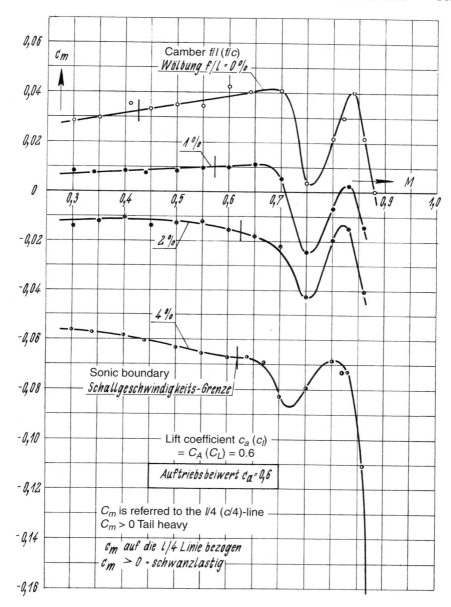

Fig. 2.74 **Pitching-moment coefficient C_m at C_A (C_L) = 0.6 dependent on *Mach* number M for differently cambered airfoils according to *Göthert* [90].**

had to try to avoid separation by boundary-layer control. It was, in addition, attempted to avoid the shock-related losses by decelerating the flow in the supersonic region to subsonic velocities by an adiabatic (isentropic) recompression instead of a shock wave.

2.3.2.7 IMPROVEMENT OF THE AIRFOIL CHARACTERISTIC AT SUPERCRITICAL CONDITIONS BY SPECIAL DESIGN AND BOUNDARY-LAYER CONTROL. Here, *H. B. Helmbold* [92] in cooperation with *Göthert* conducted a test that was reported on by *Ludwieg* [93]. As shown in Fig. 2.75, an airfoil was investigated whose contour was concave in the downstream part. The compression waves emanating from this concave surface were to decelerate the locally supersonic flow in an adiabatic process back to subsonic flow avoiding shock waves. In addition the steady pressure increase (recompression) over the downstream part of the airfoil was to avoid separation. The measured momentum-loss distribution at a *Mach* number of $M = 0.87$ was plotted in Fig. 2.75. The momentum-loss due to the boundary layer and the losses due to the shock waves were clearly distinguishable, the latter indicated by the extremely large secondary maxima. *Helmbold* pointed at still another observation. In the tests in the 2.7-m high-speed wind tunnel of the DVL, model flows were observed where, due to the extended boundary-layer separation, not only was the shock near the model surface replaced by a steady pressure rise, but nearly the entire shock, except for a small remainder, disappeared due to the separation. Concerning the reason for this, it was explained that, due to the strong flow separation, a new effective boundary was formed at the edge of the separated region such that considerably lower excess velocities with correspondingly weaker shocks occurred. This caused strong vibration in the wind tunnel because the shocks, controlled by the separation, were very unstable. As a result of such oscillating shocks, a strong shaking of the models occurred. These observations and interpretations of wind-tunnel phenomena required an excellent understanding of the flow physics!

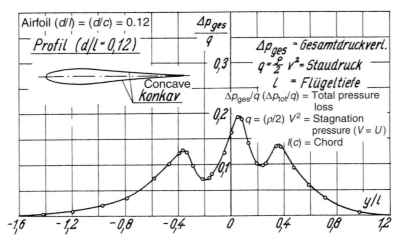

Fig. 2.75 Total pressure losses downstream of a wing with a concave downstream region ($M = 0.87$) [92]; axes: $\Delta p_{ges}/q$ ($\Delta p_{tot}/q$), y/l (y/c).

Based on the knowledge that an effective way of reducing the drag of high-speed airfoils was the reduction of the extent of separation, *Bruno Regenscheit* [94] carried out first tests with boundary-layer suction in 1940. He had been inspired to do so by remarks of *Helmbold* at the 1940 meeting of the LGL. *Regenscheit* had already gained experience concerning boundary-layer control in high-lift tests in the low-speed range at the AVA Göttingen. For these investigations, the 11- × 11-cm high-speed wind tunnel of the AVA was available (see Fig. 2.12) where *Ludwieg* had performed his first swept-wing investigations. The test setup utilized (Fig. 2.76) shows that the wing model was attached to a suction pipe, which was connected to a 40-m^3 vacuum vessel. Due to this test setup, it was not possible to employ the three-component balance. The drag had to be determined by momentum-loss measurements. For the interpretation of the measurements, a schlieren system for the visualization was routinely available. The model wing had a span of 140 mm and a chord of 60 mm. The airfoils were formed by 17%-thick circular arcs with a rounded nose. The wings were equipped with longitudinal and crosswise slots for boundary-layer suction. The investigations were only carried out at an angle of attack of $\alpha = 0$ deg but at different *Mach* numbers.

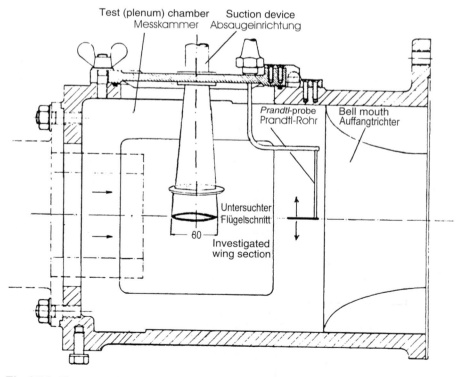

Fig. 2.76   Test setup for the high-speed airfoils with suction according to *B. Regenscheit* [94].

As shown in Fig. 2.77, considerable drag reductions were obtained dependent on the suction rate $C_Q = Q/(\rho VS)$ and the freestream *Mach* number [Q = suction rate, ρ = freestream density, V (U) = freestream velocity, and S = wing reference area]. The results showed that with increasing suction coefficient, C_Q, the drag coefficient was strongly reduced. The tests, carried out with the simplest means, clearly demonstrated the most favorable results to be obtained by longitudinal slots. *Betz* had already in 1940, immediately after the completion of the tests, applied for a patent designated "Wings for High-Speed Flight with Installations to Control Drag and Lift." *Regenscheit* was named as inventor. The patent claim was initially turned down on the grounds, among others, that the documents submitted lacked details concerning the shock location. The correspondence (Göttingen Archive) is documented up to January 1944—thereafter surely other priorities prevailed. *Regenscheit's* investigations were the first ones concerning boundary-layer suction to reduce the airfoil drag in the area of compressible subsonic flow.

In Volume E of the Göttingen "Monographs Concerning the Progress of German Aeronautical Research since 1939" many further research results

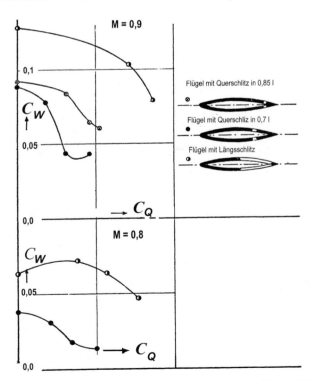

Fig. 2.77   Change of the drag coefficient C_W (C_D) for various positions of the suction slot according to B. Regenscheit [94]; axes: C_W (C_D), C_Q; legend (top to bottom): wing with lateral (cross) slot at 0.85 l (c), wing with lateral (cross) slot at 0.70 l (c), and wing with longitudinal slot.

related to the airfoil development are reported. Just the bibliography comprises 245 references, which were, however, not completely available to our enquiries. In treating our topic "High-Speed Aerodynamics," two further developments were, of course, of interest, namely 1) laminar airfoils and 2) high-lift aids (lift augmentation).

2.3.2.8 LAMINAR AIRFOILS. Keeping the boundary-layer laminar (laminarization) was a particular challenge to research because the U.S. Air Force had successfully flown an aircraft with a laminar wing, the North American Fighter Aircraft "Mustang." The question, whether laminar airfoils could be utilized at high speeds, was primarily decided by considering whether the laminar boundary layer could be maintained at the high *Reynolds* numbers of $R \approx 30 \times 10^6$. In that respect, F. W. Riegels [95], among others, obtained a negative result for laminar airfoils (see Fig. 2.78), which were theoretically described as favorable. With the aid of large model chords, *Reynolds* numbers of up to $R \approx 28 \times 10^6$ could be simulated in these tests in the Göttingen Wind Tunnel KVI. The measured airfoil drag at a lift coefficient of C_A (C_L) = 0 decreased initially at $R < 5 \times 10^6$ but increased again with increasing *Reynolds*

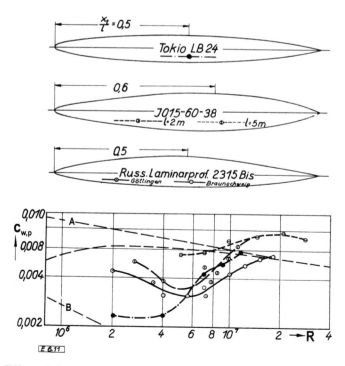

Fig. 2.78   Effect of the *Reynolds* number R on the drag C_{Wp} (C_{Dp}) of laminar airfoils according to A. Walz [95]; line A) flat plate, turbulent line B) flat plate, laminar boundary layer; legend: $x_t/l = x_t/c$ = maximum-thickness location.

number and reached at $Re > 10 \times 10^6$ already the drag-coefficient level of standard airfoils. It is interesting that large differences did not occur between the results of comparative measurements in the 8-m low-speed wind tunnel A3 of the LFA and the KVI tunnel, which pointed to similarly good flow qualities in these two facilities. In these tests at high speeds, of course, especially the comparison between the English Mustang airfoil and the optimized airfoils of *Göthert* and *Kawalki* was of interest. The test results [96] for the Mustang airfoil (1.6 50 13.6-0.825 39) could be compared to the ones for the airfoil NACA 2 35 12-0.55 40 with the DVL airfoil behaving similarly but being, however, more favorable in the compressible speed range. Because in Germany the decision concerning the airfoil selection for high-speed flight had already been made in favor of the symmetrical airfoils, which were even more superior to the Mustang airfoil regarding drag behavior, there existed in this field of research no further need for action. These investigations of laminar airfoils were supported by the great progress, which had been achieved after the basic work of *W. Tollmien*, in the area of boundary-layer stability computations by *H. Schlichting* and *A. Ulrich* [97] as well as by *J. Pretsch* [98]. With these theories, one was able to determine the limits of the laminar boundary layer no longer being stable when facing small disturbances. For the first time, one could include the pressure gradients existing on the airfoil into the stability computations. Beyond a certain "limit of stability" of the boundary layer, an amplification of the small disturbances occurred. These small disturbances were in practice, for example, individual roughness elements, such as aircraft rivets, or in wind-tunnel test the turbulence of the flow. To raise the limit of stability of the laminar boundary layer, it was tried to consider theoretically as well as experimentally boundary-layer suction. Here, *J. Pretsch* [98] had computed the "optimal suction" for a flat plate. In that regard, *Schlichting* and *Bussmann* [99] could show that in the case of suction through individual sinks the plate acquired an additional drag, because the amount of liquid sucked off transferred its momentum to the plate. However, this share was small as long as only liquid particles near the wall were sucked off. The activities concerning boundary-layer control were focused at the AVA in Göttingen because also the basic theoretical means had been developed here.

2.3.2.9 BOUNDARY-LAYER CONTROL FOR REDUCING AIRFOIL DRAG. Concerning the reduction of airfoil drag by boundary-layer control, the following alternatives had been discussed in the time period considered here (1935–1945):

- Laminarization of the boundary layer
- Turbulent boundary-layer suction over the downstream part of the airfoil (reduction of pressure drag)
- Reduction of turbulent skin-friction drag by blowing

Concerning the problem of laminarization, *H. Holstein* [100] had, based on the investigations of *O. Schrenk* [101] carried out some important basic tests. According to the theory, by means of continuously distributed suction, for example, on a flat plate at high *Reynolds* numbers, performance gains of 50% and more were possible. Because continuous suction could technically only be realized with extreme difficulties, suction through individual slots was tested in detail. As nondimensional parameter for the suction or blowing rate Q, the following coefficient was introduced

$$C_Q = Q/(VS) = Q/(US) \; [S = \text{suction area}, \; V\,(U) = \text{freestream velocity}]$$

and for the suction or blowing pressure p_A, the nondimensional pressure parameter

$$C_p = p_A/q \; [q = (\rho/2)V^2 = \text{stagnation pressure}, \; V = U]$$

Figure 2.79 shows the development of the airfoil drag C_{Wp} (C_{Dp}) and the airfoil performance coefficient $C_P = C_{Wp} + C_Q(1 - C_p)$ essential to the assessment, for the airfoil NACA 0012-64 with suction.

One may principally state that the laminarization of the boundary layer only succeeded when the suction slot was located within the laminar region of the boundary layer. Under these conditions, the laminar boundary layer could also be located in the region of increasing pressures. A relaminarization

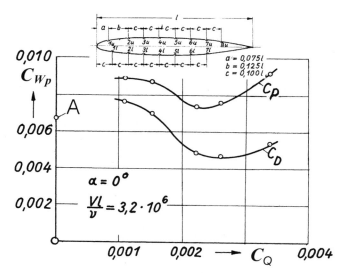

Fig. 2.79 Drag coefficient C_{Wp} (C_{Dp}) determined by momentum-loss measurements in the center section and performance coefficient $C_P = C_{Wp} + C_Q(1 - C_p)$ dependent on the suction coefficient C_Q for simultaneous suction through six different slots on the airfoil NACA 0012-64; flow condition A: all slots closed. *Holstein* [100]; legend: $Vl/\nu = Ucl/\nu$.

of the boundary layer was no longer possible once the boundary layer had become turbulent, a result that lets one conclude that through each slot an amount had to be sucked off that kept the boundary-layer laminar up to the next suction slot. This led to an extensive optimization program concerning the effect of the number of slots and the slot spacing. From the test result in Fig. 2.79, the optimum power requirement for the minimization of drag at a given flow condition could be determined.

2.3.2.10 TAKEOFF AND LANDING AIDS IN THE LEADING- AND TRAILING-EDGE REGIONS OF HIGH-SPEED AIRFOILS [102]. The maximum lift of airfoils explicitly designed for high-speed flight was, due to the relatively pointed airfoil leading edges and the early turbulent boundary-layer separation, relatively low. Because of the development of a "leading-edge (nose) flap" by *Krüger*, it was possible to considerably increase maximum lift. Because these high-lift aids, known as "*Krüger* flaps" among experts, are today still being used by many aircraft, this development will be further discussed in detail. The mode of operation of the nose flap, depicted in Fig. 2.80, was explained to be such that, at the right flap setting, the forward stagnation point of the incoming flow was forced into a position in the immediate vicinity of the leading edge of the flap. This reduced the high excess velocities that appeared at high angles of attack near the leading edge of "high-speed airfoils." The suction peak on the airfoil was thus reduced and the flow separation delayed towards higher angles of attack. The mode of operation of the leading-edge flap was demonstrated by flow visualizations in the water tunnel; see Fig. 2.81. Here, a laminar airfoil was investigated with a thickness ratio of 12%, a maximum-thickness location of 50%, and a relatively small nose radius [nose radius parameter $= (r/l)/(t/l)^2 = (r/c)/(t/c)^2 = 0.21$, with the nose radius here being designated r and not ρ!]. Extensive wind-tunnel tests [103–106] on airfoils with different nose radii led to the result that the nose radius parameter $(r/c)/(t/c)^2$ was of decisive influence on the effectiveness of the nose flap. The more pointed the airfoil investigated was, the more efficient the *Krüger*-flap proved to be. For airfoils with a standard NACA nose-radius parameter of $(r/c)/(t/c)^2 \geq 1.1$, an improvement in maximum lift $C_{A,\max}$ ($C_{L,\max}$) was no longer attainable because these airfoils already generated lower suction peaks without aids. Here, turbulent separation slowly moved towards the airfoil leading edge with increasing angle of attack; see Fig. 2.82. Further investigations showed that the lift-augmenting effect of the nose flap depended, in addition, strongly on the chord ratio l_N/l (c_N/c) (Fig. 2.83) and the nose-flap angle η_N (Fig. 2.84). In the

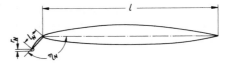

Fig. 2.80   Laminar airfoil (1 50 12-50-0.21) with nose (leading-edge) flap according to W. *Krüger* [102]; *l* = *c*.

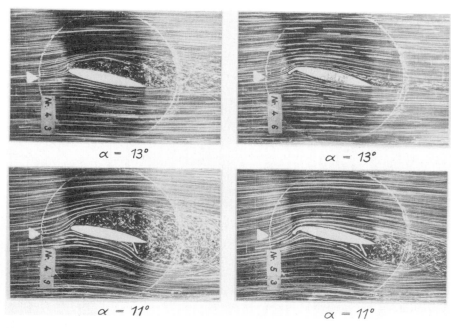

Fig. 2.81   Flow visualization on a laminar airfoil with high-lift devices in a water tunnel: *Krüger*-**flap, trailing-edge split flap [102].**

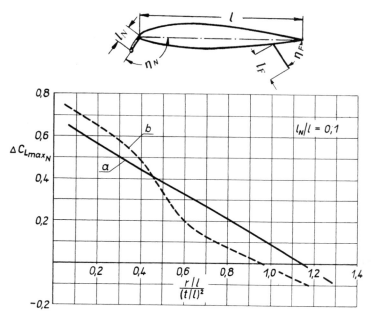

Fig. 2.82   High-lift increase due to a nose flap (leading-edge flap) dependent on the nose-radius parameter $(r/l)/(t/l)^2 = (r/c)/(t/c)^2$**, laminar airfoil (1 50 12-50-0.21) [102]; curve a = airfoil without split flap and curve b = airfoil with split flap (20% chord,** $\eta = 60$ **deg, deflected).**

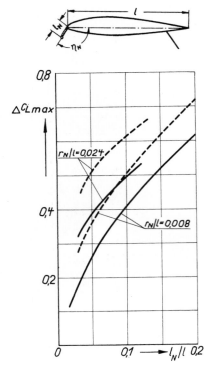

Fig. 2.83 High-lift increase due to a nose flap dependent on the nose-chord ratio l_N/l (c_N/c), laminar airfoil (1 50 12-50-0.21) [102]; without split flap (solid line), and with split flap (dashed line) [chord of the split flap $l_F = 0.2\,l$ ($c_F = 0.2\,c$)]; flap angle $\eta_F = 60$ deg).

parametric studies of the effect of the nose radius of the airfoil and the chord ratio of the nose flap on maximum lift, the effect of a so-called split flap on the lift increase was also investigated. *Doetsch* [107] had given in his report a comprehensive survey concerning the state of airfoil research and the tasks still to be solved. In his survey, he was able to correlate the results of all available tests on different airfoils with a fixed split flap of 20% chord and a flap deflection of 60 deg by a single shape parameter $(r/l)(t/l)^2/(4x_t/l) = (r/c)$ $(t/c)^2/(4x_t/c)$ (see Fig. 2.85). This way, it could be shown that the maximum lift mainly depended on the shape of the forward part of the airfoil up to the maximum-thickness location x_t. The pressure distribution over the downstream part of the airfoil was obviously so strongly dominated by the low-pressure area behind the split flap that details of the design in this part of the airfoil had no effect on the pressure distribution, and hence on maximum lift, $C_{A,\max}$ ($C_{L,\max}$). This result was also confirmed by the flow visualizations in Fig. 2.81.

2.3.2.11 EFFECT OF SURFACE ROUGHNESS ON AIRFOIL DRAG. To solve this problem, numerous investigations were carried out by *Doetsch* [108] at the DVL in Berlin and by *Holstein* [109] at the AVA in Göttingen. The reason for these activities was based on the fact that a large research effort was invested into

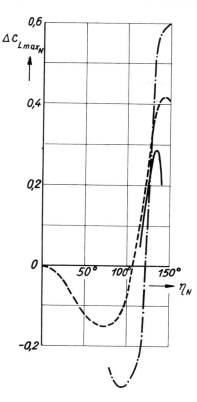

Fig. 2.84   High-lift increase due to a nose flap dependent on the flap angle η_N at different flap chords l_N/l (c_N/c). Laminar airfoil, without split flap, curvature of the flap leading edge r_N/l (r_N/c) = 0.008 [102]; A: nose-flap chord ratio $l_N/l = c_N/c = 0.05$, B: nose-flap chord ratio $l_N/l = c_N/c = 0.10$, and C: nose-flap chord ratio $l_N/l = c_N/c = 0.20$.

the development and investigation of laminar airfoils without checking their technical feasibility. The effect of technical roughness in turbulent boundary layers as an indicator of the quality of the aircraft production in wartime was largely unknown. *Doetsch* had, therefore, original wings of the aircraft types Junkers Ju 288, Heinkel He 177, Messerschmitt Me 109B, and Focke-Wulf Fw 190 investigated in the large DVL subsonic tunnel, and the results compared with results of an original wing of the North American Mustang. The table of *Doetsch* in Fig. 2.86 shows the result of the drag increase of four well-known German aircraft types and its effect on range, performance, and fuel consumption. *Doetsch* commented on these results as follows:

> From the experimental result I showed to you, it clearly follows that the German wings investigated exhibit a considerably higher drag than the ideally smooth wing and that their surface design is still greatly in need of improvement. The results show that the wing of the He 177 exhibits with 60% the largest drag difference compared to the ideally smooth wing and simultaneously the absolute highest drag coefficient. The Me 109 E which is, however, with regard to the production technique considered outdated compared to the newer aircraft, follows with 49%, the Ju 288 with 46%, and the Fw 190 with 33%.

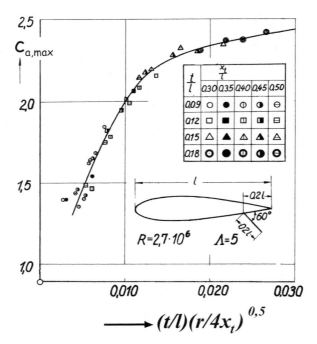

Fig. 2.85   Maximum lift coefficient $C_{A\max}$ ($C_{L\max}$) of symmetrical airfoils of different maximum-thickness locations x_t/l and a geometrically fixed split flap, H. Doetsch [107]; axes: $C_{a,\max}$ ($C_{L\max}$), $(t/l)(r/4x_t)^{0.5}$; $l = c$.

If one assumes the drag of the wing to comprise 40% of the total drag of the aircraft, it follows that for the He 177 24%, the Me 109B 20%, the Ju 288 16%, and for the Fw 190 13% of the engine power was needed for this additional drag due to surface disturbances.

The Mustang-wing stays with a drag coefficient of C_{Wp} (C_{Dp}) = 0.0072 far below the drag of all German wings investigated. It must also be taken into account that the latter was investigated without the standard surface coating. If the drag of the German wings (comprised of standard airfoils with a maximum-thickness location of 30%) is already that high compared to the ideal one, then surface disturbances will have an even stronger effect if airfoils with a larger maximum-thickness location, for example, to improve the airfoil characteristics at high *Mach* numbers, are to be utilized. If it is finally attempted to develop and employ airfoils, whose drag due to laminarization of the boundary layer is 40% or more (lower than one of the airfoils used up to now) the requirement is for a wing design of greater perfection.

After this almost disillusioning statement, one comes to the conclusion that the research and development with regard to laminar wings had, indeed, provided interesting and important contributions, but that products for the

a)

Trag-flügel	$c_{wp\ M}$ (c_a =0,2)	c_{wp} id. (c_a =0,2)	$100\ \dfrac{\Delta c_{wp}}{c_{wp}\ id.}$	v.H.Verlust bei $\frac{W_T}{W_F}$ =0,3 Motorleist., od. Reichweite.	v.H.Verlust bei $\frac{W_T}{W_F}$ =0,4 Brennstoff
He 177	0,0109	0,0068	60	18	24
Me 109 B	0,0101	0,0068	49	15	20
Ju 288	0,0102	0,0070	46	14	18
FW 190	0,0089	0,0067	33	10	13
Mustang	0,0072				

$c_{wp\ M}$ = Mittelwert im untersuchten Spannweitenbereich
c_{wp} id. = Widerstandsbeizahl des id. glatten Tragflügels
W_T = Tragflügelwiderstand
W_F = Flugzeugwiderstand

b)

$c_{wp\ M}$ = Average measured drag coefficient within the span region investigated

$C_{wp\ id.}$ = Drag coefficient of the ideally smooth wing

W_T = Wing drag

W_F = Aircraft drag

(subscript wp = Drag, a = Lift)

v.H.Verlust bei = Percent loss at; Motorleistung = Engine performance;

Brennsstoff = Fuel; Reichweite = Range

Fig. 2.86   a) Airfoil drag coefficient $c_{wp} = C_{Wp}$ ($c_{dp} = C_{Dp}$), $c_a = c_L$ (lift coeficient); design conditions: subscript id = ideally smooth, and original wing: subscript M = measurements of *Doetsch* [108], b) translation.

high-speed range could not yet be manufactured by the industry. An improvement in the production quality had to be ruled out up to the end of the war.

2.3.3   REALIZATION OF THE SWEPT-WING CONCEPT [110, 111]

2.3.3.1   CONFIGURATION STUDIES AT LOW SPEEDS.   Following *Ludwieg's* experimental confirmation of the drag reduction by sweeping the wing in the high-speed domain, corresponding investigations were, of course, in the center of interest. The disadvantage of sweepback was the movement of the wall boundary layers towards the wing tip, as indicated by the tufts attached to the wing surface in Fig. 2.87. The consequence was an early flow separation on the outer wing. From it resulted a "tail down" positive pitching moment, increasing with angle of attack, and a roll instability near stall, a lower maximum lift, a lower maximum-lift increase due to landing devices and a deterioration of the aileron effectiveness. The wing configuration in Fig. 2.87 was selected from the DLR Archive because of the good quality of the photographs. This wing, designed by *J. Weissinger* [112], had due to its special chord distribution for example compared to a trapezoidal wing, theoretically a

a)

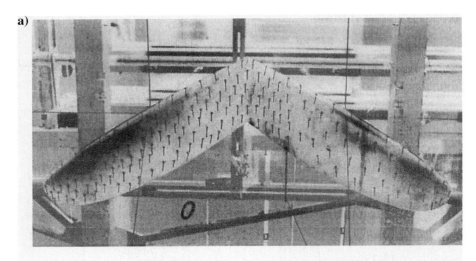

b)

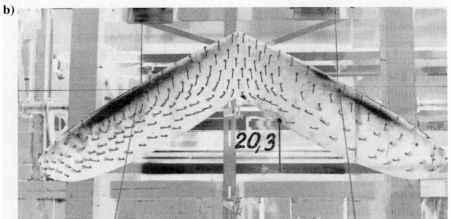

Fig. 2.87 Tufts on an "elliptical wing": a) $\alpha = 0$ **deg; b)** $\alpha = 20.3$ **deg (DLR Archive Göttingen).**

constant spanwise lift distribution. This was a purely fundamental investigation, whose experimental verification by *H. Lemme* in 1943–1944 led to better conditions with regard to the attachment of the flow along the span!

With forward sweep, the wing boundary layer moved inward (see Fig. 2.88). The boundary layer first separated on the inner wing, which also caused a tail-heaviness at large angles of attack. However, the flow at the wing tips remained attached up to the angle of attack corresponding to maximum lift. The aileron effectiveness and hence the roll stability and controllability were, therefore, maintained. The coefficient of the yaw/roll moment became smaller with increasing lift coefficient for this wing configuration, in comparison to the swept-back and unswept wings (see Chapter 6).

Fig. 2.88   M-wing flow visualization: a) angle of attack $\alpha = 0$ deg; b) angle of attack $\alpha = 20$ deg (DLR Archive Göttingen).

Next, test results at low speeds related to pure configurational investigations will be described. The initial objective of these investigations was to find optimum geometries for wings without takeoff and landing devices. A survey of the wings investigated at the Institute of Aerodynamics of the TH Braunschweig, headed by *Hermann Schlichting*, is given in Fig. 2.89. For these 15 wing configurations, extensive six-component and pressure distribution measurements on models with a maximum span of $b = 700$ mm and a wing chord of l $(c) = 150$ mm $[\Lambda = b^2/(b \; 1) = 5.18]$ were carried out in the wind tunnel of the TH Braunschweig at an air speed of 40 m/sec by *W. Jacobs* [113–117]. The conditions corresponded to a *Reynolds* number of $R = 4.2 \times 10^5$. The model wings were constructed of steel bars and brass ribs. The ribs were equipped with pressure orifices, which were more closely spaced in the area of large lift changes, that is, on the outer wing. The

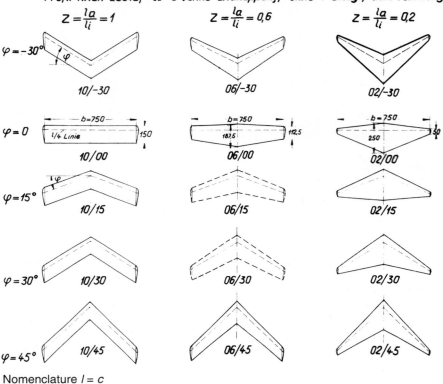

Swept Wing
Pfeilflügel

Airfoil NACA 23012; Λ = 5 (without end caps); without dihedral; without twist
Profil NACA 23012; Λ=5 (ohne Endkappen); ohne V-Stellg.; ohne Verwindg.

Nomenclature $l = c$

Fig. 2.89 Survey of the wings investigated at the Institute of Aerodynamics of the TH Braunschweig, W. Jacobs [117].

connections of the individual pressure leads were embedded. After installing the end caps, faultless force measurements on the model wing were possible. As a result of the pressure distribution measurements at different angles of attack, the normal force coefficients C_N (in Fig. 2.90, $C_N = c_n$) for instance, were plotted vs the nondimensional spanwise coordinate $\eta = 2y/b$. It was interesting that at the relatively small freestream *Reynolds* number, the lift, at a sweep angle of 30 deg, was strongly reduced at the center wing with increasing angle of attack α and towards the tip even an increase occurred (see Fig. 2.90). The latter was a result of the flow stalling in the outer region of the wing. An important task of these measurements was to establish a database that could be utilized for the validation of newly developed computational methods.

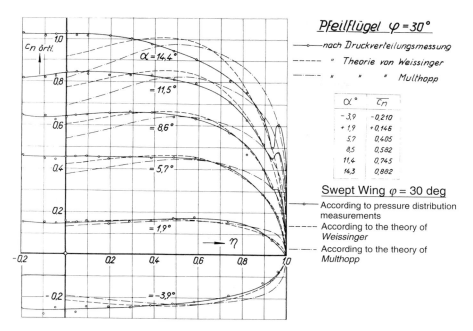

Fig. 2.90   Lift distribution on a $\varphi = 30$ **deg swept wing at different angles of attack** α**, comparison theory/experiment** ($l_i/l_a = 1$, $\Lambda = 5$), **W. Jacobs [114]; axes:** $c_{n,\text{local}}$, $\eta = 2y/b$ (b **= half-span,** c_n **= normal force coefficient).**

2.3.3.2   COMPUTATIONAL METHODS TO DETERMINE THE LIFT DISTRIBUTIONS ON SWEPT WINGS.   Three methods for the computation of lift distributions on swept wings were available at the time:

1) The method of *H. Multhopp* [118]: This method was based on *Prandtl's* lifting-line theory under consideration of a certain correction factor.
2) The lifting-wing theory according to *J. Weissinger* [119], an "F-method," large-scale computational effort: Here, a two-dimensional circulation distribution on the wing was assumed as the basis.
3) The lifting-line method of *J. Weissinger* [119]: an "L-method," about 1/3 of the computational effort of the F-method.

Weissinger was able to show that between the simpler lifting-line method and the lifting-wing theory only small differences in the results could be noticed. *Weissinger* [120] remarked in the "Yearbook of Mathematical Surveys 1985" looking back at his time at the DVL 1937–1945:

> In *Prandtl's* model the wing is replaced by a lifting line, i.e. a straight discreet vortex with the circulation $\Gamma(y)$, from which a flat string of free vortices of density $\Gamma(y)$ branches off against the flight direction, i.e. normal to the lifting line. For the description of the actual flow near the

wing this is, of course, a quite unsuitable model since the velocity in the vicinity of the "lifting line" goes, reciprocally to the distance, to infinity. All of us were greatly surprised that based on such a model one could realistically compute the demanded circulation distribution $\Gamma(y)$. Despite intensive considerations, one finally concluded that the method in an incomprehensible way emanated from *Prandtl's* congenial head, as Athena from the head of Zeus, and decided to no longer consider this miracle but to concentrate on the mathematical description, a singular integral-differential equation of second order for $\Gamma(y)$.

Weissinger's solution was based on the "lifting-wing" model where the wing is not only represented by a discreet vortex but by a continuous vortex system of density $\gamma(x,y)$ distributed over the wing-plan form area. This model had already been investigated by *Albert Betz* in his thesis for a symmetrical freestream. It was at the time, however, only of theoretical interest because the resulting two-dimensional integral equation could not be solved with the means available. *Konrad Zuse* had only started some years prior to the war to develop under contract to the Henschel Aircraft Company his first computer, especially for the solution of linear systems of equations.

The comparison between the measured lift distributions and the ones computed by the methods *Multhopp* and *Weissinger*, Fig. 2.91, allowed the following conclusions. The theory according to *Weissinger* definitely pro-vides a better agreement with the measurements than the one according to *Multhopp*. According to *Multhopp*, the multiplier between the circulation and the effective angle of attack was independent of the sweep angle, that is, ∂C_A $(C_L)/\partial\alpha$ = constant. The "lifting-wing" theory of *Weissinger* predicted a decrease in total lift, thus also a decrease in the lift-curve slope ∂C_A $(C_L)/\partial\alpha$ with increasing sweep; see Fig. 2.91. The measured lift increase ∂C_A $(C_L)/\partial\alpha$ amounted to only about 60% of the value determined according to the theory of *Weissinger*; however, the trend was correctly represented. A comparison of the lift coefficients obtained by the direct force measurements and the ones determined from the pressure distributions yielded a good agreement.

2.3.3.3 VERIFICATION OF THE CONFIGURATION STUDIES ON SWEPT WINGS AT HIGH REYNOLDS NUMBERS.

Although the measurements carried out by *Jacobs* clearly showed some characteristic influence parameters of wing sweep, due to the low *Reynolds* numbers of the investigations ($R = 4.2 \times 10^5$), they could not directly be transferred to full scale without further verification. Further inves-tigations concerning the effect of the *Reynolds* number were, therefore, carried out on swept wings with different plan forms in the large low-speed wind tunnels of the AVA, DVL, and LFA. The tests at the DVL [121] concentrated first on comparative measurements on trapezoidal wings without and with sweep ($\varphi = 0$ and 35 deg); see Fig. 2.92. The behavior of the 35-deg sweptback wing in comparison to the corresponding straight wing was determined in the

Insert: $\partial c_a / \partial \alpha (\partial c_l / \partial \alpha)$
       - - -   According to Multhopp
       –o–   According to measurements
       –•–   According to Weissinger

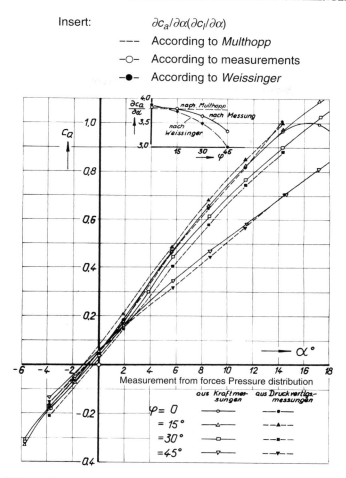

Measurement from forces   Pressure distribution

	aus Kraftmessungen	aus Druckverteilungsmessungen
$\varphi = 0$	—o—	—•—
= 15°	—△—	– –▲– –
= 30°	—□—	– –■– –
= 45°	—▽—	– –▼– –

Fig. 2.91   Lift coefficient C_A (C_L) dependent on the angle of attack α at symmetrical freestream conditions; comparison of force and pressure distribution measurements, Jacobs [114]; axes: c_a (c_l), α deg.

5 × 7 m wind tunnel of the DVL on models of 4 m span at a *Reynolds* number of $R = 2.45 \times 10^6$. The drag development at low lift coefficients (Fig. 2.93) showed that due to the maximum thickness location of 40%, a laminar effect as already discussed in Sec. 2.2.2 occurred. The drag minimum C_{Wpmin} (C_{Dpmin}) was still somewhat lower for the swept wing than for the straight wing. *Seiferth* assumed that the excess velocities were somewhat reduced due to the swept form. Concerning the drag coefficients C_{Wp} (C_{Dp}), it has to be noted that the induced drag according to the *Prandtl–Glauert* rule was, as common at the time, subtracted. The maximum lift of the swept wing was $(C_{A,max})_{\varphi = 35 \text{ deg}}$ [$(C_{L,max})_{\varphi = 35 \text{ deg}}$] = 0.96, hence substantially lower than that of the straight

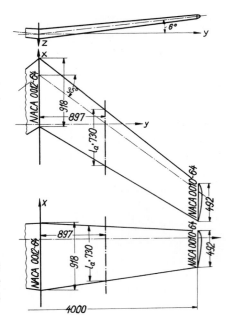

Fig. 2.92 Trapezoidal wing ($\varphi = 0$ and 35 deg, $\Lambda = 5.75$, taper ratio = 0.535). Model for the 5×5 m subsonic tunnel of the DVL at Berlin-Adlershof [110, 121].

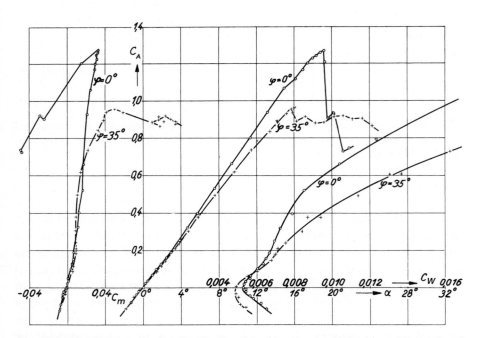

Fig. 2.93 Coefficients C_A, C_W, C_m (C_L, C_D, C_m) of two trapezoidal wings with $\varphi = 0$ and 35 deg, $\Lambda = 5.75$. Taper ratio 0.535; measurements in the 5×7 m subsonic wind tunnel of the DVL, $R = 2.45 \times 10^6$ (1942) [110, 121].

trapezoidal wing $[(C_{A,max})_{\varphi=0\ deg} = 1.27)$. However, stall was less abrupt in case of the swept wing (see Fig. 2.93) and occurred over an angle-of-attack range of $\Delta\alpha \approx 7.5$ deg. After maximum lift (increasing α) only a small decrease in the C_A (C_L) values occurred, while for the straight wing the lift coefficient was reduced from $C_L = 1.27$ to $C_L = 0.73$ within an angle-of-attack range of $\alpha \approx 2$ deg. The pitching-moment developments were almost identical for the two wings up to a lift coefficient of $C_L = 0.7$. Thereafter, a behavior in the longitudinal stability typical of the swept wing was exhibited. At higher angles of attack, the swept wing became more tail down because the outer wing areas experienced lower lift due to the boundary-layer movement. Shortly prior to and beyond maximum lift, the tail-heaviness strongly increased due to an early stall on the outer wing. On the inner wing, the flow was virtually supported by these separated regions and stayed attached longer. In case of the straight wing, the suction peaks at the leading edge disappeared nearly simultaneously along the entire span, and nose-down pitching moment was established. When exceeding maximum lift, the swept wing became, therefore, even more unstable about its lateral (cross) axis while the straight wing became stable in this flow regime. A comparison of these measurements at high *Reynolds* numbers ($R = 2.45 \times 10^6$) with measurements of *Jacobs* ($R = 4.2 \times 10^5$) and investigations of *M. Hansen* [122] and *H. Luetgebrune* [123] at relatively low *Reynolds* numbers led partly to results that could not be generalized. In the investigations of *Jacobs* as well as the ones of *Hansen* and *Luetgebrune*, maximum lift increased with the sweep angle φ, which could neither be confirmed by theory nor by experiments at high *Reynolds* numbers. Similarly contradicting results were obtained in a comparison of the drag measurements. As already noted in conjunction with the airfoil development, the drag coefficients C_{Wp} (C_{Dp}) measured at low *Reynolds* numbers were considerably larger than indicated by the results of *Puffert* at the DVL.

2.3.3.4 INVESTIGATIONS CONCERNING LONGITUDINAL STABILITY. To systematically diagnose the problem of longitudinal stability, further geometric possibilities to influence the C_m-development were studied. Here, *H. Lemme* [124] carried out measurements on a swept wing, a partly swept wing, and an M-wing; see Fig. 2.94. Prior to the massive separation of the flow and also thereafter, the M-wing (c) resulted in nose-down moments, contrary to the standard swept wing (a). For the partly swept wing (b), the pitching moment hardly changed up to stall. This wing configuration provided, in addition, the highest maximum lift at the strongest lift increase. The position of the pitching-moment reference axis was determined from the measurements according to dC_m/dC_A (C_L) = 0 for the conditions without flow separation. If one took the $\frac{1}{4}$-point of the wing section corresponding to the center of gravity of the half-wing as reference, the reference axis of wing (a) was further upstream by 0.055 l (c), the one of wing (b) by 0.047 l (c), and the one of wing (c) by 0.034 l (c). Here

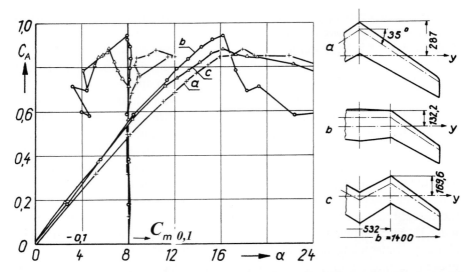

Fig. 2.94 Coefficients C_A (C_L), C_m of three swept wings with different plan forms. $\varphi = 35$ deg, reference axis determined in measurement for $\partial C_m / \partial C_A = 0$, $R = 0.45 \times 10^6$; tests (1943) [124].

l (c) = S/b is the ratio of the wing reference area to the half-span b. An interpretation of the good aerodynamic characteristics of the M-wing could be derived from the flow visualizations. (Also see the flow visualizations on the M-wing in Fig. 2.88.) The planform shows the wing to be virtually a combination of forward-swept and a swept-back wing. This investigation of different planforms of swept wings, carried out at relatively low *Reynolds* numbers, was again proof and example of the creative fundamental research still being carried out towards the end of WWII (1943–1944), even if an M-wing surely could not be realized for design and flutter reasons.

2.3.3.5 PRESSURE DISTRIBUTION MEASUREMENTS ON A SWEPT WING IN TUNNEL VI OF THE AVA. The first comprehensive swept-wing measurements had been carried out at lower or medium-high *Reynolds* numbers. Here, it became obvious that swept-wing characteristics were strongly dependent on boundary-layer effects and that, therefore, a transfer to full-scale was certainly not possible. At the end of 1944, extensive measurements initiated and supervised by *Seiferth* were, therefore, carried out by *Krüger* [125] on a swept wing of 3 m span in the large Tunnel VI of the AVA. The investigations comprised the clean wing and the wing with different types of landing devices at symmetrical and asymmetrical freestream conditions with force and pressure distribution measurements being performed. Amongst other things the pressure distribution measurements were to serve as basis for the study of the spanwise (lateral) movement of the boundary layer, which was later also to

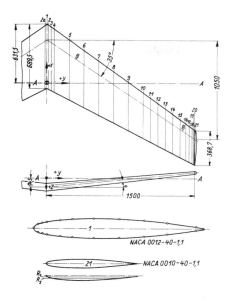

Fig. 2.95 Swept-wing model for pressure distribution measurements in Tunnel VI of the AVA; commencement of the measurements late 1944; management: *R. Seiferth* **[110].**

be surveyed. Here, we shall provide detailed information on some results obtained for the clean wing, which were new at the time and constituted valuable additions to the DVL measurements on a similarly configured model. The model wing is shown in Fig. 2.95 with numerous ribs equipped for pressure measurements. The sweep angle of the 1/4-chord line was 35 deg, the ratio of the tip and root chords was 0.535, and the dihedral angle of the center line 6 deg. The airfoils were symmetrical, at the root 12%, at the tip 10% thick. The y axis corresponded to the neutral axis determined in measurements. The line BB corresponded to the 1/4-chord points of the individual sections. R1 and R2 are the wing-tip contours. The geometric data corresponded to the design of the jet fighter that was being developed. The normal force coefficients C_{Ny} of the individual sections, determined from the pressure measurements, were plotted in Fig. 2.96. Noticeable was the sudden increase in the C_{Ny} values shortly before flow separation in most of the wing sections and the loss in lift in the outboard region of the wing at high angles of attack. Most important was, of course, detailed information concerning the longitudinal stability; see Fig. 2.97. Here, the pitching-moment coefficients C_{my} were plotted vs the angle of attack α with the reference point being the 1/4-chord location and the reference length being the airfoil chord. Noticeable was the stable position of the four inner sections with respect to the 1/4-chord point, while all other airfoil sections were unstable. In evaluating the pressure distribution measurements, the reference system was chosen such that the origin of the measured dC_m/dC_N values (gradients of the coefficients of the pitching moment and the normal force) at small angles of attack just became zero. This allowed the location of the neutral points x_n/l_y (x_n/c_y) to be

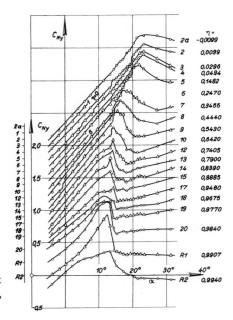

Fig. 2.96 Section normal force coefficient C_{Ny}**, determined from pressure distributions, vs angle of attack** α**, R. Seiferth [110].**

Fig. 2.97 Section pitching-moment coefficient C_{my}**, determined from pressure distributions, vs angle of attack** α**, R. Seiferth [110].**

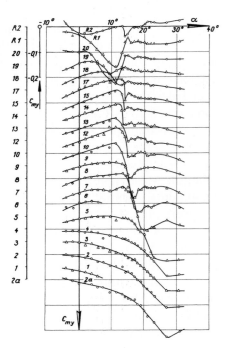

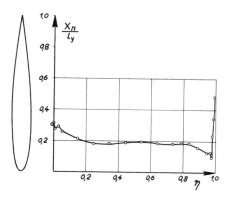

Fig. 2.98 Position of the local neutral points determined from the gradient $\partial C_{my}/\partial C_{Ny}$ at $C_{Ny} = 0$, R. Seiferth [110].

determined; see Fig. 2.98. The neutral points were for most of the wing located at about 19% of the chord, very far forward, however, at the wing center at approximately 30%. This fact required further investigations concerning the angle of yaw (yaw angle) and a statement regarding the wing-fuselage interference effects. The high lift values, determined by the normal force distributions derived from the pressure distribution measurements, and confirmed by flow visualizations, inevitably had to be changed by these two parameters, that is, the yaw and wing-fuselage interference.

2.3.3.6 SHEARED SWEPT WING. If a wing moves with an increasing angle of yaw β, a shift in the lift distribution occurs that affects all aerodynamic forces and moments which can, however, be understood as long as the flow stays attached. The aerodynamic parameters addressed up to now, such as lift, drag, and pitching moment, experienced relatively small changes. To estimate the lift changes, it was assumed that the lift was dependent on the velocity component $V \cos \beta$ normal to the wing span; see Fig. 2.99. The lift coefficient

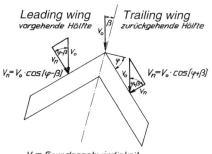

Fig. 2.99 Schematic representation of the effective freestream velocity for the sheared swept wing, W. Jacobs [116].

$V_0 = U_0 = $ Freestream velocity (speed)
$V_n = U_n = $ Effective normal component
of the velocity

C_A (C_L) thus became $C_L = C_{L0} \cos^2 \beta$ with C_{L0} being the lift at a yaw angle of $\beta = 0$. This relation also applied to the swept wing without takeoff and landing devices as was shown by *Weissinger* [126, 127].

Seiferth, who had presented a summary of the results on the sheared wing up to 1940 in the *Ring Book of Aeronautical Technology* [128], had analyzed in detail the research results that had been obtained for the sheared swept wing up to 1945 in the Göttingen monographs. Here, he also addressed the analytical solutions of *Weissinger* that led to results which at least partly confirmed the trend of the experiments. Important knowledge could be used by the design aerodynamicist, namely, that the increase of the yaw-rolling moment with lift C_L was amplified by the positive sweep of the wing while it became weaker with negative sweep. For the sheared wing, only one representative result out of the extensive investigations will be reported here. For that purpose, the lift distribution C_{Ly} at an angle of attack of $\alpha = 7.5$ deg and the range of yaw angles of $\beta = 0$ deg to $\beta = 15$ deg was plotted in Fig. 2.100. The local lift coefficients, plotted vs the half-span $\eta = -1$ to $\eta = +1$, increased on the leading-wing side. A corresponding decrease of the C_{Ly} values could be noticed on the trailing-wing side. This result agreed with fluid-physical predictions because the sweep angle of the leading-wing side of the swept

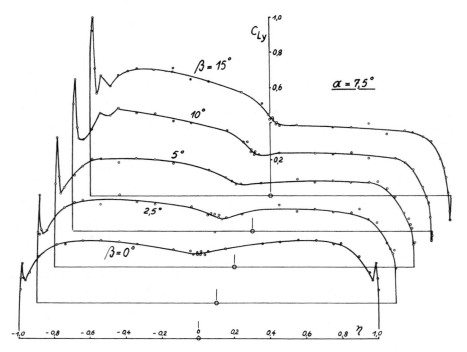

Fig. 2.100   Local lift coefficient C_{Ay} (C_{Ly}) vs the half-span $s = b/2$ dependent on the yaw angle β for $\alpha = 7.5$ deg; *Seiferth* [110].

wing decreased with yaw, contrary to the case of the straight trapezoidal wing. Consequently the lift on this wing side, compared to the corresponding trapezoidal wing, had to decrease except for the wing tip where a strong positive forward lift peak due to the flow around the tip developed. The results of the AVA investigations were only in a very limited way compatible with the measurements of *Jacobs*. In how far the differences in the measurements were purely *Reynolds* number associated could not be investigated.

2.3.3.7 Wing-Fuselage Arrangements. While the isolated swept wing had been investigated at low speeds at different research establishments, systematic measurements on swept-wing/fuselage arrangements needed for new high-speed projects were lacking. *Hermann Schlichting* was requested in 1943 by the RLM to coordinate work in the area of wing-fuselage interference. As director of the Institute of Aerodynamics of the TH Braunschweig and adviser of the LFA Braunschweig, he had, besides his institute's wind tunnel, the necessary access to other suitable test facilities and financial means. Interference between basic fuselage shapes and conventional wings had, in addition, already been determined at the Institute in an extensive research program under his supervision. Because at the institute of *Schlichting* the results of experiments on numerous swept-wing configurations (see Fig. 2.89) were available, it was obvious to task him with this activity. The fuselage selected was a 1:7 body of revolution with a main axis of 750 mm with the total span of all wing models having been $b = 770$ mm. At a wind-tunnel velocity of 40 m/s, the *Reynolds* number was $R = 0.4 \times 10^6$. As *E. Möller* [129] remarked, a prerequisite was that the stability characteristics of the wing/fuselage combinations, such as the pitching moment C_m and the rolling moment C_l, determined at the relatively low *Reynolds* numbers could be transferred with sufficient accuracy to full scale. A summarizing report of *Schlichting* [130] concerning the status of the tests in February 1945 is available as Internal Report at the Institute of Aerodynamics. The program comprised in detail the following activities:

- Pressure distribution measurements concerning fuselage and nacelle interference on a straight high-lift wing in the wind tunnel of the NLL, Amsterdam, under AVA's supervision. These investigations had been stopped in September of 1944 due to the war.
- Six component measurements, primarily on wing-fuselage combinations, were carried out in the A1 wind tunnel (2.5-m diameter) of the LFA in addition to the investigations at the Institute of Aerodynamics. The measurements had, at the time of the issuing of *Schlichting's* report, not yet been completed.
- In the 2.8-m high-speed wind tunnel of the LFA, drag measurements were planned on wing/fuselage combinations with engine nacelles under

contract to the industry. They had at the time of the report been approved by the wind-tunnel commission, but had not yet been executed.

- The complete program concerning systematic six-component measurements related to the interference of a rectangular wing with fuselage and tail unit had been concluded and documented by *Schlichting* at the end of 1942 at the Institute of Aerodynamics. A summary of the results was published in the *Yearbook of German Aeronautical Research* in 1943 [131].
- A contribution to the investigation of longitudinal stability was provided by the TH Graz at approximately the same *Reynolds* number in cooperation with the Institute of Aerodynamics in Braunschweig. The objective of this investigation was, among other things, to determine the effect of different symmetric bodies of revolution on trapezoidal wings without sweep.
- A comprehensive investigation program, in addition to the interference measurements on straight wing/fuselages, was carried out on almost all swept-wing configurations presented in Fig. 2.89 in the wind tunnel of the TH Braunschweig; see Fig. 2.101. The configurations shown in Fig. 2.89 were supplemented by three forward-swept wings of constant chord and sweep angles of $\varphi = 15$, 30, and 45 deg (Junkers Project Ju 287) and a strongly tapered swept-back trapezoidal wing with a sweep angle of $\varphi = 45$ deg (Wing Model 06/45). On all models, set up as low-, mid-, and high-wing (shoulder-wing) configurations, three- and six-component

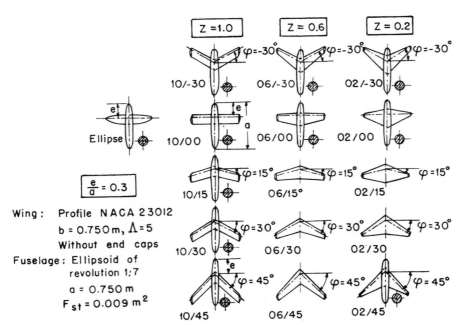

Fig. 2.101 Swept-wing model variations for interference measurements on wing-fuselage combinations at the Institute of Aerodynamics of the TH Braunschweig [130].

measurements were carried out, the results reduced, analyzed, and documented in reports of the Institute. Here, *Möller* obtained results that were of considerable importance to the design aerodynamicist.

The influence on rolling moment caused by the fuselage for different sweep angles and taper ratios $\lambda = l_a/l_i = c_a/c_i$ was relatively small. The effect of the sweep angle φ on the total rolling moment was, on the contrary, clearly recognizable; see Fig. 2.102. Another important result of the investigations of *Möller* [132] was the determination of the fuselage effect on the longitudinal stability of various swept wings; see Fig. 2.103. According to these investigations for the swept-back wing, the destabilizing shift in the neutral point due to the fuselage was substantially less, and for the forward-swept wing larger than for the unswept wing. These test results had been obtained on mid-wing configurations with the rear position of the wing, with respect to the fuselage referenced to the geometrical neutral point of the wing. In the case of the configurations investigated, the location of the neutral point for the unswept wing was shifted about 8%-chord upstream, for the forward-swept wing ($\varphi = 30$ deg) 15% upstream, and for the swept-back wing 1%-chord downstream. Further investigations concentrated on the influences of the yaw angle β in case of the wing/fuselage model with simulated rudder and different heights of the wing position relative to the fuselage. Because of the far downstream position of the rudder, its effectiveness in achieving lateral stability outweighed the effect of the position of the wing.

Supplementing the investigations on wing-fuselage configurations at relatively low *Reynolds* numbers, first systematic configurational studies on

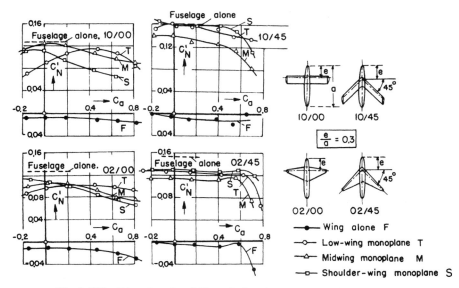

Fig. 2.102   Directional stability of wing-fuselage combinations [130].

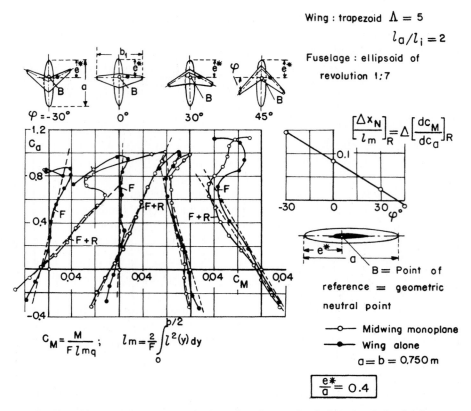

Fig. 2.103   Shift in the neutral-point location due to the influence of the fuselage on swept wings [130]. Legend: $l = c$.

small models of the swept trapezoidal Wing 8 with fuselages and nacelles at *Mach* numbers $M \leq 1$ were still important and reported by *Ludwieg* at the LGL meeting 127 in Göttingen. Under contract to the Messerschmitt Corporation (AG), he had performed three-component force measurements on models of extremely small dimensions in the 11×11-cm high-speed wind tunnel of the AVA; see Fig. 2.104. The effect of the fuselage and nacelles on the characteristics of a swept and an unswept wing was determined. The Swept-Wing 8 ($\varphi = 30$ deg, airfoil NACA 0012-64) was investigated alone,

Fig. 2.104   Models of trapezoidal wing 7 and swept-wing 8 with fuselage and nacelles for the AVA 11- × 11-cm high-speed wind tunnel according to *Ludwieg*; legend: top-trapezoidal wing 7, and bottom-swept-wing 8.

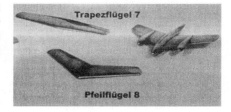

with nacelles, with fuselage and nacelles, and nacelles reduced in size. The results of drag measurements at an angle of attack of $\alpha = 0$ deg ($C_A = C_L = 0$) are plotted vs the *Mach* number in Fig. 2.105. The drag rise of the swept wing alone commenced at approximately $M = 0.84$. The fuselage affected the wing only insignificantly while due to the presence of the nacelles, the drag coefficient C_W (C_D) started to rise at $M = 0.76$. The reason for the different effects of fuselage and nacelles was probably due to the fact that the symmetrical flow in the center of the wing was only a little disturbed by the fuselage, while the nacelles were located in a flowfield where velocity components normal to the nacelle axis occurred. The drag behavior had, in addition, according to the "area rule," to deteriorate as will be shown by *Werner Heinzerling* in Sec. 2.3.4. The drag development of the unswept Wing 7 was also plotted for comparison with and without nacelles. As expected, the nacelles at high speeds disturbed the flow less than in the case of the swept wing. Nevertheless the swept wing also with simulated nacelles remained definitely superior to the unswept wing. In 1940 that was decisive for the continuation of the development of high-speed aircraft.

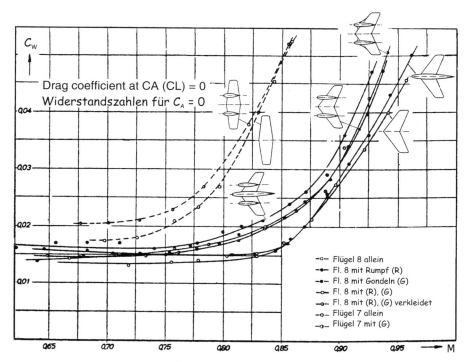

Fig. 2.105   Drag coefficients C_W (C_D) of Model 7 and Model 8 with and without fuselage and nacelles dependent on *Mach* number M according to measurements of *Ludwieg*; legend: drag coefficient at C_A (C_L) = 0; wing alone, wing 8 with fuselage (*R*), wing 8 with nacelles (*G*), wing 8 with *R* and *G*, wing 7 alone, wing 7 with *G*.

2.3.3.8 LIFT AUGMENTATION DEVICES FOR THE SWEPT WING [133]. Already during the development of high-speed airfoils, it became clear that with the swept wing, a wing unit had been created that was characterized at high-speed flight by low drag, but also by small airfoil radii, small airfoil thickness, and large maximum-thickness locations. It soon turned out that these designs, favorable with regard to high speeds, were most unfavorable with respect to maximum lift. Because maximum lift was of great importance to takeoff and landing and to flight maneuvers, this area of design aerodynamics was researched intensively and at great expense. Because of the numerous forms developed, one should in principle distinguish between two high-lift devices: 1) high-lift devices in the nose (leading-edge) region of the airfoil and 2) high-lift devices in the downstream (trailing-edge) region of the airfoil.

The total effort related to the investigations of high-lift devices becomes obvious by the compilation of *Krüger*; see Fig. 2.106. Concerning this figure, literature references are given, which are, in part, no longer traceable. An interesting contribution to these systematic high-lift investigations was provided by *G. Brennecke* [134]. He had investigated slats of different planforms without and with the so-called extendable split flap; see Fig. 2.107. The total span of the model was 1.5 m. The model size allowed to attain a maximum *Reynolds* number of $R = 0.6 \times 10^6$ in the AVA Tunnel VI. The results had to be interpreted allowing for boundary-layer effects at relatively low *Reynolds* numbers. Even if the maximum lift $C_{A\max}$ ($C_{L\max}$) achieved was not satisfactory, a favorable development of the pitching-moment coefficient C_m could be observed. On the same wind-tunnel model, a completely new idea to affect the longitudinal stability was tested, which is still today employed in modern aircraft design. *Brennecke* [135] achieved by a split flap attached to the suction side of a swept wing and being deflected in a direction opposite to the one of the split flap on the pressure side a change in zero-lift (lift at $\alpha = 0$ deg). The maximum lift was hardly affected, as is indicated by the plots in Fig. 2.108. A full-span split flap extending to the wing trailing edge served as a landing device that could be deflected. The result showed that the tail-down pitching moment could either be reduced or completely avoided—dependent on the deflection of the opposite split flap. Remarkable was, in addition, the reduction of the yawing/rolling moment. Stalling of the flow on the leading wing (half) was observed at high angles of attack. This kind of landing device reduced the danger of wing drop, for instance, in case of a lateral gust, during approach.

A good documentation of the most important basic tests, mainly carried out at the AVA, had in 1945–1946 been compiled by *Krüger* in the Göttingen monographs. On the large wind-tunnel model of a 35-deg swept wing, already shown in Fig. 2.95, high-lift devices, called landing devices by *Krüger*, were investigated concentrating on three configurations, Fig. 2.109. Besides the cambered slats, the nose flaps (leading-edge flaps) developed by *Krüger* and

Landing devices (Landehilfe)			Slat (Vorflügel)			Wing (Flügel)					Test results (Meßwerte)			
Landing devices design	l_F/l	η_F	Slat design	l_s/l	$2y_s/b$	$\varphi°$	Λ	l_i/l_a	Airfoil Inboard	Outboard	$10^{-6}R$	$\Delta c_{L_{max}}$	$\Delta\alpha_{max}$	Literaturstelle
Ohne Landeklappen / Without landing flaps														
			(design)	0,20	1,0	35,8	5,75	1,86	0012-E4	0010-64	0,30	0,43	10,0	[14]
			(design)	0,25	1,0	35,0	6,37	1,86	0012-64	0010-64	0,57	0,29	2,0	[20]
			(design)	0,13	1,0	32,5	6,00	2,16	t/l=0,136	t/l=0,085	0,72	0,30	4,0	[21]
			(design)	0,20	1,0	35,0	5,76	1,86	0012-1.1-40	0010-1.1-40	1,38	0,49	7,5	[7]
			(design)	0,20	1,0	35,8	5,76	1,86	0012-64	0010-64	2,00	0,46	8,5	[22]
			(design)	0,20	1,0	35,0	5,67	1,66	0012-64	0010-64	2,50	0,47	11,0	[23]
			(design)	0,20	1,0	35,0	5,67	1,86	0012-1.1-40	0010-1.1-40	1,91 / 5,00	0,55 / 0,54	10,3 / 5,6	[24]
Mit Landeklappen im inneren Spannweitenbereich / With inboard landing flaps														
(design)	0,25	40°	(design)	0,20	1,0	35,0	5,67	1,86	0012-64	0010-64	2,50	0,44	9,0	[23]
(design)	0,20	60°	(design)	0,20	1,0	35,0	5,00	2,00	0012	0012	1,00	0,65	12,5	[25]
(design)	0,20	60°	(design)	0,20	1,0	35,0	5,76	1,66	0012-1.1-40	0010-1.1-40	1,38	0,49	6,5	[7]
(design)	0,20	60°	(design)	0,20	1,0	35,0	5,67	1,87	0012-1.1-40	0010-1.1-40	1,98 / 4,40	0,33 / 0,54	7,0 / 7,8	[24]
(design)	0,20	60°	(design)	0,20	1,0	35,0	5,67	1,87	0012-1.1-40	0010-1.1-40	4,40	0,56	8,5	[24]
(design)	0,25	30°	(design)	0,20	1,0	35,0	5,00	2,00	0012	0012	1,00	0,80	12,8	[25]
Mit Landeklappen über die gesamte Spannweite / With full-span landing flaps														
(design)	0,20	40°	(design)	0,20	1,0	35,8	5,75	1,86	0012-64	0010-64	2,00	0,43	10,6	[22]
(design)	0,25	40°	(design)	0,20	1,0	35,0	5,67	1,86	0012-64	0010-64	2,50	0,43	10,0	[23]
(design)	0,20	60°	(design)	0,25	1,0	35,0	6,37	1,86	0012-64	0010-64	0,57	0,28	4,0	[20]
(design)	0,20	60°	(design)	0,20	1,0	35,0	5,76	1,87	0012-1.1-40	0010-1.1-40	1,38	0,47	6,5	[7]
(design)	0,20	60°	(design)	0,20	1,0	35,0	5,67	1,87	0012-1.1-40	0010-1.1-40	1,98 / 4,40	0,39 / 0,59	7,4 / 8,1	[24]
(design)	0,20	60°	(design)	0,20	1,0	35,0	5,67	1,87	0012-1.1-40	0010-1.1-40	4,40	0,61	7,9	[24]
(design)	0,30	40°	(design)	0,25	1,0	35,0	6,37	1,86	0012-64	0010-64	0,57	0,29	2,0	[20]

Fig. 2.106 High-lift devices in swept-wing investigations at the AVA according to *W. Krüger* [111], $\Delta\alpha_{max}$ = lift increase at the angle of incidence $\Delta\alpha_{max}$.

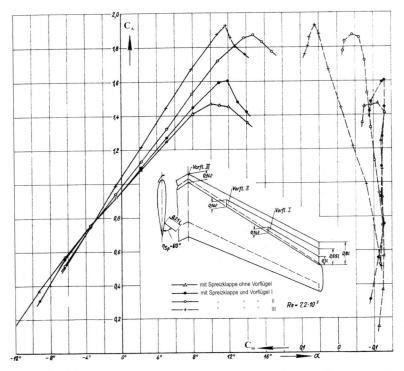

Fig. 2.107 Results of investigations on a swept wing with slats of different chord distributions and a split flap, $R = 0.7 \times 10^6$; measurements taken 1943, *G. Brennecke* [134]; axes: C_A (C_L), C_m, α; legend: with split flap without slat — △ —, with split flap and Slat I — ● —, with split flap and Slat II — ○ —, and with split flap and Slat III — + —.

already introduced in the chapter "High-Speed Airfoils" were investigated. Systematic investigations were carried out with the AVA 3-m-span wind-tunnel model and the different arrangements of slats and "*Krüger* flaps" at the nose of the wing as well as combinations of the latter with split flaps at the trailing edge. Because pretests had already shown a clear dependence of the maximum-lift coefficients attainable with these "high-lift devices," supplementary tests in a *Reynolds* number range of $R = 0.85 \times 10^6$ to 2.1×10^6 were initially conducted. The results of the measurements confirmed the earlier experience, namely, that the effectiveness of the cambered, slotted, or split flaps was substantially lower than for the unswept wing. On the other hand, the landing devices at the wing nose were generally more effective than the ones in the downstream (trailing-edge) region of the wing. This knowledge had already been gained in tests at considerably lower *Reynolds* numbers. The results had, however, to be verified and quantified at higher *Reynolds* numbers. For instance, for a nose flap of 10% chord, which *Krüger* had optimized in the context of his airfoil investigations, an increase in $\Delta C_{A\max}$ $(C_{L\max}) = 0.44$ could be achieved; see Fig. 2.110. It was remarkable that this

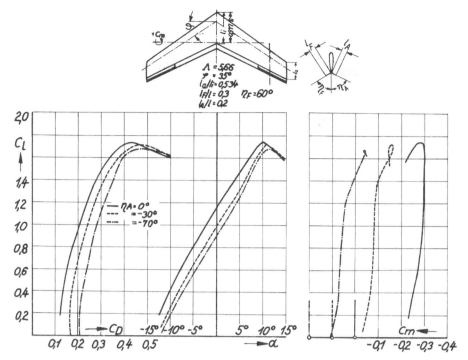

Fig. 2.108 Results of measurements on a swept wing with a 30%-chord trailing-edge split flap of opposite deflection, $R = 0.6 \times 10^6$; measurements taken 1944, *Brennecke* [135].

nose flap of 10% chord provided about the same maximum lift increase as a "20%"-chord slat. The best combination with regard to maximum lift determined in the measurements was the one with a "20%"-chord continuous split flap and the "20%"-chord nose flap. Here, the C_{Amax} (C_{Lmax}) value was increased to 2.00 as compared to 0.89 for the clean wing. For most of the flap combinations investigated the longitudinal moments (pitching-moment coefficients c_m) at stall (angle of attack too high) were nose-up. Only configurations with an outer slat or nose flap and additional split flaps behaved in a nose-down manner; see Fig. 2.111. In interpreting the test results, the visualization of the flow on the suction side of the high-lift wings was particularly helpful.

What we did not notice during our enquiries was a direct, close cooperation with the test centers, such as in Rechlin. While in England, the pilots reported their new experience immediately after the flight tests to the research institutions—the cooperation between the individual research institutions and especially with industry in Germany was scarce. For the design aerodynamicists in the development offices of industry, this meant that the excellent research potential in these areas was not always applied according to the needs. This concerns the use of scientific personnel as well as the use

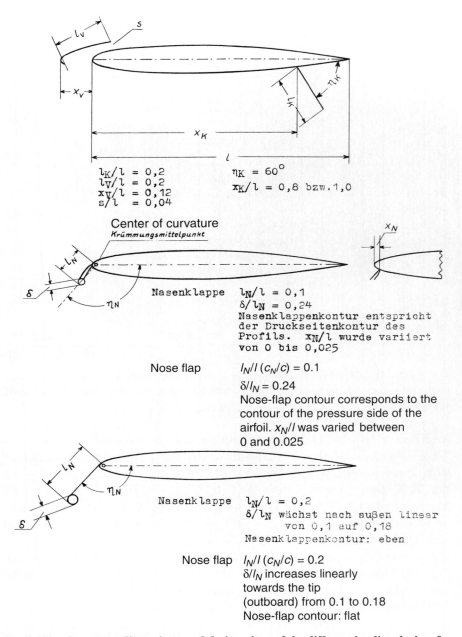

$$l_K/l = 0,2$$
$$l_V/l = 0,2$$
$$x_V/l = 0,12$$
$$s/l = 0,04$$

$$\eta_K = 60°$$
$$x_K/l = 0,8 \text{ bzw.} 1,0$$

Center of curvature
Krümmungsmittelpunkt

Nasenklappe $\quad l_N/l = 0,1$
$\delta/l_N = 0,24$
Nasenklappenkontur entspricht
der Druckseitenkontur des
Profils. x_N/l wurde variiert
von 0 bis 0,025

Nose flap $\quad l_N/l \, (c_N/c) = 0.1$

$\delta/l_N = 0.24$
Nose-flap contour corresponds to the
contour of the pressure side of the
airfoil. x_N/l was varied between
0 and 0.025

Nasenklappe $\quad l_N/l = 0,2$
δ/l_N wächst nach außen linear
von 0,1 auf 0,18
Nasenklappenkontur: eben

Nose flap $\quad l_N/l \, (c_N/c) = 0.2$
δ/l_N increases linearly
towards the tip
(outboard) from 0.1 to 0.18
Nose-flap contour: flat

Fig. 2.109 Geometry, dimensions, and designations of the different landing devices for the large AVA swept-wing model, sweep angle $\varphi = 35$ deg, $R = 2.1 \times 10^6$; W. Krüger [136].

of test facilities. All large German aircraft companies owned relatively small wind tunnels, which were in fact fully occupied, but provided only test results obtained in flows at extremely low *Reynolds* numbers and were, therefore, not always sufficiently reliable for design purposes. The use of the research

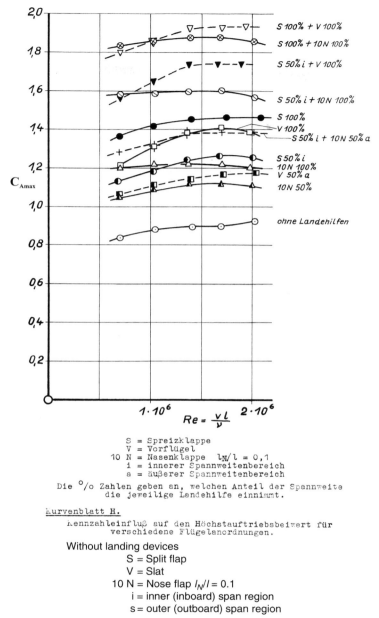

S = Spreizklappe
V = Vorflügel
10 N = Nasenklappe l_N/l = 0,1
i = innerer Spannweitenbereich
a = äußerer Spannweitenbereich
Die °/o Zahlen geben an, welchen Anteil der Spannweite
die jeweilige Landehilfe einnimmt.

Kurvenblatt H.

Kennzahleinfluß auf den Höchstauftriebsbeiwert für
verschiedene Flügelanordnungen.

Without landing devices
S = Split flap
V = Slat
10 N = Nose flap l_N/l = 0.1
i = inner (inboard) span region
s = outer (outboard) span region

The percentages (%) denote which length of the span was
covered by the respective landing devices.

Graph (Data) sheet H

Effect of characteristic parameters on the maximum lift
coefficient for different wing arrangements.

Fig. 2.110   Effect of characteristic parameters on maximum lift for different high-lift devices for the AVA swept-wing model with a sweep angle of $\varphi = 35$ deg; measurements at $R = 2.1 \times 10^6$, W. Krüger [136]; axes: C_{Amax} (C_{Lmax}), Reynolds number R.

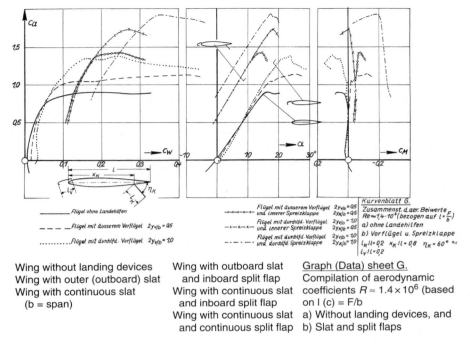

Wing without landing devices
Wing with outer (outboard) slat
Wing with continuous slat
  (b = span)

Wing with outboard slat
  and inboard split flap
Wing with continuous slat
  and inboard split flap
Wing with continuous slat
  and continuous split flap

Graph (Data) sheet G.
Compilation of aerodynamic
coefficients $R \approx 1.4 \times 10^6$ (based
on l (c) = F/b
a) Without landing devices, and
b) Slat and split flaps

**Fig. 2.111   Compilation of the aerodynamic coefficients for different high-lift devices.
AVA swept-wing model $\varphi = 35$ deg. $R = 2.1 \times 10^6$. Compilation W. *Krüger* [136]; axes:
c_a (c_l), c_w (c_d), α, c_m (continuous means over the entire span).**

results, for instance, in the field of high lift and the flight stability of swept
wings was applied after the war primarily by the Allies in their development
of high-speed aircraft. A valuable and extremely useful database was the
Göttingen Monograph F2 "Aerodynamic Coefficients," because this document contained a thorough and critical analysis and compilation of the results
of especially competent scientists.

**2.3.3.9 EMPLOYMENT OF "KRÜGER FLAPS" ON COMMERCIAL AIRCRAFT AFTER
1945.**   Like the swept wing itself, many new developments were necessary
that became an absolute prerequisite with the introduction of this new technology and were only used after 1945 in series production. This also applied
to the high-lift device, the "*Krüger* flap." Its inventor, *Werner Krüger*, was
born in Kolberg/Baltic Sea on November 23, 1910, and had studied mechanical engineering at the TH Berlin. After the completion of his diploma
examination in October 1934, he worked as an aerodynamicist at the Dornier
Company in Friedrichshafen, where *Hermann Schlichting* took on the
management of the wind tunnel in 1935. It was surely not a coincidence that
he continued his activity at the AVA Göttingen in the Wind Tunnel Branch

following *Schlichting's* call as full professor by the Institute of Fluid Mechanics of the TH Braunschweig at the end of 1937; he became head of the large Wind Tunnel VI in 1944. His scientific work concentrated first on the "improvement of flight characteristics by boundary-layer control" up to the development of a research aircraft. In 1943, *Krüger* had published a report on the first investigations on an airfoil with a nose flap [136]. A patent "High-Lift Device for Airfoils with an Arbitrarily Small Nose Radius" was applied for by the AVA on January 12, 1944, based on the results of his work; see Fig. 2.112.

a)

DW/B 57/44

9

A V A
19. FEB. 1944
an ...

Empfangsbestätigung.

Die Patentanmeldung vom 12. Januar 1944 der Aerodynamischen Versuchsanstalt Göttingen E.V., Göttingen, Bunsenstr. 10, betreffend :

"Hochauftriebsmittel für Profile mit beliebig kleinem Nasenradius "

ist am Reichspatentamt 14. 2. 1944 hier eingegangen und unter dem Aktenzeichen in den Geschäftsgang gegeben worden.

Reichspatentamt Annahmestelle

A 99248 XI, 62b

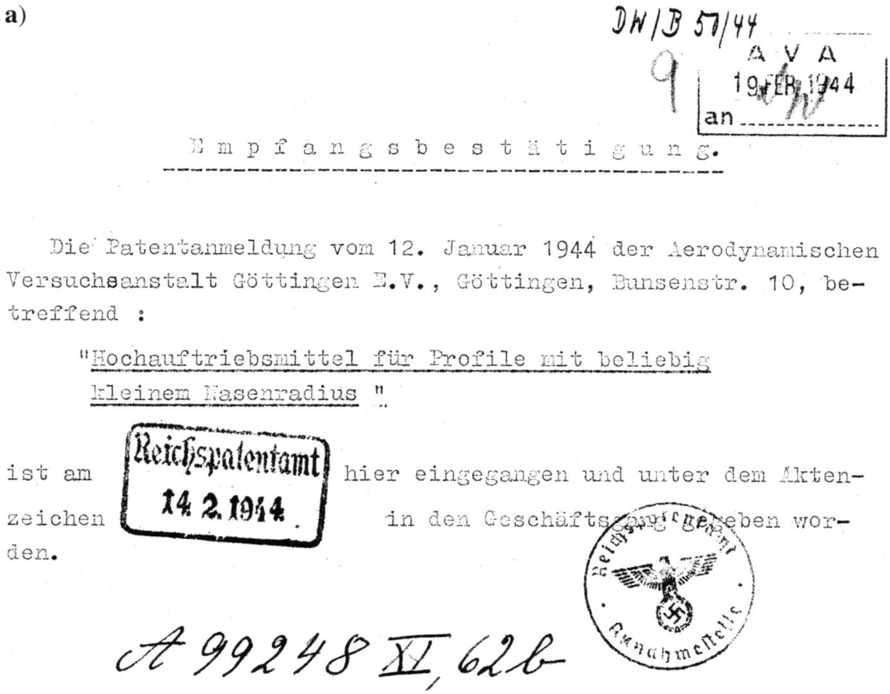

b) *Acknowledgement of Receipt*

Concerning the patent registration of January 12, 1944, of the Aerodynamic Test Establishment Göttingen E.V. (Registered Association), Göttingen, Bunsenstrasse 10:

"High-lift means for airfoils with an arbitrarily small nose radius"

was received on February 14,1944, at the State Patent Office and assigned to business procedures with Reference A 99248 XI, 62b.

State Patent Office

Fig. 2.112 a) Patent application concerning the "*Krüger* flap", 1944 (GOAR-3634); b) translation.

Fig. 2.113   Handley Page HP 80 "Victor" bomber, 1952.

From August 1945 until October 1946, he was busy preparing the Göttingen Monographs with his former Head of Institute, *Reinhold Seiferth*. Because both scientists were not taken over by the KWI, they jointly founded in 1947 an engineering office for fluid technology, which operated until 1977 extremely successfully. *Krüger* had soon lost his contact to the aerospace industry due to concentrating on his new tasks and only followed the use of "*Krüger* flaps" on commercial aircraft as somebody interested in aeronautics. A bomber of Sir Frederic Handley Page, the Handley Page HP 80 "Victor," took off on her maiden flight to the complete surprise of the entire public in December of 1952. The HP 80 "Victor" had a wing plan form similar to the "Arado" Ar 234 with crescent wings, engines buried in the wing roots as investigated by *Dieter Küchemann* at the AVA Göttingen, and novel flaps at the outer-wing leading edges; Fig. 2.113. The latter were very similar to the *Krüger* nose flaps and thus one of the first applications of his development. In December of 1968, *Krüger* was contacted by a representative of the Scientific Contact Office München of the Boeing International Corporation. Boeing wanted to contest the registration of a U. S. patent by *F. T. Watts* as follows from the letter in Fig. 2.114. *Krüger* received the offer to accept an advisory function. Here, he signed a Consulting Services Agreement directly with the representative of the headquarters office of the Boeing Company in Seattle. With expertise he explained, among other things:

> I looked at the U.S. Patent No. 3,363,859. As far as the claims refer to the spatial arrangement and principle actuation of the extended landing devices, the U.S. patent, in my opinion, does not present anything new compared to our application of 1944. Claims can at best be asserted, if need be, with respect to the construction of the actuators.

Krüger proved again that the nose flap he developed was functionally not a slat as had, for example, been developed by Handley Page. A slat provided the boundary layer with additional energy to avoid a flow separation. The "*Krüger* flap" on the other hand attempted to force the freestream stagnation point by an experimental optimization to be located in the immediate vicinity of the nose of the flap. This considerably reduced the excess velocities in the vicinity of the nose at high angles of attack.

Paul G. Kafka immediately informed *Werner Krüger* upon receipt of his written statement that he had passed this on to the patent office in Seattle and pointed out another unusual feature of the U.S. patent. Between the flap and the nose of the wing was a slot, Fig. 2.115a. In that respect he quoted an investigation of *Krüger* [137] that also contained this variant. A translation of this test report was available in the United States as NACA Technical Memorandum No. 1119 and thus represented the official state of knowledge. How explosive the positive outcome of the patent dispute was also becomes obvious by a comparison of the patented U.S.-nose flap with the *"Krüger* flap" of the Airbus A300 (Fig. 2.115b). Both companies — Airbus and Boeing — use this high-lift device developed in 1943 on many of their modern aircraft today.

2.3.3.10 CONFIGURATION INVESTIGATIONS ON SWEPT WINGS AT HIGH SUBSONIC SPEEDS. An analysis of the evaluation of the measurements shows clearly that the principle investigations of *Ludwieg* in the high-speed regime at the AVA Göttingen were supplemented by measurements under contract to the Messerschmitt Corporation (AG). Altogether 16 model configurations were tested in the 11×11-cm high-speed wind tunnel; see the table in Fig. 2.116. The results were comprehensively discussed by *Ludwieg* at the LGL Meeting 127 in 1940. Because the investigations had repeatedly shown stability problems for nearly all of the different swept-wing forms investigated, besides the important drag reductions at high subsonic speeds, the Messerschmitt Corporation tasked the AVA with the investigation of a 35-deg swept wing with a twist of 5 deg (see No. 9 in Fig. 2.116) in the high subsonic regime. *A. Roth* [138] was able to demonstrate that no essential worsening of the drag occurred at low angles of attack up to a *Mach* number of $M = 0.8$. Up to $M = 0.8$ the moments did not change at all. It could be concluded that the natural stability was unchanged by wing twist. However at a *Mach* number of $M = 0.9$ and lift coefficients between C_A $(C_L) = 0$ and 0.15, the wing became unstable due to the flow stalling in the outer wing regions. Another interesting study was added by *Roth* [139] within the context of his basic work at the AVA, namely, a study on a swept wing whose planform was built up by circular arcs; see Fig. 2.117. In our enquiries concerning the creative process, it could not be established whether *Roth* had merely carried out "internal measurements" on the order of *Betz*. The latter had already filed a patent application in September of 1939 concerning the swept wing (see Fig. 2.118) which was granted under patent 732/42 on April 9, 1941, naming as inventors *Betz* and *Busemann*. Already at the end of November 1939, *Betz* applied for a patent designated "Wing with High Sweep" supplementary to the swept-wing patent of *Betz/Busemann*. This supplementary patent was granted in June of 1942 (Fig. 2.119). The patent claim was characterized by the fact that the spanwise wing axis showed a variable angle with respect to the fuselage axis. The wing investigated proved with regard to the drag behavior even more favorable than the 30-deg swept

a) **BOEING**
INTERNATIONAL CORPORATION

WISSENSCHAFTLICHES VERBINDUNGSBÜRO MÜNCHEN
8 MÜNCHEN 2 · DIENERSTRASSE 21 · GERMANY

PHONE : 224938, 226017, 221231, 221293, 221401
CABLE : BOEMU MÜNCHEN
TELEX : VIA GENEVE-BOEINGAVION GVE 22848

IN REPLY REFER TO
1-4200M-1996
10. Dezember 1968

Herrn
Dr. Ing. Werner Krüger
34 Göttingen
Merkelstr. 30

Sehr geehrter Herr Dr. Krüger,

ich erlaube mir, Ihnen heute folgende Angelegenheit zur Kenntnis zu bringen:

Im Juni dieses Jahres wurde an einen gewissen Mr. F. T. Watts ein U.S.-Patent
erteilt, womit Prinzip und Anwendung der von Ihnen erfundenen Nasenklappe
patentiert wurde.

Wir sind natürlich der Ansicht, dass dieses Patent vollständig unberechtigt ist,
und möchten deshalb die Ansprüche des besagten "Erfinders" anfechten. Zu diesem Zwecke hätten wir gerne gewusst, welche Patente Ihnen persönlich bzw.
Firmen, für welche Sie tätig waren, verliehen wurden.

Anliegend überreiche ich Ihnen die Kopie des Patents (U.S. 3,363,859) von
Mr. Watts zu Ihrer Information.

Ich wäre Ihnen für jegliche Auskunft, die Sie mir dieshalb geben können, wirk-
lich sehr verbunden, und verbleibe inzwischen

mit freundlichen Grüssen

Dr. Paul G. Kafka
Wissenschaftlicher Vertreter

Anlage

**Fig. 2.114 a) Letter of *P. G. Kafka*, Scientific Liaison Office München, Boeing
International Corporation to *W. Krüger*, December 1968 (Estate *W. Krüger*); b) translation (courtesy of the Boeing Historical Archives, The Boeing Company).**

wing with an aspect ratio of $\Lambda = 3.36$ (Wing No. 11). The technical realization
of this wing form occurred—as already mentioned—within the "Arado" project AR 234 and after the war on the English Bomber HP 80 "Victor" with its
"crescent" wing. Another supplement to the *Betz/Busemann* swept-wing patent
was to gain even larger significance to military aircraft construction—the AVA
Patent "Aircraft with Provisions to Change Wing Sweep," which was also
granted in June of 1942 (Fig. 2.120). In the first statement on February 13, 1940,

b) Boeing
International Corporation
Scientific Liaison Office München

Dr.-Ing. Werner Krüger
3400 Göttingen

Dear Dr. Krüger,

Allow me to bring the following matter to your attention:

In June of this year, a certain U.S. patent was granted to a certain Mr. F. T. Watts, which patented the principle and application of the nose flap invented by you.

We are, naturally, of the opinion that this patent is completely unjustified and we want, therefore, to contest the demands of the said "inventor." For this purpose, we would like to know which patents were granted to you personally respectively which companies you worked for.

Attached is for your information a copy of the Patent (U.S. 3,363,859) of Mr. Watts.

I would be obliged to you for any information you could give to me in that regard and remain in the meantime

Yours faithfully
Dr. Paul G. Kafka
Scientific Representative

Attachment

Fig. 2.114 (Continued)

Fig. 2.115 a) U.S. patent concerning a nose flap, patent holder *F. T. Watts*, granted on January 16, 1968 (Estate *W. Krüger*); b) Airbus A300 "Krüger flap" (Otto Lilienthal Memorial Lecture, November 9, 1973, Berlin).

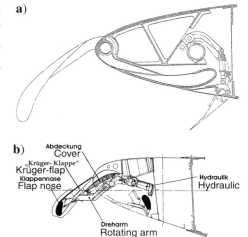

Zahlentafel 1. Hauptsächlichste Kennzeichen der untersuchten Flügelformen

Modell-Nr. Model No.	Pfeilwinkel φ Sweep angle φ	Streckung $\Lambda = \dfrac{b^2}{F}$ Aspect ratio $\Lambda = b^2/F$	Zuspitzung $\dfrac{t_a}{t_i}$ Taper ratio t_a/t_i	Profil Airfoil	Schnitt Section (see Note)	Flügelenden Wing tips (see Note)
1	0°	4,53	0,6	Gö 623	senkr. zur Mittellinie	senkr. zur Mittellinie
2	15°	4,22	0,6	Gö 623	,, ,,	,, ,,
3	30°	3,40	0,6	Gö 623	,, ,,	,, ,,
4	45°	2,14	0,6	Gö 623	,, ,,	,, ,,
5	−45°	2,14	0,6	Gö 623	,, ,,	,, ,, in Flugrichtung
6	45°	4,53	0,6	Gö 623	in Flugrichtung	in Flugrichtung
7	0°	4,53	0,6	NACA 0012—64	senkr. zur Mittellinie	,,
8	30°	3,40	0,625	NACA 0012—64	,, ,,	,,
9	35°	4,45	0,615	NACA 0012—63	in Flugrichtung	,,
10	45°	2,24	0,622	NACA 0012—64	senkr. zur Mittellinie	,,
11	30°	3,36	0,610	NACA 0012—64	in Flugrichtung	,,
12	30°	4,50	0,618	NACA 0012—64	senkr. zur Mittellinie	,,
14	−30°	4,50	0,613	NACA 0012—64	,, ,,	,,
15	45°	4,43	0,613	NACA 0012—64	,, ,,	,,
16	−45°	4,40	0,167	NACA 0012—64	,, ,,	,,

Note: senkr. zur Mittellinie = normal to the centerline; in Flugrichtung = in flight direction.

Fig. 2.116 Compilation of the investigated high-speed models of the AVA Göttingen.

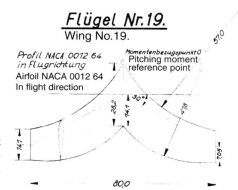

Flügel Nr.19.
Wing No.19.

Profil NACA 0012 64
in Flugrichtung
Airfoil NACA 0012 64
In flight direction

Momentenbezugspunkt 0
Pitching moment
reference point

Fig. 2.117 Swept-wing plan form built up of circular arcs (Airfoil NACA 0012-64), Λ = 3.88, measurements by *Roth* [139].

this patent application was rejected by the State Patent Office due to another similarly worded patent application. The additional patent A 90 629 XI 62b, granted on June 4, 1942, contained the following claims:

> 1. Aircraft with speeds near the speed of sound and large sweep (30 deg and more) in high-speed flight according to Patent Application A 90 203XI/62b, which is equipped with devices to change wing sweep during flight and with known landing devices, such as landing flaps, and which is characterized by the fact that the landing devices, such as landing flaps or similar, which generate a nose-down moment when activated, are arranged and adjustable in a way such that the pressure changes in case of a change in sweep can be compensated.
>
> 2. Arrangement according to Claim 1 characterized by the fact that the operation of these devices and the change in the wing sweep position are necessarily being coupled.

signed *Albert Betz*

The enormous insight into the problems associated with the introduction of the swept-wing technology, which was already available shortly after the experimental verification of *Busemann's* theory (1939), is remarkable. The application of variable wing sweep on modern fighter aircraft is today a proven technology. A change of the wing sweep during flight was intended for the project Messerschmitt Me P 1101. *Messerschmitt* insisted, however, during the construction of the experimental aircraft on trials with fixed wings of different sweep. The test aircraft had almost been completed in 1945 and was about to undergo flight testing, also see Chapter 6 of this book.

2.3.3.11 Swept-Wing Investigations in the High-Speed Wind Tunnels of the DVL (2.7-m Diameter) and the LFA (2.8-m Diameter). Measurements were carried out in the DVL high-speed wind tunnel (2.7-m diameter), following the measurements at the AVA on swept wings of different planforms, on a

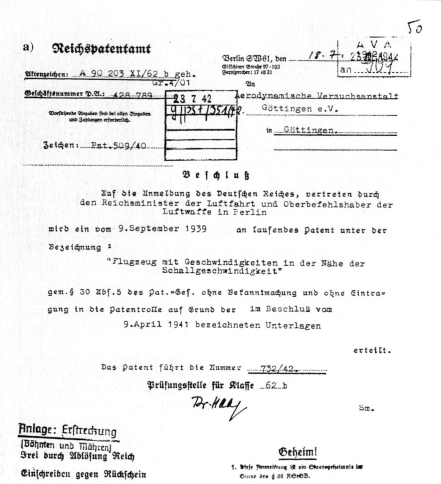

Fig. 2.118   a) Swept-wing patent 732/42 "Aircraft with speeds near the speed of sound," granted on July 18, 1942, invented by *A. Betz* and *A. Busemann* (DLR Archive Göttingen, GOAR 3674); **b)** translation.

swept wing that had been investigated as the first one of a planned, systematic series of different swept wings. Because the dimensions of the DVL models (1.2-m span) compared to the AVA models (0.08 m) were increased by a factor 15, measurements on an optimized wing form could for the first time provide information concerning the delay in the occurrence of compressibility effects due to sweep at high *Reynolds* numbers.

The model was a 35-deg swept wing (based on the l/4-chord line) with an aspect ration of $\Lambda = b^2/F = 6$ and a taper ratio $l_i/l_a = c_i/c_a = 2$ (Fig. 2.121). The airfoil, taken in the flight direction, was a standard airfoil without camber with a thickness ratio of 12% and a maximum-thickness location of 30%

b) State Patent Office
Reference:

Business number:

Reference: Pat. 509/40

To

Aerodynamic Test Establishment
Göttingen e.V.
Göttingen

Decision

Upon the registration of the German State, represented by the State Minister of Aeronautics and Commander-in-Chief of the German Air Force in Berlin, a patent, valid since September 9, 1939 under the designation:

"Aircraft with speeds near the speed of sound"

will be granted according to § 30, Section 5, of the Patent Law without announcement and enrollment into the Patent Roll based on the documents named in the decision of April 9, 1941.

The Patent has the number <u>732/42</u>
Examination Office for Class 62b

<u>Secret!</u>

Fig. 2.118 **(Continued)**

(NACA 0 00 12–1.1 30). At first, the use of an actual high-speed airfoil was deliberately rejected—according to a statement of *Göthert* [140]—to allow, among other things, a direct comparison with a standard wing, that is, a wing without sweep and with a standard maximum-thickness location also tested in the DVL high-speed wind tunnel. The *Reynolds* number based on the mean chord l_μ (c_μ) (reference chord, m.a.c.) of the swept wing ranged dependent on the *Mach* numbers investigated from $R = 1.3 \times 10^6$ ($M = 0.3$) to $R = 2.6 \times 10^6$ ($M = 0.87$). The pitching moments of the swept wing were normally based on the geometrically averaged mean chord l_μ (c_μ) and referred to an axis that corresponded in case of rectangular wings to the 1/4-chord line. Three-component force measurements were carried out at different *Mach* numbers, which in the case of especially interesting speeds and angles of attack, were complemented by total pressure-loss measurements downstream of the wing and by flow visualizations with a schlieren system.

Göthert reported in detail on important results at the LGL Meeting 156 "Questions of High Speeds," held in October of 1942 in Berlin and already mentioned:

- Up to *Mach* numbers of $M = 0.8$, perturbations did not occur in the lift and moment developments in the range of low lift coefficients while at higher lift coefficients $C_A = C_L \geq 0.4$, the slope of the lift curve C_A (C_L) $= f(\alpha)$ decreased, and, most of all, the pitching moment turned heavily tail down. As shown in Fig. 2.122, the lift-curve slope $\partial C_A/\partial \alpha$ was always positive for lift coefficients $C_A = C_L = 0$; it was positive for $C_A = C_L = 0.3$ up to

Reichspatentamt

Aktenzeichen: A 90 497 XI/62 b geh.
                            Gr. 4/01

Geschäftsnummer P.A.: 459 902

Vorstehende Angaben sind bei allen Eingaben
und Zahlungen erforderlich.

...zeichen: G/Pst/T. 319/42

Berlin SW61, den _____ 2. 9. 1942.
Glitschiner Straße 97-103
Fernsprecher: 17 18 21

An

Aerodynamische Versuchsanstalt
Göttingen e.V.

in _____ Göttingen

11 SEP. 1942

11. 9. 42

3/21

Beschluß

Auf die Anmeldung des Deutschen Reiches, vertreten durch

den Reichsminister der Luftfahrt und Oberbefehlshaber der
Luftwaffe in Berlin

wird ein vom 17.November 1939 an laufendes Patent unter der

Bezeichnung :

"Flügel mit starker Pfeilstellung"

- Zusatz zum Patent 732/42 (A 90 203 XI/62 b geh.) -

gem.§ 30 Abs.5 des Pat.=Ges. ohne Bekanntmachung und ohne Eintra=

gung in die Patentrolle auf Grund der im Beschluß vom

4.Juni 1942 bezeichneten Unterlagen

erteilt.

Das Patent führt die Nummer _____ 790/42.

Prüfungsstelle für Klasse _____ 62 b

Dr. Haa[?]   Sm.

Geheim!

1. Diese Anmeldung ist ein Staatsgeheimnis im
   Sinne des § 88 RStGB. in der Fassung des
   Gesetzes vom 24. 4. 1934 (RGBl. I S. 341 ff.).
2. Weitergabe nur verschlossen, bei Post=
   beförderung als "Einschreiben".
3. Empfänger haftet für sichere Aufbewahrung.

Anlage: Erstreckung

(Böhmen und Mähren)

Frei durch Ablösung Reich

Einschreiben gegen Rückschein

mit Merkblatt u.Gebühren=Anm.
   " Vordr. 19 a

Pat.16 a(U) Pat.=Ert.- Präf.=B.IIb
  1940.1000

Fig. 2.119   Addition to patent 732/42 "Wing with high sweep": patent 790/42 granted on September 11, 1942, invented by *A. Betz* (DLR-Archive Göttingen, GOAR 3674). [See Fig. 2.118 except for the dates and the designation, the latter being "Wing with high sweep," Addition to Patent 732/42 (A 90 203 XI/62b).]

Reichspatentamt

Aktenzeichen: A 90 629 XI/62 b geh.
Gr. 6/03

Geschäftsnummer P.A.: 475 905

Vorstehende Angaben sind bei allen Eingaben und Zahlungen erforderlich.

G/Pst/T.
293/42 g

A V A
11.SEP.1942

Berlin SW61, den   2.9. 1942.
Glisshiner Straße 97-103
Fernsprecher: 17 +8 21

An

Aerodynamische Versuchsanstalt
Göttingen e.V.

in Göttingen

11 9 42

3/22

B e s c h l u ß

Auf die Anmeldung des Deutschen Reiches, vertreten durch

den Reichsminister der Luftfahrt und Oberbefehlshaber der
Luftwaffe in Berlin

wird ein vom   17.Dezember 1939   an laufendes Patent unter der

Bezeichnung :

"Flugzeug mit Einrichtung zur Änderung der
Flügelpfeilung"

- Zusatz zum Patent 732/42 (A 90 203 XI/62 b geh.) -

gem.§ 30 Abs.5 des Pat.=Ges. ohne Bekanntmachung und ohne Eintra=

gung in die Patentrolle auf Grund der im Beschluß vom

4.Juni 1942 bezeichneten Unterlagen

erteilt.

Das Patent führt die Nummer 799/42.

Prüfungsstelle für Klasse 62 b

Dr. Haag

Sm.

Geheim!

1. Diese Anmeldung ist ein Staatsgeheimnis im
   Sinne des § 88 RStGB. in der Fassung des
   Gesetzes vom 24.4.1934 (RGBl. I S. 341 ff.).
2. Weitergabe nur verschlossen, bei Post-
   beförderung als „Einschreiben".
3. Empfänger haftet für sichere Aufbewahrung.

Anlage: Erstreckung

(Böhmen und Mähren)

Frei durch Ablösung Reich

Einschreiben gegen Rückschein

mit Merkblatt u.Gebühren=Anm.
  ″ Vordr.19 a

Pat.16 a(U) Pat.=Ert.- Präf.=B.115
  1940.1000

**Fig. 2.120 Addition to Patent No. 732/42: "Wing with device to change wing sweep."
Patent No. 799/42 granted on September 11, 1942. Inventor: *A. Betz*. (DLR-Archive
Göttingen, GOAR 3674). [See Figure 2.118 except for dates and designation, the latter
being: "Aircraft with device to change wing sweep," Addition to Patent 732/42 (A 90 203
XI/62b secret).]**

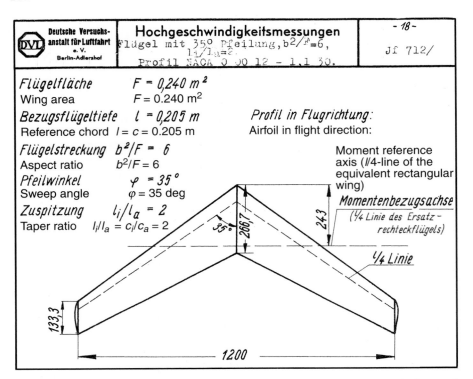

Fig. 2.121 DVL 35-deg swept-wing model for the 2.7-m high-speed wind tunnel, B. *Göthert* [140]; translation of header: high-speed measurements, wing with 35-deg sweep, $b^2/F = 6$, $l_i/l_a = 2$ airfoil NACA 0 00 12–1.130 note: $l = c$ = chord, b = span.

$M = 0.82$ and for $C_A = C_L = 0.55$ positive up to $M = 0.7$. (In Fig. 2.122 holds: $c_a \to C_A$; $c_m \to C_m$.) At $C_A(C_L) \neq 0$, the curves turned towards smaller values. This could be traced to the lift losses occurring when exceeding the critical *Mach* number, which was associated with shocks and flow separations. The lift unsteadiness observed was naturally also reflected in the development of the pitching moment $C_m = f(\alpha)$. In Fig. 2.122, the gradients $\partial C_m/\partial C_A$ (C_L) were plotted vs the *Mach* number. At the same time, the development of the gradient described the changes in the neutral-point locations at various lift coefficients C_A (C_L) especially in the range of high *Mach* numbers. Here, it becomes obvious that for lift coefficients of $C_A = C_L \leq 0.3$ with increasing speed up to a *Mach* number of $M = 0.8$ an almost unchanged neutral-point location was determined, which was followed at $M \geq 0.8$ by a strong shift of the neutral point towards a more stable position. At the higher lift coefficient of $C_A = 0.55$, a strong destabilizing shift in the neutral-point position developed, which reflected at higher *Mach* numbers the strong break in the pitching-moment development at lift coefficients above the critical one.

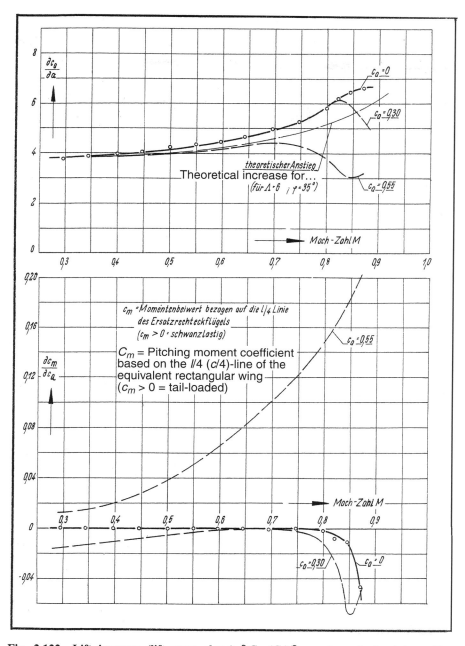

Fig. 2.122 Lift increase (lift-curve slope) ∂C_A $(C_L)/\partial \alpha$ and neutral-point position $\partial C_m/\partial C_A$ (C_L) dependent on the *Mach* number M $(C_A$ $(C_L) = 0, 0.3, 0.55)$, *B. Göthert* [140]; axes: $\partial c_a/\partial \alpha = \partial c_l/\partial \alpha$, $\partial c_m/\partial c_l$, *Mach* number M.

- For the *Mach* numbers of $M = 0.3$ and $M = 0.87$, the measured pitching moments for the DVL swept wing were plotted in Figs. 2.123 and 2.124 dependent on the lift coefficient and compared to the results for the rectangular wing, which had the same airfoil data. At the lowest *Mach* number of $M = 0.3$, the lift coefficients had as expected shown a lower, but linear increase with angle of attack in comparison to the rectangular wing. Although the sign of this deviation was in agreement with the computational methods available at the time, the reduction in the lift-curve slope ∂C_A $(C_L)/\partial \alpha$ due to sweep was, however, not proportional to $\cos\varphi$, the appropriate main element of the computational method. While the lift development for the rectangular and the swept wing was almost linear up to maximum lift, the slope of the pitching-moment coefficient at $M = 0.3$ in case of the swept wing was already slightly curved, unlike the one of the rectangular wing. It is worth mentioning that the break in the pitching-moment curves towards tail-down moments when exceeding certain high lift coefficients. These critical lift coefficients shifted to lower values with increasing *Mach*

Fig. 2.123 Lift coefficients C_A (C_L) vs pitching-moment coefficient C_m at a *Mach* number of $M = 0.3$, comparison swept wing/ rectangular wing, *Göthert* [140].

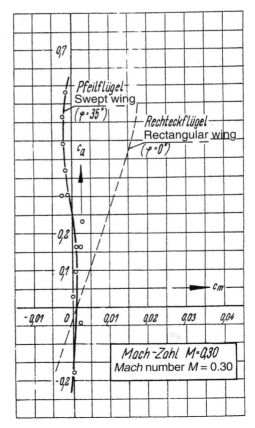

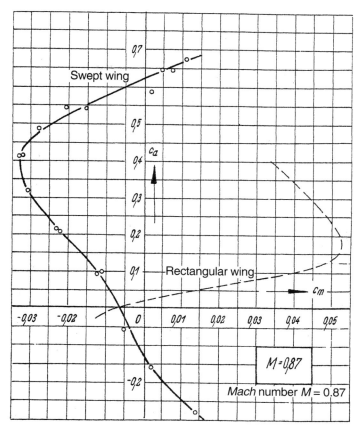

Fig. 2.124   Lift coefficients C_A (C_L) vs pitching-moment coefficient C_m at a *Mach* number of $M = 0.87$, comparison swept wing/ rectangular wing, *Göthert* [140].

number. This break in the pitching-moment development towards tail down was apparently due to a breakdown of the flow at the wing tips, which was, in the case of sweptback wings, promoted by the boundary layer flowing towards the wing tips. At the highest *Mach* number investigated, $M = 0.87$, the pitching-moment curves were at low lift coefficients turned in a nose down, that is, stable sense, which pointed to lift losses in the center region of the wing.

• A comparison of the slopes $\partial C_m/\partial C_A$ (C_L) in Fig. 2.125, as an indicator of the neutral-point location at low lift coefficients ($C_A = C_L = 0$), showed that large perturbations in the neutral-point location only occurred after exceeding the *Mach* number of $M = 0.8$. In case of the rectangular wing, a considerable shift in the neutral-point location could already be observed starting at $M = 0.73$. Noticeable here was that the neutral point of the swept wing exhibited at low lift coefficients no shift up to the onset of gross perturbations at a *Mach* number of $M \approx 0.8$. The rectangular wing

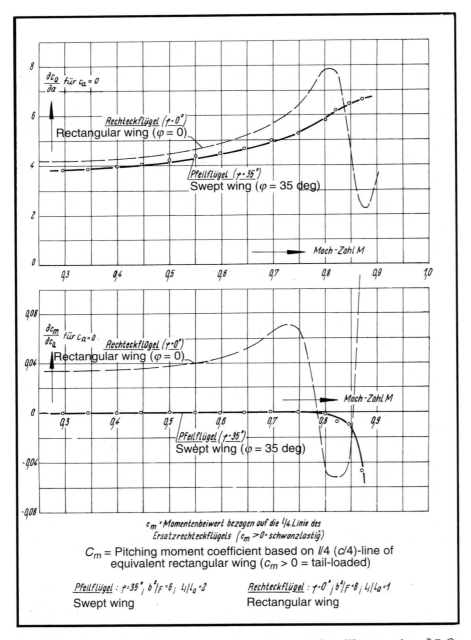

Fig. 2.125   Comparison between rectangular and swept wing: lift-curve slope $\partial C_A/\partial\alpha$ ($\partial C_L/\partial\alpha$) and neutral-point location $\partial C_m/\partial C_A$ ($\partial C_m/\partial C_L$), *Göthert* [140]; axes: $\partial c_a/\partial\alpha$ ($\partial c_l/\partial\alpha$) at c_a (c_l) = 0, $\partial c_m/\partial c_a$ ($\partial c_m/\partial c_l$) at c_a (c_l) = 0, *Mach* number.

showed, on the other hand, also at purely subsonic speeds a steady movement of the neutral point with increasing *Mach* number in an unstable sense, that is, towards the wing leading edge.

These tests clearly pointed out two flow phenomena, namely, that the swept wing showed at low-lift "cruise" a more stable behavior than the rectangular wing, but that the longitudinal stability at high angles of attack could be dramatically reduced. The limits of a conventional swept wing were thus shown, and its applicability on fighter aircraft—without the use of, for instance, stabilizing flap systems—became obvious. The solution of this so-called "tip-stall problem" after 1945 was still to cause great difficulties to the victorious powers in the East and West. *Göthert's* comparison between the rectangular and the swept wing also contained a warning to the people responsible for the flight trials because high-speed trials of conventional fighter aircraft during a dive could become very dangerous to the pilot and the aircraft due to the stability problems of conventional wings. These difficulties are described in Chapter 5.

- In Fig. 2.126 the drag coefficients $C_W(C_D)$ for the swept wing and the rectangular wing of the DVL were plotted vs *Mach* number for the lift coefficients C_A $(C_L) = 0$, 0.2, and 0.4. One observes that the drag rise for the swept wing at the lift coefficients considered was, without exception, shifted by $\Delta M \approx 0.08$ to higher *Mach* numbers. At a lift coefficient of C_A $(C_L) = 0$ the drag rise of the rectangular wing commenced at a *Mach* number of about $M = 0.72$ and the one of the swept wing at about $M = 0.8$. In addition, it became clear that the drag rise after exceeding the critical *Mach* number was obviously less steep for the swept wing than for the rectangular wing. *Göthert* explained this difference in the drag rise by the fact that in case of the swept wing shocks appeared initially only in the center region of the wing and then spread gradually across the entire span. He based his explanations on extensive schlieren observations and wake measurements, which had been carried out within the scope of this basic program at various wing sections. *Göthert* concluded that in the case of a suitable design of the fuselage side walls (junction), a wing-fuselage combination could be expected to behave with respect to drag more favorably than an isolated swept wing. (This assumption had already been confirmed by *Ludwieg's* measurements on wing-fuselage combinations; see Fig. 2.105.)

Schlichting remarked during a discussion at the LGL Meeting 156 that the flow conditions on a wing-fuselage arrangement could in principle change. As an example, he quoted results of investigations at the Institute of Fluid Mechanics of the TH Braunschweig, which had been carried out on a rotational ellipsoid (prolate spheroid) with a rectangular wing in incompressible

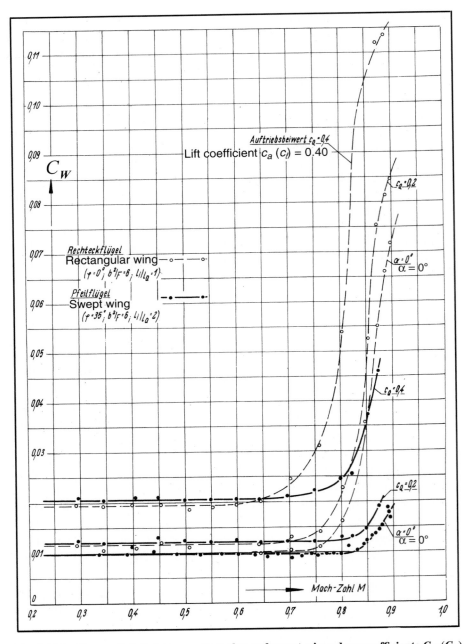

Fig. 2.126   Comparison between rectangular and swept wing: drag coefficients C_W (C_D) as a function of the *Mach* number M for various lift coeffients C_a (C_l) and angles of attack $\alpha = 0\,\text{deg}$, *Göthert* [140]; axes: C_W (C_D), *Mach* number M.

flow. Although the latter had been a rectangular sheared wing, it became clear that very high lift peaks occurred on wing-fuselage arrangements. These resulted from the lateral component of the flow about the fuselage. In the case of a high (shoulder) wing, this caused a lift increase because the leading wing half generated in the immediate vicinity of the fuselage an additional positive angle of attack. For the trailing wing half, the angles of attack were reduced and the lift decreased. An additional rolling/yawing moment, as in the case of the positive dihedral wing, was generated on the high-wing aircraft due to the influence of the fuselage. For the low-wing aircraft, the signs of the influence parameters were reversed.

• Another high-speed investigation was carried out by *G. Koch* [141] at the Institute of Gasdynamics of the LFA in Braunschweig headed by *Busemann*. These were contract measurements for the branch of *Alexander Lippisch* of the Messerschmitt Corporation to clarify how far the sheared wing at high subsonic speeds actually showed the theoretically predicted behavior. The recently completed high-speed wind tunnel A6 with a test-section cross section of 36×40 cm was available for these tests. The maximum *Mach* number attainable in the empty test section was $M = 0.94$. The wing model was supported by the solid walls of the test section. A symmetrical hyperbolic airfoil of 9% thickness, 40% maximum-thickness location, and 80-mm chord was selected as model. Figure 2.127 shows the installation of the model at a sweep (shear) angle of 40 deg. The model was, in addition, tested at a sweep (shear) angle of 20 deg and in flow normal to the wing leading edge. The pressure orifices were located in a plane normal to the wing leading edge in the center of the tunnel. The construction of this pressure distribution model required at a chord of only 80 mm a considerable effort. The flow conditions on this wing between two solid tunnel walls did, of course, not exactly correspond to those on the infinite-span sheared wing. However, it was assumed that the tunnel sidewalls had only an insignificant effect on the measurements in the

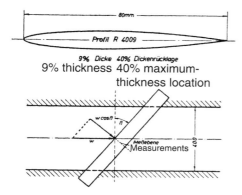

Fig. 2.127   Model installation in the LFA high-speed wind tunnel A6 (36×40 cm), hyperbolic airfoil R 4009 (40% maximum-thickness location, 9% thickness), *Koch* **[141].**

a)

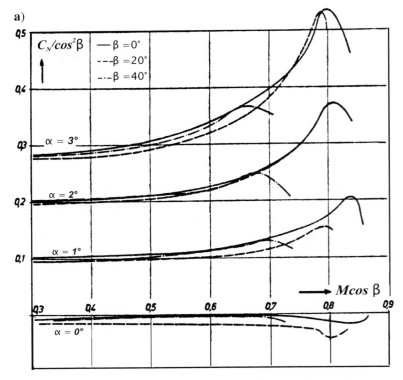

Fig. 2.128   Plots of coefficients a) Normal force coefficients C_N based on the normal velocity component; b) pitching-moment coefficient C_m based on the normal velocity component, Koch [141].

tunnel center, because the ratio of the model chord to the tunnel width was small. If one considers the possible error sources, such as the small number of pressure orifices on an 80-mm wing chord, the solid tunnel walls and possible condensation effects, as well as the deflection (bending) of the model, very interesting results have, nevertheless, been obtained. *Koch* had plotted the normal force coefficients C_N and the pitching-moment coefficients C_m in Fig. 2.128. To allow a direct comparison with the transformation factors according to the *Prandtl–Glauert* rule, $C_N/\cos^2\beta$ vs $M\cos\beta$ respectively $C_m/\cos^2\beta$ vs $M\cos\beta$ were plotted in the figure. (In the figure, $c_n \rightarrow C_N$ and $c_m \rightarrow C_m$.) According to theory, the values determined at the different yaw angles had to coincide. The comparison showed that the agreement between the results was satisfactory up to a *Mach* number of $M = 0.65$. At the yaw angles of $\beta = 0$ deg and $\beta = 20$ deg, the measured data even agreed quite well up to a *Mach* number of $M = 0.78$. That the pitching-moment coefficient at $\alpha = 0$ deg ($C_A = C_L = 0$) did not become zero was traced to the error sources associated with the measurements already mentioned.

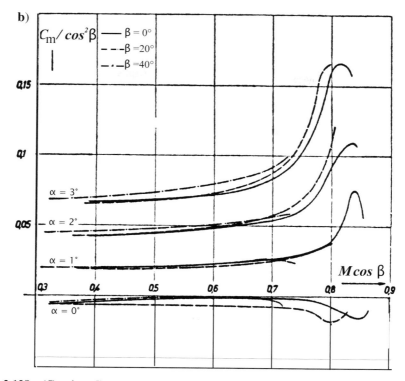

Fig. 2.128   (Continued)

The discussion of the LFA and DVL results showed clearly that the advantages of the swept wing were indisputable at high subsonic speeds; however, the demand of the RLM to build, at short notice, a fighter aircraft superior in speed and equipped with jet engines could with the current state of technology only be met by increased project investigations in the wind tunnel. This fact resulted in a greater reduction of the fundamental investigations, which was also reflected in the wind-tunnel schedules of the individual research establishments.

2.3.3.12 LOW-ASPECT-RATIO WINGS ($\Lambda \leq 3$) [142].   For aircraft that were to fly at high *Mach* numbers, swept wings with low aspect ratios were preferred, whose aerodynamic behavior had, however, first to be investigated both theoretically and experimentally. As a practical example, the project of the tailless aircraft, Delta P11, of *Alexander Lippisch* shall be mentioned in Chapter 6. But also for gliders, which were to be carried up by aircraft and then dropped, low-aspect-ratio wings were used. Subsonic measurements were carried out under contract to the Henschel (AG) Corporation at the DVL in Berlin in the 2.15- × 3.0-m wind tunnel supervised by *G. Lange* and

Wacke [143]. The project manager of the customer was *H. Voepel* of the DFS. *Voepel* had written a report after the war [144] about these basic investigations on systematically varied planforms at the AVA, the LFA, and the DVL at subsonic as well as high speeds up to $M = 2$. Concerning the application, emphasis was placed on the development of a natural stability unmanned aircraft (DFS-D2), the remotely controlled tailless flight vehicles of the Henschel Company (Hs 293F and Hs 298F), and the design of a gun rocket (Electric Ray) for supersonic speeds. (These projects are detailed in Chapter 7.) Within the entire program of the investigation, main emphasis was placed on the effect of the wing planform on the development of the pitching moment $C_m = f(\alpha)$ and $C_m = f(C_A) = f(C_L)$ whose nonlinearity with regard to the longitudinal stability and the controllability about the lateral axis posed in the case of low-aspect-ratio wings a decisive problem for the application. As an example of the experimental results, we shall here merely introduce an excerpt of the work of *Lange* and *Wacke*.

The results for wings of equal aspect ratio $\Lambda = 4/3$, but different taper ratios of $l_a/l_i = c_a/c_i = 1/2$, 1/4, 1/8, and 0 are shown in Fig. 2.129. The sweep angles of the wing leading edge were 45, 61, 66.8, and 71.1 deg. The lift-curve slope became smaller with increasing sweep. The neutral point was only at C_A $(C_L) \approx 0.3$ located at the selected reference axis l_m and was located at lower C_A (C_L) values further upstream ($l_m = F/b$ with F being the wing area and b the wing span). At a taper ratio of $l_a/l_i = c_a/c_i = 0$, it was located by 0.07 l_m upstream, and at larger $l_a/l_i = c_a/c_i$ values by 0.12 l_m downstream. The effect of sweep was relatively small. Drag increased somewhat at the same lift coefficient with sweep while maximum lift decreased slightly. Contrary to the rectangular wing and similarly to the swept wing, the flow on these trapezoidal wings—except for the wing with l_a/l_i $(c_a/c_i) = 1/2$—did not suddenly separate (stall). Today we know in detail the vortical flow on the lee-side and are able to better explain and interpret the flow behavior.

Triangular wings, today called delta wings, with different aspect ratios, that is, $\Lambda = 3$, 2, $\frac{3}{4}$, and 1, were, in addition, investigated; see Fig. 2.130. As expected the lift-curve slope $\partial C_A/\partial \alpha = \partial C_L/\partial \alpha$ and also maximum lift decreased with increasing aspect ratio and hence increasing sweep angle φ. An exception was the delta wing with $\Lambda = 3$ where a relatively early breakdown of the vortical flow could be observed. Within the range of the linear lift increase, the gradient $\partial C_m/\partial C_A$ $(\partial C_m/\partial C_L)$ was almost constant, and the neutral-point position only changed from 0.12 l_m (c_m) to 0.15 l_m (c_m). At $\Lambda = 1$, on the other hand, the neutral point shifted in a range above C_A $(C_L) \approx 0.4$ by 0.15 l_m (c_m) downstream. The shift in the neutral-point location was hence very large. There were, at the time, no pressure distribution measurements or flowfield investigations available, but *Seiferth* had already recognized in his interpretation of the test results that the leading-edge vortices in the case of

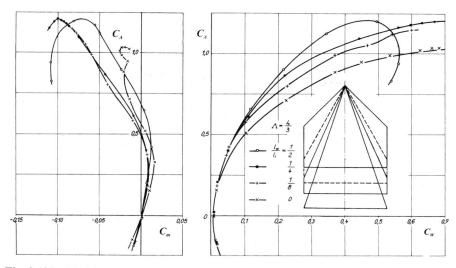

Fig. 2.129   Pitching-moment development and drag behavior dependent on lift on low-aspect-ratio trapezoidal wings; measurements in the DVL 2.15- × 3-m low-speed wind tunnel at $R = 2.1 \times 10^6$, *Lange* and *Wacke* [143]; axes: C_A (C_L), C_m; C_A (C_L), C_W (C_D); legend: $l_a/l_i = c_a/c_i$ (c = chord).

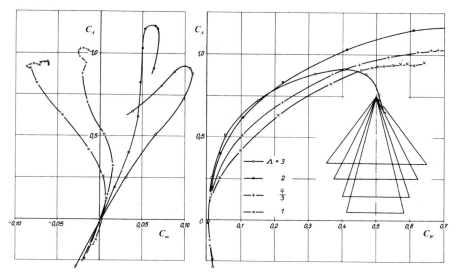

Fig. 2.130   Pitching-moment development and drag behavior dependent on lift on low-aspect-ratio delta wings; measurements in the DVL 2.15- × 3-m low-speed wind tunnel at $R = 2.1 \times 10^6$, *Lange* and *Wacke* [110, 143]; axes: C_A (C_L), C_m; C_A (C_L), C_W (C_D); legend: $l_a/l_i = c_a/c_i$ (c = chord).

low-aspect-ratio wings affected the flow in the downstream region of the suction side of the wing.

Göthert's [145] high-speed measurements on a low-aspect-ratio wing clearly showed the favorable drag development of swept wings at high *Mach* numbers. However, it also turned out that for a wing of very low aspect ratio ($\Lambda = 1$) similar effects already occurred without sweep.

In the accompanying theoretical investigations, two early estimates of the lift increase for low-aspect-ratio wings are especially interesting. In his publication [146], *Weissinger* pointed out the limits of the linear wing theory. His "Wing Method," called "F"-method for short, was based on the plane two-dimensional vortex system of the lifting wing of *Blenk* and *Wieghardt*. Here, *Weissinger* did not define the spanwise lift distribution from the start. For low aspect ratios, the "F"-method provided—under considerable computational effort—with $\partial C_A/\partial\alpha = \partial C_L/\partial\alpha = 0.55\ (\pi\Lambda)$ a somewhat too large lift increase, while his extended lifting-line theory, the L-method, resulted in a value of $\partial C_A/\partial\alpha = \partial C_L/\partial\alpha = 0.5\ (\pi\Lambda)/(\Lambda + 2) \rightarrow \pi\Lambda/2$. This value was in agreement with the so-called "theory of slender bodies" (slender-body theory), *R. T. Jones* [147].

2.3.3.13 BOUNDARY-LAYER FENCE, A BARRIER AGAINST STALL. The development of boundary-layer fences to influence flow separation on the suction side of wings goes back to investigations of *Wolfgang Liebe*, which commenced in 1937 at the DVL in Berlin-Adlershof. *Liebe* [148] reported extensively on his experiences developing boundary-layer fences. The reason was the possibility of wing drop when stalling new fighter aircraft such as the Heinkel He70 and the Messerschmitt Me 109, which could have become a serious problem for the training of inexperienced young pilots. The DVL, therefore, dealt with this problem at the Institute of Flight Mechanics. The objective was to judge the danger of rolling and especially to determine when and why a stronger rolling motion ensued. These rolling moments developed due to the asymmetrical lift decrease on the wing. To the design engineers it was especially important to obtain criteria concerning this asymmetry in the lift distribution arising when stalling the aircraft. Because a simulation of the flight processes in stationary measurements in the wind tunnel was only possible in a limited way, flight tests also had to be carried out to address this problem. In the wind tunnel, the rolling-moment coefficient C_l dependent on the yaw angle β for a fixed angle of attack was determined in preparation for the flight tests [149, 150]. Here, the following flow phenomena were observed:

- A strong dependence of the rolling moment on the yaw angle remained at maximum lift and further increasing angles of attack. The measured yaw-rolling moments $\partial C_A\ (C_L)/\partial\beta$ were larger than at small angles of attack.

- Suddenly appearing asymmetrical lift distributions and an obvious hysteresis were observed when running through angle of attack and yaw sweeps.

These results made flight tests, for instance with the Me 109, necessary because this fighter aircraft had proven to be especially "tip-prone" and could at high angles of attack, that is, especially on approach, only be flown safely with automatically operating slats. *Willy Messerschmitt* wanted to avoid the slats for cost reasons. The original wing of the Me 109 was an untwisted, unswept trapezoidal wing with the same airfoil form across the span, but with a decreasing thickness ratio of 14% inboard to 11% at the tip. The landing flap, a cambered slotted flap, extended from the wing root to 56% of the half-span. The flap angle was during landing 40 deg and the relative flap chord 20%. The wing was equipped with *Handley–Page* slats that were automatically controlled by the pressures on the wing. These slats were installed over 45% of the span at the outer wing and thus particularly endangered when the flow separated in this area. *Liebe* assumed on the basis of the flight tests that the rolling moments occurring when stalling the Me 109 did not represent an unstable flight condition, such as in the case of an autorotation. Conspicuous to him was a strange behavior of woolen tufts glued to the surface for flow visualizations during a provoked stall. In stalling the aircraft, a stronger separation initially only occurred in the vicinity of the fuselage. This perturbation then spread upstream and then suddenly across almost the entire wing up to the wing tip. The primary occurrence of flow separation in the vicinity of the fuselage was surely also caused by the 14% wing thickness. To prevent crossflow within the flow region still attached, two aluminum sheets were attached to the wing of the Me 109 (Fig. 2.131). This was the birth of the boundary-layer fence! Flight tests and flow visualizations showed that the crossflow could be reduced by these means. *Liebe* wrote about the continuation of the work:

> The tests with the boundary-layer fence were continued in Berlin–Adlershof in 1939. With the newest design of the Me 109E, it could finally be proven that the fence would have been an alternative, fully replacing the slat and making it superfluous. That way the tests suggested by Messerschmitt were after all successful. But the success was not recognized anymore. The fence was dismissed. The corresponding reports and the patent documents went into the safe. There they stayed until the end of the war.

The boundary-layer fence quickly found its application after the war due to the use of swept wings. On the swept-back wing, a spanwise pressure decrease occurred, which generated an outward boundary-layer flow. The first aircraft equipped with boundary-layer fences was the Russian fighter

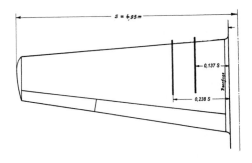

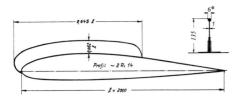

Fig. 2.131 Arrangement of the boundary-layer fence on the Me 109 according to *Liebe* [148]; top: joint, and bottom: airfoil.

aircraft MiG-15. This jet aircraft, already developed in 1947, turned out to be dangerous when stalling. With two boundary-layer fences on each wing side, flight characteristics could be considerably improved. The MiG-15 was built in large numbers. The aircraft shown in Figs. 2.132 and 2.133 were produced in Poland under license (photographed at the Military Museum at Kolberg). After the MiG models, the boundary-layer fence was used on further aircraft with swept wings, commercial as well as military. Some examples are shown in Fig. 2.134. It should still be mentioned here that the Spanish postwar design of the Me 109 and the "Hansa Jet" HFB-320 were equipped with boundary-layer fences, a late satisfaction for the inventor of the boundary-layer fence, *Wolfgang Liebe*.

Fig. 2.132 MiG-15 with boundary-layer fences, 1951 Polish construction under license, Military Museum Kolberg.

Fig. 2.133   MiG-19 with boundary-layer fences, 1955 Polish construction under license, Military Museum Kolberg.

2.3.3.14   INVESTIGATION OF BODIES OF REVOLUTION AND FUSELAGES. *Göthert* had already at the LGL Meeting 127 in 1940 in Göttingen reported on high-speed measurements on bodies of rotational symmetry and the comparison with drop tests [151]. This investigation was carried out in cooperation with the Henschel Aircraft Works in the DVL high-speed wind tunnel on bodies of rotational symmetry, which, with an axis ratio of diameter/length of <1:1.7, were of interest as basic developments for bombs rather than aircraft fuselages. This investigation that *Göthert* [152] had continued during the following years for study purposes on two bodies led to important results concerning the drag behavior of such bodies at high subsonic speeds. *Zobel* [153] had worked extensively on the control of the wake behind bodies of rotational symmetry. Here, he tested the interferometer developed for the 2.8-m high-speed wind tunnel, for instance, in investigations of the wake of cylinders simulating the crossflow on a projectile (missile). For that purpose, he compared, as in Fig. 2.135, the flowfield on a cylinder with and without controlling the wake at a *Mach* number of $M = 0.30$. The density field, which was visualized by interferometer and schlieren pictures, shows on the left an almost stationary wake due to the insertion of a flat plate. In the center picture, one observes pronounced disturbances already commencing at a downstream distance of the plate of one cylinder diameter. The right

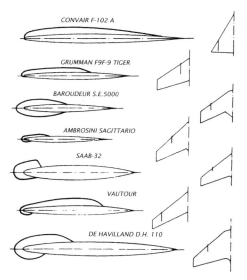

Fig. 2.134   Examples of boundary-layer fences on postwar aircraft according to *Liebe* [148].

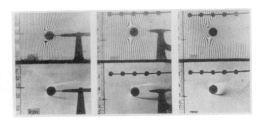

Fig. 2.135 Visualization of the flow-field (density changes) on a cylinder with wake control according to *Th. Zobel* [153]; upper row: interferometer pictures, lower row: schlieren pictures.

photograph was exposed by 1/6000 of a second but still showed only the highly unsteady wake without any pronounced structures. A comparison with pressure measurements downstream of the cylinder confirmed this effect on the wake. A direct effect on the wake downstream of bodies of rotational symmetry by changing the form of the tail of missiles was successfully investigated by *Zobel* in the 2.8-m high-speed wind tunnel; see Fig. 2.136. The results showed that the drag coefficient of the drop-shaped (ideal) body already exhibited at $M = 0.5$ a stronger drag increase that merged into a steep drag rise at $M = 0.70$. In this *Mach* number range, the drag coefficient of the body of revolution with the extreme maximum-thickness location decreased slightly and only increased steeply at $M = 0.72$. By cutting off the blunt back and introducing, instead, a drop-shaped tail, an interesting method of drag reduction was found.

Concerning the development of high-speed aircraft, important and basic tests were reported by *H. Melkus* [154]. All together eight fuselage models of

Fig. 2.136 Comparison of the drag of streamlined bodies of rotational symmetry and different plan forms; measurements in the LFA 2.8-m high-speed wind tunnel, *Th. Zobel* [153]; axes: drag coefficients C_W (C_D), *Mach* number M; legend: O = streamlined body with an extreme maximum-thickness location, nose closed with drop-shaped tail, and X = ideal fuselage.

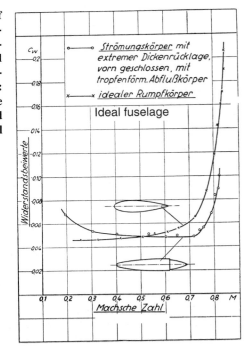

rotational symmetry were investigated in the 2.7-m high-speed wind tunnel of the DVL. Here, the first six fuselage shapes corresponded to the drop shape of the NACA airfoil systematic. At the maximum-thickness location $x_d/l = 0.45$, the models distinguished themselves by the slenderness D/L (thickness/length) that was decreased in equal intervals from $D/L = 0.4$ (Fuselage Model 1) to $D/L = 0.1$; see Fig. 2.137. The remaining two models had the same relative

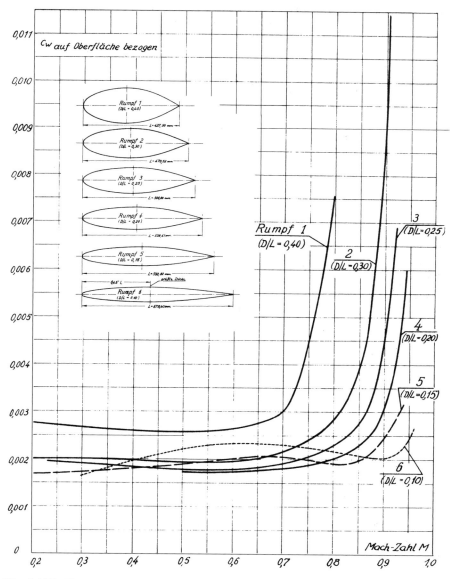

Fig. 2.137   Drag measurements on six fuselages of rotational symmetry in the 2.7-m high-speed wind tunnel of the DVL, *H. Melkus* [154]; axes: c_w (c_d) based on the surface area, *Mach* number M; legend: Rumpf = Fuselage.

thickness as Fuselage 5 ($D/L = 0.15$), but were different concerning the nose radius that was in the case of Model 5a double and in the case of Model 5b four times as large as for Model 5. In addition, cylindrical center sections were inserted in the case of Model 5b in the range between 0.3L and 0.4L.

The development of the *Reynolds* number (based on the fuselage length) was plotted in Fig. 2.138 dependent on *Mach* number. The change in the *Reynolds* numbers also allows for the temperature effects of the wind tunnel flow. The highest *Reynolds* numbers were achieved with Model No. 6, because with decreasing slenderness D/L the fuselage length L was steadily increased order to adjust the surfaces of the bodies for reasons related to the testing technique. The drag coefficient C_W (C_D) was based on the model surface. The drag of the models was determined with a special model support (Fig. 2.139), which was carefully designed. The drag of the model [including the support minus the drag of the support (alone)] was determined. The test results in Fig. 2.137 showed for the models investigated a systematic change in the drag coefficient with *Mach* number. The drag coefficients C_W (C_D) were relatively low in the incompressible flow regime. Also the steep pressure rise, which is a criterion

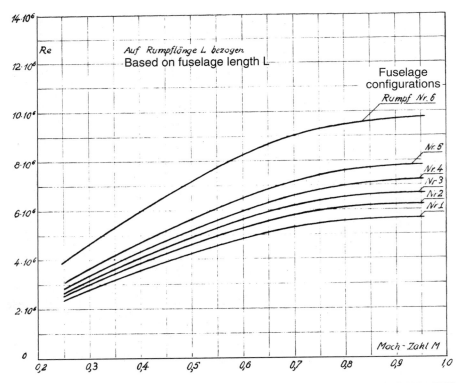

Fig. 2.138 Dependence of the *Reynolds* number *R* on the *Mach* number *M* for the DVL fuselage configurations Nos. 1–6 in the 2.7-m high-speed wind tunnel, *H. Melkus* [154].

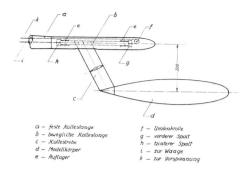

a – feste Haltestange
b – bewegliche Haltestange
c – Haltestrebe
d – Modellkörper
e – Auflager

f – Umlenkrolle
g – vorderer Spalt
h – hinterer Spalt
i – zur Waage
k – zur Vorspannung

Fig. 2.139 Schematic of the test setup for the DVL fuselage measurements in the 2.7-m high-speed wind tunnel, *H. Melkus* [154]; legend: a – fixed support rod, b – movable support rod, c – strut, d – model, e – support, f – turning reel, g – forward gap, h – rear gap, i – to the balance, k – to prestress device.

for the aerodynamic quality in the compressible flow regime, commenced for all fuselage models at relatively high *Mach* numbers. The gradually commencing drag rise in the critical *Mach* number range indicated at the same time—as already discovered by *Göthert*—that locally exceeding the speed of sound initially only caused weak shocks and no boundary-layer separation. The increase in drag in case of the really slender fuselages ($D/L = 0.15$ and 0.1) at medium-high *Mach* numbers was related to the laminar-turbulent transition of the boundary layer, which was determined by the DVL "dust deposit" method.

Especially revealing was, in that regard, the investigation of all models with a smooth and a rough surface in order to determine the effect of surface roughness on the drag development with changing *Mach* number; see Fig. 2.140.

These investigations were especially important with regard to the application of the test results to the experimental aircraft project DFS 346, which is described in detail in Chapter 6. This aircraft was to achieve maximum speeds corresponding to $M = 2.5$, propelled by liquid-fuel rocket engines, and hence provide the possibility to study characteristic free-flight data in the transonic and supersonic flow regimes. The project DFS 346 was conceived in 1944 by the DFS and the Siebel Company tasked with the construction. The fuselage measurements were carried out at the DVL Berlin-Adlershof during November/December of 1944 under the most difficult wartime conditions. All the more remarkable are the very thoroughly planned tests, the excellent results, and their documentation. To the latter also belongs the comment of *Melkus* concerning the effect of the model support on the test results. His comment on the "additional blockage" due to the model support in the rear regions of the fuselage models investigated suggests that he meant the distortion of the drag according to the "area rule" (Sec. 2.3.4). With these measurements a database was generated that can still today—after more than 60 years—be used to validate advanced computational methods. Figure 2.141 shows a Russian version of the DFS 346 that was further investigated in the USSR after 1946. It is obvious that the fuselage of rotational symmetry could be approximated by a 1:9 rotational ellipsoid corresponding to the DVL Model Configuration 6.

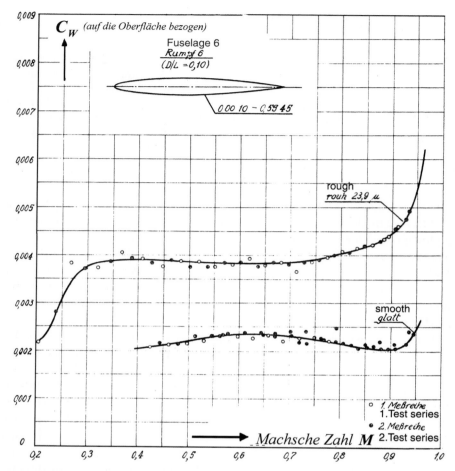

Fig. 2.140 Drag coefficients of the smooth and the rough DVL fuselage 6 dependent on *Mach* number, *H. Melkus* [154]; axes: C_W (C_D) based on the surface area, *Mach* number *M*.

Fig. 2.141 Project of the experimental supersonic aircraft DFS 346, 1944; approximation of the fuselage by a 1:9 rotational ellipsoid (prolate spheroid).

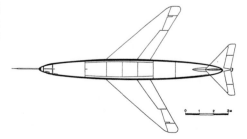

2.3.4 TRANSONIC AREA RULE, A HIGHER AERODYNAMIC DESIGN PRINCIPLE (WERNER HEINZERLING)

2.3.4.1 INTRODUCTION. The transonic area rule, formulated, patented, and applied during the years 1943–1945 by *Otto Frenzl* at the Junkers Aircraft and Engine Company (Works) in Dessau belongs—besides the swept wing, the supercritical airfoil, and the laminar airfoil—to the very few basic concepts of aircraft aerodynamics. It is, up to today, applied on high-speed aircraft in order to postpone the steep drag rise occurring when approaching the speed of sound to higher *Mach* numbers hence achieving higher flight speeds. The area rule considers in summary the overall aircraft geometry and includes the transonic speed range as well as, in a somewhat modified form, the supersonic range.

While the swept-wing concept, already clearly visible in the planform of an aircraft configuration, gives guidelines for the optimum distribution of the lift-generating surfaces, the area rule represents a guideline for the optimum axial distribution of the cross sections of an aircraft. The observation of the area rule in an aircraft design is, therefore, only recognizable on second sight. Between the area rule and the swept-wing concept there exists a direct geometrical connection, Fig. 2.172.

Probably independent of *Frenzl's* Junkers-patent of 1944, the "Area Rule" was discovered and applied for a second time in 1952 by *Richard T. Whitcomb* at the NACA and internationally published in 1955. It is surprising that in the international literature, wing sweep is, up to today, indeed correctly presented as a German development from the time period 1935–1945, but that the discovery, patenting, and first applications of the area rule are falsely ascribed to *Richard T. Whitcomb* (1952) and not to *Otto Frenzl* (1944).

2.3.4.2 AREA RULE CONCEPT. The area rule is an aerodynamic concept to minimize drag of aircraft flying near the speed of sound or at supersonic speeds by a special distribution of the cross sections of the aircraft. Here, the entire geometry of the aircraft, including the engines, is considered. An older textbook on aircraft aerodynamics of *F. Dubs* [155] introduces the concept by a comparison of an aircraft designed without and under consideration of the area rule (Fig. 2.142).

In applying the area rule to an aircraft configuration, the axial development of the cross-sectional areas is designed in a steadily increasing or decreasing way. The maximum overall cross section shall, of course, in addition be as small as possible. By a suitable design of the axial distribution of the cross sections and an appropriate axial arrangement of the aircraft components, this is to a good approximation possible. Unsteadiness in the development of the cross sections can, if necessary, be compensated by additional bulges, waistlike shaping, or separately attached displacement bodies. A typical means of improving the distribution of the cross sections

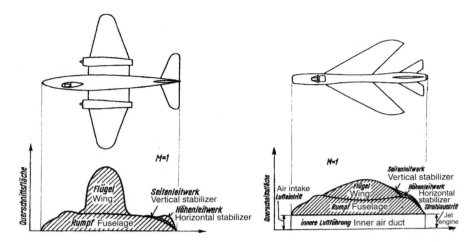

Fig. 2.142 Axial distribution of the cross-sectional displacement of an aircraft designed with and without consideration of the area rule (according to *Dubs*) [155]; axes: cross-sectional area, none.

is, for instance, the waistlike shaping of the fuselage in the region of the wing. In the case of commercial aircraft with constant passenger or cargo compartments, this is not necessarily possible. Here, the ideal continuous cross-sectional distribution must be approximated by a skillful arrangement and design of the wings and nacelles as well as additional bulges, for instance, to accommodate the landing gear, or separate displacement bodies. Typical is the placement of the jet engines below the wing in nacelles attached to forward-pointing profiled pylons.

The sweep of the wing represents, by the way, a direct geometric improvement in the sense of the area rule because the contribution of the wing to the total cross-sectional area of the aircraft is distributed over a larger distance in flow direction simultaneously reducing the maximum contribution of the wing. Especially effective in this regard is the oblique (sheared) wing where the contribution of the wing to the cross-sectional area is distributed over a particularly large distance and where always only one wing side is located in a cross-sectional plane contributing to the total cross-sectional area. (The geometric connection between the area rule and the swept-wing concept, found already some years earlier on the basis of quite a different consideration, will later be explained more precisely—see Fig. 2.172.)

In the 1940s, the practical application of the area rule in aircraft design aimed initially at the flight regime of high subsonic speeds, "Transonic Area Rule," which was just being opened up with the aid of the new jet engines. The area rule was also valid at supersonic speeds, however, in a somewhat modified form.

The area rule constitutes for the design engineer a very convenient recipe that can serve him as a lead to the design of the complete aircraft configuration with the axial development of the cross-sectional area allowing him a quick qualitative judgment of the effect of different geometrical variants with respect to transonic drag.

2.3.4.3   *OTTO FRENZL* DISCOVERS THE TRANSONIC AREA RULE IN 1943.   The first formulation, explanation, and patenting of the concept of the transonic area rule goes back to *Otto Frenzl* (Fig. 2.143). *Frenzl* (born November 12, 1909, Graz; deceased November 1, 1996, Innsbruck) was employed from 1937 as test engineer in the department Fluid Technology under *Philipp von Doepp* (1885–1967) at the Junkers Aircraft and Engine Works in Dessau. At Junkers he built and ran, among other things, the 0.30×0.30-m high-speed wind tunnel HK 900 [156, 157] (Fig. 2.144).

The small facility operated with a steam-ejector drive on a short-time basis and reached in the semi-open test section just sonic velocity. Because of the small test section, aircraft and wings were mostly investigated as half-models (Fig. 2.146). The test section was equipped with a schlieren system to visualize pressure distributions, a three-component balance for force measurements, and an advanced data-reduction technique (Fig. 2.145).

Based on many flow visualizations and test results obtained in the wind tunnel, *Frenzl* came to the conclusion that decisive for the steep drag rise of aircraft approaching the speed of sound was the magnitude and axial distribution of the flow-displacing cross sections of the aircraft. He based this correctly on the fact that pressure disturbances in the case of sonic speed

Fig. 2.143   *Otto Frenzl* **(1909–1996), the discoverer of the transonic area rule (right), with** *Theodore von Kármán***, second from right, in front of a steam-jet ejector at the SNECMA in Villaroche 1958 (Archive German Museum München).**

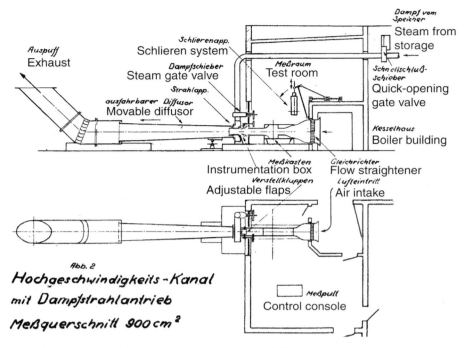

Abb. 2
Hochgeschwindigkeits-Kanal
mit Dampfstrahlantrieb
Meßquerschnitt 900 cm²

Fig. 2.144   0.30-m × 0.30-m high-speed wind tunnel HK 900 with steam-ejector drive of the Junkers Aircraft and Engine Works at Dessau (August 9, 1941) (Archive German Museum München).

Fig. 2.145   Control room with test section of the high-speed wind tunnel HK 900 at Junkers (about 1943) (Archive German Museum München).

Fig. 2.146   Typical half-models of wing-, fuselage-, and nacelle-configurations for the high-speed wind tunnel HK 900 of Junkers (1943) (Archive German Museum München).

spread most of all normal to the flow direction and that, therefore, the interference between the flow-displacing cross sections located within one plane is decisive for the drag.

An especially clear demonstration of the effect of the area rule was given by *R. T. Jones* [158] much later in 1972 by a comparison of the wave drag of aircraft flying in formation at a *Mach* number of $M = 1$ (Fig. 2.147). Two aircraft flying closely and exactly next to each other exhibit together about four times the wave drag of a single aircraft. The wave drag D of each individual aircraft of the formation is being doubled due to the interference with the other aircraft. In the case of a skillful staggering of the formation, the thickness distribution is, however, improved such that the sum of the wave drag of both aircraft corresponds approximately to the wave drag of a single aircraft.

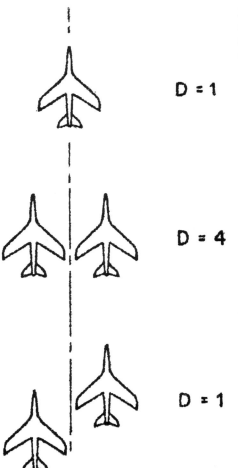

Fig. 2.147   Effect of the mutual interference on the total wave drag D of two aircraft flying in formation at $M = 1$ according to *Jones*, 1972 [158].

D = 1

D = 4

D = 1

2.3.4.4   GERMAN JUNKERS-PATENT CONCERNING THE TRANSONIC AREA RULE OF OTTO FRENZL OF 1944.   *Otto Frenzl* formulated on December 17, 1943, at the Junkers Aircraft and Engine Company [159] a report on the invention "Arrangement of Displacement Bodies in High-Speed Flight" and received 50 Marks as an inventor's bonus. The report was signed, in turn, by *Frenzl* and *W. Hempel* and according to later statements by *Frenzl* without his knowledge subsequently also by *W. Hertel*.

As an example of the effectiveness of the area rule, a test result for a forward-swept wing with nacelles related to the Project EF 122, obtained in the Junkers high-speed wind tunnel HK 900, is described in the report. By a longitudinal shift of the engine nacelles downstream away from the maximum-thickness location of the wing, the drag-rise *Mach* number of this arrangement was increased from $M = 0.72$ to $M = 0.77$, which is approximately the critical *Mach* number of the wing alone. At $M = 0.80$ the drag reduction due to the shift of the nacelles amounted to 33%.

In 1944 *Theodor Zobel* (1906–1953) of the DFL Braunschweig extensively described all results of these wind-tunnel tests of *Frenzl* to a larger audience at the German Academy of Aeronautical Research [160].

Frenzl himself described in 1957, after the area-rule patent of 1944 had been newly granted, in his publication "Engine-Nacelle-Wing Interference at High Speeds (The Area Rule)" [161] these decisive tests on a small half-model of a 27-deg forward-swept wing of an aircraft similar to the Junkers Ju 287 with a long nacelle as basis and explanation for the area rule.

Patent 932410 "Low-Drag Design of High-Speed Aircraft, also of those with Displacement Bodies Located Outside the Planform," was granted March 21, 1944 (Figs. 2.148–2.150). The inventors named were *Heinrich Hertel* (1901–1982), *Otto Frenzl*, and *Werner Hempel*. *Hertel* was at the time head of Aircraft Development at Junkers and *Hempel* was a collaborator of *Frenzl* at the high-speed wind tunnel. One can assume that *Frenzl* was the actual inventor and the driving force behind the patent.

The patent was again granted in 1955 within the territory of the Federal Republic of Germany based on the First Transfer Law (Überleitungsgesetz) of July 8, 1949. The patent is valid as of March 21, 1944, with the time period May 8, 1945–May 7, 1950, not counting towards the patent duration (law of July 15, 1951) (Fig. 2.148). The patent application was announced on March 3, 1955, and the patent document was published on September 1, 1955.

Claim 1 of 6 patent claims, seen in Fig. 2.149, reads:

> High-speed aircraft, if necessary with displacement bodies arranged outside the planform of the aircraft, characterized by the fact that its components including possible displacement bodies arranged outside the planform of the aircraft exposed to the flow are in their design and position arranged in a way that the sum of their displacing cross sections within common planes, located one after the other, normal to the flight

direction and considered from the nose to the tail of the aircraft, only increases up to a maximum and thereafter decreases to zero at the tail of the aircraft with the increase and decrease not being erratic in order not to prematurely exceed the critical *Mach* number.

In the patent claims 2–6, the special arrangement of engines and additional external aerodynamic loads to be dropped and of displacement bodies are described. In particular, for the first time an arrangement of the engine on the wing, either forward of or behind the wing, via a pylon was suggested (Figs. 2.149 and 2.150).

Considering the state of the engine development in Germany during WWII, it was the obvious objective of the aircraft development to achieve especially by a low-drag design subsonic speeds as high as possible. *Frenzl's* patent includes, therefore, also the transonic area rule.

2.3.4.5 INVESTIGATIONS CONCERNING AIRCRAFT DRAG AT HIGH SPEEDS IN THE 1940s. An important objective at the start of the 1940s in Germany was to develop faster aircraft with the aid of the novel jet engines. Besides the accelerating development of the jet engines, the airframe had to be adapted to the high speeds in order to achieve the high speeds strived for despite the massive compressibility effects occurring when approaching the speed of sound. The essential aerodynamic tasks here were the development of the swept wing and the integration of the jet engines. The latter were at the time still relatively long, thick, and heavy. For large aircraft, several of the available propulsion units were needed for each aircraft, and the optimum installation of the engines on the wing and fuselage had to be found. Because the development of the jet engines was still very much in progress and the engines were of comparatively low performance, separate nacelles in single, twin, and triple arrangements were preferred to ease the later switch to newer types of engines. The question of the aerodynamic interference between the nacelles and the air frame was extensively investigated experimentally at the aircraft manufacturers and the aeronautical research establishments [162–164]. Most of the high-speed data came from the Heinkel high-speed tunnel in Rostock, *Frenzl's* high-speed wind tunnel at Junkers in Dessau, and from the 2.7-m high-speed wind tunnel of the DVL in Berlin-Adlershof.

The results of high-speed wind-tunnel tests on components and total configurations of aircraft showed, besides the effect of the thickness ratio of individual components such as, for instance, the wing or the nacelle at higher *Mach* numbers, an additional effect of the equivalent thickness ratio of the overall configuration. The new concept of the area rule allowed for the first time to account for the effect of the mutual interference of components, which were located far apart but in the same cross-sectional plane normal to the flow direction, due to the lateral propagation of transonic disturbances. This

a)

Erteilt auf Grund des Ersten Überleitungsgesetzes vom 8. Juli 1949
(WiGBl. S. 175)

BUNDESREPUBLIK DEUTSCHLAND

AUSGEGEBEN AM
1. SEPTEMBER 1955

DEUTSCHES PATENTAMT

PATENTSCHRIFT
№ 932 410
KLASSE 62b GRUPPE 3 01
J 5040 XI / 62 b

Dr.-Ing. Heinrich Hertel, Berre Aix (Frankreich),
Dipl.-Ing. Otto Frenzel, Dessau und Werner Hempel, Dessau-Ziebigk
sind als Erfinder genannt worden

Junkers Flugzeug- und Motorenwerke AG., Dessau

Widerstandsarme Gestaltung von Hochgeschwindigkeitsflugzeugen,
auch von solchen mit außerhalb des Flugzeugumrisses liegenden
Verdrängungskörpern

Patentiert im Gebiet der Bundesrepublik Deutschland vom 21. März 1944 an
Der Zeitraum vom 8. Mai 1945 bis einschließlich 7. Mai 1950 wird auf die Patentdauer nicht angerechnet
(Ges. v. 15. 7. 1951)
Patentanmeldung bekanntgemacht am 3. März 1955
Patenterteilung bekanntgemacht am 4. August 1955

Die bisher üblichen Flugzeugbauformen weisen eine Verteilung ihrer an der Luftverdrängung beteiligten Querschnitte auf, die, über die Flugzeuglängsachse betrachtet, einen mehrfachen Wechsel von einer Zunahme zu einer Abnahme der quer zur Flugrichtung gelegenen Verdrängungsquerschnitte ergibt. Trägt man die in gemeinsamen Flugzeugquerebenen gelegenen Verdrängungsquerschnitte des Flugzeuges und etwaiger an dessen Außenseite befindlicher Verdrängungskörper, wie Abwurflasten, Motor- und Fahrgestellverkleidungen od. dgl., in einer graphischen Darstellung über der Flugzeuglängsachse auf, so ergibt sich, daß die hierdurch gebildete Kurve einen Verlauf aufweist, der vom Wert Null am Rumpfbug bis zu den Motorvorbauten an den Tragflügeln zunächst einigermaßen gleichmäßig zunimmt, dann aber infolge der sich summierenden Verdrängungsquerschnitte der Motorvorbauten, des Rumpfes, und des Tragflügels mit den Motorraum- und Fahrgestellverkleidungen und etwaigen sonstigen Ausbauten plötzlich ansteigt, um dann mehr oder weniger gleichmäßig wieder abzunehmen und am hinteren Flugzeugteil, an dem sich das Leitwerk befindet, abermals auf einen größeren Wert anzusteigen und am Flugzeugende jäh auf einen Wert Null abzufallen.

Bei Flugzeugen mit einer solcherart plötzlichen Zu- und Abnahme der Verdrängungsquerschnitte ergibt sich bei einer Steige-

Fig. 2.148 a) Title page of the German Patent No. 932410 "Low-drag design of high-speed aircraft, also those with displacement bodies located outside the aircraft planform" of March 21, 1944; b) translation (of title text only).

transonic interference goes beyond the mutual interference between neighboring components due to local excess velocities in the near field, which occur already at low speeds.

Up to 1945 even though the area rule was not yet generally accepted as a higher design principle, the principles upon which the area rule was based (due to extensive experimental experience) were more or less unconsciously practically applied—outside the Junkers Company.

2.3.4.6 RECEPTION OF THE TRANSONIC AREA RULE IN GERMANY. In March 1944, *Theodor Zobel* had shown in his already mentioned lecture "Fundamentally

b) Granted on the basis of the First Transfer Law

Federal Republic of Germany	Issued on
German Patent Office	September 1, 1955
Patent Document No. 932 410	
Class 62b Group 3	
J50 40 IX/62b	

Dr.-Ing. Heinrich Hertel, Berre Aix (France),
Dipl.-Ing. Otto Frenzl, Dessau and Werner Hempel, Dessau-Ziebigk,
were named inventors.
Junkers Aircraft and Engine Works Corporation (AG), Dessau
Low-drag design of high-speed aircraft, also those with displacement bodies located outside the aircraft planform
Patented on the territory of the German Federal Republic since March 21, 1944

Fig. 2.148 (Continued)

New Ways to a Performance Improvement of High-Speed Aircraft" at the German Academy of Aeronautical Research [165], not only results from the Heinkel high-speed tunnel, but also results concerning the interference drag of a nacelle on a swept wing, which he had used to explain the area rule (Figs. 2.151–2.153). Here, *Zobel* did, however, not mention the concept of the area rule and did not explicitly give a corresponding interpretation of the results.

In the Heinkel report "Aerodynamic Rules for the Installation of Jet-Engine Nacelles" of August 1944 by *Gerhard Schulz* [166], eight main rules are formulated for ten different parameters, based on extensive material from Heinkel and Junkers wind-tunnel tests, concerning the geometry of the nacelle

PATENTANSPRÜCHE:

1. Hochgeschwindigkeitsflugzeug, gegebenenfalls mit außerhalb des Flugzeugumrisses angeordneten Verdrängungskörpern, dadurch gekennzeichnet, daß dessen vom Fahrtwind umströmten Teile einschließlich etwaiger außerhalb des Flugzeugumrisses angeordneter Verdrängungskörper in ihrer Formgebung und ihrer Lage zueinander derart angeordnet sind, daß die Summe ihrer Verdrängungsquerschnitte in gemeinsamen, quer zur Flugrichtung hintereinanderliegenden Ebenen, vom Anfang zum Ende des Flugzeuges betrachtet, bis zum Erreichen eines Größtwertes nur zunimmt und vom Größtwert auf den Wert Null am Flugzeugende nur abnimmt, wobei die Zu- und Abnahme zur Vermeidung vorzeitiger Überschreitung der kritischen Machzahl nicht sprungartig erfolgt.

2. Hochgeschwindigkeitsflugzeug nach Anspruch 1, dadurch gekennzeichnet, daß bei Anordnung mehrerer außerhalb des Flugzeugumrisses liegender Verdrängungskörper (13, 14) diese, in Flugrichtung betrachtet, sich im wesentlichen decken.

3. Hochgeschwindigkeitsflugzeug nach einem der Ansprüche 1 und 2, dadurch gekennzeichnet, daß außerhalb des Flugzeugumrisses angeordnete Verdrängungskörper größeren Volumens in Teilkörper unterteilt sind, welche räumlich getrennt, vorzugsweise in Hintereinanderanordnung, am Flugzeug angeordnet und gegebenenfalls einzeln oder gemeinsam abwerfbar sind.

4. Hochgeschwindigkeitsflugzeug nach einem der Ansprüche 1 bis 3, dadurch gekennzeichnet, daß außerhalb des Flugzeugumrisses angeordnete Verdrängungskörper (11 bis 16) ganz oder teilweise vor und/oder hinter dem Tragflügel angeordnet sind.

5. Hochgeschwindigkeitsflugzeug nach einem der Ansprüche 1 bis 4, dadurch gekennzeichnet, daß als Triebwerke (11, 12, 15, 16) ausgebildete Verdrängungskörper an Auslegern vor und/oder hinter dem Tragflügel angeordnet sind.

6. Hochgeschwindigkeitsflugzeug nach einem der Ansprüche 1 bis 4, dadurch gekennzeichnet, daß außerhalb des Flugzeugumrisses liegende Verdrängungskörper in Nähe des Leitwerkes angeordnet sind.

Fig. 2.149 Patent claims 1–6 of Patent 932410 "Low-Drag Design of High-Speed Aircraft" of March 21, 1944, see translation on pp. 238–239.

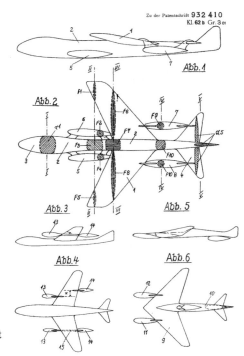

Fig. 2.150   Figures related to Patent 932410.

installation on fuselage and wing. The Heinkel report of August 1944 was, by the way, only published in 1955 in the *Journal of Flight Sciences* [167].

Also at the Junkers Company, *Erich Truckenbrodt* (1917–2010) and *Georg Backhaus* (1910–1976) wrote in August 1944 a summarizing report, "Drag Measurements on Nacelles of Special Engines and Comparison with Computations" [168], which contains *Frenzl's* results and a "list of wind-tunnel measurements carried out at Junkers Flugzeug-und Motorenwerke (JFM) (Junkers Aircraft and Engine Works) on nacelles of special engines" of 1943–1944. The test results that had inspired *Frenzl* to formulate and justify the area rule had thus already in 1944 repeatedly been published and also the patent had in 1944 already been granted.

In the "Monographs Concerning Progress in German Aeronautical Research (since 1939)," which were edited by *Albert Betz* (1885–1968) in 1946 [169 Chapter K3], in "The Installation of Jet Engines" (authored by *D. Küchemann*, *O. Conrad*, and *J. Weber*), the area rule as a higher design principle or as explanation for the measured interference drag is not mentioned. Obviously the people involved at the time also did not realize that the drag reduction due to wing sweep was based on an increased slenderness of the overall configuration in the sense of the area rule.

The extensive data compilation and the rules concerning the engine integration derived did not explicitly refer to the idea of the area rule and

Theo Zobel: Fundamentally new ways for the
improvement of high-speed aircraft

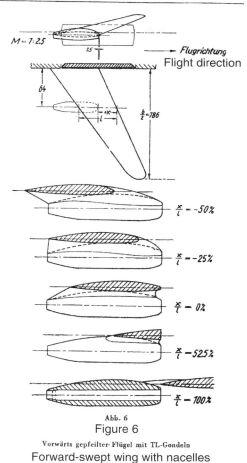

Abb. 6
Figure 6

Vorwärts gepfeilter Flügel mit TL-Gondeln
Forward-swept wing with nacelles

Fig. 2.151 Variation of the axial position of an engine nacelle on a forward-swept wing (wind-tunnel measurements on a half-model) according to *Frenzl* in [160].

seem, therefore, not always conclusive and to be generalized. The area rule obviously was, at the time, not even known to all German aerodynamicists and still not recognized as relevant to these kinds of transonic interference problems.

Especially interesting is in that regard the repeatedly made observation that the critical *Mach* number and the drag of a swept wing in the high-speed regime was not worsened by the addition of a long fuselage while short nacelles on the wing resulted in a pronounced deterioration.

Hubert Ludwieg (1912–2001) already stated this effect in 1940 in his famous lecture "Swept Wings at High Speeds" at the LGL [170]. However, he based it on the fact that the fuselage did little to distort the symmetrical flow in the center of the wing while the nacelles were located within a

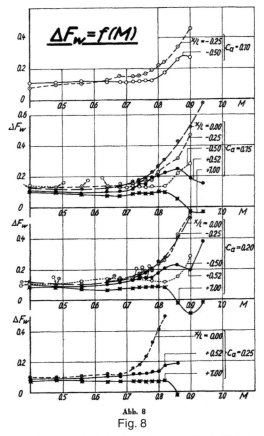

Abb. 8

Fig. 8

Schädliche Widerstandsfläche für Gondel und Interferenz über der Machschen Zahl

Parasite drag area of nacelle and interference
vs *Mach* number

Fig. 2.152 **Interference drag of an engine nacelle on a wing as function of the *Mach* number and the axial position of the nacelle according to *Frenzl* in [160]; axes: $\Delta F_W = \Delta F_D$ = interference drag area, *Mach* number; $c_a = c_l$.**

flowfield where velocity components normal to the nacelle axes existed (also see Sec. 2.3.3, Fig. 2.105).

In 1944, *H. Lindemann* [171] also noticed in the results of high-speed measurements on a model of the Ju 287 that by the addition of a fuselage to forward-swept wings the critical *Mach* number increased, while due to the attachment of nacelles, especially the relatively stocky twin- and triple-nacelles, it was noticeably decreased (Fig. 2.154), however, without further interpreting this effect in the sense of the area rule.

In wind-tunnel tests one had, therefore, also to carefully consider the effect of the displacing cross sections of the model support on the measured drag

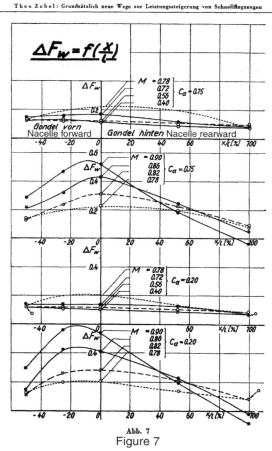

Abb. 7
Figure 7

Schädliche Widerstandsfläche für Gondel und Interferenz über der Gondelrücklage
Interference drag area of nacelle and interference
vs the nacelle location

Fig. 2.153   Interference drag of an engine nacelle on a wing dependent on the axial location of the nacelle and the *Mach* number according to *Frenzl* in [160]; axes: $\Delta F_W = \Delta F_D$ = interference drag area, x/l = nacelle location; $c_a = c_l$.

acting on the model, which is, for instance, addressed by *Lindemann* in his 1944 report mentioned previously. Also a report by *Melkus* of 1946 concerning early DVL measurements on fuselage models in the high-speed wind tunnel [172] (Fig. 2.139) contains a corresponding remark.

Because the area rule was discovered and formulated at the Junkers Company on the occasion of predevelopments for a jet-aircraft project with forward-swept wings, it is not surprising that in the case of the four-engine jet bomber Junkers Ju 287 (first flight in August 1944) the area rule was for the first time, consequently, applied to a flying aircraft (Figs. 2.155 and 2.156).

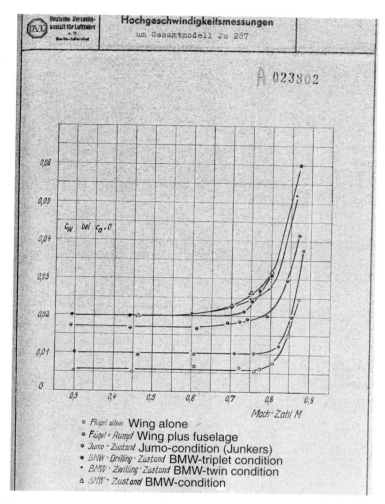

Fig. 2.154   Drag development vs *Mach* number of components and total configurations of the aircraft Junkers Ju 287 with different engines [171]; axes: c_w (c_d) at c_a (c_l) = 0, *Mach* number M; translation of header: high-speed measurements on Ju 287 full model.

Fig. 2.155   Prototype of the four-engine jet bomber Junkers Ju 287 in flight (first flight: August 16, 1944) (courtesy of Archive German Museum München).

Fig. 2.156   Model of the project EF-122, related to the swept-wing development, in the large low-speed wind tunnel of Junkers in Dessau (1943) (courtesy of Archive German Museum München).

The forward engine nacelles on the fuselage terminate as determined by the area rule at the same cross-section plane where the tips of the forward-swept wings are located. The tails of the two engines on the wing are moved as far downstream of the trailing edge as structurally possible. The strongly forward-swept wing promotes in itself a slender axial development of the aircraft cross sections. The Junkers Ju 287 was, therefore, at the end of the war considered by the Allies as one of the most interesting German aircraft projects. In Dessau, and later in the USSR, this line of development was continued for years with aircraft, material, and personnel of Junkers.

The German project proposals existing in 1945 for a jet fighter for the Heinkel He S 011 engine [173] (Fig. 2.157) still under development did, by the way, not show a clear application of the area rule, for the most part, if one disregards the employment of the swept wing.

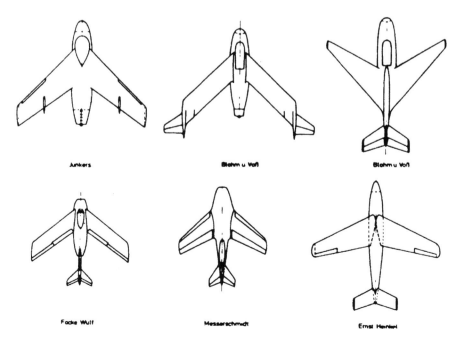

Fig. 2.157   Company designs associated with an invitation to bid for a single-engine jet fighter with the jet engine Heinkel He S 011 (1945) [173].

In 1945 the area rule, outside of Germany, was as equally unknown as the swept-wing concept. It is surprising that the knowledge of the transonic area rule did not also spread together with the more or less voluntary, or even involuntary, departure of Junkers employees to foreign countries. *Philipp von Doepp*, head of Fluid Technology, went to the United States; *Hans-Heinrich Wocke*, head of Aircraft Development, and *Georg Backhaus*, the leading aerodynamicist, ended up in the USSR; and *Heinrich Hertel*, the technical director, and *Otto Frenzl* went to France. Even after the return of most of the development engineers to Germany, a definite application of the area rule to German aircraft projects did not commence at once. In fact a patent dispute ensued relating to the extent of the application of the area rule to the first German civil aircraft after WWII, the "Hansa-Jet" HFB 320 and its relevancy to such an aircraft.

In a letter to *Siegfried Günter* (1899–1969), the chief aerodynamicist of Heinkel, of October 16, 1965 [174], as well as to *August Wilhelm Quick* (1906–1982) of February 1982 [175], *Frenzl* quotes as argument for his area-rule concept comparative measurements on half-models of a swept and a W-wing of equal sweep (Fig. 2.158). Here, an essentially higher transonic drag was obtained for the W-wing than for the corresponding swept wing. To rule out the "wing-center effect" (Mitteneffekt), at the time assumed to be the cause of the large differences in case of the swept wing, the W-wing, in addition, was tested with a strong reduction of the wing thickness in the area of the (wing) kink without, however, decisively reducing the drag. *Günter* answered *Frenzl* on October 27, 1965: "The justification of the so-called area rule by the measurements on the W-shaped wing is certainly correct and it is your personal unquestionable brainwork" [176]. Corresponding wind-tunnel measurements are discussed later.

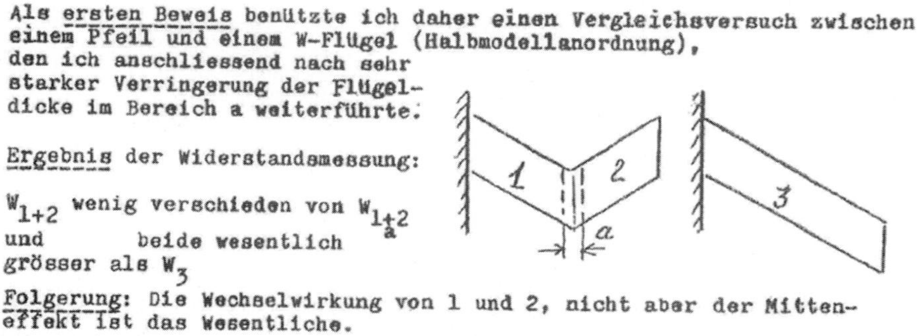

Als ersten Beweis benützte ich daher einen Vergleichsversuch zwischen einem Pfeil und einem W-Flügel (Halbmodellanordnung), den ich anschliessend nach sehr starker Verringerung der Flügeldicke im Bereich a weiterführte.

Ergebnis der Widerstandsmessung:

W_{1+2} wenig verschieden von $W_{1+2 \atop a}$ und beide wesentlich grösser als W_3

Folgerung: Die Wechselwirkung von 1 und 2, nicht aber der Mitteneffekt ist das Wesentliche.

Fig. 2.158 Argumentation of *Frenzl* with regard to the effectiveness of the area rule by a comparison of the drag of a swept wing with that of a W-wing (1982) [175] (courtesy of Archive German Museum München); translation: As first proof I, therefore, used a comparative test between a swept and a W-wing (half-model arrangement), which I subsequently continued after a very strong reduction of the wing thickness in area a. Result of the drag measurement: W_{1+2} little different from W_{1+a+2} and both considerably higher than W_3. Conclusion: Interaction between 1 and 2 and not the center effect is what matters.

A much later letter of *Ruediger Kosin* (1909–1996, chief aerodynamicist at the Arado Aircraft Company during WWII), from December 22, 1981, to *Werner Heinzerling* states [177]:

> The Junkers people, i.e. Mr. *Frenzl*, who built and ran the ejector-driven high-speed wind tunnel of Junkers, was in understanding near-sonic flow only ahead of us and all the rest of German aeronautics (industry and research institutes) by a whisker; he had, of course, also an experimental facility. Ten years prior to *Whitcomb* he established the area rule and applied for a patent in 1944. Patent No. 932410 is in the possession of the Patent Department of MBB.

2.3.4.7 AERODYNAMICIST *OTTO FRENZL* AND HIS HIGH-SPEED WIND TUNNELS. *Otto Frenzl*, an electrical engineer, educated at the TH Graz and the TH Vienna, came in 1937, at age 28, to the Junkers Aircraft and Engine Works in Dessau joining the branch "Fluid Technology" headed by *Phillip von Doepp*. Here, he got involved in the layout, design, construction, and finally the operation of the intermittently operating 0.3×0.3-m high-speed wind tunnel HK 900 (900 cm^2) well equipped with modern instrumentation [178].

The wind tunnel HK 900 was driven by a steam-jet ejector. With steam from a hot-water reservoir according to *Ruths* with a volume of 220 m^3 and a pressure of 12 atm ($\approx$174 psi), the facility could at a *Mach* number of $M = 0.95$ still be operated without interruption for some minutes. Such wind tunnels are naturally, despite the lower efficiency of an ejector drive, much simpler in their setup with lower operating costs than a comparable continuous facility with large electrically powered compressors. *Frenzl's* facility was equipped with a measuring technique conceived by him. The results of a multicomponent balance based on oil-pressure gauges were immediately displayed as aerodynamic coefficients in the usual diagram form as light dots on two large fluorescent screens. A semiautomatic *Mach*-meter based on an idea of *Frenzl* was installed, and with a schlieren system flow, visualizations in the entire test section were possible. The test section could alternatively be arranged as closed, semi-open, or three-quarter-open in order to achieve, also in the presence of models, subsonic velocities as high as possible and, at the same time, minimize wind-tunnel corrections. The half-span of typical half-models of wing-fuselage configurations was about $b/2 = 0.16$ m (Fig. 2.146).

Within the German aeronautical industry, only the Heinkel Aircraft Company in Rostock possessed a continuous 0.36×0.36-m high-speed tunnel (since 1937), a company-owned test capacity at high subsonic speeds (also see Sec. 2.3.2, Fig. 2.63). Considerably larger high-speed wind tunnels with test sections of 2.7- and 2.8-m diameter existed only at the DVL in Berlin-Adlershof and the LFA in Braunschweig.

Since 1942, *Frenzl* accompanied the current aircraft projects at Junkers with high-speed measurements in his wind tunnel and was hence in close

Institution (Establishment)	Location	Designation	Cross section (Width × Height in m)	Area in m²	Max. Mach number	Year	Remarks	Source/ References, Figures
Junkers Aircraft and Engine Works	Dessau	HK 900	0,30 × 0,30	0,09	0,95	1942	*Ruths*-steam reservoir, 220 m³, 12 at	[156, 157] Figs. 2.144, 2.145
	----	Giant wind tunnel (proposed)	20 × 13 elliptical	200,00	0,95	Memorandum 1943	Hot-water reservoir 4500 m³, 25 at	[180, 181]
	planned: Muldenstein	M3	2,34 × 2,00	4,68	0,95	1944/45	only design (Dingler Company)	[179] Figs. 2.161, 2.162
S.N.E.C.M.A.	Villaroche (France)	for engine investigations	0,40 × 0,60	0,24	1,25	1953	*Ruths*-steam reservoir 45 m³, 15 at	[182] Fig. 2.163
Institute Aérotechnique	Saint-Cyr (France)	Sigma 4	0,85 × 0,85	0,72	2,8	1962	Hot-water reservoir, 65 at	[184]
Institute of Propulsion Technology DFVLR	Trauen (Germany)	Ramjet-altitude test stand	0,70 × 0,70	0,49	4	1968	La-Mont hot-water reservoir, 200 m³, 100 at	[185]

Fig. 2.159 Survey of larger high-speed wind tunnels driven by an ejector, System *Frenzl*.

Fig. 2.160 *Frenzl's* **sketch of April 7, 1942, of a large facility with three ejector-driven high-speed wind tunnels with a common steam reservoir (courtesy of Archive German Museum München).**

immediate contact with the most advanced developments in high-speed aerodynamics. The transonic area rule could emerge within this environment due to many observations, test results, and theoretical interpretations.

In addition, *Frenzl* worked intensively on the further development and optimization of economic high-speed wind tunnels powered by ejectors (Figs. 2.159 and 2.160). Thus, he worked from 1944 up to the end of WWII together with the Dingler Company in Zweibruecken on the detailed design of a much larger ejector-driven wind tunnel, M3, with a test section of 2.34 × 2.0 m, which was to be erected for Junkers 20 km south of Dessau in Muldenstein near Bitterfeld [179] (Figs. 2.161 and 2.162).

In 1943 when large wind tunnels and towing ranges for the investigation of full-scale aircraft were being discussed in Germany, *Philipp von Doepp* and *Otto Frenzl* submitted a proposal, including economic efficiency calculations, for a "giant wind tunnel for the investigation and trial of full-scale aircraft and their components" with an elliptical test section of 20 m × 13 m and a hot-water reservoir with a volume of 4500 m^3 [180, 181]. For this facility the ejector was, as an improvement, to be supplied with hot water rather than steam.

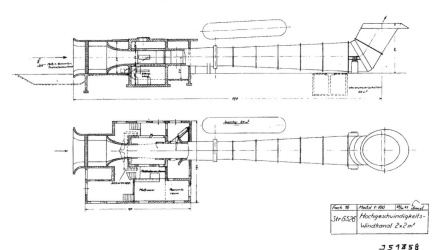

Fig. 2.161 *Frenzl's* **drawing of December 27, 1943, related to a 2- × 2-m high-speed wind-tunnel project (courtesy of Archive German Museum München).**

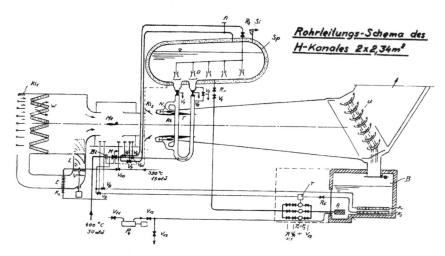

Fig. 2.162 Schematic of the pipe system of the 2.00 × 2.23-m high-speed wind tunnel, October 3, 1944, according to *Frenzl* (courtesy of Archive German Museum München).

After WWII, when *Frenzl* again found employment as a wind-tunnel engineer and aerodynamicist at the Société Nationale d'Etude et de Construction de Moteurs d'Aviation (SNECMA) in Villaroche, France, he started with the construction of a somewhat larger and improved high-speed wind tunnel compared to the Junkers HK 900 for engine and propulsion investigations. In Fig. 2.163, *Frenzl* is shown at the control console of the very elegant facility in about 1953 [182]. In 1956, *Frenzl* received a doctorate degree in engineering (Dr.-Ing.) from the TH Graz with the thesis "Flow of Evaporating Water in Nozzles" [183].

Frenzl participated in the 1960s in projecting two even larger ejector-driven supersonic wind tunnels: the Tunnel Sigma 4 at the Institut Aérotechnique in Saint-Cyr, France started operating in 1962 [184], and the Ram-Jet High-Altitude Test Stand for *Mach* numbers of up to $M = 4.0$ of the Deutsche

Fig. 2.163 *Otto Frenzl* at the control and data-reduction console in front of the test section of his 0.40 × 0.60-m high-speed wind tunnel at the SNECMA in Villaroche, France, about 1953 (courtesy of Archive German Museum München).

Forschungs-und Versuchsanstalt für Luft-und Raumfahrt (DFVLR) (German Research and Test Establishment of Aeronautics and Space) in Trauen commenced operating in 1967. In the description of the ejector facility in Trauen [185], 10 of the 18 references quoted are publications of *Frenzl*. Further reports of *Frenzl* concerning his work on steam and hot-water jet propulsion for test facilities were published between 1958 and 1963 [186–189].

In the 1970s, *Frenzl* put his experience related to steam and hot-water ejectors into the services of environmental technology at large-scale facilities concerning the desalination of brack water at Roswell, New Mexico, and the dust removal at steel mills in Lone Star, Texas, according to a method sold to the Krupp Company [190].

2.3.4.8 SECOND DISCOVERY OF THE AREA RULE IN THE UNITED STATES. In 1952, probably independent of the German investigations, the area rule was discovered as "Area Rule" in the United States by *Richard Trevis Whitcomb* (born 1921) for the second time. *Whitcomb* was employed since 1943 at the NACA, Langley Field, 8-ft high-speed wind tunnel [191]. The high-speed wind tunnel with a closed test section and solid wind-tunnel walls provided in the immediate vicinity of the speed of sound only unsatisfactory results and was, therefore, in 1950 modified by installing a test section with slotted walls, thus becoming the 8-ft transonic wind tunnel [192]. A reliable wind-tunnel simulation at transonic speeds was already at that time extremely important because the new postwar jet aircraft had their regular range of operation at very high subsonic speeds and had, at least briefly, to fly through the transonic speed range.

In 1952 *Whitcomb* summarized test results related to the transonic drag rise of fuselages with different wing configurations and a waistlike shaped fuselage in the area of the wing and in the center region thickened fuselage bodies. From the results, he concluded, among other things, that in the vicinity of the speed of sound the drag rise of the combination of a thin low-aspect-ratio wing with a slender fuselage was primarily dependent on the axial distribution of the air-displacement cross sections normal to the freestream. Waistlike shaping of the fuselage of a wing-fuselage combination, such that the distribution of the air-displacement cross sections of the combination correspond to the one of the original fuselage alone, reduces decisively the contribution of the wing to the drag rise [193] (Fig. 2.164).

In the early 1950s these findings were directly applied to the development of the Convair YF-102, a supersonic aircraft with delta wings. Only a second, strongly modified prototype with a waistlike fuselage in the area of the wing and extended at the tail, with lateral displacement bodies and a more slender nose, allowed supersonic flight. The aircraft entered series production in 1954 as F-102A and F-106 [194] (Fig. 2.165).

The findings concerning the area rule in 1952 were only communicated by the NACA to the U.S. Aircraft Industry, but otherwise kept secret up to 1955.

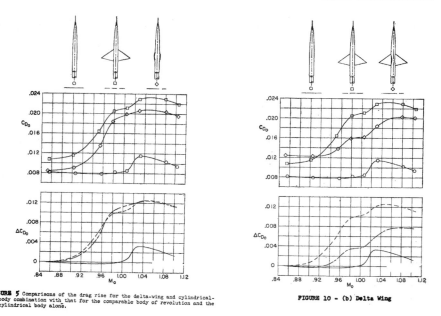

FIGURE 5 Comparisons of the drag rise for the delta-wing and cylindrical-
body combination with that for the comparable body of revolution and the
cylindrical body alone.

FIGURE 10 - (b) Delta Wing

Fig. 2.164   Drag investigations of *Whitcomb* on slender bodies at transonic speeds related to the area rule, 1952 [193].

Fig. 2.165   Application of the area rule on the Corvair F-102A; left: original prototype, and right: extended fuselage nose, waist-shaped fuselage in the wing region, and displacement bodies at the tail of the fuselage [194].

NACA Details Area Rule Breakthrough

By Richard T. Whitcomb

Conclusions

• **The shock phenomena** and drag-rise increments measured for four representative wing and central-body combinations at zero lift near the speed of sound are essentially the same as those for bodies of revolution with the same axial distributions of cross-sectional areas normal to the air stream.

• **On the basis** of these results, it is concluded that, near the speed of sound, the zero-lift drag rise of a thin, low-aspect-ratio wing-body combination˙ is primarily dependent on the axial distribution of the cross-sectional areas normal to the air stream. It follows that the drag rise for any such configuration is approximately the same as that for any other with the same distribution of cross-sectional areas.

• **Indenting the bodies** of three representative wing-body combinations, so that the axial distributions of cross-sectional areas for the combinations were the same as for the original body alone, greatly reduced or eliminated the zero-lift drag-rise increments associated with wings near the speed of sound.

AVIATION WEEK, September 19, 1955

Fig. 2.166   Summary of the first internationally available publication of *R. T. Whitcomb* concerning the area rule in *Aviation Week* of September 19, 1955 [194].

Whitcomb's publication "A Study of Zero-Lift Drag-Rise Characteristics of Wing-Body Combinations near the Speed of Sound" of September 3, 1952, was only printed in full on September 19, 1955 as "Special Service" in the noted professional journal *Aviation Week* under the title "NACA Details Area Rule Breakthrough" supplemented by photographs of the two prototypes of the Convair F-102 (Figs. 2.165 and 2.166).

Later *Whitcomb* undertook a lecture tour to Europe to introduce the concept of the area rule. Among other things, *Whitcomb* lectured in January 1958 for two days at the von Kármán Institute in Rhode-St.-Genèse, near Brussels, the International Training Center for Experimental Aerodynamics, on the area rule and took on a professional discussion with representatives of the French aircraft industry [195]. Here, *Whitcomb* named as examples of the successful application of the area rule the following especially well-known American aircraft: the carrier-based fighters Grumman F-11F and Chance Vought F8U, the interceptors Convair F-102A and F-106, the fighter bomber Republic F-105, and the bomber Convair B-58. During the discussion, representatives of the French aircraft industry described first wind-tunnel tests concerning the application of the area rule on aircraft projects of Bréguet ("Taon") and Sud-Aviation ("Vautour").

The development of the geometry of the Boeing B-52 shows that during the time from 1946 to 1954 the swept wing and the jet engines in nacelles in front of and below the wings were introduced step-by-step. Only the

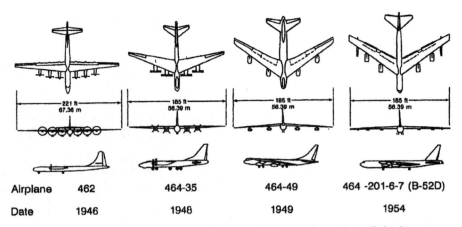

Airplane	462	464-35	464-49	464 -201-6-7 (B-52D)
Date	1946	1948	1949	1954

Fig. 2.167   The first stages in the development of the configuration of the long-range Boeing B-52, 1946–1954, according to *J. E. Steiner* [196].

configuration of 1954 allows one to recognize the application of the area rule [196] (Fig. 2.167). Aircraft developments in the Soviet Union clearly showed since the 1960s, in the case of long-range bombers (Fig. 2.168), the application of the area rule, such as, in the case of the Bomber Tupolev TU-22 "Blinder" with the waist-shaped fuselage in the wing area as well as the displacement bodies at the wing trailing edge [197]. Obviously only the

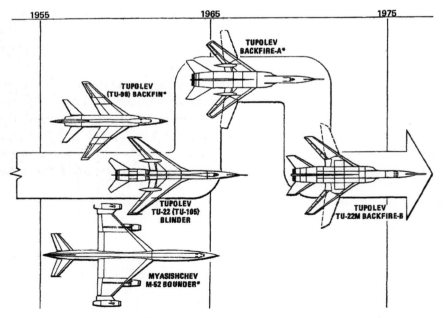

Fig. 2.168   Development of long-range bombers in the Soviet Union 1955–1975 according to *R. D. Ward* [197].

publication of the area rule in an influential American professional journal in English in 1955 had led to its general dissemination and application.

2.3.4.9 Area Rule for Supersonic Flow. The investigations regarding the area rule mentioned up to now addressed most of all the case of near-sonic speeds where pressure disturbances predominantly spread normal to the freestream direction. At higher supersonic *Mach* numbers, the configurations optimized according to the transonic area rule turned out to be unfavorable, which can be traced to the conical character of the flow at supersonic speeds. The decisive cross-sectional area distribution at supersonic speeds corresponds analogously to the average cross-sectional area, which is generated by planes inclined corresponding to the *Mach* angle $\mu = \arc \sin 1/M$ to the flow direction (Fig. 2.169).

The area rule, together with the theoretical method of the German mathematician *Wolfgang Haack* (1902–1994), already published in 1941, to determine streamlined bodies of minimum drag at arbitrary geometrical boundary conditions and their wave dag [199] (Fig. 2.170), allows now for a slender aircraft configuration, the cross-sectional area distribution to be computed theoretically at any *Mach* number for minimum supersonic drag and, in addition, the supersonic drag under arbitrary boundary conditions. These bodies of minimum wave drag are today in the United States known as "Haack–Sears–Bodies." *William S. Sears*, a pupil of *von Kármán*, had in 1948 somewhat generalized *Haack's* work for the publication in the United States.

Friedrich Keune (1908–1982) and *Klaus Oswatitsch* (1910–1993) gave in 1956 in a paper "Equivalence Law, Similarity Laws for Near-Sonic Velocities and Drag of Noninclined Bodies of Low Aspect Ratio" a good survey of the corresponding computational methods (Fig. 2.171) [200].

Robert T. Jones (1910–1999), who had already in 1945 published at the NACA the swept-wing concept, presented in his 1956 NACA-report No. 1286 "Theory of Wing-Body Drag at Supersonic Speeds" a well-founded method for the computation of the supersonic drag of slender aircraft configurations [198]. The wave drag is here computed as drag of an equivalent streamlined body optimized for a certain *Mach* number.

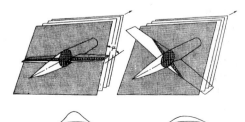

Fig. 2.169   Location of the cross sections to determine in the sense of the area rule the effective displacing cross-sectional areas at supersonic speeds according to R. T. Jones, 1956 [198].

a) ———————————— Bericht 139 der Lilienthal-Gesellschaft ————————————

Geschoßformen kleinsten Wellenwiderstandes

W. Haack, Karlsruhe

*Zusammenfassung: In der vorliegenden Arbeit werden Geschoßspitzen und Ge-
schosse kleinsten Wellenwiderstandes unter verschiedenen Nebenbedingungen be-
stimmt. Als Nebenbedingungen werden Kaliber, Volumen und Länge bzw. nur zwei
dieser Größen vorgegeben. Die Grundlage der Untersuchung bildet die Kármánsche
Näherungstheorie der Gasströmung um schlanke Rotationskörper. Für die verschiedenen optimalen Geschoßformen, bezogen auf gleiches Volumen und gleiche
Länge, wurden die Widerstandsbeiwerte näherungsweise berechnet und für U = 2 c
(c = Schallgeschwindigkeit) im Göttinger Windkanal von Walchner und Ludwieg
gemessen. Die Übereinstimmung ist über Erwarten gut, die berechneten und gemessenen Werte zeigen denselben Verlauf. Die optimalen Geschosse bei gegebenem
Kaliber und Volumen stimmen fast vollständig mit französischen Beutegeschossen
überein.*

b)   Projectile shapes of lowest wave drag

Summary: In the present work, projectile tips and
projectiles of lowest wave drag are being determined under various extra conditions. As extra conditions, caliber, volume, and length, respectively two
of those, are being prescribed. The basis for this
investigation is *von Kármán's* approximate theory of
the flow about slender bodies of revolution. For the
various optimum projectile shapes, based on the
same volume and the same length, approximate
drag coefficients were computed; they were tested
at $U = 2c$ (c = speed of sound) in the Göttingen wind
tunnel by *Walchner* and *Ludwieg*. The agreement is
unexpectedly good, the computed and measurement data show the same development. The optimized projectiles at a given caliber and volume coincide
almost completely with captured French projectiles.

Fig. 2.170   **a) Summary of the publication of *W. Haack* (1941) concerning the determination of the optimum streamline shape and the supersonic drag of projectiles [199];
b) translation.**

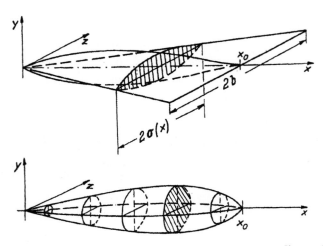

Fig. 2.171   **Slender wing and equivalent body of revolution according to *F. Keune* and
K. Oswatitsch (1956) [200].**

2.3.4.10 AREA RULE AND THE SWEPT WING. The application of the transonic area rule to the geometry of swept wings gives a clear explanation for the favorable drag behavior at near-sonic speeds as compared to the corresponding unswept wings (Fig. 2.172). The thickness ratio of the cross-sectional area distributions according to the area rule, decisive for the transonic drag, is for swept wings essentially more favorable than for the equivalent unswept wings.

The figure compares the axial development of the cross-sectional areas (displacement areas) $Q(x)$ for the sheared wing, the swept wing, and the

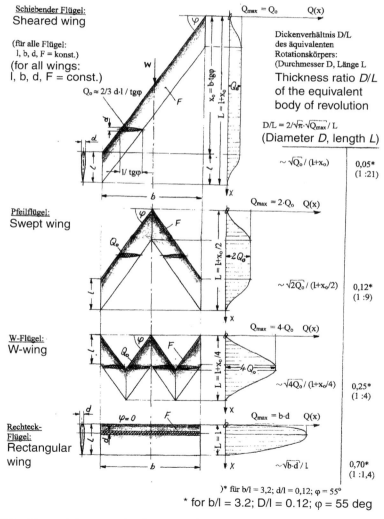

Fig. 2.172 **Geometric connection between wing sweep and area rule: axial development of the displacement cross sections and thickness ratios of equivalent bodies of revolution for the sheared wing, the swept wing, the W-wing, and the rectangular wing.**

W-wing with the corresponding rectangular wing. As a rough formal comparative parameter for the aerodynamic quality of the axial cross-sectional
area distribution of the wing, the thickness ratio of the equivalent body of
revolution with the diameter D and the aircraft length L is presented:
$D/L = (2/\sqrt{\pi}) \sqrt{Q_{max}}/L$.

The sheared wing as the cleanest form of the swept wing is only intersected
once by each cross-sectional plane and provides the flattest and longest-
stretched axial cross-sectional development. It also shows a correspondingly
favorable drag behavior. In the example of Fig. 2.172, the thickness ratio
D/L of the sheared wing is only about 7% of the one of the equivalent rectangular wing.

By the way already in 1940, *Bernhard Göthert* (1907–1988) had explained
in his lecture "Computation of the Velocity Field of Swept Wings at High
Subsonic Speeds" at the meeting of the LGL [201], among other things, the
swept-wing effect in the wind tunnel as a consequence of the lower "cross-
sectional blockage" of the sheared wing compared to the rectangular wing
(Fig. 2.173). Here, he presented beforehand part of the later argumentation
for the area rule in so far as he named the magnitude of the cross-sectional
blockage and the corresponding excess velocities especially at near-sonic
freestream speeds responsible for the additional drag.

The sheared wing was already suggested by *Richard Vogt* (1894–1979) at
Blohm and Voss during WWII for an aircraft project with variable wing
sweep [202] (Fig. 2.174). At the time, similar concepts existed at the AVA and

Fig. 2.173 *Göthert's* **model concept of the swept-wing effect as consequence of a more favorable cross-sectional blockage due to sweep (1940): a) wing without sweep between two walls, and b) swept wing between two walls [201].**

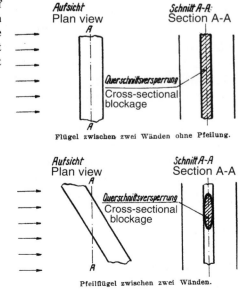

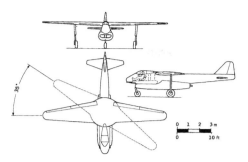

Fig. 2.174   Project P.202 of a jet fighter of Blohm and Voss with a sheared wing of a variable sweep angle of 0 to 35 deg (1943) according to *H. Pohlmann* [202].

at the Messerschmitt Company as patent proposals. *Robert T. Jones* suggested in 1972 at NASA the sheared elliptical wing ("Oblique Wing") as a high-speed wing superior to the standard mirror-image swept wing, also with, among other things, variable sweep for the optimum adaptation to the prevailing flight *Mach* number [203]. In 1972, NASA gave Boeing a study contract concerning a large transport aircraft to operate within and clearly above the sonic speed range. Besides configurations with the swept wing, the delta wing, and the variable-sweep wing, also configurations with a sheared elliptical wing were to be investigated. NASA later built a small twin-jet test aircraft, NASA AD-1, which allowed sweep angles of up to 60 deg and underwent flight trials from 1979 [204].

The standard mirror-image swept wing is naturally in its cross-sectional area development less favorable than the sheared wing because each cross section cuts the wing twice, and the cross-sectional distribution covers only less than half of the axial extent. Its thickness ratio D/L is in our example with 16% of the rectangular wing more than twice as unfavorable. The regular swept wing shows, correspondingly, at high speeds also a clearly less favorable drag behavior than the sheared wing. Because of other advantages, this classical form of the mirror-image swept wing had been intended in Germany during WWII for many projects of high-speed aircraft (see, for example, Fig. 2.157) and represents up to now the most frequent solution for transonic aircraft (Figs. 2.167 and 2.168).

2.3.4.11   "CENTER-EFFECT" ON THE SWEPT WING.   In the early 1940s, when the area rule was not yet known, the bad performance of the swept wing compared to the oblique (sheared) wing was attributed to the "center effect." In the case of the sheared wing, the streamlines close to the wing surface run, looking down, in an S-shaped pattern [205] (Fig. 2.175). For symmetry reasons this S-shaped pattern near the plane of symmetry of a swept wing cannot be very pronounced and must in the center plane, respectively at the fuselage wall, completely disappear. The common simple explanation of *Busemann* for the swept-wing effect is the splitting of the freestream velocity into a component normal to the wing leading edge, which is responsible for the pressure and

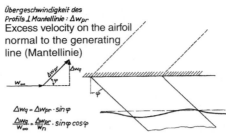

Fig. 2.175   Undisturbed streamline shape on a swept wing according to G. Backhaus (1943) [205].

velocity distribution on the wing, and a rather ineffective component in the direction of the wing leading edge. The explanation must fail at the plane of symmetry, respectively at the fuselage. *Dietrich Küchemann* at the AVA Göttingen as well as *Georg Backhaus* at the Junkers Company suggested a locally limited S-shaped contouring of the fuselage to achieve the ideal streamline shape of a sheared wing, in the root area of a swept wing, thus reducing the unfavorable "center effect." A test in the Junkers high-speed wind tunnel with a half-model of a configuration similar to the Ju 287 with a fuselage contour computed by *Backhaus* did not, by the way, show any essential drag improvement (Fig. 2.176). Later *Frenzl* himself suspected that the flow development at the wing root of the small half-model could, indeed, have been disturbed by the wind-tunnel wall boundary layer.

This fuselage contouring is limited to the immediate wing root area, and is, in no way, related to the massive waistlike shaping of the fuselage over the entire axial extent of the wing as required by the area rule; it caused, nevertheless, some discussions in Germany with the question being whether this was an early application of the area rule.

In the guidelines for the standardized determination of the drag at $M = 0.88$ to align fighter projects with the He S 011TL of 1945 [206], the "center-effect" of the swept wing is, for instance, considered by determining the drag rise with *Mach* number by computing a small defined part of the wing in the immediate vicinity of the fuselage as rectangular wing, while for the larger

Fig. 2.176   Effect of an S-shaped fuselage side wall in the area of the wing root on the drag of a swept wing according to measurements in the Junkers high-speed wind tunnel HK 900 (1943) according to G. Backhaus [205]; axes: $\Delta c_w = \Delta c_d$, *Mach* number M.

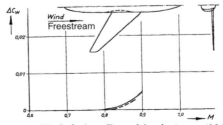

Einfluß einer Rumpfeinschnürung (1943)
Effect of waistlike shaping of the fuselage (1943)

remainder a sheared wing is assumed. An effect of the area rule on transonic drag can that way, of course, not be taken into consideration.

The W-wing is with regard to a slender cross-sectional area distribution according to the area rule still more unfavorable. In the center region of the wing, each plane normal to the flight direction cuts the wing four times, and the cross-sectional distribution stretches only over less than one-quarter of the distance of the sheared wing. The thickness ratio of the W-wing under consideration is with 36% of the rectangular wing about twice as unfavorable as that of the swept wing and about five times as unfavorable as that of the sheared wing. By the way, the thickness ratio D/L of the corresponding rectangular wing was accordingly 15 times more unfavorable than that of the sheared wing.

Before the area rule was discovered and understood, the W-wing was considered wrongly as a good compromise between a forward-swept and a swept-back wing, which most of all reduced the unfavorable boundary-layer effects in the outboard regions of a swept-back wing. This was confirmed by measurements at low speeds (see Fig. 2.94). The Blohm and Voss project of a four-engine jet bomber (P.188), for instance, had inboard a sweepback of 15 deg and outboard a forward sweep of 8 deg [202] (Fig. 2.177).

In a test in the Junkers high-speed wind tunnel on half-models of wing configurations, namely, an unswept wing, a 40-deg swept-back, and a W-wing (inboard swept back 40 deg, outboard swept forward 15 deg), an improvement in the critical *Mach* number was observed for the swept wings with the W-wing, however, exhibiting as to be expected according to the area rule the least favorable results (Fig. 2.178). A collection of the different half-models for the 0.30×0.30-m high-speed wind tunnel HK 900 in Fig 2.179 [207] shows, among other things, the three wings compared. The wing half-models had a half-span of $b/2 = 0.16$ m. The figure shows, by the way, on the right a model of a competing Heinkel-configuration with a straight wing and nacelles, which was obviously meant for comparative measurements.

The higher transonic drag of the W-wing as compared to the swept wing is one of the essential arguments for the effectiveness of the area rule

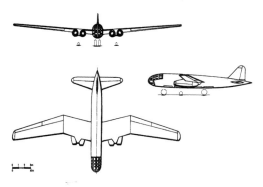

Fig. 2.177 Project P.188 of a four-engine jet bomber of Blohm and Voss with a W-wing according to H. Pohlmann (1943) [202].

Fig. 2.178 Comparison of drag of an unswept wing, a swept wing, and a W-wing measured in the Junkers high-speed wind tunnel HK 900 according to *G. Backhaus* **(1943) [205]; axes:** $\Delta c_w = \Delta c_d$**,** *Mach* **number** *M***; legend: 1) measurement unswept wing** $\varphi = 0$ **deg; 2) measurement W-wing, and 3) measurement swept-back wing** $\varphi = 40$ **deg.**

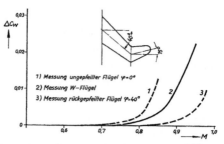

Versuchsergebnisse (1943) für einen „W"-Flügel
Test results (1943) for a W-wing

independent of the sweep angle as already shown in Fig. 2.158. In a certain sense, the area rule implicitly contains the swept-wing concept, because the most favorable attachment of a wing of a given planform, span, and absolute airfoil thickness on an aircraft for minimum drag ultimately requires a more slender axial distribution of the cross-sectional areas in the longitudinal direction and hence leads inevitably to a sweeping of the wing. An especially good adjustment of the wing sweep and the cross-sectional area distribution corresponding to the requirements of aerodynamics within a large range of flight speeds up to supersonic flight is provided by a variable-geometry wing.

2.3.4.12 SUMMARIZING REMARKS CONCERNING THE AREA RULE. The application of the aerodynamic area rule is for aircraft at high subsonic speeds and within the supersonic range decisive for a favorable drag development. The area rule and the inherent concept of wing sweep have for decades naturally and successfully been applied to high-speed aircraft.

The concept of the transonic area rule was established by *Frenzl* in 1943 at the Junkers Aircraft and Engine Works in Dessau. There it was formulated, verified, patented, further developed, and still during WWII used in individual German jet-aircraft projects without anybody outside Germany noticing or paying attention to it. Despite the fact that almost all German documents concerning these developments were available to the Allies after WWII and that complete German development teams worked with the Allies, it took 10 years and a second discovery, a theoretical formulation, and publication of the area rule by the NACA, to give it international recognition and application.

Fig. 2.179 Half-models for the Junkers high-speed wind tunnel HK 900 in Dessau (1943); front: W-wing, unswept wing and forward-swept wing [207] (courtesy of Archive German Museum München).

Because of the political situation during the post-war period, the clear priority of *Otto Frenzl's* German Junkers patent of 1944 concerning the transonic area rule was not generally known and was forgotten. The rule was then for 10 years hardly consciously applied. A corresponding international review of the history of the origin of the area rule is deemed advisable [208, 209].

2.3.5 INVESTIGATIONS ON PROPELLERS WITH SWEPT BLADE AXES

The aerodynamic advantages exhibited by the swept wing at high *Mach* numbers suggested also that the effect of sweep on the flow about propeller-blade tips at high *Mach* numbers should be investigated. *A.W. Quick* [210] had already at the 1942 LGL Meeting 156 in Berlin reported on first investigations on swept propellers at the DVL, Berlin–Adlershof. Here, three two-bladed propellers, which differed only in the design of their blade axes (Fig. 2.180), were investigated up to a maximum blade-tip *Mach* number of $M = 0.87$ at freestream speeds of up to 70 m/s. The start was propeller (b) with a straight blade axis and a diameter of 1.1 m. Propeller (a) was essentially developed on this basis by curving the blade axis backwards. The curvature was established by shifting the blade elements. Here, the same airfoil, gradient, and thickness development for propellers (a) and (b) were ensured. Curvature itself was chosen such that first at the blade tip a sweep angle of 45 deg occurred, and second the center of gravity of the entire blade was located on the extension of the blade-root axis in order to keep bending stresses at the blade root as low as possible. This requirement caused for propeller (a) an opposite curvature on its inside. Propeller (c) was generated by mirror-imaging propeller (a) at the blade axis of propeller (b).

The tests were carried out under contract of the DVL in the 3-m wind tunnel (maximum freestream velocity 70 m/s) of the Flugtechnische Versuchsanstalt Prag (FVA-Prague) (Flight-Technical Test Establishment) on a 150 kW Pendulum Dynamometer at a maximum of 600 rpm. Determined

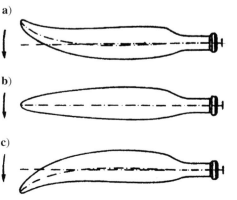

a)

b)

c)

Fig. 2.180 Swept propeller blades of the DVL, *A. W. Quick* [210].

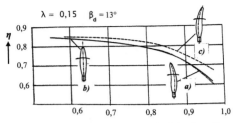

Fig. 2.181 Maximum efficiency η_{max} of the DVL swept propeller blades compared to an unswept blade, A. W. Quick [210]; axes: Efficiency η, Mach number at the blade tip.

were the thrust, the torque, the number of revolutions, and the freestream temperature. The measurements led to the following interesting results. Figure 2.181 presents a comparison of the three arrangements investigated for an advance ratio of $\lambda = 0.15$ and an angle of the airfoil pressure side (twist) of $\beta_d = 13$ deg. It can be seen that the forward sweep did not affect the efficiency at all or resulted in only a small deterioration in the efficiency compared to the unswept propeller. Here, *Quick* concluded that the gains achieved in the case of propeller (a) were not solely attributable to the later occurrence of shock waves due to sweep. He assumed that the radial pressure gradients due to different sweep, also present without the presence of shock waves, had a different effect on the boundary-layer flow, which was added to the effect of the centrifugal forces on the boundary layer. Still important to the interpretation of the results were the relatively thick sections in the blade-tip area, because during the subsequent discussion (after the presentation), *Seiferth* reported on nearly parallel propeller investigations with thinner airfoils at the AVA Göttingen. The measurements differed primarily in the selection of the airfoils especially in the blade-tip region. Swept-back and straight propeller models of the AVA are shown in Fig. 2.182. The propellers had a diameter of 0.29 m. The blade sections were designed for an advance ratio of $\lambda = 0.12$. The comparison between the propeller efficiencies of the straight and the swept blades, respectively, is plotted in Fig. 2.183. The maximum efficiency η_{max} in Fig. 2.183a allows a comparison with the DVL measurements. At a *Mach* number of $M = 0.97$, the improvement due to sweep was somewhat smaller than at the DVL, as could be expected for a thinner airfoil. For blade-tip *Mach* numbers of $M > 1$, the improvement of the efficiency was very considerable. However, for a direct comparison of the straight and the swept blade, the efficiency η at the optimum operating condition of the blade was better suited than η_{max}. The best operating conditions of the AVA propeller models were within the design range for the advance ratio $\lambda = 0.12$, where the sections had their best glide ratio. Here, sweep resulted in an improvement of the efficiency of 3 to 4% at a *Mach* number of $M \approx 1$ and at $M \geq 1$ in an improvement of 7%. However, for supersonic velocities at the blade tips the noise development increased strongly. It is probable that static tests on larger propellers with swept blades were still carried out in

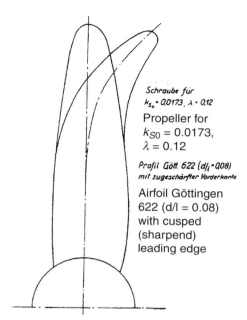

Schraube für
$k_{S_0} = 0,0173$, $\lambda = 0,12$

Propeller for
$k_{S0} = 0.0173$,
$\lambda = 0.12$

Profil Gött. 622 ($d/l = 0,08$)
mit zugeschärfter Vorderkante

Airfoil Göttingen
622 (d/l = 0.08)
with cusped
(sharpend)
leading edge

Fig. 2.182   Swept propeller blade of the AVA, cited by *A. W. Quick* in [210].

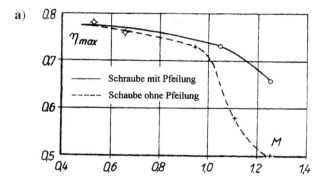

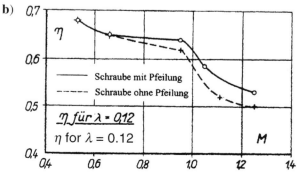

a)

η_{max}

——— Schraube mit Pfeilung
- - - - Schaube ohne Pfeilung

M

b)

η

——— Schraube mit Pfeilung
- - - - Schraube ohne Pfeilung

η für $\lambda = 0,12$
η for $\lambda = 0.12$

M

Fig. 2.183 Comparison between propeller efficience of straight and swept blades: a) Maximum efficiency η_{max}, respectively, b) efficiency η for $\lambda = 0.12$ of the AVA swept propeller blade (solid line) compared with a blade without curvature, cited by *A. W. Quick* in [210]; legend: a) propeller with sweep, propeller without sweep, and b) propeller with sweep, propeller without sweep.

Fig. 2.184 Three-blade propeller with swept-blade tips on a jacked-up Messerschmitt Me 109 (1944–1945), copy of a single frame of a cine-film, *H. Birkholz* **[211].**

1944–1945 at other locations. According to statements of *Heinz Birkholz* in 2001 in Jet & Prop [211], Flight Captain *Richard Dahm*, test pilot at the Heine-Propeller Company, has handed down in his estate, among other things, movie sequences that show provisional static-thrust measurements on a three-blade propeller with swept tips on a jacked-up Messerschmitt Me 109 (Fig. 2.184).

Quick [212] has summarized all activities in the area of propeller research in Germany up to 1945. A further short presentation was prepared by *J. Stüper* [213] for the British Military Administration.

Also the development of high-speed propellers with swept-back blade tips was only used—initially in the USSR—after 1945.

REFERENCES

[1]   A. *Bäumker*, "Zur Geschichte der deutschen Luftfahrt," Festschrift zum 70. Geburtstag von Ludwig Prandtl, Druck: C.G.Vogel, Pössneck (Thüringen), 1944, pp. 1–81.

[2]   K. *Hein, Adolf Bäumker* (1891–1976), "Einblicke in die Organisation von Luft- und Raumfahrtforschung," DLR Mitteilung 95-02, 1995, pp. 1–157.

[3]   J. C. *Rotta*, "Die Aerodynamische Versuchsanstalt in Göttingen, ein Werk Ludwig Prandtls," Verlag Vandenhoek & Ruprecht, Göttingen, 1990.

[4]   *NN.*, "25 Jahre Deutsche Forschungsanstalt für Luftfahrt e.V.," DFL, Braunschweig, 1936–1961, Festschrift.

[5]   H. *Winter*, "Der Überdruckkanal der Aerodynamischen Versuchsanstalt Göttingen," Jahrbuch der Deutschen Luftfahrtforschung, Bd. 3, 1937, pp. 574–577.

[6]   W. *Krüger*, "Der große Windkanal für Über- und Unterdruck (K VI) der Aerodynamischen Versuchsanstalt Göttingen," Monographien über Fortschritte der deutschen Luftfahrtforschung (seit 1939), ed. by A. *Betz*, AVA Göttingen, D1 Modellversuchsanlagen, Teil 1.1, 1946, pp. 1–7.

[7]   R. *Seiferth*, "Vorausberechnung und Beseitigung der Schwingungen von Freistrahl-Windkanälen," Monographien über Fortschritte der deutschen Luftfahrtforschung (seit 1939), ed. by A. *Betz*, AVA Göttingen, D1 Modellversuchsanlagen, Teil 4.4,1946, pp. 1–23.

[8]   H. *Bäuerle*, "Einfluss von Düse und Auffangtrichter auf die Strahleigenschaften," Monographien über Fortschritte der deutschen Luftfahrtforschung (seit 1939)," ed. by A. *Betz*, AVA Göttingen, D1 Modellversuchsanlagen, Teil 4.1, 1946, pp. 1–11.

[9]   H. *Schuh*, "Windschwankungsmessungen mit Hitzdrähten. Göttingen. Monographien über Fortschritte der deutschen Luftfahrtforschung (seit 1939)," ed. by A. *Betz*, AVA Göttingen, D1 Modellversuchsanlagen, Teil 4.3, 1946, pp. 1–27.

[10]   H. Seiferth, "Messungen der Kanalturbulenz mittels Kugeln und ihr Zusammenhang mit der Hitzdrahtmessung, Monographien über Fortschritte der deutschen Luftfahrtforschung (seit 1939)," ed. by A. Betz, AVA Göttingen, D1 Modellversuchsanlagen, Teil 4.2, 1946, p. 121.

[11]   M. Kramer, "Der 5 m × 7 m-Windkanal der DVL," Deutsche Luftfahrtforschung, Bd. 12, Oct. 1935, pp. 181–187.

[12]   F. Ehlers, "Der 8-m-Kanal A3 der LFA Braunschweig, Monographien über Fortschritte der deutschen Luftfahrtforschung (seit 1939)," ed. by A. Betz, AVA Göttingen, D1 Modellversuchsanlagen, Teil 1, 1946, pp. 1–7.

[13]   H. Fütterer, "Hochgeschwindigkeitswindkanäle in Göttingen 1926–2005," DLR IB 224–2005-A01, 2005.

[14]   A. Steichen, "Beiträge zur Theorie der zweidimensionalen Bewegungsvorgänge in einem Gas, das mit Überschallgeschwindigkeit strömt," Göttinger Archiv, GOAR 3652, 1909.

[15]   A. Busemann, "L. Prandtl, Näherungsverfahren zur zeichnerischen Ermittlung von ebenen Strömungen mit Überschallgeschwindigkeit," Stodola Festschrift, Zürich, 1929, pp. 499–509.

[16]   H. Ludwieg, "G. Oltmann, Der Hochgeschwindigkeitskanal der AVA mit Dampfstrahlantrieb, 0765 m Durchmesser," Monographien über Fortschritte der deutschen Luftfahrtforschung (seit 1939), ed. by A. Betz, AVA Göttingen, D1 Modellversuchsanlagen, Teil 2.3, pp. 1–6.

[17]   H. Matt, "Aufbau und Strahleigenschaften des DVL-Hochgeschwindigkeits-Windkanals," LGL-Bericht 127, 1940, pp. 9–17.

[18]   H. Matt, "Der Hochgeschwindigkeitswindkanal der DVL (2,7 m Ø), Monographien über Fortschritte der deutschen Luftfahrtforschung (seit 1939)," ed. by A. Betz, AVA Göttingen, D1 Modellversuchsanlagen, Teil 2.1, 1946, pp. 1–22.

[19]   B. Göthert, "Widerstandsbestimmungen bei hohen Unterschallgeschwindigkeiten aus Impulsverlustmessungen," LGL-Bericht 127, 1940, pp. 23–32.

[20]   F. Mirus, "Schlierenbeobachtungen im DVL-Hochgeschwindigkeitswindkanal," LGL-Bericht 127, 1940, pp. 18–23.

[21]   Th. Zobel, "Der Hochgeschwindigkeitskanal der Luftfahrtforschungsanstalt Hermann Göring, Braunschweig," Schriften der Deutschen Akademie der Luftfahrtforschung, Heft 1060/43 g, 1943.

[22]   H. Matt, "Der Hochgeschwindigkeitswindkanal A 2 der LFA-Braunschweig," Monographien über Fortschritte der deutschen Luftfahrtforschung (seit 1939), ed. by A. Betz, AVA Göttingen, D1 Modellversuchsanlagen, Teil 2.2, 1946, pp. 1–21.

[23]   H. Matt, "Hinweise zur Verbesserung der Eigenschaften des 2,8-m- Hochgeschwindigkeitswindkanal der LFA," Ministry of Aircraft Production, Bericht AGD 1021/G, 1946.

[24]   Th. Zobel, "Strömungsmessungen durch Lichtinterferenz," Deutsche Luftfahrtforschung FB1167, 1940.

[25]   W. Knackstedt, "Die Hochgeschwindigkeitskanäle A 6 und A 7 der LFA Braunschweig," Monographien über Fortschritte der deutschen Luftfahrtforschung (seit 1939), ed. by A. Betz, AVA AVA Göttingen, D1 Modellversuchsanlagen, Teil 2.5/6, 1946, pp. 1–23.

[26]   W. Knackstedt, "Die Hochgeschwindigkeitskanäle A9a und A9b der LFA Braunschweig," Monographien über Fortschritte der deutschen Luftfahrtforschung (seit 1939), ed. by A. Betz, AVA AVA Göttingen, D1 Modellversuchsanlagen, Teil 2.7, 1946, pp. 1–29.

[27]   P. Hamel, "Der Busemann Überschallkanal A-9 der LFA-Entwicklung, Nutzung und Verbleib," Deutscher Luft-und Raumfahrtkongress, Braunschweig, DGLR-2006-190/191, 2006.

[28]   J. Erichsen, B. M. Hoppe, (eds.), "Peenemünde, Mythos und Geschichte der Rakete
        1923–1989," Katalog des Museums Peenemünde, Nicolaische Verlagsbuchhandlung
        GmbH, Berlin, 2004.

[29]   F. Zwicky, "Report on Certain Phases of War Research in Germany, Vol. II," Aerojet
        Engineering Corp., Pasadena, CA, U.S., Oct. 1945.

[30]   W. Kraus, "Der Überschall-Windkanal von Peenemünde," INTERAVIA, 6. Jahrgang,
        Nr. 10, 1951, p. 558–561.

[31]   P. P. Wegener, "The Peenemünde Wind Tunnels: A Memoir," Yale Univ., New Haven,
        CT, U.S., 1996.

[32]   H. Kurzweg, "The Aerodynamic Development of the V-2," AGARD, First Guided
        Missile Seminar, München, April 1956, Verlag E. Appelhans & Co., Braunschweig,
        1957, pp. 50–69.

[33]   C. Wieselsberger, "Elektrische Anzeige von Kräften durch Änderung einer Induktivität,"
        Jahrbuch der Luftfahrtforschung, Ausg. Flugwerk, 1937, p. 592.

[34]   S. F. Erdmann, "Deutsch-Niederländische Odyssee, Autobiographischer Zeitspiegel,"
        Delft Univ. Press, Amsterdam, the Netherlands, 2001.

[35]   N. N., "Wind Tunnels of the Naval Ordnance Laboratory," NOL, 362699-56-2, 1956.

[36]   R. D. Hiscocks, J. L. Orr, J. J. Green, "High Speed Tunnels and Other Research in
        Germany," C.I.O.S. Item No. 25, File No. XXVIII-47, 1945, pp. 1–37.

[37]   R. D. Hiscocks, J. L. Orr, J. J. Green, "Luftfahrtforschungsanstalt Hermann Göring,"
        Völkenrode, Brunswick. B.I.O.S. Target Nos.: C25/385, C2/662, 26/74, 1945, pp. 1–40.

[38]   J. J. Green, R. D. Hiscocks, J. L. Orr, "Wartime Aeronautical Research & Development
        in Germany," The Engineering Journal: Part I, 1948, pp. 531–545; Part II, 1948,
        pp. 584–589; Part III, 1948, pp. 627–637; Part IV, 1949, pp. 19–25.

[39]   R. Smelt, "A Critical Review of German Research on High-Speed Airflow," The Royal
        Aeronautical Society, 696th Lecture, London, Oct. 1946, pp. 899–934.

[40]   Otto Mühlhäuser, "Über eine Forschungsanlage zur Untersuchung von Strömungs-
        vorgängen in der Nähe von Schallgeschwindigkeit," AVA-Bericht 42/G/4 + Anlage,
        1942, p. 1–50.

[41]   H. Matt, "Der Hochgeschwindigkeitskanal (3 m Ø) der Luftfahrtforschungsanstalt
        München," Monographien über Fortschritte der deutschen Luftfahrtforschung (seit
        1939), ed. by A. Betz, AVA Göttingen, D1 Modellversuchsanlagen, Teil 2.9, p. 17.

[42]   A. Bäumker, "Brief an A. Betz," vom Sept. 3, 1940. GöAR 146-1.

[43]   H. Trischler, "Luft- und Raumfahrtforschung in Deutschland," Politische Geschichte
        einer Wissenschaft, Campus Verlag, 1992.

[44]   A. Bäumker, "Briefwechsel mit dem RLM (H. Schelp)," 1943, (KPAR A25).

[45]   E. Thiel, "Von Ötztal nach Modane, Aus der Geschichte des großen
        Hochgeschwindigkeitswindkanals »Bauvorhaben 101« der Luftfahrtforschungsanstalt
        München (LFM), später Anlage S1 MA der ONERA," DGLR Jahrestagung in München,
        Sitzung der DGLR-Fachgruppe 12 (Geschichte der Luftfahrt), 1986, DGLR-Jahrbuch
        1986, Teil II, pp. 773–795.

[46]   F. L. Wattendorf, "Memorandum, Proposal to a New Development Center of the U.S.
        Airforce," Annex 8 in Ref. [35], 1987.

[47]   M. Pierre, "Création du centre d'essais de l'ONERA à Modane-Avrieux," ONERA
        Chattillon-sous-Bagneux, 1987.

[48]   See Ref. [3].

[49]   L. Prandtl, "Der Luftwiderstand von Kugeln. Nachrichten der Gesellschaft der
        Wissenschaften in Göttingen," Mathemat.-phys. Klasse, 1914, pp. 177–190.

[50]   C. Wieselsberger, "Mitteilungen der Modellversuchsanstalt (MVA)," Nr. 16, 1. Folge.

[51]   M. Kramer, H. Doetsch, "Untersuchung der Profilreihe NACA-24 im 5 m × 7 m-Wind-
        kanal der DVL," ZWB/FB 548, 1936, p. 1–93.

[52] H. Doetsch, "Ergänzungen zum Forschungsbericht »FB 548« – »Untersuchung der Profilreihe NACA-24 im 5 m × 7 m-Windkanal der DVL«, " ZWB/FB 548/2, 1936, pp. 1–45.

[53] D. D. Baals, W. R. Corliss, "Wind Tunnels of NASA," NASA SP-440, 1981, p. 16.

[54] H.-U. Meier, U. Michel, H.-P. Kreplin, "The Influence of the Wind Tunnel Turbulence on the Boundary Layer Transition," ed. by H.-U. Meier, P. Bradshaw, "Perspectives in Turbulence Studies," Springer Verlag, 1987, pp. 26–46.

[55] C. Wieselsberger, "Windkanalkorrekturen bei kompressibler Strömung," LGL-Bericht 127, 1940, pp. 3–8.

[56] H. Glauert, "Wind Tunnel Interference on Wings, Bodies and Airscrews," ARC R&M 1566, 1980, p. 5.

[57] H. Ludwieg, "Widerstandskorrektur in Hochgeschwindigkeitskanälen," AVA-Bericht 44/H08, 1944, pp. 1–31.

[58] H. Holst, "Verfahren zur Bestimmung von dreidimensionalen Windkanalinterferenzen und Wandadaptionen mit Hilfe gemessener Wanddrücke bei kompressibler Unterschallströmung," DLR-FB 90-46, 1990.

[59] B. Göthert, "Windkanalkorrekturen bei hohen Unterschallgeschwindigkeiten," LGL-Bericht 127, 1940, pp. 113–121.

[60] G. Richter, "Einfluss der Modellgröße in Hochgeschwindigkeitskanälen," LGL-Bericht 127, 1940, pp. 121–128.

[61] B. Göthert, "Windkanalkorrekturen bei hohen Unterschallgeschwindigkeiten unter besonderer Berücksichtigung des geschlossenen Kreiskanals," ZWB/FB 1216, 1940, p. 1–67.

[62] B. Göthert, G. Richter, "Messungen am NACA-Profil 0015-64 mit verschiedenen Tiefen im Hochgeschwindigkeits-Windkanal (2,7 m Ø) und Vergleich mit Ergebnissen im mittleren Windkanal der DVL," DVL Jf-700/2, 1940, pp. 1–61.

[63] B. Göthert, "Zur Frage der offenen oder geschlossenen Bauart von Windkanälen mit hohen Unterschallgeschwindigkeiten," LGL Bericht 127, 1940, pp. 32–33.

[64] A. Naumann, "Profilmessungen im kompressiblen Unterschallbereich II, Profil 23 012-64, Bericht des Aerodynamischen Instituts der T. H. Aachen, 1940.

[65] J. Ackeret, "Convegno di Fisiche, Matematiche e. Naturali. Tema: Le alta Velocita in Aviazione," Reale Accademia d'Italia, Fondazione Alessandro Volta, Sept. 30–Oct. 6, 1935, Rome, Italy, Verlag: Reale Accademia d'Italia, 1936, pp. 1–695.

[66] Kl. Oswatitsch, "Kondensationserscheinungen in Überschalldüsen. Z. angew," Math. Mech. Bd. 22, Nr. 1, 1942, pp. 1–14.

[67] R. Hermann, "Der Kondensationsstoß in Überschall-Windkanaldüsen," Deutsche Akademie der Luftfahrtforschung, 1942, pp. 201–209.

[68] Th. Zobel, "Über das Programm für Untersuchungen im Hochgeschwindigkeitskanal der Luftfahrtforschungsanstalt Hermann Göring," LGL-Bericht 127, 1940, pp. 92–93.

[69] B. Göthert, "Messprogramm des DVL-Hochgeschwindigkeits-Windkanals (2,7 m Dm.)," LGL-Bericht 127, 1940, pp. 95–96.

[70] K. H. Kawalki, "Theoretische Überlegungen über günstigste Ausbildung von Schnellflugprofilen," LGL-Bericht 127, 1940, pp. 102–112.

[71] H. Ludwieg, "Ergebnisse der experimentellen Profiluntersuchungen bei hohen Unterschallgeschwindigkeiten," Monographien über Fortschritte der deutschen Luftfahrtforschung (seit 1939), ed. by A. Betz, AVA Göttingen, Das Tragflügelprofil, Teil E8, 1946, pp. 1–28.

[72] See Ref. [19].

[73] A. Busemann, "Kompressibilitätseinfluss für dünne, wenig gekrümmte Profile bei Überschallgeschwindigkeit," Schriften der Deutsche Akademie der Luftfahrtforschung, Heft 18, 1940.

[74]   B. Göthert, "Pofilmessungen im DVL-Hochgeschwindigkeits-Windkanal (2,7 m Ø)," ZWB/FB 1490, 1941.

[75]   B. Göthert, "Druckverteilungs- und Impulsverlustschaubilder von NACA-Profilen bei hohen Unterschallgeschwindigkeiten," ZWB/FB 1505/1-5, 1941.

[76]   B. Göthert, "Hochgeschwindigkeitsuntersuchungen an symmetrischen Profilen mit verschiedenen Dickenverhältnissen im DVL-Hochgeschwindigkeits-Windkanal (2,7 m Ø) und Vergleich mit Messungen in anderen Windkanälen," ZWB/FB 1506, 1941.

[77]   See Ref. [19].

[78]   See Ref. [70], pp. 102–112.

[79]   Th. Theodorsen, J. E. Garrick, "General potential theory of arbitrary wing sections," NACA-Report No. 452, 1933.

[80]   F. Weinig, "Beiträge zur Theorie dünner, schwach gewölbter Tragflügelprofile," ZAMM 14, 1934, p. 279.

[81]   J. D. Anderson, "A History of Aerodynamics," Cambridge Univ. Press, Cambridge, U.K. 1977.

[82]   K. H. Kawalki, "Theoretische Untersuchungen von Schnellflugprofilen, die aus Ellipsenprofilen entwickelt sind. 1. Mitteilung: Symmetrische Profile bei Nullanstellung," Deutsche Luftfahrtforschung, FB 1224/1, 1940.

[83]   K. H. Kawalki, "Theoretische Untersuchungen von Schnellflugprofilen, die aus Ellipsenprofilen entwickelt sind. 2. Mitteilung: Symmetrische Profile bei Anstellung und Profile mit überlagerter Wölbung," Deutsche Luftfahrtforschung, FB 1224/2, 1940.

[84]   K. H. Kawalki, "Profile geringster Übergeschwindigkeiten bei vorgegebenen Auftriebsbeiwerten," LGL-Bericht Nr. 156, 1942, pp. 163–169.

[85]   See Ref. [60].

[86]   O. Knappe, "Schnellkanalversuche an einem symmetrischen Klappenflügel," LGL-Bericht 127, 1940, pp. 79–82.

[87]   B. Göthert, "Hochgeschwindigkeitsuntersuchungen an symmetrischen Profilen mit verschiedenen Dickenverhältnissen im DVL-Hochgeschwindigkeits-Windkanal (2,7 m Ø) und Vergleich mit Messungen in anderen Windkanälen," DVL Institut für Aerodynamik, FB 1506, 1941, pp. 1–41.

[88]   R. Heinrich, "Persönliche Mitteilungen," DLR Institut »Aerodynamik und Strömungsmesstechnik«, Braunschweig, 2005.

[89]   N. Kroll, C.-C. Rossow, D. Schwamborn, K. Becker, G. Heller, "Megaflow—A Numerical Flow Simulation Tool for Transport Aircraft Design," ICAS-2002-1.10.5, 2002.

[90]   B. Göthert, "Hochgeschwindigkeitsmessungen an Profilen gleicher Dickenverteilung mit verschiedener Krümmung im DVL-Hochgeschwindigkeits-Windkanal (2,7 m Ø), Diskussion der Messergebnisse," DVL Institut für Aerodynamik, FB 1910/6, 1944, pp. 1–30.

[91]   B. Göthert, "Widerstandsanstieg bei Profilen im Bereich hoher Unterschallgeschwindigkeiten," DVL-Institut für Aerodynamik, UM 1167, 1944, pp. 1–19.

[92]   H. B. Helmbold, "Physikalische Erscheinungen in der kompressiblen Unterschallströmung," LGL-Bericht 156, 1942, pp. 170–184.

[93]   See Ref. [71].

[94]   B. Regenscheit, "Versuche zur Widerstandsverringerung eines Flügels bei hoher Machscher Zahl durch Absaugung der hinter dem Gebiet unstetiger Verdichtung abgelöster Grenzschicht," AVA-Bericht 41/14/23, FB- 1424, 1941, pp. 1–21.

[95]   F. W. Riegels, "Russische Laminarprofile," 3. Teil, Messungen am Profil Nr. 23 15 mit AVA-Nasenklappe, AVA-Bericht 44/A/02, UM 3067, 1944, pp. 1–26.

[96]   See Ref. [91].

[97]   H. Schlichting, A. Ulrich, "Zur Berechnung des Umschlags laminar/turbulent," Jahrbuch der Deutschen Luftfahrtforschung 1942, pp. 8–35.

[98]   J. Pretsch, "Anfachung instabiler Störungen in einer laminaren Reibungsschicht," Jahrbuch der Deutschen Luftfahrtforschung 1942, pp. 54–71.

[99]   H. Schlichting, K. Bußmann, "Exakte Lösungen für die laminare Grenzschicht mit Absaugung und Ausblasen," Schriften der Deutschen Akademie der Luftfahrtforschung, Bd. 7B, Heft 2, 1943, pp. 25–69.

[100] H. Holstein, "Messungen zur Laminarhaltung der Grenzschicht durch Absaugung an einem Tragflügel," LGL-Bericht S. 10, 1941, pp. 17–27.

[101] O. Schrenk, "Grenzschichtabsaugung," Luftwissen 7, 1941, pp. 409–414.

[102] W. Krüger, "Landehilfen im vorderen Profilbereich," Monographien über Fortschritte der deutschen Luftfahrtforschung (seit 1939), ed. by A. Betz, AVA Göttingen, F2, Der räumliche Tragflügel, Teil 2.111, 1946, pp. 2–8.

[103] W. Krüger, "Über eine neue Möglichkeit der Steigerung des Höchstauftriebes von Hochgeschwindigkeitsprofilen," AVA-Bericht 43/W/64, 1943, pp. 1–13. Siehe auch ZWB/UM 3049, 1943.

[104] W. Krüger, "Systematische Windkanaluntersuchungen an einem Laminarflügel mit Nasenklappe," AVA-Bericht 44/W/19, ZWB/FB 1948,1944, pp. 1–18.

[105] W. Krüger, "Windkanaluntersuchungen an einem geänderten Mustangprofil mit Nasenklappe, Kraft- und Druckverteilungsmessungen," AVA-Bericht 44/W/44, UM 3153, 1944.

[106] F. W. Riegels, "Russische Laminarprofile," 3. Teil, Messungen am Profil Nr. 23 15 mit AVA-Nasenklappe. AVA-Bericht 44/A/02, UM 3067, 1944, pp. 1–26.

[107] H. Doetsch, "Bericht über das Fachgebiet »Profile« vor dem Sonderausschuss Windkanäle am 10.1.43 und 4.1.44," DVL Institut für Aerodynamik, DVL-Bericht Jf 196, UM 1190 (23) 2902, 1944, pp. 1–39.

[108] H. Doetsch, "Über den Einfluss von Oberflächenstörungen auf den Widerstand der Tragflügel," DVL-Jf 196, ZWB/UM 1233, 1944, pp. 1–49.

[109] H. Holstein, "Versuche an einer parallel angeströmten ebenen Platte über den Rauhigkeitseinfluss auf den Umschlag laminar/turbulent," AVA-Bericht 44/A/14, ZWB/ UM 3110, 1944.

[110] R. Seiferth, "Aerodynamische Beiwerte des Tragflügels," Monographien über Fortschritte der deutschen Luftfahrtforschung (seit 1939), ed. by A. Betz, AVA Göttingen, F2 Der räumliche Tragflügel, Teil F2 1, 1946, pp. 1–69.

[111] W. Krüger, "Der Flügel mit Landehilfen," Monographien über Fortschritte der deutschen Luftfahrtforschung (seit 1939), ed. by A. Betz, AVA Göttingen, F2 Der räumliche Tragflügel, Teil F2 2.1, 1946, pp. 1–91.

[112] J. Weissinger, "Über die Auftriebsverteilung von Pfeilflügeln," DVL- Jf. 232/1, Deutsche Luftfahrtforschung FB 1553, 1942, pp. 1–46.

[113] W. Jacobs, "Druckverteilungsmessungen an einem Pfeilflügel konstanter Tiefe (j = 15°) bei unsymmetrischer Anströmung," TH Braunschweig, Aerodyn. Inst., Bericht 43/2, ZWB/UM 2083, 1944, pp. 1–32.

[114] W. Jacobs, "Druckverteilungsmessungen an Pfeilflügeln konstanter Tiefe (j = 15°) bei symmetrischer Anströmung," TH Braunschweig, Aerodyn. Inst., Bericht 43/26, ZWB/ UM 2052, 1943, pp. 1–28.

[115] W. Jacobs, "Sechskomponentenmessungen an vier Trapezflügeln mit Pfeilstellung," TH Braunschweig, Aerodyn. Inst., Bericht 44/4, ZWB/UM 2069, 1944, p. 1–34.

[116] W. Jacobs, "Druckverteilungsmessungen an zwei Pfeilflügeln konstanter Tiefe (j = 30°, 45°) bei unsymmetrischer Anströmung," TH Braunschweig, Aerodyn. Inst., Bericht 44/12, ZWB/UM 2110, 1944, pp. 1–46.

[117] W. Jacobs, "Sechskomponentenmessungen an drei vorwärts gepfeilten Flügeln," TH Braunschweig, Aerodyn. Inst., Bericht 44/19, ZWB/UM 2103, 1944, pp. 1–42.

[118] H. Multhopp, "Die Anwendung der Tragflügeltheorie auf Fragen der Flugmechanik bei unsymmetrischer Anströmung," LGL-Bericht S. 2, 1939.

[119] J. Weissinger, "Über die Auftriebsverteilung von Pfeilflügeln," DVL- Jf.232/1, Deutsche Luftfahrtforschung FB 1553, 1942, pp. 1–46.

[120] J. Weissinger, "Erinnerungen an meine Zeit in der DVL 1937–1945," Jahrbuch Überblicke Mathematik 1985, pp. 105–129.

[121] H. Puffert, "Drei- und Sechskomponentenmessungen an gepfeilten Flügeln und ein Gesamtmodell," ZWB/FB-1726, 1942.

[122] M. Hansen, "Trapezflügel mit Pfeilstellung und Verwindung beim Schieben," AVA-Bericht 41/14/49 und ZWB/FB-1588, 1941.

[123] H. Luetgebrune, "Beiträge zur Pfeilflügeluntersuchung," TH Hannover, Inst. f. Aerodynamik und Flugtechnik, ZWB/FB-1458, 1941.

[124] H. Lemme, "Untersuchungen an einem Pfeilflügel, einem abgestumpften Pfeilflügel und einem M-Flügel," AVA-Bericht 43/W/24, und ZWB/FB 1739/1 und ZWB/FB 1739/2, 1943.

[125] W. Krüger, "Windkanalmessungen an einem 35°-Pfeilflügel mit verschiedenen Landehilfen," Teil 1: Sechskomponentenmessungen. AVA-Bericht 45/W/4, 1945. pp. 1–107.

[126] See Ref. [119].

[127] J. Weissinger, "Bemerkungen über Kräfte und Momente des schiebenden Tragflügels," DVL-Jf-232/3, siehe auch ZWB/UM 1711, 1944, pp. 1–48.

[128] R. Seiferth, "Der schiebende Tragflügel. Ringbuch der Luftfahrttechnik," I A 14, 1940.

[129] E. Möller, "Systematische Sechskomponentenmessungen an Flügel/Rumpf-Anordnungen mit Pfeilflügeln konstanter Tiefe," TH Braunschweig, Inst. f. Aerodynamik, Bericht 43/17, 1943, pp. 1–40.

[130] H. Schlichting, "Bericht über das Fachgebiet Interferenz vor dem Windkanalausschuss im Februar 1945," Aerodynamisches Institut der Technischen Hochschule Braunschweig, Bericht 45/4, 1945; published in NACA TM 1347, 1953.

[131] H. Schlichting, "W. Frenz, Systematische Sechskomponenten-Messungen über die Interferenzen eines Rechteckflügels mit Rumpf und Leitwerk," Jahrbuch der Deutschen Luftfahrtforschung, 1943.

[132] E. Möller, "Systematische Sechskomponentenmessungen an Flügel/Rumpf-Anordnungen mit Pfeilflügeln konstanter Tiefe," TH Braunschweig, Inst. f. Aerodynamik., Bericht 43/17, 1943, pp. 1–40.

[133] See Ref. [111].

[134] G. Brennecke, "Auftriebssteigerungen beim Pfeilflügel," AVA-Bericht 43/W/35, see also ZWB/FB 1876, 1943, pp. 1–28.

[135] G. Brennecke, "Untersuchungen am Pfeilflügel mit Landehilfe," AVA-Bericht 44/W/54, see also ZWB/UM 3162, 1944, pp. 1–7.

[136] See Ref. [103]

[137] See Ref. [104].

[138] A. Roth, "Untersuchung eines verwundenen Pfeilflügels bei hohen Geschwindigkeiten," AVA-Bericht 40/8/8, 1940, pp. 1–7.

[139] A. Roth, "Messungen an Pfeilflügeln mit kreisbogenförmigem Umriss bei hohen Geschwindigkeiten," AVA-Bericht 41/8/2, 1940, pp. 1–7.

[140] B. Göthert, "Hochgeschwindigkeitsmessungen an einem Pfeilflügel," DVL-Inst. für Aerodynamik, Bericht Jf-712, siehe auch LGL-Bericht 156, 1942, pp. 30–40.

[141] G. Koch, "Druckverteilungsmessungen am schiebenden Tragflügel," LGL-Bericht 156, 1942, pp. 41–46.

[142] R. Seiferth, "Aerodynamische Beiwerte des Tragflügels," Monographien über Fortschritte der deutschen Luftfahrtforschung (seit 1939), ed. by A. Betz, AVA Göttingen, F2 Der räumliche Tragflügel, Teil F2 1, 1946, pp. 1–69.

[143] G. Lange, Wacke, "Prüfbericht über 3- und 6-Komponentenmessungen an der Zuspitzungsreihe von Flügeln kleiner Streckung," Teilberichte 1943:
Trapezflügel   L = 1 + 3 → UM 1023/1
Trapezflügel mit Rumpf   L = 1 bis 3 → UM 1023/2
Ellipsenflügel   L = 1 bis 3 → UM 1023/3
(See also NACA TM 1176, 1948.)

[144] H. Voepel, "Messungen an Tragflügeln kleiner Streckung," AVA-Bericht 46/Z/18, 1946, pp. 1–368.

[145] B. Göthert, "Hochgeschwindigkeitsmessungen an einem Flügel kleiner Streckung," ZWB/FB 1846, 1943.

[146] J. Weissinger, "Über die Auftriebsverteilung von Pfeilflügeln," DVL- Jf.232/1, Deutsche Luftfahrtforschung FB 1553, 1942, pp. 1–46.

[147] R. T. Jones, "Properties of Low-Aspect-Ratio Pointed Wings at Speeds Below and Above the Speed of Sound," NACA Report No. 835, 1946.

[148] W. Liebe, "Grenzschichtzaun, Barriere gegen den Abreißvorgang," DGLR Zeitschrift Luft und Raumfahrt, Nr. 1, 1990, pp. 30–34.

[149] W. Liebe, "Ein messtechnisch einfaches Verfahren zur Bestimmung von Abkippeigenschaften," DVL-Inst. für Flugmechanik, ZWB/FB 1200/2, 1940, pp. 1–31.

[150] W. Liebe, "Ergebnisse von Abkippuntersuchungen. Teil 3: Flugmessungen am Muster Me 109 mit Störkante und Zaun," DVL-Inst. für Flugmechanik, ZWB/FB 1200/3, 1941, pp. 1–19.

[151] B. Göthert, "Hochgeschwindigkeitsmessungen an rotationsymmetrischen Körpern mit verschiedener Kopfausbildung im DVL-Hochgeschwindigkeits-Windkanal (2,7 m Durchmesser)," LGL-Bericht 127, 1940, pp. 83–89.

[152] B. Göthert, "Hochgeschwindigkeitsmessungen an rotationsymmetrischen Körpern und Vergleich mit Abwurfversuchen," LGL-Bericht 156, 1942, pp. 89–108.

[153] Th. Zobel, "Der Hochgeschwindigkeitskanal der Luftfahrtforschungsanstalt Hermann Göring, Braunschweig," Jahrbuch der Deutschen Akademie der Luftfahrtforschung, 1943–1944. 1944, pp. 120–141.

[154] H. Melkus, "Messungen an Rotationskörpern im Hochgeschwindigkeitskanal der DVL," DVL Inst. für Aerodynamik, DVL-Bericht Jf 729/3, 1944, pp. 1–33.

[155] F. Dubs, "Hochgeschwindigkeits-Aerodynamik," Basel/Stuttgart, 1961.

[156] O. Frenzl, "Projekt Hochgeschwindigkeitsanlage mit Dampfstrahlantrieb und Speicher," Masch.schr. Manuskript Junkers B. 17046/703, Dessau, April 8, 1942, Archiv Deutsches Museum München.

[157] O. Frenzl, "Hochgeschwindigkeitskanal mit Dampfstrahlantrieb," Techn. Berichte 9, 1942, H. 1, pp. 21–22.

[158] R. T. Jones, "Reduction of Wave Drag by Antisymmetric Arrangement of Wings and Bodies," AIAA Journal, Vol. 10, No. 2, Feb. 1972, pp. 171–176.

[159] O. Frenzl, "Erfindungsmitteilung. Anordnung von Verdrängungskörpern bei Hochgeschwindigkeitsflug," Dessau, Dec. 17, 1943, Archiv Deutsches Museum München.

[160] T. Zobel, "Grundsätzlich neue Wege zur Leistungssteigerung von Schnellflugzeugen (Vortrag vom 13.3.1944)," Schriften der Deutschen Akademie der Luftfahrtforschung, 1944, H. 1079/44g, pp. 1–37.

[161] O. Frenzl, "Motorgondel-Flügel-Interferenz bei hohen Geschwindigkeiten (Die Flächenregel)," Zeitschrift für Flugwissenschaften 5, 1957, H. 6, pp. 181–183.

[162] G. Schulz, "Aerodynamische Regeln für den Einbau von Strahltriebwerksgondeln," Bericht der Ernst Heinkel Flugzeugwerke GmbH, Aug. 7, 1944.

[163] G. Schulz, "Aerodynamische Regeln für den Einbau von Strahltriebwerksgondeln," Zeitschrift für Flugwissenschaften 3, 1955, H. 5, pp. 119–129.

[164] E. Truckenbrodt, G. Backhaus, "Widerstandsmessungen von Sondertriebwerken und Vergleich mit Rechnungen," Untersuchungen und Mitteilungen 7302, Aug. 8, 1944.

[165] See Ref. [160].

[166] See Ref. [162].

[167] See Ref. [163].

[168] See Ref. [164].

[169] D. Küchemann, O. Conrad, J. Weber, "Der Einbau von Strahltriebwerken," Monographien über Fortschritte der deutschen Luftfahrtforschung (seit 1939), Kapitel K3, ed. by A. Betz, Göttingen, 1946.

[170] H. Ludwieg, "Pfeilflügel bei hohen Geschwindigkeiten," LGL-Bericht 127, 1940, pp. 44–52.

[171] H. Lindemann, "Ergebnisse über Messungen am Flugzeugmodell Ju 287 aus dem DVL-Hochgeschwindigkeitswindkanal 2,7 m Ø," Untersuchungen und Mitteilungen 1345, 2.9.1944.

[172] H. Melkus, "Ergebnisse von Widerstandmessungen an Rumpfmodellen im Hochgeschwindigkeitswindkanal der DVL (1946)," Archiv Deutsches Museum München.

[173] A. W. Quick, Höhler, "Abgleich der Jäger-Projekte He 11 TL," Untersuchungen und Mitteilungen 1448, 1. und 2. Teil, 1945.

[174] O. Frenzl, "Brief an Siegfried Günter vom 16.10.1965," Archiv Deutsches Museum München.

[175] O. Frenzl, "Brief an Professor A. W. Quick vom 26.2.1982," Archiv Deutsches Museum München.

[176] S. Günter, "Brief an Otto Frenzl vom 27.10.1965," Archiv Deutsches Museum München.

[177] R. Kosin, "Brief an Werner Heinzerling vom 22.12.1981," Archiv Deutsches Museum Werner Heinzerling, München.

[178] See Ref. [157].

[179] O. Frenzl, "Beschreibung zum Junkers-Hochgeschwindigkeitskanal mit Wasserstaubstrahlantrieb und 2,00 × 2,34 m2 Halbfreistrahlstrecke. Masch.schr. Manuskript Junkers B. 18432/786, Dessau, 9.6.1944," Archiv Deutsches Museum München.

[180] Ph. von Doepp, "Hochgeschwindigkeitskanal mit Strahlantrieb für Messungen an ganzen Flugzeugen bei sehr hohen Machzahlen. Masch.schr. Manuskript Junkers B. 18065/768, Dessau, 8.4.1943," Archiv Deutsches Museum München.

[181] O. Frenzl, "Wirtschaftlichkeit großer Hochgeschwindigkeitskanäle mit Dampfstrahlantrieb. Masch.schr. Manuskript Junkers B. 18055/786, Dessau, 9.4.1943, dazu K 03.05; Sk 901," Archiv Deutsches Museum München.

[182] O. Frenzl, "La Soufflerie Transsonique de la S.N.E.C.M.A.," Société Nationale d'Etude et de Construction de Moteurs d'Aviation, S.N.E.C.M.A.-Docaero No. 23, Sept. 1953, pp. 1–18.

[183] O. Frenzl, "Strömung verdampfenden Wassers in Düsen," Maschinenbau und Wärmewirtschaft 11, 1956, H. 1. und 2 (excerpted from doctoral thesis, approved by the TH Graz, Austria).

[184] J. Brocard, "Une soufflerie transsonique et supersonique de conception moderne, la soufflerie Sigma 4 de l'Institut Aérotechnique de Saint-Cyr," Revue Docaero, Paris 1962, No. 77, pp. 3–14.

[185] *E. Riester, R. Lindemann,* "Ejektor und Heißwasserversorgung für den Staustrahltriebwerkshöhenprüfstand in Trauen," Flugwelt International 17, July 1967, H. 7.

[186] *O. Frenzl,* "Der Heißwasserstrahlapparat. Seine Bedeutung für Forschungs- und Versuchsanlagen," Luftfahrttechnik 4, 1958, H. 2, pp. 28–34.

[187] *O. Frenzl,* "Windkanal. DPB 1 120 181 vom 25.7.1959, patentiert für S.N.E.C.M.A.," Paris.

[188] *O. Frenzl,* "Eignung des Heißwasserstrahlapparats für Triebswerkswindkanäle," Luftfahrttechnik 8, 1962, H. 9, pp. 224–233.

[189] *O. Frenzl,* "Grundsätzliche Betrachtung des Heißwasserejektors mit Hilfe des Mollierdiagramms," S.N.E.C.M.A. Report No. YTO-97D, Villaroche, 30.12.1963.

[190] *O. Frenzl,* "Der Heißwassserejektor als Hilfsmittel für die Umweltverbesserung," VDI-Zeitschrift 114, 1972, H. 4, pp. 239–240.

[191] *P. Garrison,* "The Man Who Could See Air," Air and Space, Smithsonian, Washington DC, U.S., June/July 2002, pp. 68–75.

[192] *D. D. Baals, W. R. Corliss,* "Wind Tunnels of NASA," NASA SP-440, Washington, DC, U.S., 1981, pp. 60–65.

[193] *R. T. Whitcomb,* "A Study of the Zero-Lift Drag-Rise Characteristics of Wing-Body Combinations near the Speed of Sound," NACA RM L 52 H 08, 1952.

[194] *R. T. Whitcomb,* "NACA Details Area Rule Breakthrough," *Aviation Week,* Sept. 19, 1955, pp. 28–44.

[195] *J. André, P. Lhost,* "Compte-rendu des conférences présentées par Mr. Whitcomb les 21 et 22ème Janvier 1958 au Centre de Formation en Aérodynamique Expérimentale à Rhode-St.-Genèse (Belgique) sur «La loi des aires en transsonique et supersonique»," S.N.E.C.M.A., Villaroche, Jan. 31, 1958, Archiv Deutsches Museum München.

[196] *J. E. Steiner,* "Evolutionary Aspects of Large Sweptwing Aircraft," 50 Jahre Turbostrahlflug, DGLR-Bericht 89-05, Bonn, 1989, pp. 227–349.

[197] *R. D. Ward,* "Soviet Practice in Designing and Procuring Military Aircraft," Astronautics and Aeronautics, Sept. 1981, pp. 24–38.

[198] *R. T. Jones,* "Theory of Wing-Body Drag at Supersonic Speeds," NACA Report No. 1284, 1956.

[199] *W. Haack,* "Geschoßformen kleinsten Wellenwiderstandes," Bericht der Lilienthal-Gesellschaft für Luftfahrtforschung 139, 1941, pp. 14–29.

[200] *F. Keune, K. Oswatitsch,* "Äquivalenzsatz, Ähnlichkeitssätze für schallnahe Geschwindig-keiten und Widerstand nicht angestellter Körper kleiner Spannweite," Zeitschrift für Angewandte Mathematik und Physik 7, 1956, pp. 40–63.

[201] *B. Göthert,* "Berechnung des Geschwindigkeitsfeldes von Pfeilflügeln bei hohen Unterschallgeschwindigkeiten," LGL-Bericht 127, 1940, pp. 52–56.

[202] *H. Pohlmann,* "Chronik eines Flugzeugwerkes 1932–1945 (Blohm u. Voß)," Stuttgart, 1979.

[203] See Ref. [158].

[204] *J. W. R. Taylor* (ed.), "Jane's All the World's Aircraft 1980–1981," IHS Jane's, Bracknell, Berkshire, U.K. 1980.

[205] *G. Backhaus,* "Beitrag zur Ermittlung des Widerstandsanstieges von Pfeilflügeln," proceedings, 1. Polytechnische Tagung der TH Dresden, 1956.

[206] See Ref. [173].

[207] *W. Hempel,* "EF 116-mit Rümpfen und Triebwerk-Gondeln," Junkers-Versuchsbericht, Schnellbericht, pp. 43–52, Sept. 4, 1943.

[208] See Ref. [191].

[209] *John D. Anderson, Jr.,* "A History of Aerodynamics and It's Impact on Flying Machines," Cambridge Univ. Press, Cambridge, U.K. 1997.

[210] A. W. Quick, "Untersuchungen an Luftschrauben mit pfeilförmigen gekrümmten Blattachsen," LGL-Bericht 156, 1942, pp. 48–50.

[211] H. Birkholz, "So eine »109« haben Sie sicher noch nie gesehen. Die »Messer« mit dem Sichelpropeller," Jet & Prop, H. 5, 2001, p. 56.

[212] A. W. Quick, "Überblick über die drei aerodynamischen Luftschraubentagungen des Entwicklungsausschusses Luftschrauben," Reports and Translations, MAP-252 R, 1946, pp. 1–19.

[213] J. Stüper, "Modern Problems of Airscrew Aerodynamics," Reports and Translations, MAP-VG 91-50, 1946, pp. 1–7. Siehe auch: Monographien über Fortschritte der deutschen Luftfahrtforschung (seit 1939), ed. by A. Betz, AVA Göttingen, Bd. H, Luftschrauben, 1946.

Chapter 3

DEVELOPMENT OF GERMAN TURBOJET ENGINES UP TO PRODUCTION READINESS

HELMUT SCHUBERT

3.1 PROPELLER AND PISTON ENGINE REACH THEIR PERFORMANCE LIMITS

The modern, aerodynamically high-quality aircraft arose from quite different lines of development relatively independent of each other. First, the revolutionary development of the all-metal construction by *Hugo Junkers*, *Adolf K. Rohrbach*, and *Andrei Nikolajewitsch Tupolew* as well as others has to be mentioned. These designers successfully introduced metal into the air-frame construction. This essential technical and economical step in the development did not yet lead to the aerodynamically perfect high-speed aircraft. The high-speed aircraft was initially developed as a competitive flight vehicle, starting in the mid-1920s, for the classic Gordon-Bennett and Schneider Cup air races [1]. Improved high-performance versions of piston engines like the BMW VI (470 kW) or the Fiat V12 (1100 kW) and a step-by-step aerodynamic refinement yielded flight speeds of 500 to 700 km/h for seaplanes. Because of the more favorable takeoff conditions on water, all record-flight competitions of the time took place on or over lakes. In 1934, Francesco Agello was able to increase the world speed record to 681.97 km/h and on October 23, 1934, to 709.07 km/h with a Macchi-Castoldi MC 72 flying over Lake Garda (for more details see Sec. 1.3). These record-breaking aircraft are extensively descibed in the literature [2, 3] by *Ferdinand C. W. Käsmann*. According to a 1938 statement of *Ernst Heinkel*, the increase in the maximum speed between 1920 and 1930 was 75% due to an increase in the engine performance and only to 25% due to aerodynamic performance improvements.

However, according to *Ernst Heinkel*, between 1930 and 1938 the share of aerodynamics in the speed increase was 65% and that of flight engines 35% (Fig. 3.1). In the 1930s, aircraft engineers recognized that a fundamental speed increase by piston engine and propellers was no longer possible. A scientist stated at the time that an aircraft with a drive power of 2942 kW would, at the same takeoff weight, need a performance of 29,420 kW in order to achieve a speed increase from 800 to 950 km/h. *W. Voigt* concludes in a 1938 investigation [4] concerning the increase of the flight speed by increasing

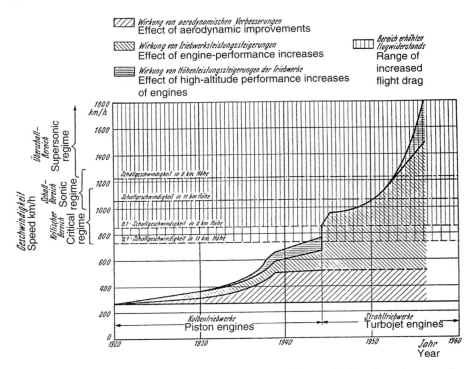

Fig. 3.1 Development of the maximum speed of high-speed aircraft and causes for performance increase [6].

the engine power of bombers and fighters that, as shown in Fig. 3.2, an increase in power above about 1800 PS (1320 kW) to increase flight speed is not worthwhile. This high engine performance was no longer achievable in a conventional way. Possibilities to improve the performance of piston engines were investigated many times: the increased use of propeller gears to increase the number of revolutions; the attachment of aerodynamic engine fairings such as Townend Rings, and since 1928 the NACA cowling [1] (Fig. 3.3); the employment of exhaust turbochargers in production engines since 1931; the use of the thrust of the engine exhaust and the increased employment of adjustable propellers since 1932; and, finally, since 1934, the improvement of the air/fuel mixture by direct fuel injection. The use of diesel flight engines to improve economic efficiency, supported by the German State Ministry of Transportation since 1930, was only adopted in some related fields, such as airships [5]. The more fuel-efficient heavy diesel engines could not gain acceptance.

A last possibility of a short-time performance improvement was then, still during WWII, the water-methanol injection (MW-50) [5] and the oxygen

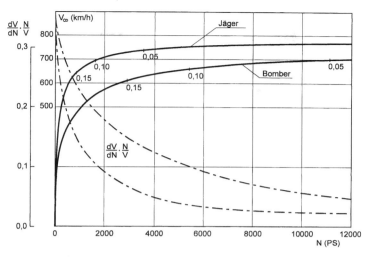

Fig. 3.2   Comparison of bomber and fighter aircraft [4] showing curves of speed vs engine performance; the curves were generated based on computations assuming a constant landing speed, engine performance, weight of the engine, amount of fuel, etc. (LGL Report No. 100, Flight Vehicles and Engines).

enrichment (GM 1 procedure) [5]. The installation of further piston engines would have led to new problems due to an increased weight, higher drag, and the necessity of larger aircraft wings. The propellers caused, in addition, in this speed range great difficulties. At the high flight speeds, the large blade diameters and the high rotational speed of the propellers placed the blade tips already within the transonic speed range. Shocks and flow separation increasingly reduced the advance efficiency of the propellers [6] (Fig. 3.4). For this reason, already during the pioneering days of aeronautics, designers started simultaneously in several countries to look for propulsion solutions to further increase the flight speed.

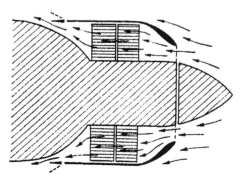

Fig. 3.3   Thrust increase by means of a controllable NACA cowling; cooling air flows through a narrow intake. Flow velocity decreases due to the increasing cross section. The heat of the cylinder heats the cooling air that exhausts with a higher velocity through an adjustable exit generating an increase in thrust.

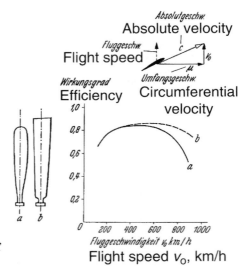

Fig. 3.4   Dependence of the efficiency of propellers [6].

3.2   DEVELOPMENT AND TESTING OF NEW PROPULSION CONCEPTS

During the pioneering days of aeronautics, the choice of efficient, reliable propulsion was already decisive for achieving longer flight distances. A good survey of the early ideas and patents concerning new propulsion systems is given in a 1944 report [7] issued on the order of the RLM (Fig. 3.5). In 1939, *W. Gohlke* introduced the possibilities of new propulsion systems known at the time in four documentary contributions to the German journal *Flugsport* [8]. The English writer *G. Geoffrey Smith* published in the professional journal *Flight* (August 28, 1941) a series concerning the story of the gas turbine and the first attempts to use the gas turbine for the propulsion of aircraft. For the first time, this series of articles referred to the suggestions and first patent (January 16, 1930) of *Frank Whittle* (Fig. 3.6). The individual contributions of *Smith* then appeared in England in book form during 1944. Several editions of this book followed in England and the United States [9].

3.2.1   ROCKET ENGINES FOR AIRCRAFT PROPULSION

The increase of the flight speed was theoretically possible with the aid of solid-fuel and liquid-fuel rockets, but first attempts in the 1920s showed that these forms of propulsion were, at first, too uneconomical to gain practical importance. Tests with rocket-propelled gliders, which had been carried out in Germany by *Fritz von Opel*, *Julius Hatry*, *Gottlob Espenlaub*, and *Fritz Stamer* at the end of the 1920s, had shown the weaknesses of this type of propulsion, most of all the missing controllability of gunpowder-driven rocket engines. On June 11, 1928, *Stamer* performed a flight on the rocket-propelled glider "Ente" (Canard) at the Wasserkuppe (Rhoen Mountains). *Von Opel* experimented

a)

Der Reichsminister der Luftfahrt
und Oberbefehlshaber der Luftwaffe
Technisches Amt
GL|C-Nr. 100 545|44 geh. (Rü IF)

Berlin, den 15. März 194.

Geheim !

Sonderdruck

Strahltriebwerke des Auslandes

I. A.

[signature]

Oberst-Ing. u. Abt.-Chef

b) The State Ministry of Aeronautics   Berlin, March 15,1944
   and Commander in Chief of the Air Force

Secret
Special Issues
Foreign Turbojet Engines

**Fig. 3.5   a) Cover of the RLM report "Foreign Turbojet Engines" of 1944 [7];
b) translation.**

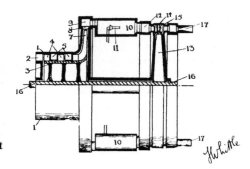

Fig. 3.6 *Frank Whittle's* **British patent
347.206, January 16, 1930.**

in 1929–1930 with gliders equipped with Sander gunpowder rockets [10]. On September 30, 1929, *von Opel* took off with his Opel-Hatry RAK I from the Frankfurt-Rebstock airfield. The E-15, designed by *Max Valier* and *Gottlob Espenlaub*, took off into the air with the aid of a rocket one month later.

In 1932, *Eugen Saenger* went public for the first time with lectures "Concerning the Construction and Performance of Rocket-Propelled Aircraft" at the Technische Hochschule (TH) (Institute of Technology) Vienna. The suggestion of a rocket-propelled aircraft that was to achieve altitudes of 60 to 70 km and flight speeds of up to ten times the speed of sound was part of his considerations.

At the end of 1935, the businessman *Ernst Heinkel*, whose interests were always in the field of high-speed flight, became acquainted with *Wernher von Braun*. *Von Braun* experimented with rockets as a collaborator of *Walter Dornberger* by order of the Heereswaffenamt (HWA) (German Army Weapons Office) at the Army Test Range in Kummersdorf near Berlin. Besides the team of the HWA, *Helmuth Walter* also conducted rocket tests in Kiel involving a new thermodynamic process, the so-called "cold" *Walter*-process. In this decomposition process, triggered by a catalyst like sodium or potassium permanganate, an oxygen-vapor mixture of about 460°C develops in the exhaust nozzle generating the thrust. *Walter* developed a further method. In the alternative "hot" *Walter*-process, hydrogen peroxide (H_2O_2) is used as oxidizer to which petrol and hydrazine hydrate are added as fuel, a mixture that reacts at a temperature of up to 2000°C in a combustion chamber [11, 12]. With a rocket engine, a Walter HWK RII 101 b with 290 to 800-kp thrust (Fig. 3.7), installed in the tail, and an additional piston engine, test pilot *Erich Warsitz* succeeded in conducting a first successful combined-propulsion flight test with a Heinkel He 112 R-V3 probably in the fall of 1937 in Neuhardenberg near Berlin (Fig. 3.8). *Heinkel* in Rostock was at the time already involved in the development of a "pure" rocket-propelled aircraft [13]. The result was the Heinkel He 176 V 1 (Fig. 3.9), which carried out her maiden flight on June 15, 1939, in Peenemünde West with a Walter HWK RI 203 rocket engine [14].

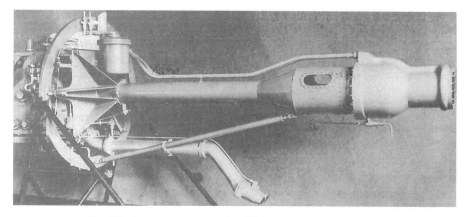

Fig. 3.7   Walter rocket-engine HWK RI 203 of the He 176.

Fig. 3.8 Heinkel experimental aircraft He 112R V-3 with Walter rocket engine; tested during autumn of 1937 with pilot *Erich Warsitz* [13].

Fig. 3.9 Heinkel experimental rocket-propelled aircraft He 176 V-1; *Ernst Heinkel* intended to penetrate with this aircraft the 1000-km/h speed range (Heinkel Archive).

The static thrust was 5.88 kN (600 kp), and the thrust at cruise was 6.76 kN at an altitude of 9000 m. The pilot was again *Warsitz*. Takeoff and a short flight proceeded smoothly. A second plane under construction at Heinkel, the He 176 V2, to be equipped with a Walter HWA RII 102 rocket engine of 7.10 kN (725 kp) thrust was probably not completed and subsequently scrapped. In how far these rocket flight tests were already meant to be an attack on the 1000-km/h speed record as requested by *Heinkel* is not clear [13].

3.2.2 FLIGHT PROPULSION WITH INTERMITTENT COMBUSTION

Prior to WWI, Frenchmen *Victor de Karavodine* and *Georges Marconnet* theoretically investigated the principle of intermittent combustion as a means of flight propulsion [5]. *Paul Schmidt* took up the idea of intermittent jet propulsion again in München in 1929–1930 and was granted a whole series of patents for it. The task was particularly difficult because neither theoretical studies nor experimental experience were available at that time to understand the workings of such an engine. Several years passed until *Adolf Busemann* of the Aerodynamische Versuchsanstalt Göttingen (AVA) (Aerodynamic Test Establishment Göttingen) in Göttingen managed to analyze the process with the laws of gas dynamics. The research was cofinanced by *Schmidt* in München with funds of the HWA starting in 1935. In 1938, he started operating a test apparatus of 120-mm diameter, which worked satisfactorily. Figure 3.10 shows the test apparatus S.R. 500 of *Schmidt*. Starting in 1935, *Schmidt's* experiments competed with the investigations of the Argus Motor Company in Berlin-Reinickendorf. The work on the new impulse engines was conducted under the supervision of *Fritz Gosslau*. The pulsejet engine Argus As 014 (Fig. 3.11) was developed and built in a large series during WWII by the Argus Motor Corporation in a special application for the Fieseler Fi 103 (V1). As propulsion for high-speed aircraft, it was, however, unsuitable.

Fig. 3.10 The experimental pulsejet device of *Paul Schmidt* (special collection of the German Museum, München).

Fig. 3.11 Argus pulsejet engine As 014 (AGARDograph 20, p. 410).

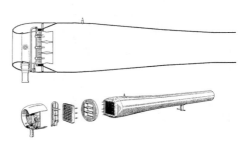

Fig. 3.12 Dornier Do 217 E-2 with *Saenger*-ramjet engine of 1000-mm diameter (AGARDograph 20, p. 342).

3.2.3 RAMJET ENGINE

In 1913, Frenchman *René Lorin* first suggested the use of the ramjet as aircraft propulsion. This type of propulsion is characterized by a particularly simple construction. It has, however, the disadvantage that it is not suitable for the takeoff of an aircraft because it requires a given freestream velocity. *Helmuth Walter* built in Kiel in 1936 an experimental device based on the Lorin-method. Theoretical investigations on ramjet engines were carried out at BMW, Junkers, and the Luftfahrtforschungsanstalt Hermann Göring, Braunschweig (LFA) (Aeronautical Research Establishment Braunschweig). Starting in 1941, *Saenger* developed several experimental ramjet devices by order of the Reichsluftfahrtministerium (RLM) (State Ministry of Aeronautics). An experimental ramjet tube of 500-mm diameter and 5.50-m length mounted on a Dornier Do17 Z was tested at the LFA in Trauen. The testing of the 1000-mm ramjet engine on a Dornier Do 217 E-2 at an altitude of 2000 m is shown in Fig. 3.12. However this type of engine was still associated with too many disadvantages for the high-speed propulsion of aircraft.

3.2.4 PISTON-DRIVEN DUCTED FAN PROPULSION

3.2.4.1 H. M. COANDA BUILDS AN AIRCRAFT WITH A HYBRID PROPULSION SYSTEM.

The aerodynamicist *Henry Marie Coanda* (Fig. 3.13), born on June 7, 1886, in Bucharest, Romania, built in the early days of aeronautics, namely in 1909, in France an aircraft with a revolutionary reaction-type propulsion [15]. In the intake of this propulsion system, a radial compressor operated that was driven via a shaft by a four-cylinder Clerget motor [16] of a performance of 36 kW (50 PS) at 1000 rpm. The air from the radial compressor (Fig. 3.14) was mixed with the exhaust of the piston engine; fuel was injected into this mixture and then ignited. This yielded a forward thrust of 215 N (484 lb) in the nozzle. This was the first hybrid propulsion consisting of a turboengine (turbocompressor) and a piston-type flight engine. The aircraft was exhibited at the Second Salon de l' Aeronautique at the Grand Palais in Paris (Fig. 3.15) in October 1910. It had a span of 10.30 m, a length of 12.50 m, and a wing area of 32.70 m². The aircraft made a flight test with, according to partly contradictory press reports, *Coanda* as a pilot on December 10, 1910, in Issy-Le Moulineaux near Paris, but the airplane was destroyed due to the flight inexperience of *Coanda*. This resulted in him

Fig. 3.13   Aerodynamicist *Henry Marie Coanda* in Paris, 1977.

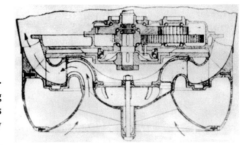

Fig. 3.14   Excerpt from the patent document of *Henry Marie Coanda* concerning the single-stage radial compressor of his propulsion system (British patent, May 26, 1911).

Fig. 3.15   Airplane of *Henry Marie Coanda* as it was exhibited in 1910 in Paris (special collection of the German Museum, München).

Fig. 3.16 1932 airplane MM 187 of *Stipa Caproni*, first flight on October 7, 1932, in Taliedo [17].

teminating related experiments for the time being. Despite this failure, the concept that *Coanda* employed in conjunction with this airplane for the first time was a milestone on the difficult way to turboflight-propulsion. The weakest part of this design was the piston engine driving the compressor. It was too inefficient and too heavy to provide the high power necessary to generate the compressor flow.

3.2.4.2 PROPULSION CONCEPTS SUGGESTED BY *STIPA CAPRONI* AND *GUSTAV KOCH*.   About 20 years later in 1929, the Italian engineer *Luigi Stipa* started with investigations concerning the aerodynamic behavior of propellers enclosed in a tube. *Stipa* published his experimental results in several professional articles [17] and tested his "Shrouded Propeller" on a *Stipa Caproni* airplane MM 187 built in Taliedo near Milan and flown by the pilot *Domenico Antonin* for the first time on October 7, 1932 (Fig. 3.16). Electrically heating the shroud yielded another reduction of the internal drag. The airplane was powered by a 90 kW (120 hp) De Havilland Gipsy III piston engine. The airplane was destroyed during further test flights at Montecelio. The Bavarian inventor *Gustav Koch* [18] born in 1843, had already made a similar suggestion in 1911 with his "turbine flying machine." The propulsion was provided by a turbine-type propeller. It is, however, not obvious how the drive of this turbine within the gondola was to be accomplished (Fig. 3.17).

3.2.4.3 EXPERIMENTAL AIRCRAFT CAPRONI CAMPINI N.1.   Italian engineer *Secundo Campini*, an employee of the Caproni Company, started in 1929 his first investigations concerning the use of jet propulsion for aircraft. In 1932, he built a boat with jet propulsion for the Italian Ministry of Aeronautics. Thereafter he developed a passenger aircraft [17] whose single-stage fan was driven by a piston-type aircraft engine (Fig. 3.18). In the case of the motorjet

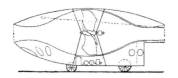

Fig. 3.17 *Koch's* "Turbine-flying-machine" of 1906 [Illust. Aeronaut. Mitteilungen X (1906)] [18].

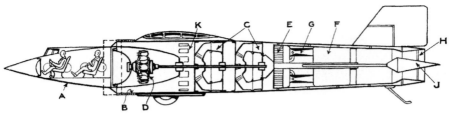

Sectional view of the original Campini design of high-altitude aircraft for operation at either sub or super-sonic speeds. A, Ovoid cabin ; B, Enshrouding cylinder ; C, Two-stage centrifugal compressor ; D, Radial engine ; E, Rectifier radiator ; F, Combustion space ; G, Annular mixing channel ; H, Discharge nozzle ; J, Cone for varying nozzle orifice ; K. Controlled lateral orifices.

Fig. 3.18 *Campini* **airplane project with hybrid propulsion (***Flight***, December 4, 1941).**

propulsion, the propeller is also included within the air-duct fairing. For that reason, it was constructed with a considerably smaller diameter than the nonshrouded propeller. To include and compress an adequate amount of air, the average speed of the air and the rpm of the propeller must be much higher. The airflow is divided inside a motorjet engine. One part constitutes the air required by the piston engine. The second part of the air bypasses the engine and exhausts via a thrust nozzle into the surroundings.

A contract with *Gianni Caproni De Taliedo* (Fig. 3.19) led in 1936 to the construction of an airframe with new hybrid propulsion with a thrust of 6.86 kN (700 kp). Two specimens of the high-speed airplane called the Campini Caproni N.1 (Fig. 3.20) made of duralumin were built in Italy within the next years. The two-stroke piston engine, Isotta Fraschini I.F.L.121 RC.40, served as a propulsion unit. The engine with a performance of 660 kW drove a three-stage fan. The overall performance was bad. It attempted to increase

Fig. 3.19 Aeronautical pioneer *Gianni Caproni de Taliedo* **(1886–1957) [17].**

Fig. 3.20   Caproni-Campini N.1 with Isotta Faschini engine [17].

the thrust by injecting and combusting additional fuel in the thrust nozzles. The first two-seater airplane, MM.487 with 14.6-m span, a length of 12.66 m, and a takeoff weight of 4440 kg made its first 10-minute flight at Forlanini (Milan) on August 27, 1940, with *Mario De Bernardi* at the controls. *De Bernardi* flew on November 30, 1941, accompanied by engineer *Giovanni Pedace*, to Guidonia near Rome. During the trials at the Italian Aeronautical Research Center in Guidonia, speeds of up to 330 km/h and with the afterburner employed, of up to 375 km/h were reached within a time period of eight months. However, the normal cruise speed was only 208 km/h. The second airplane N.1 MM.488 flew in Taliedo for the first time on April 30, 1941. This airplane survived WWII. It is now exhibited in the Italian Aeronautical Museum in Vigna di Valle.

The combination piston engine and ducted fan in the form of the piston-driven ducted fan system did not yet prove to be the suitable propulsion concept for high-speed flight. The tests of *Campini* and *Caproni* did not lead to any success as propulsive unit. The efficiency was very bad. The U.S. NACA [19] considered an "Experimental Jet Propulsion Aircraft" according to the principle of *Caproni* in the years 1939 to 1942. A Pratt & Whitney R-1830 piston engine was used for a two-stage compressor.

3.3   TURBOJET ENGINE TECHNOLOGY IS READY

The piston-driven ducted fan propulsion systems theoretically investigated and partly tested had shown that the systems using single- or multistage fans were already very strongly determined by the progress made with regard to turbojet engines. It was, therefore, no longer a large step to completely change from the piston engines equipped with turbochargers to the continuously working turbojet engines. The basic concepts of the turbojet engine in the form of the gas turbine since the first gas turbine patent of *John Barber* (1791), *Carl Gustav Patrick de Laval's* first thermic turboengine (steam turbine), and the turbocharger innovations of *Alfred Buechi*, *Auguste Rateau* (1916), and *Stanford Moss* prior to 1918 were widely very well known. The progress and granting of patents concerning the turbojet engine associated with *René Lorin*, *Francois Marconnet*, and *Charles-Edouard Guillaume* (1921) proceeded in parallel.

The theoretical fluid mechanic preparatory work of *Aurel Stodola* at the ETH-Zuerich (1925) had then, beginning in England with the computations of *Alan Arnold Griffith* and *Frank Whittle's* first patent (Fig. 3.6) of 1930, soon led also in Germany to the launch and then to the breakthrough of turbojet propulsion. *Hans Joachim Papst von Ohain* and *Herbert Wagner* were the first ones in Germany who dealt with turbojet engines as a project where the intermittent combustion of the piston engine was replaced by continuous combustion (with higher fuel consumption). In a very short time, promising projects were able to be studied in Germany at BMW, Daimler-Benz, Heinkel, and Junkers, later also with the support of the RLM. The decision concerning the most useful concept evolving from the suggestions worked out theoretically, such as a radial or axial design, was authoritatively influenced by the RLM headed by *Wolfram Eisenlohr* and *Helmut Schelp*.

A stroke of luck for the industry were the valuable contributions of the Deutsch Forschungsanstalt für Luftfahrt (German Aeronautical Research Establishments) already in existence, especially the ones of the AVA [14] that had, at that time, been addressed under the participation of *Ludwig Prandtl*, *Albert Betz*, and *Walter Encke* [20] (Fig. 3.21). *Encke* was a scientific employee at the AVA Göttingen since 1927. He owned a number of patents concerning the aerodynamic design of blade sections (profiles). At the AVA,

Fig. 3.21 *Walter Encke* (1888–1982), employee of the AVA in Göttingen up to 1945 (AVA Archive Göttingen, DLR-GOAR).

Fig. 3.22   Test bed at the AVA in Göttingen. Here, *P. Ruden* and *D. Küchemann* carried out intake tests; see Sec. 2.1.2 (courtesy of AVA Archive Göttingen, DLR-GOAR: PS44-66).

high-speed blade sections for compressors of superchargers, which had an efficiency of 70% and more, were designed and tested on AVA test beds (Fig. 3.22). The central workshop of the AVA in Göttingen had considerable experience in the mechanical treatment and particularly in the milling of compressor blades (Fig. 3.23). *Ferdinand Asmus* was the head of the workshop. The results of the DFL were summarized after the end of the war in extensive documentation for the WWII victors [21, 22].

Based on the suggestions of *Helmut Schelp* and *Hans A. Mauch,* the RLM prepared in 1938–1939 a development plan for turbojet engines that were to be used on fighter aircraft [23]. It scheduled six individual projects during a first development phase: Heinkel He S 8 (109-001), Weinrich BMW 002 (109-002), BMW 003 (109-003), Jumo 004 (109-004), Heinkel He S 30 (109-006), and Daimler-Benz DB 670 (109-007).

Sufficient progress was achieved up to 1941 for the RLM to be able in 1942 to carry out an adjustment of the turbojet development planning. It was decided to develop only engines 003 and 004 up to production readiness and to stop the development of all other jet engines. The companies involved usually did not follow the instructions of the RLM and continued the development of the turbojet projects actually terminated. First experience with the turbojet engines 003 and 004 during the test phase showed that the thrust of the engines was too low. Much higher thrust was also needed for larger fighter aircraft and most of all for bombers.

Fig. 3.23   Mechanics workshop of the AVA for the mechanical machining of turboma-chinery blades (courtesy of AVA Archive Göttingen, DLR-GOAR: NN43/FI 193).

The RLM drew up a second development program concerning higher-performance turbojet and turboprop engines in 1942. This time, engines for bombers as well as for long-range aircraft were intended. These were the turbojet engine projects Heinkel He S 011 (109-011), thrust: 12.7 kN (1300 kp), Jumo 012 (109-012) thrust: 26.4 kN (2780 kp), BMW 018 (109-018) thrust: 33.3 kN (3400 kp), and the turboprop projects Daimler-Benz/Heinkel He S 021 (109-021) performance: 1765 kW (2465 hp), Junkers 022 (109-022) performance: 3380 kW (4536 hp), and BMW 028 (109-028) performance: 5565 kW (7456 hp). Because of the progress of the war, none of these engines of the second RLM development program reached the test bed prior to the end of the war. The industrial preparatory work and thereafter the turbojet developments ordered by the RLM are discussed in the next section.

3.4 FIRST TURBOJET ENGINES ARE BUILT AND TESTED

3.4.1 HANS-JOACHIM PAPST VON OHAIN AT THE ERNST HEINKEL COMPANY

Only very few original documents are available concerning the development of turbojet [turboluftstrahl (TL)] flight engines at the Ernst Heinkel Company in Rostock-Marienehe and, from 1941 after the move of the Heinkel team, to Stuttgart-Zuffenhausen, as well as at the Hirth Motor Company after its purchase by Heinkel in April 1941. In 2000, *Volker Koos* (Rostock) took on the task of doing a chronological survey of the documents available in the special collections of the German Museum München. Included in the documents were the inventory of the Heinkel Archives, the Avicentra file collection of *Wolfgang Wagner*, and *Heinkel* assets from the National Air and Space Museum (NASM), Washington, D.C. The following presentation concerning the Heinkel turbojet development in Rostock under the supervision of *Hans-Joachim Pabst von Ohain* and later in Zuffenhausen relies very strongly on the work of *Volker Koos* [24] and an earlier publication of *Wilhelm Gundermann* [25].

After the first tests with the "Device" (Fig. 3.24) built by *Max Hahn*, on March 3, 1936 *von Ohain's* Ph.D. advisor, *Robert Pohl* of the University of Göttingen wrote a letter of recommendation to *Ernst Heinkel* in Rostock. On March 18 and April 2, 1936, after his move to Rostock *von Ohain* (Fig. 3.25) introduced his project and a reduced engine design to *Heinkel*. The preparations for the functional tests and feasibility studies within the Heinkel group "Special Development II," newly established at Heinkel, already started in April 1936. The tests with the first experimental device in Rostock lasted until December 1936 and were then abandoned. Another experimental device with hydrogen gas fuel operated successfully on the test bed in Rostock at the beginning of March 1937 (probably an F 2, VK). The construction of an engine for first flight tests was planned at Heinkel in early May 1937.

From June 1937, the group of *von Ohain* and *Gundermann* worked independently and no longer as part of the Heinkel Experimental Department

Fig. 3.24 Mechanic *Max Hahn* in his mechanics workshop in front of the first "Device-Turbojet Engine" built with *von Ohain* in Göttingen in 1935 (courtesy of special collection of the German Museum, München).

Fig. 3.25 *Hans-Joachim Pabst von Ohain*
(1911–1998, photograph taken in 1985).

headed by *Kurt Matthaeus*. In May 1937, *von Ohain* introduced "Project 2,"
a turbojet engine that was to serve as the basis for the development of an
airworthy turbojet engine. The RLM in Berlin was included in the procurement of material, and cooperation with the DVL Engine Department in
Berlin-Adlershof commenced in December 1937 at the latest. An experimental engine was shown to representatives of the RLM at Heinkel in Rostock in
May 1938. Static tests with a new engine operating on liquid fuel were
proceeding at that time. During the same month, the construction of the F 3 b
engine, which was already planned as a flight engine, commenced. Design
work on the Engine Projects F 4 (according to proposals of the DVL with
higher turbine inlet temperatures and turbine blade cooling) and F 5 (device
with an additional axial front fan to increase thrust and efficiency as compared
to the F 3 b) and an axial front engine was being carried out at the Heinkel
"Special Development II" during the summer of 1938.

First conversations with *Walter Encke* of the AVA Göttingen concerning the
development of a new engine with a combined axial/radial compressor started
in July 1938. During the same month, the first design of the Heinkel experimental aircraft He 178 by *Walter Guenter* was presented, for which a mock-up
was completed in August of 1938 by the Heinkel "Special Development I."
The RLM ordered the AVA on July 26, 1938, to carry out aerodynamic measurements on the model of the He 178 airframe. On July 29, 1938, the AVA
wrote to the RLM that the test results would be sent within the next days to

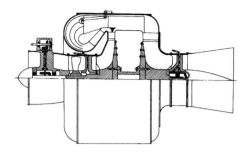

Fig. 3.26 **Sectional view of the Heinkel F 3 b, later He S 3 b.**

Heinrich Hertel respectively *Walter Kuenzel* at Heinkel in Rostock (AVA Report 38/31 of November 12, 1938). Model tests were also carried out at the DVL in Adlershof at the same time as in Göttingen. The inspection of the mock-up took place on August 29, 1938. In December of 1938, *Ernst Heinkel* asked *Ernst Udet* for the foreign release of the newly assigned Heinkel turbojet patents. In January 1939, the He 178 V-1 was to be ready at the Heinkel "Special Development I." *Gottfried Reidenbach* and *Hans M. Antz* of the RLM in Berlin inspected the F 3 b engine and the mock-up of the He 178 in Rostock-Marienehe on January 20, 1939. The F 3 b engine was essentially ready at the beginning of March 1939. An improved engine of simpler design and without an axial fan was to be built for the second experimental aircraft, the He 178 V-2, (presumably the F 6, VK). The design of a further experimental engine with an axial fan commenced in March 1939.

For the further development of the He 178, a new axial engine was also to be designed (presumably the F 9, VK). The completion of the He S 3 b engine (Fig. 3.26) and the He 178 were expected for the middle of June 1939 such that taxiing tests with the He 178 V-1 could start immediately. In May 1939, the flight tests were still planned for July 1939 in Oranienburg. Preliminary meetings concerning the further development of the experimental aircraft He 180 with two turbojet engines under the wings were scheduled for June/ July 1939. In early August 1939, the He S 8 was specified as an alternative solution for the He S 9 (axial engine) for the twin-engine fighter He 180, later the He 280. The F 8 (He S 8) engine was a further development of the F 6, which most likely was to be tested by the end of September 1939. The introduction of the abbreviation He S for the Heinkel jet engines, instead of the letter "F" used at Heinkel previously as designation of the engines, occurred in August 1939. The letter "S" stood for "special (jet) engines" as all turbojet engines were then officially being designated by the RLM.

The maiden flight of the Heinkel He 178 V-1 in Rostock-Marienehe with the He S 3 b of 4.41 kN (450 kp) thrust as engine (Fig. 3.27) occurred on August 27, 1939, early in the morning. In early August, it was expected that the first test runs of the axial engine He S 9 would be at the end of 1939. The inspection of the mock-up of the twin-engine fighter, now called Heinkel He

Fig. 3.27   Heinkel He 178 V-1 with the Heinkel engine He S 3 b. First flown on August 27, 1939, in Rostock-Marienehe—the world's first flight of an aircraft with a turbojet engine.

280, took place on September 26, 1939. Arrangements between the RLM and *Ernst Heinkel* concerning the cost sharing for previous engine and aircraft developments were carried out in October/November 1939. Here, also a demonstration flight of the He 178 V-1 was carried out for *Ernst Udet*. The second aircraft prototype of the He 178 V-2 was scrapped before the final completion, and a renewed test-bed trial was started with the engine F 3 b. The He S 6 engine was also tested in November 1939. Further in-flight tests of the He S 3 b and the He S 6 were carried out under a Heinkel He 111 and a Heinkel He 118. Figure 3.28 shows the Heinkel He 280 V-3 with two He S 8A engines, still without cowling, as propulsion after a crash landing in Rostock-Marienehe due to a turbine-blade failure.

After the takeover in 1941 of the Magdeburg Junkers-Engineering Group, headed by *Max Adolf Müller*, which worked on an axial engine, designated He S 30, it seems that the axial engine of the *von Ohain* group, the He S 9, was no longer pursued. However, because the He S 30 did not become ready as fast as expected, the *von Ohain* group worked in parallel on the He S 8A engine of a reduced diameter for the Heinkel He 280. The testing of this engine under a Heinkel He 111 is confirmed (Fig. 3.29). Altogether 27 units for the He 280 were to be manufactured.

Further details of the early engine tests at Heinkel remain unknown. This also holds for the assignment and designation of the first experimental jet engines. However, there must have been an F-2 engine because it was mentioned according to *Koos* once in a Heinkel document. The first successful

Fig. 3.28   Trial of the He S 8A on the Heinkel He 280 V-3 with pilot *Fritz Schaefer*; crash landing due to a turbine-blade fracture immediately after takeoff in Rostock-Marienehe (collection of *Walter Zuerl*).

Fig. 3.29   Trial of the Heinkel He S 8A on a Heinkel He 111 (collection of *Bruno Lange*).

test-bed runs with hydrogen combustion occurred in the spring of 1937, which confirms the recollection of *von Ohain*, while the statement "September 1937" (in *Heinkel's* memoirs) is wrong. The assertion, used again and again, that the jet-engine development at Heinkel up to the maiden flight of the He 178 V-1 on July 20, 1939, was carried out without the knowledge and consent of the RLM is not correct. The RLM was informed as of the end of 1937 at the latest. The development was, however, carried out until the autumn of 1939 at *Heinkel's* own financial risk. Compensation by the RLM was granted only thereafter. The flight testing of the He S 3 a under a He 118 prior to the maiden flight of the He 178 V-1 described in the literature is, up to now, not documented. This holds similarly for the installation of the He S 6 on the second experimental specimen He 178 V-2 that, according to documents, was scrapped prior to completion.

Further development of the jet engines after the beginning of the war at Heinkel in Stuttgart-Zuffenhausen suffered from the lack of suitable test facilities and the lack of systematic investigations because special test facilities were missing or could only be set up provisionally. *Heinkel* always tried to achieve success by shifting dates forward keeping the deployment of means and personnel as low as possible. At the same time, abilities were wasted due to the fact that too many different turbojet projects and the most different designs were being investigated at an insufficient commitment of staff and resources. The necessity of a sensible preparation of the manufacturing process for the turbojet engines was ignored, and the abilities related to the production of the jet engines in series were nonexistent at Heinkel and also not achieved by the acquisition of the Hirth Motor Works in Zuffenhausen in 1942 and the employment of *Harald Wolff* of BMW/BRAMO. Personnel disputes, particularly with or due to *Max Adolf Müller*, were added. *Heinkel* adhered to his old principle of the "Development Company," which brought the "first success," but was later not feasible under the conditions of war and the necessity to manufacture sufficient test and production-type series.

Only individual examples of the many Heinkel jet-engine projects were manufactured. Only for the type He S 11 with 12.7 kN (1300 kp) thrust and equipped with an axial/radial compressor have a larger number of engines been built. The last eight engines of the pre-series A0 were completed on U.S.

	1. Model	Project 2	F 3 b	F 3 b
Diameter mm	1500	970	930	930
Nozzle diameter mm		475	355	340
Mass kg		about 250	330	
Thrust kp		900–1000	500	
Length mm		about 1390		
Frontal area m²	1.77 (est.)	0.74	0.68 (est.)	0.68 (est.)
Specific thrust kp m⁻²		1216–1351 (est.)	735 (est.)	
Remarks		1. draft of Flight engine		
Source	Minutes of March, 18, 1936	Project of May 21, 1937	Report of June 15, 1938	Drawing of June 9, 1938

	"Device" (Unit) *Ohain*	He S 3 b	He S 3 b	F4	F5
Diameter mm	930	930 (est.)	930	630	780
Nozzle diameter mm				300	300
Mass kg	500	360	360	280	280
Thrust kp	500	500 (est.)	500	500	600
RPM min⁻¹		13,000	13,000		
Length mm					
Frontal area m²	0.68 (est.)	0.68	0.68	0.31 (est.)	0.48 (est.)
Specific thrust kp m⁻²	725 (est.)	735	735	1613 (est.)	1250 (est.)
Remarks	F 3 b		Sum of most probable data		
Source	Letter of June 30, 1938	Report of 1940		Information of June 15, 1938	Information of June 15, 1938

	He S 6	Axial proj. *Ohain*	He S 8	He S 8A
Diameter mm	920 (est.)	400	780 (est.)	770
Nozzle diameter mm				
Mass kg	420	280	380	450
Thrust kp	600 (est.)	500	670 (est.)	600
RPM min⁻¹	13,300		13,500	
Length mm				1500
Frontal area m²	0.66	0.79 (est.)	0.48	0.47
Specific thrust kp m⁻²	910	633 (est.)	1400	1277
Remarks				
Source	Report 1940	Information of June 15, 1938	Report 1940	*Ohain* Oct. 20, 1942

Fig. 3.30   Data compilation concerning the Heinkel TL engines according to investigations of *Volker Koos* (2000) (courtesy of special collection of the German Museum).

Army orders after the occupation of Germany in 1945 with the aid of BMW in Kolbermoor in Upper Bavaria, transferred to the United States, and tested on a test bed of the U.S. Navy in Trenton. Figure 3.30 shows in tabular form the jet-engine projects of Heinkel with the most important technical data as established by *Volker Koos* [24]. The information differs partly very strongly from the one given in [5]. According to the files, much is contrary to post-war statements of *Wilhelm Gundermann* [25] and *Hans von Ohain* [26].

3.4.2   HERBERT WAGNER'S TEAM DEVELOPS TURBOJET ENGINES AT JUNKERS IN MAGDEBURG

Hugo Junkers, at that time still professor for thermodynamics at the TH Aachen (Aix-la-Chapelle), founded in 1913 in Magdeburg the Junkers Motor Construction Company Gesellschaft mit beschränkter Haftung (GmbH) (limited liability company). On June 15, 1936, the Junkers Aircraft Corporation Aktiengesellschaft (AG) (Corporation) and the Junkers Motor Construction GmbH (Magdeburg) then merged into the Junkers Flugzeug- und Motorenwerke (JFM) (Junkers Aircraft and Engine Works AG) located in Dessau. At the Magdeburg location, also called Motorzweigwerk Magdeburg (MZM) (Engine Factory Branch Magdeburg), the investigation of new propulsion concepts was started in the autumn of 1935 under the supervision of *Herbert Wagner* (Fig. 3.31) and *Max Adolf Müller*.

Fig. 3.31 *Herbert Wagner* (1900–1982) [28].

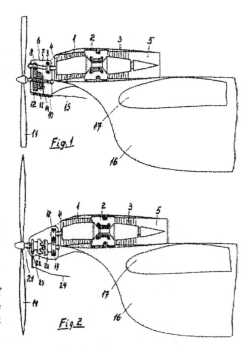

Fig. 3.32 *Herbert Wagner's* **proposal of 1935 of a shaft–performance engine (turboprop) according to British patent 495469 [28].**

Herbert Wagner [27] born on May 22, 1900, in Graz, was from 1930 to 1938 full professor at one of the two Chairs of Aircraft Construction at the TH Berlin-Charlottenburg. *Max Adolf Müller* was a chief assistant at this chair. *Wagner* had already dealt before with turboprop engines (Propeller-Turbinen-Luftstrahl-Antrieb) (PTL) (propeller-turbine-jet-propulsion) at the Offene Handelsgesellschaft (OHG) (Open Business Corporation) founded in 1934 in Berlin. A patent was filed here (Fig. 3.32) concerning a shaft-turbo engine with propeller (British patent 495469). These investigations were then, as of 1935, continued in Magdeburg at the Junkers MZM.

Herbert Wagner managed to convince Junkers general manager *Heinrich Koppenberg* of his new ideas of turbojet propulsion. Thus, according to *Rudolf Friedrich* [27], on April 1, 1936, at the Junkers MZM, later named Design Office II Kobü II, independent of the Junkers Piston Engine Development in Dessau, a group of three young engineers started to develop turbojet engines based on *Wagner's* suggestions. The closest collaborators were *Max Adolf Müller*, *Rudolf Friedrich*, and *Hans Stabernack*. As in the case of the DVL in Adlershof and BRAMO in Berlin-Spandau, piston-driven ducted fan engines based on *Smith* [9] were investigated in Magdeburg as the German Patent Document (Fig. 3.33) shows. A four-stage axial compressor was driven by a cooled two-stroke piston engine with 16 cylinders arranged in X-form. Because a basic piston engine was not available at Junkers, it was decided at a meeting between the RLM and the JFM on April 4, 1939, to

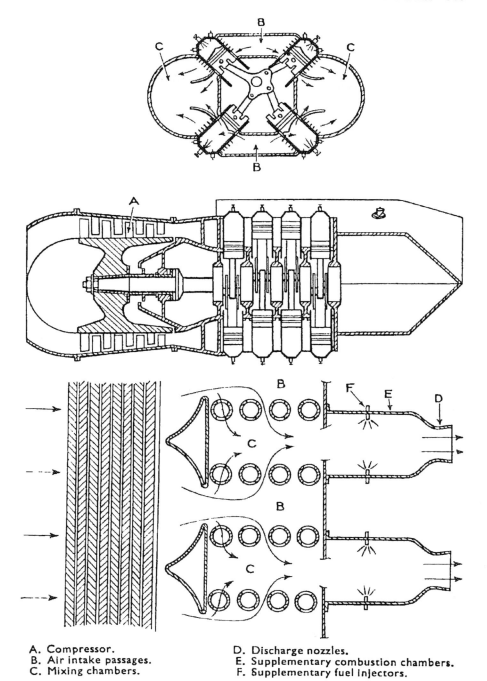

A. Compressor.
B. Air intake passages.
C. Mixing chambers.
D. Discharge nozzles.
E. Supplementary combustion chambers.
F. Supplementary fuel injectors.

Fig. 3.33   Proposal for a piston-driven ducted fan engine unit with a 16-cylinder two-stroke piston engine [9].

Fig. 3.34 "Magdeburg" experimental unit; RT 0 on a test bed at Junkers in Magdeburg, spring 1939 [28].

provide the 24-cylinder radial engine under development at Junkers, the Jumo 222, for the planned ducted fan engine.

The main attention of *Herbert Wagner* was, however, focused on the pure turbojet engine. The first experimental project was designated Rückstoss-Turbine (RT 0) (recoil or thrust turbine zero). It was an axial turbojet engine with a 14-stage axial compressor, an annular combustion chamber, and a two-stage axial turbine. During layout and design, emphasis was placed on the multiple-stage axial compressor with a reaction degree of ½ and the combustion chamber. The aerodynamic lifting-wing theory, as applied by *Curt Keller* in 1934 in his thesis about single-stage axial fans [27] at the Eidgenössische Technische Hochschule (ETH) (Swiss Institute of Technology, Zurich) supervised by *Jacob Ackeret*, was available in 1936 for the aerodynamic design of the compressor blades. The measurements at the AVA Göttingen on numerous airfoil profiles and the corresponding polar representations were another important design basis. The design considerations of *Rudolf Friedrich* regarding compressors and turbines are clearly presented in the documentation about the work of *Wagner* [28].

The experimental device RT 0 (Fig. 3.34) was installed in early 1939 on an engine test bed at the Magdeburg factory, initially for the use of propane gas as fuel. The development work at Magdeburg was stopped, at short notice and surprisingly, on orders of the RLM. Work was to be shifted to Dessau to

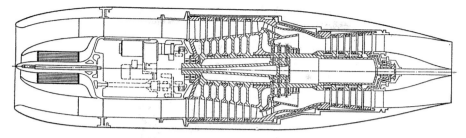

Fig. 3.35 Engine proposal of *Hellmut Weinrich* from Chemnitz of the year 1939 (courtesy of MTU Company archive).

the Junkers Motor Plant. It was, however, disbanded. *Wagner* went back to the TH Berlin at the end of 1938. In August of 1939, the strong group of 15 at Junkers of Magdeburg, followed an offer of *Heinkel* and went via Berlin to Rostock-Marienehe. Here, the Heinkel engine He S 30 (109-006) with a thrust of 7.35 kN (750 kp) was developed under the supervision of *Max Adolf Müller* based on the preliminary development work at Magdeburg.

3.4.3 HELLMUT WEINRICH STARTS HIS TURBOJET DEVELOPMENT WITH A GAS TURBINE FOR SHIPS

At about the end of 1939, the RLM arranged a contact between BMW in München and *Hellmut Weinrich*, an inventor and business man from Chemnitz, who, on orders of the German Navy High Command, was developing a gas turbine with counter-rotating compressor-blades [29] for ships. His investigations had already been running for some time, and he had some test results. Facilities for testing a small experimental engine were available in both Chemnitz and Kiel. Based on the same design thrust as the BMW P3302, namely 6 kN, the engine P3304, a counter-rotating axial turbojet engine, later designated 109-002, was designed in cooperation with BMW (Fig. 3.35). The considerable involvement of the entire BMW development department in this work to be carried out parallel to the in-house engine development led to capacity problems with regard to the BMW project P3302.

On the other hand, the ideas and the experience of *Weinrich* were also of importance to the work on this project. Already when testing the combustion chamber, the special weaknesses of the counter-rotating compressor design became obvious. Considerable casing deformations due to local overheating led to imbalances and hence to a rough rotational movement. It was particularly difficult to avoid deformations of the outer compressor casing where the counter-rotating blades were supported. Shrink joints, which were used to fasten the blades in outer rings, loosened during the engine operation. After some larger damage occurred during test runs and following insurmountable difficulties, work on the *Weinrich* engine was terminated in 1942. All development capacities at BMW in Spandau were concentrated on the turbojet project 109-003.

3.4.4 BMW AND BRAMO USE THEIR TURBOCHARGER EXPERIENCE FOR THE BMW 109-003

The modern piston engine factory of Siemens & Halske in Berlin-Spandau was in 1938 renamed Brandenburgische Motorenwerke GmBH (BRAMO) (Brandenburg Motor Company). In 1938, BRAMO started with first investigations to increase the maximum flight speed of aircraft above the limits set by propellers. Very soon, several possibilities arose from these studies, carried out by *Hermann Oestrich* (Fig. 3.36) who had already dealt with jet propulsion at the DVL in 1928, and *Hermann Hagen* (Fig. 3.37) regarding the increase of the flight speed. Also here, the piston-driven ducted fan concept as well as the turbojet concept seemed suitable to be further pursued. To demonstrate such a propulsion unit, an engine was built at BRAMO in 1938. The propulsion unit consisted of a seven-cylinder radial engine, BRAMO 325, which drove an axial four-blade shrouded fan disc that was covered by a cowling similar to a NACA cowling. This propulsion system was installed on a Focke-Wulf Fw 44 J "Stieglitz (Goldfinch)" (Fig. 3.38) [29]. The flight tests

Fig. 3.36 *Hermann Oestrich* **(1903–1973), head of the turbojet development at BMW.**

Fig. 3.37 *Hermann Hagen* **(1911–2000), the closest collaborator of** *Hermann Oestrich.*

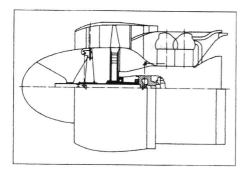

Fig. 3.38 **Piston-driven ducted fan concept for the Focke-Wulf Fw 44 J with a nine-cylinder BRAMO 325 engine.**

in Berlin-Schoenefeld in October 1938 proceeded promisingly and without disturbances. The pilot was *Hanna Reitsch* [30]. In experiments at the AVA Göttingen with a 14-cylinder twin-radial-engine BRAMO 329, respectively engine BMW 139, with motorjet propulsion the results turned out to be less favorable, and the AVA advised against further tests without having found an explanation for the heavy thrust losses that occurred [31]. As a very important by-product, BRAMO gained first experience concerning the design and the operational behavior of single axial compressor stages.

Parallel to the work on the piston-driven ducted fan project, investigations on turbojet engines (Fig. 3.39) were being performed at BRAMO in Berlin and BMW in München. It was relatively simple to establish designs for a radial-compressor engine [32]. BMW could fall back on earlier experience in the construction of aircraft-engine turbochargers and exhaust-gas turbines of radial design. Radial-wheel models with pressure ratios of 2.5:1 to 3.0:1 were available, which showed efficiencies of up to 81%. The large engine mass and the large diameter associated with this type of design remained, however, unsatisfactory.

Uncertainty also existed for a long time concerning the form of the turbines. When, in the course of 1938, newer research of the AVA concerning highly loaded axial compressors became known, BRAMO contacted this institution. Using the knowledge found there, several axial-compressor engine projects were investigated. Comparisons with the radial-compressor design showed clear advantages in diameter and the engine mass. After announcing the preparatory work to the RLM in Berlin, performance data for a military high-speed aircraft were specified. A takeoff thrust of 6 kN, a maximum engine diameter of 600 mm, and an engine mass of 600 kg were initially demanded. These were the same technical requirements that were given to the Junkers Company in parallel.

After the 1939 merger of BRAMO and BMW, already described, all aircraft-engine developments being carried out in Berlin-Spandau were coordinated with the development supervision in München and compared to the project work running in München. After the merger the initial BRAMO engine project, now

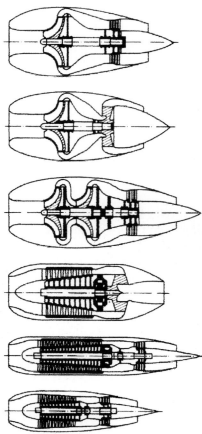

Fig. 3.39 Different radial and axial turbojet engine designs of BMW at Spandau.

designated BMW P3302, was compared to the engine project with a radial compressor being designed at that time in München (Fig. 3.39). After the first fairly useful experimental results for the axial compressor were available, *Hermann Oestrich*, head of the turbojet development, succeeded with the axial-compressor engine project [33]. The development of the turbojet engines was concentrated under *Oestrich* at the Spandau factory in September 1939. Support was subsequently given by the Piston Engine Development Branch München with regard to the design of turbines. Numerous test results concerning turbochargers were available there from the piston-engine development. BMW in München primarily possessed experience in manufacturing hollow welded turbine blades made of highly heat-resistant sheet metal and their attachment at the rotor [34].

There also existed sufficient test results for cooled turbine blades of exhaust-gas turbochargers. There was initially complete ignorance concerning the construction of individual engine components such as compressor, combustion chamber, turbine, nozzle, engine control, starter, etc. Only the location of the design point of the compressor within the working lines was fairly clearly determined. A complete design was worked out by the design department based on the project design. The construction of ten experimental turbojet engines (designation P3302 V1 to V10) was prepared (Fig. 3.40). For a number of components, possible future changes were already included in the plans. Individual tests with components commenced at the same time. Especially with regard to the design of the combustion chamber, there existed

Fig. 3.40   BMW project P3302 V1 to V10 of 1940.

little clarity. A radial-compressor test bed was set up as fast as possible at Berlin-Spandau to supply air for two individual combustion chamber test beds. The combustion chamber with its crash plate and the nozzle impacting upon it caused, as expected, considerable difficulties. The pressure losses in the combustion chamber and in the gas channels were substantially larger than assumed. The demonstration of the gas-turbine principle was, however, successful. The design of the experimental engines BMW P3302 V1 to V10 was carried out in 1940; the manufacture of the components was also completed in 1940. First trial runs started in Berlin-Spandau on February 20, 1941. After it turned out that the combustion-chamber pressure losses were appreciably higher than assumed in the design computations and the mass-flow rate was insufficient for the required thrust, the mass-flow rate was raised by means of an additional compressor stage. The pressure ratio with now seven compressor stages was increased from 2.8:1 to 3.16:1.

In the course of 1941, a static thrust of 4.5 kN was successfully achieved with the experimental engine. Already within the same year, an engine, the BMW P3302, was installed under the fuselage of a Messerschmitt Bf 110, in addition to the two Daimler-Benz piston engines DB 605 B, and tested in Berlin-Schoenefeld. Also in 1941, two of these piston engines were installed on the test aircraft Messerschmitt Me 262 V-1 in addition to the piston aero-engine Jumo 210 and on March 25, 1942, flown for the first time by the test pilot *Fritz Wendel*. The test-bed investigations with modified turbojet engines were continued in 1942 after various changes, primarily to the combustion chamber, had been carried out. In further tests of the first turbojet engine series, a thrust of 5.5 kN was reached in the summer of 1942.

It was recognized at BMW that the design thrust of 6 kN could hardly be reached, and the thrust demands of the airframe manufacturer (Messerschmitt) had in the meantime been increased; hence, a new layout and design was started and designated as BMW P3302, V11 to V14. The basic concept was, however, kept. Four experimental engines, V11 to V14, were planned. The trials commenced at the end of 1942. The seven-stage compressor of this design had a mass-flow rate increased by 30% compared to the experimental engines V1 to V10. Its efficiency corresponded approximately to that of the so-called "Göttingen" axial compressor of the AVA. However, the behavior during engine startup was very unfavorable. It necessitated the introduction of a variable (thrust) nozzle, the so-called "mushroom nozzle," which, when fully opened, facilitated the startup of the engine. Because of the good results of the tests with the experimental engines P3302, V11–V14, at the beginning of 1943, which soon delivered the required performances and came close to the specific fuel consumption specified, the production of a Series Zero was started. These first production engines received the RLM designation by which they later became known: BMW 109-003A-0. The maiden flight of a Zero-series engine was carried out during October 1943 in Berlin-Schoenefeld

Fig. 3.41   Maiden flight of a 109-003 A-0 engine below a Messerschmitt Me 110 as test bed during October 1943 (courtesy of MTU Company archive).

with an Me 110 as test bed, that is, a flying test bed where the engine was mounted under the fuselage (Fig. 3.41).

The endurance properties of the turbine blades caused the greatest difficulties at the beginning of 1944. Because of the shortage of alloys, it was necessary to introduce hollow sheet-metal turbine blades. But also other improvements, such as the introduction of an axial-roller bearing instead of a plane bearing and combustion chamber changes, were required to endure initially 20 and then 50 hours of operation. The one-hundredth "Zero-series" engine was delivered in August 1944. The engines BMW 109-003 A-0 were installed on the Arado Ar 234 C (Fig. 3.42), which was employed as reconnaissance aircraft at the Western Front and later as bomber and night fighter. In September 1944, a peak altitude of 13,000 m (an unofficial world record) was reached with an Arado Ar 234 V-8 (GK-IY) equipped with two sets of BMW engines arranged in pairs. The pilot was *Josef Bispink*. For engine investigations, two Arado Ar 234 C were available to BMW and ready at Oranienburg airfield in the north of Berlin. The first flight of a BMW 109-003 E-1 on a Heinkel He 162 A-2 "People's Fighter" took place on December 6, 1944, at Vienna-Schwechat (Fig. 3.43).

Altogether about 450 jet engines 109-003 were manufactured up to April 1945 in Berlin-Spandau, Basdorf-Zuehlsdorf, and at the central factories at Nordhausen in the Harz Mountains. December 1944 was the best production month with 100 manufactured jet engines. As total production hours per engine, BMW quoted 500 hours for this series, however, without accessories.

Fig. 3.42   Arado Ar 234 V-8 with the two-by-two installation of the BMW 109-003 A-0 engines (courtesy of special collection of the German Museum, München).

Fig. 3.43   Heinkel He 162 A "Volksjaeger (People's Fighter)" with a BMW 109-003 E-1 engine on top of the fuselage (special collection of the German Museum, München).

The individual experimental engines still needed a production time of 6000 hours per engine. Another two variants of the BMW 003 were still under development at the end of the war in 1945. The first was the new jet engine 109-003 C with a new seven-stage compressor with a pressure ratio of 3.4:1 and a longitudinally divided stator casing of the compressor designed by Brown Boveri Cie (BBC) in Mannheim. It was to deliver a thrust of 7.8 kN (800 kp). All drawings had been completed, and a prototype was being manufactured. The second variant planned was designated 109-003 D. It was to meet the RLM demand of a higher thrust of 10.8 to 11.3 kN. Here, a completely new layout was planned with an eight-stage compressor with contoured guide blades conceived by Brueckner-Kanis (Dresden) and a two-stage turbine. The engine had, nevertheless, the same main dimensions and mass as the 109-003 A. The mass-flow rate was once more to be increased by about 30%. At the end of WWII, the calculation and the design work were not yet completed. No test-bed running took place.

3.4.5   ANSELM FRANZ IS FIRST IN BRINGING THE JUMO 004 TO PRODUCTION READINESS

The Junkers Engine-Construction Company received in July 1939 an order from the RLM to develop a turbojet engine of a thrust of 5.8 kN (600 kp). It was intended for military high-speed aircraft such as the Messerschmitt Me 262. Under the supervision of *Anselm Franz* (Fig. 3.44), the development of an axial-jet engine commenced immediately at Dessau. The experience of the Junkers team of *Herbert Wagner* in Magdeburg was not used. The reason was not so much the wish of the RLM as the personal animosities of the people involved such as, for instance, *Anselm Franz.*

The sequence and the results of this development up to 1945 are described in detail in the book Aircraft Engines and Jet Engines [5]. Contrary to the engine components of *Hans von Ohain* and *Frank Whittle*, who used radial compressors, the Jumo 004 like the BMW 003 had a purely axial design which, for the high flight speeds desired, was more favorable due to the smaller frontal area (Fig. 3.45). At the beginning, the team of *Anselm Franz*

Fig. 3.44 *Anselm Franz* (1900–1994), head of Junkers turbojet development in Dessau from 1938–1945; after WWII he was head of AVCO Lycoming Aeroengine Group in Stratford, Connecticut until 1968.

had about 30 to 40 employees, three years later there were over 500. The choice of the axial compressor was, as in the case of BMW, determined by the good test results achieved with a six-stage experimental compressor with a 0.75-kg/sec mass-flow developed together with the AVA in Göttingen during 1934–1935. An eight-stage axial compressor (Fig. 3.46), based on the design and construction suggested by *Walter Encke* of the AVA, was then used as compressor. The combustion chamber was subdivided into six individual flame tubes, equipped with spin inserts to provide a good air/fuel mixing. The single-stage axial turbine, initially not cooled, was based on a design of the steam-turbine department of the company Allgemeine Elektrizitäts Gesellschaft (AEG) (General Electric Co.). The exit cross-section of the nozzle could be altered, as in the case of BMW, by means of a longitudinally adjustable mushroom nozzle.

The first Jumo 004 engine made its first run in Dessau on October 11, 1940. After some design improvements, a design thrust of 5.8 kN (600 kp) was reached on August 6, 1941. The flight testing started on March 15, 1942, with a Messerschmitt Me 110 modified as a flying test bed. At the German Air Force E-center in Rechlin, a Ju 88 A-4 was added as flight-test aircraft (Fig. 3.47). The first flight of a Messerschmitt Me 262 V-3 with two Jumo 004 A engines took place at Leipheim, Bavaria on July 18, 1942. The pilot was *Fritz Wendel*. The Me 262 had been designed without considering the knowledge regarding the advantages of the swept-wing effect. The (low) wing sweep was rather the result of a center-of-gravity shift caused by the installation of the heavy Junkers engines. Further developments of the Me 262, planned at that time, were, however, to be equipped with swept wings of up

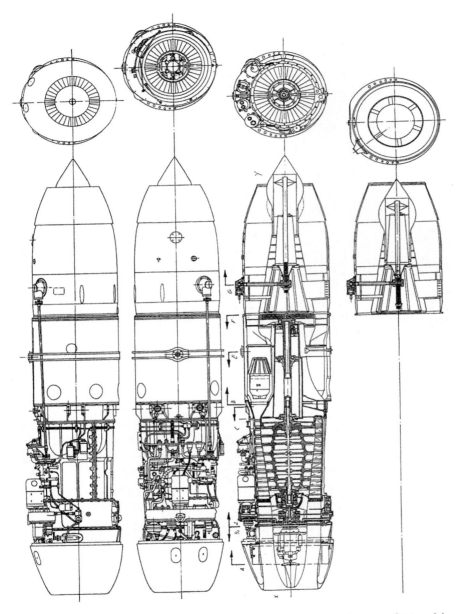

Fig. 3.45   Junkers-engine Jumo 004 B—first turbojet engine to be manufactured in a series.

to 45-deg sweep [35]. In Chapter 6, the Me 262 with swept wings will be treated in detail. The first higher performance Jumo 004 B jet engine with a thrust of 8.9 kN (900 kp) ran on the test bed in Dessau in June 1943 [36]. Engine endurance runs of 50 hours were achieved in 1943.

Fig. 3.46 Compressor blades for the Jumo 004 manufactured at the AVA, 1941 (courtesy of AVA Archive Göttingen, DLR-GOAR).

Fig. 3.47 Ju 88 A-4 as "flying test bed" for the Jumo 004 at the E-center, Erprobungsstelle (flight test center) in Rechlin. The flight testing of the TL-engines took place with the aid of several Ju 88 aircraft. The engines, with inlet cowling, but without the rest of the lining, were attached on a special support below the wing of the test-bed aircraft (Rechlin reports).

The first production jet engines of the Jumo 004 B were delivered to Messerschmitt from February 1944 (Fig. 3.48). The first Me 262 operational aircraft were delivered to the German Air Force (Luftwaffe) in April 1944. Figure 3.49 gives a survey of the state of the 109-004-development in form of the Minutes of *Anselm Franz* of August 6, 1943. The jet engines were manufactured at the factories in Dessau, Magdeburg, Muldenstein, Koethen, Leipzig, Zittau, and Prague. Altogether 6010 Jumo 004 B-1 and B-2 were manufactured between February 1944 and March 1945. The Jumo 004 was also used on long-range reconnaissance aircraft and the bomber Arado Ar 234 of which about 200 aircraft, also partly equipped with BMW 003 engines, were delivered. The heavy bomber Ju 287 V-1 (Fig. 3.50) was also under development at Junkers in Dessau; it was tested with four Jumo 004 B-1

Fig. 3.48 Single-seat fighter Messerschmitt Me 262 with Jumo 004 B-1, engine nacelle open.

a)

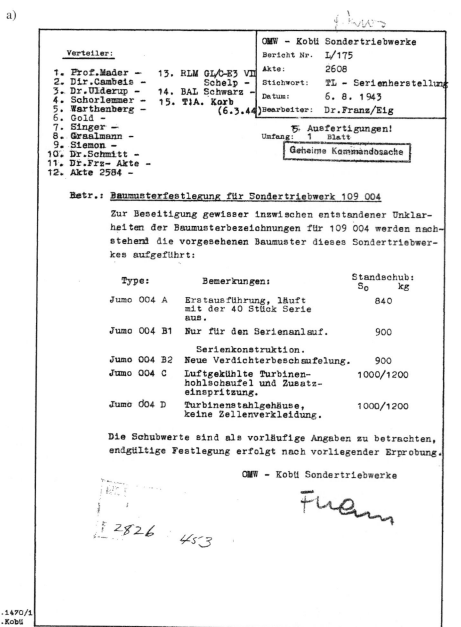

```
                                        OMW - Kobü  Sondertriebwerke
        Verteiler:                      Bericht Nr.   L/175
                                        Akte:         2608
 1. Prof.Mader –      13. RLM GL/C-E3 VII
 2. Dir.Cambeis –         Schelp –      Stichwort:    TL - Serienherstellung
 3. Dr.Ulderup –      14. BAL Schwarz –
 4. Schorlemmer –     15. TlA. Korb     Datum:        6. 8. 1943
 5. Warthenberg –         (6.3.44)      Bearbeiter:   Dr.Franz/Eig
 6. Gold –
 7. Singer –                               5. Ausfertigungen!
 8. Graalmann –                          Umfang:  1  Blatt
 9. Siemon –
10. Dr.Schmitt –                           Geheime Kommandosache
11. Dr.Frz- Akte –
12. Akte 2584 –
```

Betr.: __Baumusterfestlegung für Sondertriebwerk 109 004__

   Zur Beseitigung gewisser inzwischen entstandener Unklar-
heiten der Baumusterbezeichnungen für 109 004 werden nach-
stehend die vorgesehenen Baumuster dieses Sondertriebwer-
kes aufgeführt:

Type:	Bemerkungen:	Standschub: S_0 kg
Jumo 004 A	Erstausführung, läuft mit der 40 Stück Serie aus.	840
Jumo 004 B1	Nur für den Serienanlauf.	900
Jumo 004 B2	Serienkonstruktion. Neue Verdichterbeschaufelung.	900
Jumo 004 C	Luftgekühlte Turbinen- hohlschaufel und Zusatz- einspritzung.	1000/1200
Jumo 004 D	Turbinenstahlgehäuse, keine Zellenverkleidung.	1000/1200

   Die Schubwerte sind als vorläufige Angaben zu betrachten,
endgültige Festlegung erfolgt nach vorliegender Erprobung.

                              OMW - Kobü Sondertriebwerke

                                        Franz

```
  2826     453
```

```
.1470/1
.Kobü
```

Fig. 3.49 **a) Junkers Report No. L/175 of August 6, 1943. This report of** *Anselm Franz* **gives information concerning envisaged 109-004 engine designs (courtesy of special collection of the German Museum, München); b) translation.**

b) Distribution

OMW-Design Office Special Engines
Report No. L/175
File
Ref.: TL-series production
Date:
Responsible:
No. of pages:

Ref.: Design definition for special engine 109 004
To eliminate certain uncertainties arising in the meantime with regard to the design designations for the 109 004, the envisaged designs of this special engine are listed below

Type:	Remarks:	Ground thrust: S_o kg
Jumo 004 A	First model phases out with the 40-piece series	840
Jumo 004 B-1	Only for the series start, series design	900
Jumo 004 B-2	New compressor blades	900
Jumo 004 C	Air-cooled hollow turbine blades and additional reheat	1000/1200
Jumo 004 D	Turbine steel casing no cowel lining	1000/1200

The thrust data are to be considered as preliminary; the final definition follows after trials.

OMW-Design Office Special Engines

Fig. 3.49   (Continued)

engines as V-1 (RS + RA) prototype in Brandis near Leipzig in the summer of 1944. The aircraft had forward-swept wings, a concept that was an aerodynamic revolution at that time. Also of special interest is that the Jumo 004 B-4 was already equipped with afterburner and a shortened variable nozzle. This type of jet engine, which was still under development, had, at a gas temperature of 870°C, a static thrust of 9.8 kN (1000 kp) without afterburner, and 11.8 kN (1200 kp) with afterburner.

Fig. 3.50   Junkers Ju 287 V-1 (RS + RA) with four Jumo 004 B1 engines of 8.4 kN thrust each and four rocket engines Walter HWK 109-502 with 4.9 kN (500 kp) thrust each as takeoff aids.

3.4.6 DAIMLER-BENZ DEVELOPS A TURBOFAN ENGINE

The development of turbojet engines commenced at Daimler-Benz (DB) (Daimler-Benz Corp.) in Stuttgart relatively late. *Karl Leist* (Fig. 3.51), who came from the DVL in Berlin-Adlershof to Daimler-Benz in 1939, had initially as his main task the development of exhaust-gas turbochargers for high-altitude aircraft piston engines. At that time, the DB piston engine development department in Stuttgart-Untertuerkheim was fully occupied due to the numerous variants and special-design models of the aircraft-engine family DB 601. The personal capacity for the turbojet engine development remained, therefore, low. *Leist* describes in detail the unusual features of turbofan engines [29]. Because promising development projects of single-flow jet engines of the 7.8- to 8.8-kN category were already pursued for the RLM at BMW and Junkers, Daimler-Benz received the difficult task of the development of an engine with higher thrust and considerably lower fuel consumption. The turbofan engine suggested by *Leist*, which received later the RLM designation 109-007 (Fig. 3.52) (the in-house Daimler-Benz designation was DB 670), was designed for a thrust of 6.0 kN at a speed of 900 km/h at an altitude of 6000 m [37] and was to have, at the same time, a specific fuel consumption of 36.8 g/kNs. Converted, this resulted in a maximum static thrust of 13.7 kN and a corresponding specific fuel consumption of 22.9 g/kNs.

The actual development work started in 1941; the first of three experimental engines built entered the test bed in Stuttgart-Untertuerkheim on April 1, 1943, and reached in the autumn of the same year full operational speed of 12,600 rpm. A high compressor pressure ratio and a high turbine inlet temperature had to be chosen due to the, for the time, exceptionally low fuel

Fig. 3.51 *Karl Leist* **1901–1960, since 1939 head of Daimler-Benz Special Development in Stuttgart-Untertuerkheim.**

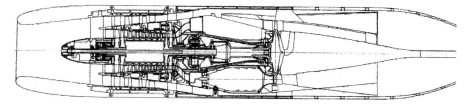

Fig. 3.52   Daimler-Benz bypass engine 109-007, in-house designation DB 670.

consumption demanded. The diameter was according to the set specifications not to exceed 900 mm and the length 5000 mm. For the required pressure ratio of 8:1, an axial compressor was, therefore, as in the case of BMW, designed based on suggestions of *Walter Encke* of the AVA Göttingen consisting of two counter-rotating bladed rotors. The production of the blade rings was also partly carried out at the AVA in Göttingen (Figs. 3.53 and 3.54). The inner cylinder had nine rows of blades, the outer cylinder on the inside had eight rows, and on the outside within the bypass another three compressor stages. The two cylinders were coupled via a planetary gear and

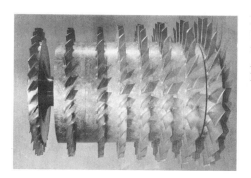

Fig. 3.53 Compressor rotor for the bypass engine DB 670; manufactured at the AVA in Göttingen (courtesy of AVA Archive Göttingen, DLR-GOAR: P41/F/537).

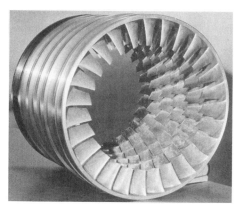

Fig. 3.54 Compressor guide-blade rings for the DB 670; manufactured at the AVA in Göttingen (courtesy of AVA Archive Göttingen, DLR-GOAR: P41/F/542).

had a maximum rotational speed of 12,600 rpm and 6200 rpm respectively. The compressor, including the counter-rotating vane rings and despite the high pressure ratio, was quite short. The static mass-flow rate of the inner flow passage was 14.8 kg/s, and the one of the outer passage was 35.8 kg/s so that a bypass ratio of 2.42:1 resulted.

A parallel compressor design was performed by the J. M. Voith Company in Heidenheim. The compressor blade rings consisted throughout the forward part of heat-resistant light metal and throughout the latter stages of steel. The combustion chamber consisted initially of four individual flame tubes; a fifth tube was added later. The turbine inlet temperature was 1373 K. The turbine blades were cooled. A partial exposure of the blades to cool air from the outer passage was chosen, which was supplied to the turbine through corresponding slots from the secondary-air annular channel. Pretests of this procedure on Daimler-Benz exhaust-gas turbine wheels had shown good results also at high gas temperatures. The development of this very complex and completely new engine concept required many tests on individual elements and components for which frequently the corresponding test facilities, such as a centrifugal test bed for turbines and a test bed for the counter-rotating gear, exposed to a performance of 2950 kW, had to be built at Daimler-Benz.

The development dragged on and was stopped at the end of 1943 by the RLM. The run time of the first experimental engine, the DB 007 V-1 (Fig. 3.55), whose first run was on April 1, 1943, amounted altogether to 152 hours at the termination of the development. The mechanical features chosen by Daimler-Benz during the engine design, such as the subdivision into three rotors with two bearings each and elastic elements arranged in between, with the application of a "floating" bearing for the engine rotor, and the separation of the load-bearing casing elements from the thermally highly stressed components, proved successful during the short trial period and provided important knowledge for future engine developments. At that time Daimler-Benz also dealt with other forms of bypass engines. The project 109-016 with compressors without counter-rotating rotors and with two one-stage turbines, and also an engine with cylindrical compressor and turbine rotors as well as cooled turbine blades, became known.

Fig. 3.55 First experimental engine DB 007 V-1 was installed on the test bed at Daimler-Benz in Untertuerkheim on April 1, 1943.

3.5   ESTABLISHMENT OF LARGE-SCALE TEST FACILITIES FOR TURBOJET ENGINES

The first high-altitude test bed setup in Germany was erected for the test of piston engines during WWI 1916–1917 by the Maybach Company on Wendelstein Mountain (upper Bavaria) [5] at an altitude of 1800 m. On the research side, a high-altitude test bed with a mass-flow rate of at first 1.2 kg/s, later of up to 3.0 kg/s, was erected in Germany at the DVL in Berlin-Adlershof in 1934 [5, 14]. Only piston engines and turbochargers were at first investigated here. In cooperation with the BBC Company in Mannheim, from 1936 a larger high-altitude test bed for piston-type engines was then established at the Test Center of the German Air Force in Rechlin, Expert Group "E3-Engines," which was used up to 1945 [38]. A high-altitude test bed was also built at the Institute of Engine Research of the LFA Voelkenrode in 1941 [21].

Experiments on piston engines started only in the middle of 1944. The first piston engine investigated in this test bed was a DB 603 of Daimler-Benz. For the investigation of turbojet engines, a test bed (Fig. 3.56) was completed in the building M1 (Fig. 3.57) of the LFA at the end of 1944. How far German turbojet engines were still investigated here could not be determined. A high-altitude test bed for flight engines was also established in 1938 at the Forschungsinstitut für Kraftfahrzeugwesen und Fahrzeugmotoren (FKFS) (Research Institute of Motor Vehicle Engines Stuttgart) in Stuttgart-Untertuerkheim headed by *Wunibald Kamm*. This test bed was to be used for large air-cooled aircraft engines of up to 3000-kW (3945-hp) performance.

Fig. 3.56   Turbojet engine test bed at the LFA in Braunschweig-Voelkenrode going into operation at the end of 1944 (courtesy of AVA-Archive Göttigen, DLR-GOAR).

Fig. 3.57 Test-bed building M1 in Branschweig-Voelkenrode where the turbojet engine test bed was housed (courtesy of AVA-Archive Göttingen, DLR-GOAR).

On orders of the RLM, a further high-altitude test bed was built at BMW in München-Milbertshofen in 1943, better equipped with a design mass-flow rate of up to 20 kg/s. Planning commenced in May 1941. The facility was again planned and built by BBC Mannheim under the supervision of *Ulrich Senger*. It carried the cover name "Herbitus" (Fig. 3.58). Although the facility was originally conceived for piston engines, it was used from mid-1944 up to the end of the war for the investigation of turbojet engines such as the BMW 003 and the Jumo 004 B. The U.S. Army occupied the BMW plant in München-Milbertshofen in May 1945. On orders of the Americans, tests with the BMW engines and the English "Derwent" engines continued until the facility was transported in mid-1946 to Tullahoma, Tennessee. In 2005, this test facility was still being repeatedly modernized, among other things equipped with DEMAG compressors still in operation. The German high-altitude test beds up to 1945 are summarized in a 1966 DVL report by *Hermann Barth* (Fig. 3.59).

Fig. 3.58 Test of a General-Electric engine in the "Herbitus" high-altitude test bed of the Arnold Engineering Development Center (AEDC) in Tullahoma, Tennessee; facility was dismantled in 1946 at BMW in München.

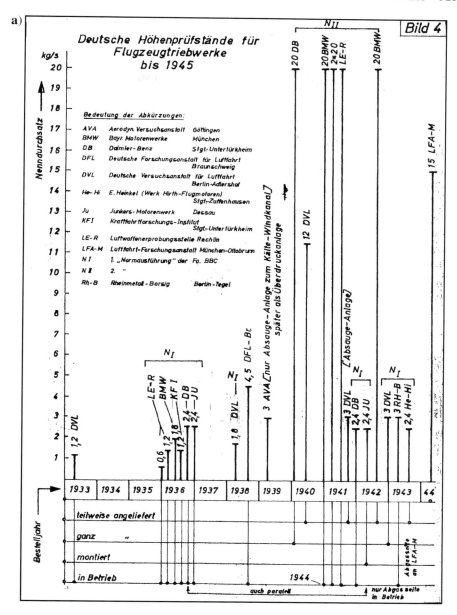

Fig. 3.59   German high-altitude test beds for aircraft engines up to 1945: a) 1966 DVL report (Historic Archive of the DLR Cologne), and b) translation.

3.6   INSTALLATION INVESTIGATIONS OF TURBOJET ENGINES

Although for the installation of single- and multipiston engines much experience was available at all German aircraft companies, the integration

b)

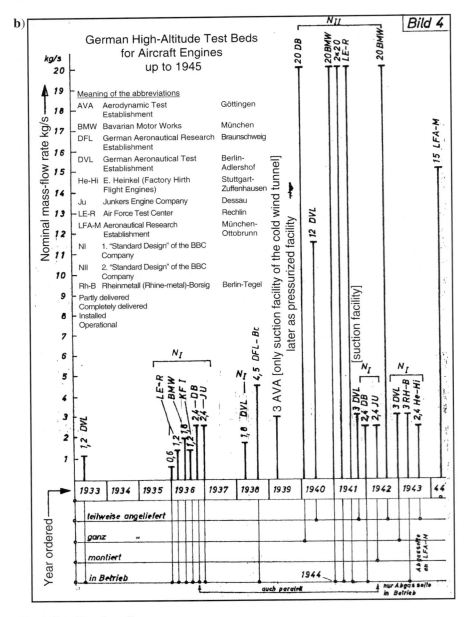

Fig. 3.59   (Continued)

and the installation of turbojet flight engines into the airframes for all aircraft companies was new.

Remedies could only be provided by extensive wind tunnel investigations. Concerning the aerodynamic arrangement of the turbojet engines, various basic installation arrangements were, in principle, possible. Firstly, the installation of the engine or engines at the wing or within the wing and, secondly, the installation of the engine or engines within or at the fuselage of the aircraft was possible. Results of extensive wind-tunnel investigations at research establishments and industry concerning the miscellaneous installation configurations are available in a variety of research reports.

In 1955, *Gerhard Schulz* [38] summarized measurements made in the Heinkel low-speed and the Heinkel high-speed wind tunnels, as well as measurements in the Messerschmitt low-speed and the Junkers high-speed wind tunnels. The measurements were divided into those with the engine nacelles arranged on the wing, the engine nacelles mounted at the fuselage, and test series with pairs and twins of nacelles. *Schulz* summarized test results according to eight main rules:

- The forward and rear positions of the nacelle shall be such that the tail of the nacelle is not located in the range between about 10% of the wing chord downstream of the wing trailing edge and 20% upstream. Supplementary information concerning the area rule can be found in Sec. 2.3.4.
- The lower nacelle position shall be as small as possible; if one is, however, for constructive reasons, forced to choose an unfavorable (nacelle) tail position, a lower nacelle position may be recommendable, but not otherwise.
- In case of a poor nacelle-tail position, a "tapered rear section" is beneficial. The latter shall not be more than 20% of the nacelle length; however, in the case of nacelle twins, this section may be extended to 40% of the nacelle length.
- Nacelles on the suction side of the wing are always worse than on the pressure side.
- The clear distance between two jet engines shall not be smaller than one to two engine diameter.
- The previously mentioned rules have to be more strictly applied at high Mach numbers.

If the nacelles are placed in the proximity of the fuselage, the following additional rules apply:

- Nacelles at the fuselage located downstream of the wing are better than upstream nacelles no matter whether above or below the wing, especially at high C_A (C_L) values. However, the flow approaching the air intakes can more easily be spoiled by boundary-layer effects if the nacelles are located downstream.
- Nacelles located upstream on the fuselage, but below the wing position must not simultaneously merge into the wing. They must have a sufficiently low position.

Many of these aerodynamic test results are still an important design basis today for the installation of jet engines. It must, however, be taken into account that the engines were, compared to today, considerably larger. The interferences in case of the flow about a wing-nacelle (engine) arrangement were hence different to today's aircraft. A low position of engines—as realized on today's commercial aircraft—could at that time not be a focal point of the investigations due to the large engine diameters and the necessary ground clearance. In the case of the American bomber project B 47 of the Boeing Company after WWII, the demanded safe distance between engine and swept wing led to a breakthrough in the drag minimization (also see Secs. 6.3 and 8.4). *Dietrich Küchemann* [39] reports primarily on model investigations for the BMW 003 and Jumo 004 engines employing various forms of the tail section and bundling of the engines with different central ribs. He also investigated fuselage forms with the engine positioned on top as in the case of the Henschel Hs 132 (Fig. 3.60) and the Heinkel He 162.

Shapes of the jet expansion and the effect of the hub in a comparison of different classes of cowling were investigated at high speeds by *Hubert Ludwig*. Also *H. Bönecke* [40] reported on measurements on a Jumo 004 with different inlet cowlings. The influence of pressure losses within the air intake as well as the influence of the hub rounding on the performance of a turbojet engine were investigated for different annular cowlings. As a basis, a cowling designed by *Paul Ruden* was employed, which was used at Junkers on the Jumo 004 and which also took the *Riedel* engine starter into account. *Jachmann* [41] showed the results of tests with the complete Jumo 004 engine with different forms of the cowling.

The results of many installation tests with motorjet and turbojet propulsion systems (Fig. 3.61) were presented by *Küchemann* in June of 1945 in a report [42] prepared at the AVA quoting many research reports. *Bäuerle* [43], *Buschner* [44], *Küchemann* [45, 46] (Fig. 3.62), *Conrad* [47], and *Bäuerle* [48, 49] reported on further engine installation and attachment investigations. Figure 3.63 shows the outlines of aerodynamic models with fuselage and nacelles for straight and swept wings. The Göttingen Monographs alone in the Archive of the AVA Göttingen contain summarizing reports on more

Fig. 3.60 Wind tunnel investigations at the AVA in Göttingen on a model of the Hs 132 with a nacelle for the BMW 003 on the upper side of the fuselage (courtesy of AVA-Archive Göttingen, DLR-GOAR: P41/F266).

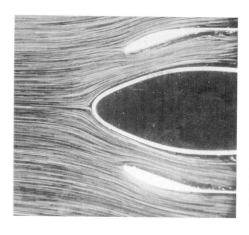

Fig. 3.61   Flow conditions on an inlet model of a turbojet engine; AVA measurements (courtesy of AVA-Archive Göttingen, DLR-GOAR: N42/A/042).

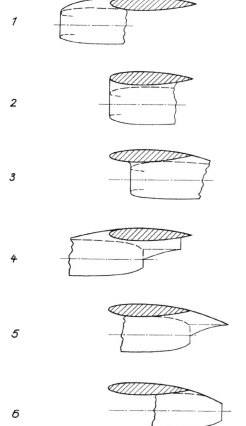

Fig. 3.62   Six basic forms of the arrangement of wing and engine installation investigated by *D. Küchemann* (courtesy of AVA-Report 44/A/42 of September 11, 1944, p. 40).

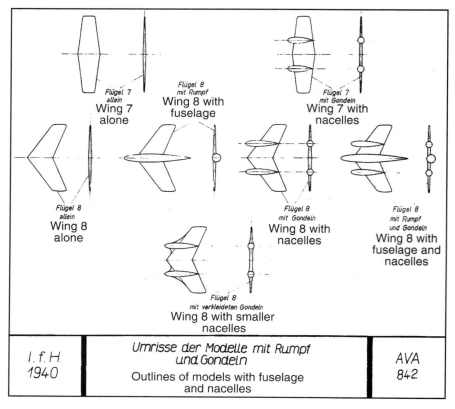

Fig. 3.63 Outlines of aerodynamic models with fuselage and nacelles for straight and swept wings; compare with the results in Fig. 2.105.

than 1000 pages concerning questions related to jet-engine integration and jet-engine inlet and tail flows. A thorough analysis of all of these results would exceed the scope of this book and not do justice to its actual subject.

3.7 TESTING OF THE TURBOJET ENGINES AT THE E-CENTER RECHLIN

The E-Center of the Air Force in Rechlin was of great importance to the testing of the new German TL-engines. In spring 1943, a special turbojet test-bed was also available in Rechlin (Fig. 3.64) in addition to the five enormous double test beds for piston engines with performance of up to 2950 kW (3940 hp). Within the Engine Department E3, a new Group E305 was created under *Johann Ruther* that dealt exclusively with the testing of the turbojet. Beginning in mid-summer 1944, the trial of the new jet aircraft types Ar 234 and Me 262 was an especially novel task for the E-Center Rechlin. The new aircraft types still had to cope during ground and flight testing with a variety of problems related to the turbojet engines BMW 003 and Jumo 004. Corresponding

Fig. 3.64 TL test bed of the E-Center Rechlin with a Jumo 004 V-11 (courtesy of special collection German Museum, München).

details are reported in "Flight-Test Centers up to 1945" by *H. Beauvais* et al. [51] and in "Turbojet History and Development" by *A. L. Kay* [52].

The test-bed trials of the TL engines commenced at Rechlin immediately after the completion of the test bed. According to information from a former employee, *Heinz Borsdorff* [51], the first tests of the Jumo 004 A V-11 occurred in February 1943. The testing of the BMW 003 started some months later. The flight testing of the turbojet engines was carried out with several Junkers Ju-88 aircraft. The engines were equipped with inlet cowling, but without any further lining and attached to a special suspension under the left wing (Fig. 3.41). The development and testing of a front test bed for the BMW 003 engines of the Heinkel He 162 (Fig. 3.65) was one of the last tasks undertaken by the E3 department of the E-Center Rechlin.

Fig. 3.65 Front test bed of the Department E3 for the test of TL engines at the E-Center Rechlin; on the test bed is the engine BMW 003 E for the Heinkel He 162 (courtesy of special collection German Museum, München).

3.8 TURBOJET ENGINE TECHNOLOGY STATUS AT THE END OF WWII

At the end of the war in 1945, three German military aircraft types with newly developed turbojet engines were in operation [30]: the Messerschmitt Me 262 (2 x Jumo 004), the Arado Ar 234 (2 x Jumo 004 or 4 x BMW 003), and the Heinkel He 162 (1 x BMW 003). The turbojet engines BMW 003 A and Jumo 004 B were in full series production; the improved version of the Jumo, the 004 E, was largely tested and ready for series production. The engine Heinkel He S 011 A was being tested and was to go into production in May 1945. The production effort for the turbojet engines was very low due to the effective production techniques of the manufacturers. However, due to the war the German turbojet flight engines had prematurely entered large-scale production. In many respects, therefore, the failure rate was correspondingly high. The Messerschmitt Me 262 reached a maximum flight speed of 870 km/h at an altitude of 8000 m. The relatively high fuel consumption can be explained by the low compression ratio and the low compressor efficiencies of the turbojet engine series. However, individual compressors with much better performances were already under development at the engine companies.

Prior to the end of the war in 1945, the multiple-jet swept-wing bomber Junkers Ju 287 was the most interesting aircraft project not yet in series production. It had a negative wing sweep. In the autumn of 1942, *Hertel*, head of development at Junkers, requested *Hans Wocke* to establish, through studies and wind tunnel tests, a scientific basis for high-speed flight with multijet large aircraft [50]. Up to the end of the war, only two experimental aircraft of the type Ju 287 were completed. As propulsion for the series-production aircraft, six BMW 003 A-1 or Jumo 004 B-1 engines were envisaged. Four Walter HWK 502 rocket engines of 4.9 kN (500 kp) thrust each, mounted under the actual engines, were used in addition to shorten the takeoff of the prototype aircraft. The maiden flight of the V-1 took place in Brandis on August 16, 1944. The pilot was *Siegfried Holzbauer*. The first flight tests with only two Jumo 004 engines proceeded successfully.

Chapter 5 addresses the flight trials of the Ju 287 V-1. *Wolfgang Wagner* reports in detail on further important German aircraft equipped with turbojet engines [30]. The Henschel Hs 132, which at the end of the war was equipped with a BMW 003 A-1 jet engine, belongs to this category. Also the Horten all-wing fighter Ho IX V-2 was equipped with two Jumo 004 B-1 engines. Another aircraft with swept wings, this time adjustable (variable sweep) and with turbojet propulsion, was the Messerschmitt experimental aircraft P 1101. The aircraft was designed to be driven by a Jumo 004 C and a He S 011. The P 1101 was a midwing installation aircraft with a 40-deg swept-wing. The aircraft did not make its maiden flight before the end of WWII. In 1945, it was taken to the United States as a study object and served at the Bell Company as the model for the development of the experimental aircraft Bell X-5 (see Chapter 6).

REFERENCES

[1] G. Wissmann, "Geschichte der Luftfahrt von Ikarus bis zur Gegenwart," Eine Darstellung der Entwicklung des Fluggedankens und der Luftfahrttechnik, 5 Auflage Berlin, Verlag Technik, 1979.

[2] F. C. W. Käsmann, "Weltrekordflugzeuge," R. Oldenbourg, München, 1989.

[3] F. C. W. Käsmann, "Die schnellsten Jets der Welt," Weltrekord-Flugzeuge, AVIATIC Verlag, Planegg, 1994.

[4] W. Voight "Flugwerk und Triebwerk," Bericht über die Sondertagung der Lilienthal-Gesellschaft für Luftfahrtforschung am 23. und 24. August 1938 in München. LGL-Bericht 100, 1938.

[5] K. von Gersdorff, K. Grasmann, H. Schubert, "Flugmotoren und Strahltriebwerke," Die deutsche Luftfahrt, Band 2. 3. Aufl., Bernard & Graefe Verlag, Bonn, 1995.

[6] O. Fuchs, W. von Gronau, R. W. Schulz (eds.), "Starten und Fliegen, Band II," Deutsche Verlags-Anstalt, Stuttgart, 1957.

[7] H. Schwencke (ed.), "Strahltriebwerke des Auslandes," Sonderdruck. Bericht des Technischen Amtes des RLM GL/C-Nr. 100545/44 geh. (Rue/F), March 15, 1944.

[8] W. Gohlke, "Heißluftstrahltriebwerke," 4 Teile: Flugsport XXXI (1939) 1, pp. 1–5, Flugsport XXXI (1939) 2, pp. 31–37, Flugsport XXXI (1939) 3, Seite 70–75, Flugsport XXXI (1939) 4, pp. 100–104.

[9] G. G. Smith, "Gas Turbines and Jet Propulsion for Aircraft, 3rd Ed.," Aerosphere, New York, U.S., 1947.

[10] J. Dressel, M. Griehl, "Die deutschen Raketenflugzeuge 1935–1945," Die Entwicklung einer umwälzenden Technik. Motorbuch Verlag, Stuttgart, 1989.

[11] E. Kruska, "Das Walter-Verfahren, ein Verfahren zur Gewinnung von Antriebsenergie," VDI-Z 97 (1955) No. 3, pp. 65–70, 97 (1955) No. 9, pp. 271–277, 97 (1955) No. 21, pp. 709–713, 97 (1955) No. 24, pp. 823–829.

[12] E. Kruska, "Raketenentwicklung bei der Firma Hellmuth Walter, Kiel," DGLR Vortrag 82- 055, 1982.

[13] V. Koos, "Ernst Heinkel Flugzeugwerke 1933–1945," Typenbücher der Deutschen Luftfahrt, Heel Verlag, Königswinter, 2003.

[14] E. H. Hirschel, H. Prem, G. Madelung, "Luftfahrtforschung in Deutschland," Reihe: Die Deutsche Luftfahrt, Band 30, Bernard & Graefe Verlag, Bonn, 2001.

[15] H. G. Stine, "The Prowling Mind of Henri Coanda," Flying 80 (1967) 3, pp. 64–68.

[16] L. E. Opdycke, "French Aeroplanes before the Great War," Schiffer Publishing, Atglen, PA, U.S., 1999.

[17] R. Abate, G. Alegi, A. Giorgio, "Aeroplani Caproni: Gianni Caproni and his Aircraft 1910–1983," Museo Caproni, Trento, 1982.

[18] "25 Jahre Geschichte des Berliner Vereins für Luftschifffahrt Teil 7," Die technische Kommission Illust. Aeronaut. Mitteilungen X (1906), H.10, pp. 345, 346.

[19] J. R. Hansen, "Engineer in Charge: A History of the Langley Aeronautical Laboratory 1917–1958," The NASA History Series, NASA, Washington, DC, U.S., 1987.

[20] A. Betz, Axiallader. Vortrag von der Lilienthal-Gesellschaft für Luftfahrtforschung am 28 Feb. 1938 in Berlin," Jahrbuch 1938 der deutschen Luftfahrtforschung Teil II, pp. 183–186.

[21] J. J. Green, R. D. Hiscocks, D. M. Holman, J. L. Orr, "Die deutsche Luftfahrtforschung im Zweiten Weltkrieg," Bericht einer kanadischen Kommission aus dem Jahr 1945 (Translation), DFVLR Historisches Archiv, Bonn, 1983.

[22] E. Schmidt, "Geschichte des Institutes für Motorenforschung der LFA 1941–1945 (mit Verzeichnis von Veröffentlichungen und Berichten des Institutes)," Teilbericht der Göttinger Monographien, Göttingen 1945.

[23] H. Schelp, "Zur Geschichte des Strahltriebwerks—Die Entwicklung bis 1945," DGLR-Jahrbuch 1981, Bd. IV. DGLR, Köln, 1981, S. 073-1 bis 073-17.

[24] V. Koos, "Verlauf der frühen TL-Entwicklung und Bau He 178 nach vorhandenen Dokumenten im DM München," Chronologische Aufstellung vom May 25, 2000, 24 Seiten; In der Sondersammlung des Deutschen Museums München.

[25] W. Gundermann, "Zur Geschichte des Strahltriebwerks—Die Entwicklungen bei Heinkel von 1936 bis 1939," DGLR-Vortrag Nr. 81-074, 1981. DGLR-Jahrestagung 12. bis 14. Mai 1981, in Aachen.

[26] H. Schubert (Redaktion), "50 Jahre Turbostrahlflug (50 Years of Jet-Powered Flight)," DGLR-Symposium, Oct. 26–27, 1989, München, DGLR-Bericht 92-05 Band I und II. DGLR, Bonn, 1992.

[27] C. Keller, "Axialgebläse vom Standpunkt der Tragflügeltheorie," Mitteilungen aus dem Institut der Aerodynamik an der ETH, No. 2, Zürich, 1934.

[28] W. Heinzerling, G. E. Knausenberger, M. Osietzki (Redaktion), H. Wagner, "Dokumentation zu Leben und Werk," Bonn-Bad Godesberg: Deutsche Gesellschaft für Luft- und Raumfahrt, 1984.

[29] A. Bäumker (ed.), "Strahltriebwerke," Schriften der Deutschen Akademie der Luftfahrtforschung,"Deutsche Akademie der Luftfahrtforschung, Berlin,1941.

[30] W. Wagner, "Die ersten Strahlflugzeuge der Welt," Die Deutsche Luftfahrt, Band 14, Koblenz, Bernard & Graefe Verlag 1989.

[31] Frerichs, Koch, Kiel, "Auslegung eines ML-Geraetes für Motor 139 und Untersuchung des Modells im Göttinger Windkanal," BMW-Flugmotorenbau, Entwicklungswerk Spandau, Bericht III, H-Nr. 38 vom, June 27, 1938.

[32] H. Schubert, "Erinnerungen. 1934–1999: Flugtriebwerkbau in München," 3 Erweiterte Auflage, AVIATIC Verlag, Oberhaching, 1999.

[33] H. Oestrich, "Die Entwicklung der Fluggasturbine bei den Bayerischen Motorenwerken während des Krieges 1939–1945," Flugwelt (1950) No. 4, pp. 69–71.

[34] A. Müller, "Der BMW-Abgasturbolader," LGL-Bericht Nr. 172 der Lilienthal-Gesellschaft, 1943, pp. 75–96.

[35] W. Radinger, W. Schick, "Messerschmitt Geheimprojekte," 3 Auflage, Aviatic Verlag, Oberhaching, 2004.

[36] A. Franz, "Zur Geschichte des Strahltriebwerks," DGLR-Vertrag No. 076-1 1981, DGLR-Jahrestagung Aachen 12, May 14, 1981.

[37] K. Leist, H. G. Wiening, "Enzyklopädische Abhandlung über ausgefährte Strahltriebwerke," Forschungsbericht des Landes Nordrhein-Westfalen No. 1128, Westdeutscher Verlag, Köln, 1963.

[38] G. Schulz, "Aerodynamische Regeln für den Einbau von Strahltriebwerksgondeln," ZFW 3 (1955) H. 5, pp. 119–129.

[39] D. Küchemann, "Bericht über das Göttinger Versuchsprogramm zum Einbau von TL-Triebwerken," ZWB U + M/Re/3125, 1944.

[40] H. Bönecke, "Prüfstand an einem Jumo-TL mit verschiedenen Einlaufhauben," ZWB U + M 3154, 1944.

[41] Jachmann, "Messungen am Junkers TL B1-176 mit verschiedenen Hauben," Bericht E-Stelle Rechlin, Nr. 684/44, 1944.

[42] D. Küchemann, "Triebwerksaerodynamik," AVA Bericht 45/A/10, June 25, 1945, In Bericht im AVA-Archiv Göttingen.

[43] C. Bäuerle, "Der Anbau von TL-Triebwerken an den Tragflügel," 3 Teilbericht, ZWB U + M 3158, 1944.

[44] R. Buschner, "Druckverteilungsmessungen an einem Pfeilflügel mit einem Triebwerk," 4 Teilbericht, ZWB U + M 3176, 1944.

[45] D. Küchemann, "A. Scherer, Zur Frage des Triebwerkeinbaus bei Ein-TL-Jägern," ZWB U + M 3205, 1945.

[46] D. Küchemann, "Der Anbau von TL-Triebwerken an den Tragflügel," 2 Teilbericht: Druckverteilungsmessungen an einem Rechteckflügel mit einem Triebwerk, AVA-Bericht 44/A/42, 1944.

[47] O. Conrad, "Übergänge zwischen Flügel und Triebwerk," Göttingen Monographien, Abschnitt K3.2.4, S. 146–206, Göttingen, 1945.

[48] C. Bäuerle, "Untersuchungen an dem Modell eines Strahltriebwerkes im Windkanal," 1 Teilbericht: 3-Komponenten-Messungen an einer Strahlgondel, AVA-Bericht 44/W/17, 1944.

[49] H. Bäuerle, "Widerstands- und Schubänderung bei nachträglichem Anbau eines TL-Triebwerkes unter dem Rumpf der He 219," ZWB U + M 3041, 1943.

[50] W. Wagner, "Hugo Junkers, Pionier der Luftfahrt—seine Flugzeuge," Die deutsche Luftfahrt, Band 24, Bernard & Graefe Verlag, Bonn, 1996.

[51] H. Beauvais, K. Kössler, M. Mayeri, C. Regel, "Flugerprobungsstellen bis 1945," Die deutsche Luftfahrt, Band 27, Bernard & Graefe Verlag, Bonn, 1998.

[52] A. L. Kay, "Turbojet History and Development 1930–1960," Vol. 1, Great Britain and Germany, The Crowood Press, Ramsbury, U.K. 2007.

Chapter 4

AEROELASTICITY PROBLEMS IN COMPRESSIBLE SUBSONIC AND TRANSONIC FLOW

HANS FÖRSCHING

4.1 INTRODUCTION

An aircraft is an extremely lightweight construction and thus relatively flexible. Therefore, it necessarily deforms appreciably under load. Such static as well as time-dependent elastic deformations (natural vibrations)—especially of lift-generating and control surfaces—change the distribution of the aerodynamic load, which in turn change the deformations of the elastic aircraft structure. This interacting "aeroelastic" feedback process is the cause of many flight-related phenomena with far-reaching consequences. They significantly affect the flight-mechanical behavior of the aircraft and the stress of the aircraft structure. Under certain conditions, such aeroelastic interactions may lead statically and dynamically to structural instabilities with destructive events—the most dangerous is flutter, a self-excited oscillation of the lifting surfaces mostly together with the airframe. Extensive investigations must be carried out in the design stage of a new aircraft development based on specifications to avoid such disastrous failures and to prove the aeroelastic safety of the aircraft.

Aeroelasticity has influenced the evolution of aircraft since the earliest days of flight, gaining increased importance with rapidly progressing technical development and flight speed. New aeroelastic problems were constantly added and confronted the aircraft engineers with growing scientific-technical challenges. This was also the situation when reaching the sound barrier and entering the transonic speed range, initially with unswept aircraft. Pilots repeatedly reported newly occurring serious flight mechanical problems. The flight stability was noticeably reduced, the control surfaces became increasingly ineffective up to the reversal of the aileron effect, the wings and the controls started vibrating intensely, and the entire aircraft was shaken by shocks. One was rather helplessly confronted by these phenomena, and the fundamental question arose as to the effect of the compressibility of the flow—and after the introduction of the swept-wing concept also as to the effect of wing sweep—on the aeroelastic behavior in near-sonic and transonic flow. It became a certainty that mastering these problems and understanding the prevailing physical relations were essential prerequisites for the further increase of the flight speed up into the supersonic area. Basic and guiding development work was performed in Germany up to the end of WWII from 1935 to 1945, that is, at the threshold of a new age in aeronautical technology, as shown in this chapter.

To fundamentally understand the aeroelastic behavior of swept-wing high-speed aircraft, the typical elasto-mechanical properties and design features of the swept wing are first described. Then, as a consequence, the important effect of wing sweep on the "static" aeroelastic stability (bending-torsion divergence), on the control-surface effectiveness, and on the lift distribution on the elastic wing is shown. Subject of further considerations is then the "dynamic" aeroelastic behavior of unswept- and swept-wing aircraft in compressible subsonic and transonic flow with emphasis on the problem of flutter stability. The various dynamic aeroelastic problems caused by compressibility and swept-wing effects at high speeds—especially within the transonic speed range—will be outlined, and the scientific-technological progress achieved in Germany by the end of WWII to cope with these problems will be presented and discussed. The outstanding contributions to further develop the scientific foundations of aeroelasticity—today a fundamental field of knowledge in modern aeronautics—achieved towards the end of the war under increasingly more difficult conditions—will be highlighted.

4.2   DESIGN FEATURES AND ELASTO-MECHANICAL BEHAVIOR OF SWEPT WINGS

4.2.1   LARGE-ASPECT-RATIO SWEPT WINGS

As a rule, the wings of subsonic aircraft—unswept and swept—have an aspect ratio of $\Lambda > 5$ and, therefore, possess to a good approximation a typically one-dimensional beamlike elasto-mechanical behavior. Thus, the deformation of the elastic wing caused by the aerodynamic lift can be described by a vertical bending and a torsion about an "elastic axis" (idealized line of the shear centers of the wing cross sections) as reference axis in the spanwise direction. In the case of an unswept wing in steady flow, concerning aeroelastic and flight-mechanical behavior, only the torsional deformation is of importance, because it results in a change in the angle of attack and hence in the aerodynamic lift distribution on the wing. Bending is not a factor.

The large-aspect-ratio swept wing can be idealized by a rotation φ of the elastic axis of the straight wing about the fuselage attachment location (root) ($y = 0$); see Fig. 4.1. Thereby, the characteristic beamlike elasto-mechanical behavior under the effect of the aerodynamic lift remains essentially unchanged. However, aeroelastically now also the wing bending is of fundamental significance, because a change in the angle of attack occurs kinematically in the freestream direction due to the bending of the swept wing along

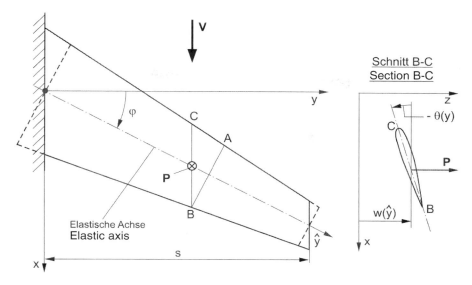

Fig. 4.1   Static bending deformation behavior of a large-aspect-ratio sweptback wing.

the elastic $\hat{y}$ axis as reference axis. This is clearly shown in Fig. 4.1. One recognizes that in the case of the swept-back wing, exposed to a discrete static lift force P on the elastic reference axis $\hat{y}$, the points A and B on a line normal to this axis have approximately the same deflection $w(\hat{y})$, while the deflection at the point C is considerably smaller. The wing segment BC of the (positively) swept-back wing in the freestream direction experiences a reduction in the angle of attack of $-\theta(y)$ by this kinematic effect. It is opposite for the forward-swept wing (negative sweep) $\theta(y) > 0$ as can easily be verified. Hence, due to the bending, for the sweptback elastic wing the local freestream angles of attack $\alpha(y)$ along the wing span become smaller, whereas for the forward-swept wing $\alpha(y)$ they increase, as is shown in Fig. 4.2. These changes in the freestream angle of attack, kinematically caused by the bending of a large-aspect-ratio swept wing, affect to a large degree its static and—incorporated in the natural vibration modes—also its dynamic aeroelastic behavior, with many important flight-technological consequences.

However, the elasto-mechanical behavior of the large-aspect-ratio swept wing is also in a certain way affected by its structural design, that is, the

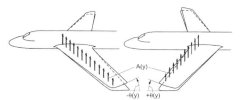

Fig. 4.2   Angle-of-attack change in the flow direction in case of a bending deformation of a forward-swept and a swept-back wing, left: sweepback φ, and right: forward sweep $-\varphi$.

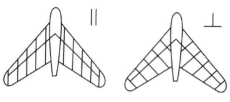

Fig. 4.3 Constructive arrangement of the wing ribs relative to the spar of a swept wing (original sketch of *P. Jordan*).

constructive arrangement of the wing ribs relative to the spar (box). Both arrangements outlined in Fig. 4.3 are applied practically. The arrangement of the ribs parallel to the freestream direction (||) is applied for aerodynamic reasons on low-aspect-ratio swept wings to minimize elastic airfoil-section (profile) cambering. The arrangement normal to the wing spar (⊥) is preferred for constructive and production-related reasons. This can, for instance, be seen in Fig. 4.4 in the wing design of the Boeing 747 as a typical representative of today's large-aspect-ratio swept-back wings of commercial aircraft [1]. Also illustrated in Fig. 4.4 are details of the constructively similar arrangements (⊥) for the swept horizontal and vertical stabilizers of the empennage and moreover, due to aeroelasticity requirements, the nonconventional design

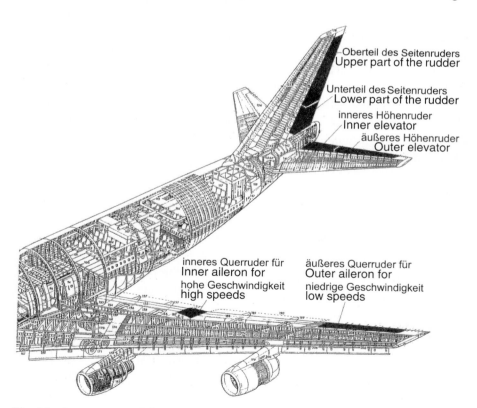

Fig. 4.4   Arrangement of the wing ribs and primary control surfaces of the Boeing 747.

and positioning of the primary control-surfaces of this large swept-wing aircraft. This will be addressed in more detail in Section 4.3.2. For swept wings of large aspect ratio, the elastic airfoil section cambering in the freestream direction, caused by bending and torsion, is according to experience negligibly small and, therefore, in general aerodynamically insignificant for the aeroelastic behavior.

All of these constructive and elasto-mechanical details are of fundamental importance to the understanding of the characteristic aeroelastic behavior of swept wings. Historically, it should be mentioned that a first detailed account of these correlations has been documented by *P. Jordan* (AVA Göttingen). He was not to publish these in a report prior to the end of the war, but did so in 1946 in his guiding contribution to the "Göttingen Monographs Concerning Progress in German Aeronautical Research (since 1939)" [2], instigated by the Allies after the end of the war. He presented a summary of the state of knowledge at that time concerning the aeroelastic behavior of swept wings and explained the corresponding theoretical foundations.

4.2.2   SWEPT LIFTING SURFACES OF LOW ASPECT RATIO

Swept empennages (stabilizers), but also the slender delta wings of supersonic aircraft, possess as a rule an aspect ratio of $\Lambda < 5$ and therefore show, at partly considerably different structural designs, a more two-dimensional plate-like elasto-mechanical behavior. The deformations in the spanwise and chordwise directions are approximately of the same order of magnitude, that is, the resulting cambering of such lifting surfaces in the flight direction now attains aerodynamic importance and can no longer be neglected. On the other hand, such swept and unswept plate-like lifting surfaces of low aspect ratio have by far a larger specific stiffness. Therefore their static aeroelastic behavior, also in the interaction with control surfaces, is far less problematic than in the case of large-aspect-ratio wings. The dynamic aeroelastic behavior of low-aspect-ratio lifting surfaces, however, still requires full attention.

4.3   EFFECT OF WING SWEEP ON THE STATIC AEROELASTIC BEHAVIOR

Static aeroelastic problems result, by definition, from the interactions of static deformations of the elastic aircraft with the hereby induced changes of the steady aerodynamic forces. With regard to the structural reaction of the aircraft, there arises primarily the question to the static aeroelastic stability of the wing—the so-called "wing torsional divergence." The bending and torsional deformations of the swept and unswept wing, however, also lead to significant flight-mechanical effects due to changes in the lift distribution and, aeroelastically interacting with a control surface, also to a reduction of the control effectiveness. With flight speeds increasing up into the near-sonic range, first flight experience showed that the overall static aerolastic problem becomes seriously more important, as will be discussed next. Here, wing sweep plays a very important role.

4.3.1 WING TORSIONAL DIVERGENCE

Crashes due to static wing-torsion fractures had already occured in the early days of aircraft engineering and increased with the emergence of the mono-wing airplanes and with the further increase of flight speeds. The torsional failure of the wings was initially regarded as a strength problem due to a too low torsional stiffness of the wing spars—a mostly correct but insufficient conclusion. It was initially not recognized that possibly a static aeroelastic stability problem is the real cause, later called "wing torsional divergence," which was for the first time analyzed as such in 1926 by H. Reissner Technische Hochschule (TH) (today Technical University) Aachen and formulated for a single-spar beam-idealized wing [3].

The static aeroelastic divergence problem can easily be explained with the aid of an idealized two-dimensional model, Fig. 4.5. It is assumed that in case of a rotation-elastically supported (rigid) wing segment the axis of rotation is located at a distance e downstream of the aerodynamic neutral point (located in subsonic flow at about the ¼-chord wing position). Looking first at the unswept wing, at a very large spring (torsional) stiffness and a low freestream speed V, the twist of the wing towards larger angles of attack α by the aerodynamic pitching moment $Ae\cos\alpha$ about the elastic axis would be very small and practically unproblematic. However, for actual elastic wings, at large

Fig. 4.5 Rotation-elastically supported wing segment (top) and angle-of-attack change dependent on the freestream velocity (bottom).

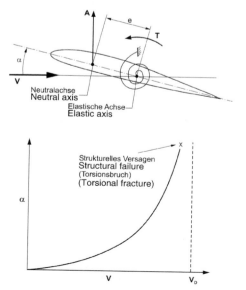

flight speeds, this twist effect can quickly lead to reaching the torsional-fracture limit of the wing because the aerodynamic moment increases with the speed V squared while the elastic restoring moment (torque) T is independent of V. The latter is, therefore, only up to a "critical" speed V_D able to maintain stable aeroelastic equilibrium. For $V > V_D$, the angle of attack α goes to infinity ($\alpha \to \infty$), and a torsional fracture will generally occur prior to that; see Fig. 4.5. The structural requirement in case of the unswept wing to avoid torsional divergence is a sufficient torsional stiffness at the maximum flight speed together with the position of the elastic axis as close as possible to the aerodynamic neutral axis ($e \to 0$).

The static aeroelastic divergence behavior in case of the swept wing is also quite substantially affected by the wing bending. As shown in Fig. 4.2, the swept-back wing experiences, in the case of a bending deflection, kinematically caused in the freestream direction, a reduction in the angle of attack. This restoring effect $-\theta(y)$, increasing rapidly with the sweep angle φ and the wing aspect ratio, has the consequence that already at a small sweepback of only about 10 deg the static aeroelastic "bending-torsion divergence" problem becomes irrelevant. Conversely the forward-swept wing achieves in case of bending an enlargement of the angle of attack in the flow direction. By this twist effect $+\theta(y)$ the static divergence problem for forward-swept wings becomes a cardinal problem. This is clearly shown in Fig. 4.6 where the

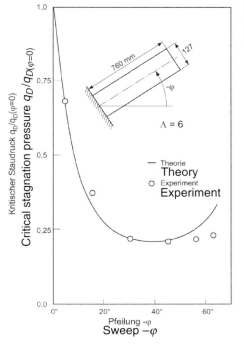

Fig. 4.6 Static bending-torsion divergence behavior of a forward-swept elastic homogeneous wing.

dramatic decrease of the critical stagnation pressure q_D with increasing forward sweep for an idealized beam-like wing becomes quite obvious.

In the case of the world's first aircraft with forward-swept wings, the Ju 287 with four jet engines, which was already projected in 1942 as prototype and flight tested for the first time on August 16, 1944, this characteristic static aeroelastic divergence behavior became a serious problem, whose importance was already fully recognized at that time. The leading-edge sweep of the Ju 287 was 23 deg (see Fig. 4.7), the wing aspect ratio was 6.9, and the flight tests showed, as reported, that at a flown *Mach* number of $M = 0.7$ "the values for the wing stiffness and the twist-up of the wing still just corresponded" [4]. With regard to further details of the Ju 287 and its further development after the end of the war by a Junkers team in the Soviet Union, reference is made to Sec. 6.3.

For forward-swept wings of conventional metal construction, the problem of the aeroelastic bending-torsion divergence is a fundamental limiting technological factor and an essential reason that high-speed forward-swept (negatively swept) aircraft with sweep angles larger than 15 deg were, to date, never developed up to series readiness. However, with the innovative application of multidirectional carbon-fiber compound constructions for a strength-optimized nonisotropic stiffness distribution ("Aeroelastic Tailoring"), new

Fig. 4.7 Ju 287 V1 with four Jumo 004 B-1 jet engines.

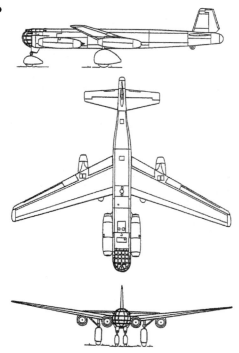

promising constructive possibilities to reduce the kinematic twist effect have risen. They were, for the first time, advantageously applied in the United States (1984) in the case of the forward-swept wing of the Grumman experimental aircraft X 29 and in Russia (1997) on the forward-swept demonstrator Suchoj S-37. For both military-oriented prototypes, the aspect ratio was relatively small ($\Lambda \approx 5$); however, they did not go into series production. Finally, it should be mentioned that the static aeroelastic divergence problem in the case of large-aspect-ratio wings is a typical subsonic phenomenon. The reason is that, after reaching a pure supersonic flow, the local aerodynamic neutral point moves to the wing-section center, that is, to the immediate vicinity of the elastic axis ($e \rightarrow 0$).

4.3.2 CONTROL-SURFACE EFFECTIVENESS

The flexibility of the wing also has an important effect on the effectiveness of control surfaces. The physical connections existing here are illustrated in Fig. 4.8 for an idealized torsion-elastically supported rigid airfoil segment with control surface. If the latter at a stable aeroelastic equilibrium of the acting moments about the axis, at an effective angle of attack α and a freestream velocity V, is deflected downwards by an angle η, an additional lift A_R is generated whose resultant is located near the control-surface axis of rotation thus generating a dynamic pitching moment M_R that rotates the torsion-elastically supported wing counterclockwise by an amount of $-\Delta\alpha$. This pitching moment increases with the square of the freestream velocity V, while the elastic restoring moment (torque) M_R remains independent of V. Thus, the control-surface effectiveness becomes increasingly smaller with increasing V, until a critical velocity V_R is finally reached where the control surface is completely ineffective. This velocity is called "control-surface reversal velocity." Here, additional lift is no longer generated, but only a change in the lift distribution on the elastically supported wing occurs. The control-surface effect turns into the opposite direction when exceeding V_R. Analogous relations exist, as can easily be verified, at an upwards deflection of the control surface. This static aeroelastic phenomenon is primarily of importance to the effectiveness of the ailerons located at the wing tip where the change in the angle of attack due to the torsional elasticity of the wing is largest.

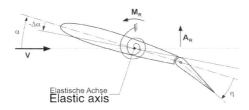

Fig. 4.8 Angle-of-attack change of a rotation-elastically supported two-dimensional wing segment caused by a control-surface (flap) deflection.

The possibility of a negative influence on the control-surface effectiveness due to the torsional elasticity of the wing was already recognized for unswept wings. But, at the still moderate flight speeds, the problem was of minor flight-technical importance, because with the specified static flight load strength requirements the torsional stiffness of the wings was sufficiently large. Typically the first scientific publications [5] on the static aeroelastic problem of control-surface effectiveness appeared only in 1943 and 1944 in England, still for unswept aircraft. These dealt especially with the efficiency of ailerons, but also with the additional effect of the bending and torsion elasticity of the fuselage on the effectiveness of the elevator and rudder controls, all in conjunction with the development of large military aircraft.

The increase in the flight speed with fighter aircraft into the near-sonic range, and most of all the introduction of the swept-wing concept then led to a drastic change of the situation. The first flight experiences in the near-sonic range showed a distinct reduction in the control-surface effectiveness, in case of ailerons often up to control reversal, and the fundamental question arose concerning the effect of the compressibility of the flow and of the wing sweep on the effectiveness of the control surfaces as a whole. The effect of the wing sweep can easily be explained. In case of the swept-back elastic wing, there arises in the freestream direction due to the downward control-surface deflection η—in addition to the angle-of-attack reduction $-\Delta\alpha$ due to the aerodynamic pitching moment M_R (see Fig. 4.8)—still another angle of attack reduction due to the wing bending, kinematically caused by the lift A_R. Thus, the control-surface effectiveness, primarily that of the ailerons, is in addition adversely affected. The transonic compressibility effects, occurring after exceeding the critical *Mach* number, have still another important effect on the control-surface effectiveness on large-aspect-ratio elastic wings. Here, the aeroelastic relations become exceptionally complicated and difficult to analyze.

In nondimensional form, the aileron effectiveness is expressed by the change in the angle of attack at the wing tip, induced by the rolling motion, $\omega_x s/V$, per unit aileron deflection η, with ω_x being the angular rolling velocity and s the wing half-span. Figure 4.9 shows as a typical example the aeroelastic reduction in the aileron effectiveness at sea level as function of the flight *Mach* number M_∞ for the wing of the Boeing XB-47, that is, the prototype of the first American large-scale swept bomber B 47 (span 2 s = 35.4 m, sweep angle of the aerodynamic neutral axis at $^1/4$-wing-chord $\varphi = 35$ deg, aspect ratio $\Lambda = 9.43$) [6]. The dramatic reduction of the aileron effectiveness at the wing tips with increasing *Mach* number and the critical speed of the aileron control reversal at $M_\infty \approx 0.8$ can be seen.

This serious swept-wing effect due to aeroelasticity with all its consequences was already well-known to the German aircraft engineers at the end of the war. However, in conjunction with the development of the first aircraft

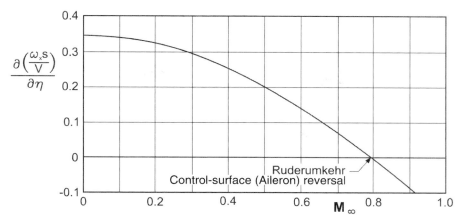

Fig. 4.9 **Aileron effectiveness dependent on the flight *Mach* number at sea level for the XB-47.**

projects with sweep angles larger than 20 deg, this was nowhere reported in detail. In particular, during the development of today's large swept-wing passenger aircraft great importance has been attached to the realization of effective roll control. For static aeroelastic reasons, unconventional design measures became necessary, namely, additional inner ailerons at the very stiff inboard wing for all flight speeds, and the use of the conventional outer ailerons at the wing tip for low flight speeds and during landing and takeoff; see Fig. 4.4. These are today, together with spoilers as secondary aerodynamic control aids, typical characteristics of large-aspect-ratio sweptback aircraft for control about the longitudinal axis. Sometimes, the outer ailerons are even completely deleted, for instance, in case of the Airbus A340. Also due to the same aeroelastic requirements, the elevators and rudder are split into inner and outer parts, as is also illustrated in Fig. 4.4.

In the case of forward-swept wings, the conditions are exactly reversed. The now positive change $+\Delta\alpha$ of the angle of attack of the elastic wing in streamwise direction, kinematically caused by the wing bending, increases with increasing forward sweep, and quickly compensates the torsion-elastic nose-down twist effect $-\Delta\alpha$ due to M_R when deflecting a control surface (flap); see Fig. 4.8. This has a positive effect on the aileron effectiveness at the wing tips and considerably improves the control effectiveness compared to the unswept wing. This was also shown by experience gained during the flight testing of the forward-swept Ju 287. It was reported [4] that the controllability about the longitudinal axis "was felt as especially pleasant," which was solely explained by the "healthier" flow at the wing tips in the case of the forward-swept wing. But certainly here also the positive effect of the nose-up twist of the forward-swept wing, resulting from the wing bending, was of at least the same importance.

4.3.3 STATIC LIFT DISTRIBUTION ON AN ELASTIC SWEPT WING

In the case of the swept elastic wing, the bending deformation also has an important effect on the static lift distribution with flight-mechanical consequences, particularly in the event of flight-load changes. Figure 4.10 shows a typical example of the lift distribution on the forward-swept wing of the XB-47 at a 3-*g* pull-out maneuver at a *Mach* number of 0.8 and a flight altitude of 27,000 ft [6]. A significant shift of the lift resultant in the direction of the wing root occurs. This leads not only to a favorable reduction of the static wing-root bending moment, but also to a shift of the steady lift resultant forward in flight direction, with repercussions on the static longitudinal flight stability behavior of the aircraft. Conversely the change in the steady lift distribution on the forward-swept wing leads to a shift in the steady lift force resultant towards the wing tips and, therefore, also forward in the direction of flight. The consequence is, as shown by the flight tests of the Ju 287, a noticeably more sensitive longitudinal flight stability that increases with increasing flight speed. In tests with rigid models carried out in various wind tunnels during the development of the Ju 287, the three-dimensional boundary-layer behavior on the forward-swept wing was also closely investigated. Contrary to the swept-back wing, the boundary layer flows on the forward-swept wing towards the inner wing and merges with the fuselage boundary layer. Together they then flow downstream along the fuselage. In this way, the ailerons at the wing tips are indeed located in a "healthier" flow, which has a positive influence on their effectiveness. It has thus been reported [4] "that the spanwise

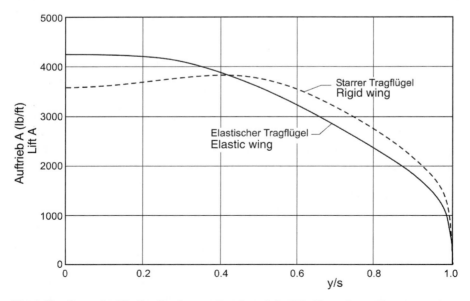

Fig. 4.10   Spanwise lift distribution on the wing of the XB-47 at a 3-*g* pull-out maneuver at a *Mach* number of 0.8 and an altitude of 27,000 ft [6].

extent for the ailerons could be reduced, so that the spanwise extent of the landing flaps could advantageously be enlarged."

4.3.4 ANALYTICAL INVESTIGATIONS

Wind-tunnel tests with elastic swept-wing models to experimentally investigate the static aeroelastic behavior of swept wings, with and without control surfaces, were probably not carried out in Germany until 1945. Whether project-oriented theoretical investigations were carried out is also unclear. Anyway, there are no records. At that time such analytical investigations with the computational aids available (mechanical electrically operated desk calculators) were only possible with a very large expenditure in time and personnel and had, presumably, no priority. However, the theoretical fundamentals, also for compressible flow up to the critical *Mach* number, were already available. The wing could be structurally modeled as an inhomogeneous beam in the form of differential and integral equations, and the steady aerodynamic forces on large-aspect-ratio swept (three-dimensional) wings could be computed by the simple strip-theory with the application of the cos φ-law [7] as well as the extended lifting-line theory of *J. Weissinger* [8] with the application of the quadrature method of *H. Multhopp* [9]. For steady flow, compressibility could simply be accounted for by the application of the *Prandtl–Glauert* transformation [10]. Applying these theoretical fundamentals, all established in Germany prior to 1945, the static aeroelastic behavior of large-aspect-ratio swept wings was then systematically researched in England and the United States with great effort after the end of the war. All relevant influence parameters were taken into consideration, corresponding analytical methods were worked out, and the results published in a number of scientific reports. In all of these, one looks in vain for a reference to the basic and guiding first findings of *P. Jordan* [2], which were certainly known to the scientists in England and the United States.

4.4 EFFECT OF COMPRESSIBILITY AND WING SWEEP ON DYNAMIC AEROELASTIC BEHAVIOR

4.4.1 INTRODUCTION

Dynamic aeroelastic problems arise from the interactions of time-dependent deformations of the elastic aircraft and the resulting induced "unsteady" aerodynamic reactions. The time-dependent deformations—externally or self-excited—occur as structural vibrations. Here, neglecting structural dampening, the following system forces are interactively involved at every instant t:

- Elastic forces
- Vibration-induced unsteady aerodynamic forces $A(t)$
- Inertial forces

Vibrations of the elastic aircraft occur, as a rule, excited by external or also system-immanent time-dependent disturbance forces, such as by gusts when flying through atmospheric turbulence, in the transonic flow regime by pressure fluctuations and shocks within the basically steady flow, or of stabilizers (the empennage) by the turbulent wakes of the wing or the jet engines. Dynamic aeroelastic stability exists when the externally excited vibrations of the aircraft, or individual components, occur with finite amplitudes and decay again—aerodynamically damped by unsteady aerodynamic forces $A(t)$ and the structural damping—after the disturbance disappears. Here, the question of the structural fatigue strength arises. In the case of a dynamically aeroelastic instability—under certain conditions during an undisturbed flight when reaching a critical speed—after a small disturbance, spontaneously self-excited vibrations of a lifting surface, and most frequently of the complete aircraft, may occur with exponentially increasing amplitudes and disastrous consequences within a few seconds. Such dangerous destructive aeroelastically self-excited wing and empennage vibrations, called "flutter," already appeared in the beginnings of aircraft technology. They accumulated with the transition to monoplanes, the application of new advanced aerodynamic and structural designs, and increasing flight speeds. The reliable forecast of the "critical flutter speed" and the proof of the flutter safety during the design stage, demanded by the design specifications, confronted the aircraft engineers with great new challenges and problems at increasing flight speeds up into the supersonic range and the introduction of the swept-wing concept.

The effect of the compressibility of the flow and the wing sweep on the dynamic aeroelastic behavior—and here especially the flutter behavior—are now discussed. The variety of the problems to be solved is shown, and the fundamental contributions to the aeroelasticity of modern high-speed aircraft acquired in Germany through 1945 are demonstrated.

4.4.2 NATURAL VIBRATION PARAMETERS OF AIRCRAFT AND VIBRATION-INDUCED UNSTEADY AERODYNAMIC FORCES

For the analytical investigations of dynamic aeroelastic problems in aeronautical engineering, most accurate knowledge of the characteristic vibration behavior of the aircraft—in particular of the lifting surfaces—as well as of the vibration-induced unsteady aerodynamic forces $A(t)$ is a basic prerequisite. In the continued efforts to improve the reliability of flutter predictions, a permanent further development of corresponding theoretical and experimental methods and procedures to determine the required parameters of the natural vibrations of the aircraft, and the unsteady vibration-induced unsteady aerodynamic pressure distributions, was an absolute necessity. The respective elementary research and development work performed in Germany towards the end of WWII under increasingly worse conditions is commented on in the following.

4.4.2.1 PARAMETERS OF NATURAL VIBRATIONS OF AIRCRAFT. A theoretical determination of the parameters of the natural vibrations of complete aircraft was not yet possible with the computational aids available up to the end of the war. Such computations could, at the most, only be performed for aircraft components, primarily for the wing and the tailplane. At two special meetings concerning wing vibrations (1938 and 1941), the existing analytical possibilities and experiences were discussed. The application of the energy method and the Galerkin procedure with assumed displacement functions to derive the equations of motion—also to formulate the flutter equations by taking into account the aerodynamic forces $A(t)$—and the solution of the resulting eigenvalue problem by determining the zero points of the characteristic equation for the calculation of the eigenfrequencies and the corresponding natural vibration mode shapes became common practice [7, 11].

In consideration of these computational limitations, the experimental determination of the required vibration parameters (natural frequencies and corresponding mode shapes), possibly for the complete aircraft, became an early attractive alternative. Since 1930 in Germany, such vibration measurements were being carried out at the DVL in Berlin-Adlershof on original aircraft, elastically suspended to simulate free-floating flight conditions, so called "shake tests." They were the beginning of today's "ground vibration tests," an indispensable tool in conjunction with the aeroelastic qualification not only of high-speed aircraft.

Ground vibration measurements on elastically suspended, later on elastically supported large aircraft, were then also carried out in Germany by industry. Thereby, systematic further developments of the testing technique were primarily being driven by *G. de Vries* at Dornier in Friedrichshafen [12]. To determine the natural vibration modes and the relating frequencies, he introduced the "phase resonance criterion," developed electrodynamic vibration exciters and frequency generators, and special measurement techniques—also still used today as standard techniques. Figure 4.11 shows the high standard of the ground-vibration test technique in 1942. The six lowest symmetrical natural vibration modes measured on the Do 17 aircraft are presented here [12]. They show that for flutter investigations not only the vibration modes of individual aircraft components (wings and empennage), but the vibration modes of the complete aircraft as individual degrees of freedom must be taken into account, at that time an unrealistic demand. The ground-vibration test technique was further successfully developed after the war by *de Vries* in France at the ONERA.

4.4.2.2 VIBRATION-INDUCED UNSTEADY AERODYNAMIC FORCES. By far the more difficult task, compared to the determination of the natural vibration mode parameters, is the theoretical determination of the vibration-induced unsteady aerodynamic forces on the oscillating lifting surfaces required for flutter

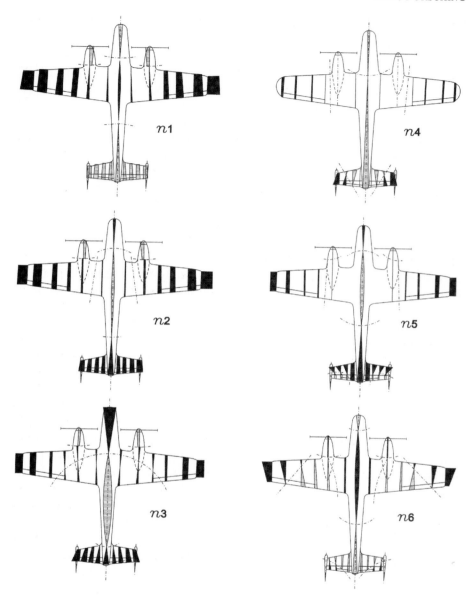

Fig. 4.11 Six lowest symmetrical natural vibration modes of the Do 17 measured in ground-vibration test by *G. de Vries*.

computations. This was and remained up to today the key problem in the effort concerning the continual further developments in aeroelasticity—beyond the compressible subsonic speed range up into the supersonic regime. Here, in unsteady aerodynamics, new analytical ways had to be followed. For compressible unsteady flow, the *Prandtl–Glauert* transformation could no longer be applied. Pioneering scientific work in this field was being performed from 1920 up to the end of the war in Germany, primarily at the AVA in Göttingen, where *Ludwig Prandtl* laid the foundation for incompressible flow with his singularity concept of time-dependent bound and free vortices on the oscillating wing [13].

In 1922 *Prandtl's* pupil, *W. Birnbaum*, published the world's first unsteady aerodynamic theory of a harmonically oscillating two-dimensional flat-plate airfoil having two degrees of freedom, heaving and pitching in inviscid incompressible flow [14]. Here, basic physical knowledge was gained, and the flutter problem was, for the first time, proven and understood to be an aerolastic feedback stability problem. In 1929, *H. G. Küssner* published an extended theory of the two-dimensional flat-plate airfoil, oscillating harmonically in inviscid incompressible flow, adding as a third degree of freedom an oscillating control surface (flap) [15]. This theory was later (1940) extended by *H. G. Küssner* and *L. Schwarz* for the harmonically oscillating two-dimensional airfoil, taking into account an aerodynamically balanced control surface with an auxiliary control flap as fourth degree of freedom [16]. In all of these theories, the consequent application of *Prandtl's* ingenious singularity concept of bound and free vortices was the physical basis.

With his "Theory of the Aircraft Wing in a Compressible Medium," published in 1936, *Prandtl* then laid—virtually as a culmination of his impressive scientific work—the foundation for the further development of the unsteady aerodynamic lifting surface theory in compressible subsonic flow [17]. In his brilliant publication, he showed not only the validity of the application of his classical potential-theoretical singularity method also in the domain of gas dynamics, but with the introduction of the concept of an "acceleration potential" also a promising way for further development of the aerodynamic theory of the oscillating wing in compressible flow, urgently required for dynamic aeroelastic investigations, in particular with regard to the flutter problem. In his work, he showed that the pressure jump on the oscillating lifting surface is directly proportional to the change in the acceleration potential, formulated this for the corresponding dipole pressure singularities obtained from the solution of the classic linearized acoustical wave equation, and transferred it to the wing moving at a constant speed V of the order of the speed of sound by applying a Galilei transformation. *Prandtl* explained explicitly that the limit of his extended linearized wing theory is given in the subsonic range by the occurrence of shock waves and also that the validity in the supersonic domain is restricted to a purely supersonic flow. So he was fully aware of the special problem of crossing the sound barrier where a linearized potential theory loses its validity.

The advantages and possibilities of *Prandtl's* concept of an acceleration potential for an advanced aerodynamic theory of the nonstationary oscillating wing in compressible subsonic flow were already soon recognized. In 1938,

C. *Possio* [18] in Italy published a theory of the harmonically oscillating two-dimensional flat-plate airfoil in inviscid compressible subsonic flow based on the acceleration potential. At the end of 1940 *Küssner* published his "General Wing Theory" [19] in the field of aerodynamics—a first-class pioneering work. This linearized inviscid three-dimensional lifting surface theory comprised all (linearized) steady and unsteady aerodynamic wing theories known at that time. First solutions of the "*Possio* Integral Equation," relating the unknown load distribution over the lifting surface and the known (prescribed by the vibration) normal velocity to the surface, the downwash, for wing heaving and pitching, were published in 1943 by F. *Dietze* [20].

As of 1942, the AVA Institute of Nonstationary Phenomena in Göttingen, headed by *H. G. Küssner*, was already working on the development of a two-dimensional linearized aerodynamic theory of the harmonically oscillating wing at supersonic speeds. This was achieved by applying *Pandtl's* singularity concept, derived from the acoustical wave equation, the basis of which is the velocity potential of a source moving at supersonic speed. For harmonically heaving and pitching motions L. *Schwarz* [21] published in 1943 the theoretical formulations that were then employed by P. *Jordan* [22] shortly before the end of the war to calculate—then also with addition of a control surface—the corresponding two-dimensional aerodynamic force coefficients. These were the first approaches to a theory concerning the harmonically oscillating wing with control surface in supersonic flow.

4.4.3 FLUTTER STABILITY IN COMPRESSIBLE SUBSONIC FLOW

The statements presented have shown that the essential theoretical aerodynamic fundamentals, as a prerequisite for the performance of flutter computations for wings of an arbitrary three-dimensional planform in inviscid compressible subsonic flow, had already been worked out scientifically in Germany by the end of WWII. Also first attempts to compute the unsteady aerodynamic forces on harmonically oscillating wings with control surfaces in supersonic flow were accomplished. Numerical solutions for *Küssner's* integral equation [19] were not possible until the emergence of large computers nearly two decades later. In other words, it was not possible to compute unsteady aerodynamic pressure distributions in compressible subsonic flow on harmonically oscillating three-dimensional lifting surfaces with arbitrary planform and control flaps, including that for swept wings. Up to that time the aerodynamic strip theory for unswept and swept wings with stripwise calculation of the aerodynamic loads applying the two-dimensional unsteady aerodynamic lift coefficients for wing heaving and pitching and also the control-surface rotation remained the standard procedure in flutter calculations.

When increasing the flight speed into the high subsonic speed regime, the question of the influence of compressibility on the flutter behavior attained increasing importance. For lack of the necessary unsteady aerodynamic coefficients in compressible subsonic flow, applying the strip theory, one simply extrapolated in routine flutter computations the results obtained with incompressible aerodynamic forces into the corresponding compressible *Mach* number range, a risky procedure. However, the unsteady two-dimensional aerodynamic coefficients for rigid wing heaving and pitching up to $M \approx 0.75$ computed by *F. Dietze* [20] in 1943 showed, indeed, a certain justification for this approach, since the vibration-induced unsteady aerodynamic forces responsible for the flutter onset or aerodynamic damping within the frequency range of interest exhibit only a relatively low dependence on the *Mach* number.

Concerning now the practical possibilities and experience in the execution of such flutter computations at that time, a presentation [23] given by *A. Teichmann* (DVL) in 1941 has some quite remarkable passages. He states:

> It is still necessary today to account for the aeroelastic interaction of a larger number of degrees-of-freedom in such a way, that numerous partial systems of the aircraft (wing and empennage, Note of the Author), each with two or three degrees-of-freedom, are taken separately into consideration in the expectation that one of them suffices to essentially describe the flutter characteristics of the complete system. This might in some cases be more or less justified by evaluating the presently existing coupling relations. However, in many cases uncertainties remain, so that such a procedure can only be regarded as a "stop-gap." This situation can only be improved by considerably reducing the computing time. With suggestions for a reduction of this time by 20% or even 50%, practically nothing at all is gained; much more, it is necessary to cut back the computing time to 1/10 or 1/20. This is only possible with the aid of automatic calculators... *Zuse* is presently working on such a device which is supposed to be suitable here. His device distinguishes itself from the ones known up to now by expressing all numbers in yes/no combinations of telephone relays; the course of the desired computations is then controlled by a prepared (punched) paper tape. The DVL is sponsoring the development of this promising device as far as possible.

Teichmann clearly addressed the state of the art and the possibilities in carrying out flutter computations. With the available computational aids, flutter analyses were already extremely time consuming in the case of taking into account only three simultaneous degrees of freedom (natural vibration modes) because the critical flutter velocity—if at all existent—cannot be computed explicitly from the characteristic flutter determinant of the complex mathematical eigenvalue problem. It must be laboriously determined implicitly in an iterative process by a parameter variation. The computing time could later only drastically be reduced with the availability of large computing facilities and the application of an iteration procedure for the solution of the non-self-adjunct complex eigenvalue problem. Just before the end of the war, such a solution procedure was developed by *H. Wielandt* [24] at the AVA in Göttingen. It should be mentioned that during the evacuation from Berlin in March 1945 the last computer development Z4 of *Konrad Zuse* was stationed for three weeks at the AVA in Göttingen where the world's first program-controlled aerodynamic computations were carried out.

4.4.4 WIND-TUNNEL FLUTTER-MODEL INVESTIGATIONS

The execution of flutter model investigations in the wind tunnel to experimentally check the theoretical results had been a tradition in Germany since the 1930s. They were, at first, phenomenon-oriented and applied primarily to the investigation of specific aeroelastic problems. A focal point of the activities was at the DVL in Berlin-Adlershof. Here, numerous systematic flutter model investigations with different objectives were carried out after 1936 by *H. Voigt*, whose results were summarized at the LGL meeting "Wing and Empennage Oscillations" on March 6–8, 1941 in München [25]. From 1939 on, he investigated intensively the flutter behavior of control surfaces with auxiliary flaps. This problem became acute with the introduction of aerodynamically balanced control surfaces with (motion-controlled) auxiliary flaps to reduce the manual control forces in the case of large aircraft and at high flight speeds. *Voigt* conducted a whole number of fundamental experimental wind tunnel investigations with different control-surface/auxiliary-flap configurations.

Since about 1940, aircraft companies increasingly also conducted project-oriented experimental flutter investigations on dynamically similar complete and component models in the large wind tunnels now available. This was a helpful necessity in the evaluation of the accuracy of flutter computations considering the aforementioned insufficiencies. The construction of such dynamically similar complete models is very expensive. It must be carried out observing numerous model laws, and the investigations must be conducted on the elastically suspended model at the real flight *Mach* number to simulate flight conditions. For *Mach* numbers up into the near-sonic range, these demands can only be realized with sufficiently large high-speed wind tunnels. Because these were not available, the test results obtained in the large low-speed wind tunnels had to be extrapolated with the corresponding velocity scaling into the high-speed range, a questionable procedure already applied in flutter computations. Such flutter investigations on dynamically similar complete models and components of new developments were up to the end of the war mainly carried out in the large wind tunnels of the LFA in Braunschweig and the AVA in Göttingen, on the He 177 (wing, empennage, and complete model), the He 219 (complete model), the He 162 (complete

Fig. 4.12 Dynamically similar flutter model of the Ju 288 in the LFA large wind tunnel A3 in Braunschweig.

model), the Me 264 (wing with engines), the Ar 234 (empennage), the Ju 288 (complete model with 7-m span), and the forward-swept Ju 287 (complete model) amongst others. Apparently no specific investigations with a dynamically similar strongly swept-back complete model were carried out up to the end of the war. These were the world's first flutter tests conducted in a wind tunnel on free-flying suspended dynamically similar complete models. They attracted great attention from the Allies after the end of the war.

Figure 4.12 shows the elastically suspended flutter model of the Ju 288 in the large wind tunnel A3 of the LFA in Braunschweig (1940), and Fig. 4.13 shows the related schematic details of the elastic suspension in the test section

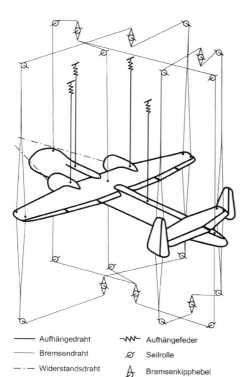

Fig. 4.13 Sketch of the elastic suspension of the Ju 288 flutter model to simulate free-flight conditions and of the mechanical flutter brakes.

——— Aufhängedraht	⟶⟍⟍⟍ Aufhängefeder
——— Bremsendraht	⊘ Seilrolle
— · — Widerstandsdraht	⚡ Bremsenkipphebel

Legend (top to bottom, left to right):

Suspension wire	**Suspension spring**
Brake wire	**Coil**
Drag wire	**Brake lever**

Fig. 4.14 Flutter model of the Ju 287 in the AVA large wind tunnel VI in Göttingen, with the engines on the wing shifted upstream (compare to Fig. 4.7).

together with the "flutter brakes." In the case of a flutter onset, these were mechanically operated by hand through wires fixed on the wing, the fuselage and empennage applying a rocking lever. Figure 4.14 shows the dynamically similar "free-flying" flutter model of the forward-swept Ju 287 using this testing technique in the large wind tunnel VI of the AVA in Göttingen (1944). This model had previously also been flutter tested in the large wind tunnel A3 of the LFA in Braunschweig. For a better agreement between model and the full-scale aircraft, plastic material "Vinidur" was used as model material for the first time, with the following advantages compared to wood, the model material normally used—the material is homogeneous, the ratio of elasticity modulus to shear modulus is the same as of "Dural," and the elasticity is lower. Thus, for a geometrically and dynamically similar true to scale reproduction of the aircraft, one obtains a model with the same elasto-mechanical characteristics as the full-scale aircraft and with the model scaled to the size of the wind-tunnel test section, a sufficiently large velocity scale.

The flutter tests carried out by *P. Jordan* in 1941 with the complete model of the twin-engine He 219 in the large wind tunnel VI of the AVA in Göttingen were documented in an instructive movie. This movie, still up to date, shows in several variants the flutter of the wing with ailerons, the empennage, and the elevators with auxiliary flaps, and explains clearly the physical relationships of the complex aeroelastic interaction of the contributing system forces, and the coupling of natural vibration modes participating in the flutter process. The effectiveness of a control-surface mass balance about the axis of rotation is also impressively demonstrated.

All of these experimental flutter model investigations could, of course, not provide a reliable proof of the quality of the unsteady aerodynamic forces employed in the flutter computations. Additional measurements of the unsteady aerodynamic coefficients, as well as the unsteady aerodynamic pressure distributions on realistically profiled oscillating wing models, were also required. Because of the lack of necessary measuring techniques, the experiment was for a long time not able to keep up with the theory. For the first time a paper appeared in 1941 by *W. Albring* [26] (Institute of Aeromechanics at the Technical University Hannover) reporting measurements of unsteady aerodynamic coefficients. In the minutes of a meeting of the "Specialists Committee Wind Tunnels" within the "Expert Office Vibrations" of the RLM, held on November 14, 1944 in Bad Eilsen [27] concerning tests to determine the global aerodynamic forces and moments on oscillating wings in a wind tunnel, several such activities already conducted in low-speed wind tunnels are mentioned at the DVL, the AVA, and the LVA on two-dimensional wings with flaps with the Goe 409 profiles, as well as at Focke-Wulf and Messerschmitt with project-specific airfoils. Corresponding measurements in high-speed wind tunnels were allegedly planned. Under contract and supervision of the AVA, the NLR in Amsterdam had worked since 1941 on the measurement of unsteady aerodynamic coefficients on harmonically oscillating two-dimensional wings with control surfaces (flaps). These activities did not bring about any results up to the end of the war; however, they were continued thereafter and published by the NLR in 1952 and 1955.

H. Drescher (AVA) reported in 1939 for the first time on unsteady pressure distribution measurements on a two-dimensional airfoil with a slotted flap at rapid changes in the flap deflection. In two other reports [28], he presented the results of measurements on two-dimensional wings with Goe 409 airfoil sections (13% thickness), and harmonically and impulsively excited flaps. These two-dimensional unsteady pressure distribution measurements, carried out in a water tunnel, provided for the first time an impressive insight into the complex unsteady fluid-mechanical processes on oscillating wings with control surfaces (phase relations of the pressure distributions with regard to the motion, formation of the oscillatory vortex sheet within the wing wake, pressure singularities). Figure 4.15 shows a representative result of these measurements. The diagrams on the left of the photographs exhibit the real part of the nondimensional pressure distribution π' in-phase with the flap motion and the imaginary part π'' with a 90-deg phase shift with respect to the flap motion, for two oscillatory circular frequencies ω. Here, the agreement between theory and experiment is surprisingly good. It still took more than a decade before such measurements also became routine for three-dimensional wings.

In conclusion, it should be remarked that in 1940 the construction of a special large wind tunnel for aeroelastic investigations was also seriously being taken into consideration. This situation was discussed at the AVA Göttingen during two meetings (November 9 and December 19, 1940). In interesting memoranda [11] one finds, among other things, the reasons for this. It is stated that the available wind tunnels were already fully occupied by steady routine tests and that "these conditions will become unbearable should the majority of the German aeronautical industry begin to carry out the desirable wind tunnel flutter investigations." It is further concluded that it is very uneconomical "to employ the available wind tunnels, which are equipped for steady measurements with balances, tunnel-air cooling, pressurization, and vacuum systems, for flutter investigations since this equipment is here not needed." However, the projected world's first large-scale wind tunnel for aeroelastic investigations with an envisaged test section width of 8 m, a flat rectangular cross section, and a velocity of 100 m/s was not realized. The construction of such a transonic wind tunnel for aeroelastic investigations,

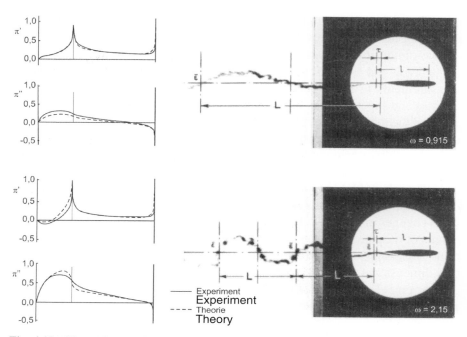

Fig. 4.15 Unsteady aerodynamic pressure distributions on the two-dimensional wing with a harmonically oscillating control surface according to *H. Drescher*: on the right, the vortex sheet emanating from the control-surface (flap) trailing edge; on the left, the components of the pressure-difference distributions π′ in phase with the control-surface motion and π″ with a 90-deg phase shift.

operating for the achievement of high *Reynolds* numbers with the flow medium Freon-12 and variable pressure, with a square 4.87-m (16-ft) test section and a *Mach* number of up to 1.2, then was started in the United States at the NACA in Langley, Virginia, in 1955. This "Langley Transonic Dynamics Tunnel" was put into operation in 1960 and proves today as an excellent, unique test facility, especially in research activities and in the dynamic aeroelastic qualification of modern high-speed aircraft up into the supersonic flow regime.

4.4.5 SWEPT-WING FLUTTER STABILITY IN SUBSONIC AND SUPERSONIC FLOW

With the increase in the flight speed up into the near-sonic range—beside the question concerning the effect of compressibility of the flow—there arose with the introduction of the swept-wing concept also the fundamental question of the effect of the wing sweep on the dynamic aeroelastic behavior of this new generation of high-speed aircraft, in particular concerning the flutter behavior. A conclusive answer was, ad hoc, not possible because hereto no experience was available. The following event best describes the state of knowledge at the end of the war.

In a secret letter, dated September 12, 1944, without giving reasons the Institute of Nonstationary Phenomena of the AVA received from the "Electromechanical Works GmbH (Limited Liability Company), Karlshagen (Pomerania)," the order to conduct the following aeroelastic investigations:

- Flutter of swept wings with control surfaces at supersonic speeds
- Flutter of thin, flat sheet-metal panels

These were for the A4 and the A9 rockets, respectively. This meant in plain language to carry out various flutter investigations on orders of Peenemünde on the V2 rocket. With the knowledge and possibilities at that time this was a new great scientific challenge. The problem here was presumably the flutter ("bursting") of the planking occurring several times during the trial of the A4 as well as the possibility of flutter of a 45-deg swept-back wing at *Mach* numbers of up to 5 in case of a winged version of the A4, the A9, projected to increase the range. Work on this contract was taken on by *P. Jordan*. Regarding the flutter of thin, flat sheet-metal panels, he already reported to the Peenemünde customer on November 7, 1944, in a first interim report the following findings:

> The question of the flutter of thin, flat sheet-metal panels (diaphragms) is theoretically and through tests—the latter within the subsonic range—basically solved. The danger of flutter exists if the freestream velocity V is higher than the wave velocity $v_W = (\sigma/\rho)^{1/2}$ within the sheet-metal, where σ is the tension in the flow direction and ρ the material density.

Jordan describes in an AVA report [29], published shortly before the end of the war, how he came to these findings. He investigated this problem, following his teachers *L. Prandtl* and *A. Betz*, with a congenially simple model of a sheet-metal strip in parallel flow. Figure 4.16 shows a corresponding principle sketch of his model for the comparative experimental investigations in a Göttingen small subsonic wind tunnel. With the formulation of the participating

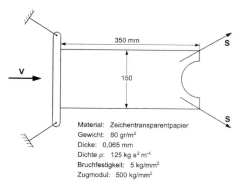

Fig. 4.16 Principle sketch of *P. Jordan's* wind tunnel model for the investigation of panel flutter.

Legend (top to bottom):
material: transparent drawing paper
weight
thickness
density ρ
fracture strength
tensile modulus

350 mm

v

150

Material: Zeichentransparentpapier
Gewicht: 80 gr/m²
Dicke: 0,065 mm
Dichte ρ: 125 kg s² m⁻⁴
Bruchfestigkeit: 5 kg/mm²
Zugmodul: 500 kg/mm²

forces—inertia force, elastic reaction force due to bending and diaphragm tensile-force S, and the unsteady aerodynamic force $A(t)$ normal to the sheet-metal skin—he derived a partial differential equation, which he solved, with a mathematical separation approach for a traveling wave, for various parameter variations. With physically correct foresight he formulated the aerodynamic force quasi-steady applying the supersonic theory *of J. Ackeret* [30]—a concept which formed ten years later the linearized base for the nonlinear supersonic "Piston Theory" developed in the United States [31]. These were the first guiding investigations concerning the aeroelastic problem of panel (skin) flutter, a typical supersonic phenomenon. This was then systematically further researched in the 1950s and 1960s in the United States within the context of the Apollo program for all practically relevant parameters confirming the basic findings of *Jordan.*

He informed his Peenemünde customers in a further interim report, dated January 8, 1945, regarding the swept A9 wing, quoting, "the flutter computations were carried out according to the method developed here for swept wings." Regarding the knowledge gained so far, he remarked:

1. Contrary to the subsonic range, the supersonic range shows a fundamental effect of wing sweep on the flutter behavior such that the assertion of the limitation of flutter to certain frequency domains is no longer valid for swept wings.

2. Individual test computations indicate that in the case of $M \approx 1$ an appreciably higher flutter tendency may occur, up to now only noticed in the range $n_T/n_B < 1$. (Here, $1/n_T$ is the circular frequency of the torsional and $1/n_B$ that of the fundamental bending vibration mode.)

This shows that at the end of the war at the AVA—and worldwide probably only there—a computational method was already available for the investigation of the flutter behavior of large-aspect-ratio swept wings in compressible sub- and supersonic flow. Thereby, with the application of the aerodynamic strip theory, the two-dimensional solutions for the unsteady aerodynamic coefficients of the harmonically oscillating two-dimensional wing in supersonic flow could be applied. This also follows from the report by *Jordan* [32] published at the end of 1944. Further details are explained in his contribution "Aircraft Flutter" to the "Göttingen Monographs" [2] in which he mentions another essential result of his parametric flutter investigations, carried out for the first time for swept-back wings:

Also the very torsion-stiff swept wing may flutter. The statement that sufficient torsional stiffness prevents flutter, valid for the normal wing, is not necessarily true for the swept wing; the kinematical change in angle of attack due to bending suffices in the case of heavy wings (at large flight altitudes and, therefore, large mass parameters) and near-sonic *Mach* numbers to cause self-excited one degree-of-freedom flutter [of the fundamental natural wing bending vibration mode].

With these explanations to Peenemünde, that in the range $M \approx 1$ a higher flutter tendency exists, and that there is a possibility of a (nonclassical) one degree-of-freedom wing bending flutter in near-sonic flow, *Jordan* was probably the first to recognize the significant changes in the flutter behavior of the swept wing existing in this flow regime. An appreciative reference to these important findings, which with the "Göttingen Monographs" were certainly known in the United States and England at all aeronautical research establishments and also in industrial firms after the end of the war, is not found in the Anglo-American literature.

A revealing attachment to a memorandum of *Jordan* from January 24, 1945 [33], concerning the present orders of his working group at the AVA Institute of Nonstationary Phenomena in Göttingen shows, that in January 1945—in addition to the investigations already demanded by Peenemünde—an order was placed by Heinkel. It concerned the execution of flutter computations for the projected further development of the "People's Fighter" He 162 with a swept wing at various sweep angles. The latter was supposed to reach high subsonic speeds ($M \approx 0.9$). Here, Heinkel intended to provide desk calculating machines and two assistants. Finally, a further order followed in early January 1945 from the Forschungsgemeinschaft Halle (Research Association Halle) (FH) to carry out flutter computations for the DFS 346, a supersonic experimental fighter ($M \approx 2$) with a 45-deg sweepback and two rocket engines (for further details see Sec. 6.5). During the contract placement, *Jordan* explicitly pointed out the unknown difficulties when passing the sound barrier, where no experience and no computational possibilities existed. The FH intended to provide personnel capacity for the computation of the supersonic unsteady aerodynamic forces on the ailerons, so far only being known by their theoretical formulations [22], for which they promised the support of the Mathematical Seminar of the University of Halle as well as of the Academy of Technology in Chemnitz.

The special problem of control-surface flutter (aileron flutter) in the supersonic range up to $M \approx 1.4$—and here the additional effect of wing sweep—was apparently already known to *Jordan*. It should be remarked here that within his working group only five assistants were at his disposal in early 1945 to cope with this extensive and new research work; an assistant with an academic degree (a secondary-school teacher, available for three months), a female high school graduate (with an advanced technical education) as assistant for simple tasks, an advanced female assistant (after courses for technical assistants), and two female assistants without mathematical know-how. Considering the difficulty and the extent of the tasks to be solved, this was a practically hopeless venture, typical of the personnel situation towards the end of the war within German aeronautical research institutions, at the end of an extremely successful era with fundamental contributions to aeronautical technology. The execution of this work, marked with the highest priority and to be completed by the end of April 1945, was probably not started at all, because the provision of the necessary design documents was delayed as well. The theoretical documents for additional aid in the computation of the urgently needed unsteady aerodynamic control-surface force coefficients at supersonic speeds were sent to another two outside offices at the end of February 1945, namely, to the training center of the Fieseler Plant in Kassel and to the TH Darmstadt. Also this work was probably not seriously started during the short time remaining before the end of the war.

P. *Jordan* went to the RAE in Farnborough, England, in 1946 where he completed, among other things, the computation of the two-dimensional aerodynamic force coefficients for the harmonically oscillating control surface in supersonic flow [34]. In 1953 he moved to the United States and worked there for the Martin Marietta Corporation in Baltimore. At his retirement in 1976, the company issued the "Selected Published Works of Peter Friedrich Jordan" [35], which provides a comprehensive survey of his impressive scientific lifework.

Little is known about the results of flutter investigations carried out until the end of the war by industry during the development of the first swept-wing aircraft. Results of such investigations for the forward-swept wing of the Ju 287 are presented in an internal report of the Junkers Company of May 20, 1944 [36]. An essential finding here was that a shift of the center of gravity of the wing-mounted engines to a position upstream of the elastic axis, other than originally conceived (see Fig. 4.7), considerably increased the critical flutter speed. This was a fundamental finding. Today it is a generally valid design rule for the arrangement of the engines on unswept wings, and forms the flight silhouette of all swept-wing aircraft with pylon-mounted engines under the wing. In addition, theoretical and experimental flutter investigations are known [37] that were carried out in cooperation between the DVL and the Henschel Aircraft Company on a swept-back wing with a 38-deg leading-edge sweep. The computations were carried out with the strip theory, the measurements in one of the medium-speed wind tunnels of the DVL. The test results were allegedly in good agreement with the theory. The characteristic flutter behavior of swept wings in high subsonic and in transonic flow could, of course, not generally be judged from these measurements.

At the beginning of WWII, nearly all aircraft companies in Germany had their own officials in charge, and working groups concerned with the execution of aeroelastic investigations related to their manifold new developments. The fundamentals were provided by competent research institutions, primarily the AVA Institute of Nonstationary Phenomena in Göttingen and the working group around A. *Teichmann* at the DVL in Berlin-Adlershof. There were as yet no special lectures on aeroelastics at technical universities. Therefore, new personnel in charge in the aircraft industry, working in this field, were delegated to the two aforementioned research institutions at the instigation of the Technological Office of the RLM for a brief training, which constituted for the institutions an additional burden. *Jordan* has commented critically on the questionable success of this measure in his already mentioned memorandum [33] by writing:

> Concerning the question of the industry orders, we were initially of the opinion that we are willing to assist in the training of the industry employees, but that the prototype certification itself is left to the companies, while we wanted to devote our activities to undisturbed research work. This opinion has proved to be wrong. The education of industry employees is a burden without producing the desired extent of mutual fertilization. On the other hand, the permanent collaborators required for the necessary research work could not be obtained.... The peculiar feature of flutter problems requires a long time to get acquainted, temporary help is harmful.

Jordan expressed with these statements that the cooperation with outside organizations and short-term support lacking professional competence was of little help. He pleaded for a concentration of the professional experts— which steadily diminished towards the end of the war as technological challenges grew—at institutions, where the specialized knowledge was located. He describes the situation resulting from these insufficiencies in his memorandum as follows:

> Due to newly arising urgent problems, the final completion of reports— swept wing, supersonic flow—was delayed for a long time. Extensive material, which had already frequently been proven valuable at visits and meetings, thus awaits its utilization. But not only reports were left uncompleted, also urgent tasks in flutter research, primarily experimental investigations could not be carried out—for instance the aeroelastic behavior of swept wings with advanced near-sonic profiles and aerodynamically balanced flaps, semi-empirical determination of unsteady aerodynamic forces and comparison of observed flutter cases with theoretical computations.

This was a disillusioning conclusion of an outstanding scientist, who had contributed a lot to the theory of wing flutter as well as to the aeroelasticity of the swept wing and hence to modern aircraft engineering.

Figure 4.17 shows again, in summary, the tendency in the static and dynamic aeroelastic behavior of a conventional metallic wing without engines (nacelles) in compressible subsonic flow up to about $M = 0.75$, with the grave

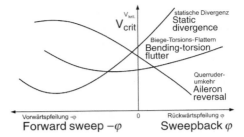

Fig. 4.17 Tendential effects of wing sweep on the critical speeds in the compressible subsonic speed range up to $M \approx 0.75$.

effect of the wing sweep being clearly recognizable. Presented are the relative magnitudes of the critical speeds for the static divergence, the aileron reversal, and the (classical) bending-torsion flutter. Their knowledge is of fundamental importance to an optimum wing design for swept high-speed aircraft. The critical speeds for the torsion divergence and the aileron reversal are in case of the unswept wing generally close to each other while the bending-torsion flutter speed is considerably lower and requires the highest attention. The relations are, however, fundamentally different in the case of swept wings. For the forward-swept wing, the critical speed of the bending-torsion divergence, rapidly decreasing with increasing negative sweep, quickly becomes a technically limiting factor while the aileron effectiveness is advantageously affected. On the other hand, in case of the swept-back wing, the drastic decrease in the critical speed for the aileron reversal with increasing sweep gains primary importance while the static divergence is unproblematic. One also recognizes in Fig. 4.17 that the effect of the wing sweep on the classic bending-torsion flutter behavior is in the case of the conventional wing without engines (nacelles) relatively small, and in the case of sweepback due to the kinematical coupling of bending and change in angle of attack even positive. As experience shows, control-surface flutter on swept wings and stabilizers (empennages) can also in compressible subsonic flow up to $M \approx 0.75$—just as in the case of the unswept wing—relatively easily be avoided by the well-tried method of a control-surface mass balance. Other laws apply, however, to swept-wing aircraft with pylon-mounted jet engines under the wing with regard to an aeroelastic wing optimization.

4.4.6 DYNAMIC AEROELASTIC PROBLEMS IN TRANSONIC FLOW

Despite the fact that the German aircraft companies had already worked on swept supersonic aircraft at the end of WWII, the flight-technical problems to be expected when "overcoming the sound barrier" were largely unknown, the crossing of the transonic range as fast as possible being considered essential. The first experience of test pilots when briefly exceeding the critical *Mach* number with still unswept, later slightly swept, aircraft showed that serious aeroelastic problems were to be expected. Strong vibrations of the wing and control surfaces occurred, and the entire aircraft was subjected to more or less irregular buffeting oscillations. Because of the lack of the necessary facilities for corresponding experimental and analytical aeroelastic investigations, such as large transonic wind tunnels as well as large electronic computers for a numerical solution of the basic nonlinear fluid mechanic equations—the Euler and Navier–Stokes equations, already known for a long time—the actual causes of these transonic aeroelastic phenomena, just like many others, could only be surmised. Thus, the sound barrier could only be conquered step by step by risky flight tests thereby using the gained experience. The American test pilot *Charles Yeager* finally succeeded in performing the first supersonic flight at a *Mach* number of $M = 1.06$ on October 14, 1947, with the unswept rocket-propelled experimental aircraft Bell XS1. However, the transonic range remained a largely unexplored "island," its scientific development still a long-time vision.

Because of these insufficiencies, for the flutter calculations of the first supersonic aircraft, the transonic range was simply bridged by an interpolation of the results obtained with application of the strip theory and the linearized two-dimensional aerodynamic unsteady force coefficients for the high subsonic and low supersonic speed ranges. Then in the 1960s with the availability of powerful computers, mathematical collocation methods for the numerical solution of "*Küssner's* integral equation" [19] of his linearized three-dimensional lifting surface theory were developed to compute the unsteady pressure distributions on harmonically oscillating wings with control surfaces in subsonic flow—so-called panel- and *Mach*-box methods with a decomposition of the wing planform into a number of finite elements [7]. The basis for the supersonic flow was the analogous integral equation for the supersonic range, developed in 1947 in the United States by *I. E. Garrick* and *S. I. Rubinow* [38], for which the two-dimensional solution of *L. Schwarz* [21] set the example.

With these linearized three-dimensional numerical lifting surface methods, which constituted great progress for the computation of the unsteady aerodynamic forces on oscillating wings of arbitrary geometry and control surfaces, there was still no possibility of realistic analytical solutions for the specific dynamic aeroelastic problems occurring in transonic flow and, in the meantime, known from wind tunnel investigations. However, these were urgently needed because the proof of flutter safety of swept subsonic aircraft, operating close to the critical *Mach* number, must be provided up into the transonic range. To understand the characteristic compressibility effects occurring here, one has to go back to the basic nonlinear fluid-mechanical equations already mentioned. Only the introduction of a new generation of supercomputers from the beginning of the 1980s presented for the first time possibilities for the solution of these basic equations and hence for the execution of realistic flutter computations for swept wings also in transonic flow. Here, a large amount of the pioneering work was performed mainly in the United States. Because of the complexity of the dynamic aeroelastic behavior of wings in transonic flow and the multitude of influence parameters, these tasks still possess a high scientific actuality.

In the following, based on the present state of knowledge, first the multitude of flutter problems occurring in transonic flow will be presented, and then the "buffeting" behavior of the unswept- and swept-wing aircraft when entering the transonic speed range will be discussed.

4.4.6.1 FLUTTER BEHAVIOR. The main reasons for the typical (nonclassic) flutter phenomena in transonic flow are the specific flow processes occurring on the steady wing. These are characterized by shock waves and oscillating shock boundary-layer interactions with associated flow separations. These nonlinear aerodynamic processes occur on the oscillating wing interactively with the oscillatory motion. The induced unsteady aerodynamic forces are, therefore, nonlinearly dependent on the oscillatory amplitude; thus, transonic flutter happens in finite-amplitude limit cycles. This is a peculiar characteristic feature of the flutter phenomena in transonic flow.

Figure 4.18 shows, as an instructive example, the typical features of high-speed, low angle-of-attack flutter of a swept wing. In compressible flow, the dynamic pressure at flutter tends to decrease with increasing *Mach* number to

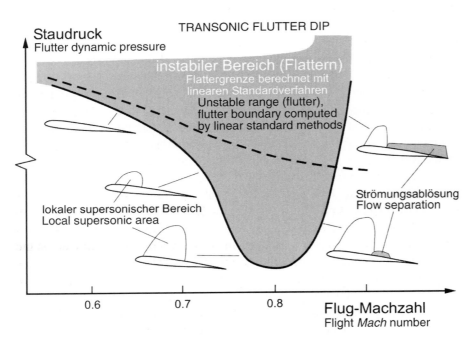

Fig. 4.18   Typical transonic flutter characteristics of a swept-back wing.

a minimum "critical flutter point" value in the transonic speed range. Thereby, the flutter boundary is directly connected to the dramatic changes in the flow characteristics. When entering into the transonic flow regime, with the increasingly larger local supersonic regions on the upper (suction) side of the wing and the associated greater increasing pressure gradients and vibration-induced unsteady aerodynamic forces, at first a reduction of the flutter dynamic pressure down to a minimum occurs. With further increasing *Mach* number and the occurrence of now massive shock boundary-layer interactions with flow separations finally reaching the wing trailing edge, a rapid increase in the flutter dynamic pressure up to the supersonic range then follows at now decreasing unsteady aerodynamic forces due to energy losses within the flow. This typical feature of transonic flutter, first discovered with the emergence of large transonic wind tunnels, is described as "Transonic Flutter Dip." Figure 4.18 shows that the "considerably higher flutter sensitivity in transonic flow," already forecast by *P. Jordan*, actually exists. It is also indicated that the flutter boundary can be reasonably well computed by linearized inviscid standard methods only for subsonic attached flow.

Flutter in transonic flow with limit-cycle amplitudes may occur in several forms. In the classical form it may occur by an aerodynamic coupling of several degrees of freedom (natural vibration modes), but also in the non-classical form as one-degree-of-freedom wing bending and wing torsion flutter as well as control-surface flutter. Thereby, often an immediate transition of one form of flutter into another may be possible already at very small changes in the flight *Mach* number or the angle of attack. The transonic one degree-of-freedom flutter of control surfaces is called "control-surface buzz." It depends on many parameters, but primarily on the flight *Mach* number M_∞ and the angle of attack α but also on the wing sweep, the natural rotational frequency, airfoil thickness and on the control-surface chord and aspect ratio. As shown in Fig. 4.19, it may occur in three variants. Buzz-type A occurs in the range $0.75 < M_\infty < 0.9$ with the shock wave on the wing upper surface being positioned upstream of the control surface axis of rotation (hinge line). This control-surface flutter may occur spontaneously when exceeding the critical *Mach* number, as was often reported by the pilots. The shock characteristic changes here in the interaction with the control-surface movements and the nonlinear flutter vibrations occur quasi harmonically with finite amplitude. Buzz vibrations of Type B may then occur with increasing *Mach* number in the range $0.9 < M_\infty < 1$ with the shock wave now positioned on the control surface itself. The oscillations stimulated by the pressure pulsations of the flow separations occur primarily more or less irregular in time and are, therefore, also described as "transonic control-surface buffeting." Finally there exists in pure supersonic flow up to $M_\infty \approx 1.4$ the possibility of a classical control-surface flutter with one degree of freedom, following linearized potential theory, which is also described as Buzz Type C. Obviously *Jordan*

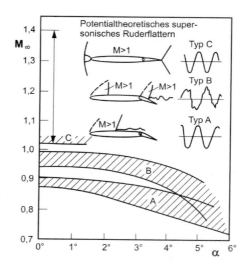

Fig. 4.19 Typical buzz behavior of a control surface in transonic flow; legend: supersonic control-surface flutter according to potential theory ("typ" means "type").

had already clearly recognized this dangerous flutter possibility with theoretically unlimited amplitudes. Presumably this was the reason why he was intensively looking for help in the computation of the corresponding force coefficients for his intended flutter investigations on the DFS 346 at supersonic speeds; see Sec. 4.4.5. Sometimes there is a small stable *Mach* number range between the individual types of control-surface buzz at low angles of attack. During the 1950s and 1960s, the transonic control-surface buzz behavior was experimentally investigated [39] most of all in England in systematic wind tunnel tests. Here, it was shown that a control-surface mass balance about the rotation axis—effective in subsonic flow—remained completely ineffective. Sufficiently high natural rotational stiffness of the control-surface joint and a correspondingly high control-surface natural rotational frequency to avoid control-surface buzz could only be achieved with the introduction of hydraulic control-surface actuators.

4.4.6.2 TRANSONIC BUFFETING. When flying close to the critical *Mach* number and entering the transonic speed range, pilots also reported the sudden appearance of an impulse-like onset of irregular oscillations of the wing and the complete aircraft. Conclusive physical explanations for these mysterious unsteady aeroelastic oscillation phenomena were impossible, due to the ignorance of the real fluid-mechanical processes occurring when entering the transonic flow regime. However, they were a clear symptom for the difficulties to be expected when trying to pass the sonic barrier and were later described as "transonic buffeting." It should be remarked that the term buffeting was introduced for the first time in the early 1930s in England to describe the forced vibrations occurring on empennages in the turbulent wakes of wings and propellers (empennage buffeting).

It is a characteristic feature of the transonic buffeting oscillations that the exciting unsteady aerodynamic forces are already intrinsically induced in the flow on the steady wing. The shock waves appearing on the upper side of the wing when entering the transonic speed range are associated with extreme pressure gradients and broadband pressure fluctuations, whereby the elastic wing—and hence the entire aircraft—is excited to carry out more or less irregular structural vibrations in their low-frequency vibration modes. The aeroelastic interactions between the exciting and the vibration-induced unsteady aerodynamic forces on the elastic wing are extremely complex and hardly accessible to an analytical prediction to determine the real vibration characteristics [40]. The occurrence of these aeroelastic vibration phenomena, typical when entering the transonic *Mach* number range, is described by aerodynamicists as "buffet boundary (in German schuettelgrenze means shaking boundary). The latter defines the upper limit of the flight speed of today's subsonic (swept) aircraft as indicated in Fig. 4.20. A sufficiently safe distance to the buffet boundary under consideration of minor maneuvers (1.3 *g*) as well as a brief crossing of the buffet boundary due to a vertical gust (1.6 *g* or 12.5 m/sec) limits the maximum cruise flight speed and constitutes an important structural load assumption for civil passenger aircraft. Here, *g* denotes the gravity constant.

For the contour of the transonic buffet boundary and the intensity of the buffet oscillations, the flight *Mach* number M_∞ and the flight angle of attack α are also dominant parameters. This is clearly shown in Fig. 4.21 for a typical example, where the results of wind-tunnel tests on a torsion-elastically supported rigid swept half-wing model are illustrated [40]. The buffeting intensity increases rapidly after exceeding the buffet boundary and in a limited M_∞/α-range torsional flutter occurs. Because the buffeting and flutter oscillations occur each with finite amplitudes and a drastic change in the vibration behavior already occurs for very small changes in the *Mach* number or the

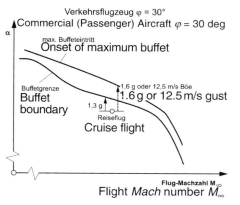

Fig. 4.20   Buffet boundary and buffeting criteria of a large subsonic commercial aircraft.

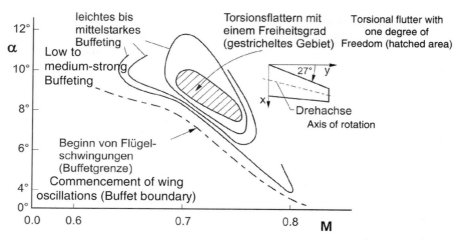

Fig. 4.21 Experimentally determined buffeting and flutter contours of a torsion-elastically supported rigid swept-wing model.

angle of attack, the vibration mode characteristics are often hard to distinguish. The wing-section geometry (profile) also plays an important role.

With the introduction of the swept-wing concept, it was possible to significantly increase the buffet boundary (critical *Mach* number) and also to reduce the buffet intensity with the introduction of the new (transonic) high-speed wing sections (supercritical profiles). For a swept-back wing, the buffeting intensity is considerably lower than for the unswept wing of the same wing area, stiffness, aspect ratio, and wing load. Moreover, for a swept-back wing, the kinematical reduction of the angles of attack in flow direction due to wing bending is of importance. As a consequence, this effect results in a reduction of the buffeting sensitivity and also in the gust sensivity in the case of atmospheric turbulences. Concerning the transonic buffeting behavior of forward-swept wings, no (published) information exists up to now. However, this should be far more problematic—just like the flutter behavior of such wings in transonic flow.

4.5 CONCLUSION

The explanations presented have shown that with the rapid development of high-speed aircraft and the introduction of the swept-wing concept, new ground was entered in the still young flight-technical area of aeroelasticity. For the solution of the newly arising problems, enormous efforts in theory and experiment were required. Germany performed in the time period between 1935 and the end of the war in 1945—at the threshold of a new age in modern aeronautics—globally acknowledged pioneering work. Not only in consideration of the activities described, but also in the opinion of many well-known foreign experts, one may state in summarizing that at the end of WWII, Germany played a leading role in the field of aeroelasticity and contributed essentially to the further development of this fundamental scientific field of modern aircraft engineering [41]. This concerns particularly the first fundamental contributions to the aerodynamic theory of the nonstationary oscillating wing in compressible subsonic and supersonic flow, the contributions to the theory of wing flutter and to the technology of ground vibration testing. Pioneering work was also performed in the urgently needed development of advanced testing techniques for the pursuance of aeroelastic investigations on oscillating models in the wind tunnel. Here, the first flutter tests with dynamically similar complete free-flight aircraft models, as well as the experimental determination of unsteady aerodynamic coefficients of oscillating wings and the first measurements of unsteady aerodynamic pressure distributions, were outstanding innovative achievements. But also the basic contributions to the aeroelasticity of swept-wing aircraft elaborated towards the end of WWII under difficult conditions deserve high recognition. All of these scientific-technical activities have substantially contributed to the achievement of high-speed flight up into the supersonic range.

REFERENCES

[1] *P. Allez-Fernandez* (ed.), "Die berühmtesten Flugzeuge der Welt," Bernhard und Gräfe Verlag, Koblenz, 1986, pp. 215–222.

[2] *A. Betz* (ed.), "Monographien über Fortschritte der Luftfahrtforschung (seit 1939)," Band G, Instationäre Vorgänge, 1946.

[3] *H. Reissner*, "Neuere Probleme der Flugzeugstatik," Zeitschrift für Flugtechnik und Motorluft-schiffahrt, 17 Jahrgang, Heft 7, April 1926, pp. 137–146.

[4] *W. Wagner*, "Die deutsche Luftfahrt," Band 14, Die ersten Strahlflugzeuge der Welt, Bernhard & Gräfe Verlag Bonn, 1989, pp. 174–187.

[5] *A. R. Collar*, "The Expanding Domain of Aeroelasticity," Journal of the Royal Aeronautical Society, Vol. 50, Aug. 1946, pp. 613–636.

[6] *T. H. Pian, H. Lin*, "Effect of Structural Flexibility on Aircraft Loading," U.S. Air Force, Technical Report 6358, 1951.

[7] *H. W. Försching*, "Grundlagen der Aeroelastik," Springer Verlag, Berlin, 1974.

[8] *J. Weissinger*, "Über eine Erweiterung der Prandtlschen Theorie der tragenden Linie," Math. Nachrichten, Band 2, 1949, pp. 45–106, auch ZWB FB 1553, 1942 und NACA TM 1120, 1947.

[9] *H. Multhopp*, "Die Berechnung der Auftriebsverteilung von Tragflügeln," Luftfahrt-Forschung 15, 1938, pp. 153–169.

[10] *L. Prandtl*, "Über Strömungen, deren Geschwindigkeiten mit der Schallgeschwindigkeit vergleichbar sind," Journal of the Aeronautical Research Institute, Tokyo Imperial Univ., 5, No. 65, 1930, pp. 25–34.

[11] *P. Bublitz*, "Geschichte der Entwicklung der Aeroelastik in Deutschland von den Anfängen bis 1945," DFVLR-Mitteilungen 86-25, 1986.

[12] *G. de Vries*, "Zur Technik des Standschwingungsversuchs," Jahrbuch 1942 der Deutschen Luftfahrtforschung, pp. 590–595.

[13] H. W. Försching, "Ludwig Prandt's grundlegende Beiträge zur instationären Aerodynamik schwingender Auftriebsflächen"; In Ludwig Prandtl, ein Führer in der Strömungslehre, ed. by G. E. A. Meier, Vieweg Verlag Braunschweig/Wiesbaden, 2000, pp. 147–171.

[14] W. Birnbaum, "Das ebene Problem des schlagenden Flügels," Zeitschrift für angewandte Mathematik und Mechanik (ZAMM), 4, 1924, pp. 277–292.

[15] H. G. Küssner, "Schwingungen von Flugzeugtragflügeln," Luftfahrt-Forschung, 4, 1929, pp. 41–62.

[16] H. G. Küssner, L. Schwarz, "Der schwingende Flügel mit aerodynamisch ausgeglichenem Ruder," Luftfahrt-Forschung, 17, 1940, pp. 337–354.

[17] L. Prandtl, "Theorie des Flugzeugtragflügels im zusammendrückbaren Medium," Luftfahrt-Forschung 13, 1936, pp. 313–319.

[18] C. Possio, "L'Azione aerodinamica sul profilo oscillante in uno fluido compressibile a velocita Iposonara," L'Aerotecnica 18, 1938, pp. 441–458.

[19] H. G. Küssner, "Allgemeine Tragflächentheorie," Luftfahrt-Forschung 17, 1940, pp. 370–378. Auch NACA Techn. Memo. 979, 1941.

[20] F. Dietze, "Die Luftkräfte des harmonisch schwingenden Flügels im kompressiblen Medium bei Unterschallgeschwindigkeit," Zentrale für wissenschaftliches Berichtswesen der Luftfahrtforschung (ZWB), FB 1733, 1943.

[21] L. Schwarz, "Ebene instationäre Theorie der Tragfläche bei Überschallgeschwindigkeit," Jahrbuch 1943 der Deutschen Luftfahrtforschung, Beitrag I A 010.

[22] P. Jordan, "Instationäre Luftkraftbeiwerte bei Überschall," AVA-Bericht 45/J/08, 1945.

[23] A. Teichmann, "Stand und Entwicklung der Flatterberechnung," Sitzung, Flügel- und Leitwerksschwingungen, 6. bis 8. März 1941, LGL-Tagungsbericht 135, pp. 11–18.

[24] H. Wielandt, "Beiträge zur mathematischen Behandlung komplexer Eigenwertprobleme," AVA-Berichte 43/J/9 (1943), 43/J/21 (1943), 44/J/21 (1944), 44/J/37 (1944).

[25] H. Voigt, "Übersicht über bisherige in der DVL durchgeführte Windkanalversuche an Rudern mit aerodynamischem Ausgleich," Sitzung. Flügel- und Leitwerkschwingungen, München 6.-8. März 1941, LGL-Tagungsbericht 135, pp. 78–83.

[26] W. Albring, "Kraftmessung am schwingenden Tragflügel," Jahrbuch 1941 der Deutschen Luftfahrtforschung, pp. 324–337.

[27] H. Roos, "Kurzer Bericht des Fachamts Schwingungen im 'Sonderausschuß Windkanäle' über die im letzten Jahr durchgeführten Kanalversuche (nach den bis zum 10.9.1944 eingelaufenen Berichten)," Nov. 11, 1944.

[28] H. Drescher, "Messungen an einem Tragflügel mit harmonisch bewegtem Ruder in ebener Strömung, AVA-Bericht B/44/J/10, 1944 und Untersuchungen an einem symmetrischen Tragflügel mit spaltlos angeschlossenem Ruder bei raschen Änderungen des Ruderausschlags," AVA-Bericht B/44/J/41, 1944.

[29] P. Jordan, "Über das Flattern von Beplankungen," AVA-Bericht 45/J/3, 1945.

[30] J. Ackeret, "Luftkräfte auf Flügel, die mit größerer als Schallgeschwindigkeit bewegt werden," Zeitschrift für Flugtechnik und Motorluftschiffahrt (ZFM), 16, 1925, pp. 72–74.

[31] H. Ashley, G. Zartarian, "Piston Theory—A New Aerodynamic Tool for the Aeroelastician," Journal Aeron. Sci., Vol. 23, 1956, pp. 1109–1118.

[32] P. Jordan, "Flatterberechnung im Überschall, 1. Mitteilung: Instationäre Luftkraftbeiwerte des Flügels ohne Ruder," AVA-Bericht 44/J/43, 1944.

[33] P. Jordan, "Aktenvermerk über den augenblicklichen Auftragsbestand (mit Anlage) der Arbeitsgruppe Jordan," Jan. 24, 1945.

[34] P. Jordan, "Aerodynamic Flutter Coefficients for Subsonic, Sonic and Supersonic Flow (Linear Two-Dimensional Theory)," ARC R.&M. 3057, 1955.

[35] Martin Marietta Corp. (ed.), "Selected Published Works of Peter Friedrich Jordan," Baltimore Maryland, U.S., June 1976.

[36] *J. Dörr*, "Flattersicherheit der Ju 287," Interner Bericht der Junkers Flugzeug- und Motorenwerke, Dessau, 1944.

[37] *W. Heger, F. Walter, W. Quessel*, "Experimentelle und theoretische Untersuchungen über das Flatterverhalten des Pfeilflügels," ZWB-UM (Untersuchungen und Mitteilungen) 1501, 1945.

[38] *I. E. Garrick, S. I. Rubinow*, "Theoretical Study of Air Forces on an Oscillating or Steady Thin Wing in a Supersonic Main Stream," NACA Report No. 872, 1947.

[39] *N. C. Lambourne*, "Control Surface Buzz," ARC R.&M. 3364, 1962.

[40] *H. W. Försching*, "Aeroelastic Buffeting Prediction Techniques—A General Review," DFVLR-FB 81-15, 1981.

[41] *I. E. Garrick, W. H. Reed*, "Historical Development of Aircraft Flutter," AIAA Journal of Aircraft, Vol.18, Nov. 1981, pp. 897–912.

Chapter 5

EFFECT OF HIGH *MACH* NUMBERS ON HIGH-SPEED AIRCRAFT

HANS GALLEITHNER

5.1 INTRODUCTION: THE PROBLEM

Prior to the beginning of WWII, success had been achieved in increasing the flight speed of fighter aircraft through improved aerodynamics and propulsion performance together with the metal material used for the fuselage and wing structures. The maximum speeds in horizontal flight, and during a dive and the corresponding *Mach* numbers to be flown during in-flight structural loads testing, were, however, sufficiently far below the values where the compressibility of the air could have an effect on the stability and controllability of the aircraft configurations and designs of the time.

The development of jet propulsion and its use as aircraft propulsion from 1939 opened up the possibility for a further increase in flight altitude and flight speed. With unchanged conventional aerodynamics, a speed could be reached in horizontal flight that was previously hardly possible during a dive. With the flight speed approaching the speed of sound, compressibility effects had to be taken into account. In principle these had already been verified in 1928 in Germany [1] and in 1929–1930 in investigations of the National Advisory Committee for Aeronautics (NACA) by measurements. In Germany, *Prandtl* and *Busemann* dealt with fundamentals of high-speed flow [2]. Compressibility effects did not yet play a conscious role in practical flight trials and in-flight operations of fighter aircraft. But soon there were already indications of unusual events that occurred in flights with high dynamic pressures such as, for instance, trim changes that could not be explained solely by aeroelastic deformations. Already in 1937 a fatal accident occurred with a Messerschmitt Bf 109 where compressibility could have played a role. In 1939, during combat operations, there were increasing reports of crashes without enemy encounter where fighter aircraft could not be pulled out of high-dynamic-pressure steep dives. The Bf 109 was particularly susceptible, but the Fw 190 was also criticized. The flight testing of the first jet aircraft, such as the He 280, Me 262, and Ar 234 and also of the rocket-propelled aircraft Me 163 commenced at the same time in approximately 1942–1943. The rapid increase in the flight speed achieved was just as desired as it was alarming. During trials, some of the new aircraft showed unexpected and dangerous flight characteristics, such as the loss of controllability at high dynamic pressures and flight speeds when approaching the speed of sound. No definite knowledge existed as to why this was the case and how this behavior could favorably be influenced.

5.1.1 AERODYNAMIC RESEARCH

The design offices of the aircraft companies were quite conscious that aerodynamics and flight mechanics would cross new frontiers when the flight speed approached the speed of sound. Theories and experimental results existed that pointed to the phenomena. So the aerodynamicists and designers of the aircraft industry had some idea of the effects that could occur when approaching the speed of sound, at least with regard to conventional airfoils and the wing and fuselage design. Uncertainty existed on the one hand, because aeroelastic effects could not easily be separated from *Mach* number effects. On the other hand, it was not clear how characteristic features of a configuration, such as the airfoil shape, the relative airfoil thickness, and the wing and fuselage geometries, would influence the aerodynamics in the near-sonic range. If at all, solutions had been insufficiently investigated let alone proven in practice. Aerodynamic research in the field of compressibility effects was at the beginning of the WWII just 10 years old (see Chapter 2: New Test Facilities). The required large wind-tunnel facilities were only just being built, and flight research had not yet really started.

K. *Tank* described in his lecture "Aeronautical Research and Aircraft Design" on May 8, 1942, the uncertainty of the load assumptions accompanying the increased flight speed of aircraft [3]. Initially, prior to the occurrence of shock waves, it was possible with the help of the *Prandtl–Glauret* rule to transfer aerodynamic research results also to speeds where the volume change (compressibility) already started to play a role. Among other things, he pointed to flight accidents that could have possibly been avoided with a more accurate knowledge of the effects associated with the occurrence of shock waves on the aerodynamics and hence on the structural layout. He appreciated the discovery of the swept-wing effect, however, and appealed to aeronautical research to provide the designers with data for the fuselage layout of the complete aircraft when the increase in the critical *Mach* number was of concern.

In the report on the meeting "Flight Characteristics" of the Lilienthal-Gesellschaft für Luftfahrtforschung (LGL) (Lilienthal Society for Aeronautical Research), held April 8–9, 1943, *W. Friebel*, Reichsluftfahrtministerium (RLM) (State Ministry of Aeronautics), formulated in his introductory lecture 11 main concerns and wishes of the RLM. His last point addressed his concern about the "flight characteristics at high *Mach* numbers" [4]. In the meeting paper of *E. G. Friedrichs*, Deutsche Versuchsanstalt für Luftfohrt (DVL)

(German-Aeronautical Test Establishment), concerning the "examination of basic properties" as part of a plan for a revised form of Flying Qualities Specifications, presented by the DVL (*K. Doetsch*, DVL, 1943), the questions of the controllability at high speeds when approaching the speed of sound were discussed [5]. Serious difficulties had arisen during pull-out maneuvers where the elevator-control forces had increased above any known levels. A kind of overcompensation was observed in case of the rudder where the control effectiveness remained unchanged despite fading control forces so that any feeling of the pilot for the loads on his aircraft had been lost. The only way to protect the operators from possible accidents would be the limitation of the speed, better, even of the *Mach* number. However, here a *Mach* number indicator would be required, which had already been developed at the DVL. Particularly unpleasant was the fact that a fighter aircraft behaved quite normally, for example, during a dive at $M = 0.8$, but was no longer fully controllable during a pull-out. The danger of the associated flight testing was pointed out as well.

Thus it is clear that the phenomena of high flight speed, as far as they were related to the flight speed approaching the speed of sound, that is, concerning compressibility, were attentively pursued by both flight research and the RLM. The need for a systematic investigation of the effects and their explanation was expressed as well as the desire to point out the dangerous situation to the operational pilots.

5.1.2 MACH NUMBER EFFECTS

It is known that with the flight speed approaching the speed of sound, a gradual rearrangement of the airflow about the contour of the aircraft from a pure subsonic to supersonic flow conditions takes place. At first, some local flow velocities dependent on the outer form of the aircraft components immersed within the flow, may become equal or higher than the speed of sound. At the same time, zones of pure subsonic flow remain with shock waves occurring at the transition from supersonic to subsonic velocities (Fig. 5.1). The transition between the first appearance of shock waves and the flow turning completely supersonic is described as the transonic range. It may in the case of today's fighter aircraft assumed to be located between the *Mach* numbers of $M = 0.8$ and $M = 1.1$. The pressure distributions on the lifting surfaces, the fuselage, and the empennage are determined by the form of the airfoil, and its relative thickness, the planform and the sweep angle of the wings as well as the elevator and rudder configurations, including the geometry of the control surfaces, trailing-edge flaps, and the trim flaps. The pressure distribution is dependent on the *Mach* number, the angle of attack, and the dynamic pressure of the airflow and is responsible for the aerodynamic forces and moments acting on the aircraft within the transonic range. The pressure

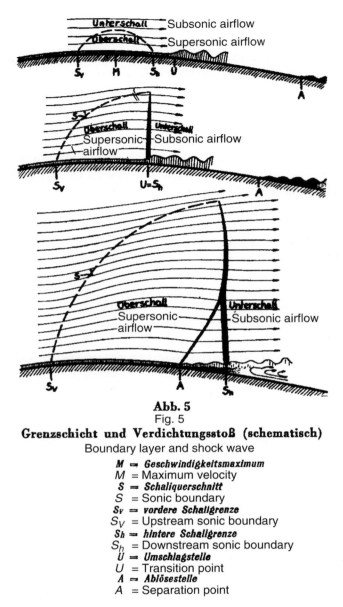

Abb. 5
Fig. 5

Grenzschicht und Verdichtungsstoß (schematisch)
Boundary layer and shock wave

M = *Geschwindigkeitsmaximum*
M = Maximum velocity
S = *Schaliquerschnitt*
S = Sonic boundary
S_v = *vordere Schallgrenze*
S_V = Upstream sonic boundary
S_h = *hintere Schallgrenze*
S_h = Downstream sonic boundary
U = *Umschlagstelle*
U = Transition point
A = *Ablösestelle*
A = Separation point

Fig. 5.1   **Boundary-layer and shock-wave interaction (*H. Helmbold*, 1942).**

distribution on the aircraft, changed by shock waves and due to the occurrence of pure supersonic flow, causes a downstream shift in the center of pressure and the neutral-point locations whose travel is configuration dependent. At a fixed center-of-gravity location, nose-down moments generally arise here. The uncontrolled motions of the aircraft about the pitch, roll, and yaw axes caused by an irregular pressure distribution must be compensated by control-surface deflections initiated by the pilot. With the *Mach* number dependent pressure distributions on fins, control surfaces, and trim flaps, the effectiveness of the control surfaces, and their corresponding hinge moments also change. So-called "buffeting" may, in addition, occur due to flow separations that form between the contour ("wetted area") and the local shock waves. The expressions, such as wing drop, nose slice, snaking, pitch down, pitch up, or *Mach* buffet describe aircraft reactions that may arise from such effects. The flight characteristics, primarily during air combat and weapons delivery, may considerably be affected, and the pilot has to pay full attention to keep the aircraft under control. So the elimination or compensation of the *Mach* number effects is of fundamental importance for a fighter aircraft, which is to operate within the near-sonic speed range.

The complex relations of the flow changes can only be determined in a limited way and even with today's computational methods and wind-tunnel investigations cannot be completely described and predicted in all details for a specific aircraft configuration. The aerodynamics of the most modern fighter aircraft within the transonic range may, therefore, provide surprises which only show up during flight testing.

The use of a swept wing or empennage cannot avoid these phenomena. But the associated *Mach* numbers are decisively increased and the aircraft reactions delayed and possibly attenuated. After the end of the war, numerous German fighter and bomber projects became known where the swept wing found extensive applications [6]. Some projects, like the Messerschmitt Me P1101, were in an advanced stage of development (see Chapter 6, Experimental Aircraft Me P1101).

The investigations of captured German jet aircraft in the United States, Britain, and France after the war were, among other things, aimed at determining the maximum *Mach* numbers to which they were safely controllable, and tactical maneuvers could be flown effectively during air combat encounters and which flying qualities they had as weapons platforms in this range. The swept wing, introduced for the first time on operational combat aircraft, namely, the Me 262 und Me 163, was of special interest in comparison to the jet aircraft Gloster "Meteor I" and de Havilland "Vampire," developed in Britain, as well as the Lockheed XP-80 and the Republic XP-84 in the United States, which were still developed, built, and tested with their unswept wings. The American jet-aircraft projects North American XP-86 and Boeing XB-47, and also the MiG 15, developed in the former Soviet Union, benefited especially as the first ones from the German wind-tunnel results concerning the aerodynamics of the swept wing.

5.1.3 PITOT-STATIC SYSTEM

One aspect that plays an important role when approaching the transonic range concerns the instrumentation of an aircraft to determine the total and static pressures (pitot-static system). The measurement of the static and the total pressures on the aircraft is also subject to considerable disturbances in the transonic range. The altitude indication, which is derived from the static pressure, as well as the indicated airspeed, which is derived from the difference between the measured total pressure and the static pressure on the aircraft, is also distorted depending on the *Mach* number. The correction of the displayed pressure readings is a prerequisite of the determination of the true *Mach* number of the aircraft. When using combined pitot-static probes (so-called *Prandtl* probes) mounted on a nose boom or strut-mounted pitot probe on the side of the fuselage or on the wing, one may generally assume that the indicated *Mach* number is below the true value in the transonic range due to a dynamic-pressure error of up to $M = 0.05$. This means that, for instance, at an indicated *Mach* number of $M = 0.97$, the aircraft is already flying within the supersonic speed range at a true *Mach* number of about $M = 1.02$. The well-known "jump" of the needle of the altimeter is a sure sign that one has now "gone supersonic." The jump occurs when the pressure rise due to the shock after reaching the speed of sound moves across the static pressure orifice to a position upstream of the pitot probe. A normal shock is now positioned upstream of the pitot probe at the nose of the aircraft. The altitude indication may jump here to a value up to 300 m lower than the one prior to the shock movement. Anyway, the altimeter indicates a lower value than the real one, because the static pressure in the bow region is, at least in the case of the usual extent of a boom-mounted probe, at subsonic velocities generally higher than the static pressure of the undisturbed atmosphere. When approaching the speed of sound, the static pressure decreases, dependent on the location of the static probe to increase in the transition shortly before and after $M = 1$ jumping to a substantially higher pressure than before. The static pressure measured by a bow-mounted probe and the true static pressure agree at supersonic flight with $M > 1$ quite well so that both altitude and speed indications are now nearly error free (Fig. 5.2).

The German fighter aircraft of WWII did not have any *Mach* number indicator on their instrument panel. Instead, they had partly so-called altitude-compensated airspeed indicators, which also showed, in addition to the speed derived from the dynamic pressure, at higher dynamic pressures and pressure altitudes the true airspeed relative to the surrounding air mass. Knowing the speed of sound (1225 km/h at sea level at a standard temperature of 15°C and 1062 km/h at an altitude of 11 km and −56.5°C) the pilots would, at least, have a idea concerning the margin of the actual airspeed of the aircraft with respect to the speed of sound or of their own flight *Mach* number. The speed limits in the operating instructions were given as values of the true airspeed (such as for the Me 262 A-1) [7]. At the DVL, equipment for the measurement and display of the *Mach* number [8] existed. Whether a *Mach* number

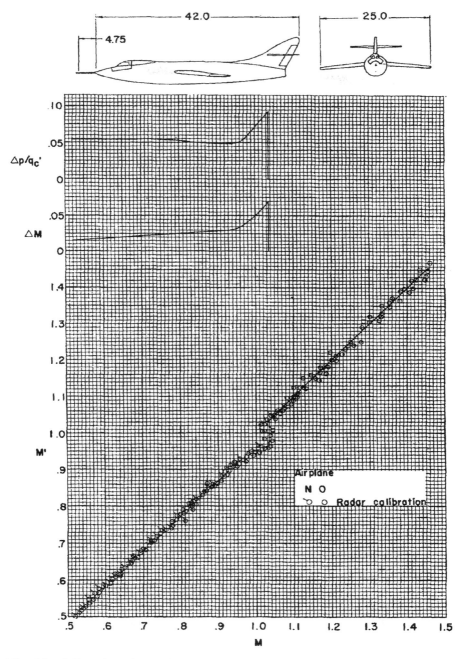

Fig. 5.2   Calibration of the pitot-static error on the D 558-II (NACA RM 57A02): Legend: *M'* = indicated *Mach* number, *M* = true *Mach* number.

warning device was evolved, which was later to be installed by way of a trial on the Me 163 and also on the Me 262, is not verifiable.

The pilot could not satisfactorily have controlled the approach to the stalling speed, for instance, solely by the altitude-compensated airspeed indication. If this would have been, for instance, 150 km/h at ground level, then the stalling speed at an altitude of 11 km would correspond to an indication of about 260 km/h on an altitude-compensated airspeed indicator. For flight control reasons, it was not possible to do without a speed indication, which evaluated the pressure difference between the measured total and static pressures. To stay with the example of the stalling speed, the latter would at ground level as well as at an altitude of 11 km provide nearly the same value for the stalling speed. Also the airspeed reading for best climb rate, maximum range, or maximum endurance does not change with the flight altitude; however, the associated true airspeed does. This corresponds to the information a pilot normally requires for the supervision of the flight condition and the speed limits of his aircraft. On the other hand, however, the indication of the true airspeed was indispensable for the flight planning and aerial navigation. The maximum flutter-safety speed was also defined according to the German, Bauvorschrifte für Flugzeuge (BVF) (design regulations for aircraft) as true airspeed. (Up to 1945, the BVF referred only to the terms "dynamic pressure" and "speed." The indication of the dynamic pressure on a scale calibrated in speed units was called "Fahrt," which means "indicated airspeed" in German). Two scale sectors were envisaged for the altitude-compensated airspeed indicator, which were used for high speeds and flight altitudes. The "stagnation pressure range" in case of the display of the Me 262 went up to 400 km/h. The second section, whose scale ranged from 400 to 1000 km/h, commenced at altitudes where the true airspeed exceeded 400 km/h (see Appendix 3, Figs. 5.15 and 5.16).

In the high-speed tests of the Royal Aeronautical Establishment (RAE) in Britain, carried out in 1943–1944, a so-called "*Mach* meter" was already employed as an instrument, developed at the RAE laboratories, and used in "Spitfire" high-speed dive tests [9]. This was claimed to be the first *Mach* meter in the world installed on an aircraft. With a speedometer indicating true flight speed one would, however, also be able to establish and monitor the true airspeeds representing critical *Mach* number values. All limiting speeds, which also corresponded to the critical values of the *Mach* number, were, indeed, as mentioned in the manuals of the German aircraft, given as values of the true airspeed. The speed indicator of the Me 262 had at 950 km/h a red mark on the scale, which corresponded on the altitude-compensated speed indicator also to the critical, for example, the maximum permissible *Mach* number, of the aircraft. Since 1956, German aeronautics took over the concept, common in the United States, Britain, and France, where the so-called calibrated airspeed and the *Mach* number are mostly read from a single instrument inside the cockpit whereas the true airspeed is generally not indicated.

5.1.4 CRITICAL MACH NUMBER

A very general definition of the notion of a critical *Mach* number is found in the well-known book of *Schlichting* and *Truckenbrodt* titled *Aircraft Aerodynamics*, which states "As critical *Mach* number one understands the *Mach* number, formed with the freestream speed and the speed of sound at the freestream conditions, where locally on the airfoil the speed of sound is reached. . ... shock waves occur and, as a result, also airflow separations and a strong drag increase" [10]. This definition was already the basis of statements and publications that prior to and during WWII, German wind-tunnel and flight-test results regarding *Mach* number contained dependent airfoil data (as in the book of *F. W. Riegels* [11]). This definition is basically also used in the NACA reports that were evaluated for this contribution.

In a report concerning the questioning of *Lippisch* by the Combined Intelligence Objectives Subcommittee (CIOS) as remarked by *Ch. Burnett* [12], the value may also be regarded as critical *Mach* number where compressibility effects (buffet onset, trim change) were first felt by the pilot during flight (see Appendix 1 of this chapter).

5.1.4.1 DRAG RISE. In conjunction with the airflow development in the transonic regime, a pronounced drag increase of the aircraft is observed ("drag rise"), which is dependent on the angle of attack (respectively the lift coefficient) and the *Mach* number. The onset of the drag rise can clearly be related to a particular *Mach* number of an aircraft, which could then directly be described as the "critical" *Mach* number for the drag rise. This relationship is, however, not clear in publications about corresponding measurements. Rather the definition mentioned previously, namely, the first attainment of $M = 1$ on the contour immersed within the airflow, is taken as basis. To a certain extent, the drag rise protected the early jet aircraft from penetrating the range of high subsonic *Mach* numbers with an aerodynamic design unsuitable for flying in the transonic range. The *Mach* number of a subsonic aircraft, which is diving at its maximum possible dynamic pressure, decreases with decreasing altitude because the static pressure increases, and hence the ratio of dynamic pressure to static pressure, which determines the *Mach* number, rapidly decreases. During the dive, the aircraft flies automatically as it were, out of the range of the critical *Mach* numbers.

On the other hand, the drag rise associated with high *Mach* numbers had to be overcome by a corresponding increase of engine thrust. At the attempts to reach and exceed $M = 1$, therefore, the aircraft had to dive at a more or less steep flight-path angle due to the poor performance of the engines. This meant, however, that at the same (critical) *Mach* number of the aircraft, the corresponding changes in the pressure distributions were shifted with decreasing altitude to higher levels of dynamic pressures, that is, to higher effective aerodynamic forces. It is no wonder that the first pilots confronted with *Mach* number effects had the impression of breaking through an invisible barrier, the "sound barrier." Aeroelastic effects may, in addition, superpose compressibility effects. This was the result of an investigation of the Me 262, where the compressibility effects were cushioned by the softness of the rear fuselage and the elevator without this having been realized by the designers (see Sec. 6.2.1 on the Me 262). If the propulsion performance at the time had been sufficient to allow the aircraft to break through the "sound barrier" at high altitudes, that is, at lower impact pressure, the expression "sound barrier" would possibly never have been created. Because of the lower density of the air, the dynamic pressure, responsible for the forces acting on the aircraft, is considerably lower at high altitudes and the same *Mach* number than at low altitude and hence less effective regarding the impact of the changes in the pressure distributions. Figure 5.3 presents a survey of the drag increase due to compressibility for a number of propeller and jet aircraft of the first generation.

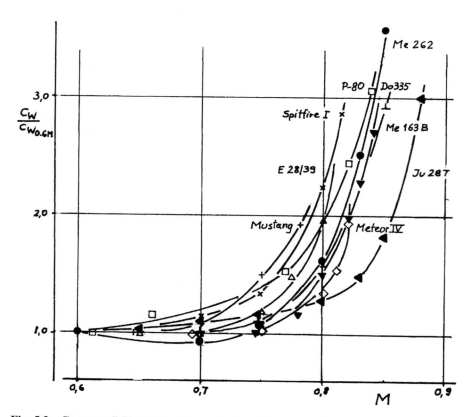

Fig. 5.3   Compressibility-related drag rise for various propeller and jet aircraft of the first generation.

5.1.4.2 TRIM CHANGES. In addition to the drag rise, a prominent change also occurs in the transonic speed range in the lift gradient. For thicker airfoils, the additional lift caused by an angle-of-attack change is, after a small increase, quickly reduced to less than half of its initial value within a few hundredths change in the *Mach* number. The neutral point as point of action of the change in lift caused by the change in the angle of attack shifts downstream when shock waves appear on the upper and lower surfaces of the airfoils. The pitching moment tends towards nose-heaviness at a fixed center of gravity. If the elevator gets into the turbulent wing wake thickened by shock waves, it may become completely ineffective. The local shock waves on the wing may, moreover, affect the downwash and hence the local angle of attack at the elevator. If the pressure distributions on the fin/control surface combination change due to shock waves, the control surface may feel as if it is locked. If it is, however, located in the area of the shock-induced separated region of the horizontal stabilizer, it may lose its effectiveness and is felt to be "loose," that is, the hinge moment goes towards zero. The *Mach* number, which is characteristic for the associated effects, such as the trim change and the controllability, can be regarded as another critical *Mach* number this time related to the erratic stability and the control reactions. It is also dependent on the lift coefficient tending here to lower values the higher the angle of attack becomes. It does generally not coincide with the critical *Mach* number defined for the drag rise (the drag rise *Mach* number). Figure 5.4 shows for various aircraft the change in the elevator angle due to compressibility required to maintain a trimmed lift coefficient of 0 to 0.1 when the *Mach* number exceeds critical values. The rapidly changing stability and maneuverability behavior of the aircraft is difficult to judge by the pilot and requires all of his attention in controlling the aircraft. When approaching the critical *Mach* numbers, the sudden tendency of the aircraft to pitch down, which could not effectively be countermanded, must have been surprising and at the same time alarming to the pilot. Even more so the fact that the aircraft resisted a pull-out by unexpectedly building up almost insurmountable elevator control forces at increasing load factors or completely blocking the control stick. The pilot could have been prepared here by an intensive briefing. In most instances, as, for instance, in case of the Me 262, the pilot had to retrain, by familiarizing himself with flying the aircraft in Mach critical regimes. There were only a few Me 262 twin-seat aircraft that could have been used for familiarization and training flights.

5.1.5 PROJECTS

As described by *D. Fiecke* [13], a look at German jet-aircraft projects resulting from RLM contracts, being designed and built at the end of 1944 and the beginning of 1945 during the last hectic months of the war at the

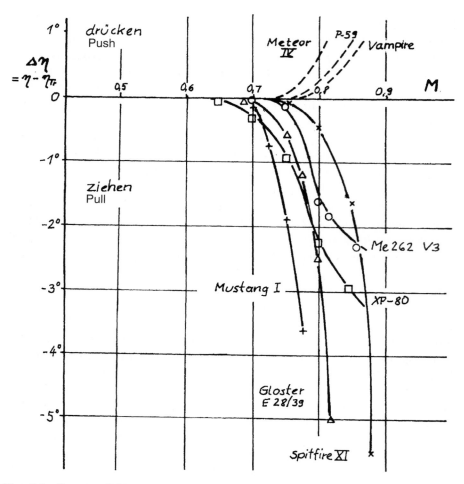

Fig. 5.4 **Compressibility-related change of the elevator angle for pitching-moment equilibrium at lift coefficients between 0 to 0.1 and altitudes of 5 to 7 km for different propeller and jet aircraft of the first generation.**

well-known German aircraft companies, is, therefore, interesting because it shows the status of the applied aerodynamic research results achieved and how practical solutions might look. On instructions from the RLM, the designs presented were subjected to a comparative evaluation by the DVL which was, however, confined exclusively to aerodynamic and flight mechanics questions. The objective was the selection of an efficient jet fighter, which could quickly be achieved and whose design and manufacture had, in addition, to consider the prevailing shortage of high-quality material and an efficient series production despite using unskilled personnel who were supported, for instance, by companies outside the aircraft industry (for example, furniture factories producing wooden aircraft components). The involvement of the

DVL in the assessment of the aerodynamic/flight mechanics aspects had, among other things, become necessary because the computational methods employed by the companies were not coordinated and consistent aerodynamic assumptions were missing as well as corresponding extended design specifications suitably adjusted to the extension of the flight regime into the immediate vicinity of the speed of sound. Even under the care of the DVL, the computational methods and considerations within the tasks were still strongly determined by the familiar incompressible aerodynamics of the propeller era. The report of *Fiecke*, published at the DVL eight years after the war, is probably a revised and adjusted version and does, therefore, not quite reflect the situation prevailing in January 1945.

The flight characteristics to be expected when approaching the speed of sound could only vaguely be assessed due largely to missing confirmation of the wind tunnel and computational results by flight tests. The wing sweep of the suggested project designs, partly going up to 35 deg, would have meant a clear increase in the critical drag-rise *Mach* number. Sooner or later one would, in all probability, have been confronted during test flights with the subject of the "transonic pitch up" as well as with the "pitch-up" effect at high angles of attack. As the development of the swept wing in Britain, the United States, France, and the Soviet Union has shown, one would not have been able to avoid a long period of intensive flight research and trials to finally validate the swept wing for operational aircraft. The advanced project aircraft Me P 1101, found after the end of the war by the U.S. troops at the Messerschmitt plant at Oberammergau, was nearly complete and could have still in 1945 undergone flight trials (for details see Chapter 6 Experimental Aircraft P 1101). The continuation of the P 1101 project under U.S. management resulted in the X-5 research aircraft developed by the Bell Company. It took, typically enough, up to about 1955 before the results of the NACA concerning the flight trials of the X-5, which possessed a variable-wing sweep which could be adjusted between 20–59 deg during flight, were completely reduced and presented in form of reports. To be fair, it must be mentioned that the wing sweep of the original Me P 1101, different from the X-5, was only adjustable on the ground. Among other things pursued by the Messerschmitt Company was the objective to find a wing sweep that would constitute the most favorable compromise between the expected increase in the drag-rise *Mach* number and the unfavorable flight characteristics present at high angles of attack.

In the following sections we will report on the experience gained during the flight trials and in combat missions with compressibility effects prior to and during WWII with aircraft generations developed in Germany. At first, these were fighter aircraft, then also light bombers that were able to reach very high speeds. Comparisons with the experience made in the opposing foreign countries, primarily in the United States and Britain, at the same time with regard to compressibility effects can be drawn by considering the aircraft developments carried out there. Suitable examples of propeller and jet aircraft are discussed. Also revealing are reports on captured aircraft found in Germany after the war, which were partly flown and tested in the United States, Britain, and France up to 1948. Projects started in Germany and later continued in the victorious countries as far as the documents can contribute to the subjects within this chapter are also included.

5.2   PROPELLER AIRCRAFT

At first, the occurrence of compressibility effects on propeller aircraft may be surprising. The available shaft horse power of reciprocating flight engines strongly increased at the end of the 1930s and reached up to 2500 hp towards the end of the war in Germany for serial-production engines. The development abroad proceeded similarly. This performance as such would have already been sufficient for a single-engine fighter aircraft to fly close to the sound barrier, that is, to overcome the drag rise. The transformation of the shaft horse power into thrust proved, however, to increase insufficiently when approaching high *Mach* numbers. The main reason here is the propeller efficiency, which decreases rapidly with the occurrence of shock waves at the tips of the rotating propeller blades. References to the effect of compressibility on the propeller profiles had already shown up in 1933 as follows from a report by *J. Stack* [14]. In 1937–38, papers appeared in Germany that investigated the potential of propeller propulsion to increase flight speed [15, 16]. The propeller was ruled out for propulsion reaching high *Mach* numbers within the transonic regime and at supersonic speeds.

The improvements in the drag characteristics were the reason, however, that also propeller aircraft could reach flight speeds in a shallow dive high enough for compressibility effects to occur on wings and stabilizers. Significant examples of German and Allied propeller-driven fighter aircraft are addressed in the following sections.

5.2.1   GERMANY

5.2.1.1   MESSERSCHMITT ME 109.   During an in-flight demonstration of the Bf 109 B with a Jumo 210 engine, an inexplicable crash occurred in 1937 when the test pilot of the German Air Force Flight Test Center at Redlin (Erprobungsstelle der Luftwaffe), *K. Jodlbauer*, was not able to pull his aircraft out of a high-speed dive, starting at an altitude of 5000 m, and plunging nearly vertically into Lake Mueritz near Rechlin. *H. Beauvais* of the E-Stelle had the case of *Jodlbauer* on his mind and was warned, when in 1941 during a dive test he was barely able to safely pull out his Me 109 E using both hands and all his strength. The RLM had ordered the E-Stelle to investigate the dive behavior of the Me 109 after the German Air Force had reported difficulties during the high-speed dive of the aircraft and the subsequent pull-out. About one year later, fatal crashes with the Me 109 G had accumulated during air combat missions where the pilots, partly in the heat of a fighter engagement during air combat, obviously had become too fast in diving their aircraft and were no longer able to recover. The Messerschmitt Company, therefore, carried out a program named "Final Dive Speed Me 109 G." In 28 flights test pilot *L. Schmid* systematically increased the dive speed and initial altitudes and noticed that the manual control force erratically changed during a dive repeatedly from push to pull depending on the trim position set on the elevator [17]. He finally found a most favorable trim position, which had to be set at the start of the dive in order to be able to counteract the tendency of the Me 109 G to pitch down, beginning at about $M = 0.78$ to 0.80, with still tolerable manual forces thus allowing the pilot a safe pull-out. In addition at approximately $M = 0.8$, uncontrolled rolling motions occurred, which would today be classified as "wing drop." By trimming the aileron trailing edges, they could be rendered to some extent controllable. During the tests of the Me 109 G, a maximum *Mach* number of $M = 0.805$ was reached at an altitude of 7 km in a dive with flight-path angles of 70 to 80 deg.

Figure 5.5 shows the speed and *Mach* number development during a dive by an example taken from the Me 109 test series and a comparison with a test, which is described in Sec. 5.2.2.2.

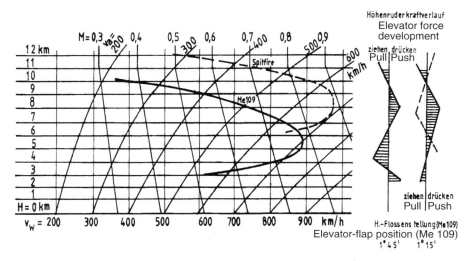

Fig. 5.5   Development of the indicated and true airspeed and *Mach* number as well as the required elevator deflection during dives of the Me 109 and "Spitfire"; legend: $v_a =$ indicated airspeed, M = *Mach* number, $v_w =$ true airspeed.

5.2.1.2 FOCKE-WULF FW 190 A, Fw 190 D-9, TA 152. In the summer of 1943, the
Fw 190 gave increasing reasons for complaints as follows from reports of
pilots involved in flight testing and at the front. From high altitudes it would
not have been possible to pull the aircraft out of dives at high and medium-
high altitudes; however, success was achieved in some instances at low altitudes. In addition to the Fw 190, concern was most of all with the Me 109.
Compressibility was named as a possibility because the phenomena suggested the formation of local shock waves on wings and empennages, possibly
superposed by effects of elastic deformations as a result of high dynamic
pressures and angles of attack. The Fw 190 had, of course, within the framework of strength testing undergone high-speed dives. Here, speeds between
930 and 955 km/h at an altitude of 5 to 6 km, corresponding to $M = 0.82$ to
0.84, were said to have been reached.

In 1944 the DVL, commissioned by the RLM, carried out investigations
concerning the behavior of the Fw 190 A equipped with the BMW 801
engine at high speeds (similar tests followed later also with the Me 262 V-3;
see Sec. 6.2.1). Included were quasi-stationary inclined-trajectory flights
with load factors around $n \approx 1$. Higher angles of attack were achieved during
dive recoveries. Starting at an altitude of about 1 km, the recoveries from
dives were performed at an altitude of 7 km and $M = 0.70$, starting at 9.5 km
and 6 km at $M = 0.75$, and starting at 5 km and 3 km at $M = 0.65$ to investigate the effect of the angle of attack. To fly at the highest attainable *Mach*
numbers, flight-path angles of up to 90 deg were selected. At a dive from an
altitude of 10-km *Mach* numbers of $M = 0.76$ were reached prior to pulling
out at an altitude of 7 km [18]. The report states that the deviations in the
measured elevator angles from the expected values had to be attributed to
elastic deformations and propeller-wake effects. Assumptions that the perturbations in the balance of the pitching moment could have been caused by
shock waves occurring within the flight regime were not confirmed
(Fig. 5.6).

The reports of the chief designer of Focke-Wulf, *K. Tank*, who used to fly
his aircraft personally, show that he reached with the Fw 190 A-7 in inclined-
trajectory flights from an altitude of 6 km an indicated airspeed of 700 km/h
[19]. This corresponds to a *Mach* number of about $M = 0.8$ and to a true flight
speed of 910 km/h (assuming a standard atmosphere without errors in the
pressure measurements but with compressibility corrections applied). The
load factor at pull-out was $n = 7$. *Tank* reported that during the dive "white
flakes flew off the wings." This was very likely due to the formation of con-
densate, as may be observed in areas of low static pressures upstream of local
shock waves after having exceeded the speed of sound, if the humidity of the
air is sufficiently high. Nearly the same values of the airspeed and *Mach*
number were achieved with the Fw 190 D-9 equipped with the V-12 engine
Jumo 213 A. Remarkable is that *Tank* again did not mention any of the known

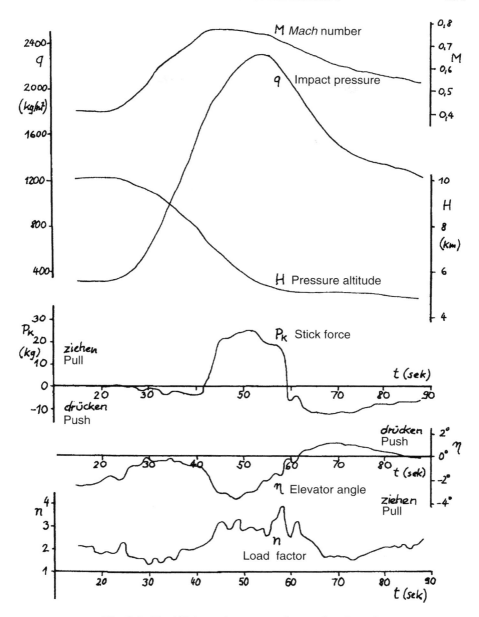

Fig. 5.6   Fw 190 A, testing process for $n = 2$ and $n = 3$.

compressibility effects, such as buffet, vibrations, or loss of controllability, except for a distinct increase in the control forces, which were otherwise generally slight and agreeable. The pull-out that was carried out at high lift coefficients was also flawless. So the topic "*Mach* number effects" hardly seemed to play a role during the practical operation of the Fw 190. From individual field reports one may conclude, however, that also in the case of the Fw 190 *Mach* effects could have played a role, but were, however, not as pronounced as in the case of the Me 109 [20].

Aircraft of the Ta 152 series, which evolved from the Fw 190 D series equipped with in-line engines (Jumo 213, DB 603), dived at the beginning of 1945, according to reports [21], during introductory and training flights of the Luftwaffe pilots with speeds of up to more than 1000 km/h. Here, only strong buffeting of the wings was observed each time. The pull-out was faultlessly possible with the electrically operated elevator trimming device. For the Ta 152 H, only a maximum speed of 800 km/h had been recommended. The E-Stelle in Rechlin even had doubts whether the aircraft should dive with more than 600 km/h, because "strong instabilities" about the vertical (yaw) and cross (pitch) axes were supposed to have been observed [22].

Whether the aerodynamic behavior with regard to *Mach* number effects of the FW 190 was more favorable than that of the Me 109 and also of the Me 262 cannot be decided without exact test results. The difficulty is that at high dynamic pressures a superposition of *Mach* number effects and aeroelastic effects is unavoidable. It was known that the British "Spitfire" showed a considerably higher critical *Mach* number (regarding loads) than the first British jet aircraft Gloster E28 "Meteor I" or de Havilland "Vampire." In the case of his aircraft, *Tank* had obviously instinctively done the right thing at high *Mach* numbers with regard to the selection of the wing and empennage airfoil sections and the geometry and the arrangement of the stabilizers as also had the Vickers Supermarine Company with the "Spitfire." Anyway, in his lecture at the Academy of German Aeronautical Research in May of 1942 [23], he had revealed that he and his development engineers were well aware of the problem of *Mach* number effects and the associated open questions. His conclusions and recommendations on aeronautical research activities fitted well to what *W. Messerschmitt* had reported in conjunction with the high-speed accidents of the Me 109 (possibly also the Me 210) [24]. The wings of the Focke-Wulf fighters had a relative airfoil thickness of nearly 16% at the root and 13 to 9% in the area of the aileron respectively at the wing tips and used airfoils of the NACA 24 series [25]. Compared to the Messerschmitt fighter, the Fw 190 had an important difference in the empennage design possibly important to the load behavior in compressible flow. The horizontal stabilizer of the Me 109, and also that of the Me 262, was mounted at approximately half the height of the vertical stabilizer. Especially at higher angles of attack, flow separations due to shock waves on the wings could lead to a loss in the effectiveness of the control surfaces. Shock formations on the control surfaces themselves may, in addition, contribute. Aerodynamic corrective actions on the intersection of the vertical tail with the elevator of jet aircraft of the first postwar generation, for instance, on the British Gloster "Meteor" or the Hawker "Sea Hawk," suggest similar problems. The horizontal stabilizer of the Fw 190 and the Ta 152 were, on the contrary, mounted lower and at the fuselage itself. There, the consequences of shock-induced separations seemed to be less compared to the empennage typical of the Messerschmitt aircraft. The "Spitfire" also had the horizontal stabilizer attached to the fuselage and, as is well known, could safely be dived well above $M = 0.8$. In case of the American and English supersonic fighter aircraft of the 1950s, horizontal stabilizers attached in low positions were preferred to keep the elevators out of the wing wakes at high angles of attack and to guarantee a sufficiently high nose-heavy pitching moment also in the case of the airflow stalled on the wing. Remarkably one has not deviated from these considerations since the beginning of the 1970s in the case of the American aircraft F-14, F-15, F-16, and F-18 up to today's F-22.

5.2.1.3 DORNIER DO 335. In 1944, the Dornier Company entered the market for heavy fighter aircraft with the development of the Do 335. Up until that time, the name *C. Dornier* was associated with seaplanes (flying boats), amphibious aircraft, bombers, and night fighters derived from the latter. Although the Do 335 was described as the fastest reciprocating-engine driven fighter of the war, there is little information in the literature (for example, *K.-H. Regnat* [26]) noting any deficits in the flight characteristics due to compressibility effects. It is suspected that the Do 335, similar to the Fw 190 and Ta 152, profited from a favorable choice of airfoil sections and stabilizer (empennage) geometry. At the *Mach* numbers ($M = 0.76$) reached during the flight trials, there did not seem to have been any serious *Mach* number effects. Only *W. Green* [27] points out that at the end of 1943 during high-speed tests with the Do 335 V-1 at Rechlin yaw movements ("snaking") and nose up and down motions ("porpoising") about the pitch axis had been observed. However, these observations could not challenge the generally positive impression that the test pilots of the E-Stelle apparently had of the flight characteristics of the Do 335. A model of the aircraft was investigated in the high-speed wind tunnel of the DVL in 1944. The Do 335 distinguishes itself with regard to the drag rise as function of the *Mach* number remarkably little compared to that measured for the Me 262 or Me 163 [28], although the wings of the Do 335 showed a relative airfoil thickness of only 12% (airfoil NACA 23012-635). The performance data of the Do 335 A [29] taken from the information in the aircraft manual show an authorized maximum horizontal speed of $V_h = 600$ km/h at ground level and 835 km/h at an altitude of 8700 m. The latter corresponds at standard atmospheric conditions to a *Mach* number of $M = 0.76$. The authorized maximum stagnation pressure during a glide corresponded to a true airspeed of $V_W = 900$ km/h near the ground ($M = 0.735$ at standard atmospheric conditions). Although further details are not available, one may assume that at the maximum permissible true airspeed, corresponding to $M = 0.76$, still no noticeable trim changes affecting the flight characteristics occurred. From the reports available, it cannot be determined whether the Do 335 completed the dives and pull-out testing required by the Construction Group H5 of the BVF of that time without encountering noticeable *Mach* number effects. According to the BVF, a dive from a maximum altitude of 6000 m was sufficient to prove flutter safety; here, the highest dynamic pressure was probably reached but hardly also the maximum, that is, the critical, *Mach* number of the aircraft. This had not yet been considered in the BVF effective in 1936 through 1945.

5.2.2 GREAT BRITAIN

5.2.2.1 ROYAL AERONAUTICAL ESTABLISHMENT. An analysis of the contemporary propeller aircraft of the countries opposing Germany during WWII offers itself for comparison. It quickly became clear that they too had to cope with the phenomenon of compressibility and the resulting *Mach* number effects. Because an important limit for the tactical use of a particular type of aircraft was reached with the maximum "flyable" *Mach* number, the RAE in Britain devoted considerable effort, since 1941, to investigate the phenomena and, if possible, to remedy the situation be it by limiting the airspeed or by aerodynamic means. It could be assumed that the German aircraft, whose aerodynamic design was not fundamentally different from the British high-performance aircraft, were subject to quite similar effects. The lack of suitable test facilities for high-speed investigations, for example, on airfoils for fast rotating propeller blades, led to the construction of a new powerful wind-tunnel facility at the RAE in Farnborough where an airflow velocity of 600 mph (966 km/h) was supposed to be reached. The wind tunnel was commissioned in November 1942. Because of the war situation, it was inevitable that applied research rather than basic research had priority to be able to give answers to current questions. In 1941, a critical situation arose for Fighter Command when the German Fw 190, an aircraft, arrived at the front whose performance surpassed that of the Bristish fighters "Spitfire" and "Hurricane." Comparative flights carried out by the Royal Air Force (RAF) with the "Spitfire VB" and the Fw 190 A-3, mistakenly landed by *Otl Faber* in 1942 in England, showed that the Fw 190 was faster than the "Spitfire" at altitudes of up to 3 km by 40–56 km/h and at 7.5 km by 32 km/h and also had a higher climb rate. The resulting increased engine performance of the "Spitfire VC" (as "Spitfire IX" the most built version of the British fighter aircraft), only allowed at altitudes above 7.6 km considerably higher speeds than the A-3 to be achieved [30].

   In May 1943, the RAE started a test series with the objective of comparing wind-tunnel results of high-speed model measurements available up to then with in-flight measurements and to analyze the differences. The questions to be answered ranged from drag rise and trim changes due to compressibility, to the confirmation of the load assumptions for high subsonic speeds. The flight tests, which consisted of dives at flight-path angles up to 45 deg from high altitudes, were initially carried out with the U.S. fighter aircraft North American P-51 "Mustang." However, the "Mustang" with its single-stage supercharger was only able to reach an altitude of about 28,000 ft (8534 m), and at least another 10,000 ft (3048 m) was required to reach high *Mach* numbers, which only a "Spitfire IX," which could climb as photo-reconnaissance aircraft to an altitude of 40,000 ft (12,192 m), was likely to achieve. The aircraft was equipped with cameras to register airspeed, pressure altitude, *Mach* number, and load factors. Later, further fighter aircraft, among them the "Typhoon," "Tempest," American Republic P-47 "Thunderbolt," and Bell P-39 "Airacobra," were evaluated. Already at the time of the tests in 1943, occasional crashes of British and American fighter aircraft had occurred because they could not be pulled out of steep high-speed dives. The exact circumstances were, however, not clear, and measurement data were lacking almost completely. The tests gained further importance due to the fact that under controlled test conditions valuable experience concerning the behavior in the presence of compressibility effects and how to overcome them were gained. The findings were immediately passed on to the operational pilots thus certainly contributing considerably to the situation awareness of the pilots so that they did not encounter *Mach* number effects unprepared and knew how to cope with them [31].

Here, one recognizes the parallels to the experiments of the DVL on the Fw 190 or of the Messerschmitt Company on the 109 regarding both the cause as well as the phenomena occurring and their evaluation. Whether and how knowledge concerning compressibility effects reached the operational pilots of the German Air Force can no longer be reconstructed. The training of the German fighter pilots offered, not the least due to time and fuel shortages, too little opportunity to treat the topic of *Mach* number effects in detail. The availability of validated information from research or industry data, which could go via the RLM to the operational units, seems to have been only poor.

5.2.2.2 VICKERS SUPERMARINE "SPITFIRE." The tests run by the RAE in 1943–1944 at Farnborough showed that the "Spitfire XI" could dive at *Mach* numbers far higher than $M = 0.80$. In dives from an altitude of over 10 km, the aircraft started at $M = 0.75$ to become slightly nose heavy. The control stick had to be pulled to maintain the flight-path angle. The nose-heaviness increased up to $M = 0.83$, remained, however, controllable. At *Mach* numbers above $M = 0.85$, tail-heavy trimming did not suffice to end the pitch-down trend, and finally at $M = 0.86$ also a pull force on the stick of 70 daN (152 lb) did not suffice to recover the aircraft [32]. On one of these flights, during a dive from an altitude of 40,000 ft (12,200 m) and a flight-path angle of 47 deg and with the stick full aft, the test aircraft reached a *Mach* number of $M = 0.89$

at altitudes between 31,000 ft (9500 m) and 27,000 ft (8200 m). The *Mach* number dependent drag rise prevented a further increase in the *Mach* number. The controllability problems automatically disappeared with decreasing altitude at about $M = 0.83$. Caution was required during the pull-out process where between $M = 0.82$ and 0.78 after an initial pull at a nose-heavy trim, the stick had to be strongly pushed forward to avoid a strong pitch-up and hence a large load factor increase (see Fig. 5.5). The "Spitfire" was even said to have reached $M = 0.92$, where the aircraft must have been, however, in a nearly vertical flight trajectory and completely out of control. On one of these high-speed dives, the "Spitfire PR XI" (Serial No. EN.409) used for the tests lost propeller and gearbox because of overspeeding the engine at $M = 0.9$. The aircraft could, however, be recovered and safely landed at the Boscombe Down test center [33]. The *Mach* numbers achieved and the *Mach* number limit of $M = 0.85$ finally laid down in the pilot's manual were substantially higher than the ones of the English jet aircraft of the time such as the Gloster "Meteor" ($M = 0.78$) or the de Havilland "Vampire" ($M = 0.75$). The "Spitfire" could — if only during a steep dive — at the *Mach* numbers achieved by the Me 262 safely and also in a tactically meaningful way compete well keeping her own full controllability. In this way the aircraft was able to surprise the Me 262 during air combat also at flights within the high-speed regime.

The results of the high-speed tests were subjected to an assessment during a meeting in 1946 between the RAE and experts of the Vickers Supermarine Company. Altogether 15 aircraft were investigated up to their *Mach* number limits. The "Spitfire" was the only aircraft where the limiting factor was the drag rise. Others were, however, limited by oscillations in pitch, uncontrollable nose-heaviness, and buffeting. In the case of the American aircraft, investigated *Mach*-number-related "aileron buzz" was the limiting factor, especially with regard to structural loads aspects. Some results are compiled in Table 5.1.

5.2.3 UNITED STATES OF AMERICA

5.2.3.1 LOCKHEED P-38 "LIGHTNING." Compressibility effects also troubled the American Lockheed P-38 at the beginning of the development trials. The aerodynamicists and the development team of the Lockheed Company in about 1937 were conscious of possible compressibility effects; however, there was no practical flight-test experience. In 1942, the company test pilot *R. Virden* was not able to pull the aircraft P-38 E out of a dive, which he wanted to demonstrate to an observer team of the U.S. Army. The P-38 was probably the first U.S. fighter aircraft where compressibility effects were proven to be the cause of a crash [34]. Besides the wing, the *Mach* number effects affected primarily the twin tail unit (empennage). When approaching the critical *Mach* number, a strong nose-heavy pitching moment developed.

TABLE 5.1 LIMITING FACTORS

Aircraft	Propeller = P Jet = S	Maximum test *Mach* number	Limiting factors
Mustang I	P	0.80	Pitch oscillations
Mustang III	P	0.82	Nose-heaviness, porpoising
Spitfire XI	P	0.9 +	Propeller efficiency, trim changes
Spitfire IX	P	0.85	Drag rise
Spitfire XXI (with guns)	P	0.85	Buffeting
Spitfire XXI (without guns)	P	0.88	Drag rise, propeller efficiency
Tempest V	P	0.85	No information, probably nose-heaviness
E28/39	S	0.82	Buffeting
Vampires I	S	0.80	Pitch oscillations, tail-heavy
Meteor I	S	0.80	Buffeting, tail-heavy
Meteor I (long nozzle.)	S	0.84	Trim changes, tail-heavy

This could only be eliminated by an electrically operated "dive flap" later installed under each wing. The resulting change in the pressure distribution reduced the nose-heavy moment. In 1944, Lockheed test pilot *Kelsley* achieved an indicated flight speed of 750 mph (1207 km/h) with a P-38J that resulted after a correction of the static pressure source still in an indicated airspeed of 550 mph (885 km/h) [35]. This would have corresponded at an altitude of 4 km to a *Mach* number of approximately $M = 0.9$. During combat, it was advisable to pull out at an altitude of 30,000 ft (9144 m) with no more than 290 mph (467 km/h)/3 g, at 20,000 ft (6100 m) with no more than 360 mph (580 km/h)/4.5 g, and at 10,000 ft (3050 m) with no more than 430 mph (692 km/h)/6 g if one wanted to avoid buffeting and a "pitch down" [36]. The P-38 was, by the way, the first fighter aircraft that was equipped with hydraulically supported ailerons to cope with the large aileron forces at flights with high dynamic pressures.

5.2.3.2 REPUBLIC P-47 "THUNDERBOLT." During high-speed tests in 1941, airspeeds had already been reached during dives with the heavy and stocky P-47 where local shock waves with airflow separation (shock-induced separation) occurred on the wings. This led to *Mach* number related buffeting and strongly impaired the controllability about the pitch axis at high dynamic pressures and *Mach* numbers. The aircraft developed in dives above $M = 0.75$ a strong tendency to independently increase the flight-path angle up into the vertical and further. Even with a 40- to 50-lb (18- to 23-daN) fully pulled elevator, this could not be avoided ("frozen stick dive"). In an almost vertical dive, a maximum *Mach* number of $M = 0.86$ was reached. The controllability about the pitch axis was regained only below an altitude of approximately 20,000 ft (6100 m) after a more or less uncontrolled dive of 15,000 ft (4570 m) with a decrease in the *Mach* number. If the elevator (or the trimming if initially set to a tail-heavy position) was not released quickly enough, enormous *g*-forces could build up in a short time (up to $n = 8$), which would render the pilot unconscious ("black out") and lead to bent wings [37]. As an immediate measure, as in the case of the P-38, electrically operated "dive recovery flaps" were mounted on the lower sides of both wings. Also with the latter, one was not supposed to pull out at altitudes above 25,000 ft (7620 m), at speeds of more than 400 mph (644 km/h), and at lower altitudes at less than 500 mph (805 km/h). When exceeding these values, the use of the elevator trim for a pull-out recovery was recommended in case of an emergency [38].

5.2.3.3 NORTH AMERICAN P-51 "MUSTANG." Results of measurements on a wind-tunnel model of the P-51 carried out at the RAE in Britain in April 1945 compared to flight-test and wind-tunnel data of the NACA [39] showed that the critical drag-rise *Mach* number of the P-51 amounted to $M = 0.70$ to 0.74. The smaller value dated from results of the NACA wind-tunnel and flight tests. It deviates from the RAE value, which is mentioned second. The values have been determined during flights at conditions not closely specified and may already differ due to different lift coefficients. Similarly different are the critical *Mach* numbers determining the trim positions of the elevator. The value given by the NACA for the loads-related critical *Mach* number here is $M = 0.75$ while the value determined by the RAE amounts to $M = 0.7$. Information concerning the associated lift coefficients is missing here too. Results of model measurements of the pitching moment as a function of the *Mach* number show that at a lift coefficient of 0.6 the aircraft starts already at $M = 0.65$, and at 0.8 already at $M = 0.63$ to develop a strong nose-down pitching moment. The drag characteristic is quite comparable to that of the Me 109 and other German aircraft.

Probably little known are tests that the NACA carried out in 1944 with a P-51B without propeller whose results were to be used for a validation of high-speed drag data that were obtained on a 1/3-model in tests in the 16-ft tunnel of the Ames Aeronautical Laboratories. To exclude the influence of the propeller effects in the comparison with the wind-tunnel model data, the propeller of the P-51B was removed. The in-flight measurements were carried out in high-speed dives at lift coefficients near zero after the P-51 had been towed by a P-61A "Black Widow" to altitudes of 24,000 to 28,000 ft (7300 to 8500 m). The in-flight *Mach* numbers were increased stepwise from $M = 0.71$ to 0.76 during three flights. The compressibility-related increase in drag occurred at approximately $M = 0.68$ to 0.70. The drag coefficients obtained with the 1/3-model agreed surprisingly well with the data obtained from the in-flight measurements [40]. At these *Mach* numbers, controllability problems or trim changes were, similar to the case of the Fw 190 (see Sec. 5.2.1.2), not definitely identified. The flights of the propeller-less P-51B ended with an emergency landing during the fourth flight after the towing cable broke at a low altitude and partly wrapped itself around the P-51. The pilot, *J. M. Nissen*, escaped the emergency crash landing with only slight injuries. The aircraft was destroyed; the instrumentation could, however, be recovered undamaged. The tests were not resumed.

The available literature does not mention any complaints concerning anomalies in the flight characteristics of the P-51 at high speeds other than the ones already known from similar aircraft. The pilot training manual of the P-51 specifies maximum permissible flight speeds assigned to certain flight altitudes: at 40,000 ft (12,200 m) 260 mph (418 km/h) corresponding to $M = 0.74$, at 20,000 ft (6100 m) 400 mph (644 km/h) corresponding to $M = 0.73$, and at 5000 ft (1500 m) still 505 mph (813 km/h) corresponding to $M = 0.72$ were safely possible; g-loads of up to 8 g were allowed [41]. The manual reminds the pilots: "Never exceed these speeds. If you do, you are asking for trouble." With high probability, *Mach* number effects were decisive.

5.2.3.4 BELL P-39 "AIRACOBRA." The Ames Aeronautical Laboratory of the NACA at Moffet Field, California, carried out in 1945–1946 high-speed tests on a fighter aircraft P-39 "Airacobra." They showed that at *Mach* numbers above $M = 0.55$ to 0.75 a tail-heavy pitching moment developed which had to be compensated by pushing the control stick requiring up to 70-lb (32-daN) stick force. At a lift coefficient of 0.2, the stick force became less beginning at about $M = 0.77$ to 0.78 as the tail-heavy pitching moment decreased. The drag, whose rise due to compressibility effects commenced at $M = 0.7$, became so high that even the full engine power in a vertical dive did not suffice to exceed $M = 0.8$.

5.3 JET AIRCRAFT

Compared to the propeller, the thrust of the jet-turbine engines decreases initially relative to the static or takeoff thrust with increasing dynamic pressure and at the same altitude; however, thereafter it increases noticeably with increasing airspeed. Together with the wing-fuselage designs chosen for the early jet aircraft and despite the conventional aerodynamics, a further increase in the flight speed was achieved, which could tactically be used. The (hidden) objective of flying faster than the speed of sound was now within reach. Towards the end of the 1930s in Germany, the Heinkel Company had performed pioneering work and had as a first company not only developed a jet engine, but had also successfully demonstrated it in flight on an aircraft, the

He 178. Arado, Junkers, and Messerschmitt also developed subsequently, in addition to Heinkel, jet-propelled aircraft that were to go into operation as combat aircraft, fighters, bombers, and reconnaissance aircraft. The experiences of Germany, Britain, and the United States at high flight speeds with jet aircraft are described next.

5.3.1 GERMANY

The first flight of an aircraft with a turbojet engine took place on August 27, 1939, at the airfield of the Heinkel Company in Rostock-Marienehe. Both the jet engine and aircraft were developed by Heinkel. The aircraft was the He 178, and with it, the test pilot, *E. Warsitz*, performed altogether about 15 flights. The airspeeds achieved were at the most 600 km/h, which does not quite correspond to $M = 0.5$ at nearly sea level. This was still far away from speeds and *Mach* numbers that could have led to compressible effects.

5.3.1.1 ARADO AR 234A. At about the same time as work on the Me 262 started at the Messerschmitt Company, the Arado Company commenced studies for an aircraft with jet engines. At the end of 1941, the design of a single-seat long-range reconnaissance aircraft, the Ar 234A, existed with a straight wing and two Jumo 004B engines attached to the lower sides of the wing. The takeoff took place from a droppable three-wheel takeoff cart, and the landing was achieved on a retractable skid under the fuselage and supporting skids under each of the engine nacelles. The flight mass was 7.75 t, and the wing area was 26.4 m². At an altitude of 8 km, a flight speed of 765 km/h was to be achieved. The range amounted to 1940 km [42]. *Mach* numbers of up to $M = 0.82$ were to be attained (also see Chapter 6, Arado 234 with Swept Wings).

During the first flights in August of 1943, the V-1 was flown at speeds up to $V_a = 650$ km/h. Stability and controllability characteristics were surprisingly good. Despite control forces that were not yet optimized, an outstanding agility existed that met the fighter demands. The V-type [Versuch (test)] aircraft Ar 234A (up to V-8) had not been flown at airspeeds above 700 km/h by the end of 1943. The V-5 was, as well as the V-7 (both with Jumo 004B-0 engines), deployed in tests with the Luftwaffe (1st Squadron/Test Unit, Supreme Command of the Air Force). A memorandum of September 11, 1944, concerning combat deployment [43], states that when the V-5 was pushed into a dive from an altitude of 10 km and as an altitude of somewhat below 9 km was reached and a speed of $V_a = 550$ to 600 km/h ($M = 0.80$ to 0.83), the aircraft could suddenly no longer be controlled by any of the three control surfaces. The controllability, however, reestablished itself again at lower altitudes.

5.3.1.2 ARADO AR 234B. The bomber version of the Ar 234 was equipped with a fully retractable three-wheel landing gear instead of the skid. The flight mass was increased to 9.5 t. An amount of 3.8 t of fuel could be carried internally. Bombs (1000 kg) could be carried under the fuselage and on racks underneath the engine nacelles. The maximum speed when carrying bombs decreased in horizontal flight to 680 km/h.

With the V-9, the first aircraft with the retractable landing gear, the highest speed attained up to that time with an Ar 234, 920 km/h ($M = 0.79$), was reached in April 1941 with Jumo 004B-1 engines in the 23rd flight at an altitude of 4 km (test pilot was *G. Eheim*). Large aileron forces in the presence of relatively good elevator and rudder controllability were noted [44]. The V-9 was thereafter also used for performance measurements with bomb loads during which indicated airspeeds of up to 750 km/h were reached. During a dive maneuver with the V-10 starting at an altitude of 9 km, the terminal velocity of the aircraft was reached at an altitude of 5 km; the indicated airspeed was 780 km/h corresponding to a *Mach* number of $M = 0.83$. However, no *Mach* effects were observed [45].

After the enlargement of the elevator of the V-9 in preparation for the Ar 234C series, a number of steep dive tests was carried out to reach the projected true airspeed of $V_w = 1000$ km/h. The true airspeed actually achieved by then was $V_w = 960$ km/h at an altitude of 7 km. Here, a sudden elevator vibration occurred, which grew with increasing airspeed. Simultaneously, the impression of a slowly emerging control-force reversal occurred, which was attributed to *Mach* number effects as well as to structural elasticity due to air loads. In case of the Ar 234B production aircraft, elevator buffeting was already observed at speeds above $V_w = 850$ km/h. Contour deviations (waviness) on the elevator flap were diagnosed as the cause [46].

After the war, the Ar 234B was thoroughly tested at the British RAE with regard to *Mach* number effects [47]. During dive maneuvers started from an altitude of 30,000 ft (9144 m), the flight-path angle had to be increased up to 30 deg to approach the range of compressibility effects. In addition, a nose-heavy trim had to be used because otherwise the push force would have become too large for the pilot. At $M = 0.76$, nose-heaviness commenced, and the elevator started to feel "sloppy." The effects increased at $M = 0.82$, and the elevator now had to be pulled to hold the trajectory angle. The density increased during the descent leading to a steady reduction in the *Mach* number, and recovery could be initiated with quite normal stick forces and satisfactory control-surface effectiveness. The RAE engineers established $M = 0.75$ as a tactically still useful *Mach* number of the Ar 234B. The latter was in their assessment too close to $M = 0.72$, which was already reached during horizontal flight, so that unwanted *Mach* number effects could already occur during only slightly inclined flight-path angles.

5.3.1.3 ARADO 234 C. To increase the flight speed, twin engines were mounted on the C-version under the wings instead of single engines. By this means, airspeeds of up to 890 km/h were to be achieved. The V-13 and V-19, intended as preproduction aircraft, were equipped with BMW 003 A-0 respectively A-1 engines. The flight trials of the V-19 led already during the second flight to a speed of $V_w = 790$ km/h near the ground without the engines having reached full thrust. Towards the end of 1944, control forces and control effects were determined in flights with the V-21 at speeds of up to $V_a = 900$ km/h and the aircraft judged to be controllable by the pilot. *Mach* number effects had not explicitly occurred apart from the known vibration of the elevator in case of the B-series. It was, however, planned to investigate whether consequences of the *Mach* number related neutral-point movement at 1000 km/h could be ascertained (flight report of January 4, 1945 [48]). Further tests were, however, no longer carried out.

5.3.1.4 HEINKEL HE 162. Hardly any flight-characteristic features that could have been directly attributed to *Mach* number effects have become known from the He 162 trials. The maximum permissible airspeed authorized in the model description was $V_{hor} = 800$ km/h during horizontal flights and $V_{glide} = 1000$ km/h at an altitude of 1 km. The aircraft showed, however, instabilities about the pitch axis at high speeds, the reason why the airspeed was limited to 600 km/h when deployed with the Luftwaffe. During trials at Leck airfield from March 1945, this dangerous limit was frequently exceeded, and speeds of up to 950 km/h were reached. The aircraft "bucked" at these speeds without getting out of control. At the same time a strange noise was noticed by the pilots as if somebody was hitting the roof of the cockpit with a blunt object [49].

5.3.1.5 HEINKEL HE 280. The trials of the He 280 commenced with He S8 turbojet engines whose thrust just sufficed to reach a speed of about 700 km/h during level flight. The He 280 V-2, later equipped with Jumo 004 engines, was able to reach substantially higher airspeeds. Some of the V-models of the He 280 were employed for high-speed research as well as for the investigation of the stall behavior and as a test carrier of jet engines [50, 51]. The He 280 V-7 without engines was transferred to the DFS at Linz-Hoersching airfield in April 1943 for test purposes. Here, the DFS carried out high-speed dive tests, after the aircraft had been towed to high altitudes, where true airspeeds of up to 770 km/h were reached without compressibility effects having been reported [52].

5.3.1.6 JUNKERS JU 287. The design characteristics and the aerodynamics of the forward-swept wing configuration will be dealt with in detail in Chapter 6 concerning the Ju 287. The flight testing of the Ju 287 started with a concept demonstrator for the new wing design designated Ju 287 V-1. The fuselage was "borrowed" from a He 177, the tail from a Ju 188, and the landing gear from a U.S. bomber of the type Convair B-24, which was shot down during a bombing mission over Germany. The maximum airspeed was not to exceed 650 km/h due to the landing gear, which was mounted in a fixed-down position. The objective was primarily to investigate the stability and controllability of the forward-swept wing in the low-speed regime and the stall behavior under symmetrical and asymmetrical yaw conditions as well as at various landing-flap positions. Also the asymmetries occurring in case of an asymmetrical thrust after an engine failure played an important role. The aircraft was equipped with wool tufts for airflow visualization. Pictures taken by a camera, mounted in front of the vertical stabilizer, were used to document the airflow characteristics. The results of the flight tests largely confirmed the forecasts concerning the aerodynamic behavior of the forward-swept wing. The stall characteristics during flight at various sideslip angles were repeatedly demonstrated without running into really dangerous situations.

The question concerning the critical *Mach* number of the configuration remained unanswered during 16 or 17 flights with the Ju 287 V-1, carried out at Brandis airfield. A *Mach* number of $M = 0.70$ was supposed to have been reached, which would correspond to about $V_a = 430$ km/h at an altitude of 10 km. This is, however, not credible for flights of the V-1 with the fixed landing gear. The wing, however, lets one expect critical *Mach* numbers of about $M = 0.85$. It can be read that later flight tests had shown that the aircraft could have been safely flown up to $M = 0.92$ [53]. A Ju 287 V-2 was also stationed at Brandis airfield, which had, however, as far as is known never been flown.

The prototype Ju 287 V-3, which was also intended for high-speed tests and considered as a zero-series model of the Ju 287, was not completed any more due to the war [54]. It may be taken from more recent investigations [55, 56] that after the war a Ju 287 V-3 in an advanced state of construction had been present at the Junkers Company in Dresden occupied by the Soviet Army. The occupational forces who were very interested in the project, decided, however, to have a Junkers EF 131, which was largely similar to the Ju 287 V-3, with triple nacelles virtually developed from scratch. Under Russian supervision this aircraft reached ground testing status at Dessau. For secrecy reasons, the flight testing was, however, transferred to Ramenskoje in the Soviet Union. There, a German crew carried out about 15 test flights between May and October of 1947. Primarily the oscillatory behavior of the wing conveyed here a bad impression. The maximum permissible airspeed at ground level was set to 600 km/h, at an altitude of 6 km to 800 km/h, at 9 km to 960 km/h, and at an altitude of 11 km to a maximum of 900 km/h [57]. According to *H. Walter*, only 780 km/h was reached [58]. Despite otherwise satisfactory characteristics, the EF 131 program was stopped in June of 1948 by the Soviet Union (see also Chapter 8).

The development of the forward-swept wing technology under the participation of the German Junkers engineers was, nevertheless, continued. The airframe of the EF 131 was equipped with modified wings and more powerful engines mounted on the lower sides of the wings, which were, in the meantime, also available in the Soviet Union. The aircraft received the new designation EF 140. The flight tests with the EF 140 V-1 took place from September 1948 at Tjoplistan (south of Moscow) with a German crew onboard. A maximum speed of 904 km/h was said to have been reached (presumably at an altitude of 9 km, $M = 0.83$). Unreliable engines and wing oscillations resulted in a request for major redesign work. To increase the range, fuel tanks were attached to the wing tips as they have become known later from the HFB 320. The aircraft EF 140R and EF 140B/R resulted. After four flights of the EF 140R, beginning in October 1949 where a maximum true airspeed of $V_a = 837$ km/h, corresponding to $M = 0.78$ at an altitude of 9 km had been reached, the program was finally abandoned in June 1950.

5.3.1.7 MESSERSCHMITT ME 262. The Messerschmitt Company started the specific testing of the Me 262 at high speeds during the late summer of 1943. These tests were to prove structural integrity of the aircraft as well as to investigate compressibility effects. The flights were initially carried out with the modified Me 262 V-3. The V-3 was the first of altogether three V-models that were mainly used for high-speed tests. With the V-3, Messerschmitt test pilot *G. Linder* reached on October 3, 1943, a true airspeed of 950 km/h at an altitude of 5 km, corresponding to $M = 0.82$, in a shallow dive maneuver. The aircraft was later (in 1944) transferred to the DVL for investigations in the high-speed flight regime.

The DVL had already started in 1943–1944 at Berlin-Adlershof to carry out measurements concerning the behavior at very high flight speeds with the Fw 190, however, without being able to establish distinct *Mach* number effects (see Sec. 5.2.1.2 regarding the Fw 190). The high-speed flights with the Me 262 were, inspired by the results of the trials at Messerschmitt, therefore, a logical continuation of the DVL measurements, because substantially higher *Mach* numbers than with the Fw 190 had become attainable [59]. Compared to the Fw 190, primarily the propeller influence was eliminated in the case of the Me 262. However, pitching moments due to the engine thrust as well as interference effects between the jet and the elevator had to be accounted for. The question of how far the results of wind-tunnel tests at high speeds would be confirmed by the results from flight-test measurements played an important role. In the wind tunnel, the DVL had already obtained such results on a model of the Me 262 whose geometry was, however, characterized by a straight unswept wing with engine nacelles (without internal airflow effects) attached to the wing [60].

The flight tests of the DVL initially started with various comparatively flat trajectory-inclination flights at $n \approx 1$ at altitudes of 1, 3, 5, and 7 km. Later followed pull-out tests at maximum speed at an altitude of 3 km, each with constant load factors of $n = 2$, 3, and 4, respectively. For each test maneuver a constant trim setting was to be selected, which rendered the stick force during the high-speed flight nearly zero. By means of a DVL double-trace recorder the total and static pressures (to determine airspeed, altitude, and *Mach* number), the load factor, the horizontal stabilizer trim and elevator angles, and hinge moment (elevator control force) were measured and registered. To be able to determine the elastic deformations of the elevator during flight, DVL elevator-deflection sensors were attached near the fuselage and at the tip. From plots of the elevator angle and the elevator hinge moments vs the dynamic pressure and *Mach* number, it was realized that at high dynamic pressures deviations from the expected dependence of these parameters existed. In addition to the expected *Mach* number effect, also a dynamic pressure effect occurred that was mainly explainable by the "softness" of the structure of stabilizer and fuselage because it only dominated at low altitudes. With increasing dynamic pressure, a tail-heavy moment arose that could only be controlled by a manual push force on the stick. The control forces became so large that they reached the performance limit of a pilot. With an increase in the *Mach* number, however, an increasingly nose-heavy moment developed, caused by the compressibility effects, which had now to be compensated by a pull force on the stick. This critical *Mach* number amounted to $M = 0.78$. The nose-heavy pitching moment, resulting from the *Mach* number effect, was at low altitudes and very high dynamic pressures much more pronounced than at high altitudes and correspondingly lower dynamic pressures. The flight behavior at high dynamic pressures and the same *Mach* number turned out to be less severe because the tail-heavy dynamic pressure related "softness effect" counteracted the nose-heavy moment due to compressibility effects. At conditions where both effects were in balance, noticeably higher *Mach* numbers (up to $M = 0.86$) could be achieved than the critical *Mach* number of about $M = 0.78$, solely related to the trim changes, would have allowed. This meant that the *Mach* number effects favorably overlapped with regard to the overall behavior and were delayed due to the structural softness of the rear fuselage and the elevator attachment. The trim changes at high altitudes at the same *Mach* number but lower dynamic pressures, hence lower stick forces, were less critical. With increasing load factors (at a pull-up maneuver) the occurrence of shock waves shifted to lower *Mach* numbers. The manual stick forces increased up to about 70 daN (152 lb) due to the *Mach* number effect at $M = 0.78$ and 4 g while at 2 g the same *Mach* number and the same dynamic pressure, actually a push force of 1 daN (2.2 lb), were still required (see Figs. 5.7, 5.8, and 5.9). The softness of the tail region of the V-3 was treated separately in a second report of the DVL. Here it was shown that mainly the twist of the stabilizer fin and the deformation of the elevator attachment were responsible for the "softness effects" which could, moreover, be identified as a special characteristic of the V-3, because the corresponding series aircraft showed deviating

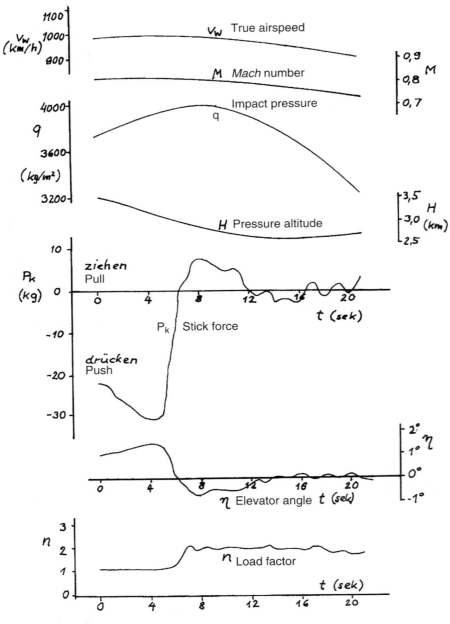

Fig. 5.7   **Me 262 V-3, testing process for** *n* = 1 **and** *n* = 2.

characteristics. The report comes to the conclusion that it must be generally possible by a skillful structural design to considerably reduce or delay *Mach* number effects, thus making the attendant symptoms at high flight speeds controllable for the pilots [61]. The Me 262 V-3 was destroyed on September

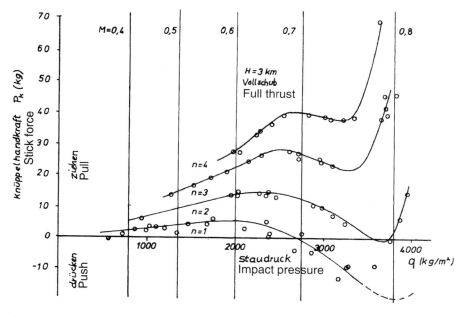

Fig. 5.8 Me 262 V-3, control-surface hinge moments and stick forces dependent on dynamic pressure and *Mach* number at load factors $n = 4$.

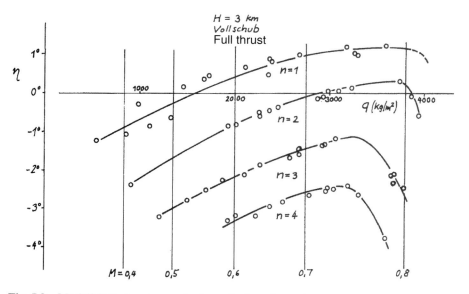

Fig. 5.9 Me 262 V-3, elevator angle dependent on dynamic pressure and *Mach* number at four load factors.

12, 1944, during a bomb raid. This prevented further flights still scheduled to recalibrate the pitot-static system within the extended operational *Mach* number regime.

The Me 262 S-2 (VI+ AG) was also used for high-speed tests. After the maiden flight on March 28, 1944, the test pilot *Herlitzius* achieved about three months later in a shallow dive maneuver across a calibrated test course a measured true airspeed of 1004 km/h at an altitude of 5 km corresponding to $M = 0.86$. The S-2 was destroyed on July 19, 1944, after 47 flights, at the Messerschmitt flight-test center at Lager Lechfeld airfield during an air raid.

The experience gained especially from the tests with the V-3, but also during test flights of the Me 262 series production aircraft, which had started in the meantime, inspired W. *Messerschmitt* to investigate within a high-speed program (Hochgeschwindigkeitsmessungen) open questions concerning, for instance, the unresolved conditions of the control-surface effectiveness and load distributions on wings and empennage, as well as the possibility to reduce by a systematic adaptation of the configuration the *Mach* number effects respectively to delay them to higher *Mach* numbers. The use of swept wings played a central role. This is discussed in detail in Sec. 6.2.1.

The high speeds attainable in dive maneuvers with the Me 262 were associated with phenomena little known to the Air Force up to then. According to information "leaked out," the young operational Luftwaffe pilots were mostly ignorant of these phenomena. A careful briefing of the pilots with these problems would have been urgently needed. However, due to the lack of validated information and also lack of time, this was no longer possible. The war situation and the aircraft readiness as well as the supply situation regarding the jet fuel J2 added to the problems. The maximum horizontal flight speeds of the Me 262 A-1 free of *Mach* number effects were about 850 to 870 km/h at an altitude of 6 km ($M = 0.77$); however, a speed increase in a dive maneuver could rapidly lead to critical *Mach* numbers being encountered. At high airspeeds it was noticed that first the stick forces about the aircraft roll and pitch axes dramatically increased especially in the case of increased load factors. Above approximately $M = 0.80$ heavy buffeting and a deep, alarmingly loud roaring noise occurred that seemed to emanate from the rear part of the cockpit canopy. At $M = 0.83$, the aircraft wanted to pitch down rather abruptly and in irregular jerks. At the same time for the aircraft trimmed at $M = 0.78$, a pull force of 15 daN (33 lb) was required to counteract a further increase of the flight-path angle. With increasing speed the nose-heavy pitching moment finally became so large that, at the elevator trim position set for level flight, the pilot could barely manually control the aircraft.

After about $M = 0.78$, in addition to the trim changes, some disturbance about the vertical axis ("snaking") set in, and at $M = 0.83$ an abrupt roll-off of the wing ("wing drop") occurred. At $M = 0.86$ an unacceptably high 50 daN (109 lb) pull force was finally required to at least prevent the flight-path angle from becoming any steeper. The pilot, surprised by such a behavior during air combat, could only let the aircraft dive with an apparently blocked elevator at a steadily increasing flight-path angle. Only with the adjustment of the electrically operated elevator trimming device could the pilot influence the flight trajectory during this phase. The pitch trim position at the beginning of the dive really seemed to have a large influence on the high-speed effects. Parallel to the behavior of the Me 109 G within this range are apparent (see Sec. 5.2.1.1). Because the *Mach* number, as already described, decreases quickly at a constant dynamic pressure, that is, a constant airspeed indication, with decreasing altitude, the critical *Mach* number range was fortunately automatically excited so that it was primarily important to control the airspeed increase during a dive by reducing the thrust. However, this was not easily managed because the jet engines could flame out altogether while idling. Below the critical *Mach* number for controllability and load factors of about $M = 0.82$, the aircraft started almost immediately to react to control inputs in a normal way again.

The speed limits listed in Table 5.2 applied in 1945 to the production aircraft delivered to the Air Force. Speeds of up to 1000 km/h were authorized during a dive corresponding at an altitude of 4 km to a *Mach* number of $M = 0.86$ (standard atmosphere). The values given in the table for the speed achieved during level flight are average values for the Me 262 A-1 of so-called multiple measurements. They consisted of 125 data points, which were evaluated by the Bauaufsicht (program office) of the Luftwaffe at Messerschmitt (BAL-Mtt) up to December 1, 1944, during test flights with the Me 262 A-1 with correctly adjusted production jet engines Jumo 004B (average fuel injection pressure $p_k = 62$ kg/cm^2 at $N = 8700$ rpm) [62]. The highest *Mach* number achieved during level flight at a correctly adjusted engine performance output was accordingly $M = 0.78$ at an altitude of 8 km.

During series-production aircraft acceptance test flights, the aircraft had in level flight to reach at least a true airspeed of 830 km/h and in a 30-deg dive

TABLE 5.2   SPEED LIMITS

Altitude (km)	Max. permissible flight speed (km/h)[a]	Maximum *Mach* number	True airspeed in level flight (km/h)	*Mach* number in level flight
0	950	0.78	845	0.70
2	950	0.79	852	0.715
4	950	0.81	864	0.74
6	950	0.84	870	0.77
8–12	900	0.86	860–770A	0.78–0.73

[a]Conversion: 1 km/h = 0.62 mph

between 950 and 1050 km/h ($M =$ to 0.83 to 0.78 at an altitude of 1000 m) to be declared accepted. Because of the considerable manufacturing inaccuracies of the production aircraft (waviness, sagging sheet-metal panels, skin steps, inaccurately closing fuselage, and landing-gear doors) and due to the frequently large deviations in the engine thrust performance (mainly caused by incorrect fuel injection pressure), a large scatter of the in-flight performance data resulted. However, the acceptance flights hardly went above altitudes of 4 to 5 km. During these flights, *Mach* number effects, therefore, only rarely occurred because critical *Mach* numbers up to or above $M = 0.78$ were hardly reached during dives from this relatively low altitude [63]. Nevertheless, also during test-flight operations there were incidents due to *Mach* number effects as the following example shows. During a dive from high altitudes, the chief test pilot of the test-flight operations of the production facility at Hessenthal/ Schwaebisch Hall airfield, *H. Wagner*, did not always have the airspeed indicator in his sight when his aircraft at $V_w > 1000$ km/h suddenly entered a range where "local *Mach* number perturbations" occurred. The engine cowlings tore off, and the aircraft also suffered structural deformations [64].

It happened frequently that aircraft during combat unintentionally encountered *Mach* number effects. During dive attacks or escape maneuvers, the aircraft accelerated very quickly especially at "zero *g*" conditions when making the transition into steep dive maneuvers. Quickly, the known tendency to pitch down then showed up, and the aircraft could, as explained, at high load factors only be pulled out by applying large stick forces. Intuitive use of the electrically operated horizontal stabilizer trim devices could help recover the aircraft preventing disaster. The pilot could support the pull-out from a dive within the critical *Mach* number range applying nose-up trim. Caution was, however, advised when the aircraft began to respond normally again to elevator control inputs after leaving the critical *Mach* range resulting in a nose-up movement according to the existing tail-heavy trim. High load factors could develop very quickly if the pilot did not anticipate the high push forces required on the stick with a relatively slow trim device operation not being able to move trim fast enough in a nose-heavy direction (the American XP-80 aircraft had similar problems, which are addressed in Sec. 5.3.3).

Especially in the case of the roll control (there was no support by a hydraulic system), it was tried to make the stick forces at high dynamic pressures controllable for the pilot by an extension of the stick length. The warnings to the pilots not to exceed the maximum airspeed values, fixed at 900 km/h at altitudes of >8 km ($M = 82$) and at 950 km/h below 8 km, were not sufficient to prevent accidents whose victims were not only inexperienced pilots. During air combat, hardly any pilot could afford to keep the airspeed indicator of his Me 262 in sight for long enough to monitor airspeed limits. A *Mach* number warning device still under development at the DVL had supposedly been installed for test purposes on only a few Me 262 aircraft [65]. In the book by

M. Boehme [66] about the JG 7 examples of how the flight operations with the Me 262 were arranged in practice are described.

In the assessment of the tactical benefit and the effectiveness of the Me 262 during air combat, next to pure technical aspects one should not ignore the fact that the operational Luftwaffe pilots had not been sufficiently familiarized with the special characteristics of their aircraft in the high-speed regime. Many obviously did not know what to do with the set speed limit of 950 km/h. In the case of propeller aircraft, they could "do anything" if combat success or the saving of their life was concerned. Broken rivets and buckled wings from pull-outs with much more than 7 g did not represent too great a risk because the aircraft—especially the Fw 190—were known to be very robust and could withstand "quite something." There existed a sufficiently large strength-related safety margin thanks to the safety factor $j = 1.8$. For the aircraft of the Stress Group H5, BVF, a load factor of $n = 7$ was the maximum at pull-out, which was, as a rule, demonstrated in support of the in-flight proof load testing. It was generally not realized that in case of the Me 262 and some other jet aircraft at the permitted true airspeeds structural integrity was less of a concern than the avoidance of critical flight characteristics encountered in the near-sonic speed range. Effective air brakes, repeatedly demanded for the 262, were not available although their lack had, for instance, already been criticized in the report on the maiden flight of the V-3 written by the test pilot *F. Wendel* [67].

The first real air combat missions with the Me 262, starting in about July 1944, were aimed at intercepting enemy reconnaissance aircraft (RAF "Mosquito" or "Spitfire" aircraft) against whom the Me 262 Test Command *Thierfelder*, operating from Lager Lechfeld airfield, was employed. There was soon an accumulation of unexplained cases where the pilots could not pull out of high-speed dive maneuvers anymore resulting in fatal crashes. Lt. *Schnoerrer* survived one of these accidents by a lucky coincidence and was able to report what had happened [68]. He was about to intercept a U.S. reconnaissance aircraft, which tried, however, to escape in a dive. When *Schnoerrer* decided to follow, his Me 262 went out of control. When throttling back the engines, both flamed out. Activating the elevator did not have any effect. The pilot decided to bail out and jettisoned the cockpit canopy. Surprisingly before bailing out he found that his Me was controllable again. The open cockpit must, after canopy jettison, have acted as an air brake slowing down the aircraft. He was able to recover control of the aircraft and, after having successfully restarted the engines, landed safely at Lechfeld. The aircraft, however, suffered considerable damage and had to be repaired.

After four retraining flights transferring from the Me 110 to the Me 262 (this altogether needing about two flight hours), the former Me 262 pilot *H. G. Mutke* had been on his first high-altitude flight, which led him according to plan to an altitude of 12 km [69]. In the Innsbruck region (Tyrol, Austria), he overheard on the radio the warnings to a colleague in a Me 262 who was obviously attacked in the vicinity of Lager Lechfeld airfield by "Mustangs." Without hesitation *Mutke* initiated a steep dive maneuver to aid the pilot under attack. The aircraft dived with a 40-deg flight-path angle and at full thrust. It soon started to vibrate, shake, and oscillate sideways becoming increasingly more nose heavy. The needle of the altitude-compensated air-speed indicator finally jammed at a stop speed indication of 1100 km/h. The aircraft became uncontrollable, and the stick could be freely moved without the least reaction. Solely with the electrical stabilizer trim was *Mutke* able to carry out slight trajectory corrections. In an attempt to reduce the thrust, both engines flamed out. With decreasing speed and *Mach* number normal con-trollability returned, *Mutke* succeeded in pulling the aircraft out of the dive and restarted the engines. However, he noticed that the aircraft structure was considerably deformed and out of trim, and so he decided to carry out the approach to Lechfeld airfield not at the usual 250 km/h, but for safety rea-sons at 350 km/h. The landing succeeded without crashing the aircraft. The inspection of the aircraft showed that the wings were twisted and rivets had been torn off. The assumption, later expressed by *Mutke*, to have exceeded the speed of sound during the reported dive, is not realistic. According to his descriptions, he had experienced with his Me 262 the already well-known compressibility effects. Unusual features in the behavior, for instance, after reaching the critical *Mach* number that the controls were completely "freely movable," are within the bandwidth of possible *Mach*-related phenomena, which may occur depending on the stabilizer trim position and the manufac-turing deviations primarily seen in the skin of the control surfaces (waviness, deformations). Depending on the angle of attack and stabilizer trim position, the boundary-layer thickening or even separation caused by the shock waves on the wing surfaces or the stabilizer may have caused the freely movable control surfaces.

Also *W. Späte* repeatedly reported *Mach* number effects that he had expe-rienced with the Me 163 as well as with the Me 262 [70]. After several combat missions, which he carried out at the JG 7 at the end of April 1945 taking off from Prague-Rusin airfield, in an attack against B 17 bomber formations in the area of Dresden-Chemnitz he had pushed his aircraft to over 960 km/h. When he tried to get into an attack position, at the slightest attempt to pull positive *g*s (load factor), as he described it, he got into the "dreaded high-speed flow separation regime" as a result of reaching the critical *Mach* num-ber. The aircraft shook alarmingly. However, he was able to carefully stabilize the aircraft, but lost contact to his Me 262 formation.

5.3.1.8 ME 262 IN BRITAIN. After the war, the characteristics of the Me 262 at high speeds were thoroughly flight tested at the RAE in Farnborough. Of particular interest were the characteristics of the swept-wing configuration of the Me 262 and the behavior as a weapons platform. Compared to the Me 262, the contemporary British Gloster "Meteor I" and the de Havilland "Vampire" aircraft were both still with conventional straight wings. *E. Brown* stated in his book [71] that at $M = 0.8$ the "Meteor I" showed a tail-heaviness ("pitch up") that became so strong that the speed was automatically reduced and the aircraft recovered from the critical *Mach* number range on its own accord. In the case of the "Vampire," he observed that trim changes already occurred at $M = 0.74$ that grew to an intense pitching about the pitch axis at $M = 78$. His flight-test results obtained with the Me 262 essentially confirmed the results of the Messerschmitt test series (test pilots *Linder, Herlitzius, Baur*). At the RAE *Mach* numbers of up to $M = 0.84$ were, according to *Brown*, reached with the Me 262. As tactically still usable, a maximum *Mach* number for the aircraft of $M = 0.82$ was determined by the RAE. The latter was considerably higher than that for the Ar 234 determined by the RAE to be $M = 0.72$ (see Sec. 5.3.1.1).

5.3.1.9 ME 262 IN THE UNITED STATES. In the United States (Freeman Field, Wright Field), captured Me 262 aircraft were compared with the performance of the Lockheed XP-80 and P-80A-1 as well as the Republic XP-84. This first generation of jet fighters of the United States had conventional straight wings. The drag-rise *Mach* numbers were determined to be for the XP-80 about $M = 0.78$ and for the XP-84 $M = 0.80$. For the Me 262, it was determined to be $M = 0.81$ to 0.82. This higher value and the considerably better flight performance of the Me 262 were ascribed to the sweep of the wings and the thinner, symmetrical airfoil sections [11% to 9% MAC compared to 13% mean aerodynamic cord (MAC) for the XP-80 and the XP-84] [72]. An Me 262 aircraft flown by the test pilot *H. Fay* in April 1945 from the already occupied Frankfurt airfield (serial number 111711, U.S. designation T-2-711) and another Me 262 with the U.S. designation T-2-4012 participated in the comparative tests. The tests comprised, primarily, measurements of level flight speeds and rate of climb. In detail, the following results were obtained. (Data in tables obtained from U.S. Report TSFTE-2008, pp. 5–8.)

Tables 5.3 and 5.4 show that the Me 262 without exception comes off quite well considering the poor manufacturing quality of the series-production aircraft and the poor condition of captured aircraft transferred to the United States [73]. It can be seen that the true airspeed of 568 mph (914 km/h) of the Me 262 #711 at an altitude of 5000 ft (1500 m) is substantially higher than the speed to be expected from the aforementioned "multiple measurements," which was at that altitude 850 km/h. Equally the airspeed of 539 mph (867 km/h, $M = 0.82$) of the Me 262 #711 at an altitude of 35,000 ft (10,700 m) is distinctly higher than the 810 km/h obtained by the "multiple measurements."

TABLE 5.3   LEVEL FLIGHT SPEEDS V_h, MPH (KM/H)

Pressure Altitude in kft (km)	Me 262 #711 13,500 lb (6140 kg)	XP 84 #1 14,000 lb (6360 kg)	P 80 A-1 # 44-35044 11,600 lb (5270 kg)	P 80 A-1 # 44-85121 11,600 lb (5270 kg)	XP 80 A 11,600 lb (5270 kg)
5 (1.5)	568 (914)	596 (959)	530 (853)	514 (827)	554 (891)
20 (6.1)	566 (911)	565 (909)	536 (862)	521 (838)	534 (529)
35 (10.7)	539 (867)	528 (850)	515 (829)	508 (817)	497 (800)

5.3.1.10 ME 262 IN FRANCE. Three Me 262 were also intensively tested in France at the flight-test center in Bretigny near Paris [74, 75]. Altogether seven Me 262, among them a double seater, were secured for France, most of them, however, incomplete and in bad condition partly caused by sabotage. Two of the aircraft could be restored by the Société Nationate de Constructions Aeronautiques de Sud Ovest (SNCASO) to an airworthy condition. They were designated "Number 1" and "Number 3." One further aircraft taken over from the U.S. Army Air Force in a ready-to-fly condition was designated "Number 2." This was a Me 262, which had been seized at Lager Lechfeld airfield after the occupation by the U.S. Army Air Force 54th Air Disarmament Squadron (ADS, "Feudin' 54th"). In 1947, high-speed tests were carried out at the test center of the French Air Force [Centre d'Essais en Vol (CEV)] in Bretigny airfield with the Messerschmitt Me 262 "Number 2." The French test pilots *Receveau* and *Cabaret* of the CEV also had their first experience with compressibility effects here. Like their colleagues of the RAF and the USAF, they stated that their Me 262, beginning between $M = 0.78$ and 0.82, increasingly developed intense yaw, roll, and primarily porpoising motions about the pitch axis, which became harder and harder to control with increasing speed. Because of the lack of air brakes and the danger of engine flame out when reducing the throttles, the further exploration of the *Mach* number limits was abandoned and the tests not

TABLE 5.4   CLIMB RATES w, FT/MIN (M/S)

Pressure Altitude in kft (km)	Me 262 # 711 # 4012 # 44-85121 13,100 lb (5960 kg)	XP 84 #1 14,000 lb (6360 kg)	P 80 A-1 # 44-35044 11,600 lb (5270 kg)	P 80 A-1 # 44-85121 11,600 lb (5270 kg)	Me 262 S 20 Report 13,000 lb (5900 kg)
10 (3.05)	3700 (18.5)	3700 (18.5)	3700 (18.5)	3200 (16)	3200 (16)
20 (6.1)	2900 (14.5)	2900 (14.5)	2900 (14.5)	2100 (10.5)	2100 (10.5)
35 (10.7)	1500 (7.5)	1500 (7.5)	1500 (7.5)	800 (4)	400 (2)

Fig. 5.10 Me 262 A-1 with nose boom with pitot-static pressure sensor, France, 1948.

continued above $M = 0.82$. This *Mach* number was apparently already reached during level flights [76]. This seems surprising because the German aircraft reached an average of only about $M = 0.78$ during level flight. The aircraft tested at the CEV had the gun and canon ports covered by tape for these tests. The quality of the pitot-static calibration may possibly also have played a role. The French Messerschmitt Me 262, Number 2, had been equipped with pitot-static probes on a nose boom installation (see Fig. 5.10). Possibly different from the standard pitot-static probe at the left wing tip of the series production aircraft, after calibration the boom installation provided fairly exact values of the in-flight static and the total pressure measurements. This allowed the determination of reliable values for the calibrated airspeed and the *Mach* number. However, the difference in the *Mach* numbers seems too large to be solely explained in that way. The Me 262 No. 711 of the U.S. Air Force showed, however, similarly good level flight performance (see Table 5.3). The aircraft were kept in flying status up to about 1948. The last flight of a Me 262 in France took place on October 7, 1948. The other aircraft still existing were not airworthy and were used to provide spare parts until the final termination of all flights [77].

5.3.1.11 ME 262 IN FORMER CZECHOSLOVAKIA AND IN THE SOVIET UNION. At the end of the war, numerous Me 262 had been captured by Czech and Soviet troops. Up to the last days of the war, many aircraft of the Luftwaffe, especially of the JG 7, the KG (J)6, and the KG (J)54, had operated against Czech and approaching Soviet troops from the Luftwaffe airfields at Saaz, Pisek, and Prague-Ruzine [78]. Many off-site production plants, which had produced parts for the Me 262 series production in Germany, were located at the end of the war in Czechoslovakia. A final assembly plant and test-flight operations were located at Eger (Cheb) airfield, which were, however, completely destroyed by a heavy U.S. bomb raid shortly before the end of the hostilities. From parts still in storage within Czechoslovakia, 17 aircraft were assembled and made ready for flight. The Czech Air Force had thus already in 1946 access to the most modern jet fighter at that time and an immediate entry into the jet age. The first Czech Me 262 flew on August 27, 1946. The tests confirmed the earlier German experience. The planned large-scale series production could, however, not be achieved due to a Soviet intervention in favor of their own products. In May 1951 the available Me 262, designated

S-92, were finally scrapped except for two which are today still preserved in Prague's museums [79]. *A. Alexandrov* and *G. Petrov* reported [80] that a number of intact Me 262 aircraft had supposedly been taken over and intensively tested in flight by the Soviets. Although a planned resumption of a series production of the aircraft did not materialize due to the "bad flight characteristics" and because of the implications for independent Soviet designs, valuable know-how and technology assets could be gained.

5.3.1.12 SUMMARY. Table 5.5 provides a summary of known flight-test results obtained with German propeller- and turbojet-fighter aircraft and light bombers in the high *Mach* numbers range.

The flights of the first jet aircraft, the He 178, and the ones of the rocket-propelled aircraft He 176 were only of a pioneering character and thus need not to be mentioned here. The trials of these aircraft are frequently documented

TABLE 5.5   HIGH *Mach* RANGE FLIGHT-TEST RESULTS

Model	P = Propeller S = Jet R = Rocket Propelled	Achieved maximum *Mach* numbers	Limiting factors
Me 109 G8	P	0.805	Nose-heaviness, elevator ineffective or blocked
Me 109 F	P	0.80 (?)	Nose-heaviness
Do 335	P	0.76	Yawing, porpoising
Fw 190 A	P	0.82	Trimming, drag rise
Fw 190 D 9	P	0.80 (?)	Buffeting
Ta 152	P	0.80	Buffeting
Are 234 A	S	0.75	Buffeting
Are 234 B	S	0.80	Buffeting, loss of control effectiveness
Are 234 C	S	0.75 (?)	Drag rise, buffeting
Me 262	S	0.86	Buffeting, nose-heaviness, elevator ineffective or blocked, yaw, roll
He 162	S	0.80	Buffeting, oscillations
He 178	S	< 0.5	No *Mach* number effects (600 km/h) observed
He 280 V 7	S	0.75	Buffeting
Ju 287 V 1	S	< 0.70	Aeroelasticity, no *Mach* number effects
He 176	R	0.66	No *Mach* number effects (800 km/h) observed
Me 163	R	0.80	Buffeting, nose-heaviness, roll
DFS 346	R	0.9+ (?)	Trim changes, loss of control-surface effectiveness

elsewhere, for instance, in Vol. 33 of [81]. The flight speeds achieved were still too low to lead to noticeable *Mach* number effects.

5.3.2 GREAT BRITAIN

5.3.2.1 GLOSTER E 28/39 W 4041. This aircraft was the first jet-propelled aircraft that flew in Britain. Like the He 178, it was only designed and built as a flying test bed to test the jet engine developed by *F. Whittle* completely independently of *Hans Joachim Pabst von Ohain* at Heinkel. As a part of the flight-test program, which was conducted in 1945 at the British RAE to investigate phenomena at near-sonic speeds, an aircraft was conceived where any interference of the wing with propeller or jet-engine nacelles was not present. The Gloster E28/29 met this requirement in a perfect way [82]. The aircraft had its air intake located in the aircraft nose and the nozzle exit at its tail (interestingly enough this is just like the Heinkel He 178), so that no effect of the propulsion on the airflow about the wing was to be expected. The aircraft could be equipped with two different wing sets for the tests. Both wing sets were without sweep and had an airfoil-section thickness of 12.5% MAC at the root and 10% MAC at the wing tip. The airfoil types differed being an NACA 23012 and an EC 1240/0640 "high-speed" airfoil, the latter being a British development. It possessed a flat upper surface ("flat top") to slow down the velocity increase thus delaying the onset of critical *Mach* numbers. The right wing of the "flat-top" wing set that was attached first was equipped with pressure orifices to measure static pressure distributions, and with a pitot rake at the trailing edge of the wing. The latter was located downstream of the wing covering the wing wake and was intended to allow airflow momentum-loss measurements and hence the determination of the airfoil-section (wing) drag. The pressure data (30 pressure orifices) were led to three instrument panels with airspeed indicators. These had been installed together with the absolute pressure displays of the pitot rake in the radio compartment under the pilot seat inside the fuselage. The displays of the instruments were continuously photographed during flight by cameras. The measurements successfully commenced with the "flat-top" wing, dives from an altitude of 37,000 to 40,000 ft (11,280 to 12,200 m) being carried out. The *Mach* numbers were increased, starting at $M = 0.65$, up to the maximum *Mach* number attainable, which was $M = 0.816$. Already at $M = 0.72$, uncontrollable pitch oscillations occurred (frequency 1 Hz, amplitude 0 to 4 g). The fabric-covered elevators whose covers inflated and influenced the pressure distributions were identified as the reason. The attachment of 5-mm high strips at the trailing edges on the upper and lower sides of the control surfaces remedied the problem, but considerably increased the stick forces. At $M = 0.73$ and 1 g, the aircraft started to become highly nose heavy. The trim change could only be controlled by the pilot with difficulties. The stick forces changed correspondingly; the 10-daN (22-lb) push force required at a $-g$ flight at $M = 0.75$, already with a tail-heavy trim as a precautionary measure, increased to about 20 daN (44 lb) at $M = 0.81$. For the pull-out with 3 g at $M = 0.81$, almost 30 daN (66 lb) pull force was then required. Approaching $M = 0.816$, buffeting as well as intense uncontrollable yawing, pitching, and rolling commenced. Despite the "flat-top" airfoil sections, the trim changes and the drag rise observed here were considerably larger than the ones for the "Spitfire XI" whose wing had only about 10% MAC relative airfoil-section thickness. The cause of the trim changes was assumed to be the strong shock waves forming on the wing upper and lower surfaces. This was confirmed by pressure distribution measurements. This also confirmed later accessible German wind-tunnel results, which predicted for so-called high-speed airfoils of a similar form, but with a higher critical *Mach* number (= first local attainment of the speed of sound) later but higher drag rises than for conventional airfoils. Unfortunately the tests with the second wing set, composed of conventional NACA 23012 airfoil sections, could not be carried out anymore.

5.3.2.2 GLOSTER "METEOR I-IV." The experience gained at the British RAE in 1944 from a series of dives with the propeller-driven fighter aircraft "Spitfire," "Mustang," and "Thunderbolt" with respect to procedures to determine the critical drag-rise *Mach number* raised considerable doubts concerning the attained measurement accuracy. The usefulness of the results seemed restricted by the inaccuracies of the pressure measurements needed to determine pressure altitude, airspeed, and *Mach* number in the dive maneuvers. Progress was made at the same time in the technique of near-sonic wind-tunnel measurements so that it seemed justified to achieve by new measurements a drag determination as well as the validation of the latter by flight-test results. In 1947, this was attempted with aircraft of the type "Meteor IV" but also employing an aircraft with shorter wing span ("clipped wing") [83]. In addition to the cockpit instrumentation, the aircraft were equipped with a *Mach* meter, a longitudinal and a normal accelerometer, as well as a second precision altimeter. The latter was to contribute to determining and correcting the lag effects in the pressure lines when transmitting the total and static pressure information. The scale readings/displays of the instruments were recorded by a robotic camera system installed in the cockpit and either triggered by the pilot or operating automatically if set that way by the pilot.

The flight tests were initially carried out in level flights at an altitude of 25,000 ft (7620 m) reaching *Mach* numbers of up to $M = 0.811$ with the "Meteor IV." To reach higher *Mach* numbers, dives were executed with different engine power settings up to full engine thrust. They were started at altitudes of up to 33,000 ft (10,060 m) to reach the maximum *Mach* number at about 25,000 ft (7620 m) in order to be able to use these data together with the ones obtained at this altitude in level flight. To be able to establish the dependence of the drag-rise *Mach* number on the lift coefficient, in addition to the drag coefficients determined at near-zero lift during the dive, also drag coefficients at higher lift coefficients (0.24, 0.33) were determined. For the "Meteor I," the onset of the drag rise was found to be $M = 0.73$, and for the "Meteor IV" $M = 0.75$. In a comparison of the *Mach* numbers achieved with the "Meteor IV" (with standard wings) during level flight and in a dive, it was surprising that the maximum *Mach* number of $M = 0.82$ obtained in a full-power dive (lift coefficient = 0) was only insignificantly higher than the one achieved during level flight, that is, $M = 0.811$. This showed that the drag rise must have been quite considerable. The "Meteor IV" with "clipped wings" also showed hardly any measurable differences compared to the results with the "Meteor I." The agreement of these drag measurements with wind-tunnel results, which were obtained with a 1/12-scale model of the "Meteor IV," was surprisingly good. The *Mach* number limit was then fixed at $M = 0.8$. With the high engine thrust installed and the still low drag, this value was very easily reached. The trim changes on the "Meteor IV" when approaching the drag-related critical *Mach* number remained easily controlled. The aircraft showed up to $M = 0.8$ an increasingly tail-heavy moment. This meant that the aircraft pulled out automatically and was, therefore, able to quickly recover from the critical *Mach* number regime again. This behavior was in conjunction with the strong drag rise desirable both as a warning and as a safety factor. Mainly at low altitudes the developing load factors could strongly increase during the automatic pull-out due to the higher dynamic pressures. They had, therefore, to be strictly observed and controlled by the pilot in order not to overload the aircraft. An air brake, effective at 400 mph (644 km/h) with up to -0.8 *g*, could be used within the entire operational range of the aircraft and hence contributed to a fast deceleration of the aircraft [84].

5.3.2.3 DE HAVILLAND "VAMPIRE." A first assessment of the flight characteristics of the DH100 "Vampire" aircraft was carried out in April 1944 by the British Aeroplane and Armament Experimental Establishment (A&AEE) in Boscombe Down [85]. The maximum permitted flight speed at that time was an indicated 510 mph (821 km/h). Apart from an increase of the stick forces, *Mach* number effects were not observed up to this airspeed. The dynamic stability about the vertical axis was criticized. Even in calm air, the aircraft developed oscillations about the vertical axis that were unpleasantly amplified in turbulent air and also with control-surface deflections not completely controllable. The requirements on this aircraft as a weapons platform were, therefore, not met. Acceleration and deceleration behavior was a further matter of complaint. Higher thrust and air brakes were supposed to remedy the shortcomings. During test flights between July and October 1945, the first series-production aircraft reached a maximum level speed of 526 mph (846 km/h) at an altitude of 25,500 ft (7770 m). The critical *Mach* number, given as $M = 0.76$, could thus already be reached during level flight, which had also to be taken into account by the pilot during tactical flight maneuvers. The later "Vampire" versions 5 to 9 showed during high-speed flight already from $M = 0.71$ to 0.76 an increasingly diving-pitching-like (porpoising) motion about the pitch axis and buffeting. At $M = 0.78$ to 0.79, abrupt tail-heavy trim changes occurred (sometimes also nose-heavy) partly accompanied by "wing drop," which provided the impression of a snatched roll [86]. The automatic nose-up movement at near-sonic speeds was judged to be an additional safety factor not given in the case of aircraft with nose-heaviness within that range. The intense "porpoising" motion about the pitch axis commencing at $M = 0.78$ rendered any tactical use of the aircraft within this *Mach* number regime impossible. The "Vampire" production aircraft were for this reason limited at the RAF to 525 mph (845 km/h) true airspeed indication and a maximum *Mach* number of $M = 0.78$. The later "Vampire" training aircraft (double seater) was, on the other hand, limited to 875 km/h and $M = 0.82$. This aircraft also showed within this limiting *Mach* number range the favorable safety-related tail-heaviness.

5.3.3 UNITED STATES OF AMERICA

5.3.3.1 BELL P-59. With the Bell P-59A, the first American aircraft with jet propulsion took off on October 1, 1942, from the Army Air Force Muroc Bombing and Gunnery Range, today's Edwards Air Force Base (EAFB) in California. The flight testing of the P-59 aircraft soon showed, however, that this first jet aircraft would hardly have the potential of an operational fighter aircraft. It was hopelessly inferior to the standard piston engine fighters. The high-speed testing of the P-59 was carried out starting in June 1944. During the fourth high-speed dive, the landing gear came out unintentionally. The aircraft could, however, be recovered and belly landed by test pilot *J. Woolams*. On the fifth flight during November 1944, the complete empennage broke away at an altitude of 25,000 ft (7620 m). The trials were discontinued, and the aircraft was limited to a maximum speed of only 400 mph (644 km/h) and a *Mach* number of $M = 0.70$. As a safety factor when unintentionally exceeding the *Mach* number limit, the fact was considered that the aircraft started at $M = 0.77$ to develop a tail-heavy pitching moment, which made it practically impossible for the pilot to continue an initiated dive at speeds above this *Mach* number due to the large push forces required on the control stick. The aircraft thus behaved similarly to the British "Meteor" and "Vampire" aircraft.

5.3.3.2 LOCKHEED XP-80. During flight tests that the NACA carried out in 1947 to determine the characteristics of the new Lockheed XP-80 aircraft at high flight speeds [87], problems arose that seriously threatened the stability and controllability within this speed range. Especially serious was the "pitch down" from $M = 0.7$ with a subsequent intense bucking of the aircraft during the pull-out from dives at high speeds and *Mach* numbers. During a test after reaching a maximum *Mach* number of $M = 0.866$ at a pull-out and a subsequent reduction of the *Mach* number to $M = 0.85$, the positive load factor of 4 g held by the pilot had increased within about one second to 7 g without pilot control inputs. The flight altitude was approximately 27,000 ft and the indicated airspeed 400 mph at an average gross weight of 10,220 lb (4645 kg) and a normal center-of-gravity location (27.5% MAC). *Mach* number and flight speed reduced during that second to $M < 0.8$ and 380 kts (704 km/h) respectively. The immediate easing and pushing of the stick, at 4 g still held with a pull force of about 70 lb (32 daN), led to the abrupt decrease of the positive load factor from 7 g to a negative 1 g (Fig. 5.11). In wind-tunnel measurements it had been noticed that from $M = 0.7$ a nose-heavy pitching moment started to develop. This was traced back to the fact that the zero-lift angle of attack decreased due to *Mach* number effects with increasing *Mach* number, and hence the incidence or the downwash, respectively, decreased at the location of the elevator. This resulted in a nose-heavy moment. This by itself did not, however, explain the subsequent abrupt "pitch up." The cause for this behavior was suspected to be the interaction between the wing and the elevator in the presence of shock waves on the wing, which could finally also be proven.

In the evaluation, the contributions of the fuselage, wings, and empennage on the pitching moment were examined step-by-step, and the conclusion was reached that the momentary loss of the equilibrium of the aircraft was primarily caused by the changing elevator effectiveness. The angle of attack for zero lift, increasing with decreasing *Mach* number thus increasing the downwash at the elevator, and the fast increase ("return") of the control-surface effectiveness caused the surprising "pitch-up" effect. It was, among other things, recommended as a remedy to provide the wing with symmetrical airfoil sections and the horizontal stabilizer unit with a distinct sweep, which slowed down the increase in the downwash at the horizontal stabilizer. The suggestion of a swept stabilizer unit was incorporated on the Lockheed F-94C "Starfire" aircraft developed in 1950 and essentially derived from the P-80; in the case of the F-94C "Starfire," the Lockheed Company further developed the otherwise almost unchanged P-80 configuration as a double-seat fighter aircraft with a radar nose and integrated rocket armament. Furthermore, the Lockheed F-94C received as one of the first fighter aircraft of the U.S. Army Air Force a jet engine with an afterburner.

5.3.3.3 LOCKHEED XP-80 AND REPUBLIC XP-84.

In the case of the prototypes of the U.S. jet aircraft Lockheed XP-80 and Republic XP-84 (Fig. 5.12), it was noticeable that they exhibited when approaching load-related critical *Mach* numbers different or even opposite characteristics with respect to trim changes

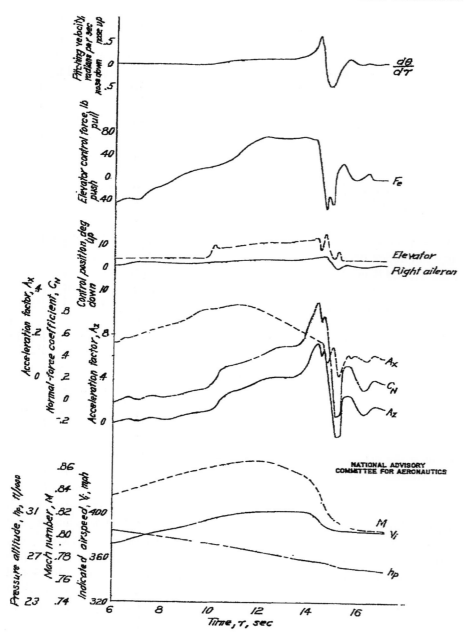

Fig. 5.11   XP-80, parameter time history during pull-out process with pitch up.

about the pitch axis. Both aircraft had conventional, straight trapezoidal wings that had, however, considerably different airfoil-section characteristics. The P-80 (used as a single-seat fighter during the Korean War, later known as jet trainer Lockheed T-33) became from $M = 0.7$ to 0.75 more and more nose

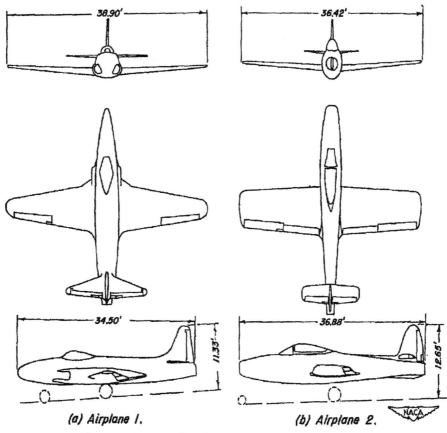

(a) Airplane 1.   (b) Airplane 2.

Figure 1.- Three - view drawings of airplanes 1 and 2.

Fig. 5.12   XP-80 (Aircraft 1), XP-84 (Aircraft 2), side views.

heavy. Similar to the P-80, the P-84, after about $M = 0.755$, became steadily nose heavy, but then tended to become at $M = 0.82$ at first abruptly tail heavy (pitch up) with this trend being amplified up to $M = 0.86$. A just as abrupt nose-heavy load reversal ("nose down") occurred at $M > 0.86$. The controllability of the P-84 was thus strongly impaired in the high *Mach* number regime; therefore, the aircraft was tactically hardly useful to U.S. Air Force pilots.

The facts were more accurately investigated and thoroughly analyzed by the NACA with the knowledge of experience already gained during dives with the XP-80 [88]. As a cause for the initially inexplicable difference in the behavior, the different pressure distributions on the suction and pressure sides of the airfoil sections selected for the XP-80 and the XP-84, obtained during flight measurements, were identified. The local shock waves on the wings of the two aircraft, already noticeable above $M = 0.70$ to 0.75, exhibited different forms and developments for the wing sections selected. The thickness and the maximum-thickness location of the XP-84 were larger than those of the XP-80. The symmetrical airfoil sections of the XP-80 wing showed within the critical *Mach* number range a steady development and progression of the shock waves on the upper and lower surfaces, so that a steadier downstream shift of the neutral point resulted. On the other hand, the local shock waves on the upper and lower sides of the XP-84 wing sections developed irregularly and initially generated pressure distributions and downwash angles, which caused the observed tail-heavy pitching moments. When further increasing the *Mach* number, the movements of the shock waves on the upper and lower wing surfaces of the XP-84 aligned themselves again, thus causing the nose-heavy trim change. Figure 5.13 shows the dependence of the parameters related to the controllability about the pitch axis as function of the *Mach* number during the pull-out process from a high-speed dive maneuver.

5.4   ROCKET-PROPELLED AIRCRAFT

Even prior to the development and application of the jet engine, rocket propulsion as a means of thrust for aircraft had been tried. Because of the short operating time of a rocket engine, it was clear from the beginning that its use for aircraft propulsion would only be possible for very special in-flight operations and that the return to the airfield and the landing were only possible in a glide. The Heinkel Company was a forerunner and had been able to gain experience with a He 112 propeller-driven fighter aircraft with a rocket engine additionally installed on the aircraft.

The development of rocket-propelled aircraft commenced in Germany with the pioneering achievement of *E. Heinkel*. In 1936, *W. von. Braun* was able to convince him to demonstrate this type of aircraft propulsion by way of trial. The test carrier He 112R was followed by a research and experimental aircraft especially designed for rocket propulsion, the He 176, which took off on June 20, 1939, for the first time. The maximum speed achieved was about 850 km/h corresponding to about $M = 0.7$ at mean sea level (MSL) conditions. *Mach* number effects did not, however, as far as is documented, play a role during the flight tests.

5.4.1   SIEBEL DFS 346

The technical development of the DFS 346 research aircraft up to the end of WWII in Germany and the continuation under Soviet supervision, at first at Siebel in Halle (as OKB 3) (OKB refers to experimental project design office) and from October 1946 at the OKB 2 in Podberesje (Dubna, 120 km north of Moscow) is described in detail in Sec. 6.5. After its transfer to the

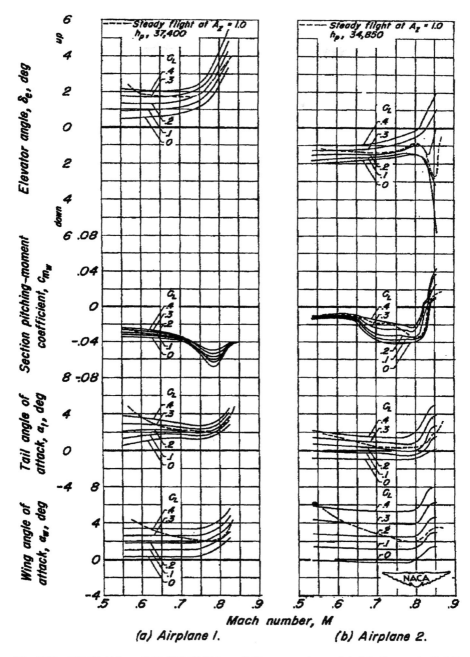

Fig. 5.13   XP-80 (Aircraft 1), XP-84 (Aircraft 2), parameters related to the controllability about the pitch axis as a function of the *Mach* number.

Soviet Union, the first prototype of the DFS 346, still completed at Halle, was investigated intensively in the large wind tunnel of the Zentralinstitut für Aero- und Hydramechanik, Zhukovsky, Moskau (ZAGI) (Central Aero- and Hydrodynamics Institute). After numerous aerodynamic improvements, a glider version without propulsion evolved at the OKB 2 designated 346 P (also named 346A and 346-I [89]). The glider aircraft was equipped with ballast instead of the rocket engine weight including the additional fuel load.

Flight testing of the 346 P was transferred to Tjoplystan, an airfield located 8 km south of Moscow. It was prepared by personnel of the Siebel Company who had been brought at the end of 1946 to the Soviet Union. German test pilot W. Ziese carried out the test flights with the 346 P. Training flights had preceded where he practiced piloting in a prone position with the gliders "Crane" and "Baby," which were still modified in Germany for this purpose. There followed flights with the 346 B towed by an American B-25 aircraft that had been confiscated by the Soviet Union. The 346 P was said to have been positioned on a takeoff cart and, after takeoff, release, and a short flight, had landed on a skid similar to the Ar 234 or Me 163. The cockpit of the 346 p, partitioned off, was also dropped from the B-25 to test the rescue system. After the landing technique had been satisfactorily tested, the next phase starting in spring 1948 was prepared, which was to consist of glides from high altitudes. The airplane was now attached between the two engines under the right wing of a B-29 aircraft also retained by the Soviet Union. With the 346 p carried up to altitude, Ziese now examined the technique of separating from the carrier aircraft as well as the special requirements of the pilot's prone position with regard to the handling qualities and the controllability within a larger speed range including the landing approach and landing using the extended skid. During this phase, about 30 flights were carried out supposedly reaching repeatedly speeds of more than 900 km/h without compressibility effects being in evidence [90].

Of the three aircraft intended for rocket-propelled flight, the first one, the prototype designated 346-1 (also 346 V-1 or 346-I), became ready in May of 1949. The rocket engines were not yet available. They were, as in the case of the 346 P, replaced by weight dummies. The first flight occurred on September 30, 1949, with Ziese in a 346-1 under the wing of the B-29 carried up to an altitude of almost 10 km. The flight mass amounted to 3145 kg. After the unproblematic release, the 346-1 must during the subsequent shallow dive already have attained high airspeeds of about the 900 km/h because a controllability loss and buffeting occurred indicating compressibility effects. The flight ended with a hard landing with Ziese being injured. The skid was apparently not completely extended and hence not locked. The skid and the underside of the fuselage were considerably damaged. After repair, the first flight of the 346-1 was carried out by the

Russian test pilot *P. I. Kasmin*, but also ended with a crash due to the skid not being locked. However, the damage was minor because snow covered the landing ground. Also *Kasmin's* landing after the subsequent flight failed when, after the release at an altitude of 2000 m, he missed the landing strip. After the renewed repair, *Ziese* was tasked with the further test flights that were transferred to Luchowizy airfield southeast of Moscow in 1950. Because of the *Mach* number effects, the airspeed of the repaired 346-1 was limited to $M = 0.8$. Also for the 346-2 and 346-3, which had been completed in the meantime, the *Mach* number limit was set to $M = 0.9$, only slightly higher. Further flight testing with the 346 P, the 346-1, and 346-3, the latter with an installed (Soviet modified), but not yet operational Walter rocket engine, lasted until 1951.

Flights with the 346-2 are not documented. The flight tests with the 346-3 with operational rocket propulsion commenced in August of 1951 in Luchowizy airfield. *Ziese* carried out three flights with the 346-3 [91]. The first one had been performed on August 15, 1951, however, due to the *Mach* number restrictions with only one ignited rocket chamber. The previously mentioned compressibility effects, that is, the loss of control-surface effectiveness and "wing drop," recurred but still seemed controllable by the pilot. During the second flight on September 2, 1951, 900 km/h was exceeded without problems. During the third flight on September 14, 1951, after the release at an altitude of 9300 m and the ignition of the rocket engines at an altitude of 8500 m, the aircraft climbed to an altitude of 12 km accelerating simultaneously to more than 900 km/h ($M > 0.85$). Here, controllability loss occurred again, and the aircraft turned this time into an uncontrolled dive (also a spin after stalling may have developed [92]). The controllability was lost once more at an altitude of 7000 m after control-surface effectiveness was regained for a short period of time. Consequently, *Ziese* jettisoned his rescue capsule at an altitude of about 6500 m and landed safely with the parachute. The aircraft, including the instrumentation, was completely destroyed. After an investigation of the crash, whose causes could not completely be explained, the entire program was abandoned, and in 1953 the design office "OKB 2" in Podberesje was dissolved.

Parallel to the work on the DFS 346 research aircraft, the development of jet-propelled fighter aircraft in the Soviet Union had progressed. Already the MiG 15, equipped with swept wings, reached in 1950 tactically usable subsonic *Mach* numbers of up to $M = 0.92$ and had proven to be a success during the Korean War starting in 1951. The ambitious research program, which started at the Siebel Factory in Halle with the DFS 346, did not bring the lead hoped for in the race to achieve high supersonic *Mach* numbers. With their programs X-1, X-2, and D-558-II, the Americans were already in 1951 much further along so that a success as a matter of prestige seemed no longer possible.

5.4.2 MESSERSCHMITT ME 163

The development and technology of the Me 163 aircraft will be described in Chapter 6.2.3. The flight testing commenced at the Messerschmitt airfield in Augsburg with nonpowered towed flights and glides because the delivery of the Walter Company rocket engines for the Me 163, whose thrust was adjustable between 150 and 750 daN, was delayed. In shallow dive maneuvers the Me 163 V-1 already attained speeds of up to 850 km/h.

On August 13, 1941, the first rocket-propelled flight of the V-1 took place from the Peenemünde-West airfield where the trials with ignited rocket engines had been transferred. In several flights, the speed along a fixed test range equipped with cine-theodolites was increased to 800, 880, and finally 920 km/h. The Me 163 V-4, loaded with fuel and with *H. Dietmar* at the controls, was towed by a Bf 110 flown by *R. Opitz* to an altitude of 4000 m on October 2, 1941. The rocket engine was ignited immediately after the release. Within a short time, the aircraft reached a speed of 1004 km/h at an altitude of 3000 m measured along the established cine-theodolite basis, which corresponded to a *Mach* number of $M = 0.84$.

Lippisch's suggestion to carry out high-speed research with the Me 163 did not go down well. Contrary to the warnings of *Lippisch*, the RLM was determined to turn the Me 163 into an armed rocket-powered fighter aircraft designated Me 163 B. The testing of the V-models of the B-series commenced on June 26, 1942, at Lager Lechfeld airfield. Rocket-propelled flights, however, took place again at Peenemünde-West airfield. The maximum speed at all altitudes was limited to 900 km/h due to compressibility effects. The operational ceiling amounted to 12 km. The maximum safe *Mach* number of the Me 163 was reached at approximately $M = 0.85$. It must be considered that the tail-heavy trim changes may lead to extremely high load factors if the critical *Mach* number is reached at low altitudes, that is, at high dynamic pressures.

The test pilot *H. Peters* reported that when flying at more than 900 km/h, $M = 0.80$, a slight lifting of the forward fuselage became noticeable despite holding the control stick in a fixed position. Subsequently a fast quivering roll to both sides and an equally fast nose-down pitch followed. The rocket engine also shut down abruptly. The process usually proceeded so fast that the individual motions of the aircraft could not exactly be identified [93].

From operational reports of the Me 163 it becomes clear that the danger of entering the critical *Mach* number regime was especially high if the rocket propulsion was not shut off in time, i.e. was "forgotten" when leveling off out of a climb into level flight or even entering a shallow dive maneuver. This was probably not always to be avoided during the stress of air combat. Intense, mostly uncontrollable reactions, like "heavy buffeting," sudden "wing drop," or "pitch down" with high negative g-loads, were the consequence. In one instance, even the complete rudder on the vertical tail was lost. Already prior to the experience with the Me 262 during high-speed trials, there existed, therefore, considerations to install a *Mach* number warning device on the Me 163 (also see Me 262).

W. Späte reported that in 1944 he had just caught up in his Me 163 with a formation of USAF P-47 escort fighters and was about to go into an attack position when the left wing suddenly and without warning dropped. Counter control measures remained without reaction. Instead, the nose dropped, and the aircraft buffeted and felt like flying over a washboard. A kind of porpoising with severe negative g-loads occurred. The altitude-compensated airspeed indicator showed $V_W = 960$ km/h true airspeed. *Späte* found himself within the critical speed range. When he reduced thrust, everything returned to normal again [94].

During an attack on three Lockheed P-38 "Lightning" aircraft on April 9, 1945, the Luftwaffe pilot, *Roesle*, and his squadron companion, *Bott*, both of the JG 400 at an altitude of 7300 m had just agreed on their attack tactics and assigned their targets when they unintentionally entered the *Mach* number range critical to their Me 163. Both had failed to throttle back their rocket engines during the transition into level flight so that the aircraft further quickly accelerated. *Bott's* Me 163 started to slowly roll to the left followed by a pitch up. At the same moment, the wing surfaces buffeted and a strong nose-heavy moment was noticeable so that *Bott* tried to prevent by a hard pull at the control stick. His airspeed indicator reading was 960 km/h. The rocket engine shut down as a result of the negative g-loads. Almost at the same moment the same thing happened with *Roesle's* aircraft. When control was regained, both were positioned far below their prey who disappeared undamaged on the horizon [95].

To reduce the principal faults of the Me 163 aircraft, such as the short flight duration, the droppable landing gear, and the skid landing, a further development of the Me 163 at the Junkers Aircraft Company was suggested. The aircraft, designated Ju 248 V-1 (later Me 263 V-1) without a rocket engine and with a fixed landing gear, carried out from February 1945 some towed gliding flights at the Dessau and Brandis airfields with test pilot *H. Peters* at the controls. Following the experience with the Me 163, a maximum *Mach* number limit of $M = 0.80$ and a maximum speed of 1000 km/h were planned for the trials. Flights at high speeds without or with rocket propulsion are, however, not documented. Planned series production of the Me 263 aircraft was not achieved [96].

REFERENCES

[1] A. *Busemann*, "Profilmessungen bei Geschwindigkeiten nahe der Schallgeschwindigkeit," Jahrbuch der WGL 1928, p. 95; quoted in *F. W. Riegels*, Aerodynamische Profile, R. Oldenbourg, München 1958, p. 48.

[2] L. *Prandtl*, "Die Rolle der Zusammendrückbarkeit bei der strömenden Bewegung der Luft," *A. Busemann*, Aufgaben der Hochgeschwindigkeitstechnik, Schriften der

Deutschen Akademie der Luftfahrtforschung, Heft 30, Vorträge der 2, Wissenschaftssitzung, Nov. 12, 1937.

[3] "Jahrbuch der Deutschen Akademie der Luftfahrtforschung 1942/43," Berlin, 1943, pp. 292–295.

[4] W. Friebel, "Die Bedeutung der Flugeigenschaften für die Frontbewährung der Flugzeuge," Lilienthal-Gesellschaft für Luftfahrtforschung, Bericht 163 (Geheim), Berlin, 1943, pp. 3, 4.

[5] E. G. Friedrichs, "Grundeigenschaftsprüfung," Lilienthal-Gesellschaft für Luftfahrtforschung, Bericht 163 (Geheim), Berlin, 1943, p. 11.

[6] "German Aircraft, New and Projected Types," U.K. Air Ministry, A I 2(G) Report No. 2383, Jan. 1946.

[7] "Me 262 A-1 Bedienungsvorschrift-Fl," Stand Aug. 1944, Messerschmitt A. G. Augsburg, p. 10.

[8] H. Friedrichs, "Messung und Anzeige der *Mach*'schen Zahl im Fluge," DVL, UM 1028, 1944.

[9] Ch. Burnet, "Three Centuries to Concorde," Mechanical Engineering Publications Ltd., London, 1979, p. 37.

[10] H. Schlichting, E. Truckenbrodt, "Aerodynamik des Flugzeugs," Springer–Verlag, Berlin 1960, pp. 127–130.

[11] F. W. Riegels, "Aerodynamische Profile," R. Oldenbourg, München, 1958, pp. 35–49.

[12] Ch. Burnett, "Three Centuries to Concorde," Mechanical Engineering Publications Ltd., London, 1979, p. 82.

[13] D. Fiecke, "Stand der deutschen Jagdflugzeugentwicklung zu Kriegsende," Zeitschrift Flugwelt 6/1953, pp. 165–168, and 7/1953, pp. 203–208.

[14] J. Stack, "The NACA High-Speed Wind Tunnel and Tests of Six Propeller Sections," NACA Report No. 463, 1933; see also Chap. 3 of this book.

[15] F. Weinig, "Luftschrauben für schnelle Flugzeuge," DVL, Berlin-Adlershof, Jahrbuch 1937, pp. 49–55.

[16] G. Bock, R. Nikodemus, "Die Aussichten des Luftschraubenantriebs für hohe Fluggeschwindigkeiten," DVL, Berlin-Adlershof, Jahrbuch, 1938, pp. 50–55.

[17] W. Späte, "Testpiloten," Aviatic Verlag, Planegg, 1993, pp. 38–44.

[18] E. Kuhle, "Flugmessung der Eigenschaften um die Querachse bei hohen Mach'sche Zahlen am Muster Fw 190," DVL UM 1077, Feb. 1945.

[19] W. Wagner, "Die Deutsche Luftfahrt, Kurt Tank—Konstrukteur und Testpilot bei Focke-Wulf," Bernard & Graefe Verlag, München, 1980, p. 139.

[20] J. Prien, P. Rodeike, "Jagdgeschwader 1 und 11," Teil 3, 1944–1945, o. V., p. 1517.

[21] J. L. Ethell, "Ta 152, monogram close-up 24," Monogram Aviation Publications, (Eagle Editions Ltd.), Hamilton, MT, U.S., 1990, pp. 24, 25.

[22] D. Hermann, "Focke-Wulf Ta 152," Der Weg zum Höhenjäger, Aviatic Verlag, Oberhaching, 1998, (from TAT report from Focke-Wulf, section flying qualities), p. 93.

[23] K. Tank, "Luftfahrtforschung und Flugzeugbau," Jahrbuch der DAL 1942–1943, pp. 289–295.

[24] W. Messerschmitt, correspondance with Udet, 1941; quoted in W. Späte, "Testpiloten," Aviatic Verlag, Planegg, 1993, p. 40.

[25] F. W. Riegels, "Aerodynamische Profile," R. Oldenbourg München, 1958, p. 51 (Fig. 4.2).

[26] K.- H. Regnat, "Dornier Do 335," Aviatic Verlag, Oberhaching, 2000.

[27] W. Green, "The Warplanes of the Third Reich," Galahad Books, New York, 1990, p. 157.

[28] G. Bock, "Neue Wege im Deutschen Flugzeugbau," Schriften der Deutschen Akademie der Luftfahrtforschung, Berlin 1945; see also Fig. 6.1.

[29] K.-H. Regnat, "Dornier Do 335," Aviatic Verlag, Oberhaching, 2000, p. 148.

[30] W. Wagner, "Die Deutsche Luftfahrt," Band 1, Bernard & Graefe Verlag, 1980, pp. 111, 113–115.

[31] Ch. Burnet, "Three Centuries to Concorde," Mechanical Engineering Publications Ltd., London, 1979, pp. 29–37.

[32] A. Bullen, B. Rivas, "John Derry: The Story of Britain's First Supersonic Pilot," W. Kimber, London, 1982, p. 179.

[33] Ch. Burnett, "Three Centuries to Concorde," Mechanical Engineering Publications Ltd., London, 1979, pp. 38, 39.

[34] Jay Miller, "The X-Planes X-1 to X-31," Aerofax, Arlington, TX, U.S., 1988, p. 9.

[35] W. Green, "Famous Fighters of the Second World War," Macdonald & Co. Ltd., London, 1959, p. 76.

[36] "Pilot's Flight Operating Instructions for Army Models P-38H Series, P-38J-5, and F-6B-1," T. O. No. 01-75FF-1, Fig. 25 (dive placard), Sep. 25, 1943.

[37] R. Whitford, "Fundamentals of Fighter Design, Part 1-Stability & Control," Air International Magazine, June 1996, p. 344.

[38] "Pilot's Flight Operating Instructions for P-47B, -C, -D and –G Airplanes," T. O. No. 01-65BC-1, Sec. II, Para.17, Jan. 20, 1943.

[39] W. A. Mair (ed.), "High-speed Wind-tunnel Tests on Models of Four Single-Engined Fighters (Spitfire, Spiteful, Attacker and Mustang)," A.R.C. Technical Report, R&M No. 2535, April 1945, Part 5, pp. 78, 79.

[40] J. M. Nissen, B. L. Gadeberg, W. T. Hamilton, "Correlation of the Drag Characteristics of a P-51B Airplane obtained from High-Speed Wind-Tunnel and Flight Tests," NACA ACR No. 4K02, 1944.

[41] "Pilot's Flight Operating Instructions for Army Models P-51-D-5, British Model Mustang IV Airplanes," AN 01-60JE-1, Fig. 5.2, Apr. 5, 1944.

[42] G. Bock, "Neue Wege im Deutschen Flugzeugbau," Schriften der Deutschen Akademie der Luftfahrtforschung, Berlin, 1945.

[43] K. R. Pawlas, "Arado Ar 234 Der erste Strahlbomber der Welt," Luftfahrtdokumente LD 21, 1976, p. 279.

[44] See Ref. [43], p. 103.

[45] See Ref. [43], p. 119.

[46] See Ref. [43], p. 180.

[47] E. Brown, "Wings of the Luftwaffe," McDonald and Jane's Publishers Ltd., London, 1977, p. 92.

[48] See Ref. [43], p. 407.

[49] G. Hanf, "Ich flog die He 162," Zeitschrift Flugzeug, 6/92, p. 17.

[50] W. Späte, "Testpiloten," Aviatic Verlag, Planegg 1993, pp. 56, 57.

[51] J. R. Smith, E. J. Creek, "Jet Planes of the Third Reich," Monogram Aviation Publications, (Eagle Editions Ltd.), Hamilton, MT, U.S., 1982.

[52] H. Lommel, "Geheimprojekte der DFS, Vom Höhenaufklärer zum Raumgleiter 1935–1945," Motorbuch, 2000, p. 10.

[53] "So kam es zum roten Stratojet Typ 150," Zeitschrift Flugwelt, Vol. 5, 1961, p. 357.

[54] S. Holzbaur, "Vorwärts gepfeilte Flügel," Interavia Magazine, No. 7, 1950, pp. 380–382.

[55] H. Walter, "Deutsche Luftfahrtspezialisten für die Sowjetunion, part 2," Flugzeug Classic Magazine, Vol. 4, 2003, pp. 45–49.

[56] H. Lommel, "Junkers Ju 287," Aviatic Verlag, Oberhaching, 2003.

[57] "Junkers EF 131 Bedienvorschrift," Report No. S-317, August 1946, p. II–35; reprint Dr. Korrell.

[58] H. Walter, "Deutsche Luftfahrtspezialisten für die Sowjetunion," Teil 3, Flugzeug Classic Magazine, Vol. 5, 2003, p. 56.

[59] E. Kuhle, "Fluguntersuchungen über Eigenschaften um die Querachse bei hohen Machschen Zahlen am Muster Me 262 V-3," UM 1418/1, DVL Berlin-Adlershof, Nov. 1944.

[60] H. Lindemann, "Hochgeschwindigkeitsmessungen am Flugzeugmodell Me 262 im HWK der DVL," Industriebericht J 900/4, DVL 1943.

[61] E. Kuhle, "Fluguntersuchungen über Eigenschaften um die Querachse bei hohen Machzahlen am Muster Me 262 V-3," UM 1418/2, DVL Berlin-Adlershof, Feb. 1945.

[62] H. Horn, "Figures of Mtt-AG, compiled for U.S. Report," July 10–11, 1945, pp. 330–331.

[63] E. Englander, "Interrogation Report of H. Fay, 1945"; in "Messerschmitt Me 262: A Pictorial and Design Study Including the Pilot Handbook," Aviation Publications, Appleton, WI, U.S., 1978.

[64] M. Jurleit, "Strahljäger Me 262," Die Technikgeschichte, Transpress Verlagsgesellschaft, Berlin, 1992, p. 112.

[65] E. G. Friedrichs, "Messung und Anzeige der Machschen Zahl im Fluge," DVL, UM 1028, 1944.

[66] M. Boehme, "Jagdgeschwader 7," Motorbuch Verlag, Stuttgart, 1983, pp. 62, 63.

[67] F. Wendel, "Mtt A.G., Me 262 V3 PC+UC, 1st and 2nd flight," Flight Report No. 783/1, July 22, 1942, p. 3.

[68] J. Foreman, S. E. Harvey, "The Me 262 Combat Diary," Air Research Publications, Walton-on-Thames, U.K., 1995, p. 42.

[69] H.-G. Mutke, "The Story of My First Supersonic Flight on 9 April 1945 over Innsbruck," http://mach1.luftarchiv.de/mach1.htm, accessed Nov. 2009, no longer available online; contact H. Galleithner for hard copy.

[70] W. Späte, "Top Secret Bird: The Luftwaffe's Me-163 Comet," Pictorial Histories Publishing Co., Missoula, MT, U.S., 1989, pp. 259–261.

[71] E. Brown, "Wings of the Luftwaffe," Macdonalds and Jane's Publishers Ltd., London, 1977.

[72] "Memorandum Report on ME-262 Airplanes, Nos. T-2-711 and T-2-4012," TSFTE-2008, Head Quarters Air Material Command, Wright Field, Dayton, OH, U.S., Sept. 3, 1946.

[73] P. Butler, "War Prizes," Midland Counties Publications, Leicester, U.K., 1994, p. 218.

[74] A. Marchand, "Les Messerschmitt 262 de Bretigny," L'Album du Fanatique de l'Aviation, No. 41, 42, Editions Lariviere, Paris, France, Feb./March 1973, p. 23.

[75] J.-C. Fayer, "Vols d'Essais, le centre d'essais en vol de 1945 a 1960," E-T-A-I, Boulogne-Billancourt, France, 2001, pp. 20, 21, 32, 51.

[76] A. Marchand, "Les Messerschmitt 262 de Bretigny," L'Album du Fanatique de l'Aviation, No. 41, 42, Editions Lariviere, Paris, France, Feb./March 1973, p. 25.

[77] P. Butler, "War Prizes," Midland Counties Publications, Leicester, U.K., 1994, p. 270.

[78] M. Jurleit, "Strahljäger Me 262 im Einsatz," Transpress Verlagsgesellschaft, Berlin, 1993, pp. 137–147.

[79] "The Postwar Messerschmitt," Air Enthusiast Quarterly Four, pp. 167–172.

[80] A. Alexandrov, G. Petrov, "Die deutschen Flugzeuge in russischen Diensten 1914–1951," Band 2, Flugzeug Publikations GmbH, p. 223.

[81] W. Wagner, "Die ersten Strahlflugzeuge der Welt," Buchreihe Die deutsche Luftfahrt, Band 14, Bernard & Graefe Verlag, Koblenz, 1989, He 178: pp. 15–23; He 176: pp. 48–52.

[82] A. W. Thom, F. Smith, J. Brotherton, "Flight Tests at High Mach Number on E28/39 W 4041 (Single-engined Jet-propelled Aircraft)," ARC R&M No. 2264, Oct. 1945.

[83] D. J. Higton, R. H. Plascott, D. A. Clarke, "The Measurement of the Overall Drag of an Aircraft at High Mach Numbers," A.R.C. Technical Report, R.&M. No. 2748, Jan. 1949.

[84] J. J. Partridge, "The Gloster Meteor F.IV," No. 78, Profile Publications Ltd., Leatherhead, U.K.

[85] B. Johnson, T. Heffernan, "Boscombe Down 1939–1945: A Most Secret Place," IHS Jane's, Bracknell, Berkshire, U.K., pp. 158–166.

[86] F. K. Mason, "The de Havilland Vampire Mk. 5 & 9," No. 48, Profile Publications Ltd., Leatherhead, U.K.

[87] H. H. Brown, L. S. Rolls, L. A. Clousing, "An Analysis of Longitudinal Control Problems Encountered in Flight at Transonic Speeds with a Jet-propelled Airplane," NACA RM No. A7G03, Ames Aeronautical Lab., Moffett Field, CA, U.S., Sept. 1947.

[88] NACA Report, RM A51E14, Oct. 1951.

[89] "SKYROCKETS mit dem roten Stern," in "Edition 02 E International" der Fliegerrevue, 1993, p. 10.

[90] H. Lommel, "Geheimprojekte der DFS—Vom Höhenaufklärer bis zum Raumgleiter 1935–1945," Motorbuch-Verlag Stuttgart, 2000, pp. 107–109.

[91] "SKYROCKETS mit dem roten Stern," in "Edition 02 E International" der Fliegerrevue, 1993, p. 14.

[92] H. Walther, "Deutsche Spezialisten für die Sowjetunion," Teil 7, Flugzeug Classic 10/2003, p. 57.

[93] H. J. Ebert, J. B. Kaiser, K. Peters, "Willy Messerschmitt—Pionier der Luftfahrt und des Leichtbaus," Die Deutsche Luftfahrt, Band 17, Bernard & Graefe Verlag, Bonn, 1992, p. 264.

[94] J. L. Ethell, "Messerschmitt Komet," Motorbuch Verlag, Stuttgart, 1980, pp. 142, 143.

[95] J. L. Ethell, "Messerschmitt Komet," Motorbuch Verlag, Stuttgart, 1980, p. 206; W. Späte, "Top Secret Bird," Pictorial Histories Publishing Co., Missoula, MT, U.S., 1989, p. 253.

[96] H. J. Ebert, J. B. Kaiser, K. Peters, "Willy Messerschmitt—Pionier der Luftfahrt und des Leichtbaus," Die Deutsche Luftfahrt, Band 17, Bernard & Graefe Verlag, Bonn, 1992, p. 267.

[97] H. Schlichting, E. Truckenbrodt, "Aerodynamik des Flugzeugs," Vol. 2, Springer–Verlag, Berlin, 1960, pp. 126, 168.

[98] C-T. E. Lan, J. Roskam, "Airplane Aerodynamics and Performance," Univ. of Kansas, Lawrence, KS, U.S., 1981, pp. 68, 140.

[99] "Bauvorschriften für Flugzeuge," Heft 1, Vorschriften für die Festigkeit von Flugzeugen, Reichsluftfahrtministerium, Technisches Amt, Dec. 1936.

[100] Bände I-V des "Ringbuch der Deutschen Luftfahrttechnik," RLM, 1940.

[101] G. Siegel, "Angewandte Lastannahmen," C.J.E. Volckmann Nachf. E. Wette, Berlin, 1938.

[102] H. Wenke, "Praktische Theorie in der Flugtechnik," Band 8, Flugleistungsermittlung, Verlag Dr. Max Gehlen, Leipzig/Berlin, 1938.

[103] H. Wenke, "Praktische Theorie in der Flugtechnik," Band 9, Flugeigenschaften, Verlag Dr. M. Mathiesen & Co., Berlin, 1940.

[104] H. Koppe, "Fahrtmessung," V E 5, Ringbuch der Luftfahrttechnik, RLM, Berlin, 1941, p. 2.

[105] G. Kiel, "Die Messung des Flugstaudrucks und statischen Drucks bzw. der Fluggeschwindigkeit und Höhe," I B 7, Ringbuch der Luftfahrttechnik, RLM, Berlin, 1941, pp. 10, 11.

[106] H. Koppe, "Fahrtmessung," V E 5, Ringbuch der Luftfahrttechnik, RLM, Berlin, 1941, p. 3, fig. 2.

[107] D.(Luft) T. 5005, "Fahrtmesser mit Höhenausgleich," Geräte-Handbuch, RLM Technisches Amt, Part D, T. 5005, Berlin, Sept. 1942.

[108] *D. Nijboer, D. Patterson*, "Im Cockpit, Jagd- und Kampfflugzeuge des II. Weltkriegs," Motorbuch Verlag, Stuttgart, pp. 167, 171.

[109] *K. A. Merrick*, "German Aircraft Interiors 1935–1945," Vol. 1, Monogram Aviation Publications, (Eagle Editions Ltd.), Hamilton, MT, U.S., p. 186.

[110] *M. Griehl*, "Strahlflugzeug Arado Ar 234 Blitz," Motorbuch Verlag, Stuttgart, 2003, p. 190.

[111] *K. Kracheel*, "Flugführungssysteme, Blindfluginstrumente, Autopiloten, Flugsteuerungen," Die Deutsche Luftfahrt Band 20, Bernard & Graefe Verlag, Bonn, 1993, pp. 18–20, 42.

[112] *K. A. Merrick*, "German Aircraft Interiors 1935–45," Vol. 1, Monogram Aviation Publications, (Eagle Editions Ltd.), Hamilton, MT, U.S., pp. 14, 15.

[113] *R. Ahlers, G. Sauerbeck*, "Geschichte des Forschungsstandortes Braunschweig-Völkenrode," Braunschweig, 2003, p. 25.

[114] *B. Göthert*, "Hochgeschwindigkeits-Untersuchungen an symmetrischen Profilen mit verschiedenen Dickenverhältnissen im DVL-Hochgeschwindigkeits-Windkanal (2.7 m) und Vergleich mit Messungen in anderen Windkanälen," Berlin-Adlershof, Jf 701/3, FB 1505, Dec. 17, 1941.

[115] *F. W. Riegels*, Aerodynamische Profile, R. Oldenbourg München, 1958.

APPENDIX 1:   COMMENTS ABOUT THE CRITICAL MACH NUMBER AT FLIGHT TEST

In the literature, different interpretations are assigned to the idea of a critical *Mach* number. Most frequently one associates this *Mach* number with the one where the near-sonic drag rise commences. But also the trim changes due to compressibility that must be compensated by corresponding control-surface deflections using trim or control stick inputs by the pilot may serve as an indication or reference point. *Mach* number induced buffeting (for instance, "aileron buzz") or uncontrolled movements of the aircraft ("wing drop" or "nose slice"), caused by shock-induced flow separations, which influence above a certain *Mach* number the controllability of the aircraft, are additional characteristic features that may be associated with a critical *Mach* number.

An aerodynamic definition can be found in the well-known German textbook "Aerodynamik des Flugzeugs" (Aerodynamics of the Aircraft) of *H. Schlichting* and *E. Truckenbrodt* [97]. The following wording (slightly modified) is taken from that book: The critical *Mach* number, M_{crit}, is the *Mach* number M_∞, composed of the freestream speed and the speed of sound of the freestream, at which sonic velocity first occurs locally on the wing. Shock waves and as a result also flow separation and strong drag increases may generally occur at higher *Mach* numbers.

In the United States, for practical purposes, in the evaluation of flight data, definitions are used that are based on the compressibility-related increase in the drag coefficient ("drag-rise or drag-divergence *Mach* numbers"). The methods used by the aircraft companies Boeing and Douglas became known.

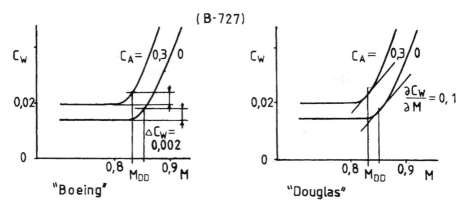

Fig. 5.14 **Definition of the *Mach* number at "drag divergence,"** M_{DD} **(drag-rise *Mach* number),** $C_W = C_D$, $C_A = C_L$.

According to Boeing, M_{DD} is the *Mach* number where $C_D = C_{Dincompressible} + 0.002$, but according to Douglas $M_{DD} = M_{crit} + 008$ (average value). M_{DD} stands here for the "drag divergence" *Mach* number, and M_{crit} is the *Mach* number of the aircraft where shock waves first occur on the wing, that is, where $M = 1$ is reached locally. The results of both methods lie in practice close together and may be considered equivalent [98]. Figure 5.14 shows the application of the definitions of M_{DD} to the Boeing B-727 as example.

Mach number effects on the trim change, or other effects such as "buffeting," "wing drop," or "snaking," may each become a criterion for the critical *Mach* number limiting the use of the aircraft if the effects, when reaching this *Mach* number, unacceptably limit the tactical use of the aircraft or even become a risk to flight safety. The lowest of these values may then be considered the critical *Mach* number of the aircraft where one or more of the above effects become so limiting that a safe and effective operation of the aircraft is no longer possible and even the structural or fatigue strength is affected.

A typical phenomenon is the *Mach* number related aileron buffeting as it may, for instance, be observed on the AMD "Falcon 20E." Near the maximum operating *Mach* number M_{Mo}, a fine humming noise initially develops that is also felt at the control stick when small aileron deflections are input. If M_{Mo} is exceeded for testing purposes, pronounced aileron buffeting with higher amplitudes may quickly be the result. (The Falcon does not reach its M_{Mo} of 0.86 during level flight.)

Vortex generators upstream of the ailerons on the upper side of the outer wing of an aircraft generally indicate that buffeting has already occurred at relatively low *Mach* numbers and that the situation has to be remedied in order to be able to use a larger speed respectively *Mach* number range. Vortex generators on the swept wing at low speeds and high angles of attack, as well as in the case of shock-induced separation, delay the loss of lift on the outer wing thus counteracting the "pitch-up" effect.

Vortex generators are frequently used as an aid on vertical tails (Panavia "Tornado") or rear fuselages (Fiat G-91) against *Mach* number induced buffeting or yawing moments due to local airflow separations. New energy is supplied by vortex generators to the partially separated boundary layer downstream of the shock wave from the undisturbed outer flow so that a separation with large velocity fluctuations is prevented.

APPENDIX 2: GERMAN CONSTRUCTION REGULATIONS FOR AIRCRAFT

The aircraft developed and manufactured in Germany during the war were based on the so-called Bauvorschriften für Flugzeuge ("Design Regulations for Aircraft") (Vols. 1, 3, 4) of December 1936 ("BVF 36"). Understandably no load cases are included which could be related to the phenomena occurring when reaching the critical *Mach* number at near-sonic speeds. From about 1938, this was recognized, however, and preparations were made to account in later issues of the regulations for *Mach* number effects (airload assumptions, flight characteristics, verification). However up to 1945, exact formulations and requirements were not published. However the valid calculation methods and the proof of individual requirements were unwittingly partially responsible for *Mach* number effects, as a rule, being able to be ignored pending the approval of a new model. The *Mach* number effects corresponding to the data and observations of the test pilots available after the evaluation of the test results were effectively "eliminated" by suitable airspeed limits for the operational use of the aircraft. This, however, did not prevent effects occurring increasingly more often during combat—especially in the case of jet aircraft—endangering flight safety. The investigations of the DVL within the high-speed range just described, for example Fw 190, Me 262 V-3 were also an expression of the efforts to cope with these problems. The BVF designated user groups, which were identified by capital letters (G for transportation of goods; P for the transportation of passengers; R for journeys; S for training and sports; KI, KII for aerobatics; and H for high performance and testing), and stress-related groups 1–5. The fighter aircraft, including the jet aircraft, were generally assigned to the group H5.

From today's point of view, the BVF can be described as a very well-proven design basis. On this basis about 130,000 German aircraft, primarily of military design and application, were developed, manufactured, tested, and certified between 1928 and 1945. The aircraft were successfully flown in peace and war, the latter under conditions hardly conceivable today. The German BVF was equally applicable to military and civilian aircraft. The following remarks provide a compressed survey [99].

Decisive for the dimensioning of the components and the structure are the loads due to the dynamic pressure, the angle of attack, and the aerodynamic forces expressed by the load factor $n = P/G$ (P = airload, G = weight). Here, the safe load conditions have to be considered with respect to the loads resulting from the following:

- Aerodynamic forces occurring in a dive (maximum dynamic pressure), a pull-out (maximum load factor) and composite load conditions as resulting from gust loads, control-surface deflection as well as from propeller effects
- Ground forces, such as those occurring during landing, taxiing and towing, a headstand, or turnover
- Water forces occurring on floats of seaplanes
- Manual stick and pedal forces, machinery and other forces

Some important requirements essential to understanding the remarks regarding the proof of strength according to the BVF are explained next.

BVF, Vol. 1, Sec. 1020: Critical speed was understood as the speed where flutter occurred or where wing or stabilizer areas reached the limit of their static stability. The speed was allowed to just reach the 1.3 times the maximum speed. During the proof by flight tests, it was sufficient to show in case of the Load Group H5 that the aircraft was free of flutter at the speed reached in a dive from the operational ceiling respectively a maximum altitude of 6000 m. The propeller drag of the engine, or engines, throttled to idle motion should be effective. This was not possible in the case of the new jet aircraft; however, the jet engines could be throttled back to idle operation. (See Sec. 1125, issued 1936.)

BVF, Vol. 1, Sec. 1125: For the Loads Group H5 during a dive (the so-called Case C), the final dynamic pressure was decisive, which was reached at c_a (c_1) = 0 and hence $q_{end} = (G/F_{Tr})/c_{wg}$ (G = weight, F_{Tr} = wing area, $c_w = c_d$). The proof had to be furnished according to Number 1020 in a dive that was started at the operational ceiling of the aircraft, but not higher than 6000 m. (The values of the maximum stagnation pressures of the other load groups followed from Table 2 of the BVF, Vol. 1.)

BVF, Vol. 1, Sec. 1126: The pull-out proof at a positive lift coefficient was such that for the Load Group H5 the following had to be shown:

For $q_h \geq 7.09\ G_{max}/F_{Tr}$:

$n_{Tr} = 7.5 - (4.25\ G_{max}/F_{Tr})/(q_h - 4.25\ G_{max}/F_{Tr})$, but ≤ 7.0;

for $q_h \leq 7.09\ G_{max}/F_{Tr}$:

$n_{Tr} = 6.0$

Here, q_h was the maximum flight dynamic pressure (not the maximum flight speed) attainable in level flight. Subscript Tr indicates that the values of the wing had to be taken. The proof by computations and flight tests had to be furnished at the maximum gross weight.

Section number 1141: The additional gust load factor for level flight conditions was calculated with $n = 1$ at $v = v_h$. If a gust velocity v_b of $+10$ m/s or -10 m/s prevailed, the following applied [η is a coefficient ≤ 1 (for details see BVF pp. 31, 32)]:

$$n_{Tr} = 1 + q(F_{Tr}/G)\ (v_b/v)\ (dc_{aTr}/d\alpha)\eta$$

Information concerning the verification in flight trials according to the BVF, can be found in a number of documents [100–103], which were available up to 1945.

APPENDIX 3:   INDICATED AIRSPEED AND SPEED MEASUREMENTS

The following definitions apply to the presentation of the flight speed customary in Germany up to 1945 and its measurement or display on an aircraft. The increasing altitudes and flight speeds becoming closer and closer to the speed of sound required more refined measurements and corrections to be provided. The following definitions were valid [104]:

- Speed (km/h; m/s)
- Speed through the air (flight speed, true airspeed, airspeed)
- Velocity of the air (wind speed)
- Speed over ground (ground speed)
- Dynamic pressure (kg/m², mm water)
- Indicated airspeed (km/h) (with the indication of the airspeed based on dynamic pressure measurements and, therefore, associated with certain errors)

The knowledge of the ground speed is important to flight planning and navigation. The true airspeed (flight speed) can be derived from the pitot-static onboard system, applying a density correction and, if necessary, a compressibility correction. The true airspeed may also directly be measured by means of the so-called air log on the aircraft. The wind speed, respectively the velocity of the air mass over ground in which the aircraft moves, can be determined by external means, for instance, by meteorological measurements. On board, the wind speed may be determined via the "wind triangle" with the aid of the ground speed obtained by aerial navigation (range/time measurement, dead reckoning) or radio navigation and the true airspeed.

A distinction between "indicated airspeed" and true airspeed" proved to be useful for flight operations. "Indicated airspeed" derived from a dynamic pressure measurement represents a measure for the ability of an aircraft to glide. The setting up, change, and supervision of a flight condition can be guaranteed independently of the altitude, by the corresponding speed readings. The true airspeed, which is important to aerial navigation, such as during a long-range flight (flight time, range, and navigation), and also to the adherence to the limits of the flight regime (flutter), can only be obtained, corresponding to the flight altitude, after applying a density correction if the dynamic pressure measurement is available. As long as the flight altitudes and dynamic pressures are low, as for the aircraft designs of up to the mid-1930s, the differences between the indicated airspeed and the true airspeed remain low. (Deviations are acceptable up to about 400 km/h and altitudes well below 3000 m.) With increasing speed and especially pressure altitudes the differences become, however, so large that they can no longer be neglected.

The dynamic pressure q in an incompressible air environment is defined as the difference between the total (stagnation) pressure and the static pressure at altitude:

$$q = p_{\text{total}} - p_{\text{static}} = 1/2\ \rho v^2$$

Compressibility effects were taken into account by a factor ß, that is, $q_{\text{comp}} = 1/2\ \rho v^2 ß$, with the factor approximated by the expression [105] $ß = 1 + 1/4(v/a)^2$ with a = speed of sound. Because the speed of sound changes with the static temperature T, ß is not only dependent on v but also on the pressure altitude. The compressibility correction of the display was based on the temperature distribution of the Internationale Normalatmospäre (INA) (international standard atmosphere), which resulted in small errors compared to having the correction based on the actual temperature of the atmosphere. Altogether the uncertainty in such a speed indication (let alone errors of the static pressure pickup and the instrumentation errors) above 400 km/h amounted to about ± 1% of the actual value up to an altitude of about 12 km and increasing up to an altitude of 14 km insignificantly [106].

The instrumentation industry in Germany had already in about 1938 suggested a separation of "indicated airspeed" and true "airspeed" on the same indicator on the cockpit instrument panel. One display design of the *R. Fuess* Company, Berlin-Steglitz, had two display sections separated by a divider (see Fig. 5.15). Average speeds in a range of 70 to 250 km/h were shown as "indicated airspeed." The indicator needle disappeared at 250 km/h below the divider. At 250 km/h, another needle emerged from under the divider that now took over indicating the true airspeed (flight speed) of the aircraft up to 750 km/h being the maximum indication on the airspeed scale [107]. A similar type was used for larger speed ranges showing a range between 100 and

Fig. 5.15 Central cockpit instrument panel of a Messerschmitt Me 262 A-1; the airspeed indicator is visible in the left upper corner.

400 km/h as "indicated airspeed" for the purpose of monitoring the flight condition. At the 400 km/h mark, the needle appeared located under the divider covering the range of the scale to the right of the divider that indicated the true airspeed up to 1000 km/h on the scale. This display was not only altitude compensated, but also contained a compressibility correction and was suited for *Mach* numbers up to approximately $M = 0.85$. Such altitude compensated and compressibility corrected airspeed indicators were installed on aircraft such as the Me 262A, Ar 234B, Me 163B, and Do 335 A-1 [108–110]. (See Fig 5.15.)

The display of the so-called INA altitude-compensated airspeed indicator (Höhen-Fahrtmesser) lacked the separation of the scale into an indicated airspeed (Fahrt) and a true airspeed range, and used a mechanical compensation dependent on the static air pressure, and an altitude-dependent temperature scale function representing the INA atmosphere. The speed indication, thus obtained, hardly differs from values that would have been achieved with the actual static air temperature instead of the INA temperatures. If the difficulties of an accurate derivation of the static temperature from a stagnation-temperature measurement are taken into account, the INA temperature is an acceptable approximation that, furthermore, foregoes the need for an additional sensor. Such an altitude-compensated airspeed indicator was also developed by the *R. Fuess* Company [111]. For speeds up to flight *Mach* numbers of $M = 0.45$, the compressibility correction may be neglected. For takeoff, approach, and landing at low altitudes, the information corresponded with sufficient accuracy to the indicated airspeed that is necessary for the control of the flight condition (gliding ability, distance to the stall speed). The altitude compensation at higher altitudes is based on the INA atmosphere, and the true airspeed is indicated as pressure altitude increases. The scale ranges normally up to 750 km/h or with a small inner-scale extension up to 900 km/h [112] for faster aircraft. The latter may, for instance, be found on the cockpit instrument panels of the Me 109 G-6, Fw 190 D-9 or Ta 152 H-0, and also of the He 162 aircraft (see Fig. 5.16).

Fig. 5.16 Altitude compensated airspeed indicator (Höhen-Fahrtmesser) up to 900 km/h installed, for instance, on the Messerschmitt Me 109 G, Focke-Wulf Fw 190D, or Ta 152 H.

APPENDIX 4: AIRFOIL NOMENCLATURE

A large part of the aircraft characteristics at transonic speeds is determined by the pressure distributions on the wing or empennage airfoil sections. The selection of an airfoil section is an extensive process, decisive for the flight performance and flight characteristics, which must be completed long before the start of aircraft design work. The airfoil characteristics were determined in wind-tunnel measurements. The Luftfahrt forschungsanstalt Hermann Göring, Braunschweig (LFA) Aeronautical Research Establishment, Braunschweig) possessed already from late 1941 a very efficient high-speed wind-tunnel (A2) in Voelkenrode near Braunschweig, Germany, with a nozzle diameter of 2.8 m and *Mach* numbers of up to $M = 0.95$ in a continuous operation [113]. Numerous project-supporting measurements on aircraft models, such as Me 262, Ju 287, Do 335, but also basic research measurements were carried out in this wind tunnel.

The first part of the following explanations is taken from an original report of the Institute of Aerodynamics of the DVL [114] that deals with *Mach* number effects and also compares results from different wind-tunnel measurements. The nomenclature for airfoil sections of empennages and wings employed throughout the DVL not only for airfoil sections developed in-house is explained by the example of the airfoil section designated NACA 1 30 12-1.1 40. The numerals have, in sequence, the following meaning:

- Height (camber) of the mean line, $f/t = 1\%$
- Maximum-camber location, $x_f/t = 30\%$
- Thickness ratio, $d/t = 12\%$

- Nose radius, $(\rho/t)/(d/t)^2 = 1.1$
- Maximum thickness location, $x_d/t = 40\%$
- $t = \text{chord}$

The letters NACA indicate that the contour was determined according to NACA standards. NACA used airfoil designations consisting of four- and five-digit numeral strings.

The four numerals of the airfoil section NACA 2415 have the following meaning:

- Camber, 2%
- Maximum-camber location, 40% downstream of the leading edge
- Thickness ratio, 15%

The investigated four-digit airfoils started with the numerals 0, 2, 4, and 6 and possessed thickness ratios between 6 and 21% (in steps of 3 %).

The numerals of the five-digit NACA airfoil series have, as shown by the example NACA 23012, the following meaning:

- 1st numeral—Twenty third (20/3) of the lift coefficient for smooth incoming and trailing flow (= "no flow" about the mean line, shock-free entrance) $c_a = (c_1) = 0.3$ $(20/3 \times 0.3 = 2$, or $2{:}20/3 = 3{:}10 = 0.3)$. This represents a measure for the curvature. (Investigated values: 2, 3, 4, 6.)
- 2nd and 3rd numeral combined—Twice the maximum camber location; $30{:}2 = 15 \%$
- 4th and 5th numeral combined—Thickness ratio = 12%

The NACA used still more designations that will, however, not be further addressed here. They are described in detail by *F. W. Riegels* [115].

Chapter 6

EXPERIENCE GAINED DURING DEVELOPMENT AND TESTING OF THE FIRST SWEPT-WING JET AIRCRAFT

BERND KRAG

6.1 INTRODUCTION

A special conference of the Lilienthal-Gesellschaft für Luftfahrtforschung (LGL) (Lilienthal Society for Aeronautical Research) took place in München on August 23 and 24, 1938. The topic of this conference was "Airframe and Engine" for high-speed flight [1]. At this conference, it became clear that aircraft with reciprocating engines and propellers had reached their performance limit and that substantial increases of the flight performance could only be expected by so-called "special engines," like rocket or turbine-jet engines. Still within the same year, the Reichsluftfahrtministerium (RLM) (State Ministry of Aeronautics) asked the engine industry to deal with the development of jet engines. It is stated in a memorandum of the RLM of October 1938 that one must also work on the airframe to find optimum configurations for high-speed flight and for the adaptation of the new jet engines [2]. The memorandum addresses both aeronautical research organizations with the request to provide high-speed wind tunnels and test beds and the industry to develop suitable aircraft configurations. Research should be concerned with high speed, however, not forgetting flight at high lift coefficients. Concerning the airframe, this report envisaged slender fuselages and thin wings with low aspect ratio respectively tailless configurations as favored by *Lippisch*. Nobody thought at that time of the swept wing suggested three years before by *Busemann*.

The development of the first aircraft with swept wings and the difficulties that had to be overcome in the application of the new concept are described in the following chapters. The first generation of jet aircraft did not yet have a swept wing. An exception was the rocket fighter Messerschmitt Me 163, which required such a wing for stability reasons. We shall, therefore, not elaborate on this first generation of jet aircraft because there are sufficiently good descriptions in the relevant literature. However, these aircraft already possessed considerable flight performance due to their high-quality aerodynamics. The flight testing of these aircraft is covered in Chapter 5. These aircraft will be introduced together with the most important technical data within an introductory paragraph.

Up to the end of the war, only two aircraft with a real swept wing were actually completed in Germany. These were the experimental aircraft Messerschmitt P 1101 and the jet bomber Junkers Ju 287. The research aircraft DFS 346 was in an advanced state of construction when the Siebel plants were occupied by the Americans. The designs of *Lippisch* and the utilization on antiaircraft rockets showed that low-aspect-ratio wings were also of interest to high-speed flight. Merely a glider version of the *Lippisch* P 13 was built, but not tested in Germany anymore. Some of these aircraft served as examples for designs in Britain and the United States, which were partly quite successful. The Soviet Union also had benefited to a considerable degree from the German know-how, however, attached the greatest importance to independent designs. Because, except for the Junkers 287, no flight tests with swept-wing aircraft were carried out in Germany, the experience gained by the Americans and British with follow-up models is of great interest. The flight testing of this pioneer generation of aircraft revealed the problems that the German engineers would have had to cope with had one of the planned swept-wing projects actually been achieved.

The role of German aeronautical research in the development of these new aircraft will also be addressed in the following chapters. The aerodynamics of the swept wing were exemplarily investigated within all speed regimes by the aeronautical research establishments. A very good survey of the work performed by German aeronautical research establishments concerning swept wings and the application in industrial projects is given by *P. Hamel* [3]. Sufficient data were, correspondingly, available to industry, which could be used in the wing development. Despite these data, the industry carried out its own wind-tunnel measurements to investigate the behavior of the swept wing in the high-lift regime. In the opinion of the Arado Company, the aeronautical research establishments concentrated too much in their investigations on symmetrical airfoils within the high-speed range and neglected, therefore, the need for cambered high-speed airfoil sections [4]. This statement was according to the state of affairs not correct and revealed that the cooperation between research establishments and industry was not as close as often claimed. The investigations by aeronautical research also extended to airfoils with low camber (<2%) and hence covered the area of industrial applications in the high-speed range. The need for airfoils with higher camber was, indeed, recognized and corresponding investigations started [5].

The Arado and Junkers Companies, which had turned since 1942 to the development of jet bombers with large-aspect-ratio swept wings, needed airfoils that could also provide sufficiently high lift coefficients at low speeds without the occurrence of the dreaded "tip stall," that is, the early flow separation at the outer wing due to the spanwise boundary-layer flow. Despite results available from aeronautical research, much work was invested at Arado and Junkers into the development of suitable cambered airfoils which also exhibited characteristics at high speeds which were hardly inferior to those of symmetrical airfoils. It was recognized by aeronautical research rather late that high-speed aircraft with strongly swept wings had special flying qualities that differed from the known ones and could possibly cause considerable problems to the pilots. On request of the RLM, research started in mid-1944 on a program that addressed this question. The numerous aircraft projects, which the German aeronautical industry dealt with during the last months of the war, will not be further considered here, because there are no reliable documents available that allow a sensible judgment. These projects, with partly fantastic configurations, were presented in the relevant literature mostly in a more simplistic way; due to the lack of data, a serious analysis is scarcely possible. Most of these projects were paper studies with superficial performance calculations. Only some of the configurations found their way into the wind tunnel or reached mock-up status. This also held for efforts to equip already established jet aircraft belatedly with a swept wing. All of these attempts did not get beyond the design stage and configurational studies in the wind tunnel.

After the war these designs stimulated the imagination of numerous engineers abroad and led there to equally fantastic layouts. Some of these designs found their way into the design offices of the aeronautical industry and grew into useful aircraft. In most instances it is not possible to establish anymore in how far a new design originated from German ideas. Therefore, we shall introduce here only those projects where the German origin can be proven beyond question.

6.2   FIRST JET AIRCRAFT WITH SWEPT WINGS

6.2.1   JET AIRCRAFT OF MESSERSCHMITT

6.2.1.1   MESSERSCHMITT ME 262.   The fighter aircraft Messerschmitt Me 262 was the world's first jet-propelled aircraft that was manufactured in series. Compared to a contemporary, modern propeller-driven, fighter (Messerschmitt Me 109 K with DB 605 D reciprocating engine), the Messerschmitt Me 262 provided an almost 200-km/h higher maximum speed [6] and was thus with regard to flight performance superior to any allied fighter aircraft. The flight test center of the German Air Force in Rechlin released the Me 262 as a fighter for combat missions in September 1944 [7]. Despite its superiority, this aircraft did not present any serious obstacle to the Allies in their fight to suppress the national-socialist regime in Europe.

With the development of the Me 262, new technical ground was entered with regard to aerodynamics and jet propulsion. The consistent application of the latest results in high-speed aerodynamics also provided the Me 262 with advantages relative to the contemporary aircraft in Britain and the United States. With 760 km/h at an altitude of 9000 m, the Gloster "Meteor" Mk 3 [8] was considerably slower than the Me 262, which achieved a maximum speed of 830 km/h at the same altitude and almost the same engine thrust. The United States' first jet, the Bell XP-59 A "Aeracomet," could not compete with the Messerschmitt Me 262. The maximum speed of the experimental YP-59 amounted to merely 660 km/h at an altitude of 10,000 m [9]. Only the Lockheed XP-80 reached with 805 km/h at an altitude of 6500 m a comparable performance [10]. The first series version of the P-80, the P-80 A, which became available in the spring of 1945 was considerably more efficient than the prototype XP-80. This aircraft reached a maximum speed of 900 km/h near the ground and still achieved 850 km/h at an altitude of 8000 m. The P-80 A would have been a serious opponent for the Me 262 if this aircraft would still have been introduced during the war.

Comparative flights of the Messerschmitt Me 262 and the Lockheed P-80 A, carried out after the war at Wright Field, showed that the Me 262 reached higher speeds, accelerated faster, and had about the same climb performances as the P-80 A. The flight characteristics of the P-80 A were judged by the test pilots to be better [11]. The Lockheed XP-80 is the best example of the status of the American high-speed research in 1943 and its implementation by the Lockheed chief designer *Clarence "Kelly" Johnson*.

Extensive literature exists concerning the development and operation of the Messerschmitt Me 262. Therefore, we shall not go into the development of this remarkable aircraft. We shall rather report on the behavior of the aircraft in limiting areas. In addition, the question will be addressed as to which technologies of the Me 262 have entered aircraft developments outside Germany after the war. The effects of compressibility on the flight characteristics are treated in more detail in Sec. 5.3.

The Messerschmitt Me 262 represented a successful combination of advanced aerodynamics and the most modern engine technology. The technical data listed next provide a good survey. Decisive for the good flight performance of this model was the wing. The wing thickness amounted to 12% at the wing root decreasing to 9% at the wing tip. (These data apply to sections normal to the 25% chord line. In the freestream direction, the relative thickness was 11% and 8% respectively at the wing tip.) At the time, *Ludwig Bölkow*, an employee at the Messerschmitt project office since March 1939, was responsible for the wing aerodynamics. As airfoil section, he selected a symmetrical airfoil of the type 0012 E4 (wing root) with a maximum thickness at 40% of the wing chord [12]. The forward part of the airfoil section consisted of an ellipse whose minor diameter corresponded to the airfoil thickness at 40% of the wing chord. The airfoil contour followed down to the trailing edge the National Advisory Committee for Aeronautics (NACA) procedure for the computation of airfoil contours. According to a statement by *Waldemar Voigt*, chief designer at Messerschmitt and responsible for the development of the Me 262, it was important that the trailing-edge angles of the airfoil sections remained ≤15 deg [13]. Trailing-edge angles that were too large caused undesirable fluctuating neutral-point locations (see [5], p. 5). *Bölkow* essentially followed the guidelines for high-speed airfoil sections shortly before being specified by *K. H. Kawalki*, Deutsche Versuchsanstalt für Luftfahrt (DVL) (German Test Establishment of Aeronautics) and considered favorably by German aeronautical research (see Sec. 2.3.2). Dependent on the wing thickness and maximum-thickness location, these airfoils possessed critical *Mach* numbers of $0.8 < M < 0.9$. The nose radius was determined by an ellipse and was relatively large, which was considered advantageous regarding the high-lift regime. Although not thought of with respect to the reduction of the transonic drag, the wing possessed a sweep of 18.5 deg (wing leading edge), which also had a favorable effect on the critical *Mach* number due to the effect of the area rule (see Sec. 2.3.4). The wing had an automatic "slat," extending over the entire span, characteristic of the Messerschmitt aircraft. This slat ensured sufficient lift during landing and obviously also prevented the feared "tip-stall" problem in the case of swept wings. All control surfaces were equipped with geared tabs to reduce the control forces at high speeds. *Messerschmitt* was not a friend of hydraulically actuated control surfaces because he feared the complexity and the additional weight of such systems. Like his chief designer, *Voigt*, he approved of a good aerodynamic balance and tabs although they caused more problems with regard to the flutter safety (see [13], p. 23). The elevator was electrically adjustable via a trimming button, which was a great relief to the pilot when compressibility effects commenced. The airfoil sections of the vertical and horizontal stabilizers were similar but had, however, only a thickness of 8%.

Because of the higher sweep of the leading edges and the thinner airfoil sections, compressibility effects at the empennage occurred only at speeds beyond 1000 km/h at altitudes above 6000 m [14]. Pressure distribution measurements on the wing and the horizontal stabilizer of the Me 262 V3 showed that up to a *Mach* number of $M = 0.855$ controllability was not yet impaired [15]. The critical *Mach* number of $M = 0.81$ for the transonic drag rise was determined in the United States by comparative flight measurements with the Lockheed P-80 A and the Republic XP-84 [16]. These values were measured during level flights. According to wind-tunnel measurements of the DVL, the transonic drag rise already occurred, however, at the smaller *Mach* numbers of $M = 0.75$ (Fig. 6.1) [17].

The occurrence of compressibility effects is given for the Messerschmitt Me 262 at $M = 0.86$. This is also based on the evaluation of comparative flight measurements with the Lockheed P-80 A and the Republic XP-84 (see [11], p. 108). This coincides with the information from *Bölkow* for the onset of compressibility effects at speeds >1000 km/h. This speed could, however,

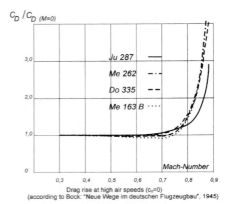

Fig. 6.1 Critical *Mach* number for some German fighter aircraft (see [17], p. 15).

only be achieved in a dive at zero lift ($C_A = C_L = 0$). For lift coefficients $C_A = C_L > 0$, compressibility effects already occur at lower *Mach* numbers. Operational experience showed that the Me 262 could still be flown up to *Mach* numbers of $M = 0.83$ without problems [18]. At a *Mach* number of $M = 0.86$, serious handling problems occurred, which could not be mastered anymore by regular pilots.

Production defects and aeroelastic effects were responsible, however, for this flight performance not being reached by many series-production aircraft. At high dynamic pressures, the slats showed a tendency to slowly open due to a wing deformation [19]. This caused additional drag for the aircraft. The torsion of the outer wing at large speeds caused control problems about the longitudinal axis. The cause was insufficient fastening of the maintenance covers in the wings so that the wing skin was no longer able to correctly take up the torsional forces [20]. Furthermore the fabric-covered control surfaces caused problems. At high speeds, the fabric, covered control surfaces inflated ("ballooned") or started to flutter [21]. In addition, it was difficult to pull the Me 262 out of a dive at high speeds. Above a certain speed ($M > 0.86$), a pull at the control column led to an increase in the trajectory inclination. In this case, the pull out from the dive was only possible with the aid of the elevator trimming. The Me 262 was really dangerous in flight with one engine. The performance of one engine was hardly sufficient to keep the aircraft in the air during flight with large side-slip angles [22]. For these and other problems, Rechlin Erprobungsstelle der Luftwaffe (E-Stelle) (E-Center) had first to find solutions before the aircraft could be delivered to the troops. The flight testing of the Me 262 is described in detail in Sec. 5.3.

The Me 262 (Table 6.1) developed pronounced yawing oscillations ("snaking") at high speeds that were hardly dampened at higher altitudes. For the pilot, considerable difficulties arose here during the approach to firing. To remedy the problem in a simple way, the size of the vertical stabilizer of the Me 262 was successively reduced (see [7], p. 60). This was, at first sight, an

TABLE 6.1 IMPORTANT TECHNICAL DATA OF THE MESSERSCHMITT ME 262

Wing area	21.7 m²
Aspect ratio	7.37
Airfoil (root)	0012-E4 with 40% maximum-thickness location
Airfoil (wing tip)	0009-E4 with 40% maximum-thickness location
Sweep	18.5 deg (leading edge)
Critical *Mach* number	0.75 ("drag rise") according to DVL measurements Respectively 0.81 from U.S. measurements
Critical *Mach* number	0.83 ("buffeting onset")
Critical *Mach* number	0.86 (operational limit)
Maximum permitted speed	850 km/h at an altitude of 8000 m
Permitted speed during a dive	1000 km/h

absurd step because the dampening of the yawing oscillations was thus reduced. At the same time this led, however, also to a reduction in the oscillation frequency making it, therefore, easier for the pilot to control this movement by the rudder. The investigations carried out by *K. H. Doetsch* at the DVL were, however, considerably more promising. The installation of a DVL yaw damper that acted on a separate auxiliary flap or tab (trailing-edge spoiler) provided the desired dampening and stabilized at the same time the Dutch roll in gusty air [23]. This yaw damper was not installed on the series-production aircraft anymore.

The flight tests of the Messerschmitt Me 262 at very high speeds had revealed that a considerable need for clarification with regard to the influence of the *Mach* number on the flight characteristics existed. At the Project Office Oberammergau (Oberbayerische Forschungsanstalt Oberammergau) (OFO) (Upper-Bavarian Research Establishment), a special high-speed program was established in February 1944 for the Me 262 (see [6], p. 251). Although the swept wing had been systematically investigated at the AVA, DVL, and LFA since 1940, it had never occurred to the Research Directorate to develop a special experimental aircraft to investigate the influence of the *Mach* number on the flight behavior. The opinion was that wind-tunnel investigations were completely sufficient to settle all questions. Only during the summer of 1944 when industry had already started the development of a new generation of high-performance aircraft with swept wings did the Research Directorate (Forschungsführung) invite the experts from research and industry to discuss the development of a "research aircraft for high speeds" [24]. The result of this meeting was finally the research aircraft DFS 346 (see Sec. 6.5.1), which was, however, not completed by the end of the war. Almost at the same time, the NACA also had the idea of developing a research aircraft for high speeds [25]. This idea resulted in the research aircraft Bell XS-1, which broke the "sound barrier" in October 1947 for the first time (see Sec. 6.5.2).

The Messerschmitt high-speed program consisted of three steps. During the first step (Project HG I), the Me 262 V-9 was equipped with a swept horizontal stabilizer and a flat cockpit hood. Stability problems arose during the flight testing in January 1945. Whether the changes actually led to higher critical *Mach* numbers is not known.

It can be gathered from the interrogation protocols of *W. Voigt* at Oberammergau that the Messerschmitt HG II program planned a whole string of measures to increase the speed (see [13]: Interrogation of *W. Voigt*). In a first step, the outer wing of the Me 262 was to be installed without alteration with a sweep of 35 deg (25% chord line) by the insertion of a "wedge" at the engine. Alternatively it was planned to sweep the entire wing by 35 deg by inserting a "wedge" at the fuselage. The sweep would have required a rearward shift of the engines to keep the center-of-gravity location within a stable range. It is not known whether the experimental aircraft envisaged here could still be completed.

During a second step, several new wings were to be built and investigated:

1) Wing 1: 35-deg sweep (only the outer wing), 8% thickness at the root (seen in the flow direction), 6% thickness at the wing tip. The airfoil was taken unchanged from the original wing Me 262 wing.

2) Wing 2: 35-deg sweep, 8% thickness constant over the entire span, symmetrical airfoil with elliptical nose down to 40% of the wing chord. This airfoil had changed coordinates compared to the original airfoil in the downstream area. The trailing-edge angle was 13 deg.

3) Wing 3: 35-deg sweep, 8.25% thickness at the wing root (seen in the flow direction) and 12% thickness at the wing tip. The wing aspect ratio was 4.3. Whether the latter statement referred to the complete wing or only to the outer wing is not clear.

The program HG II served to find a suitable wing shape for the project P 1101. The new wing was to provide good high-lift flight characteristics as well as high critical *Mach* numbers. *Voigt* hoped especially in the case of the latter wing for an improvement of the "tip-stall" problem. In addition, the wing-fuselage interference was to be reduced by the low thickness at the wing root. This is astonishing because already the first measurements initiated by *Messerschmitt* had shown that the influence of the fuselage was only small (see Sec. 2.3.3 and Fig. 2.105). Landing flaps, ailerons, and slats had a constant chord. Because of the taper of the wing, this led to a relatively large relative flap chord, that is, from 16% at the root to 25% at the wing tip (Sec. 2.3.2). None of these wings were built. Wind-tunnel measurements on the 35-deg swept wing, carried out at the DVL, showed that the critical *Mach* number (drag rise) could be increased by 0.1 M compared to the normal wing [26]. Drop tests with models of the HG II configuration were, in addition, carried out (see [7], p. 79) in cooperation between the DVL (*Göthert*) and the Deutsches Forschungsinstitut für Segelflug (DFS-Ainring) (German Research

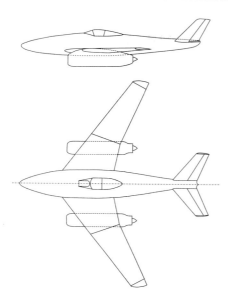

Fig. 6.2   Plan view of the Messerschmitt 262 HG II configuration (see [6], p. 252).

Institute for Gliding) (*Ruden*). These tests had the purpose of clarifying the stability characteristics of the aircraft at high *Mach* numbers. Figure 6.2 shows views of one of the HG II configurations with a 35-deg swept wing.

Messerschmitt extended the high-speed program in December 1944 by another step, the Project HG III. Within this program, the Me 262 was to receive a wing with a sweep of 45 deg using a series-production wing. The jet engines were positioned at the wing root. As a tail unit the swept empennage of the HG I version was envisioned (Fig. 6.3). Wind-tunnel tests with this configuration were carried out in 1945 at the AVA Göettingen (Fig. 6.4). Special attention was given here to the design of the air intake at the wing root.

The strong interest of *Messerschmitt* in the clarification of questions concerning high-speed flight is shown by the fact that, besides the high-speed programs HG I to HG III on the basis of the Me 262, he still planned another experimental aircraft to gather specific knowledge about the optimum wing sweep of fighter aircraft. The program resulted in the P 1101 project, already mentioned, this being a single-jet experimental aircraft with variable-sweep wing (see Sec. 6.2.1.2).

The Messerschmitt Me 262 did not have any direct successors except, maybe, the Russian Suchoj Su 9 [27]. This aircraft was, however, not a success and offered no improvements compared to the Me 262. Also the first layout of the MiG-9 envisaged an engine mounting analogous to the Me 262 under the wings. *Mikojan* changed the design and placed the jet engines into the fuselage thus obtaining an aerodynamically clean wing. In addition, the trimming problem at flights with one engine was simplified [28]. The design of the twin-engine night fighter Douglas F3D would have almost looked like

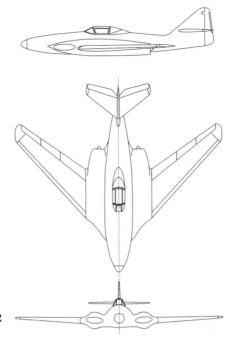

**Fig. 6.3   Plan view of the Messerschmitt 262
HG III configuration (see [4], p. 154).**

an Me 262 had not *Ed Heinemann* placed the jet engines into the fuselage,
probably for the same reasons as in the case of the MiG-9 [29].

   In Britain and the United States at the end of the war, jet aircraft were
already under development, which could be used as a basis. The North
American Aviation (NAA) Company was only interested in the wing and the
slat mechanism of the Me 262. The findings in high-speed aerodynamics
regarding the wing design and the optimum sweep angle were more important. Also the considerations of *Waldemar Voigt* concerning the solutions of
the "tip-stall" problem should have been of great interest. It can be assumed
that, besides the research reports, the "interrogation reports" were also read.
Engineers of the U.S. aerospace industry were also involved in these interrogations so that the knowledge could directly be introduced into current

**Fig. 6.4   Wind-tunnel model of the Me 262
HG III at the AVA Göttingen, 1945 (DLR
Archive Göttingen).**

developments. How far the Messerschmitt projects played a role in the development of the NAA XP-86 will be discussed later (see Sec. 8.2).

6.2.1.2 EXPERIMENTAL AIRCRAFT MESSERSCHMITT P 1101. Besides the Junkers Ju 287, the experimental aircraft Messerschmitt P 1101 was the only aircraft that was built in Germany with a swept wing to reduce transonic drag. The P 1101 already showed with the swept wing and empennage all features of a modern fighter-aircraft configuration as known from the North American F-86 and the MiG 15. With the development of the P 1101, *Messerschmitt* pursued two objectives. Firstly he reacted with this design to an invitation by the RLM in July 1944 to bid on a high-speed single-engine jet fighter, and secondly this aircraft was to serve the investigation of the stability and control behavior at high *Mach* numbers. The P 1101 was later still to play an important role in the development of the experimental aircraft Bell X-5 and the North American XP-86. The development of this aircraft is well described in the quoted literature so that there is no need to once again address this. Here, we are primarily concerned with the advanced aerodynamics of the P 1101 and design details, which later entered other aircraft designs.

The first layout of *Hans Hornung* from August 1944 shows an aircraft with a 40-deg swept wing and equally swept V-type empennage. The jet engine was integrated into the fuselage, and air was supplied by lateral air intakes at the wing root (Fig. 6.5). The outer wing of the Me 262 was to be selected for simplicity and attached swept (see [6], p. 277). The shape of the fuselage with the lateral air intakes was possibly selected due to investigations that had been carried out at the AVA Göttingen in 1944 on rotationally symmetric fuselage bodies (see Fig. 6.6) [30]. The design was changed on request by *Messerschmitt* in September 1944. The lateral air intakes were replaced by a central air intake at the bow of the fuselage, and instead of the V-type stabilizers, conventional swept tail surfaces were selected. *Messerschmitt* decided at the end of November 1944 to build an experimental aircraft parallel to the military version of the P 1101 aimed at investigating flight performance and flight characteristics at high speeds. In the case of this aircraft, the wing could be rotated about the spar-attachment point so that the sweep angle could be adjusted between 35 and 45 deg. Simultaneously the dihedral of the wing could be changed from 2 deg (for the 35-deg sweep) to −3 deg (for the 45-deg sweep). The rolling moment due to yaw, increasing with sweep, was thus accounted for. The wing corresponded largely to the outer wing of the Me 262. The aileron was extended to the wing tip. The inner part of the aileron was replaced by a landing flap. The slat remained unchanged. Figure 6.7 shows views of the experimental aircraft. The idea of an adjustable wing sweep had already been patented by *A. Betz* of the Aerodynamische Versuchsanstalt Göttingen (AVA) (Aerodynamic Test Establishment, Göttingen) in 1942 (see Sec. 2.3.3). For the investigation of the effectiveness

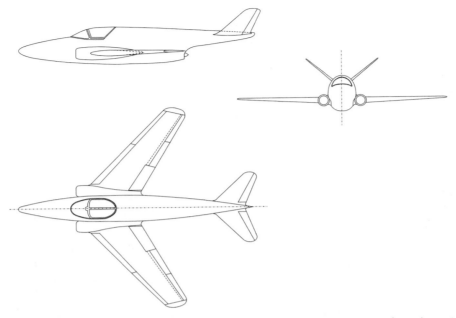

Fig. 6.5   First design of the Messerschmitt P 1101 with lateral air intakes from August 30, 1944 (see [6], p. 277).

of the control surfaces, slats and landing flaps wind-tunnel tests on a model of almost 2-m span were carried out at the DVL. Figure 6.8 shows a wind-tunnel model of this magnitude, which could possibly be associated with the P 1101. The photo was taken by *George S. Schairer* in Göttingen in May 1945. Entries in the wind-tunnel-occupancy log for March 1945 also mention investigations for the OFO on swept wings [31]. It may be that the DVL model was still shipped to Göttingen and again investigated there shortly before the end of the war. The wind-tunnel model shown has, however, a wing with a sweep of 35 deg.

A second set of wings was, in addition, prepared for the experimental aircraft [32]. These wings corresponded in many respects to the wing design

Fig. 6.6   Measurements on bodies of revolution with lateral air intakes at the AVA Göettingen (courtesy of DLR Archives Göttingen, GOAR: PS44-101/5) [30].

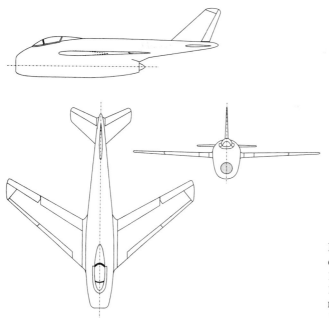

Fig. 6.7   Plan view of the experimental aircraft P 1101 from November 8, 1944 [Luftfahrt International, Heft 7 (Issue 7) Jan.–Feb. 1975].

Fig. 6.8   Wind-tunnel models at the AVA Göttingen; the swept-wing model in the background could be a model of the P 1101 without engines (photograph taken by *G. Schairer* May 1945, Boeing Archives).

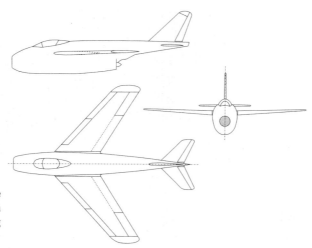

Fig. 6.9 Plan view of the planned series-production design with the second wing set (see [6], p. 279).

No. 3, which was envisaged for the project Messerschmitt HG II. The relative thickness was on average 10%, and the wing aspect ratio was 4.29. In *Voigt's* mind these wings promised improvements compared to the previous wing of the Me 262 at higher speeds (higher critical *Mach* numbers) as well as in the high angle-of-attack regime ("tip-stall" problem). The second set of wings was completed and delivered to Oberammergau. The three-view drawing of February 1945 shows clearly (Fig. 6.9) that the series-production aircraft was to be manufactured with the second wing set.

The construction of the experimental aircraft commenced at the OFO on November 10, 1944. By the end of the war, the aircraft was up to 80% completed and then fell into the hands of the Americans during the occupation. Together with the documentation and the second wing set, the aircraft was shipped to Wright Field, Ohio in May 1945. There it was extensively investigated and the documents made accessible to the U.S. aeronautical industry (Fig. 6.10). Especially the Bell Company was strongly interested in the design and suggested to the U.S. Air Force the development of a fighter aircraft on the basis of the Messerschmitt P 1101. This resulted finally in the experimental aircraft Bell X-5 that was later used by NACA for the investigation of the effect of wing sweep on the flight characteristics.

Whether the P 1101 would have met the requirements of an experimental aircraft is difficult to say. The sense and purpose of such an experimental aircraft could at that time hardly be disputed. There was a large need for clarification with regard to control-surface effectiveness, high-lift behavior ("tip stall"), and flight characteristics. From wind-tunnel measurements at the DVL, it was known that the effectiveness of the ailerons, landing flaps, and slats decreases with increasing sweep angle. With regard to the planned development of a night fighter, based on the HG III configuration, these were

Fig. 6.10 Experimental aircraft Messerschmitt P 1101 at Wright Field, OH, United States (http://users.dbscorp_net/jmustain/p1101_1.jpg).

problems that definitely had to be clarified. The research aircraft DFS 346 was not suitable because it was designed for investigations at supersonic speeds. However, some conclusions concerning the flight behavior of the P 1101 may be drawn from the flight testing of the Bell X-5 and the North American XP-86 in Table 6.2.

6.2.1.3 MESSERSCHMITT PROJECT P 1106. An aircraft derived from the Messerschmitt P 1101 was the P 1106. The P 1106 was envisaged as a light fighter aircraft. The dimensions were smaller than those of the P 1101; with 40 deg the wing sweep was the same. A version of the aircraft, the P 1106 R, equipped with rocket propulsion was even supposed to reach supersonic speeds. Figure 6.11 shows the plan view of the aircraft. Unusual was the aft location of the cockpit in front of the horizontal stabilizer. Whether wind-tunnel measurements were still carried out on the P 1101 is not verified. However, it can be assume that a set of design data existed, which were used for the computation of flight performance and flight characteristics.

TABLE 6.2   TECHNICAL DATA OF THE SECOND WING SET OF THE
MESSERSCHMITT P 1101[a, b]

Airfoil section (wing root)	0008 with the maximum-thickness at 40% chord
Airfoil section (wing tip)	0012 with the maximum-thickness at 40% chord
Sweep	40 deg (25% chord line)
Slats	Constant chord (24% chord at the tip and 13% chord at the root)
Wing area	15.84 m^2
Aspect ratio	4.29

[a]See [13], p. 5, and Chap. 2.
[b]Airfoil data correspond to the HG II-wing, version No. 3. The nose radius is larger at the wing tip than at the root. This has a positive effect on C_{Amax} (C_{Lmax}).

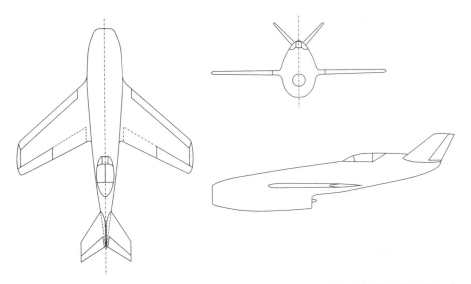

Fig. 6.11   Plan view of the Messerschmitt P 1106 (www.luft46.com/mess/mep1106.html).

Because of the unusual configuration, *K. H. Doetsch* of the Flight Department of the DVL suspected also unusual flight characteristics [33]. The available data were sufficient, indeed, for an analysis of the flight characteristics. The very high position of the empennage and the arrangement of the jet engine at the front fuselage provided at high speeds a large and positive deviation moment (coupling between rolling and pitching moments of inertia). Because of the mass distribution, this would have led to strong pitching and yawing motions when initiating a rolling motion. The high rudder position produced, due to the large leverage with respect to the roll (longitudinal) axis, a relatively large roll-induced yawing moment. The result of *Doetsch's* analysis is summarized in Fig. 6.12. Because of the strong coupling between rolling and yawing motion, the Dutch-roll oscillation at high speeds (small lift coefficient C_L) was unstable. The roll-yaw ratio reached values of 6 at sea level and 8 at high altitudes. The analysis of the causes for this instability shows clearly that, at that time, *Doetsch* had already recognized the phenomenon of inertia coupling. It was thus clear that the P 1106 would have had at high speeds severe problems concerning flight characteristics. Whether this was the reason for *Messerschmitt* to withdraw the P 1106 at the end of February 1945 in favor of newer projects is not known.

6.2.2   EXPERIMENTAL AIRCRAFT BELL X-5

The experimental aircraft Messerschmitt P 1101 had a direct successor in the Bell X-5 research aircraft. Among the specialists of the Combined Advances Field Team (CAFT), arriving on May 7, 1945, at the OFO, was the

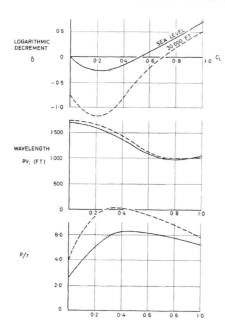

Fig. 6.12 Result of the stability analysis of the P 1106 of *K. H. Doetsch*. Upper figure: Damping of the Dutch-roll oscillation. The logarithmic decrement is a measure of the damping. Lower figure: Roll-yaw ratio *p/r* (*p* = velocity of roll, *r* = velocity of yaw) (see [33], Fig. 18).

chief engineer of the Bell Aircraft Corporation in Buffalo (see [6], p. 281), *Robert J. Woods*. *Woods* was very interested in this aircraft especially in the possibility of being able to change the sweep of the wing. However, before the practical realization on the X-5 some years would still pass that were devoted to becoming familiar with the special features of the swept wing and to studying the valuable prize from Germany, the P 1101, in detail. Bell also considered at that time the design impact of changing the sweep of the wing during flight. Variable sweep seemed extremely attractive because the wing position (sweep) could be adjusted to the respective flight condition avoiding the disadvantages of a highly swept wing during the landing approach. An important problem was the location of the neutral point that had, independent of the wing sweep, always to be positioned within the stable range.

In 1948 Bell submitted a proposal to the U.S. Army Air Force to develop a fighter aircraft on the basis of the Messerschmitt P 1101. Unlike the P 1101, this aircraft should be able to adjust the wing sweep during flight. Because the Air Force showed little interest, Bell suggested in February 1949 to use the design as a test aircraft for variable wing sweep. In July of the same year, Bell received a contract to build two prototypes designated X-5. The Bell X-5 was to accomplish three important tasks:

1) Validation of wind-tunnel measurements on swept wings with different sweep angles between 20 and 60 deg by flight tests.

2) Development and construction of a usable mechanism to adjust the wing position during flight.

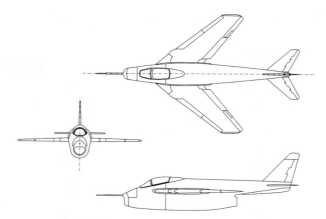

Fig. 6.13 Plan view of the experimental aircraft Bell X-5 (see [34], p. 56).

3) Determine whether variable sweep is advantageous for future fighter aircraft [34].

In its external appearance, the Bell X-5 was very similar to the P 1101 (Fig. 6.13). The wing also corresponded to the layout of the P 1101. As airfoil section for the wing, a NACA 64 A011 section was selected inboard (symmetrical airfoil with 11% thickness) that tapered off to an NACA 64 A00828 airfoil section (8% thickness) at the wing tip. The wing sweep was continuously adjustable in a range between 20 and 60 deg. The hinges shifted longitudinally when sweeping so that the neutral point always remained within the stable range. The wing possessed normal landing flaps, which could be deflected at all sweep angles. The wing leading edge had full-span slats. A special fairing took care that the openings in the fuselage at the wing root, required for the movement of the wing, always remained covered during the sweeping process. The dihedral of the wing remained the same and could not be altered.

The first of the two X-5 took off on June 20, 1951. After the company tests, where the entire range of wing sweep was covered, the X-5 was transferred to the NACA "High-Speed Flight-Research Station" at the Edwards Flight Test Center, where the flight testing was continued under the supervision of the NACA (Fig. 6.14). The flight-test program was extremely successful and showed that the aircraft possessed excellent short takeoff qualities and very good climb performance. The X-5 served, in addition, to investigate the "tip-stall" problem for which the U.S. aeronautical industry urgently needed a solution. Similar to the Convair XF-92A, the X-5 exhibited considerable "pitch-up" problems during maneuver flights with higher wing loads and higher wing sweep. Another serious problem, which gave the pilots trouble, was the dangerous stall behavior [35].

Fig. 6.14   Experimental aircraft Bell X-5 during flight (courtesy of NASA).

To be able to cope with the pitch-up problem, investigations were already being carried out at the NACA on other research aircraft. Slats, boundary-layer fences, and extensions of the wing nose were tested. A recommendation also concerned the position of the horizontal stabilizer. A low positioned elevator remains at higher angles of attack and separated flow on the wing within the range of undisturbed "healthy" flow and keeps its effectiveness. In case of a high positioned empennage, the danger exists that during a pitch up it gets into the downwash of the wing and loses its effectiveness. This defect could not be remedied in the case of the X-5 due to the layout of the aircraft. During stall, this led, together with the high vertical-tail position (stabilizer), to an intense spinning motion from which the aircraft could be scarcely recovered. This characteristic was probably responsible for the death of Air Force test pilot *Raymond Popson*, who had an accident with the first X-5. *Popson* was not able to recover from a spin at a 60-deg sweep angle. An extract from a pilot's report gives an impression of the flight behavior of the X-5 when stalling [35]:

> As the aircraft pitches, it yaws to the right and causes the aircraft to roll to the right. At this stage aileron reversal occurs; the stick jerks to the right and kicks back and forth from neutral to full right deflection if not restrained. It seems that the airplane goes longitudinally, directionally, and laterally unstable, in that order.

This description matches the result of the investigation that *K. H. Doetsch* had carried out on the P 1106. The high vertical-tail position and the mass distribution were responsible for the strong coupling between roll and yawing motions and led at high lift coefficients to stability problems, which could only be controlled with difficulty.

However, the Bell X-5 had not been built as a fighter aircraft where good flight characteristics at the limits of the flight envelope could be demanded, but as a research aircraft for variable sweep. In this role the X-5 has succeeded excellently and provided valuable information to aircraft design. If the flight characteristics of the Messerschmitt P 1101 can be inferred from the Bell X-5, the P 1101 may well have a similar behavior. The P 1101 would certainly have rendered good service as an experimental aircraft. Turning the P 1101 into a fighter aircraft would certainly have required many changes due to the problematic flight characteristics.

6.2.3 TAILLESS AIRCRAFT

6.2.3.1 MESSERSCHMITT ME 163, THE FIRST OPERATIONAL ROCKET AIRCRAFT. After the war, hardly another German aircraft fascinated the engineers of the victorious powers more than the rocket-propelled aircraft Messerschmitt Me 163. This aircraft, a creation of *Alexander Lippisch*, apparently simple in design and combined with, for the time, fantastic flight performance and good flight characteristics, seemed the ideal of a small and light interceptor for object protection. Not the least *Lippisch's* lecture on his delta developments, held at St. Germain in May 1945 before representatives of the Allies, contributed to this assessment. Test flights with the Me 163, carried out in the United States in spring of 1946 within the framework of "the Foreign Aircraft Evaluation," confirmed the good flight characteristics of this aircraft [36]. The flights were, however, carried out without propulsion so that the speed range near the critical *Mach* number could not be covered. It can be assumed that information concerning the operational flight behavior had been available to the Americans from pilot interviews.

The history of the development of the Me 163 is well documented in the relevant literature so that here emphasis is placed on the special features of the aircraft with the focus on the aerodynamics of the swept wing and the associated flight characteristics as far as sources permit corresponding statements. The tailless aircraft Me 163 had, as already explained elsewhere, for stability reasons a swept wing. The advantage of wing sweep in the higher *Mach* number range was well known to *Lippisch*. In the case of the Me 163, he had always pointed to this fact in reports and lectures.

Starting point for the development of the Me 163 was the DFS 39 of 1937, also known under the *Lippisch* designation "Delta IV c," because this aircraft possessed acceptable flying qualities [37]. The DFS 39 was a tailless mid-wing aircraft with standard propeller propulsion. *Lippisch* could in 1937 already build on ten years of experience in the design of tailless aircraft. The essential questions concerning the airfoil selection, the most useful arrangements of flaps and control surfaces, as well as the flying qualities had been settled during that time. However, up to that time, none of his designs had advanced into the high-speed regime where compressibility effects were noticeable.

The DFS 39 had a wing sweep of about 23 deg. The wing had a relatively strong dihedral; the wing tips were drooped downwards and served as rudder. The airfoil layout followed *Lippisch's* philosophy for airfoils with constant

Fig. 6.15 Wind-tunnel model DFS 39d, Configuration I, in the large wind tunnel of the AVA Göttingen (1937) (courtesy of DLR Archive Göttingen).

center-of-pressure locations. Systematic investigations at the DFS had shown that airfoils with an S-shaped chord line, whose maximum-camber location was far upstream, yielded favorable lift-to-drag ratios (see [37], p.16). To achieve a center-of-pressure position for the wing within the entire speed range, the wing was twisted and the airfoil sections changed towards the wing tip into symmetrical sections. A stable center-of-pressure location was required because the all-wing aircraft had only limited trimming possibilities about the pitch axis.

At the beginning of the development of jet engines in January 1938, clarity about the most favorable shape for an aircraft with jet propulsion existed neither in industry nor at the RLM. Because of the compressibility effects to be expected at near-sonic speeds and its influence on the flow about the empennage, the RLM saw a tailless aircraft of *Lippisch's* design as a reasonable alternative to a conventional layout with wing and separate empennage [2]. The research department of the RLM had already in January 1937 tasked the DFS with the development of a tailless aircraft for testing rocket engines. The Delta IV d was derived from the Delta IV c, later designated "Project X." On the request of the DFS, extensive wind-tunnel measurements were carried out in 1938 at the AVA on two different configurations of the Delta IV Configuration I (Fig. 6.15) essentially corresponded to the Delta IV c without an engine but with a midwing layout. With regard to the flow development, the wing-fuselage junctions were more favorably shaped than in the case of the Delta IV c. On Configuration II (Fig. 6.16) the drooped wing tips were replaced by a central vertical stabilizer fin.

The two configurations showed hardly any difference in the longitudinal stability [38]. The directional stability of Configuration I was indifferent; however, the model with the central vertical stabilizer was clearly stable. The

Fig. 6.16 Wind-tunnel model DFS 39d, Configuration II, in the large wind tunnel of the AVA Göttingen (1937) (courtesy of DLR Archive Göttingen).

rolling moment due to sideslip ("dihedral effect") for both configurations at positive angles of attack was stable ($C_{l\beta} < 0$), but changed sign in the case of Configuration I at negative angles of attack. The rudder effectiveness was in the case of Configuration II four times higher than for Configuration I. Configuration I of the Delta IV d surely would have caused considerable problems in flying the aircraft. This was finally decisive for the removal of the wing-tip droop in favor of a central vertical stabilizer.

The construction of a high-speed aircraft could not be carried out anymore in the workshops of the DFS, which were set up for building gliders. On instructions from the RLM, *Lippisch* moved in January 1939 with a number of employees to the Messerschmitt Aktiengesellschaft (AG) (Corporation) in Augsburg where he worked within "Department L" on the further development of the Delta IV d, now called Me 163. *Lippisch* emphasized that a considerable part of the aerodynamic design had already been carried out at the DFS and that *W. Georgii* should take credit because he had always unselfishly supported the tailless project despite many setbacks.

To quickly allow flight tests, the airframe of the DFS 194, an experimental aircraft with a pusher propeller already under construction, was modified based on results from wind-tunnel measurements on the Delta IV d. The changes concerned: 1) reduction of the wing sweep from 35 deg (25% chord line) to 23 deg, 2) removal of the dihedral of the wing, 3) removal of wing-tip droop, 4) installation of an extendable slat, and 5) installation of a central vertical stabilizer.

The wing section remained unchanged. A Walter-rocket engine was installed on the DFS 194 and the aircraft tested at the Flight Test Center of the Air Force Peenemünde-West. Because of the positive test results *Lippisch* was allowed to further develop the rocket fighter Me 163. Two airframes of the Me 163 were completed in spring 1941 (Fig. 6.17). Compared to the Delta IV d (Configuration II), the wing had no dihedral, and instead of the movable slat, a fixed slot was now provided, and split flaps were attached on the wing lower side. The split flaps were positioned close to the center of gravity so that when extending the flaps, large pitching-moment changes did not occur. The inner flaps at the wing trailing edge were not landing flaps, but served for trimming (trim flaps) when activating the split flaps. The wing had a relative thickness of 14% close to the fuselage that reduced to 9% towards the wing tip. The wing sections were similar to the ones of the Delta IV d with airfoil sections having zero lift pitching moment equal to zero ($C_{m_0} = 0$). The wing had a twist of 6 deg, an aspect ratio of $\Lambda = 5$, and a maximum-lift coefficient of $C_L = 2.2$. The aircraft was normally flown with a center-of-gravity location of 21% MAC (mean aerodynamic chord), the trimming range extended, however, from 17% up to 22% MAC.

Special care was taken by *Lippisch* in the design of the fixed slot. The air was not to flow through the slot at normal flight conditions so that only little

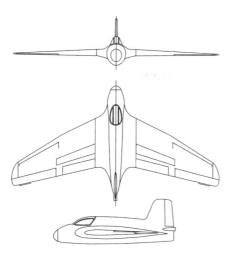

Fig. 6.17 Three-view drawing of the Messerschmitt Me 163 A (1941).

additional drag was incurred. *Lippisch* was of the opinion that a well-designed slot, built with close tolerances, would cause additional drag similar to an extendable slat [19]. (Interrogation of *Lippisch* and *Latscher* on June 10, 1945, at St. Germain by *Tsien*). The fixed slot covered 40% of the span. With this extent of slot, the aircraft could not go into a spin.

Also the combined ailerons/elevators ("elevons") were carefully designed. There was 26.3% of the control surface located forward of the hinge line, but it did not cause any instability (unstable control-surface characteristic). There was a gap of 1.2% of the control-surface chord between the control-surface leading edge and the wing trailing edge. The wing trailing edge was rounded. The overall arrangement served to achieve good control-surface effectiveness during stall. Obviously, there were no wind-tunnel measurements at high speeds with this configuration, although the design envisaged a speed of 1000 km/h at an altitude of 4000 m ($M = 0.85$).

The first aircraft came to Peenemünde-West in summer 1941 where the Walter-rocket engine was installed. On October 2, 1941, test pilot *H. Dittmar* obtained with the aircraft, now designated Me 163 A, a speed of 1003 km/h at an altitude of 3600 m corresponding to a *Mach* number of $M = 0.84$ (see Chap. 5). At this speed *Dittmar* already had to fight with severe stability problems. The aircraft became unstable about all axes, and a strong pitch-down moment occurred. As cause for the pitch-down moment, a downwash at the outer wing due to compressibility effects was assumed. *Lippisch* associated this with the strong wing twist, which at high flight speeds caused negative angles of attack on the outer wing (see [37], p. 33). It is, however, more likely that shock waves in conjunction with flow separations occurred first on the inner wing in the area of the thick airfoil sections, which led to a strong rearwards shift of the center of pressure.

The success of the Me 163 A caused the RLM to task the Messerschmitt Company with the development of a rocket-propelled interceptor. Investigations concerning such an aircraft were already being carried out for some time by Department L. However, besides the rocket engine *Messerschmitt* also considered reciprocating and jet engines as propulsion. Numerous variants, designated "P-01" of tailless configurations, were investigated in the wind tunnel of the AVA Göttingen in 1941. What relation these measurements had to the Me 163 is not clear. Three fuselage designs combined with two wings and three wing locations (high-wing, midwing, and low-wing aircraft) were investigated. Fuselage I ended in a point where one could imagine a pusher propeller, fuselage IIa corresponded more or less to the design of the Me 163, and fuselage IIb was simply somewhat longer. All fuselages could, in addition, still be combined with a small fixed horizontal tail. The wings had different spans but the same wing area. The aspect ratios were correspondingly 4.4 and 6.0. With a 32-deg sweep angle, based on the 25% chord line, the sweep was almost 10 deg higher than that of the Me 163 A. Wing I had near the fuselage a maximum thickness of 10%, wing II a thickness of 13.5%. The thickness of both wings tapered down to 8% at the wing tip. The wings served simultaneously to investigate suitable landing aids. In addition to a split flap on the lower side of the wing, positioned at 50% chord, inner flaps with a chord of 40% could alternatively be deployed as landing flaps. The enormous chord was necessary to keep the additional pitching moment as small as possible. Figure 6.18 shows one of the configurations with fuselage I and wing I [39]. Without dealing with the large number of measurements in detail, it can be said that the findings of this test program entered into the design of the Me 163 B.

The design of the Me 163 B commenced in December 1941. The aircraft was basically an enlarged version of the Me 163 A with a more voluminous fuselage to carry the required fuel (Fig. 6.19). The technical data of the Me 163 B are listed next. The wing geometry remained the same as for the Me 163 A. The airfoil sections in the center region of the wing were, however, replaced by sections with a positive zero moment, and the wing twist was reduced to 5.7 deg. The Me 163 B configuration was investigated in the high-speed wind tunnel of the DVL. However, it is not clear whether this happened

Fig. 6.18   Wind-tunnel model Project P-01, Order J 1279, as high-wing aircraft (1941) (courtesy of DLR Archives Göttingen).

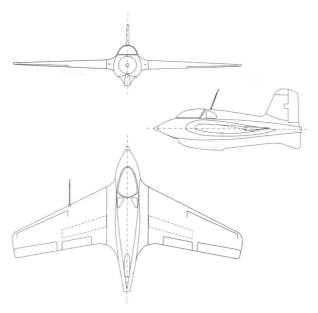

Fig. 6.19   Plan view of the rocket-propelled fighter Messerschmitt Me 163 B.

to accompany the development of the aircraft in 1941. The maiden flight of the Me 163 B took place, still without engine, on June 26, 1942. Because of the lack of engines, part of the flight testing was carried out without propulsion. Here, the Me 163 B showed the good flight characteristics already known of the Me 163 A. The first flight with rocket propulsion was carried out on February 21, 1943, by test pilot *Rudolf Opitz*.

The weaknesses of the aircraft only showed themselves during combat. As well as engine problems, the armament and the complicated organization on the ground, the aircraft showed considerable flying-qualities problems at high speeds. The critical *Mach* number for the drag rise amounted to $M = 0.75$. The flying qualities deteriorated at higher *Mach* numbers. Measurements in the DVL high-speed wind tunnel showed that the important requirement for tailless aircraft, namely, that the neutral point and the pitching moment at zero lift (zero moment) should only change slightly with *Mach* number, held only up to $M = 0.75$ for the Me 163 B. During operations, it was difficult to fully use the advantages of high flight speeds. Above $M = 0.8$ the aircraft reacted unexpectedly with a one-sided lift loss ("wing drop") or with a strong nose-heavy pitching moment, thus ruining the target approach [40]. The operational flight limit was at $M = 0.85$.

The negative effect of wing twist on the critical *Mach* number, already suspected in case of the Me 163 A, was to be settled at the DVL with a wind-tunnel model without wing twist. The measurements, carried out in December 1943 in the high-speed wind tunnel, showed, however, that the wing twist

Fig. 6.20 Me 163 as low-wing aircraft (Project Me 334 with reciprocating engine, 1943) (courtesy of DLR Archives Göttingen).

was of only secondary importance. The twist had hardly any effect on the pitching-moment balance and thus on the stability about the pitch axis important to tailless aircraft [41]. The DVL saw the problem rather in airfoil sections unfavorable for high-speed aircraft and referred to measurements on the Henschel Hs 117 missile with a 12%-thick symmetrical airfoil and a cylindrical fuselage that showed that up to $M = 0.85$ shifts in the neutral-point location did not occur.

At the AVA Göttingen even further variants of the Me 163 B were investigated on order of Messerschmitt. Figure 6.20 shows the wind-tunnel model of a Me 163 B with landing gear and the vertical stabilizer located on the underside. It is presumably a configuration derived from the P-01 with reciprocating engine and a pusher propeller (Project Me 334). Further variants investigated were a Me 163 B with an extended fuselage and a version with a V-type empennage [42]. Also models of the Me 163 with underwing stores were tested at Göttingen (Fig. 6.21). The Me 163 C was a further developed Me 163 B with an additional cruise engine and an untwisted wing. The Me 163 C was supposed to have a limiting *Mach* number of $M = 0.92$ (operational limit). A further version, the Me 163 D, was a more slender and aerodynamically improved version of the Me 163 B. These two designs no longer originated under the supervision of *Lippisch* but were designed at the Project Office Oberammergau of the Messerschmitt Aircraft Company.

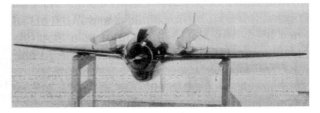

Fig. 6.21 Complete model of the Me 163 B with bombs (1944) (courtesy of DLR Archives Göttingen).

Fig. 6.22 Wind-tunnel model of the swept-wing aircraft M 263 (Design P-03) in the large wind tunnel of the AVA (1941) (courtesy of DLR Archives Göttingen).

In spring 1941, an aerodynamically improved version of the Me 163, the Project P-03 or Me 263, was developed. With this aircraft, higher speeds than with the Me 163 A were to be reached. Wind-tunnel measurements on this configuration were carried out in Göttingen in Tunnel K VI in autumn 1941. A comparison with some versions of the P-01 shows similarities so that it can be assumed that also the Me 263 evolved from the P-01 design. The designation Me 263 (P-01) is found in some of the test records supporting this statement. Figure 6.22 shows a wind-tunnel model of the Me 263 at Göttingen. Noticeable is the slender and longer fuselage as well as the higher sweep of the wing. The sweep of the 25% chord line was 32.1 deg. The wing possessed a slight negative dihedral of -0.7 deg. The airfoil-section thickness was 11% at the inner and 9.5% at the outer wing. The wing was not twisted and had the typical *Lippisch* airfoil sections with an S-shaped chord line. The maximum camber was 1.4% at 25% of the wing chord. The airfoil sections were the same over the entire span. The wing leading edge in the area of the aileron/elevator was taken up by an adjustable slat [43]. A split flap was located on the lower wing surface at about half-chord. The wing aspect ratio was 4.41. The test results showed at subsonic speeds without exception stability about all axes and a somewhat lower zero-drag than in the case of the Me 163 B. *Lippisch* writes that the work on this aircraft was stopped after the Me 163 A had exceeded 1000 km/h. It was thus clear that the Me 163 had sufficient performance potential making the development of an improved version superfluous.

An example of the successor model, the Me 163 C, was not built. The development work on the Me 163 was entrusted to the Junkers Company in summer 1944. It was decided in favor of a larger aircraft, designated Me 263, later renamed Ju 248. A first experimental model, designated Me 163 B V-18, was tested in December 1944 (see [6], p. 266). *Lippisch* named this experimental model Me 163 D hereby claiming copyright [44]. Regarding the form of the fuselage, similarities with the original Me 263 design are unmistakable. The Ju 248 had, however, a completely new fuselage with a

TABLE 6.3   IMPORTANT TECHNICAL DATA OF THE MESSERSCHMITT ME 163 B [46]

Span	9.30 m
Area	19.60 m²
Aspect ratio	4.5
Wing section rib 1	*Lippisch* airfoil with 14.0% thickness, maximum-thickness location [41] 30%
Camber	0.8%
Maximum-camber location	25%, mean line S-shaped
Airfoil at wing tip	8.0% thickness, 25% maximum-thickness location, symmetrical
Sweep	23.28 deg at 25% chord line
Twist	5.7 deg
Presetting of the wing	3 deg
Slat	fixed as of 60% of the span
Maximum speed	950 km/h
Critical *Mach* number (drag rise)	0.75
Critical *Mach* number (operational)	0.85

larger thickness (diameter) so that the ratio length to fuselage diameter corresponded closely to that of the Me 163 B. Unfortunately the details concerning the Ju 248 are extremely sparse. The wing was essentially similar to that of the Me 163 B, but possessed a somewhat larger aspect ratio. Thus, profile, airfoil-section thickness, and twist must have been similar to those of the Me 163 B (Table 6.3). This, too, corresponded to the philosophy of Junkers to fall back during new developments on as many of the Me 163 components as possible. Also different data exist concerning flight performance. The cruise speed at an altitude of 12,000 m was supposed to be 980 km/h, corresponding to $M = 0.94$ (see [4], p. 170). According to another source, the maximum speed at an altitude of 10,000 m was limited to 950 km/h, corresponding to $M = 0.87$ (see [13], p. 68). The latter value seems more credible considering that hardly any change in the wing-fuselage aerodynamics compared to the Me 163 occurred.

The first prototype was flown during February 1945 at Dessau, still without propulsion, as a glider. Three V-models of the Ju 248 had been completed up to March 1945 and delivered to the airfield at Brandis. The flight testing was allegedly continued here under Soviet supervision with one Junkers pilot killed in an accident in June 1946 (see [6], p. 268). This information seems doubtful because already during the first half of 1945, the Soviet Union had all still fairly operable aircraft transported to Russia. Only a single Me 163 B of these aircraft was tested as a glider for lack of suitable fuel in the Soviet Union [45].

The aircraft model Me 163 B was not able to fulfill its intended role as a fighter (interceptor). The reasons for this are sufficiently presented in the relevant literature. *Messerschmitt* and his chief designer *Voigt* were critical towards the idea of a tailless aircraft. As already reported elsewhere, they considered, due to the trimming problems to be expected, a separate tail-unit as necessary, which was possibly to be positioned outside the wing wake likely to be disturbed by compressibility effects. Besides his later tailless designs, such as P 1111, *Messerschmitt* consequently investigated as an alternative an aircraft with a separate empennage, such as P 1110. Just within the speed range important to this aircraft, the problems occurred. The concept of the Me 163 as a high-speed aircraft of the simplest design found after the war, nevertheless, quite a number of supporters abroad.

6.2.3.2 EXPERIMENTAL AIRCRAFT DE HAVILLAND D.H. 108.   The de Havilland D.H. 108 was the first aircraft with swept wings built in Britain. It was a private initiative of the de Havilland Company with the objective to obtain information about the behavior of swept wings at low, medium, and high speeds. After the evaluation of German documents, it was concluded that a repetition of the wind-tunnel measurements would provide little new knowledge. It was hoped to obtain this from test flights with a manned experimental aircraft. An experimental aircraft was envisaged for each of the three speed ranges. The flight test data were to provide important information regarding the wing development of the D.H. 106 "Comet." The tailless configuration appeared to be the simplest way to build a high-speed aircraft without additional drag sources such as an empennage. After all the Me 163 had shown that this simple concept could achieve good flight performance. The design was possibly also inspired by the project Messerschmitt P 1111, a tailless aircraft with a 45-deg wing sweep. The development of the three experimental aircraft commenced in October 1945 after studying German documents about swept wings and all-wing aircraft [47].

To quickly obtain results, the newly developed wings were mated to the fuselage of a D.H. 100 "Vampire" (Fig. 6.23). The wing had a sweep of 40 deg, the wing thickness was 12% at the wing root remaining constant at 10% from outside the air intake to the wing tip. As in the case of the Me 163, the ailerons were also used as elevators ("elevons"). Trim flaps were attached to both trailing edges. The leading edge of the wing in the area of the aileron was equipped with a slat, which was in the case of the first aircraft fixed in an extended position. The wing trailing edge between elevon and fuselage consisted of a so-called "trim flap." On the lower side of this trim flap, a split

Fig. 6.23 Tailless aircraft de Havilland D.H. 108 (1946).

flap was attached as a landing flap. This "trim flap" fulfilled the same function as in the case of the Me 163, namely, to balance the additional pitching moments when activating the landing flap. It served at the same time to compensate the pitching moments that occur at transonic speeds due to the motion of shock waves. Without the careful use of this trim flap, the entire high-speed trials could not have been carried out at all.

Without exact information about the wing airfoil section being available, one may assume that a suitable high-speed airfoil with a lesser thickness had been chosen. It was probably the same laminar airfoil type E.C. 1240, which had also been employed on the "Vampire." According to the statements of R. Smelt (see [26], p. 912), the British at that time were well aware of the good high-speed characteristics of thin symmetrical airfoils.

The first prototype took off for its maiden flight on May 15, 1946, crashed, however, after some flights on May 1, 1950, with the pilot being killed. The second prototype was designed for higher speeds. The slats could be manually extended and locked during high-speed flight. The second prototype flew in June 1946 for the first time. The high-speed testing brought out a problem that could not be solved by simple means. Above $M = 0.875$ fast pitch oscillations commenced that were further amplified when the pilot tried to dampen these oscillations by control inputs. During a training flight on September 27, 1946, the aircraft broke up in midair during an attempt to establish a new speed record. The test pilot *Geoffrey de Havilland* was killed. The analysis of the cause of the accident yielded that the pitching oscillations resulted in the permitted structural loads being exceeded and the structure failed [48].

The third prototype received an aerodynamically refined fuselage, a more powerful engine as well as automatically extendable slats. It was planned to strengthen the wing structure and to equip the aircraft with powered control surfaces as soon as they became available. This aircraft was, in addition, equipped with an ejector seat although nobody could tell whether this rescue device would work at speeds higher than 450 mph (720 km/h). Test pilot *John Derry* approached cautiously step-by-step the speed of sound. It was important that at each step sufficient data could be collected. By a sensitive and careful use of the trim flap, he was able to accelerate the aircraft in a dive up to close to the speed of sound. At each *Mach* number step the safe pull out from the dive was investigated. He also managed to prevent the pitch oscillations from getting more intense with increasing *Mach* number by employing the trim flap.

On April 12, 1948, *Derry* established with this aircraft a new world speed record over a distance of 100 km with an average speed of 974 km/h. Some months later, on September 6, 1948, *Derry* reached the sound barrier in a dive at an altitude between 12,000 and 9000 m thus being the first Englishman flying faster than the speed of sound. This aircraft, too, finally crashed in February 1950 with the pilot killed.

That all three aircraft crashed with a fatal outcome for the pilots revealed serious problems regarding flying qualities at high speeds. The aircraft entered into dangerous pitch oscillations when approaching the speed of sound (see [35], p. 37). In addition, an increasing instability about the vertical and longitudinal axes arose, which rendered the aircraft almost uncontrollable. Only an experienced test pilot like *Derry* could manage to control this aircraft at high speeds. After the experience with the D.H. 108, no further tailless aircraft of this configuration were built in Britain. Only an artificial pitch damping could have overcome the oscillatory problem. Such systems were, however, at the end of the 1940s not yet available. De Havilland learned from the flight testing of the D.H. 108 and equipped the supersonic fighter aircraft D.H. 110 "Vixen" with a separate horizontal stabilizer, as in the case of the "Vampire," to obtain sufficient natural dampening about the pitch axis.

6.2.3.3 FIGHTER AIRCRAFT NORTHROP XP-56. The tailless aircraft Northrop XP-56, although not a rocket-propelled aircraft, may nevertheless be compared with the version of the Me 163 with a reciprocating engine projected by *Messerschmitt*. Many of the problems encountered by Northrop with regard to this aircraft would certainly also have made life difficult for *Messerschmitt*.

In September 1939 a commission ("Kilner Board") established by the U.S. Air Corps published its report regarding the future role of the air forces in the defense policy of the country and what had to be done to meet the demands associated with this role. The report recommended a special research and development program that would secure the United States a lead in aeronautical engineering relative to future adversaries. Also included was a list of requirements to be met by the next generation of fighter aircraft. Among other things, a new fighter aircraft was demanded that was to reach a speed of 850 km/h at an altitude 5000 m.

Thirteen interested parties were invited in February 1940 to present their designs and data for a new fighter aircraft. The Northrop Aircraft Inc., which suggested a flying-wing aircraft, also belonged to the bidders. A reason why *Northrop* received a development contract for this unconventional aircraft designated XP-56 was its construction of magnesium, which was not considered a strategic material. The design of the aircraft was based on the prior experience *Northrop* had gathered with the all-wing aircraft N-1M. *Northrop* was quite familiar with the developments of the *Horten* brothers and *Lippisch* in Germany but pursued an independent way. The all-wing aircraft Northrop XP-56 had an air-cooled Pratt & Whitney R-2800 radial engine that acted on a counter-rotating propeller at the stern of the aircraft. The XP-56 possessed a central vertical stabilizer below the fuselage, similar to the one envisaged for the Me 163B with pusher propeller (see Fig. 6.24). Unlike the Me 163B, this stabilizer did not have any rudder that was located at the wing tips. The wing had a leading-edge sweep of 27 deg and an aspect ratio of 5.91. The

Fig. 6.24 Northrop XP-56 (1943) [48].

wing section at the root was a NACA 66, 2-0191 (19% relative thickness) and at the wing tip a NACA 66, 2-0167. The wing had no twist but a slight dihedral. The wing tips were slightly deflected downwards. It is interesting that Northrop in flight trials with the N-1M had the same experience as *Lippisch* in case of the Delta IV d, namely, that the deflected wing tips hardly contributed to stability. In the case of the XP-56, they were kept for safety reasons. Similar to the *Horten* aircraft, the rudders, which were designed as drag rudders, were located in the area of the deflected wing tips. The rudders opened in the center deflecting up and downwards (clamshells). There was, in addition, on the outer wing a small airbrake that could be deployed as a yaw damper. The elevons, which were operated via a boost tab, were located at the trailing edge. The landing flap was located at the inner wing. The flight tests with the N-1M had shown that this configuration had hardly any tip-stall problems. In tests in the NACA spin tunnel the XP-56 spin model showed no tendencies of going into a spin (see [48], pp. 55–69).

The maiden flight of the XP-56 took place at the test ground Muroc Dry Lake of the Army Air Force in September 1943. Already during the second flight the weak directional stability was criticized. As a remedy, a second vertical stabilizer was attached to the top of the fuselage. The third flight ended with an accident during landing where the aircraft was destroyed. Because there were still problems with the directional stability, *Northrop* decided to change the rudder actuation of the second prototype. The actuator bellows of the split flaps were supplied with air via air scoops at the wing tips. The pilot operated valves controlling the air supply to the actuator bellows with the pedals. The flight tests resumed in January 1944, and the flight-test program proceeded satisfactorily, although the demanded maximum speed could not be achieved. It turned out that the yawing motion became increasingly unstable at higher speeds.

Concerned about the stability behavior of the XP-56, *Northrop* decided on an investigation in the 40 x 80 ft wind tunnel of the NACA Ames Research Center, Moffet Field. Because the XP-56 did not receive any high priority at the NACA, it was stored on the Muroc Salt Lake. In December 1945 it was decided at Wright Field that the XP-56 offered no advantages over newer fighter aircraft with jet propulsion. The wind-tunnel tests were cancelled, and the aircraft returned to Northrop. Thus this interesting attempt to build a high-performance flying-wing aircraft finally failed due to stability problems that could not be mastered.

6.2.3.4 ROCKET-PROPELLED AIRCRAFT NORTHROP MX-324/334 (XP-79). *John K. Northrop* would not have been himself if he had not tried the flying-wing principle for all applications in aeronautics. He still wanted to build a rocket-propelled interceptor in addition to the bomber XB-35 and the fighter aircraft XP-56. Whether he heard rumors about the tests of *Alexander Lippisch* with rocket-propelled all-wing aircraft in Germany is not confirmed.

He decided in the summer of 1942 to develop a light interceptor with rocket propulsion. For this interceptor the all-wing principle was to be logically applied, that is, a flying wing without any empennage was to be built. Because of the high loads to be expected, the pilot was to be accommodated in a prone position. This allowed at the same time a thinner airfoil section to be selected. The development of this aircraft was to be carried out jointly by the Daniel Guggenheim School of Aeronautics at the California Institute of Technology (Caltech) and the Aerojet Corporation. At this time *Theodore von Kármán* was engaged in cooperation with Aerojet with the development of rocket engines as a takeoff aid for aircraft. *Von Kármán* had worked already during the development of the XB-35 as an adviser for *Northrop* and also supported the development of the rocket fighter aircraft.

The Air Materíel Command (AMC) at Wright Field recognized the potential of such an interceptor and approved funds for the development of three prototypes designated XP-79. For testing the flight characteristics of this configuration, the AMC wanted a glider without propulsion. New project designations were provided, namely, MX-324 for the glider with an installed rocket propulsion and MX-334 for the pure glider. The MX-334 had originally only an aileron/elevator combination (elevon) as control surface. The natural stability of the slightly swept wing was considered as sufficient so that *Northrop* did without a vertical stabilizer. However, stability computations showed that a vertical stabilizer was necessary to achieve sufficient lateral stability. Therefore, a simple wooden stabilizer without rudder was installed. As wing section an NACA 66, 2-018 airfoil section with a constant thickness of 18% was used along the span. The wing aspect ratio was 5.2. The aircraft was to take off by means of a landing gear that it could be jettisoned and then land on skids (Fig. 6.25).

The flight tests commenced in August 1943 on the Muroc Dry Lake in the Mojave Desert. After some unsuccessful towing tests behind a car, a Lockheed P-38 was ordered as a towing plane. Because the skids were hardly suitable for these tests, the MX-334 was fitted with a rigid three-legged landing gear. The flight testing showed some deficits, which resulted in a number of changes. The most important change concerned the control that was

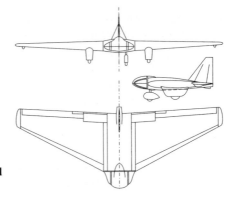

Fig. 6.25   Plan view of the rocket-propelled interceptor MX-3.24 (1944).

supplemented by a drag-rudder as it had already been implemented on earlier Northrop flying-wing aircraft in similar form. The MX-324, at first with a provisional rocket engine, was ready for its maiden flight in July 1944. The aircraft was towed to an altitude of 8000 ft (2400 m) and then released. The flight testing proceeded without problems and gave an idea of how the aircraft could perform with a stronger rocket engine.

Parallel to the flight tests the production of the desired prototypes XP-79 was initiated. While the experimental aircraft MX-324/334 had been manufactured from tubular steel and wood, the XP-79 was to be built as an all-metal aircraft largely using magnesium. The AMC decided in March 1943 to equip the third prototype with two Westinghouse 19B jet engines. Long delays during the development of the prototypes as well as the admission of Aerojet that the rocket engine could not be delivered in the foreseeable future led in September 1944 to the termination of work on the rocket-propelled interceptor. Only the configuration with jet engines was still completed and delivered for flight testing to Muroc in June 1945. The XP-79 took off for the maiden flight on September 12. Because of the failure of the electrical trimming device, the aircraft started a roll that changed into an increasingly steep spiral dive. The pilot was not able to compensate the trimming failure. The aircraft crashed killing the pilot. This was also the end for the XP-79 (see [48], pp. 93–103).

Neither the MX-324 nor the XP-79 had an opportunity to prove their qualities at high speeds, a point of great interest to the Army Air Force. Concerning the flight characteristics at high speeds, one can only speculate. Because wind-tunnel tests within this range are not available, it can be doubted whether the XP-79 with an airfoil section of 18% thickness would have reached the speed of 540 mph at an altitude 40,000 ft (860 km/h in 12,000 m) without any problems. It may be concluded, therefore, that the Messerschmitt Me 163 was the more advanced design.

6.2.3.5 RESEARCH AIRCRAFT NORTHROP X-4. After the war, after a wealth of information concerning German aircraft had reached the United States, the U.S. Army Air Force planned to examine more closely promising results of German aeronautical technology with regard to their usefulness. Also of interest here was the employment of tailless aircraft at high speeds. In the case of tailless aircraft, the interests of the Air Force in the Me 163 and *Jack Northrop's* preference for this design met. In February 1946, *Northrop* got another chance and received the order to build two prototypes of a flying-wing aircraft designated Project MX-810. *Northrop* and the Air Force hoped that the tailless aircraft was a reasonable configuration for a high-speed aircraft. The flight tests with the XP-79 had not provided any indication because the aircraft had not entered the critical speed range. The NACA feared, however, serious pitch-up problems at near-sonic speeds due to the experience with the D.H. 108, but was interested in the stability and controllability of such configurations at transonic speeds. The purpose of the Project MX-810 was thus clear, namely, to investigate the stability and controllability of tailless aircraft.

A small tailless aircraft, designated XS-4, later X-4, with a span of 8.20 m evolved, driven by two Westinghouse J30 jet engines (Fig. 6.26). Probably influenced by the Me 163, *Northrop* deviated from the proven concept of the flying wing without empennage. While all his tailless aircraft were based on the original model N-M1, he was now breaking with the tradition and chose a highly swept wing with a thin airfoil section. The aircraft now possessed a fuselage with a central vertical stabilizer. Instead of the drag rudder there was now a normal rudder.

The wing sweep of the X-4 at the leading edge was 41 deg, and the airfoil was an NACA 00010-64 with a constant thickness of 10% across the entire span. The combination aileron/elevator (elevon) took up almost two-thirds of the trailing edge. The rest was occupied by a split flap with a large chord whose upper or lower side could be deflected. This split flap also served as an air brake. Concerning the high-lift range, there was no slat but only a small boundary-layer fence close to the wing tip.

The maiden flight of the Northrop X-4 took place on December 15, 1948. The flight testing at high speeds showed a similar flight behavior as already

Fig. 6.26 Experimental aircraft Northrop X-4 (1949) (courtesy of NASA).

observed in the case of the Me 163 and the D.H. 108. Above $M = 0.76$ a constant roll and yaw motion developed with the effectiveness of the elevons decreasing. At $M = 0.88$, pitch oscillations commenced that intensified with increasing *Mach* number. The roll and yaw oscillations became so strong that they forced the pilot to constantly stabilize the aircraft about all three axes. Above $M = 0.9$ the pitch oscillation became almost uncontrollable.

The NACA investigated different remedies. Some success was brought by a thickening of the wing and elevon trailing edges. The roll rate was increased by 25%, and the controllability noticeably improved. The pitch, roll, and yaw motions commenced, however, again at $M = 0.9$ and became so strong at $M = 0.94$ that vertical accelerations of ± 1.5 g occurred. As a result, the NACA concluded that this configuration (Me 163, D.H. 108) was unsuitable for a high-speed aircraft that was supposed to fly within the transonic regime. The Northrop X-4 had simply too little damping area, horizontally and vertically. The thick control-surface (flap) trailing edges had proven themselves, however, and were introduced on further aircraft designs (see [35], pp. 49–62).

6.2.3.6 ROCKET-PROPELLED AIRCRAFT MIG I-270. The retesting of the Me 163 B in the Soviet Union, although without propulsion, led to the opinion at the Soviet Aeronautical Research Establishment (TsAGI) (Central Aero- and Hydrodynamics Institute) that the configuration of the Me 163 was unsuitable for a high-speed aircraft that was to operate near the speed of sound. The thick wing, the low sweep, and the insufficient strength of the design were criticized (see [45], p. 8).

Despite the negative assessment, the design office of *Mikojan* and *Gurewitsch* decided in 1946 to develop a rocket-propelled interceptor. If at all, the Me 263 with takeoff and cruise combustion chambers could have served as an example. The experimental aircraft MiG I-270, built in 1946, was not a tailless aircraft but had a low-aspect-ratio straight wing. At that time, TsAGI was already intensively busy with the swept wing and was not yet able to provide any specifications. The wing had a thin symmetrical laminar airfoil, and the empennage was constructed as T-tail.

The maiden flight with rocket propulsion was carried out in October 1947. The flight testing was, however, terminated after a couple of workshop flights because no advantage was seen in this concept compared to the guided missiles under development. Flights at the projected maximum speed of 940 km/h at an altitude of 15,000 m were no longer carried out [49].

6.3 AIRCRAFT WITH FORWARD-SWEPT WINGS

6.3.1 JET BOMBER JUNKERS JU 287

The jet bomber Junkers Ju 287 with its forward-swept wings belongs to the most interesting aircraft designs developed during the last war in Germany

Table 6.4   Aerodynamic Specifications for the Ju 287 [51]

Wingspan	19.4 m
Wing area	58.4 m ^{2}
Sweep	−19.8 deg (25% chord line)
Dihedral	8.5 deg
Twist	+2 deg at the wing root and −1.5 deg at the beginning of the ailerons
Airfoil (root)	1 23,3 12,5 0,825 40 0,178
Airfoil (tip)	1 25,3 10,5 0,825 40 0,150
Aspect ratio	6.4
Taper	0.38
Horizontal and vertical stabilizers:	
Airfoil	0 00 10 0,825 40 0,125

All control surfaces were equipped with an inner aerodynamic balance and tabs.

Takeoff weight	21,500 kg
Maximum speed	855 km/h at an altitude of 6000 m

(Table 6.4). This is also the reason why it has already been described in numerous publications. A detailed description of the Ju 287 is given in [50]. We shall, therefore, only discuss in this chapter special features related to the application of the swept wing. For the history of the development, reference can be made to the existing literature.

The Messerschmitt Me 262 took off from the Leipheim airfield in July 1942 for the first flight with pure jet propulsion. Although the jet engines used were still experimental engines, it could be foreseen that this new type of engine would soon be ready to go into series production and would soon be available to the aeronautical industry in larger numbers. Jet propulsion promised a considerable increase in altitude and speed compared to the propeller propulsion, an advantage that would compensate the air superiority of the Allies. Also jet propulsion offered advantages for long-range bombers. Because of the higher altitudes, higher cruise speeds and longer ranges were possible. Because of the speed of sound decreasing with altitude, attention had to be paid to the occurrence of *Mach* number effects.

During the fall of 1942, *Heinrich Hertel*, chief of development of the Junkers Aircraft Company, ordered his colleague, *Hans Wocke*, to establish the scientific prerequisites for high-speed flight with multijet aircraft [50]. *Wocke* quickly realized that a combination of jet engines with swept wings represented the most promising solution for high-speed flight. Except for wind-tunnel measurements on individual wings, no experience existed regarding the design of jet aircraft that he could reference. Neither was the question answered of whether a swept-back wing would be better than a forward-swept wing, nor was it clear where would be the best position to attach the jet engines. All of these questions had first to be answered before the design of a specific aircraft could be started.

Wocke investigated numerous model versions in the wind tunnel that distinguished themselves by wing sweep, wing arrangement, engine attachment, and fuselage form. The Junkers Company possessed a number of wind tunnels, among them a high-speed tunnel, and was very well prepared for these investigations. The swept-back as well as the forward-swept wings provide a reduction in the transonic drag at high speeds due to the "cosφ-effect." At lower flight speeds, especially during the landing approach at high lift coefficients, important differences exist between the two wing forms. Corresponding explanations are presented in the Technical Appendix. Because of the lift distribution on the forward-swept wing, the wing tips are not as strongly loaded as in the case of the positively swept wing. For a correct wing design the ailerons are located within a healthy flow and keep their effectiveness even at high lift coefficients. For the controllability during the landing approach, this is of utmost importance. The lift maximum which is located further inboard and the boundary layer moving towards the fuselage cause, during stall, the flow to first separate on the inner wing, a desired behavior. Against this the unfavorable aeroelastic qualities of the wing (see Chap. 4) require a particularly stiff and hence heavy design.

For the planned jet bomber, the forward-swept wing promised, however, some advantages that were finally decisive for *Wocke* to select this wing form. The forward-swept wing permitted a continuous bomb bay. Because of the low wing arrangement, the attachment of the landing gear was not a problem. A relatively short landing gear with a wide track and low-pressure tires that could be retracted into the fuselage could be installed. A high-wing aircraft with a swept-back wing would have also allowed a continuous bomb bay; however, problems would have arisen with regard to the accommodation of the landing gear. The far aft wing locations required a negative sweep due to the center-of-gravity location. A positive side effect was the higher critical *Mach* number compared to the straight wing. Wind-tunnel measurements at the DVL showed, however, that the steep drag rise shifted by almost 0.08 M to higher *Mach* numbers (see Fig. 6.1), although the onset in the case of the Ju 287 occurred earlier than for comparable aircraft (Me 262, Me 163).

The result of the optimization was a wing with a moderate sweep of about −20 deg and an aspect ratio of 6.4. The aerodynamicist *Hans Gropler* was responsible for the aerodynamic design of the wing. He recommended a symmetrical high-speed airfoil with no more than 12% thickness and a maximum-thickness location of 40–45% adapted to a cambered mean line. The forward part of the airfoil up to the maximum thickness consisted of half an ellipse whose small axis corresponded to the maximum thickness (see Sec. 2.3.2). Important to the high-lift behavior was also the trailing-edge angle that should not be larger than 15 deg. Small nose radii are advantageous for high-speed flight, but have, however, disadvantages in the high-lift range. As a favorable compromise *Gropler* recommended, therefore, relative nose radii of 0.55 to 0.88.

Junkers finally selected an airfoil designated NACA 0012 0,55 40 0,125 with the airfoil contour corresponding to the layout previously described. This airfoil possessed, according to measurements in the DVL high-speed wind tunnel, a favorable $C_m/C_A = C_m/C_L$ dependence. During high-speed flight, the larger nose radius would have led to an unfavorable pressure distribution. To correct this, a slightly cambered mean line with a maximum camber location of 20–25% of the chord was introduced. The mean line had a slight S-shape such that the zero moment (C_m at zero lift) at high speeds became as small as possible (see [13], *H. Gropler*, Aerodynamic Wing Sections for High Speed). These considerations led, among other things, to the quoted airfoil data, which are taken from the original design specifications of the Ju 287 of January 1944 [51]. The last digits mark the trailing-edge angle. Accordingly the trailing-edge angle would have been >15 deg at the wing root as well as at the wing tip. The wing had a slight twist that meant that the ailerons kept their effectiveness also at maximum lift. The lift maximum was located near the wing root and caused there at higher angles of attack a premature flow separation. To remedy this, a small slat was attached to the inner wing. Contrary to the swept-back wing, the forward-swept wing has a destabilizing effect on the directional stability. This can only be remedied by a sufficiently large vertical stabilizer or an appropriately large lever arm. The changed yaw-induced rolling moment was accounted for by an enlarged dihedral of the wing. All of this had been carefully investigated in the wind tunnels so that no problems were expected with regard to the lateral stability of the Ju 287 [52]. By these means a design was achieved that satisfied both the high-lift and the high-speed regimes.

The correct attachment of the engines presented a far bigger problem. To solve this problem, numerous combinations were investigated not only in the Junkers wind tunnels, but also at the DVL and the Luftfahrtforschungsanstalt Hermann Göring, Braunschweig (LFA) (Aeronautical Research Establishment, Braunschweig) (see [50], p. 527) (Fig. 6.27). Measurements on engine nacelles in the high-speed wind tunnels of Junkers and the Ernst Heinkel Aircraft Company showed that the interference drag between wing and engine

Fig. 6.27 High-speed measurements on the wing of the Ju 287 with aileron in the A2 wind tunnel of the LFA (1945) (Hamel).

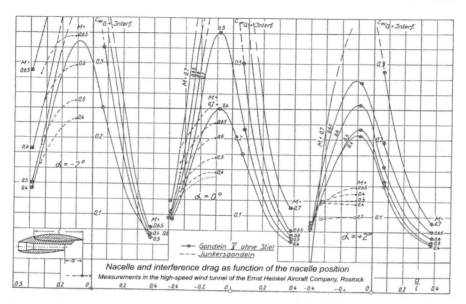

Fig. 6.28 **Interference drag of different nacelle arrangements on the wing [53]; $C_W = C_D$; legend (top to bottom): Nacelle V with pylon, Junkers nacelles.**

nacelles reaches a maximum if the nacelle is attached directly below the wing. The interference drag decreases with an increasingly forward or rearward position of the engine nacelles (Fig. 6.28) [53]. The interference drag becomes a minimum for extreme forward or aft locations (see Sec. 2.3.4 and Chap. 3). A comparison with Junkers jet engine nacelles is, in addition, shown in Fig. 6.28. The lower total drag of the Junkers nacelles was attributed to the more slender design with a long cylindrical center section. Junkers had also considered in the context of the development of the Ju 287 pylon-supported nacelles; however, they were not investigated in the high-speed wind tunnel.

Figure 6.29 shows a design with nacelle forward and aft locations of 50% where this principle was realized in an extreme way. However, this model possessed another interesting characteristic. The Junkers employees *Otto Frenzl* and *Werner Hempel*, entrusted with the wind-tunnel measurements, noticed that the favorable drag behavior of this configuration was not only attributable to the relative position of the engines with respect to the wing, but also to the optimum distribution of the cross-sectional areas along the fuselage thus minimizing local excess speeds [54] (see Sec. 2.3.4). The plan view of the first prototype, the Ju 287 V-1, shows a corresponding arrangement of the jet engines (Fig. 6.30). It is an interesting fact that during the interrogation by Allied experts Junkers engineers had repeatedly pointed out this drag-reducing effect without it having been recognized, at that time, as important, let alone having found its way into the designs of newer fighter aircraft

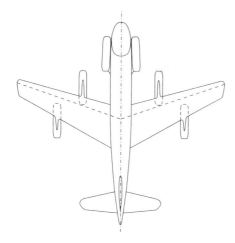

Fig. 6.29 Minimum-drag configuration of the Ju 287.

Fig. 6.30 Prototype Junkers Ju 287 V-1 (1944) (see [50], p. 525).

(see [19], interrogation of *P. v. Doepp* and *Frenzl* by *G. Schairer, H. L. Dryden, H. S. Tsien,* and *A. M. O. Smith*). Measurements at Junkers have shown that the critical *Mach* number can be increased from $M = 0.75$ to $M = 0.78$ by waisting the fuselage in the area of the wing junction (see [13], German High-Speed Airplanes and Design Developments, p. 36).

The decision for the forward-swept wing had consequences with regard to strength. Because of its geometry, the forward-swept wing in bending is subjected to torsion that leads to an increase in the local angle of attack (see Chapter 4). This can lead to a fracture of the wing if the strength is insufficient. Junkers was conscious of this fact, and extensive flutter investigations were, therefore, carried out on an elastically similar flutter model in different wind tunnels (AVA, Forschungsinstitut für Kraftfahrzeugwesen und Fahrzeugmotoren (FKFS), LFA) (Fig. 6.31). These investigations showed that aft engine positions had an unfavorable effect on the torsional divergence.

Fig. 6.31 Junkers 287 flutter model in the A3 wind tunnel of the LFA Braunschweig-Völkenrode (Hamel).

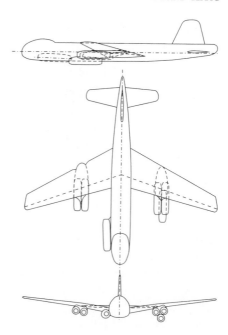

Fig. 6.32 Engine arrangements on the series design of the Ju 287 V-3.

This was confirmed by the flight tests with the Ju 287 V-1 (see Chapter 5) [55]. The wing torsion also had a destabilizing effect on the yawing motion. It furthermore turned out that the engines attached to the forward part of the fuselage led to thrust losses due to the exhaust jet impinging on the landing flaps (see [19], Interrogation of *Hertel, Gropler,* and *Zindel* of June 19, 1945). These results were probably the reason why the idea of the front-fuselage engine was abandoned in case of the Ju 287 V3 series and in favor of a three-engine cluster on each wing. With the weight of these three engines the torsional divergence could be counteracted (Fig. 6.32, right wing).

For the horizontal stabilizer, unswept, forward-swept, and swept-back forms were investigated. The stabilizer airfoils were symmetrical, but of a smaller thickness. The decision for an unswept horizontal tail resulted from the fact that a thick boundary layer draining away from the wing root would have led on a swept-back wing quickly to flow separations. A forward-swept stabilizer would have shortened the effective distance (lever) to the center of gravity, disregarding the aeroelastic problems. The unswept stabilizer was a compromise that could easily be achieved. All control surfaces were operated via auxiliary tabs. Ailerons and rudders were equipped with "spring tabs," and the elevator was adjusted via a geared tab. The ailerons during takeoff and landing, together with the landing flaps, were deflected symmetrically downwards to increase the effectiveness (see [51], p. 95). The configuration of the Ju 287 V3 was extensively tested in the high-speed wind tunnel of the DVL and in the 8-m low-speed wind tunnel A3 of the LFA (see [3], p. 9) (Fig. 6.33).

Fig. 6.33   Wind-tunnel model of the Junkers V-3 with open bomb bay in the A3 wind tunnel of the LFA Braunschweig-Völkenrode (1944) (Hamel).

6.3.2 *CONTINUATION OF DEVELOPMENT WORK IN THE SOVIET UNION—EF 131 AND EF 140*

When the Americans occupied the Junkers Plants in Dessau in April 1945, they got hold of the design documents of the Ju 287. Several jet bomber projects were at that time in an advanced stage of development in the United States. As mentioned in the chapter about the Boeing XB-47 later (see Sec. 8.4), the concept of the Ju 287 met there with only little interest. To the Soviet Union whose troops occupied the Junkers Factories in July 1945, this aircraft, on the contrary, was of great importance because the Soviet aeronautical industry had no experience in the design of jet aircraft. The further development of the Ju 287 under Soviet supervision to the EF 131 and further to the EF 140 is described in detail in Chapter 8. In Chapter 5, something is said about the flight testing as far as the sources permit reliable statements. According to this, in the operational speed range up to 780 km/h, there was no indication of a change in the flight characteristics and controllability due to a wing torsion divergence.

This changed after the further development of the EF 140, designated EF 140 R, had been redesigned into a long-range reconnaissance aircraft. The project manager for this redesign was *Hans Wocke* who had already decisively participated in the development of the Ju 287. *Wocke* increased the wing aspect ratio by an enlargement of the span to 21.9 m. He counteracted the growing danger of the wing-torsion divergence with additional fuel tanks at the wing tips, which projected far upstream (Fig. 6.34). During flight tests, wing vibrations occurred, which, after an investigation at the TsAGI, could be explained by the additional tanks. Because the problem could not be remedied, the flight tests were terminated in March 1950. The EF 140 experience left the impression at TsAGI that the forward-swept wing was associated with too many disadvantages and was not to be recommended for aircraft design.

Fig. 6.34   Experimental aircraft EF 140R in the Soviet Union, 1949 (Korrell).

Only with the emergence of new materials and methods of construction was it finally possible to also build a forward-swept wing similar to a swept-back wing with regard to weight and stiffness. Thus it became finally possible to fully use the potential of this wing without the disadvantages. On September 25, 1997, the Suchoi Su-47, a highly maneuverable fighter aircraft with a forward-swept wing and "closed coupled canards," took off for its maiden flight. Up until now, this aircraft has not gone into series production.

6.3.3 TYPE 150

With the development of the Type EF 140 B/R, the work of the "Special Design Office Baade" in the Soviet Union was not yet completed. Already during 1948, the layout of a new medium jet bomber designated "Type 150" had started (see [45], pp. 45–47). This aircraft was a high-wing design with a 35-deg swept-back wing. Also the tail surfaces were correspondingly swept. The aircraft was to be equipped with two jet engines in pylon-supported nacelles under the wings. As landing gear, a tandem design was chosen similar to the Boeing B-47 design. Because the development of the Type 150 took place under the participation of Soviet experts, it is possible that information about the Boeing XB-47 was available. The maiden flight of the XB-47 took place on December 17, 1947, and was, at the time, reported in the press. After extensive wind-tunnel measurements at TsAGI, the horizontal stabilizer was positioned as T-tail on top of the vertical stabilizer. Large boundary-layer fences were mounted on the wings to avoid an early flow separation at the wing tips (tip-stall problem). The aircraft received an irreversible hydraulic control with an artificial control-force generation. In 1951, two examples of the aircraft were completed. The Type 150 took off for its maiden flight on October 5, 1952. The flight tests were continued until May 1953. The first aircraft crashed due to a pilot error from a low altitude on May 9 killing the crew. Although the Type 150 met all demands of the ministry, there was no more interest in this aircraft. In 1952, the Tupolew Tu-88 was being flight tested; this aircraft, which had been developed according to the same specifications as the Type 150, surpassed the flight performance of the latter, however, in all respects. With the development of the Type 150 the activity of the German experts of the former Junkers Company in the Soviet Union ended. None of the German aircraft had a direct successor in the Soviet Union. However, Soviet aeronautics had greatly benefited from the work of the German experts (see Chapter 8).

6.3.4 EARLY POSTWAR DEVELOPMENT IN THE FEDERAL REPUBLIC OF GERMANY—HFB 320

With the Type EF 140 the history of the forward-swept wing was not yet over. In 1957, *Hans Wocke* came to West Germany after a short stay in East

Germany and became head of the development office at the Hamburger Flugzeugbau (HFB) (Hamburg Aircraft Construction Company). After the efforts of HFB to find competent partners for the Project "HFB 314" failed, they turned in 1960 to a new task, the development of a small business jet aircraft designated HFB 320 "Hansa Jet" [56].

It is not surprising that *Wocke* in designing this aircraft was guided by his experience with the forward-swept wing and also planned such a wing for the HFB 320. The arguments for the choice of this wing were the same as at the time for the Ju 287, that is, the advantage of a continuous cabin with standing space and without the disturbing wing box crossing the cabin. The forward-swept wing offered so many advantages to HFB that they were willing to accept the disadvantage of a heavier design. The wing was designed for a cruise speed of 850 km/h at an altitude of 9000 m ($M = 0.78$). The wing sweep was at first −20 deg with respect to the 25% chord line, as in the case of the Ju 287. To compensate the yaw-induced rolling moment which increased strongly with the lift coefficient, the wing had a dihedral of 8 deg. The vertical stabilizer was relatively large to improve the directional stability. Several proposals regarding the jet engine attachment were investigated by the project aero-dynamicists. It was important that the engines should not be affected by a flow separation on the inner wing during stall. As already in the case of the Ju 287, a slat was attached in the inner-wing region to prevent a premature flow sepa-ration. The empennage of standard sweep had been designed as T-tail. The wing was redesigned in mid-1961. The sweep was reduced to −15 deg and the dihedral of the wing accordingly also reduced to 6 deg. The profiles were NACA 63-A -1,8 11 (11% thickness and 30% maximum-thickness location) for the outer wing and an NACA 65-A -1,5 13 (13% thickness and 50% maximum-thickness location) for the inner wing. The mean line was slightly cambered and the airfoil nose somewhat drooped. High-speed measurements at Göttingen gave a critical *Mach* number of $M = 0.76$. All control surfaces were directly activated by the pilot without any support by auxiliary tabs or hydraulics.

The landing flaps were a specialty. In the case of the forward-swept wing the trailing edge has a higher sweep than the leading edge. The landing flap has correspondingly also a higher sweep that leads at flight with extended flaps to an unwanted amplification of the yaw-induced rolling moment. By a simultaneous rotation of the landing flap about a point at the inner flap, the sweep during a deflection of the landing flap was reduced, and the negative effects on the yaw-induced rolling moment could be avoided.

As already mentioned, the aeroelastic divergence was accounted for by an especially stiff construction. The smaller aspect-ratio wing and the lower sweep made, in addition, the HFB wing less sensitive than the wing of the EF 140. To relieve the wing at higher loads as for the EF 140, forward-positioned tip tanks were attached to the wing tips. To dampen the wing oscillations observed on the EF 140, the "tip tanks" had small fins.

Fig. 6.35 HFB 320 V2 (D-CARA) operated by the DLR as Flying Simulator (1980) (courtesy of DLR Archives Braunschweig).

The maiden flight of the HFB 320 took place from Hamburg-Finkenwerder on April 21, 1964. The flight tests provided good results with regard to the flight characteristics and flight performance. Only the spin tests in Spain uncovered a problem, which was before largely unknown. During stall tests, where angles of attack of $\alpha > 45$ deg were flown, the aircraft got into a so-called "super stall" from which the aircraft could not be recovered. During a super stall, the T-tail gets into the separated flow downstream of the wing and becomes ineffective. A lower horizontal stabilizer would already have left the wing wake again at these angles of attack, and the controllability would have been sustained. As a remedy, a so-called "anti-stall device" was installed on the HFB 320. Wind vanes at the nose of the aircraft measured the angle of attack and caused a "shake" of the control column as soon as the critical value was reached.

The HFB 320 received its certification as a commercial aircraft in February 1967. The HFB 320 found many friends at home and abroad. Altogether only 45 aircraft were built. The Federal Armed Forces also acquired three aircraft of this type that were deployed as electronic countermeasures (ECM) training aircraft. The second prototype, D-CARA, served until 1982 as a "flying simulator" at the DLR Research Center Braunschweig (Fig. 6.35).

6.3.5 CONVAIR XB-53/XA-44

Information concerning the swept wing also reached Convair in the spring of 1945. At that time, the jet bomber XB-46 was in an advanced stage of development (see Fig. 8.14). Parallel to the XB-46, Convair started with the development of a high-speed attack aircraft designated XA-44. The project name was altered in 1948 for a bomber aircraft into XB-53. Convair's entrance into the swept-wing technology envisaged a forward-swept wing with a sweep of 35 deg for the XB-53 designed as an all-wing configuration. Whether this design had been inspired by the Ju 287 or was based on wind-tunnel measurements on different swept-wing configurations cannot be settled anymore. The low-speed characteristics of the forward-swept wing as well as the tip-stall problem were certainly the reasons why Convair chose this wing. The plan view in Fig. 6.36 provides an impression of the design of this project [57].

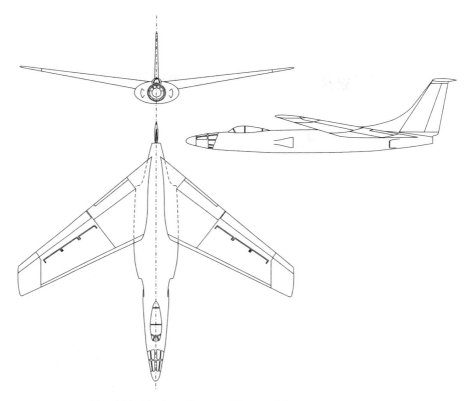

Fig. 6.36   Project Convair XB-53 (1946) (see [57], p. 181).

For reasons already mentioned the wing had a strong positive dihedral of 8 deg. The elevators were located at the rear inner wing. The wing tips could be rotated with respect to the wing by 20 deg, which provided, on the one hand, always sufficient aileron effectiveness and, on the other hand, allowed the wing torsion to be influenced. The ailerons were supported by several spoilers ("plug spoilers"), which were attached in front of the landing flap. As propulsion, three jet engines were considered, all located within the fuselage. The maximum speed was supposed to be 583 mph (940 km/h), a little less than that of the Boeing XB-47. The project XB-53 was discontinued in 1948 probably due to the lead Boeing had achieved with the XB-47.

6.4   FLYING WINGS

6.4.1   TAILLESS AIRCRAFT HORTEN HO IX

In early 1942, the brothers *Reimar* and *Walter Horten* could look back on almost 10 years of experience in the design of tailless aircraft. Among the designs were not only gliders with outstanding flight performance, but also powered aircraft. The usefulness of this type of aircraft for many applications was thus proven.

The design of the jet-propelled fighter aircraft Horten Ho IX was realized in spring 1942. Propelled by two jet engines BMW 003, *R. Horten* hoped to reach a maximum speed of 1000 km/h in a shallow dive to meet a requirement of the RLM. The design followed the rules established by the Horten brothers for natural-stability flying-wing aircraft without vertical stabilizer. As in the case of the Messerschmitt Me 163, for stability reasons the wing was swept, this also being advantageous for high-speed flight. By means of wing twist and the airfoil selection, a special lift distribution (bell-type distribution) was generated that provided an unproblematic flight behavior within the high-lift domain. Because of the bell-type distribution, maximum lift was shifted to the wing center thus unloading the wing tips. The wing was constructed of wood and possessed on each side three trailing-edge flaps functioning as aileron and elevator. Following the Horten design methods, two air brakes/drag rudder flaps were installed on each side. When operating the individual pedals, the flaps extended simultaneously up and down thus generating a yawing moment or, when simultaneously activating the pedals, an additional drag. The outer control surfaces were smaller and served as high-speed control surfaces with the larger control surfaces positioned further inboard being blocked. Figure 6.37 shows a plan view of this aircraft (see [58], p. 142).

The Horten Ho IX V-1 was completed in the spring of 1944 as a glider without engines serving to investigate the flying qualities at low speeds. The aircraft was equipped with DVL instrumentation to be able to register important data that were needed for an analysis of the flight behavior. Data were obtained up to a flight speed of 300 km/h. The flight tests yielded too large control forces and a weakly dampened yaw oscillation whose control required

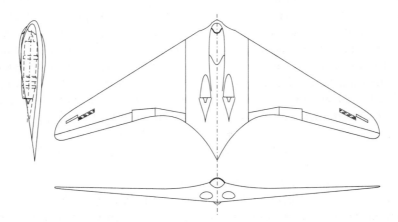

Fig. 6.37   Three-view drawing of the Horten Ho IX V-2 (see [58], p. 142).

great effort from the pilot [59]. According to the DVL flying-qualities guidelines, an undisturbed target approach under these conditions would not have been possible. By a simultaneous actuation of both air brakes, a distinct improvement in the yaw dampening to the level recommended by the DVL could, however, be achieved. The measurements showed that this dampening effect increased with increasing flight speed. The DVL, therefore, recommended in the case of tactical operations the additional deployment of the large air brakes while with the small drag rudder flaps necessary course corrections could be carried out. Low aileron effectiveness and an opposing yawing moment due to aileron deflection that counteracted banked flight were criticized. The effectiveness of the rudder was judged to be sufficient also in the case of a one-engine flight. The DVL pointed out, however, that the latter still had to be investigated more closely.

The question is how the Horten Ho IX would have behaved at high flight speeds. The DVL predicted a slight deterioration of the flying qualities. The experience with other tailless aircraft showed that the instability about the yaw (vertical) axis would have increased with increasing speed especially when approaching the critical *Mach* number. A smooth target approach run would thus hardly have been possible. Activating the air brakes on both sides to dampen the yaw oscillations, as recommended by *Horten*, would have certainly introduced disadvantages for the pilot during air combat. It is interesting that *Northrop* had suggested a similar solution to dampen the yaw oscillations of the all-wing aircraft XP-56. *Siegfried Günter*, aerodynamicist and head of the development office of the Heinkel Aircraft Company, attested the Horten Ho IX an uncontrollable yaw-rolling motion [60]. Because in the case of a tailless aircraft the directional stability and the yaw damping respectively can only be improved by additional drag, *Günter* suggested the use of a DVL yaw damper for such aircraft to keep the drag as low as possible.

Wind-tunnel tests to determine the flight behavior in the high-speed range were not carried out. The wing had a sweep of about 28 deg at the 25%-chord line. The wing airfoil section had a relative thickness of 13% at 30% of the wing chord. The airfoil was with 1.8% slightly cambered and had an S-shaped mean-line. The airfoil developed into an 8%-thick symmetrical airfoil up to the wing tip. Because of the larger diameter of the Jumo 004 jet engines that were installed on the Ho IX V-2 instead of the planned BMW 003 engines, the airfoil thickness at the wing root had to be enlarged to 14.5%. This meant a reduction of the critical *Mach* number to $M = 0.75$ as estimated by the *Horten* brothers based on DVL documents.

The maiden flight of the Horten Ho IX V-2 was probably carried out in December 1944; the flight tests were then only continued in February 1945. The literature gives little information about how far the flight tests had progressed and whether they had entered the high-speed range at all. The aircraft crashed during a descent from an altitude of 4000 m on February 18, 1945. The cause was a one-sided engine failure at low altitude that could not be compensated in time. Because the pilot was also not able to lower the landing gear, the aircraft crashed on impact with the ground. According to another account [61], an engine fire was the cause. According to a third source (see [4], p. 205), the flight on February 18, 1945, was actually not planned, and the aircraft flew with a lowered landing gear at a low altitude. After the engine failure, the aircraft turned into a spiral dive from which it could not recover any more. Why the pilot was not able to still stabilize the aircraft one can only speculate. The rudders were possibly insufficient to compensate the yawing moment at asymmetric thrust conditions after a one-sided jet engine failure.

The wing airfoil section and wing thickness allow the conclusion that the Ho IX V-2 would hardly have achieved the planned flight performance (see Figs. 2.55 and 2.56). The wing data correspond in many respects to those of the Messerschmitt Me 163 B so that a comparison with this aircraft is permitted. Above $M = 0.75$, the flight characteristics of the Me 163 B deteriorated drastically (see [41], p. 4). This was attributed to the wing thickness and the S-shaped airfoil section selected. Because the airfoil section at the root of the Ho IX V-2 had a larger thickness than that of the Me 163 B, compressibility effects would certainly have already occurred earlier. One can only guess at the effect when suddenly deploying the large brake flaps during air combat at high speeds. Compressibility effects and a lack of directional stability would certainly have led to considerable problems.

6.4.2 NORTHROP YB-49

Another aircraft can be used for comparison. A configuration similar to the Horten Ho IX was represented by the Northrop YB-49. Like the *Horten* brothers, *John K. Northrop* was an advocate of the pure flying-wing aircraft. The culmination of his flying-wing developments was the giant XB-35, which took off on its maiden flight on June 25, 1946. Developed as a competitor to the Convair XB-36, the aircraft was, however, unconvincing with regard to its flight performance. Problems with the long drive shaft of the counter-rotating propellers as well as with the reduction gear had constantly delayed the flight test program. This was the reason why already in the fall of 1944 a retrofit of the XB-35 with jet engines was considered. The decision was made in May 1945 by the Air Technical Service Commands to rebuild two of the available YB-35 airframes to install jet engines (see [48], pp. 149–167). The aircraft, having been built under the designation YB-49, became now a competitor for other jet-propelled bomber projects of the U.S. Air Force, such as the North American XB-45, the Convair XB-46, the Boeing XB-47, and the Martin XB-48.

Northrop's chief aerodynamicist, *William R. Sears*, suggested installing instead of the propellers and gear boxes that had contributed to the directional stability but now being forgone, four smaller fixed vertical fins with large boundary-layer fences to reestablish the directional stability and to stop the crossflow of the boundary layer towards the outer wing (Fig. 6.38). Without weapons platforms and other attachments *Northrop* had promised a maximum speed of more than 500 mph (>800 km/h). He realized that the thick wing of the XB-35 with 19% thickness at the wing root had to become thinner in order to increase the critical *Mach* number. The wing was also regarding strength not designed for high speeds. Wind-tunnel measurements had shown that already at $M = 0.65$ compressibility effects with flow separations in the aileron area occurred.

The YB-49 took off for the first flight on October 21, 1947. Flight tests close to the stall limit showed at first an unproblematic stall behavior. The aircraft changed into a stable descent at a large pitch attitude. However on further deceleration the aircraft flipped over backwards and could only be brought back into a controlled flight condition with effort. Prior tests in the spin tunnel had actually predicted a docile spin behavior that could actually easily be controlled by aileron inputs.

The YB-49 was weakly stable about the yaw (vertical) axis and exhibited a distinct Dutch-roll oscillation. The pilot needed a very long time to stabilize the permanent yawing motion. To overcome this flying quality problem, a roll/yaw damper was installed [62]. The additional installation of a Honeywell autopilot in 1949 also did not produce any convincing improvement (see [48], p. 167). Corresponding statements are, however, controversial. Like the Boeing XB-47, the YB-49 was also equipped with an irreversible hydraulic control system. The large control forces here were not the reason, but unstable control-force characteristics caused by flow separation at the wing trailing edge at high angles of attack. The Northrop YB-49 was thus the first aircraft

Fig. 6.38 Flying-wing aircraft Northrop YB-49 during flight (courtesy of www.wpafb.af.mil/museum).

where it was attempted to solve the flying-quality problems by a consequent application of, at that time, the most modern technical methods of flight control. The Horten Ho IX would possibly also have become a usable high-performance aircraft with the aid of a modern automatic flight control, which at that time would definitely have been possible in Germany.

As in the case of the initial model XB-35, the YB-49 was also not success-ful. The competing model Boeing XB-47, equipped with a thin swept wing, was more than 100 mph faster than the YB-49. The wing sweep did not suffice to compensate the compressibility effects caused by the thick wing. Also the possibilities of the flight control systems of the time were too limited to com-pletely remedy the stability problems of the YB-49. (See Table 6.7 for techni-cal data about this aircraft.)

Still another aircraft failed due to the fact that the wing was not optimized for high-speed flight. The Convair YB-60, developed out of the YB-36, was considered a competitor to the Boeing B-52. For the YB-60, the fuselage of the B-36 was equipped with a swept empennage, and the old B-36 wing was swept by inserting a "wedge" at the wing root [63]. Despite the same engine performance the aircraft was 80 mph slower than the Boeing XB-52 and was, therefore, eliminated from the race. Wing sweep alone was not sufficient for high-speed flight. The entire aircraft design and particularly the wing airfoil sections had to be adapted to this speed range to really benefit from the effect of the wing sweep.

6.4.3 HORTEN PROJECT HO XIIIB

The *Horten* brothers apparently worked on the design of a supersonic fighter, designated Horten Ho XIIIb, starting in 1944 (see [58], pp. 150–153). The plan view of this design (Fig. 6.39) shows a strongly swept tailless aircraft with a central vertical stabilizer. The *Horten* brothers regarded the introduc-tion of a central stabilizer as necessary because the drag rudders were suspected to have only little effectiveness due to the small span and, correspondingly, too short a lever arm. The jet engine Heinkel He S 011 was envisaged as propul-sion. With a wing sweep of 70 deg (leading edge) and an airfoil section with a maximum thickness of 7% at 45% of the wing chord the *Horten* brothers hoped for a maximum speed corresponding to $M = 1.4$. A glider model of this aircraft was built and obviously also tested in 1945. Wind-tunnel measure-ments at high speeds to support this *Mach* number were not carried out.

High-speed flights with similar configurations (Northrop X-4 and de Havilland D.H. 108) showed serious stability problems when approaching the speed of sound that could be attributed to the too low pitch and yaw damping of this configuration. None of these aircraft has entered the supersonic speed range. It is doubtful whether the Horten XIIIb (Tables 6.5 and 6.6) would have had this success. Although only a preliminary design, this aircraft appears in

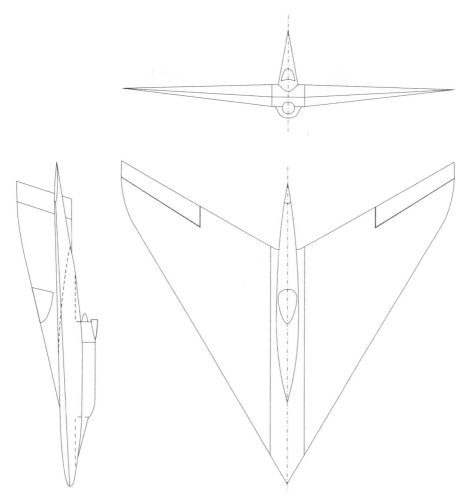

Fig. 6.39   Three-view drawing of the Horten Ho XIIIb (see [58], p. 153).

many publications as an example of the imagination of German aircraft
designers aimed at still achieving a success by advanced designs against the
Allied air superiority towards the end of the war.

6.5   RESEARCH AIRCRAFT

6.5.1   RESEARCH AIRCRAFT DFS 346

The deployment of the first jet aircraft Messerschmitt Me 262 and Me
163 showed that when approaching the speed of sound, massive flying-quality
problems occurred which had not been predictable based on wind-tunnel
measurements alone. During measurements with aircraft models in high-speed

TABLE 6.5   TECHNICAL DATA HO IX V-25[a]

Span	16.8 m
Sweep	32-deg leading edge
Sweep	about 28 deg (25% chord line)
Area	52.8 m^2
Aspect ratio	5.35
Airfoil (wing root)	Horten airfoil section with 14.5% thickness and 1.8% camber at 30% of the wing chord
Airfoil (wing tip)	Symmetrical airfoil with 8% thickness
Critical *Mach* number	0.75 (predicted)
Root chord	4.7 m
Taper	1:6
Twist	3 deg
Maximum flight mass	6876 kg
Surface load	130 kg/m^2
Maximum speed	960 km/h
Dive speed	1000 km/h
Design load multiple	7

[a]See [58], pp. 135–154.

TABLE 6.6   TECHNICAL DATA HORTEN HO XIIIB, SUPERSONIC DELTA

Propulsion	Heinkel He S 011
Airfoil	Symmetrical high-speed airfoil with 45% maximum-thickness location
Airfoil thickness	7%
Sweep	70 deg (leading edge)
Span	12 m
Aspect ratio	1.52
Maximum speed	corresponding to $M = 1.4$

TABLE 6.7   TECHNICAL DATA NORTHROP YB –49[a]

Span	172 ft (52.4 m)
Wing area	4500 ft^2 (418 m^2)
Aspect ratio	7.4
Wing airfoil section	NACA 65, 3019 (root)
Wing airfoil section	NACA 65, 3018 (tip)
Average flight mass	213,552 lb (9685 kg)
Engines	8 Allison J-35
Maximum speed	>520 mph [57] (830 km/h)

[a]See [48], p. 253.

wind tunnels, it was realized that reasonable data at transonic speeds could not be obtained due to blockage effects. The transonic speed range of $0.85 \leq M \leq 1.2$ constituted a gap where reliable data were missing. From early 1944, a new generation of fighter aircraft was being developed in the project departments of the aeronautical industry, which were equipped with swept wings and jet engines and were supposed to enter this speed range. The Research Directorate of the RLM proposed, therefore, to build a high-speed manned research aircraft to collect data within this critical speed regime. Looking at it today, it is hard to understand that German aeronautical research that had devoted itself to high-speed research for four years with great experimental effort only now realized that there were flying quality problems that had to be urgently solved. Already at the end of 1938 the Department LC 7/III of the RLM (Droppable Weapons) had pointed out in a report that in view of the progress in jet-engine design it was urgently necessary to do something about the airframe to find favorable aircraft configurations for high-speed flight. In addition to the development of high-speed wind tunnels LC 7, they especially recommended drop tests with instrumented models from high altitudes to quickly obtain useful data [64]. Drop tests would have been well suited to fill the measurement gap between $M = 0.8$ and $M = 1.2$ in a simple way. This test method was adopted by the DVL only three years later to compare wind-tunnel measurements on bomb-shaped models with corresponding free-flight tests.

In 1945, the results of flight tests with a research aircraft would have been in any case too late for the new jet aircraft under development. This late reaction to a pressing problem can today only be explained by the poor cooperation between Air Force, industry, and research. With the development of the experimental aircraft P 1101, *Messerschmitt* had already reacted in his own way.

On June 14, 1944, *Walter Georgii*, chairman of the Research Directorate, called leading representatives of industry and research to a meeting at the DFS-Ainring to learn about their views concerning a planned research aircraft [65]. In this meeting the leading minds of research and industry participated, among them *A. Busemann* (LFA), *B. Göthert* (DVL), *W. Voigt* (OFO), *S. Günter* (Heinkel), and *H. Multhopp* (Focke-Wulf).

At this meeting, wind-tunnel experts (DVL, LFA) explained that one could still test up to $M = 0.87$ with models of reasonable size. Near $M = 1$ wind-tunnel measurements failed because already prior to that sonic velocity was reached at the tunnel walls leading to shock waves and blockage effects. At transonic speeds, shock waves formed on the upper and lower wing surfaces with flow separation possibly occurring downstream. These flow separations had an influence on the effectiveness of the control surfaces thus possibly leading to controllability problems. However, these problems would disappear as soon as the shock waves had moved to the wing trailing edges.

From $M = 1.2$ the drag coefficients and lift-to-drag ratios would again reach values that would allow flight at supersonic speeds to be near at hand. To close this gap, the DFS suggested a manned rocket-propelled research aircraft which was to fly, at altitudes between 20 and 25 km. The flight tests were to concentrate on longitudinal stability, controllability, and the pressurized cabin [66].

The industry representatives were of the opinion that the construction of such an aircraft only made sense if speeds up to $M = 1$ were reached. The flight altitude should not exceed 10 to 12 km to avoid the need for a pressurized cabin. They suggested as takeoff procedure a piggyback start from a Heinkel He 177. They considered as tasks for the new research aircraft the airfoil-section development and investigations concerning wing shapes, empennages, boundary-layer control in conjunction with shock waves, longitudinal stability, and controllability. *Günter* (Heinkel) and *Schubert* (Blohm & Voss) pleaded for additional drop and wind-tunnel tests to obtain correct data.

As a result of the meeting at Ainring, three different test procedures were suggested:

1) Wind-tunnel measurements attempting to approach the critical range of $M \approx 1$ from below ($M > 0.8$) and from above ($M < 1.2$) and to cover the range around $M = 1$ by interpolation.

2) Drop tests with instrumented models to investigate the various possibilities with regard to fuselage, wing, and empennage.

3) Flight tests with manned models up to supersonic speeds to explore flight characteristics still unknown up to now, the stability behavior and the controllability.

The main difficulty in the construction of manned models was seen in the availability of workshop capacity. The Siebel Company was to work on the designs supplied by industry and build models. The research establishments would have been responsible for the experiments and supplying the test crews. For the flight tests, different aircraft were to be built:

1) Construction of an aircraft according to the present state of the art (wing sweep 35 to 45 deg, high-speed airfoil with 50% maximum-thickness location and reduced nose radius, etc.) within the shortest time, at the latest, however, within three months.

2) Construction of two aircraft to break new ground (extreme sweep, wing with a small aspect ratio, oblique wing according to Messerschmitt) within another half-year.

3) Construction of a pure research aircraft with possibilities to study all questions arising when passing through $M = 1$; completion within the next two years.

For the accomplishment of these tasks, three subcommittees were formed, which were responsible for the execution of all work on schedule. Members of the subcommittee "Manned Models" were *W. Georgii, S. Günter, W. Voigt, H. Multhopp*, and *F. Kracht* (DFS). *Busemann* (LFA) functioned as an adviser. Regarding the drop models, it was considered a priority to first advance the development of suitable instrumentation [67].

One could foresee that this very ambitious program could not be carried out in the anticipated form due to the difficult war situation and the complete overburdening of the industry. No reference is found to tests with instrumented drop models in the available documents. However, a well-prepared proposal from October 20, 1944, exists for a wind-tunnel test program for high-speed measurements that was to be presented to the Special Wind-Tunnel Committee in November 1944 [68]. In this wind-tunnel test program, no reference is made to the research aircraft, but all topics that were of importance to its development were addressed, that is, sweep, airfoil shape, fuselage form, wing-fuselage junction, wing-engine nacelles, and wing aspect ratio. The measurements were to be carried out in the high-speed wind tunnel of the DVL and in the tunnels A2, A3, A6, and A9 of the LFA. It was also important to answer the question whether the wing-nacelle combinations tested by *D. Küchemann* in the AVA low-speed wind tunnel would also be favorable in the high-speed range. Because the measurements were of a considerable urgency and the workshop capacity of the research establishments was not sufficient, the industry was to be tasked with the model construction. The DFS did not appear on the distribution list for this wind-tunnel test program.

The test program was not completed. Fuselages of rotational symmetry, derived from an NACA drop-shaped airfoil, were still tested in the high-speed wind tunnel of the DVL between the end of November and mid-December 1944 [69]. From these measurements it followed that fuselages of thickness ratios <0.15 possess critical *Mach* numbers $M > 0.85$ and are, therefore, favorable for flights in the transonic speed regime (see Fig. 2.137). The fuselage shape selected for the DFS 346 allows the conclusion that these measurements had been taken into account in the design of the aircraft. The measurements carried out at the DVL say nothing, however, about the drag in the supersonic speed range where the aircraft was to cruise. A favorable fuselage shape for supersonic flight had already been found years before at Peenemünde for the long-range missile A4. *Busemann* also had recommended this fuselage form for the antiaircraft rocket "Feuerlilie" of the LFA (see Sec. 7.2). There were, in addition, extensive investigations concerning the optimum fuselage shape for the flight at supersonic speeds [70]. It is, therefore, not understandable why *Busemann* did not recommend this shape for the DFS 346.

It seems that research and industry were not particularly interested in this aircraft. Anyway, the DFS 346 does not carry the signature of *Voigt, Multhopp*, or *Pabst* (Focke-Wulf). *Messerschmitt* pursued his own objectives with the

P 1101. In how far the consultation of the DVL and the LFA influenced the design remains vague. Research was also engaged at the time in a variety of "ventures essential to the war effort" so that staff was simply not available for such an ambitious project as a research aircraft for the investigation of the flight behavior at supersonic speeds. Merely the idea of the divided control surface seems to be attributable to *Busemann* who pointed out at an interrogation by the British in May 1945 that split ailerons were advantageous for supersonic flight [71].

The design work could only have started at the end of 1944 after first wind-tunnel results of measurements on fuselages and wings were available. *Felix Kracht* who could fall back on experience with the DFS 228 was responsible for the design [72]. *Kracht* has described his design in a report that is, unfortunately, no longer available [73]. The design work was carried out at the Siebel Company at Halle, which had, up to that time, still not been able to gain any experience in the development of high-performance aircraft.

Figure 6.40 shows views of the DFS 346. The fuselage consisted in its forward part of a body of revolution, which was formed out of an NACA 00121 0,66 50 airfoil section [74]. The maximum thickness of the fuselage was, therefore, located at 50% of the length. That fuselage shapes with a large maximum-thickness location were particularly favorable at transonic speeds had already been found before by *Th. Zobel* (LFA) in measurements on bodies of revolution in the A2 wind tunnel [75]. This result had also been taken up again by the DVL measurements already quoted. Fuselage shapes of this kind are also found later on aircraft that operate within the transonic regime such as, for instance, the North American F 100 "Super Sabre" and the Martin XB-51. The wings, arranged as midwings, had a sweep of 45 deg referred to the 25%-chord line. The airfoil section was an NACA 0012 0,55 50 0.125 with 12% relative thickness along the entire

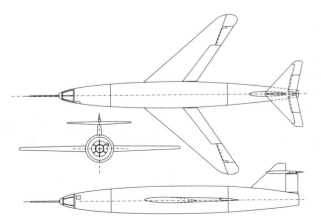

Fig. 6.40   Three-view drawing of the research aircraft DFS 346 (1945).

span and a maximum-thickness location of 50%. The only more accurate drawing from a Russian source seems to confirm this information. This would then correspond to the demands on the first of the research aircraft that was to be built according to the present state of knowledge. Fitting here is the statement of *Hans Gropler*, who recommended such an airfoil section, although with only 10% thickness, for high-speed aircraft with sweep angles of 40 to 45 deg (see [13], *H. Gropler*, "Aerodynamic Wing Section for High Speeds"). The last numerals in the airfoil designation indicate the trailing-edge angle that was to be 14 deg. The relatively large nose radius was to have a favorable effect on the outer-wing stall characteristics at high lift. Unusual was the relatively large taper of the wing of $\lambda = 3$ which had, at that time, already been recognized as being unfavorable with regard to the high-lift behavior. Also unusual was that apart from a landing flap, no further high-lift devices, for example, slats, were intended. The ailerons were divided into two parts. The inner part had an aerodynamic nose balance while the outer part extended over more than 50% of the wing chord. Both parts could be operated simultaneously or separately with the outer part having been designed for high-speed flight at supersonic speed. The horizontal stabilizer was fitted on top of the vertical stabilizer as a so-called "T-tail." This arrangement was regarded as favorable to keep the horizontal stabilizer out of the turbulence behind the shock waves on the wing. Nothing is known about the airfoil. The stabilizer could be adjusted as a whole for trimming purposes. Similar to the ailerons, the elevator was also divided into two parts. The inner elevators had an inner aerodynamic balance, the outer ones a small horn compensation. The rudder was relatively small compared to other supersonic aircraft. A Walter HWK 109 509 twin-engine with a total thrust of 40 kN was to serve as propulsion.

Whether wind-tunnel measurements were still carried before the end of the war is uncertain. No measurements on the DFS 346 are listed in the wind-tunnel logs of the AVA and LFA for 1944 and 1945. The tunnels were, at the time, fully occupied by industry orders. Besides the knowledge of the aerodynamic forces throughout the entire speed regime, most of all the flutter safety was important. The Siebel Company had, therefore, already contacted the Forschungsgemeinschaft Halle (FH) (Research Foundation Halle) regarding flutter computations. Because there already existed experience with flutter computations for swept wings at the Institute of Unsteady Aerodynamics of the AVA Göttingen, the FH turned at the end of 1944 to *P. Jordan* inquiring whether flutter computations for the DFS 346 could be carried out at the AVA [76] (see Chap. 4). These flutter investigations were to be completed by the end of April 1945 because flight tests were to commence in July 1945. Accompanying wind-tunnel tests at a *Mach* number of $M = 1$ were to be carried out on a wing model. At the end of January 1945 the AVA received from the FH a set of drawings of the DFS 346 that were needed for the flutter computations. Details on mass and stiffness distributions were missing and were requested by *Jordan* in February. The Mathematical Seminar of the University of Halle had already before been charged by the AVA to carry out numerical computations [77]. From a letter of February 26, 1945, to *Heinrich Dietz*, it follows that also the Institute of Technology Darmstadt had received a contract for numerical computations [78]. Whether flutter computations were actually still carried out under the chaotic conditions during the last months of the war cannot be ascertained any more.

When the Americans occupied the DFS-Ainring at the beginning of May 1945, they found a lot of documents, among them also documents about the "8-346," however, no completed components of the aircraft. A wind-tunnel model also did not seem to have existed because such a find would have been mentioned in the CIOS report [79]. The Americans obtained more information about this aircraft after the occupation of the Siebel Company in Halle [80]. The most valuable find at Halle was the design documents of the DFS 346, which showed that the work was already very far advanced. Completed components of the DFS 346, as described in newer reports, are not mentioned in the corresponding CIOS report (see [45], p. 23). However, according to other sources a wooden mock-up of the DFS 346 existed, if also not complete (see [72], pp. 84–87).

The DFS 346 presented itself accordingly as a research aircraft that was to reach a maximum speed of 2000 km/h ($M = 2$) at altitudes of up to 20 km. The aircraft could alternatively be equipped with different empennages. It was planned to instrument the DFS 346 such that all motions (accelerations and rotations) as well as the load on the airframe could be recorded. The attachments of the wings and the empennage were designed for the adaptation of DVL Ritz strain gauges to determine the load on the wing and the location of the center of pressure when passing through $M = 1$. The airframe was designed to carry structural loads of up to 14.4 gs. The CIOS report explicitly emphasizes the great importance of this aircraft to American aeronautical research and industry. It urgently recommended to have the DFS 346 built under American supervision, especially because the rocket engines, some material and the complete development teams were available. Under these conditions it would have been possible to build the aircraft in a relatively short time. It is thus certain that the Americans were well informed about the DFS 346 and that the design documents had obviously reached the United States.

After the occupation of Central Germany by the troops of the Soviet Union, the entering specialists recognized immediately the significance of this aircraft to Soviet aeronautics. At the Siebel Company, a Special Design Office (OKB 3) (Experimental Projects Design Office) was established with the objective to build and test the research aircraft DFS 346 (see also

Sec. 8.5.4.2 and [45], p. 24). At the end of September 1946, the first example of the DFS 346 was completed. Because of lack of fuel for the rocket engines, flight trials were not carried out in Germany anymore. The aircraft was shipped in the autumn to the Soviet Union and first extensively investigated in the large wind tunnel of TsAGI. There, it was found that the flow already separated at moderate angles of attack on the outer wings ("tip-stall"). As a remedy, the Soviet engineers attached two large boundary-layer fences on the wings, one close to the fuselage and a further one in front of the ailerons. The boundary-layer fences later became a trademark of Soviet aircraft with swept wings. This way a simpler solution had been found for the "tip-stall" problem than the mechanically more elaborate and heavier slat system. Altogether three aircraft were built with the third prototype "346-3" scheduled for rocket-propelled flights at high speeds. This aircraft had a modified horizontal stabilizer with a thinner airfoil section. Whether this change had been initiated by the wind-tunnel measurements at TsAGI or was still based on German recommendations cannot be ascertained anymore. Concerning the flight testing of the DFS 346, reference is made to Chapter 5.4.

A wind-tunnel model for measurements at high subsonic speeds had also been built at TsAGI in the meantime. The wind-tunnel measurements showed that the elevator as well as the aileron effectiveness strongly decreased at high speeds. A similar phenomenon had already been found in wind-tunnel measurements with the XP-86 model. As a cause, flow separations at the wing trailing edge because of a too large trailing-edge angle had been recognized. This also could have been the cause for the poor control-surface effectiveness of the DFS 346 at high speeds which has, however, not been verified for certain. The selected airfoil section with 12% thickness and 50% maximum-thickness location would have exceeded with its trailing-edge angle the recommended limit of 15 deg had the contour not been adapted. This can, however, not be discerned from the available documents. The wing was, however, not changed at TsAGI (except maybe by a larger chord), but an operational flight-envelope restriction to $M = 0.9$ was ordered for the DFS 346-3. Flights with rocket propulsion (Fig. 6.41) took place in late summer of 1951 nearly simultaneously with the trials of the American "competitor,"

Fig. 6.41 DFS 346 during flight trials in the Soviet Union (1950) (http://en. wikipedia.org/wiki/ File:Aircraft_346-P_ Circa_pre-1950.jpg).

the Douglas 558-II "Skyrocket." During the third flight with only one engine operating, the aircraft climbed after the separation from the carrier aircraft at an altitude of 9300 m in a rapid ascent to an altitude of 12,000 m. At this altitude the aircraft became uncontrollable at a speed of over 900 km/h. The aircraft went into a dive and could only be pulled out at an altitude of 7000 m. After further problems with the controllability, test pilot *Ziese* ejected with the cabin and landed safely by parachute. The flight testing of the DFS 346 was, thereafter, terminated because the Soviet engineers did not expect any further new information from the flight tests.

Unfortunately, no exact analysis of the cause for the loss of the control at high altitudes exists. *Ziese* had presumably exceeded the limit of $M = 0.9$, and the control-surface effectiveness decreased dramatically. There was, in addition, the low lateral stability, already established during the glide tests which could hardly be compensated by *Ziese*. It may also be possible that the unfavorable fuselage shape caused problems at near-sonic speeds. However, the tests allow the conclusion that the thrust of both engines would have been enough to accelerate the DFS 346 to supersonic speeds.

6.5.2   RESEARCH AIRCRAFT IN THE UNITED STATES—BELL XS-1, DOUGLAS 558-I, AND 558-II

At the beginning of the war, the newest generation of fighter aircraft in the United States could also achieve transonic speeds during dives. The accident with a Lockheed P-38 (see Chapter 5) in November 1941 caused the NACA to address the topic "compressibility effects" in detail. The head of the department High-Speed Wind Tunnels, *John Stack*, reacted with an extensive wind-tunnel test program in the NACA 8-ft High Speed Tunnel (HST) to clarify the problems with the P-38. After a four-month test campaign the so-called "NACA Dive Recovery Flap," a simple solution that enabled the pilots to safely pull out of a high-speed dive, was proposed. Stating with the Lockheed P-38 and the Republic P-47 and up to the Lockheed P-80, different high-performance aircraft series were equipped with the "Dive-Recovery Flap."

Not only the problems with the newest generation of fighter aircraft, but also the upcoming development of rockets and guided missiles required a deeper understanding of the phenomena occurring at flight speeds close to the speed of sound. In 1942, the first supersonic wind tunnel was put into operation at the NACA Langley Research Center with the objective to investigate flow problems occurring when passing through $M = 1$ (see [25], pp. 249–279). Tests there and in other high-speed wind tunnels showed that investigations in a *Mach* number range $0.8 \leq M \leq 1.2$ were not possible due to blockage effects. *Stack* considered a research aircraft that could fly within the transonic speed range the only possibility to remedy this dilemma. As a useful interim solution, the NACA started drop tests with instrumented models, respectively fired off models mounted on rockets.

Stack in his approach to Director *George Lewis*, responsible for basic research at NACA, remained initially unsuccessful because military research had absolute priority. *Lewis* allowed *Stack*, however, to deal with the research aircraft as far as his time permitted. During the summer of 1943, a first design of a research aircraft was developed that was to fly in a *Mach* number range of $M = 0.8$ to $M = 1.0$. As propulsion, a jet engine was proposed. Despite the fact that due to the problems that had arisen during flight at high speeds, the U.S. Army as well as the Navy showed interest in a high-speed research aircraft, it took until March 1944 before the NACA and the military got together to talk about it. The discussion, however, proceeded disappointingly because the Army and the Navy could not agree on a common approach.

At the instigation of *John Stack* a further meeting was held at Wright Field with representatives of the Air Material Command in May 1944. Head of the delegation from Wright Field was *Ezra Kotcher* who suggested rocket propulsion for the research aircraft because the experience with the Bell XP-59 had shown that the transonic speed range could not be penetrated with jet engines. This suggestion met with intense opposition by NACA. *Stack's* arguments were that rocket propulsion was completely untested for aircraft and that no pilot would get into a rocket-propelled aircraft. In addition due to the enormous fuel consumption, the flight duration of such an aircraft would be too short to collect meaningful data. *Stack* advocated instead an aircraft with a jet engine because it possessed a longer flight duration and he was, anyway, only interested in the speed range up to a maximum of $M = 1$. At a further meeting at Langley in December 1944, the Army turned down the suggestion of the NACA for a research aircraft with a jet engine as being too conservative. The Army desired an aircraft that was able to fly at supersonic speeds up to $M = 1.2$. Already one week after the meeting at Langley the Army charged the Bell Aircraft Company with the development of a rocket-propelled research aircraft designated MX-524 "Experimental Sonic 1" or for short XS-1.

Stack did not abandon his idea of a research aircraft with a jet engine and tried to interest the Navy in it. In September 1944, the Navy decided to acquire a research aircraft that met both military and NACA requirements. The aircraft was to be able to take off from the ground and have good flight characteristics at low speeds with regard to the deployment of jet aircraft from aircraft carriers. In December 1944, the Navy asked the Douglas Aircraft Corporation whether they could build such an aircraft. Douglas answered in January 1945 with the proposal "Douglas Model 558, High-Speed Test Airplane," for short D-558. Almost simultaneously with the development of the DFS 346 in Germany the development of two research aircraft in the United States of which the Bell XS-1 was to penetrate the supersonic speed range commenced at the beginning of 1945.

Of course the NACA, as consultant, was involved in the development of both aircraft. There were discussions concerning the design of the wing. *Robert Gilruth* of the NACA Flight Test Department advocated the choice of a wing as thin as possible with no more than 5% relative thickness to penetrate the supersonic range without any problems. *John Stack* held the opinion that it was more practical to select for the studies of transonic effects a 12%-thick wing, because here the phenomena would be more pronounced than in the case of an extremely thin wing. In the spring of 1945, the NACA decided to build two wings, one with 8% thickness and the other one with 10% thickness. The fuselage nose of the Bell XS-1 was pointed bulletlike followed by a cylindrical center section where the wings were attached as midwings. As an airfoil section, the slightly cambered laminar airfoil NACA 65-110 (for the 10%-thick wing) was chosen [81]. Already, during the summer of 1944, the NACA had received important information about German and Italian wind-tunnel measurements on high-speed airfoils up to $M = 0.94$ by the aerodynamicist *Antonio Ferri* from Italy (see [81], p. 37). Thereafter, the NACA must have known that symmetrical airfoils were better suited for high-speed flight than cambered airfoil sections. However, the camber was with 1% low, thus possibly having hardly any effect within the transonic speed regime. The horizontal stabilizer was located in the high position to stay clear of the perturbations caused by shock waves on the wing. It was, as a whole, adjustable for trimming purposes so that sufficient control possibilities remained for the pilot within the critical speed range. *Stack* and *Gilruth* agreed that thinner airfoil sections were more necessary for the empennage than for the wing, so that compressibility effects on the empennage commenced later than on the wing.

The first example XS-1, still without rocket engine, was ready in December 1945. The aircraft was initially to be tested as a glider up to $M = 0.80$ before it was allowed to fly with rocket propulsion. The flight testing at Pinecastle Field, Florida, from January until March 1946 showed that the aerodynamic design was healthy and that the aircraft possessed good flight characteristics at low speeds. For the flight testing with rocket propulsion the second XS-1 was brought to the flight test center of the USAF Muroc Dry Lake. The flight tests with Bell test pilot *Chalmers H. Goodlin* commenced in October 1946. During flights at altitudes of 36,000 ft (11,000 m) up to $M = 0.8$ and pull-out maneuvers of up to 8 g, the contractual agreements were met, and the actual high-speed trials could commence. The flight tests anticipated a stepwise increase in the flight speed from $M = 0.8$ up to $M = 1$. In August 1947, *Charles E. Yeager* of the Air Force took over the flight testing because the NACA pilots were not yet prepared. At the beginning of October, he reached a *Mach* number of $M = 0.94$. Here, he noticed that the elevator was nearly ineffective. As an experienced test pilot, he did the right thing and descended to lower altitudes where the elevator effectiveness returned again. At $M = 0.94$, the shock wave had moved to the hinge line and had caused flow separations on the elevator (see [35], p. 14).

Before his epoch-making flight on October 14, 1947, the NACA advised him not to exceed $M = 0.97$ because the data from the previous flight had not yet been evaluated. After the release from the carrier aircraft, *Yeager* started a climb up to an altitude of 35,000 ft (10,700 m). *Yeager* continued to climb to an altitude of 42,000 ft (12,800 m) with the thrust of only two of the four combustion chambers. There, he ignited the third combustion chamber and increased the speed (*Mach* number M) M steps of 0.02 M until it reached $M = 0.98$. At this speed, the needle of the *Mach*-meter jumped above $M = 1$. The evaluation yielded later that *Yeager* had reached with the XS-1 a *Mach* number of $M = 1.06$ thus being the first human being to have flown at supersonic speeds (Fig. 6.42).

The flight showed that sufficient control-surface effectiveness about all axes was available. Between $M = 0.88$ and $M = 0.90$ "wing drop" occurred that was, however, controllable. "Buffeting" occurred only at *Mach* numbers of $M > 0.92$. Large trim changes at transonic speeds were also not required. The flight trials of the Bell XS-1 showed that also with a conventional design (straight wing of small thickness, favorable fuselage shape), a safe flight up into the supersonic domain was possible.

Already before the Navy had contacted Douglas with regard to the planned research aircraft *Ed Heinemann*, Technical Director of Douglas, had thought about the development of a research aircraft to investigate the problems arising at high subsonic speeds (see [29], p. 28). So the wish of the Navy fell at Douglas on fertile ground. *Heinemann* had, however, thought of a research aircraft that could easily be changed into a fighter aircraft, hoping to obtain a lucrative contract for a jet-propelled fighter aircraft. The NACA, however, turned down a corresponding suggestion of Douglas and insisted on a pure research aircraft. With the support of the Navy, Douglas changed the proposal according to the wishes of the NACA. In April 1945, Douglas introduced the design D-558, which was accepted by the NACA. Representatives of the NACA and the Navy were able to inspect a mock-up of the D-558 at Douglas in early July 1945. Already prior to this inspection *John Stack* had proposed to equip one of the planned six aircraft with swept wings to validate *R. T. Jones'* swept-wing theory. The Navy turned down this proposal arguing that

Fig. 6.42   NACA research aircraft Bell XS-1 (1947) (courtesy of NASA).

there still would be too little reliable information about the usefulness of the swept wing. Army and Navy, however, approved the request of the NACA to investigate during the wind-tunnel test program of the D-558 in the 8-ft high-speed tunnel also a model of the D-558 with a 35-deg swept wing. The first documents from Germany, which showed the advantages of wing sweep for high-speed flight, arrived at the Navy Bureau of Aeronautics and at Douglas in August 1945. The evaluation of these documents resulted in a request to Douglas to study a D-558 with sweptback wings.

In the meantime, the development of the D-558 progressed. *A. M. O. Smith* dealt at Douglas with the aerodynamic design of the D-558. He writes in his memoirs that only a few documents were made available to him by the NACA and that he relied more on German measurements on a wing model of 8-cm length [82] that he had "dug up somewhere." The objective was to build the smallest possible airframe around the General Electric TG 180 engine. The demand of the NACA to provide sufficient space for the accommodation of the instrumentation was important. The aircraft possessed a cylindrical fuselage with a central air intake. The wing with an aspect ratio of 4.15 was composed of an NACA airfoil section with a 10% relative thickness. Similar to the Bell XS-1 the horizontal stabilizer, adjustable as a whole, was high. With this aircraft the speed range corresponding to $M = 0.75$ to $M = 0.85$ at altitudes between sea level and 40,000 ft (12,000 m) was to be investigated.

During April 1947, Douglas test pilot *Eugene F. May* carried out the first flight from Muroc Dry Lake (Fig. 6.43). The flight testing up to $M = 0.85$ proceeded without problems. The flight characteristics were so confidence-inspiring that the Navy decided to go for the world speed record. Navy Cdr. *Turner Caldwell* flew on August 20, 1947, along a prescribed course with an average speed of 640.663 mph (1031 km/h) and broke the record held by a Lockheed P-80 R up to that time [83]. In the context of investigations with vortex generators, the D-558-I "Skystreak" reached in June 1950 its performance limit at $M = 0.99$.

Based on the documents concerning swept wings that *A. M. O. Smith* had brought along from Göttingen in August of 1945, Douglas first investigated a version of the D-558 with swept wings. This aircraft was to advance up into the supersonic range. Because the thrust of contemporary jet engines did not suffice for flight at supersonic speed, the Douglas D-558 had to be equipped in addition with a rocket engine. Because the D-558 did not have sufficient

Fig. 6.43 Research aircraft Douglas D-558-I "Skystreak" during flight (1947) (courtesy of NASA).

space to accommodate the tanks for the rocket fuel, Douglas suggested a modified fuselage with more volume. Douglas received in January 1946 the permission from the Navy to develop a new aircraft designated D-558-II.

Like the Bell XS-1, the D-558-II had a bullet-like pointed fuselage ("Ogive") that connected to a cylindrical center section (Fig. 6.44). The wings were arranged as midwings and had a sweep of 35 deg referred to the 25%-chord line. The airfoil was a symmetrical NACA design with 10% thickness at the wing root and 12% thickness at the wing tip. For flight at high lift the wing was equipped with a full-span automatic slat and additional boundary-layer fences on the upper surface. A slightly negative dihedral of the wing was to improve the lateral stability. The empennage had a sweep of 45 deg. In how far the design of the wing was based on German documents can today no longer be determined. It is not known whether the Douglas engineers had any knowledge of the DFS 346. The choice of a wing with only little taper and the lowest thickness at the wing root reminds one of the Messerschmitt HG II wing design No. 3 that the chief designer *W. Voigt* had considered as optimum. Afterwards it is hard to ascertain which information was used for the development of this wing in the lack of available evidence. The reports related to the interrogation of German engineers might possibly have played a role here. Slats and boundary-layer fences had already, within the context of investigations on flying-wing aircraft, been determined by the NACA as favorable for flight at high lift.

The first aircraft, still without rocket engine, took off for its maiden flight on February 4, 1948, from the Flight Test Center Muroc Dry Lake. The restricted view of the pilot and the unsatisfactory lateral stability were criticized. The aircraft was fitted as a consequence with a modified cockpit and an enlarged vertical stabilizer. Figure 6.45 shows views of the aircraft after the modifications. The first "pitch-up" problems arose in August 1949 when maneuvers at higher loads and *Mach* numbers of $M > 0.6$ were performed. The NACA test pilot *John Griffith* encountered severe difficulties during a 4-g turn when the aircraft, under intense rolling motions, went out of control. *Griffith* was able to recover the aircraft and then tested the stall behavior with extended landing flaps and landing gear. The D-558-II suddenly pitched up and went into a spin. The aircraft dropped 2000 m before *Griffith* recovered

Fig. 6.44 Research aircraft Douglas D-558-II "Skyrocket" (1951) (courtesy of Boeing Historical Archives, The Boeing Company).

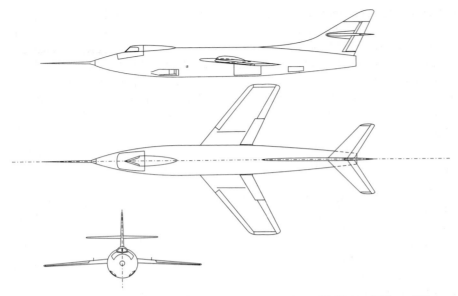

Fig. 6.45   Plan view of the research aircraft Douglas D-558-II (see [83], p. 451).

from the spin (see [35], p. 37). The NACA recognized that a severe problem existed that had to be urgently solved. The "Skyrocket" was used for pitch-up investigations up to 1953 testing different wing leading-edge designs.

Only after the second and third aircraft had been equipped with a rocket engine and prepared for a drop from a carrier aircraft could the step towards the flight at $M = 2$ be taken. During the flight tests at supersonic speeds, the "Skyrocket" demonstrated a dangerous roll behavior that was attributed to a poor lateral stability. The test pilot *Bill Bridgeman* stated that during the transition from climb to horizontal flight, lateral stability decreased dramatically. The NACA limited, thereafter, the maximum speed to $M = 1.88$. After some changes in the empennage area, which, in particular, removed the interference of the exhaust jet with the rudder, the NACA dared to approach $M = 2$. On November 20, 1953, test pilot *Scott Crossfield* reached an altitude of 18,900 m at $M = 2.005$; thus, he became the first human to fly at twice the speed of sound.

The flight characteristics, established in the flight tests of the Douglas D-558-II, are similar to the flight characteristics of the DFS 346, which allows a comparison between the two aircraft. The DFS 346 as well as the D-558-II showed with increasing *Mach* number a poor lateral stability, which the pilots had constantly to correct. There was, in addition, a reduction in the elevator and rudder effectiveness when approaching the speed of sound. Both aircraft had a wing with 12% thickness with the DFS 346 having had a wing sweep of 45 deg. Without major changes having been carried out on the aircraft, the

D-558-II has been flown up to a *Mach* numbers of $M = 2.0$. The flight testing of the "Skyrocket" has shown that the aircraft could be controlled by the experienced test pilots of the NACA despite its stability problems. The few documents concerning the flight tests of the DFS 346 show that the flight characteristics of this aircraft could not have been much worse than those of the D-558-II. This allows the conclusion that it was probably the lack of experience of the test pilot *Ziese* that caused the failure of the DFS 346 project.

6.5.3 BRITAIN—MILES M52

After the usefulness of the jet engine for high-speed flight had been proven during the flight tests of the experimental aircraft Gloster E 28/29, flight at supersonic speeds drew nearer. As in the other countries the Royal Aeronautical Establishment (RAE) recognized that a large gap existed in the knowledge about transonic flow, which had first to be closed. The RAE also thought of a research aircraft to investigate the flow phenomena occurring when crossing the sound barrier because the wind-tunnel technology did, at that time, not allow measurements in the area around $M = 1$. In 1942, two years prior to similar considerations in Germany, the Air Ministry contracted the Miles Aircraft Company with the development of an aircraft that was to achieve a speed of 1000 mph (1600 km/h) at an altitude of 36,000 ft (11,000 m).

Under the company designation M52, a configuration was developed with a cylindrical fuselage and midwings. A Whittle W2/700 jet engine equipped with augmentor, an early form of turbofan, and with an afterburner was proposed for propulsion. In the center of the air intake there was a cone housing the cockpit for the pilot (Fig. 6.46). The wings had a thin symmetrical biconvex airfoil with a sharp leading edge. The horizontal stabilizer was a flying tail while the vertical stabilizer was equipped with a standard rudder. The design was about 90% complete in February 1946 when the project was cancelled. The reasons for it were the urgently necessary economic savings after the end of the war, and the fact that the ideas of the RAE regarding the form of supersonic aircraft had changed after the evaluation of German documents related to the swept wing [84]. After the termination of the project, all information about the M52 was made available to the Americans.

Fig. 6.46 Model of the research aircraft Miles M.52 (1946).

After the termination of the M52 project, there was no aircraft available able to generate the data urgently needed for flight near the sonic barrier. The flight testing of the de Havilland D.H. 108 had not yet progressed far enough for it to be a possibility for that purpose. In view of this situation the British government decided to develop an unmanned aircraft on the basis of the M52. The Vickers-Armstrong Aircraft Company started with the development of a smaller and unmanned version of the M52 with rocket propulsion. Data were to be sent to the ground by telemetry. The aircraft, controlled by an autopilot, was to achieve supersonic speeds during horizontal flight. It was planned to test six different wing configurations with different sweep angles. The aircraft was suspended from the fuselage of a de Havilland "Mosquito" and dropped from an altitude of 12,000 m. The first attempt failed. In October 1948 the second aircraft, still with straight wings, reached a speed of 930 mph (1400 km/h) corresponding to $M = 1.4$ (see [47], p. 66). It was thus proven that this configuration of the M52, similar to the Bell XS-1, was able to enter the supersonic regime without any problems. For cost reasons, further flight tests with the M52 model were suspended in favor of rocket-propelled models that could be launched from the ground.

6.6 DELTA AIRCRAFT

6.6.1 LIPPISCH P 13 AND GLIDER DM-1

The term "Delta" wing is inseparable from the name of *Alexander Lippisch*. He created this expression for his tailless aircraft that had a wing of a triangular planform. During his time at the Messerschmitt Aircraft Company, he had already carried out investigations into the design of a supersonic aircraft thought of as a successor to the Messerschmitt Me 163 (see [44], pp. 93–106). For this aircraft he had envisaged a delta wing with a 60-deg leading-edge sweep and a small thickness ratio. *Lippisch* had recognized that the induced drag of a low-aspect-ratio delta wing at supersonic speeds caused less problems than at subsonic speeds (see [26], p. 918). For the landing approach the high induced drag was in his opinion tolerable, although under acceptance of very large angles of attack. He did not see any problems with the flow separation at the wing tips if the aspect ratio was less than 2.5. *Lippisch* considered the slender delta wing as the ideal configuration for a supersonic jet. Because of its length, the slender delta wing did not cause any trim problems when crossing the transonic range; it offered sufficient volume for fuel tanks and allowed a compact and torsion-proof design.

Wind-tunnel measurements at sub- and supersonic speeds on a wing model were carried out in early 1944 at the AVA Göttingen. The polar plots of the Göttingen measurements, presented in his book, show the drag rise at a *Mach* number of 0.8 (see [44], pp. 93–134). Later measurements on the wing of the "Zitterrochen," with the same leading-edge sweep, also carried out at

Göttingen show the drag rise only at a *Mach* number > 0.85 [85]. This can possibly be attributed to the airfoil and the smaller thickness ratio of the wing of the "Zitterrochen." Whether *Lippisch* was aware of the measurements on low-aspect-ratio wings, which had been carried out at the AVA and the LFA and which had also included the 60-deg delta wing, is not known.

Lippisch further pursued the idea of a supersonic jet at the Luftfahrt Forschungsanstalt Wien (LFW) (Aeronautical Research Establishment Vienna). In March 1944, he started the design of an experimental aircraft designated P 12. The P 12 was a tailless aircraft with a 60-deg delta wing. As propulsion, a ramjet was envisaged that was to be directly integrated into the design. Because the jet engines in 1944 – 1945 did not yet have sufficient thrust to penetrate the supersonic regime, the ramjet engine was considered as the only possibility to accomplish flight at supersonic speeds for a longer period of time with acceptable fuel consumption. In contrast to his first wing design, the P 12 had a wing of considerable thickness that was probably required for the accommodation of the propulsion unit.

After a short time the P 13 resulted from the P 12 design. The P 13 was very similar to the P 12, but was to have a solid-fuel ramjet engine. Corresponding investigations were at the time, being carried out at the LFA as well as at the LFW. The P 13 had a 60-deg delta wing with an aspect ratio of $\Lambda = 1.8$. *Lippisch* selected an airfoil section with his favored elliptical airfoil with a 45% maximum-thickness location. The wing thickness ratio was 13.6% at the wing root and probably constant over the entire span. Despite this relatively thick wing the P 13 configuration had according to his statements a critical *Mach* number of $M = 0.9$. The combined aileron/elevator (elevon) arrangement at the wing trailing edge was separated from the wing by a gap similar to the Me 163. The control surfaces were aerodynamically balanced, and the gap was to ensure that control-surface reversal did not occur as a result of compressibility effects when shock waves developed at the leading edge (see [13], p. 24). One-third of the wing volume was provided for the ramjet. Figure 6.47 shows a plan view of this aircraft. A so-called two-dimensional nozzle was envisaged as thrust nozzle. The thrust direction could be changed by flaps above and below the thrust nozzle. The pilot sat above the engine in front of the giant vertical stabilizer. The aircraft was supposed to be able to fly at speeds up to 1900 km/h. Whether this would have been at all possible with a 13.6%-thick wing is doubtful.

Lippisch pursued his own ideas with regard to the optimum configuration of a supersonic aircraft that partly contradicted the views of the industry. He was an advocate of the low-aspect-ratio tailless aircraft. Focke-Wulf and Messerschmitt worked during the last year of the war on designs with swept wings and separate horizontal stabilizers. In their opinion, due to the shorter layout, a tailless aircraft possessed too little pitch damping and less control effectiveness to be able to successfully compensate the neutral-point

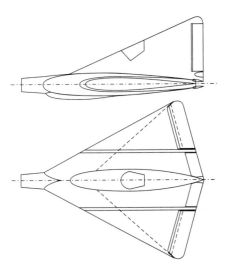

Fig. 6.47   Lippisch project P 13 for a super-sonic fighter aircraft (1944) (see [44], p. 99).

movement when crossing the critical *Mach* number range (see [13], p. 32). Because of the completely uncertain flow conditions within the critical *Mach* number range, *Lippisch*, however, did not believe that it would be possible to design an empennage that would dampen intense oscillations as well as ensure good controllability. He was of the opinion that the pitching-moment problem could be avoided within the critical *Mach* number range by crossing it in a dive at zero lift instead of in horizontal flight. The prerequisite was, however, a symmetrical wing airfoil section. In his opinion, in case of the slender delta wing only small center-of-pressure movements were to be expected within the transonic range. With regard to the directional stability, *Lippisch* also saw an advantage for the tailless aircraft. A short fuselage with a relatively large vertical stabilizer would provide an almost constant center-of-pressure location within the entire yaw angle range. He based this on the flight characteristics of the Messerschmitt Me 163 which, unlike the Me 262, did not have any "Snaking" problem within the range of small yaw angles. Because of the completely three-dimensional flow, *Lippisch* also thought that his low-aspect-ratio delta wing would have no problems at high angles of attack thus avoiding the flow separation at the wing tip.

Wind-tunnel measurements on the P 13 configuration showed that the directional stability of the aircraft was low. The reason was the short distance between the vertical stabilizer center-of-pressure location and the center of gravity. Because of its small aspect ratio, the vertical stabilizer produced considerable drag during yawed flight and contributed considerably to the yaw-induced rolling moment. Why *Lippisch* deviated from the more favor-able P 12 configuration with a conventional vertical stabilizer located further back is not clear.

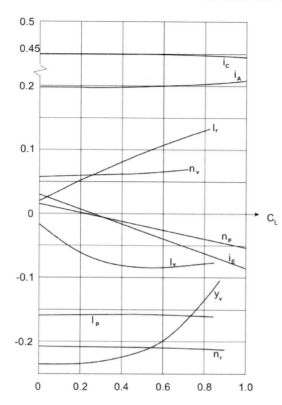

Fig. 6.48 Development of the most important derivatives and moments of inertia of the P 13 dependent on the lift coefficient; nomenclature follows the British notations for stability derivatives: i_y are the radii of inertia, l_x the roll derivatives, n_x the yaw derivatives, and y_x the side-force derivatives, $x = p$, r, v means the rotational velocity in roll, in yaw, and the lateral speed (sideslip) ([33], Fig. 15).

Apparently investigations concerning flight characteristics were still being carried out on the P 13 configuration at the DVL. The results of these investigations were subject of the lectures of *K. H. Doetsch* regarding flight control at the Institute of Flight Guidance Braunschweig. To produce the high-lift coefficients required for landing, the P 13 had to fly at high angles of attack. In the case of a low-aspect-ratio wing, this results in not only high drag but also the derivatives of lateral motion change with incidence (see [33], p. 25). Especially the rolling moment due to sideslip, the rolling moment due to yaw and the side force due to sideslip are shown to be strongly dependent on the angle of attack (see Fig. 6.48). This meant that the already small roll damping almost disappeared at higher lift coefficients and that the ratio of roll to yaw of the Dutch-roll oscillation, 0.5 at low-lift coefficients, increased to values of 2.7 at high-lift coefficients. During the landing approach, this would have meant a serious controllability problem for the pilot!

It was also clear to *Lippisch* that there was a need for clarification regarding the flight characteristics of this unusual configuration. He had an airworthy glider version of the P 13 built by students designated DM-1 (see Fig. 6.49). The airplane possessed essential features of the P 13. It was a tailless airplane with a 60-deg delta wing. The airfoil was the same as for the P 13 and was

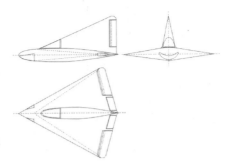

Fig 6.49 Experimental glider Lippisch DM-1.

later declared in the United States to be similar to the NACA 0015 64 [86]. Figure 6.50 shows two versions of the DM-1 with modified vertical stabilizers (see [44], pp. 93–100). These configurations are more like the P 12 than the P 13. Why *Lippisch* finally decided on the less favorable configuration remains unexplained. The DM-1 was to be carried piggyback by a carrier aircraft to altitude and, after release, return in a glide. Prior to its completion it fell into the hands of the entering American troops. It was *Theodore von Kármán* who spoke for a continuation of the design and flight testing in the United States [87]. In November 1945, the DM-1 was shipped to the United States where it arrived in January of 1946. There, the DM-1 was prepared for measurements in the 30 × 60 "full scale" wind tunnel of the NACA at Langley Field. A smaller model of the wing was, however, built prior to these measurements and tested in another wind tunnel to obtain a feeling for the magnitude of the aerodynamic forces to be expected. The first measurements took place at the end of 1946.

These tests yielded a lower maximum lift coefficient compared to the measurements in Germany and measurements on the small wing model. *Lippisch*, asked for advice, recognized this as a *Reynolds* number effect. To not overstress the model, completely built of plywood, the freestream speed was set at too low a value. After attaching transition strips to the wing nose, lift coefficients of C_A (C_L) > 1.0 were reached [86].

Fig. 6.50 Alternative configuration of the DM-1 with a modified vertical stabilizer (see [44], p. 100).

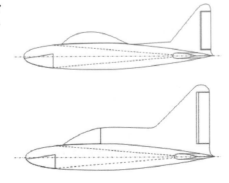

During a second test campaign, the DM-1 was systematically modified. These investigations were to provide information about the flight characteristics. The measurements on the original configuration yielded an inadequate directional stability and the drag too high. The NACA considered the airplane without substantial improvements as not flyable [88]. In succession a sharp-edged wing nose was then attached, the giant vertical stabilizer fin replaced by a smaller and thinner vertical stabilizer, the slots between control surfaces and wings were closed, and finally the DM-1 was fitted with a F-80 cockpit hood [89]. Figure 6.51 shows a plan view of the modified DM-1. All measures together led to an improved directional stability, a higher maximum lift, and produced a positive rolling moment due to sideslip (stable according to the NACA notation of the time) up to lift coefficients of 1.0. The NACA concluded here that such an aircraft with 60-deg sweep and a small nose radius and with acceptable flight characteristics at subsonic speeds could be built. As mentioned, this knowledge concerning the characteristics of low-aspect-ratio delta wings was already available two years earlier in Germany. The reason why *Lippisch* did not use this knowledge is possibly due to the poor communication between the research establishments caused by the exaggerated secrecy regulations.

The delta wing was also not unknown at NACA. In the context of reports from Germany concerning swept and delta wings, measurements that had been carried out already years before in the United States were recalled. However, these measurements concerned only the subsonic speed regime and the high-lift behavior of swept and delta wings, respectively wings of low aspect ratios [90]. The conclusions of the NACA at that time were interesting. They were of the opinion that a horizontal stabilizer could have an additional destabilizing effect because of the stability problems of swept wings at high lift. Therefore, a natural-stability tailless aircraft would possibly be better suited for high-speed flight.

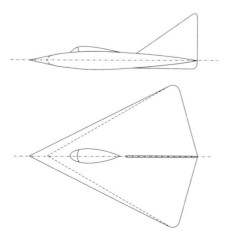

Fig. 6.51   Lippisch experimental glider DM-1 after the modification by NACA.

Even before the 1/15-scale model of the DM-1 wing was tested at the NACA, measurements took place in May 1946 in the Guggenheim Aeronautical Laboratory at the California Institute of Technology (GALCIT) 10-ft wind tunnel on a delta-aircraft model designated "Space Waif." The customer was the U.S. Marine Corps, which had these measurements carried out in the context of a "Pilotless Aircraft Program." "Space Waif" had a delta wing with 53-deg leading-edge sweep and a relative thickness of 16%. This model is now at the Western Museum of Flight in California. This design is probably based on a contact with *Lippisch*, although he does not mention this anywhere in his memoirs. The aircraft itself was never built.

6.6.2   CONVAIR *XP-92* AND *XF-92 A*

After the war, no other aircraft configuration fascinated the development engineers in the United States, Britain, and France more than *Alexander Lippisch's* delta projects. His idea of a 60-deg delta wing as the most suitable wing for supersonic flight could no longer be achieved in Germany. It was left up to the Americans to turn this idea into an actual design and to determine the qualities of this revolutionary concept in flight tests.

The "delta wing" was, of course, known in the United States, although not under this name. When a wealth of information and results of wind-tunnel measurements arrived from Germany confirming the advantages of the swept wing at high speeds, NACA started its own extensive test program to explore the characteristics of the swept wing. Important here was the stability behavior of this wing configuration at low speeds. Measurements of the U.S. Navy in 1943 on a 65-deg delta wing showed a stable behavior up to high angles of attack (see [90], Fig. 25). In 1941 *Michael Gluhareff*, chief engineer at Vought-Sikorsky, suggested an aircraft with a delta-shaped wing and 56-deg leading-edge sweep (see [88], p. 4). The aircraft possessed conventional propulsion with a pusher propeller. *Gluhareff* claimed that this configuration would shift the onset of compressibility effects to higher speeds and was, therefore, an ideal configuration for very high flight speeds. However *Gluhareff* never justified his claim. *A. Lippisch* justified this at the end of 1943 with wind-tunnel measurements on a 60-deg delta wing model in Göttingen. In the spring of 1945, before results of German high-speed research became known in the United States, the aerodynamicist *Robert T. Jones* of the NACA published his "Slender-Wing Theory" [91]. An essential statement of his theory concerns the lift distribution, which for wings with very low aspect ratios becomes independent of the *Mach* number. He develops his theory considering the example of the slender delta wing ("pointed wing"), but in no way suggests this wing as the ideal form for flight at high speeds. This was also true for *A. Busemann* when he introduced his "infinitesimal conical supersonic flow" in 1942 employing as an example the "lifting triangle" [92].

On May 10, 1945, just after the war in Europe, *Lippisch* was taken by the Air Technical Intelligence of the U.S. Army Air Force (USAF) to St. Germain in France to be questioned by U.S. experts. There, on May 23, he gave a lecture on his delta-wing developments to representatives of the Allies (see [44], pp. 93–107). This shows the interest in his design at that time. Besides *Theodore von Kármán*, there were also representatives of the American aeronautical industry in the audience. In January 1946, he went together with other aeronautical scientists to Wright Field in Dayton, Ohio. During the same year on request of the USAF, he went on a lecture tour through California where he also met engineers of Convair.

In September 1945, the USAF asked the industry to make proposals for a high-speed interceptor [93]. The aircraft was to attain a speed of 700 mph (1130 km/h) at an altitude of 50,000 ft (15,000 m). The Consolidated Vultee Aircraft Corporation, later renamed "Convair," won the contract for the development of a prototype. Convair proposed an aircraft with a 45-deg (leading-edge) swept wing and a swept V-tail. Figure 6.52 shows an illustration of this design. Similar to the Lippisch P 13 a ramjet engine was envisaged as propulsion. A number of smaller rocket engines were to ensure the takeoff from the ground and ignite the ramjet engine [94]. The takeoff was to be made from a takeoff trolley similar to that of the Messerschmitt Me 163. The landing gear was designed for landing only. The Convair engineers *T. M. Hemphill*, *R. H. Shick*, *F. W. Davis*, and *A. Burstein* were responsible for the development of this new aircraft; see Fig. 6.53.

In the spring of 1946, after wind-tunnel measurements, it was found that in the case of the swept-wing configuration there were problems in the high-lift range and the controllability about the roll and yaw axes was inadequate. In July 1946 the design was changed into a 60-deg delta wing configuration, and it was decided to further develop the aircraft into a supersonic fighter. Because the expression "delta wing" was used here at Convair for the first time, it may be assumed that this had something to do with the already mentioned visit of *A. Lippisch* to Convair in Downey for an exchange with the local engineers [95]. *Lippisch* was obviously able to convince the engineers of the advantage of a 60-deg delta wing. The result of these consultations was the design of a

Fig. 6.52 First design of Convair for an interceptor (1945) (see [93], p. 244).

Fig. 6.53 Convair engineers *T. M. Hemphill, R. H. Shick, F. W. Davis, and A. Burstein* (Interavia) (see [99], p. 30).

project, now designated XP-92, with delta wings and a similar vertical stabilizer. The affinity of the Convair XP-92 with the P 13 and the DM-1 is clearly visible in the plan view of the Convair XP-92; see Fig. 6.54. The wind-tunnel tests with the new delta configuration were very promising so that there was interest in a more intensive exchange of experience with *Lippisch*.

In the same month, *R. Shick* visited *Lippisch* in Dayton by arrangement of the U.S. Air Force to talk to him about the problems of this aircraft (see [88], p. 7). *Lippisch* could not gain acceptance from the engineers for his 15%-thick airfoil. Convair had rightly selected the thin symmetrical airfoil NACA 651 006,5 with a thickness ratio of 6.5%, which was considerably better for high-speed flight. The aerodynamic compensation (balance) of the control surfaces had now been located at the outer edge of the wing. Different

Fig. 6.54   Convair project XP-92 of a supersonic fighter aircraft (1946).

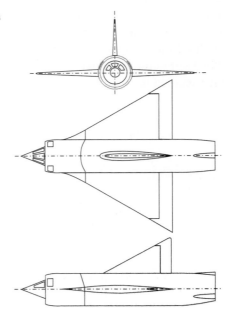

drawings of the XP-92 show both a fully moving vertical stabilizer and a version with a conventional rudder. The pilot was housed within a cone that protruded from the intake of the ramjet engine. This cone was thought of by Convair as a supersonic diffuser with an external compression to decelerate the flow upstream of the air intake by an oblique shock.

In conjunction with the XP-92 this diffuser is called a *"Ferri*-Diffuser." *A. Ferri* was an employee of the Italian Aeronautical Research Center Guidonia and, after the occupation by the Americans in September 1944, came to the NACA (see [25], p. 319). He brought with him a multitude of reports and information concerning German high-speed research. The cone as supersonic diffuser had already been suggested in 1941 by *A. Busemann* and *K. Oswatitsch* for projectiles with ramjet propulsion [96]. This concept had been extensively investigated at the time in the Göttingen and Braunschweig wind tunnels.

During the transition to the delta configuration, Convair recognized that in the case of the XP-92 they now had to deal with two critical technologies challenging engineers: on the one hand the completely new ramjet propulsion and on the other hand the delta wing untested until now. It was, therefore, decided not to test both technologies together on one test bed but to investigate the aerodynamics of the delta wing on a separate experimental aircraft ("Demonstrator").

The unusual configuration of the XP-92 met with great interest at the NACA and was extensively investigated at both subsonic and supersonic speeds. Several variants of the basic design, among them a version with vertical fins at the wing tips and a somewhat more conservative design without the inlet cone but with the cockpit on the upper side of the fuselage, were tested in the wind tunnel. The wind-tunnel models had adjustable control surfaces. Two of the configurations investigated are shown in Figs. 6.55 and 6.56. The configuration BsW60V1, Fig. 6.55, resembles the basic design of the XP-92 while the configuration B12W60V1K, Fig. 6.56, already corresponds to the later XF-92A.

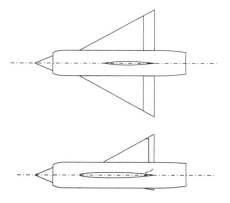

Fig. 6.55 Position of the neutral point, XP-92 wind-tunnel model BsW60V1.

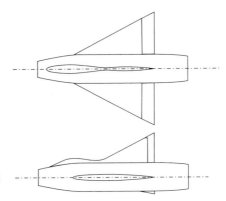

Fig. 6.56   Drag rise, XP-92 wind-tunnel model B12W60V1K.

The first measurements were carried out in 1947 at a *Mach* number of 1.53 [97]. The largest difference between the two configurations concerned the drag. Schlieren pictures showed that the intake cone of the BsW60V1 was poorly designed. The oblique shock waves coming from the tip of the cone did not hit the lip of the air intake thus causing a higher drag when the flow was passing through the model. Both configurations were stable with respect to the pitching-moment reference point selected. The location of the neutral point proved, however, to be strongly dependent on the angle of attack. This was interpreted as the influence of the fuselage, the latter being relatively thick in comparison to the wing. The directional stability was marginal in the case of both models. The control-surface effectiveness was, however, good and hardly dependent on the angle of attack.

The models were investigated during a second test series within a *Mach* number range of 0.5 to 1.5 [98]. Also during these measurements, where the dependence of the stability parameters on the *Mach* number was to be investigated, no serious defects were exhibited by the basic configuration. The control effectiveness was good throughout the entire *Mach* number range. The longitudinal stability improved with increasing *Mach* number. The reason here was the downstream movement of the center of pressure.

In the two reports quoted [97, 98], there is no reference to German sources. American reports only are quoted that once refer to the report of *A. Busemann* on conical supersonic flow [92]. Convair spent about 5000 wind-tunnel hours in various tunnels to obtain information concerning the stability and controllability of this aircraft [99].

Meanwhile the USAF and Convair agreed that the planned ramjet engine could not be realized. Work on the XP-92 was, therefore, discontinued. On recommendation of the USAF the work on the "demonstrator" for the delta configuration was continued. As propulsion, a standard jet engine was envisaged. Under the designation XF-92A the first aircraft evolved with a 60-deg delta wing. The aerodynamic layout corresponded largely to the model

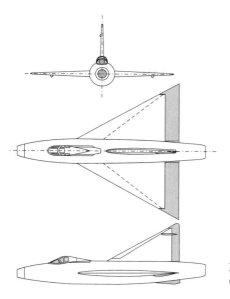

Fig. 6.57 Experimental aircraft Convair XF-92A.

B12W60V1K, although with a substantially more slender fuselage (see Fig. 6.57) and a larger vertical stabilizer (see [95], p. 178). All control surfaces were driven for the first time by irreversible hydraulic actuators. The complete aircraft was tested in the 40×80 "full-scale" tunnel of the NACA Ames Aeronautical Laboratory in December 1947. On December 18, 1948, the XF-92 A took off for its maiden flight from the Flight Test Center of the USAF Muroc Dry Lake (Fig. 6.58).

Phase 1 of the flight-test program was dedicated to flight performance. The USAF took over the aircraft in the autumn of the 1949 for Phase 2 of the flight test program, that is, the determination of the flight characteristics according to U.S. regulations. No less a person than test pilot *Charles "Chuck" Yeager* was to find out whether the XF-92 A was suitable as a

Fig. 6.58 Experimental aircraft Convair XF-92A at the Flight Test Center Muroc Dry Lake.

Fig. 6.59 USAF test pilot *Charles "Chuck" Yeager* mounts the XF-92A.

fighter aircraft (Fig. 6.59). After he had run the aircraft through all maneuvers without problems, he stated that the XF-92 A was a step in the right direction. An interesting experience was to him that the XF-92 A was spin free. At a speed of 108 km/h and an angle of attack of 44 deg, the aircraft was still controllable; however, horizontal flight was no longer possible. He found the ability to fly abrupt and fast rolls particularly impressive. "This would make one heck of a fighter plane" was his comment (see [63], p. 30). The flight tests confirmed the wind-tunnel measurements, namely, that the stability increased with increasing *Mach* number. The aircraft could be flown without problems within the transonic speed regime and showed no "buffeting" problems, which had been the case during flight testing the F-86 (see [95], p. 180). Stability and controllability were also sufficient without a separate empennage. *Lippisch* had exactly predicted these characteristics, which were here confirmed.

After the completion of Phase 2 of the flight-test program by the USAF, where the speed of sound was repeatedly exceeded briefly during a dive, the aircraft was taken over by the NACA for further flight tests up to its limits. The NACA test pilot was *Scott Crossfield*, who had, like *Chuck Yeager*, made a name for himself in testing experimental aircraft. During these flight trials, a problem occurred that had not been predicted by *Lippisch* and that had also not yet been indicated in wind-tunnel measurements—"transonic pitch-up." During "wind-up turns" at a *Mach* number of 0.87, an intense pitch-up occurred accompanied by vertical accelerations of up to 6 g. A breakdown of the lift in the outer-wing area was determined as cause, an effect which had already been observed on swept wings. The stalling flow caused a rapid upstream shift of the center of pressure and thus a strong tail-heavy pitching moment. The NACA suggested as a remedy boundary-layer fences that were to be attached to the wing leading edge at 60% of the span and extended chordwise up to the combined elevator/aileron installation (elevons). The boundary-layer fences proved to be effective at moderate but not at high

Mach numbers. The continuation of the boundary-layer fences on the lower side of the wing provided small improvements, primarily producing a better directional stability at subsonic speeds. The pitch-up problem at high *Mach* numbers ($M > 0.85$) was, however, never satisfactorily solved. At the end of 1953, the flight tests were discontinued and the aircraft put out of operation.

The XF-92 A proved the usefulness of the slender delta wing for high-performance fighter aircraft. The XF-92 A was the father of a whole line of very successful aircraft developments at Convair. The good flight-test results encouraged other aircraft manufacturers, such as Dassault, Avro, or Fairey in Great Britain, to develop their own delta configurations. The supersonic transport aircraft "Concorde" also belongs to the heirs of the XF-92 A and thus also of *Lippisch*.

6.6.3 LIPPISCH P 11 DELTA VI

After his retirement from the Messerschmitt Company, *Alexander Lippisch* took over as head of the LFW in May 1943. One of his first tasks was the further development of the project P 11-121 to a twin-engine "turbo destroyer (fighter)" designated P 11 Delta VI (see [44], p. 93). Like the design of the P 13, this project attracted the attention of British and American engineers when they found, after the war, a wind-tunnel model of it at the AVA Göttingen.

The *Lippisch* Delta VI (Table 6.8) was a tailless aircraft of delta-wing configuration with a leading-edge sweep of 40 deg. The aircraft was to be propelled by two jet engines Junkers Jumo 004C with 10.1 kN thrust each. The wing was according to *Lippisch's* type of design very thick so that the jet engines could be completely integrated into the wing with still sufficient space for large fuel tanks. The rudders and elevators were located at the wing trailing edge, and split flaps (landing flaps) were attached on the lower side. The entire

TABLE 6.8   TECHNICAL DATA *Lippisch* DELTA VI

Span	10.80 m
Length	7.485 m
Wing area	50 m²
Aspect ratio	2.36
Airfoil (root)	0 0016-37-1,036
Airfoil (tip)	0 0009-30-1,625
Dihedral	− 0.8 deg, no twist
Taper ratio	0.372
Flight mass	7300 kg
Propulsion	2 Jumo 004C with 10.1 kN thrust each
Maximum speed	1040 km/h at an altitude of 6000 m
Cruise speed	850 km/h

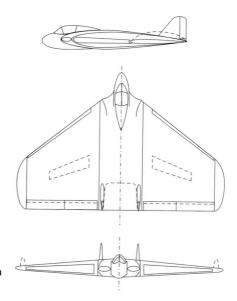

Fig. 6.60   Plan view of the Lippisch P11 Delta VI (1944).

wing leading edge was taken up by movable slats, Fig. 6.60. The symmetrical wing airfoil sections had been designed according to *Lippisch's* development formula for airfoils (see [13], p. 33). The wing-tip segments could be positioned vertically to increase the lateral stability. The following technical data of the project P 11 Delta VI were taken from various documents [100, 101].

Wind-tunnel measurements with this configuration, designated Lippisch Delta VI V2, were carried out at the AVA Göttingen. Here, two wind-tunnel models were available: a wooden model of 3-m span (Fig. 6.61), and a smaller metal model with 1.20-m span that could accommodate model engines (Fig. 6.62). This model was also prepared for pressure distribution measurements. The wind-tunnel measurements served essentially to clarify configuration-related questions.

During these measurements, the control-surface chords were varied, different split flaps on the lower wing surface investigated, and the advantages of a slat were compared to those of a split flap at the nose (Krüger flap). The

Fig. 6.61 Large wind-tunnel model of 3-m span of the Lippisch Delta VI V2 (courtesy of DLR Archives Göttingen).

Fig. 6.62 Small wind-tunnel model with model engines of the Lippisch Delta VI V2 (courtesy of DLR Archives Göttingen).

wing shape and the airfoil sections stayed the same. Because of the very large wing area and the large chord, the AVA suspected a larger influence of the ground effect on the lift and the stability behavior. Additional measurements were, therefore, carried out with a so-called "ground board" to simulate the effect of ground on the aircraft. The last measurements in Tunnel I with running engines and with the large model in Tunnel IV were still listed for March 1945 [31]. Whether all measurements were still carried out as planned is not known because the relevant documents have been lost. Anyway, this configuration raised not only the interest of the Americans but also of the British. In October 1945 the head of the Tunnel K I, *Seiferth*, received an order from the British "Resident Scientific Officer," *J. R. Ewans*, to test the model, including propulsion, in the K I [102]. It was, however, required to first repair the model. In a note of a British officer, it can be read that the Americans took this model apart during the occupation trying to remove the model engines [103]. However, in 1945 there were seemingly not any measurements because a "crash program" of January 5, 1946, scheduled extensive measurements in Tunnel I for a period of two to three months. Whether these measurements on behalf of the British actually took place is not certain. The results were shipped to England without leaving a trace. There are no documents about this subject at the DLR Göttingen.

The drag polar measured at subsonic speeds showed the expected behavior of a low aspect ratio wing. The lift dependence on the angle of attack ($C_A = C_L$ vs. α) was linear up to the maximum lift coefficient of $C_A = C_L = 1.0$ and then broke off abruptly [104]. This indicated a problematic stall behavior near the stall region. The pitching-moment dependence up to maximum lift was also linear and stable. With a slat, covering over 80% of the span, lift coefficients of up to $C_A = C_L = 1.3$ were possible (Fig. 6.63). With the split flap at the nose, lift coefficients of $C_A = C_L = 1.25$ were reached. Neither slat nor split flap at the nose had, however, an effect on the stall behavior. Regarding the lateral stability, the aircraft proved to be stable about the vertical axis. As expected the rolling moment due to sideslip was strongly dependent on the angle of attack. For large angles of attack, therefore, the danger of increased crosswind sensitivity existed as well as of a large roll/yaw ratio with the resulting handling problems for the pilot.

Unfortunately, there are no measurements in the high-speed domain so that not much can be said about the critical *Mach* number of this design. The

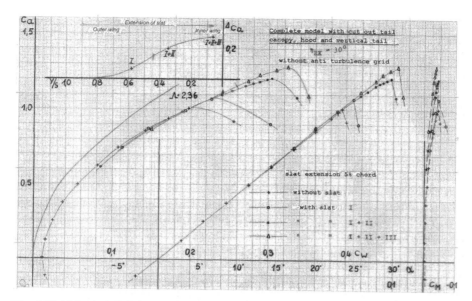

Fig. 6.63   Lippisch Delta VI V2 wind-tunnel measurements with slats (1945) [104], C_A **(C_L) = lift coefficient,** C_W **(C_D) = drag coefficient,** C_M **(C_n) = pitching-moment coefficient,** α **= angle of attack.**

available measurements at subsonic speeds showed, however, that the *Lippisch* Delta VI configuration had the potential for a successful aircraft design. The British and American engineers must also have recognized this when they found a wind-tunnel model of this aircraft at the AVA Göttingen.

6.6.4   DOUGLAS XF4D-1 "SKYRAY"

The Douglas XF4D-1 "Skyray" was the first fighter aircraft of the U.S. Navy capable of supersonic flight that was designed to operate from an aircraft carrier. The aircraft was the answer of the U.S. Navy to the North American F 100 "Super Sabre" of the U.S. Air Force. The design of this aircraft was at the time revolutionary and convincing with outstanding flight performance. The technical head of the Douglas Aircraft Company, *Edward "Ed" Heinemann* received the sought after "Collier Trophy" in 1953 for the development of this aircraft [105]. As in the case of the Convair XF-92 (see Sec. 6.6.2), the Douglas XF4D can also be traced back to a design of *Alexander Lippisch*. In 1947 there was a request for proposals by the U.S. Navy for a fighter aircraft that could operate off aircraft carriers and achieve a speed of 600 mph (970 km/h) at an altitude of 40,000 ft (12,200 m). The evaluation of German wind-tunnel measurements on low-aspect-ratio aircraft had raised the interest of the Navy in *Lippisch's* delta configurations. In the request for proposals, the Navy particularly indicated that the designs to be submitted should orientate themselves on *Lippisch's* delta configurations [106]. The Douglas Company responded to this request with their design D-571. On June 17, 1947, the Navy Bureau of Aeronautics selected the Douglas Company to carry out further studies on the basis of the D-571 design.

It was not a coincidence that the Douglas Company won this request for proposals. Already prior to the Navy request *A. M. O. Smith* [107], at that time aerodynamicist at the Douglas El Segundo Division, had tried to convince his boss, *Ed Heinemann*, of the advantages of the delta configuration with a low-aspect-ratio wing. *Smith* had worked on tailless aircraft prior to joining the Douglas Company and, since then, had never lost his interest in this aircraft configuration (see [29], p. 22). He was sent in May 1945, together with his supervisor, *Gene Root*, as a member of a U.S. Navy delegation to Europe to look for the latest results of German aeronautical research. In Paris, hundreds of results of wind-tunnel measurements and technical reports were available that had been recovered from a well shaft and were spread out for drying. Among the documents were reports on *Lippisch's* delta configurations and the Messerschmitt Me 163, which immediately raised great interest with *Smith*. The trip to St. Germain offered, in addition, the opportunity to meet *Lippisch* personally on the occasion of his lecture on his delta developments [108].

After Paris, *Smith* went first to Göttingen where he visited the AVA and the University. Of course here interesting documents were sought. Although not mentioned in his memoirs, it is unlikely that he would have missed the *Lippisch* Delta VI wind-tunnel models at the AVA. After his visit to Göttingen, he visited the LFA at Braunschweig where he was very impressed by the wind tunnels. After an excursion to Switzerland to the Swiss Institute of Technology Zurich to meet *J. Ackeret*, he returned to the United States.

Back from Germany, he wrote a memorandum to *Ed Heinemann* where he suggested that a tailless configuration of low aspect ratio should be investigated. In his opinion such a configuration would be far less problematic than the flying-wing aircraft of Northrop with large aspect ratios. Considering the short range of the first jet aircraft, the tailless configuration would have an additional advantage due to the large wing volume that could be used for large fuel tanks. To reduce drag, he suggested boundary-layer suction.

The first wind-tunnel investigations commenced at the California Institute of Technology (Caltech) on a wing configuration with 45-deg leading-edge sweep and a taper ratio of 0.33 (Fig. 6.64) in the autumn of 1946. The planform

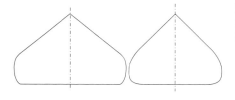

Fig. 6.64 First all-wing design of *A. M. O. Smith* (1946, right) and planform of the *Lippisch* Delta VI (1944, left).

of the *Lippisch* Delta VI wing is also shown in this figure for comparison. The latter wing had a leading-edge sweep of 40 deg. The similarity with *A. M. O. Smith's* first design is unmistakable. He had presumably seen the Göttingen wind-tunnel measurements on the Delta VI configuration, however, in how far he made use of it is not known. Because of the higher leading-edge sweep, the aspect ratio of 2.36 (Delta VI) was automatically reduced to 2, which was what *Smith* had in mind. The change to a higher leading-edge sweep had possibly somewhat defused the problematic stall behavior of the *Lippisch* configuration. The wind-tunnel measurements at Caltech proved the usefulness of the concept and confirmed *Lippisch's* data (see [105], p. 186).

Investigations concerning the installation of the jet engines and the equipment led during the next step to a configuration with a 50-deg leading-edge sweep and a slightly swept trailing edge. With this configuration, wind-tunnel investigations were carried out at low as well as at high speeds. The main objective of the investigations was the correct airfoil choice. After initial investigations, the idea of boundary-layer suction was dropped again because this would have led in practice to considerable effort. This configuration led to the Model D-571, which was submitted to the U.S. Navy in the spring of 1947 (Fig. 6.65). The Model D-571 was a tailless aircraft with a wing area of 700 ft² (65 m²); it was to be propelled by two Westinghouse J-34 jet engines with afterburner. For the entire wing, a modified NACA 0012 airfoil section had been chosen. The flight mass was 15,000 lb (6800 kg), and the flight speed was supposed to be >600 mph (965 km/h). With 12% thickness at the root, the wing was considerably thinner than that of the *Lippisch* Delta VI

Fig. 6.65 Douglas model D-571, first layout of the fighter aircraft for the U.S. Navy (January 1947).

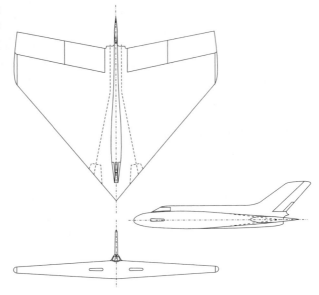

(16%). *A. M. O. Smith* had, in addition, decided on a central vertical stabilizer instead of the twin vertical stabilizer fins of the Delta VI. The model D-571 became the starting point of a whole series of designs where the wing thickness was gradually reduced finally leading to the XF4D-1.

After the presentation of the tailless design, the Navy expressed doubts regarding the spin characteristics. It was feared that this aircraft could turn over backwards as actually happened in the case of the Northrop YB-49 (see [48], p. 154). Flight tests with correctly scaled balsawood models showed no tendency to spin. Especially the low-aspect-ratio models were extremely stable. The doubts of the Navy could thus be dispelled to the great relief of *Ed Heinemann*.

The model D-571-1 was the next design step. The wing remained essentially unchanged. As wing airfoil section a NACA 0012-63/.30-12 deg was selected for the inner wing and at the wing tip a NACA 0010-63/.30-12 deg. The fuselage was extended upstream beyond the wing leading edge (Fig. 6.66) and the jet engines arranged differently. The separate landing flaps were omitted, and instead the span of the elevons was enlarged. The landing gear was changed to one with a tail wheel due to the large angle of attack during takeoff.

During the next step in the development, that is, the model D-571-2, the Westinghouse engines were replaced by a single Rolls-Royce "Nene" jet engine (Fig. 6.67). This required different air intakes. The inner wing thickness was now reduced to 10% and the outer wing thickness to 7.5% at otherwise the same airfoil section parameters. The wing area was simultaneously reduced to 620 ft^2 (57 m^2) thus also reducing the weight to 13,750 lb (6240 kg). As maximum speed, 700 mph (1125 km/h) was to be reached.

Fig. 6.66   Design of the model D-571-1.

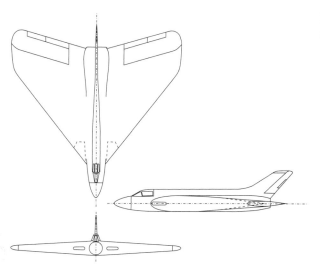

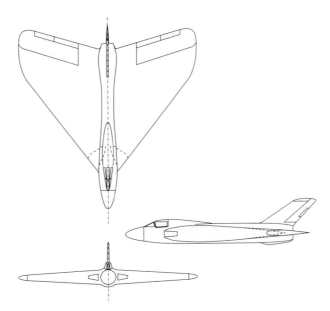

Fig. 6.67   Design of the model D-571-2 (1947).

In the spring of 1948, the last modification prior to the final solution occurred. This was the Model D-571-4 for which a Westinghouse XJ-40 engine with afterburner was envisaged (Fig. 6.68). The wing design was quite similar to the final design. The wing thickness was once more reduced to now 7.5% at the inner wing and to 4.5% at the outer wing. The air intakes were no longer integrated into the wing leading edge, but were positioned normal to the fuselage center axis. Obviously with the design of the D-571-4 a suitable high-speed configuration had been found. Until August 1949 the design was still repeatedly revised and further refined.

Ed Heinemann was confronted with problems due to the trimming of the aircraft throughout a large *Mach* number range. The solution was to design the inner part of the wing trailing edge as a special trim surface. Figure 6.69 shows a three-view drawing of the final version of the XF4D-1. Compared to the model D-571-4 the fuselage was once more extended, and the wing trailing edge near the thrust nozzle was extended backwards. In addition, slats were now scheduled to stabilize the flow at maximum lift in the elevon area. The wing area was 557 ft^2 (51.7 m^2), and the flight weight was 17,000 lb (7710 kg). The aircraft was equipped with an irreversible hydraulic control system that could, if necessary, in case of an emergency be switched over to "manual." The aircraft was equipped with a yaw damper to guarantee sufficient stability about the vertical axis and to dampen the strong Dutch-roll oscillations occurring in the case of highly swept aircraft. The landing gear with the tail wheel had been abandoned and returned to the landing gear with the nose wheel.

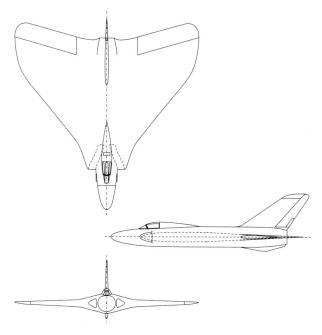

Fig. 6.68 Design of the model D-571-4 (March 1948).

Fig. 6.69 Final design of the Douglas XF4D-1 (August 1949).

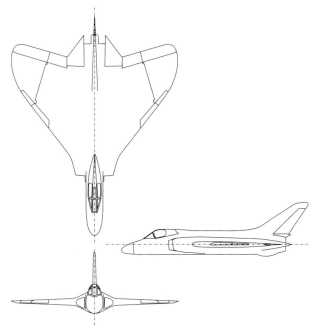

The flight testing of the Douglas XF4D-1 commenced on January 21, 1951, at the flight-test center of the U.S. Air Force, Edwards Air Base (Fig. 6.70). Besides flying-quality investigations, flight at supersonic speeds was part of the program. Serious handling-quality problems were not encountered at subsonic speeds. The stability about all axes was sufficient but not especially high. The large angles of attack, typical of delta-wing aircraft at takeoff and landing, were hard to get used to by the pilots. Flights at speeds close to the speed of sound revealed, however, a problem that could only be solved with greater effort—"rudder buzz and tail buffeting." Both are phenomena that are related to flow separations downstream of shock waves. In the area of the rear fuselage alone, 14 modifications were carried out to solve the problem of "tail buffeting."

Spin tests were extensive. The U.S. Navy required all forms of spin and corresponding pull-out maneuvers to be investigated. The aircraft did not prove to be spin free but showed unusual spin characteristics, which required a closer investigation. Before the spin trials could be continued, the NACA examined the spin behavior extensively in the spin tunnel. The pull out from some forms of spin proved to be difficult and required the full skill of the test pilots. The spin tests were successfully completed in mid-1952.

An outstanding characteristic of the XF4D-1 was the "buffet boundary" at flights with large g-loads. "Wind-up turns" could be flown at an altitude of 40,000 ft (12,000 m) with up to 3.4 g at $M = 0.90$ without problems and without buffeting. During these maneuvers the accompanying F-86E chase plane was never able to keep pace (see [108], p. 262). With the installation of the Westinghouse XJ40-WE-8 jet engine with afterburner in September 1953, the time had come for an attack on the world speed record. On October 3, 1953, *James Verdin* set up a new world speed record with an average speed of 753.4 mph (1212 km/h).

Fig. 6.70 Douglas F4D-1 on the ground (courtesy of Boeing Historical Archives, The Boeing Company).

A test team of the U.S. Navy that was tasked with familiarizing itself with the aircraft judged it, without exception, positively and certified its great developmental potential. The aircraft showed itself to be superior to all other aircraft that the test pilots had flown up to that time. The enormous climb performance resulting from the low wing loading was impressive. During the flight testing of the prototype XF4D-1, the Douglas test pilot *Bob Rahn* said about the aircraft that it was the best fighter aircraft that he had flown since the days of the "Spitfire." The XF4D was the dream of every fighter-aircraft pilot because it possessed an excellent maneuverability at high altitudes, a fantastic climb performance, short takeoff and landing paths, and excellent flight characteristics at lower flight speeds.

The success of the Douglas XF4D-1 resulted from the uncompromising use of a wing with a low aspect ratio and a high leading-edge sweep. Because of the large wing chord, airfoil sections could be used with very low thickness ratios, and due to the longitudinal extent of the wing control surfaces and trim flaps exhibited a high effectiveness. Together with the slender fuselage, compressibility effects could largely be avoided, and sufficient stability about all axes was taken care of. This aircraft best typified *Lippisch's* idea of a tailless aircraft for high speeds because it avoided the disadvantages of similar designs such as the Messerschmitt Me 163, Northrop X-4, and de Havilland D.H. 108. The Douglas XF4D-1 can be considered as a successful compromise between a pure delta-wing aircraft and a tailless configuration like the Messerschmitt Me 163.

6.7 FAST RECONNAISSANCE AIRCRAFT ARADO AR 234 WITH SWEPT WINGS

The jet aircraft Arado Ar 234 served as a fast reconnaissance aircraft and was also employed as a bomber. The development of this aircraft has already been sufficiently documented in the series "Die Deutsche Luftfahrt" (German Aeronautics) [109] so that only the planned versions with swept wings are addressed herein.

As a fast jet aircraft, the Ar 234 was to reach *Mach* numbers of up to approximately $M = 0.82$. For this reason, a wing thickness as small as possible was aimed at and was selected to be 13% at the root decreasing to 10% towards the tip. A weakly cambered airfoil section with a 30% maximum-thickness location was used. To achieve a high surface quality, developable panels were used for the skin and joined without steps normal to the flow direction. To obtain information about the aerodynamic quality of this wing, wind-tunnel measurements were carried out on an original wing at the DVL. The local drag coefficient was determined by the momentum-loss (wake-rake) method. It was found that control-surface gaps and the skin steps increased the drag coefficient by 10% compared to a smooth wing. Even the smooth wing still had a higher drag coefficient than the untreated wing of the P 51 "Mustang" equipped with a laminar airfoil (see [17], p. 13). A new wing geometry with different airfoil sections had, therefore, to be employed for the flight at higher speeds. The flight testing had shown (see Chapter 5) that the aircraft could only be deployed up to a *Mach* number of $M = 0.75$.

In 1942 Arado became interested in the application of the swept wing (see [4], p. 94). In the opinion of the chief aerodynamicist of Arado, *R. Kosin*, decisive for its use on the Arado Ar 234 was the overcoming of the stability problems of the swept wing in the high-lift regime. Aeronautical research was, at that time, mainly concerned with the flight behavior of the swept wing at high speeds. Only later did high-lift regime become of interest to scientists. Arado as well as Junkers had to acquire the needed data through their own investigations.

Kosin saw the solution to the problems within the high-lift regime in a wing whose sweep decreased from inboard to outboard. The lesser sweep of the outer wing provided sufficient aileron effectiveness at high angles of attack and, simultaneously, defused the so-called "tip-stall" problem. Arado registered this idea in 1942 for a patent, probably without knowing that *A. Betz* of the AVA had already patented the same idea in 1939 (see Sec. 2.3.3). This fact once more illustrates the poor communication between research and industry at that time. The characteristics of the new wing were investigated in the wind tunnels of the DVL, AVA, and LFA. A so-called "variable model" was used that could be equipped with different wings and engine arrange-ments. Figure 6.71 shows the Arado "variable model" in the A1 wind tunnel of the LFA Braunschweig-Völkenrode. Figure 6.72 shows a three-view draw-ing of the model scheduled for the Göttingen wind tunnels. As obvious in this figure, the wing possessed a slat in the area of the aileron. Figure 6.73 shows the wind-tunnel model equipped with two jet engines.

Besides these wind-tunnel models Arado ordered the design of four experi-mental wings. The latter possessed different airfoil sections with different

Fig. 6.71 Arado variable model in the low-speed wind tunnel A1 of the LFA Braunschweig.

Bericht: 44/W/49

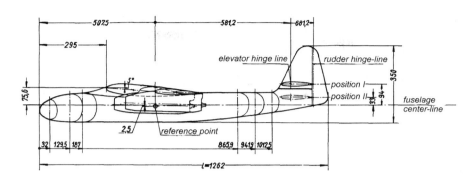

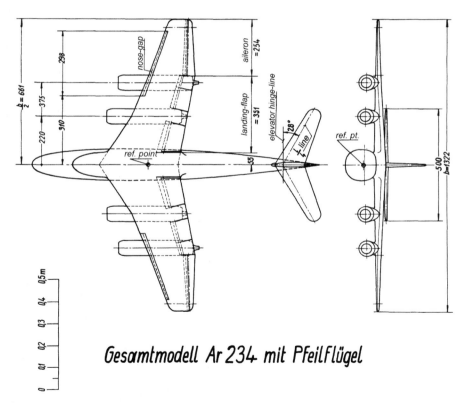

Gesamtmodell Ar 234 mit Pfeilflügel

Fig. 6.72 Three-view drawing of the variable model (G. Brennecke, W. Keydell, Investigation of the Ar 234 model with swept wing, AVA Report 44/W49, 1944); legend: complete model Ar 234 with swept wing.

Fig. 6.73 Wind-tunnel
model Arado Ar 234 with
swept wings at Göttingen
(courtesy of DLR Archives
Göttingen).

thickness distributions along the span. All airfoils had slightly cambered mean lines that caused the flow separations in the high-lift regime to be shifted to higher lift coefficients. Wings 1 and 3 had the "sickle form," while the second wing corresponded to the basic design (see Fig. 6.72). The fourth experimental wing had a constant sweep of 35 deg on the 25% chord line. Unlike the other wings this wing had a symmetrical high-speed airfoil with 12% thickness and a 50% maximum-thickness location. It is not known which of these wings Arado had finally selected for the further development of the Ar 234. None of the Arado experimental wings had been flight tested up to the end of the war. However, it may be assumed that the British company Handley Page was inspired during the development of the V-bomber "Victor" by the Arado "sickle wing."

REFERENCES

[1] "Flugwerk und Triebwerk," Bericht über die Sondertagung, München, Lilienthal-Gesellschaft für die Luftfahrtforschung, Bericht 100, Aug. 23–24, 1938.
[2] "Gegenwärtiger Stand und künftige Entwicklungsarbeit auf dem Gebiete des Schnellfluges mit Strahltriebwerk," Bericht RLM LC 7/III, Bundesarchiv-Militärarchiv, Freiburg, Oct. 14, 1938.
[3] P. Hamel, "The Birth of Sweepback—Related Research at LFA-Germany," SAE Aerospace Control and Guidance Systems, Committee 90 Meeting, Kill Devil Hills, NC, U.S., Oct. 16–18, 2002.
[4] W. Wagner, "Die ersten Strahlflugzeuge der Welt," Die deutsche Luftfahrt, Band 14, Bernard & Graefe Verlag, Koblenz, 1989, p. 94.
[5] H. Doetsch, "Bericht über das Fachgebiet Profile 'vor dem Sonderausschu Windkanäle am 10.1.43 und 4.1.44'," Deutsche Luftfahrtforschung UM 1190, Feb. 25, 1944, p. 6.
[6] H. J. Ebert, J. B. Kaiser, K. Peters, "Willy Messerschmitt—Pionier der Luftfahrt und des Leichtbaus," Die deutsche Luftfahrt, Band 17, Bernard & Graefe Verlag, Bonn, 1992, p. 245.
[7] W. Radinger, W. Schick, "Me 262, Entwicklung, Erprobung und Fertigung des ersten einsatzfähigen Düsenjägers der Welt," AVIATIC Verlag, Planegg 1992, p. 51.
[8] N. J. Butcher, "The Gloster Meteors," Model Aircraft, Vol. 19, No. 227, May 1960, p. 147.
[9] R. D. Neal, "The Bell XP-59A Airacomet: The United States' First Jet Aircraft," Journal American Aviation Historical Society, Vol. 11, No. 3, 1966, p. 177.

[10] E. T. Wooldridge, "The P-80 Shooting Star, Evolution of a Jet Fighter," "Famous Aircraft of the NASM," Vol. 3, Smithsonian Institution Press, Washington, DC, U.S., 1979, p. 104.

[11] W. J. Boyne, "Messerschmitt Me 262, Arrow to the Future," Smithsonian Institution Press, Washington, DC, U.S., 1980, p. 139.

[12] L. Bölkow, "Mit dem Pfeilflügel zum Hochgeschwindigkeitsflug," 50 Jahre Turbostrahlflug, DGLR Symposium, Deutschen Museum, München, Oct. 26–27, 1989, p. 248.

[13] R. W. Kluge, C. L. Fay, "German High Speed Airplanes and Design Development," U.S. Technical Industrial Investigation Committee, Aeronautical Subcommittee, CIOS-Report, Aug. 1945.

[14] Statement of Ludwig Bölkow, 50 Jahre Turbostrahlflug, DGLR Symposium, Deutschen Museum, München, Oct. 26–27, 1989.

[15] Tilch, "Druckverteilungsmessung Me 262 Fläche und Höhenleitwerk," Messerschmitt AG Augsburg, Versuchs-Bericht Nr. 262 02 E44 vom 10.3.44, Bundesarchiv-Militärarchiv Freiburg, RL 39/674.

[16] M. S. Koziol, Rüstung, Krieg und Sklaverei: der Fliegerhorst Schwäbisch Hall-Hessental und das Konzentrationslager, Sigmaringen: Thorbecke; Schwäbisch Hall: Hist. Verein für Württemberg, Franken, 1989, Anhang; "Memorandum Report on Me-262 Airplanes, Nos. T-2-711 and T-2-4012," Headquarters Air Materiel Command Wright Field, Dayton, OH, U.S., Sept. 3, 1946, p. 207.

[17] G. Bock, "Neue Wege im deutschen Flugzeugbau," Deutsche Akademie der Luftfahrtforschung, Schr. 1091/45 gKdos, 1945, p. 15.

[18] J. Ethell, A. Price, "The German Jets in Combat," Jane's Publishing Company, London, 1979, p. 28.

[19] G. S. Schairer, "Miscellaneous German Interrogations (Multhopp)," Boeing Doc. No. D-7076, Oct. 1945, p. 5.

[20] H. Müller, "Testflüge mit hohem Risiko," Klassiker der Luftfahrt, Heft 3/2004, p. 40.

[21] M. Jurleit, "Strahljäger Me 262 im Einsatz," Transpress Verlagsgesellschaft mbH, Berlin, 1993, p. 10.

[22] 50 Jahre Turbostrahlflug, DGLR Symposium, Deutschen Museum, München, Oct. 26–27, 1989, DGLR-Bericht 89–5, 1989.

[23] K. H. Doetsch, "Deutsche Luftfahrtforscher nach 1945 in England," Die Tätigkeit deutscher Luftfahrtingenieure und wissenschaftler im Ausland nach 1945, Blätter zur Geschichte der Deutschen Luft- und Raumfahrt V, DGLR Bonn Bad-Godesberg, 1992, p. 183.

[24] W. Georgii, "Einladung der Forschungsführung an Vertreter aus Forschung und Industrie vom 14.6.44," Bundesarchiv-Militärarchiv Freiburg, RL 39/466.

[25] J. R. Hansen, "Engineer in Charge, A History of the Langley Aeronautical Laboratory 1917–1958," NASA Scientific and Technical Information Office, Washington, DC, U.S., 1987, p. 271.

[26] R. Smelt, "A Critical Review of German Research on High-Speed Airflow," Journal of the Royal Aeronautical Society, No. 432, Vol. L (50), Dec. 1946, p. 914.

[27] W. Bergholz, "Russlands große Flugzeugbauer," AVIATIC Verlag Oberhaching, 1992, p. 167.

[28] B. A. Belyakov, J. Marmain, "MiG—Fifty Years of Secret Aircraft Design, Airlife Publishing Ltd., Shrewsbury, 1994, p. 81.

[29] T. Cebeci, "Legacy of a Gentle Genius, The Life of A. M. O. Smith," Horizons Publishing, Long Beach, CA, U.S., 1999, p. 35.

[30] A. Scherer, D. Küchemann, "Integration eines Triebwerks (LA 1014) in einen Rumpf (LA 1014-c) mit zentralem Triebwerk u. zwei seitlichen horizontal liegenden Einläufen (LA 1014-c-III)," ZWB/UM-3205 bzw, AVA-45/AW/06, 25.1.45.

[31] "Belegung der Windkanäle ab 15.3.45," Aerodynamische Versuchsanstalt Göttingen, 1945, Bundesarchiv-Militärarchiv Freiburg, RL39/350.

[32] "Messerschmitt P 1101," Luftfahrt International, No. 7, Jan.–Feb. 1975, Publizistisches Archiv R. Pawlas, Nürnberg, 1975, p. 1060.

[33] K.H. Doetsch, "The Time-Vector Method for Lateral Stability Investigations," Royal Aircraft Establishment, Technical Report No. 67200, Aug. 1967, pp. 26, 27.

[34] J. Miller, "The X-Planes X-1 to X-29," Specialty Press Publishers and Wholesalers, Inc., Marine on St. Croix, MN, U.S., 1983, p. 57.

[35] R. P. Hallion, "On the Frontier, Flight Research at Dryden, 1946–1981," NASA Scientific and Technical Information Branch, Washington, DC, U.S., 1984, pp. 48–54.

[36] K. Chilstrom, P. Leary, "Test Flying at Old Wright Field," Westchester House Publications, Omaha, NE, U.S., 1995, p. 189.

[37] A. Lippisch, "Über die Entwicklung der schwanzlosen Flugzeuge," Schriften der Deutschen Akademie der Luftfahrtforschung, Heft 1064/43, Berlin, 1943, p. 28.

[38] Winter, "Kraft- und Druckverteilungsmessungen an einem schwanzlosen Flugzeug," AVA Bericht 39/18, April 4, 1939.

[39] R. Seiferth, "Untersuchungen eines Gesamtmodells mit Pfeilflügeln (P-01), AVA-Bericht 41/14/22," Sept. 23, 1941.

[40] W. Späte, P. Bateson, "Messerschmitt Me 163 Komet, Profile No. 225," Profile Publications Ltd., Windsor, p. 68.

[41] H. Lindemann, "Hochgeschwindigkeitsmessungen am Modell des schwanzlosen Flugzeuges Me 163B mit ungeschränkten Flügeln," Deutsche Luftfahrtforschung UM 1398, Oct. 30, 1944.

[42] "Beauftragung der AVA durch die Messerschmitt AG vom 16.4.1943 (Auftrag J1304V) und Messplan vom," June 19, 1943, DLR-Archiv Göttingen.

[43] Th. Schwenk, "Windkanalmessungen an einem Pfeilflugzeug Me 263," AVA Bericht 41/14/32 vom Nov. 28, 1942.

[44] A. Lippisch, "Ein Dreieck fliegt. Die Entwicklung der Delta-Flugzeuge bis 1945," Motorbuch Verlag, Stuttgart, 1976, p. 91.

[45] D. A. Sobolew, "Deutsche Spuren in der Geschichte der sowjetischen Luftfahrt," Aviatik Verlag, Moskau, 1996, p. 8.

[46] "Luftfahrt International, No. 9, May/June 1975, p. 1325.

[47] C. Burnet, "Three Centuries to Concorde," Mech. Eng. Publ. Ltd., London, 1979, p. 99.

[48] G. R. Pape, J. M. Campbell, "Northrop Flying Wings: A History of Jack Northrop's Visionary Aircraft," Schiffer Publishing Ltd., Atglen, PA, U.S., 1995, p. 196.

[49] R. A. Belyakov, J. Marmain, "MiG Fifty Years of Secret Aircraft Design," Airlife Publishing Ltd., 1994, p. 107.

[50] W. Wagner, "Hugo Junkers - Pionier der Luftfahrt-seine Flugzeuge," Die deutsche Luftfahrt, Band 24, Bernard & Graefe Verlag, Bonn, 1996, p. 525.

[51] H. Lommel, "Junkers Ju 287, Der erste Jet-Bomber der Welt und weitere Pfeilflügelprojekte," AVIATIC Verlag GmbH, Oberhaching, 2003, p. 96.

[52] W. Eisenmann, "Das Seitenstabilitätsverhalten eines Flugzeugs mit Pfeilflügeln," MAP Völkenrode, MAP-V. 589R, AGD Nr. 1039/G, April 1946.

[53] Th. Zobel, "Grundsätzlich neue Wege zur Leistungssteigerung von Schnellflugzeugen," Schriften der Deutschen Akademie der Luftfahrtforschung, Nr. 1079/44, 1944.

[54] W. Heinzerling, "Flügelpfeilung und Flächenregel, zwei grundlegende deutsche Patente der Flugzeugaerodynamik," Technische Universität Darmstadt, Festkolloquium zur 6, Verleihung des August-Euler-Luftfahrtpreises, May 29, 2002.

[55] S. Holzbauer, "Vorwärts gepfeilte Flügel," INTERAVIA 5, Jahrgang, No. 7, 1950, p. 381.

[56] H. Neppert, "Die Hansa-Jet-Story: Erinnerungen eines Aerodynamikers," Verlag Peter GmbH, Jesteburg, 1984.

[57] *L. S. Jones*, "U.S. Bombers 1928 to 1980, 3rd ed.," Aero Publishers Inc., 1980, p. 180.

[58] *R. Horten, P. F. Selinger*, "Nurflügel, Die Geschichte der Horten-Flugzeuge 1933–1960," H. Weishaupt Verlag, Graz, 1987; Dieses Buch liefert einen guten Überblick über die Entwicklung aller Horten-Nurflügelflugzeuge.

[59] *H. Lugner, W. Pinsker*, "Kurzbericht über Flugeigenschaftsmessungen am Muster Horten IX V-1," DVL G 167/2, July 7, 1944 (Deutsches Museum München).

[60] *Siegfried Günter*, "The Influence of Fuselage and Wing Sweepback on the Range and Speed of Turbo-Jet Airplanes," Monografie, Landsberg, July 1945, p. 27.

[61] *G. W. Heumann*, "Der fliegende Flügel," Flugrevue Heft, Feb. 2, 1959, pp. 27–30.

[62] *D. McRuer, D. Graham*, "Eighty Years of Flight Control: Triumphs and Pitfalls of the Systems Approach," Journal of Guidance and Control, Vol. 4, No. 4, July/August, 1981, p. 358.

[63] *F. Stoliker, B. Hoey, J. Armstrong*, "Flight Testing at Edwards," Flight Test Historical Foundation, 1996, p. 57.

[64] "Gegenwärtiger Stand und künftige Entwicklungsarbeit auf dem Gebiete des Schnellfluges mit Strahlantrieb," RLM LC 7/III, Oct. 10, 1938, Bundesarchiv-Militärarchiv Freiburg.

[65] *W. Georgii*, "Forschungsflugzeug für Hochgeschwindigkeit," Schreiben Fo-Fü, July 14, 1944, Bundesarchiv-Militärarchiv Freiburg RL39/466.

[66] *R. Schubert*, "Reisebericht über Besprechungen bei der DFS, OFO und DVL14," July 14, 1944. Bundesarchiv-Militärarchiv Freiburg RL39/466.

[67] *W. Georgii*, "Niederschrift über die Besprechung Flug bei großen Machzahlen am 30.6.1944 in Ainring," Bundesarchiv-Militärarchiv Freiburg, RL39/466.

[68] *B. Göthert*, "Messprogramm für Hochgeschwindigkeitsuntersuchungen an Pfeilflügeln sowie Pfeilflügeln mit Rumpf und Gondeln," Berlin-Adlershof, Oct. 28, 1944, Bundesarchiv-Militärarchiv Freiburg RL39/466.

[69] *H. Melkus*, "Messungen an Rotationskörpern im Hochgeschwindigkeitskanal der DVL," AVA 46/2/11, DLR Archiv Göttingen, GO-R 2202.

[70] *W. Haack*, "Geschoßformen kleinsten Wellenwiderstandes, Bericht über die Sitzung Widerstand und Stabilität von Geschoßkörpern," Oct. 9–10, 1941, Peenemünde, Lilienthal-Gesellschaft für Luftfahrtforschung, Bericht 139.

[71] *J. J. Green, J. L. Orr, R. D. Hiscocks*, "Bericht über den Besuch bei der Luftfahrtforschungsanstalt Hermann Göring," Völkenrode, 1945, Übersetzung im DLR Historischen Archiv, 1983, p. 102.

[72] *B. Ciesla*, "Top Secret, DFS 346," Überschall-Forschungsprojekt aus Deutschland, Flug Revue, Dec. 1997, pp. 84–87.

[73] *F. Kracht*, Entwurf eines Versuchsflugzeugs zur Untersuchung der flugmechanischen Vorgänge beim Durchgang durch $M = 1$, DFS, 1944 (Dieser Bericht ist nicht mehr verfügbar).

[74] *K. J. Kosminkow*, "Skyrockets mit dem roten Stern," Flieger Revue Edition No. 2, 1993, pp. 8–9.

[75] *Th. Zobel*, "Hochgeschwindigkeitskanal der Luftfahrtforschungsanstalt Hermann Göring, Braunschweig," Schriften der Deutschen Akademie der Luftfahrtforschung, Heft 1060/43.

[76] *P. Jordan*, "Aktenvermerk über den Besuch von Herrn Thielemann," Forschungs-Gemeinschaft Halle, Dec. 28, 1944, DLR Archiv Göttingen, GOAR 3131.

[77] *G. Küssner*, Brief, Mathematische Seminar der Universität Halle, Jan. 19, 1945, DLR-Archiv Göttingen, GOAR 3131.

[78] *P. Jordan*, "Berechnung der Luftkraftbeiwerte des Flügels mit Ruder," Brief an *Dr. Dietz*, TH Darmstadt, Feb. 26, 1945, DLR-Archiv Göttingen, GOAR 3131.

[79] *Harkinson, Ewans, Bratt*, eds., "Deutsche Forschungsanstalt für Segelflug," CIOS Target No. 25/7, Evaluation Report 92, Air Force Historical Research Agency, Maxwell Air Force Base, AL, U.S., June, 13, 1945.

[80] *P. Fischer, J. A. Davison,* "CIOS Evaluation Report," C.I.C. 75/145, Air Force Historical Research Agency, Maxwell Air Force Base, AL, U.S., July 8, 1945.

[81] *J. V. Becker,* "The High-Speed Frontier, Case Histories of Four NACA Programs, 1925–1950," NASA Scientific and Technical Information Branch, Washington, DC, U.S., 1980, p. 43.

[82] *H. Ludwieg,* "Pfeilflügel bei hohen Geschwindigkeiten," AVA, LGL-Bericht 127, 1940, pp. 44–52.

[83] *R. J. Francillon,* "McDonnell Douglas Aircraft since 1920," Putnam & Company Ltd., 1979, p. 438.

[84] *L. Bridgeman,* "Jane's all the Worlds Airfcraft 1947," Sampson Low, Marston & Company, Ltd., 1047, p. 63c.

[85] *O. Walchner,* "Windkanalmessungen am Flügel des Henschelgerätes 'Zitterrochen' bei Unter- und Überschallgeschwindigkeiten," AVA Bericht: 44/H/22, Oct. 24, 1944.

[86] *H. A. Wilson, J. C. Lovell,* "Full-Scale Investigation of the Maximum Lift and Flow Characteristics of an Airplane Having Approximately Triangular Plan Form," NACA RM No. L6K20, Feb. 12, 1947, p. 10.

[87] *H.-P. Dabrowski,* "Überschalljäger Lippisch P 13a und Versuchsgleiter DM-1," Podzun-Pallas-Verlag, Friedberg, 1986, p. 17.

[88] *R. P. Hallion,* "Lippisch, Gluhareff, Jones: The Emergence of the Delta Planform and the Origins of the Sweptwing in the United States," Aerospace Historian, March 1979.

[89] *J. C. Lovell, H. A. Wilson,* "Langley Full-Scale-Tunnel Investigation of Maximum Lift and Stability Characteristics of an Airplane Having Approximately Triangular Plan Form (DM-1 Glider)," NACA RM No. L7F16, Aug. 1947.

[90] *J. A. Shortal, B. Maggin,* "Effect of Sweepback and Aspect Ratio on Longitudinal Stability Characteristics of Wings at Low Speeds," NACA TN No. 1093, July 1946.

[91] *R. T. Jones,* "Properties of Low-Aspect-Ratio Pointed Wings at Speeds below and above the Speed of Sound," NACA Report No. 835, May 1945.

[92] *A. Busemann,* "Infinitesimale kegelige Überschallströmung," Schriften der Deutschen Akademie der Luftfahrtforschung, Band 7B, Heft 3, 1943, p. 116.

[93] *R. E. Bradley,* "The Birth of the Delta Wing," AAHS Journal, Vol. 48, No. 4, Winter 2003, pp. 242–260, this report contains detailed documentation of the development of the first American delta-winged aircraft Convair XF-92; see also *R. P. Hallion,* "Convair's Delta Alpha, Air Enthusiast Quarterly, No. 2, pp. 177–185).

[94] *F. A. Johnsen,* "The Green Book Explains X-Planes," Airpower, Nov. 2003, p. 56.

[95] *R. P. Hallion,* "Convair's Delta Alpha," Air Enthusiast Quarterly, No. 2, p. 178.

[96] *W. Trommsdorff,* "High-Velocity Free-Flying Ram-Jet Units (TR-Missiles)," ed. by *Th. Benecke, A. W. Quick,* "History of German Guided Missiles Development," AGARDOGraph No. 20, Verlag Appelhans & Co., Braunschweig, 1957, pp. 352–374.

[97] *R. Scherrer, W. R. Wimbrow,* "Wind-Tunnel Investigation at a *Mach* Number of 1.53 of an Airplane with a Triangular Wing," NACA RM No. A7J05, Jan. 1948.

[98] *L. F. Lawrence, J. L. Summers,* "Wind-Tunnel Investigation of a Tailless Triangular-Wing Fighter Aircraft at *Mach* Numbers from 0.5 to 1.5," NACA RM No. A9B16, June 1949.

[99] "Zukunfts–Luftfahrt im Zeichen des Dreiecks," Interavia, 8 Jahrgang, No. 1, 1953, p. 29–31.

[100] *D. Herwig, H. Rode,* "Luftwaffe Secret Projects, Ground Attack & Special Purpose Aircraft," Motorbuch verlag, Stuttgart, 2002, p. 165.

[101] "Lippisch Delta VI V2," Auftrag J 7087, Modellzeichnungen, DLR-Archiv Göttingen, GOAR 374.

[102] *J. R. Ewans,* "Brief an *Prof. Seifert*," Oct. 23, 1945, DLR-Archiv Göttingen, GOAR 331, Schriftwechsel Lippisch Delta IV.

[103] "Notiz über Modellzustand Lippisch Delta IV," DLR-Archiv Göttingen, GOAR 1997, Auftrag J 1311.

[104] "Messungen am Nurflügelflugzeug Delta IV-V2," AVA Bericht 45/W/3 (Auftrag J 1311), DLR-Archiv Göttingen, GOAR 2479.

[105] *E. H. Heinemann, R. Rausa, Ed Heinemann,* "Combat Aircraft Designer," United States Naval Institute, Annapolis, MD, U.S., 1980.

[106] *R. J. Francillon,* "McDonnell Douglas Aircraft since 1920, Putnam & Company, London, 1979, 474.

[107] *Apollo Milton Olin Smith,* abgekürzt AMO Smith.

[108] *N. M. Williams,* "The X-Rays: Douglas XF4D-1 Skyrays," Journal of the American Aviation Historical Society, Winter 1977, p. 243. In diesem Bericht wird die Entwicklung der XF4D ausführlich dargestellt.

[109] *J. A. Kranzhoff,* "Arado-Flugzeuge," Die deutsche Luftfahrt, Band 31, Bernard & Graefe Verlag, Bonn, 2002.

Chapter 7

SPECIAL FEATURES OF ANTIAIRCRAFT ROCKETS WITH SWEPT OR LOW-ASPECT-RATIO WINGS

BERND KRAG

7.1 INTRODUCTION

An antiaircraft rocket, or in today's language, a surface-to-air missile, is designed according to aspects other than an aircraft. The antiaircraft rocket takes off from a launcher with a high initial acceleration. A separable launch stage or "booster" accelerates the rocket to a speed where the aerodynamic control becomes effective. The rocket either follows a guide beam or receives the control signals via onboard radar. During flight, the rocket must also be able to follow a maneuvering high-speed target and needs, therefore, an effective control. In following the target, the rocket is exposed to very high loads.

7.1.1 ROCKET DESIGN

In contrast to an aircraft, the antiaircraft rocket is of a symmetric design. Wings and horizontal stabilizer are in the case of a "horizontal-wing configuration" located within one plane. The rocket is only controlled by ailerons and elevators ("control in polar coordinates"). In case of a "cross-wing configuration" wings and stabilizers are arranged crosswise. The cross-wing configuration has the advantage that the rocket needs for a change of direction not to be rotated about the roll axis to use the lift vector for lateral accelerations ("control in Cartesian coordinates"). Thanks to its wing and stabilizer geometry, unnecessary delays due to activating the rolling degree of freedom will be avoided thus reducing the response time in case of course commands. Direct lift and side force control ensure that the rocket can quickly be guided to its target. As in the case of an aircraft, the swept wing results for rockets that operate at high subsonic speeds in a reduction of the transonic drag. For rockets operating in the supersonic speed regime, which is today exclusively the case, low-aspect-ratio wings (delta wings or trapezoidal wings) are best suited. Although for the development of antiaircraft rockets the swept wing was less important than for aircraft, rockets will be discussed here because only little has been reported in the relevant literature concerning aerodynamic layout, stabilization, and controllability up to now.

Everything that we know today had to be compiled by the engineers at the beginning of the 1940s. For the antiaircraft rockets, as they were called at the time, no models existed. This concerned aerodynamics, stabilization, propulsion, remote control, and automatic control by target-seeking sensors respectively (radio, acoustics, or infrared sensors).

The guarantee of adequate stability within all operational regimes was one of the hardest problems the engineers had to cope with. The adherence to natural stability is in the case of rockets difficult because

- They have to cruise within a large *Mach* number range.
- The center-of-gravity location changes rapidly due to the high fuel consumption.
- The configuration changes after the separation of the booster.

There existed no automatic trim or stabilization systems for antiaircraft rockets 60 years ago. Also the aimer operating the remote control system was not able to trim the rocket via remote control. Therefore, the greatest importance was, at that time, attached to configurations that were within the complete operational range insensitive to *Mach* number effects and caused only little change in the moments when crossing the transonic regime.

The controllability of the rocket was of the same importance as the stability. The aerodynamic forces and moments, necessary for the pursuit of targets, were produced by the control surfaces. The latter thus had to keep their effectiveness within the entire speed regime. The stability question was often more important than the minimization of the drag for the aerodynamic layout. The majority of the wind-tunnel measurements were dedicated to the solution of this problem. Of concern were the choice of the wing, the correct wing arrangement, the movement of the center of gravity relative to the center of pressure, and the arrangement of the control surfaces. Because of the urgency and because the problem was hardly amenable to a theoretical treatment, many measurements and experiments were carried out. During flight trials, numerous rockets were lost due to stability problems. These difficulties were never satisfactorily solved for any of the rockets built up to the end of the war. No single rocket was tested against an airborne target.

Even more problematic than the stability question were the guidance procedures that were to bring the rocket into the proximity of the target. It was quickly recognized that automatic guidance by a radio beam was the best method to steer the rocket into the target. The problem was in principle not the generation of a radio beam, but the evaluation of the signals within the rocket and the extraction of the control commands necessary to stabilize the rocket on the guide beam. Here, a start has been made, but, most of all, the theoretical fundamentals had been worked out. At the end of the war, it was still far away from a realization.

The guidance procedure most easily achieved is the manual control by sight. This control method was the only procedure that could be used at the time in testing antiaircraft rockets. It was already clear to the development engineers that they would not go far with the manual guidance based on the "line-of-sight" method. The complete sequence of operation from the target recognition to the launch of the rocket and the guidance into the target had to proceed fully automatic. All automatic guidance procedures under development were, at the end of the war, still far away from a practical application. An excellent survey concerning the efforts of the German radio industry and research to develop practicable radio-guidance methods is provided by *Fritz Trenkle* [1]. The report describes how extremely complex the field of automatic guidance was and the effort that was required for the design and development. For all of the tasks the German armament industry had to deal with during the last years of war, the experts required for the development and manufacture of such complex systems were simply not available. The multitude of the possibilities investigated at the time and the solutions, nevertheless, is astonishing from today's point of view.

7.1.2 ANTIAIRCRAFT ARTILLERY PROGRAM—INITIAL DEVELOPMENT

Soon after the start of the war in 1941, the German General Staff posed the question whether a remotely controlled antiaircraft rocket was to be preferred for defense purposes [2]. It was clearly recognized that aircraft would fly at increasingly higher altitudes and at higher speeds. This meant for the artillery a larger lead time, longer projectile flight durations, and hence reduced certainty of hitting the target.

A new antiaircraft artillery program, set up in April 1942, was, however, only approved by the Commander in Chief of the Air Force, *H. Göring*, on September 1, 1942. Within this antiaircraft artillery program, the guidelines for the antiaircraft artillery rocket development were fixed [3]. The antiaircraft artillery development office (GL/Flak E) of the Reichsluftfahrtministerium (RLM) (German Air Ministry) was entrusted with the supervision of the developments ([2], p. 176).

Regarding the development of antiaircraft rockets, this program envisaged:

1) Development of an (uncontrolled) solid-fuel rocket suitable for use with established command equipment (predevelopment stage).

2) Development of a rocket controlled by sight (interim solution within the overall development; final solution for the use with the antiaircraft field artillery).

3) Development of an interference-free remote-controlled antiaircraft rocket with the capability to automatically home in on a target during final approach and detonate itself in the vicinity of the target.

Concerning the tactical/technical requirements on the development of these antiaircraft artillery rockets of September 25, 1942, four projects in the development were envisaged [4]:

- Project 1—The uncontrolled antiaircraft rocket.
- Project 2—The target-seeking antiaircraft rocket.
- Project 3—The antiaircraft rocket remotely (radio) controlled on sight.
- Project 4—The antiaircraft rocket remotely controlled by electrically measured data.

The uncontrolled antiaircraft artillery rocket (R-Fla 42) was to target aircraft at altitudes from 700 to 7000 m. Emphasis was placed on simple handling qualities and the use of established launchers and gun directors. The development of this rocket had already commenced in early 1942 at the companies Rheinmetall and Krupp.

The target-seeking antiaircraft rocket was supposed to be able to aim at targets at slant ranges of 7000 m up to altitudes of 20,000 m at horizontal distances of 20,000 m and target speeds of up to 300 m/s. Initially, a rocket with solid-fuel propulsion was considered. At ranges of 3000–4000 m, the rocket was automatically guided by an automatic target homing device. A proximity fuse was to detonate the warhead in the proximity of the target. Later, this was changed to a radio-controlled remote fuse. Based on radio readings (radar readings) a remote ignition was to take place when rocket and target were at the same distance.

The rocket took off from a rotatable launcher, which was remotely controlled by means of a command device and lined up towards the target. A reloading time of 60–75 seconds was planned! The command device obtained its initial data by radar. The rocket, remotely controlled by sight, was envisaged for operation at medium altitudes between 1000 to 7000 m. Targets were to be hit up to a slant range of 10,000 m and target speeds of 300 m/s. A proximity fuse was also envisaged for this rocket [4]. The launch was carried out from a launcher, which had to be aligned manually. The remote control by sight was carried out via a control stick similar to the control already used in the case of the glide bomb Henschel Hs 293. As a director to pursue the target and the rocket, a double telescope was envisaged that was adjusted by radar. The tactical/technical specifications mentioned in this context "Project Henschel" is a reference to the antiaircraft rocket Hs 117 "Schmetterling (Butterfly)."

The last project, the rocket remotely controlled according to electrical readings, was to be used as object protection against attacks from high-speed, high-altitude aircraft. A liquid-fuel rocket was planned. It was to hit targets at slant ranges above 10,000 m, at altitudes of up to 20,000 m, and at horizontal distances of up to 50,000 m. This rocket was to be controlled by a radio beam directed towards the target. In the vicinity of the target, it switched to an automatic target-seeking control. The guide beam of the radio transmitter of the rocket was to be automatically aimed at the target by radar after the launch.

7.1.3 DEVELOPMENT CONTINUES

The responsibility for the antiaircraft rocket development was taken over by the Department L, Artillery, at the RLM. The Army Weapons Office (Wa A) at the Oberkommando des Heeres (OKH) (Supreme Command of the Army) was to look after Projects 1, 2, and 4, while the departments G.L./C-E at the RLM were to take care of the development of the radio beam and the director equipment as well as of Project 3.

After examining the tactical/technical requirements, the Head of the Army Armament (WaPrüf 11), the division responsible for rocket research and development, came to the conclusion that Project 2, the target-seeking anti-aircraft rocket (a predevelopment step to Project 4), was far too expensive, contained considerable risks, and should, therefore, not be further pursued [5]. Also, Project 4 would be associated with considerable effort and only achievable in several steps. The end of the development would then be a liquid-fuel rocket with a thrust of 6 to 8 tons.

As an intermediate step, Wa Pruef 11 in particular suggested, in order to test the control procedures, to examine a guide-beam controlled two-stage solid-fuel rocket with 100-kg payload, which needed, at first, not to have the performance of the final device. This rocket was to be launched from an adjustable launcher and then be controlled by means of a "Würzburg C" radar into the target. The utilization of a booster permitted high initial accelerations, which greatly eased the diagonal launch from a launcher. A liquid rocket had, however, to take off vertically and would have required a second radar to direct the projectile onto the guide beam proper.

7.1.4 AVAILABLE TECHNOLOGY IN GERMANY

As can be seen, the Army Weapons Office expensed, in a very short time, considerable thought on the development and the technical possibilities of rockets. The beginning of the development of this new weapon in early 1943 fell into a time, however, when the aeronautical industry was already employed to full capacity and had no more resources for the development of completely new equipment. The initially demanded 600 experts, required for the development, had to be brought back from the front, which caused the greatest difficulties ([2], p. 176). Only 80% of the needed personnel had arrived in the autumn of 1943. Another difficulty arose from the provision of the required material. We shall here, however, not describe the history of the development of the antiaircraft artillery rocket, but rather deal with the technical problems with which the engineers were, at the time, confronted.

What was available at the end of 1942? At that time, the radar technology already had a high development status. The standard radar equipment of the type "Würzburg" could locate a target and deliver good data to the antiaircraft artillery, but the accuracy did not suffice to guide a rocket on a radio beam. The radar device FuSE 65 "Würzburg Riese" had an angular accuracy of ± 0.2 deg, which corresponded at a distance of 10 km to a deviation of ± 35 m [6]. The distance error was of the same order of magnitude. Because there existed at the time on the German side no automatic adjustment of the radar antenna—an important prerequisite for an automatic tracking—the radar antennas for the determination of the target and rocket locations had to be adjusted manually, which led to another enlargement of the measurement error. Only by the centimetric wavelength technology, which was in Germany only available towards the end of the war, could the measurement errors be reduced to a few meters.

For remote control, one could fall back on the radio-control method, "Kehl/Strassburg," developed by the German radio industry and the research establishments. This radio-control method was intended for the glide bombs PC 1400X ("Fritz X") of the DVL and the Henschel Hs 293. The radio-control system consisted of the transmitter "Kehl," the receiver "Strassburg", and a command module with a control stick, ([1], pp. 30–53). The radio-control method "Kehl/Strassburg" permitted the control about two axes according to the "line-of-sight method." For this method the aimer had to guide the rocket (missile) so that it was always located on the line of sight between aimer and target. The control of the projectile was carried out either in "Cartesian coordinates" (rudder and elevator) or in "polar coordinates" (aileron and elevator). So "Kehl/Strassburg" could serve as a starting point for the manual antiaircraft rocket control. Liquid- and solid-fuel rocket propulsion for different applications was under development at the time. The most powerful rocket engine was being developed at the Heeresversuchsanstalt (HVA) (Army Research Establishment) in Peenemünde for the long-range missile A4. Sufficient know-how was hence available and had the usual exaggerated secrecy not obstructed a free exchange of information.

For aerodynamics, there were measurements on projectiles up to high supersonic speeds being carried out. For antiaircraft artillery rockets, which were to operate within the supersonic speed range, guidelines, therefore, existed with regard to the design of rocket bodies. In this field, many investigations had been carried out at Peenemünde for the A4, which could be used. Also of importance was the knowledge of the aerodynamicists, namely, that wing sweep reduces drag at high speeds or shifts the occurrence of undesirable compressibility effects to higher *Mach* numbers. Wings of a very low aspect ratio have a similar effect and are, therefore, well suited for antiaircraft rockets. Low-aspect-ratio wings have the advantage that they can be built with a very high stiffness and are not as flutter-prone as swept wings.

7.1.5 AVAILABLE TECHNOLOGY INTERNATIONALLY

Germany's former adversaries, the United States and Great Britain, have since 1943 a considerable lead over Germany in the development of the radar technology. This lead also concerned the automatic guidance of missiles by means of passive and active target homing devices and proximity fuses. Also with regard to the design of missiles for supersonic flight, they were not necessarily dependent on German knowledge. During WWII, the United States had started with the development of advanced antiaircraft rockets. At the end of the war, work had already progressed far enough so that information about German rockets could not have exerted any influence worth mentioning on the developments. This explains perhaps why the German surface-to-air missiles and air-to-air rockets did not become the starting points of the developments in the United States.

7.1.6 CHAPTER SCOPE

This chapter does not describe the general development of individual antiaircraft artillery rockets, as pictured in Fig. 7.1; for that, see [8, 9]. Instead, this chapter examines the influence of high-speed aerodynamics on the design of these antiaircraft artillery rockets. We do not consider the air-to-air rockets, at that time called "pursuit rockets." There is too little information concerning the aerodynamic design of these. There were at end of the war only two projects whose development had reached the trial stage: Henschel Hs 298 and Ruhrstahl-Kramer X-4. Both rockets possessed swept wings and were manually controlled according to the "line-of-sight" method. Whether a target could effectively be hit under operational conditions is unknown. Homing devices, which allowed the rocket to automatically guide itself to the target, were at the initial stages of the development. It is not known whether these rockets served as a model for developments abroad.

7.2 ANTIAIRCRAFT ROCKETS "FEUERLILIE (FIRE LILY)" F-25 AND F-55 OF THE LFA BRAUNSCHWEIG

Because with the development of the antiaircraft rockets new technical ground was broken, the Technical Office of the RLM expected considerable problems during the development. To be able to answer basic questions concerning the aerodynamic design, remote control, stabilization, and launch procedures, the Research Department LC 1 on May 13, 1941, contracted the Luftfahrtforschungsanstalt Hermann Göring, Braunschweig (LFA) (Aeronautical Research Establishment) to develop and test an experimental rocket and to collect data for the antiaircraft rocket development [10]. The Department of Flight Mechanics of the Institute of Aerodynamics of the LFA was responsible for carrying out the work. The project manager for the development of the experimental rocket F 22 was *Gerhard Braun*. This work

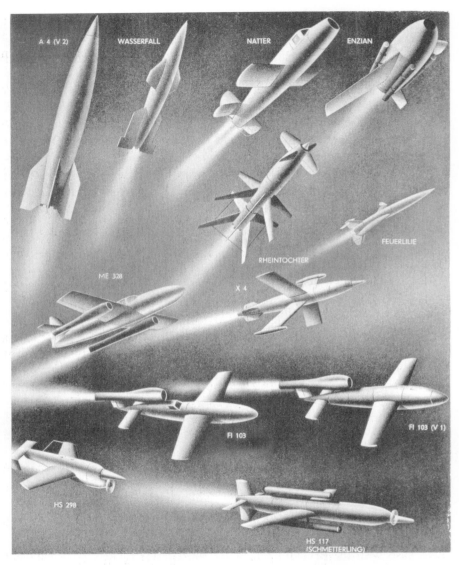

Fig. 7.1   German artillery rockets and missile developments.

proceeded in parallel to industrial developments that had partly been undertaken on their own initiative during the same year. *W. Wernitz* reports later
that the LFA development had been performed to be able to compare wind-
tunnel measurements with original high-speed test data [11]. With regard to
the purpose of the LFA development, there were obviously discrepancies
at the RLM. In a letter of February 2, 1943, the Research Directorate stated
that the LFA did not have the task of developing an operational antiaircraft

rocket [12]. Rather questions of control were to be examined, and problems of flight mechanics were to be solved. At least the Rheinmetall-Borsig Company had accompanied this development and was later, also involved in the evaluation of the test results.

It is not obvious from the report of G. *Braun* and P. *Retert* [10] on which specifications the experimental rocket F 22 was based. The highest demands with regard to the speed represented the self target-seeking rocket (Project 2). Supersonic speeds were demanded by such a rocket to be able to reach a target at the desired distance and altitude. It was, nevertheless, the wish of the LFA to venture with the projectile into the supersonic speed regime to be able to collect important data.

The layout of the F 22 (fuselage diameter 0.22 m) envisaged correspondingly a slender fuselage with a bullet-like nose. The nose formed an ogive with a 10-caliber curvature; it was copied off the long-range missile A4. It is likely that the missile "Hecht (Pike)," which was at the time under development at the LFA (see Fig. 7.2), served as a further model. Basic investigations were to be carried out with the "Hecht" concerning the stabilization and control of glide bombs.

No reference to wings is found in the description of the F 22. Because however, large side forces were required to keep the F 22 on the guide beam, it may be assumed that *Braun* had also planned wings for the projectile. The relatively simple stabilizers at the tail were dimensioned so that the static stability was ensured within the entire speed range. For propulsion, a solid-fuel rocket of Rheinmetall-Borsig was to be employed. The takeoff was carried out vertically from a launch tower. The rocket was guided by a 20-m long ramp. When leaving the launch tower, the rocket was expected to reach a speed of 4 m/s, sufficient for an aerodynamic stabilization by stabilizer fins.

In 1941, further considerations and configuration studies led to the model F 25 with a fuselage diameter of 0.25 m and conventional wings. The name "Feuerlilie (Fire Lily)" was mentioned for the first time in the last quarterly report of the Institute of Aerodynamics that year. The work commenced in the fall of 1941 with wind-tunnel measurements. The determination of the main load cases and the design of an airworthy version were completed in the summer of 1942. The manufacture of the rocket was carried out at the Ardelt Works in Breslau.

It was clear from the start that the selected configuration must reach supersonic speeds. Therefore, a swept wing with a 45-deg sweep angle (at the

Fig. 7.2 Wingless drop model of the missile "Hecht (Pike)" (1942).

25% chord line) was chosen. This is probably the first practical application of
a swept wing with the objective to reduce transonic drag. As airfoil a
symmetrical airfoil section with 9% thickness and 30% maximum-thickness
location was selected. A cross wing was deliberately foregone to save weight
and reduce drag. As stabilizers (empennage), double horizontal and vertical
stabilizer fins were selected. The rocket was built up completely symmetric
and had, therefore, at positive and negative angles of attack respectively yaw
angles, the same flight characteristics. During the vertical ascent, the rocket
was to be stabilized about the roll axis. In this way, a roll angle was prevented
from building up due to manufacturing asymmetries or perturbations, which
would have led to deviations from a straight trajectory. As a roll stabilizer, the
same device that had been developed for the missile "Hecht" was used [13].
The stabilization gyroscope was together with a pressure vessel contained
within the bow; the supporting ailerons for the roll stabilizer were at the outer
wing. The associated magnetic actuators were contained in an elliptical
housing at the wing tips. The first experimental versions of the "Feuerlilie"
F 25, which were still tested without remote control, did not have any rudder,
but only the upper horizontal stabilizer was equipped with a control surface.
To be able to adhere without remote control to a straight trajectory, some of
the rockets were equipped with an elevator control program. As propulsion, a
solid-fuel rocket, Type Rheinmetall RI 502, was chosen. This rocket had been
developed as a takeoff aid (booster) for large aircraft. It had two exhaust
nozzles with a pressure-compensation valve in between. The pressure-
compensation valve was to compensate pressure fluctuations due to an irregular combustion. Figure 7.3 shows the layout of the "Feuerlilie" F 25, and
Table 7.1 lists technical data.

**Fig. 7.3   Three-view arrangement of
the "Feuerlilie" F 25.**

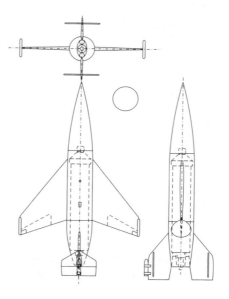

TABLE 7.1   "Feuerlilie" F 25 Technical Data

Length	2.0 m
Span	1.20 m
Fuselage diameter	0.25 m
Mass	115 kg
Propulsion	Rheinmetall RI 502 with 5000 N thrust
Range/altitude	10,000 m (set point)
Wing area	0.37 m²
Aspect ratio	1.05
Sweep	45 deg (25% chord line)
Airfoil section (wing)	NACA 0009 with 30% maximum-thickness location
Airfoil sections (stabilizers)	NACA 0006

Parallel to the development of the missile, investigations were carried out concerning the guide-beam control. These investigations commenced with the simplest case, the vertical flight path (ascent). Important here were questions concerning the stability, controllability, and the correct control-surface signals to keep the rocket on the guide beam. Computations showed that a horizontal-wing configuration can easily be stabilized and controlled by a guide beam [14]. This result certainly encouraged *Braun* to also design the further development of the F 25 not as cross-wing configuration but as horizontal-wing configuration.

As it was during the development of the "Hecht," in the case of the "Feuerlilie," flight tests with small models were to provide information about the stability behavior to be expected at different center-of-gravity positions. Small models with a fuselage diameter of 4.4 cm were manufactured of wood and cardboard and equipped with small powder-rockets. Figure 7.4 shows the construction of these miniature rockets. The first launches in the spring of 1942 did not produce any results because the rockets exploded prematurely. Thereafter further rockets were ordered and installed in metal casings. The launches now carried out during the summer of 1942 were successful and showed negligible effect on stability for changes in the center-of-gravity positions within ±10%.

During the last quarter of 1942, all design documents were ready and submitted to the Ardelt Works. The roll stabilizer was tested, and the elevator-control program and its drive (two antiaircraft artillery ignition clockwork motors) were ready. Furthermore the recovery system had been tested by the dropping of two F 25 fuselages. Three missiles were completed during the spring of 1943.

A higher priority (DE) was assigned to the project "Feuerlilie" by the RLM in the summer of 1943. This project consequently claimed the largest portion of the staff of the Department AF. The first completed F 25 was tested in the

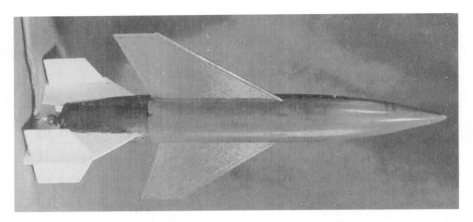

Fig. 7.4   F 25 model rocket for investigation of the stability behavior.

windtunnel A1 (see Fig. 7.5) to check the roll stabilizer, which showed a faultless operation, so that nothing stood in the way of the first launch. Interesting here is the testing technique applied. The model was suspended within the test section to freely rotate about the roll axis. After applying a perturbation, the roll stabilizer was to bring the model back into the initial position.

The first two launches on June 1, 1943, at the Test Center Peenemünde West were failures because the roll stabilizer malfunctioned. After increasing the spin of the stabilizer gyroscope and its range of precession, the stabilizer worked during the third launch faultlessly. All launches were carried out vertically from a launch tower. The rockets themselves were not instrumented. The flight trajectories were recorded by means of cine-theodolites. To facilitate the tracking of the rockets, a magnesium flare was mounted on the vertical stabilizer and ignited during the launch. The evaluation was carried out in cooperation with the Rheinmetall-Borsig Company, which had experience

Fig.   7.5 Wind-tunnel model of the "Feuerlilie" F 25 in the subsonic wind tunnel A1 of the LFA Braunschweig (1942).

with this method from dropping bomb-shaped bodies. To be able to determine the desired time-dependent drag and lift, the speed and acceleration dependence had to be generated from the trajectory by differentiation. Another four launches were carried out with an additional damping of the roll gyroscope on July 22, 1943. Further investigations dealt with the effect of the burn time and the momentum on the climb performance. The best climb performance yielded a burn time as short as possible at a constant total momentum. Eight wingless F 25 missiles were prepared to prove the validity of these considerations.

The flight tests were continued during the second half of 1943 with another two launches. The last launch was performed with an activated elevator-control program. Another three F 25 missiles were ready to be dropped by an aircraft. With these, *Mach* 1 was to be reached. The bad weather during the winter of 1943–1944 prevented, however, any test activity. During spring 1944, another three program-controlled F 25 missiles were ready to be launched from the ground and, furthermore, four wingless F 25 missiles for burn-time tests. The successful launches of the last year were analyzed by the evaluation of cine-theodolite measurements as described. The latter had, in the meantime, been simplified and improved. It had been hoped to reach altitudes of up to 10,000 m. The results of the data evaluations showed, however, that the rockets had reached maximum altitudes of only 4800 m at the most (information is uncertain).

The launches of the program-controlled F 25 missiles from the ground occurred during May–June 1944 at the test center Leba in Pomerania. All three launches were failures: at the first launch the roll stabilizers; at the second launch the elevator-control program failed; and at the third launch, the control program started too early, and the missile overturned. With the launches of four wingless F 25s during May 11–13, 1944, the theoretical considerations of the effect of the burn time could be confirmed. The three missiles that were provided for the high-speed tests were also dropped from a Junkers Ju 86 on June 19, 1944. The drops occurred from altitudes between 2000 and 6000 m at a flight speed of 440 km/h. The elevators were set for a lift coefficient of $C_L = 0.06$, which corresponded to a very steep trajectory angle. At the first drop, the rocket did not ignite. The two other drops were successful; the roll stabilizer was able to keep the rockets on a straight trajectory. The tracking with the cine-theodolite proved, however, to be problematic because the flare could hardly be recognized [15]. A brief evaluation showed that probably not more than 220 m/s had been reached. Not all F 25 rockets were launched. One specimen was discovered by British troops in a shed near Bad Harzburg in June 1945 and shipped to Britain [16]. This rocket is now at the Manchester Museum of Science and Industry.

The tests with the "Feuerlilie (Fire Lily)" F 25 were discontinued in autumn 1944 because an improved version of the "Feuerlilie," the F 55 (Table 7.2),

TABLE 7.2   "FEUERLILIE" F 55 TECHNICAL DATA

Length	4.80 m
Length with booster	6.115 m
Span	2.37 m
Fuselage diameter	0.55 m
Wing sweep	55 deg (leading edge)
Wing airfoil section	NACA 0009 with 30% maximum-thickness location
Wing area	1.7 6 m²
Aspect ratio	1.0
Mass	500 to 650 kg dependent on design
Propulsion	4 Rheinmetall RI 502 solid-fuel rockets with a total thrust of 60,000 N, or a liquid-fuel rocket *Lutz/Noeggerath* with a thrust of 64,000 N
Booster	Rheinmetall "Pirat," thrust 100,000 N
Maximum *Mach* number	$M = 0.85$ to $M = 1.25$
Range/altitude	10,000 m (planned)

was already under development. The designation F 55 derives again from the fuselage diameter of 55 cm. The rocket was substantially larger than the F 25 and was to reach supersonic speeds. As propulsion, four solid-fuel rockets, Type Rheinmetall RI 503, or one liquid-fuel rocket, developed by *O. Lutz* and *W. Noeggerath*, were considered.

Figure 7.6 shows a side view of the "Feuerlilie" F 55. The rocket was a "tailless" configuration, meaning there was no separate horizontal stabilizer. The wing had a leading-edge sweep of 55 deg with the airfoil section corresponding to a NACA 0009 airfoil with a maximum-thickness location of 30% chord. The wing trailing-edge flaps served on the inner wing as elevators and on the outer wing as ailerons. During the launch tests, the inner flaps were fixed. Similar to the F 25, the vertical stabilizers were attached at the wing tips. These had a small toe-in angle with respect to the fuselage, and each contained a magnesium flare. A Rheinmetall "Pirat" booster could, in addition, be attached. As a launcher, a modified 8.8-cm antiaircraft gun mount was used. The rocket was completely built at the workshops of the LFA.

Work commenced in autumn 1942 on the construction of wind-tunnel models. Different wing locations were tested in the A1. On a smaller, 1:25 scale model, three-component measurements up to $M = 0.95$ were carried out in the high-speed wind-tunnel A2 of the LFA. Further measurements were carried out on a model with boosters in the tunnel A1 during summer 1943 (see Fig. 7.7). After load assumptions had been established and strength calculations were available, design work on the full-scale configuration commenced in summer 1943. In addition to further wind-tunnel measurements and design work, the investigations concerning the guide-beam control were

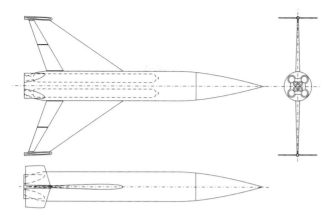

Fig. 7.6 Three-view drawing of the "Feuerlilie" F 55.

extended to the F 55 with the choice of the control parameters again having priority.

The first launch of the "Feuerlilie" F 55 occurred on May 12, 1944, at the test center Leba in Pomerania still without roll stabilizer and elevator program. Figure 7.8 shows the "Feuerlilie" on the launch ramp. The evaluation of the trajectory yielded an altitude of 4800 m (4550 m had been predicted) and a speed of 412 m/s corresponding to $M = 1.25$. The determination of the lift and drag coefficients from the measurements yielded a satisfactory agreement with calculations. Figure 7.9 shows the calculated and measured drag coefficients C_D dependent on the *Mach* number [17]. A second launch with liquid-fuel propulsion and boosters took place on October 19, 1944, at the antiaircraft artillery test ground Karlshagen, Usedom (island of Greifswald Oie) [18]. For this test, the roll stabilizer was activated, the elevator program, however, still turned off. The launch was carried out under an angle of 80 deg.

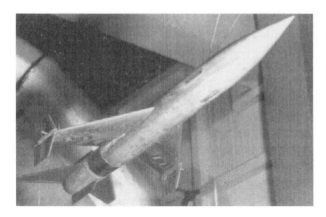

Fig. 7.7 Wind-tunnel model of the "Feuerlilie" F 55 with booster in the A1 wind-tunnel of the LFA (1943).

Fig. 7.8 "Feuerlilie" F 55 on launcher at
Leba, Pomerania in 1944 (Griehl).

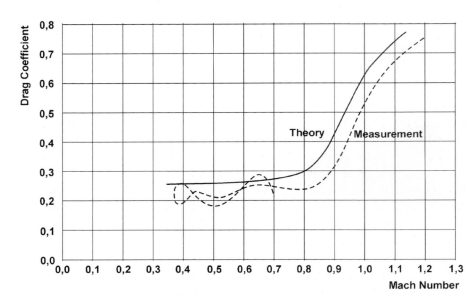

Fig. 7.9   Results of the first flight of the F 55 of May 12, 1944 [1].

The rocket crashed shortly after leaving the launch ramp. In the questioning by British experts of the Combined Intelligence Objectives Subcommittee (CIOS) [19], *Braun* names the static instability of the rocket in the configuration with boosters as cause. The mass of the booster led to a very strong rearward shift of the center of gravity. In addition the booster did not have any stabilizing fins as pictures of the launch pad at Greifswalder Oie show. *Braun* was, however, of the opinion that this instability would not have any effect during the very short launch phase with booster. The instability and the relatively large inclination of 20 deg with respect to the vertical at the launch were, according to *Braun*, the reason why the rocket could not stabilize itself anymore after the separation of the booster. A third test, scheduled for the beginning of 1945, was not carried out, because the rocket had been destroyed during an air raid on Peenemünde. Still during summer 1944 a missile, designated F 55 E, with a horizontal fin located near the nose was prepared for wind-tunnel measurements. The forward horizontal stabilizer ("canard") was to lead at supersonic speeds to an improved controllability because separations on the wing at high speeds flow due to compressibility effects were feared. In the meantime, it was probably also clear to *Braun* that an elevator in case of a low-aspect-ratio tailless configuration and small lever arm has to generate considerably larger moments than a control surface further away from the center of gravity. Either a biconvex airfoil section or the NACA 0009 airfoil with a larger maximum-thickness location were planned for the wing. The program "Feuerlilie" F 55 was discontinued in February 1945 on the instruction of the Research Directorate.

Because all results of wind-tunnel measurements, flight tests, and theoretical investigations concerning stability and guide-beam control had been published, these results were presumably also known to industry and the sister establishments. Rheinmetall had surely profited most from the cooperation. It is not clear, however, which of the Rheinmetall developments were affected by the results achieved at the LFA (see Sec. 7.3.1).

Looking back, it can be said that the development of such a complex device as an antiaircraft rocket was too much for the LFA. There was no experience that could be built on, and the staff level was far too low. At the "Flight Mechanics" Department of the Institute of Aerodynamics, no more then seven employees had worked simultaneously on this task. There was hardly any support by the industry. Equipment, like transmitters and receivers for the remote control or gyroscope stabilization equipment, was not available. These were reserved for the industry. The LFA was forced to develop and build all important equipment itself, which required an enormous amount of time. The "Feuerlilie" led as an antiaircraft rocket to a dead end, as an experimental means to correlate wind-tunnel data and flight-test results, the "Feuerlilie" had, however, met its purpose. One of the "Feuerlilie" F 55 survived the war. It was found in the A3 by the British and shipped to Britain

Fig. 7.10 "Feuerlilie" F 55 at the Royal Air Force Museum at Cosford (courtesy of RAF Museum Cosford).

and is now part of the collection of the Royal Air Force Museum at Cosford (Fig. 7.10).

7.3 ANTIAIRCRAFT ROCKET DEVELOPMENT OF INDUSTRY

7.3.1 RHEINMETALL-BORSIG "RHEINTOCHTER (RHINE DAUGHTER)"

The antiaircraft artillery rocket "Rheintochter" is one of the rockets that were to operate, controlled by sight, at medium altitudes of up to 8000 m. The development of the antiaircraft artillery rocket "Rheintochter," has already been reported in detail (see [8], Chap. "Rheintochter"). The development is only here presented in a condensed form that will be complemented where new documents will provide additional insight. Emphasis will be placed, as in the case of the other rockets, on aerodynamics, stability, and control.

In a memorandum of the HVA Peenemünde of October 15, 1942, the design of an antiaircraft rocket with solid-fuel propulsion is sketched for the first time, which must be seen in conjunction with the "Rheintochter" [21]. Figure 7.11 shows a single-stage missile whose forward part carries adjustable control fins. With a thrust of 14.3 tons, distances of up to 15 km were to be reached at maximum flight speeds of 300 m/s. The sketch shows with the control surfaces located forward already essential features of the basic layout of the "Rheintochter."

On December 7, 1942, the Heereswaffenamt (WaPrüf 11, rocket research and development division) (HWA) (Army Weapons Office) contracted the Rheinmetall-Borsig Company in Berlin-Marienfelde with the development of an antiaircraft rocket with solid-fuel propulsion of high specific momentum. The starting point was to be the design suggested by the HVA of a

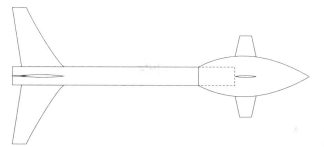

Fig. 7.11 First sketch of a solid-fuel rocket-driven antiaircraft rocket [2].

cross-wing configuration with a payload of 100 kg [22]. The antiaircraft rocket was to be designed for control by sight. Important questions concerning the control were, however, still open when placing the order. It was expected of the contractor that he should work out suitable proposals for a solution on his own. Of the four antiaircraft artillery rockets developed up to the end of the war in Germany, the "Rheintochter" was the only rocket with solid-fuel propulsion.

According to the design philosophy of Rheinmetall-Borsig, existing construction elements and methods in the design of the rocket should be used to reduce the development effort. This also concerned the launch procedure using equipment already introduced in the field, such as radar, gun directors, and gun mounts.

The rocket was to be manually guided into the proximity of the target (line-of-sight method) and then exploded by means of a distance-controlled fuse, that is, the fuse detonates the rocket when rocket and target have arrived at the same distance (coincidence) according to radar data. The Rheinmetall-Borsig Company employed a complete team of scientists for the development of the "Rheintochter," which was headed by *W. Fricke*.

For the rocket to reach the demanded combat altitude of 900 m in a vertical ascent, a launch stage with a very high specific momentum was envisaged. This launch stage was to accelerate the rocket to a speed of 300 m/s in only 0.6 s, which corresponded to an initial acceleration of 50 *g*. After dropping the launch stage, the second stage accelerated the rocket to a maximum speed of more than 400 m/s. Burnout of the second stage and hence the maximum *Mach* number of 1.3 were reached at an altitude of 3500 m. When reaching the target altitude, the speed dropped to 200 m/s. This procedure was necessary so that such altitudes could be reached despite the very short combustion time of the contemporary solid-fuel propulsion systems. This meant that the "Rheintochter" covered part of its flight range at supersonic speeds. This fact had, of course, a large effect on the aerodynamic design. Of all of the antiaircraft rockets, the "Rheintochter" had the highest initial acceleration. This was

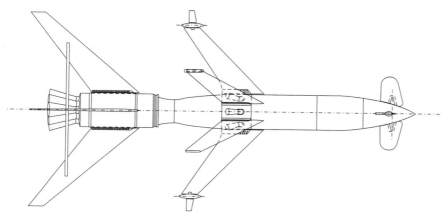

Fig. 7.12   Plan view of the antiaircraf t rocket "Rheintochter 1."

the reason why it was preferred for the testing of components of the antiair-craft rocket control (see [11], *K. H. Schirrmacher*, "Guidance of Surface-to-Air Missiles by Means of Radar").

The rocket "Rheintochter 1" (first development stage) consisted of a slender cylindrical fuselage with a long bullet-like tip. The six thrust nozzles of the cruise engine projected laterally from the rocket body (Fig. 7.12). The control surfaces were mounted at the tip of the rocket to keep them out of the exhaust jet. For flight within the transonic speed range, this arrangement had at the same time the advantage that the control surfaces could not be disturbed by compressibility effects on the wing. The control surfaces were vertically slightly staggered and controlled the pitch and yaw axes. Because of the large effective lever arms relative to the center of gravity, the control surfaces could be kept small. The control surfaces were adjustable as a whole with the axes of rotation having been located such that the control forces could be kept low throughout the entire speed range [23].

Decisive for the complete aerodynamic design was the guarantee of sufficient stability throughout the entire operational range. Here, wind-tunnel measurements at the AVA Göttingen were carried out up to $M = 1.45$. These measurements showed for the prescribed moment reference point a stable behavior within the complete *Mach* number range [24]. This meant for the design of the rocket that the center-of-gravity location was not allowed to change substantially during the combustion of the rocket fuel. Because of the high flight speeds, swept wings were chosen to avoid, if possible, compressibility effects. The sweep was 45 deg at the leading edge. Because the rocket crossed the transonic regime relatively quickly, Rheinmetall did not pay too much attention to the design of the wings. Six wooden swept wings were attached, evenly distributed about the circumfer-ence, close to the center of gravity. They did not have any particular airfoil sections. Because of the small chord, the aspect ratio was relatively large. The shift in the neutral-point location could, thus be kept within limits when crossing the transonic regime. The six wings provided, independent of the roll angle, sufficiently high lateral forces during maneuvering flight.

The "Rheintochter 1" was roll stabilized. The "ailerons" were attached at two opposite wing tips. They were small control surfaces that could be adjusted by electromagnets. The complete magnetic actuation system was accommodated in a streamlined housing. The roll stabilization was controlled by an attitude and a rate gyroscope with the latter providing roll damping. Rudder and elevator were activated by electric motors via gears that were directly attached below the control-surface shafts.

The first flight tests took place at the test center Leba in Pomerania in November 1943. In addition to the investigation of components and the remote control, the tests served to validate wind-tunnel measurements. The rockets went up to altitudes of 5000 m, reaching speeds of up to $M = 1.1$. To determine the lift and the normal force respectively, the rocket was launched on a helical trajectory. The normal force then resulted from the equilibrium with the centrifugal force. Because the rockets were not instrumented, the parameters drag and normal force coefficients had to be computed from cine-theodolite measurements. The values thus determined agreed in their *Mach* number dependence with the ones measured in the wind tunnel, although not in their order of magnitude. The difference can certainly be explained by the different *Reynolds* numbers of the wind-tunnel measurements and the flight test [25].

The flight tests had also shown that the effectiveness of the ailerons was not sufficient to stabilize the missile about the longitudinal axis. Some of the rockets were, therefore, equipped with a large aileron at one of the wings instead of the control surfaces at the wing tips. However, the simple wooden wings now had to be discarded and replaced with a new metal wing. This led at Rheinmetall-Borsig to considerations to completely forgo the roll stabilization. Anyway, the further development of the "Rheintochter 3" was already planned without roll stabilization. This was also the reason why the "Rheintochter 1" was investigated without roll stabilization. Wind-tunnel measurements at AVA Göttingen were to provide information about the range where the rocket could be considered as "roll-angle insensitive" (Fig. 7.13). There was now the problem that the control surfaces had to be deflected dependent on the roll angle to generate a lateral force as well as a moment in the desired direction. This problem too was investigated at Göttingen.

The measurements at Göttingen showed that the side force was indeed independent of the roll angle for small angles of attack ($\alpha < 10$ deg). Rolling moments already occurred, however, without empennage due to small asymmetries on the model. Further measurements were carried out at different stabilizer positions with both stabilizers, dependent on the roll angle, being

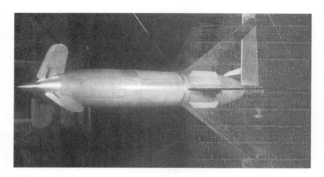

Fig. 7.13 Wind-tunnel model of the "Rheintochter 1" at AVA Göttingen.

deflected simultaneously to generate pitching moments in only one space-oriented fixed direction. Up to angles of attack of $\alpha \leq 5$ deg, the values were fairly independent of the roll angle. For larger angles of attack, in addition to the rolling moment, a yawing moment simultaneously occurred about an axis normal to the pitch axis. This coupling was strongly dependent on the roll angle φ (Fig. 7.14) [26]. With the configuration of the "Rheintochter 1," it was not possible to produce forces or moments in only one direction. At moderate angles of attack, a coupling of the degrees of freedom roll, pitch, and yaw occurred. The reasons for this asymmetry were the wakes downstream of the deflected stabilizers that led to perturbations of the lift of the wings. After these measurements, it was expected that it would be difficult to guide the "Rheintochter" to a target without roll stabilization.

During the trials of the first rockets of the development stage 1, Rheinmetall-Borsig considered a further development of the "Rheintochter" designated "Rheintochter R 3f" or "Rheintochter R 3p." The letters f and p, respectively, designate the propulsion by liquid and solid-fuel rockets respectively. Decisive for the new development was the too low combat altitude of the "Rheintochter 1," which was located between 5 and 6 km. The new rocket was to be able to cover a combat area up to an altitude of 15 km. This required primarily an engine with a longer burn time. Although this was, at that time, easily possible with liquid-fuel rockets, suitable solid-fuel propulsion with a sufficiently long burn time first had to be developed. The new rocket was supposed to be equipped with a proximity fuse from the start. A roll stabilization was forgone because it was thought possible to be able to guide the rocket without any problems with the aid of the "Rheinmetall-Borsig-K-control" into the vicinity of the target. "K" stood here for "onboard roll-angle-dependent-coordinate-converter," which was needed for the correct transmission of the control commands onto the control surfaces. Alternatively, also a ground-based coordinate converter, developed by the Telefunken Company, was to be tested; this obtained the information about the roll angle of the rocket from the rotation of the polarization plane of a transponder, "Rüse," within the rocket. This transponder was also provided for the exact determination of the

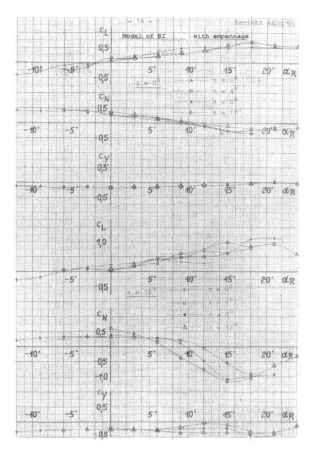

Fig. 7.14 Results of model measurements of the "Rheintochter 1" at AVA Göttingen. (Nomenclature: C_L = rolling-moment coefficient, C_N = yawing-moment coefficient, C_Y = side-force coefficient, α_R = angle of attack, φ = roll angle, η = control-surface deflection.)

range. The transponder "Rüse" was attached within a streamlined housing at a wing tip [27].

The "Rheintochter R 3" was externally not substantially different from its predecessor. The complete forward part with the stabilizers remained unchanged. Instead of the six wings, four wings in a cruciform arrangement were now envisaged, which in design and sweep were not different from the wings of the R 1. The most obvious difference was two boosters that were laterally attached to the fuselage. Less obvious was the explosive charge, whose effectiveness had been considerably enlarged. The reason for it was the inaccuracy in the determination of the distance between target and rocket by means of the contemporary radar instrumentation, which, in case of a manual tracking, did not permit a more accurate distance determination than

40 m under the most favorable conditions. Also in the structure of the rocket, changes had been carried out. For instance, the fuel tanks of the R 3f were part of the load-bearing structure. Cables and pipes had, therefore, to be placed in special channels along the outer skin. Further changes concerned the ground-based equipment for remote control and tracking. Figure 7.15 shows a side view of the "Rheintochter R 3f." The solid-fuel propelled R 3p was to have the same dimensions as the R 3f. The engine now had only four nozzles that protruded sideways from the fuselage below the wings.

The efforts of Rheinmetall-Borsig to keep the rocket design as simple as possible and to uncouple the pitch and yaw control as far as possible from the roll angle led at the AVA to wind-tunnel measurements on various R3 configurations (Figs. 7.16 and 7.17). During these measurements, numerous variants were investigated that differed in the arrangement of the control surfaces with respect to the wings. None of these measurements led to satisfactory results regarding the decoupling from the roll angle [28].

The information on the first flight tests of the "Rheintochter" varies. The first successful launch of an R 3p probably took place on December 18, 1944 (see [9], p. 21). The propulsion was to accelerate the rocket nominally to a speed of 400 m/s at an altitude of 12 km. Only a few examples of the R 3 had been launched before the suspension of the program in February 1945. The poor controllability probably motivated Rheinmetall-Borsig to also build a configuration of the R 3 with ailerons. This rocket was equipped with ailerons on one of wing pairs. Unfortunately, nothing is known about flight tests. Concerning the concept, the "Rheintochter" was more advanced than its competitors of Henschel and Messerschmitt. Compared to today's ground-to-air missiles, the "Rheintochter" shows many similarities. The deletion of roll stabilization led, however, to problems, which could, at the time, not be remedied immediately. It was not possible to guide the rocket to a particular

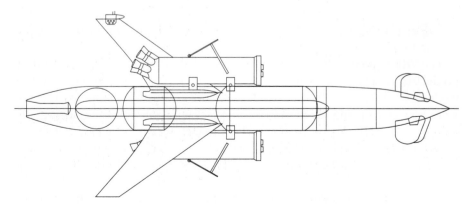

Fig. 7.15   Plan view of the antiaircraft rocket Rheinmetall-Borsig "Rheintochter 3" [4].

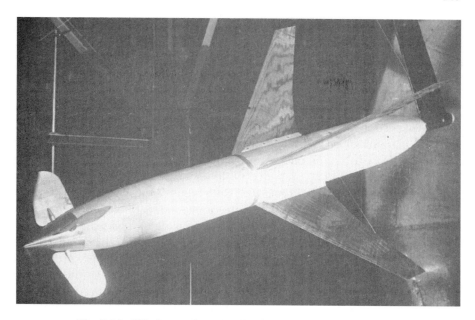

Fig. 7.16   Wind-tunnel model "Rheintochter" with four wings.

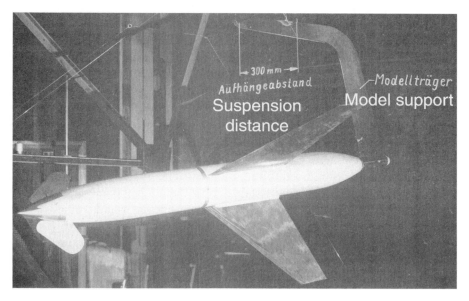

Fig. 7.17   Wind-tunnel model "Rheintochter" with four wings and three swept-back control surfaces.

point without problems. With modern instrumentation and computer technologies, the concept pursued by Rheinmetall-Borsig would be easily achievable today.

After the occupation of Central Germany by Soviet troops and the foundation of the institute "Berlin," the "Rheintochter" could be reproduced and tested. The institute "Berlin" was founded in 1946 in Berlin by the Soviet forces of occupation. The objective of this institute was the acquisition of German antiaircraft rocket technology. The difficulties in acquiring design documents as well as components for the control system were the reasons why only one rocket could be completed up to October 1946 [29]. It is not clear whether the latter was an R 1 or an R 3. Both types of rockets were thoroughly examined by the Soviet experts based on corresponding documents. However, a reproduction in the Soviet Union and flight tests did not take place.

Some specimens of the "Rheintochter 1" have survived the war and found their way to Great Britain, France, and the United States. It is not verified whether the "Rheintochter 1" in these countries became the starting point for any of their developments. For the United States, this can even be excluded. These rockets have, nevertheless, been thoroughly investigated, and, thereafter, found their way into technical museums.

7.3.2 HENSCHEL HS117 "SCHMETTERLING (BUTTERFLY)"

In 1941 the Henschel Flugzeugwerke AG (HS) (Henschel Aircraft Company AG) had suggested an antiaircraft rocket, designated Hs 297, to the RLM, which was based on the technique of the missile Hs 293 [30]. The Hs 293 was a rocket-propelled glide bomb that was remotely controlled according to the line of sight procedure. *Herbert Wagner* was responsible for the development of the Hs 293. The Hs 293 was solely controlled via ailerons and elevator. The ailerons were actuated by electromagnets working intermittently; the elevator was driven by an electric servo motor for adjustment. An automatic stabilization about the roll (longitudinal) axis was not envisaged. The airframe was designed such that sideslip-induced rolling moments were negligible. This was to prevent a yawing of the carrier aircraft during a drop leading to an unwanted sideslip-induced rolling moment. The missile was aerodynamically designed to be directionally stable about the pitch and yaw axes [31]. In manufacturing wings and fuselage, the greatest value was attached to the avoidance of asymmetries, which, in the long run, would have led to a roll-angle deviation and hence to deviations in the desired flight trajectory. Wings and horizontal stabilizer possessed thin symmetrical airfoils. A *Mach* number limiter kept the flight speed below a *Mach* number of 0.84 to avoid compressibility effects and the corresponding shifts in the neutral-point location.

Following the order of the Army Weapons Office in the autumn of 1942 to develop an antiaircraft artillery rocket, work on the antiaircraft rocket resumed at the Henschel Aircraft Company. The project manager for this development program was *Julius Henrici*. In the technical/tactical requirements of September 25, 1942, the "project Henschel" is already mentioned in conjunction with an antiaircraft rocket, remotely controlled by sight, for altitudes of 1000 to 7000 m and distances of up to 10 km. The rocket was now designated Hs 117 "Schmetterling (Butterfly)" (known as Project Henschel S).

The development work commenced in March 1943. There were no technically detailed specifications for the design of the rocket given by the customer. That is why *Henrici* during the layout of the rocket followed the philosophy already practiced in the case of the Hs 293. The rocket was to have the following features: 1) sufficiently high speeds compared to the target (Boeing B 17); 2) superior lateral acceleration; 3) low sideslip-induced rolling moments during launch; 4) natural stability about cross and longitudinal axes; and 5) low power requirements of the control devices.

The rocket had a horizontal-wing configuration with ailerons and elevator. Investigations of *Herbert Wagner* had shown that control by ailerons and elevator (polar-coordinates control) were the most favorable for the desired purpose. As propulsion, a liquid-fuel rocket was chosen from the beginning due to the possibility to regulate the thrust as function of *Mach* number. The speed was initially limited to $M = 0.75$ to avoid *Mach* number related compressibility effects. Computations showed that a lateral acceleration of 3 to 3.5 g was sufficient for attacking a target. Considering the high *Mach* number, a swept wing with 38-deg sweep and a symmetrical NACA 0012 1, 1-30 airfoil was chosen. The advantage of a larger maximum-thickness location was known to *Henrici* but, however, due to structural reasons could not be realized. Because of the positive experience made in case of the Hs 293 with low-aspect-ratio wings, a similar aspect ratio of $\Lambda = 3.5$ was selected for the Hs 117 [32].

Instead of control-surface flaps, spoiler-type control surfaces already introduced by *Wagner* on later versions of the Hs 293 were used. The spoilers were located at the wing trailing edge and, like a standard flap, were deflected up or down. Wind-tunnel measurements at the DVL showed that the trailing-edge spoilers exhibited a constant effectiveness up to $M = 0.88$. Because of their arrangement, they did not produce any moments when being adjusted and could, therefore, be operated by simple electromagnets. Different from the Hs 293, the elevator of the Hs 117 did not have a servo drive but also possessed a spoiler control surface with an intermittent actuator. The launch was carried out from a gun mount that had to be aligned with the target by a director device. After dropping the booster, the cruise engine took over the propulsion. For the remote control by sight, the well-tried "Kehl/Strassburg" equipment was used.

The first version of the Hs 117 did not yet have a cylindrical fuselage, but the fuselage diameter was largest in the area of the wing. The horizontal stabilizer had the same airfoil section as the wing, and the vertical stabilizer was designed as a double-fin stabilizer to not obstruct the exhaust jet of the boosters. The propeller of the electric generator was located below the proximity fuse extending far forward.

In the aerodynamic development of the Hs 117, the Henschel Company was supported by the aeronautical research establishments. The DVL as well as the AVA and the LFA acted as consultants. The wind-tunnel tests were carried out at the DVL in Berlin and the AVA in Göttingen. During the development of the cruise engine, the fuselage received a long cylindrical center section to house the fuel tanks. In addition, the booster rockets became shorter so that a central vertical stabilizer could be used. The first flight tests were carried out with this version, which was designated Hs 117 A. Based on the flight tests, the forward fuselage was rotated by 90 deg to achieve a better symmetry with respect to the lateral axis and hence a steadier pitching-moment dependence. A relocation of the fuel tanks allowed a rearward shift of the wings and thus a relocation of the spar and the use of the wing airfoil section NACA 0012 0,825 40 (40% maximum-thickness location). The configurations of the "Schmetterling (Butterfly)," thus modified, bore the designations Hs 117 B, C, D, and A1.

The demand not to allow any *Mach* number dependent center-of-pressure movements, which would have required a continuous trimming, led at the beginning of 1944 to measurements at Göttingen where special attention was given to just this question. The model investigated corresponded to the Configuration III of the beginning of that year (see [8], p. 156). The measurements were carried out up to $M = 0.9$ at an angle-of-attack range of $\alpha \pm 12$ deg once with and once without the boosters [33]. The center of pressure shifted, indeed, only insignificantly within this speed range (see Figs. 7.18–7.20). According to statements of the DVL, the cylindrical fuselage and the thin wing with a symmetrical airfoil section were responsible. The stability parameter dC_m/dC_A (dC_m/dC_L) indicated sufficient stability at $M = 0.6$ for the configurations with and without boosters. At $M = 0.80$ the stability became indifferent without boosters, and at $M = 0.90$ the rocket was unstable with respect to the given moment reference point. The stability behavior became, in addition, strongly dependent on the angle of attack. Because the stability depends on the center-of-gravity location and the center-of-gravity shifts in the case of a rocket as the fuel tanks are being emptied, the complete center-of-gravity range has to be considered during the aerodynamic design. In an almost vertical ascent where the wing provides hardly any lift, the consequences are minor. During the pursuit of a target where maneuvers with high load factors are likely, serious problems are, however, to be expected. Whether the Henschel Company had found a solution for these problems, which guaranteed satisfactory flying qualities within the entire flight envelope, remains doubtful.

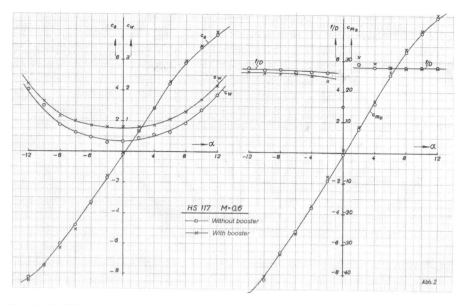

Fig. 7.18 Wind-tunnel measurements of a Hs 117 model at a *Mach* number of $M = 0.6$ [5]. (α = angle of attack, C_w = drag coefficient, C_a = lift coefficient, c_{m0} = pitching-moment coefficient, f/D = pressure-point location based on the diameter.)

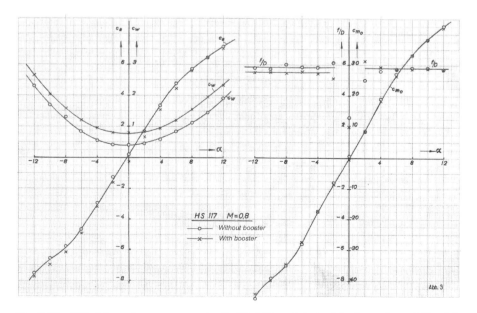

Fig. 7.19 Wind-tunnel measurements of a Hs 117 model at a *Mach* number of $M = 0.8$.

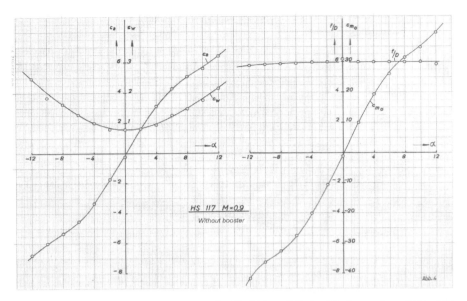

Fig. 7.20 Wind-tunnel measurements of a Hs 117 model at a *Mach* number of *M* = 0.9.

The evaluation of high-speed measurements, which had in the meantime been carried out at the DVL, showed that *Mach* number dependent perturbations occurred already in a *Mach* number range of *M* = 0.70 to 0.75 (see Fig. 7.21). As a cause, the unfavorable design of the horizontal stabilizer was identified. Sweeping the horizontal stabilizer and reducing the airfoil thickness were for structural reasons rejected. So a trapezoidal stabilizer with a straight trailing edge was finally selected. In this way, the critical *Mach* number could be raised to *M* = 0.85. This rocket received the designation Hs 117 A2. Table 7.3 lists the technical data.

The flight testing took place at Peenemünde-West. Three drops from an aircraft were carried out up to August 1944 to prove the airworthiness. Wooden models of the Hs 117 were used. Subsequently, six launches from a launcher were carried out confirming the launch procedure (Fig. 7.22). Also a launch with remote control was successfully carried out by August 1944 [18]. The time until December was employed to improve the launch procedure and the loading of the launcher. Further tests in 1945 confirmed the desired control and stability behavior. The fact that, up to the end, an optimum setting of the horizontal stabilizer was not found showed that the Hs 117 had a trimming problem. Possibly this problem was related to the quite considerable configurational change when dropping the boosters. The performance of the rocket engines also did not meet the expectations. Cine-theodolites were employed for tracking. Because of the lack of data registration or telemetry, it was only infrequently possible to correlate the flight trajectory with

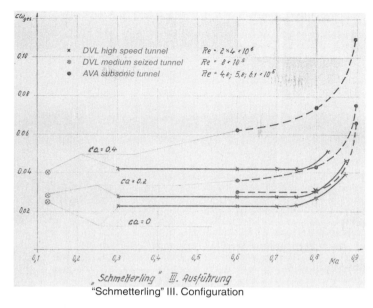

Fig. 7.21   Critical *Mach* number of the Hs 117 obtained in different wind tunnels [6]; axes: c_{wges} (c_{dtotal}), M.

the movement of the stick. This shows how difficult it was at that time to obtain a complete data set during the tests that were carried out under enormous pressure. Figure 7.23 shows plan views of the Hs 117 A1, which was built in larger numbers.

Altogether 49 rockets were launched off the ground up to February 1945 with 47 of them having had remote control. Of these, more than half failed for various reasons. Twenty-one rockets were dropped from an aircraft. Altitudes

TABLE 7.3   Hs 117 A2 TECHNICAL DATA

Length	4.30 m
Span	2.00 m
Fuselage diameter	0.35 m
Wing sweep	38 deg (25% chord line)
Wing airfoil section	NACA 0012 0,825 40 with a 40% maximum-thickness location
Wing area	0.75 m²
Aspect ratio	3.5
Takeoff (launch) mass	460 kg
Propulsion	BMW 109-558 liquid-fuel engine with 3.7 kN maximum thrust
Boosters	2 Schmidding 109-553 with 17 kN thrust each
Maximum speed	$M = 0.85$ limited
Range/altitude	1000 m/7000 m

Fig. 7.22   Antiaircraft rocket Henschel Hs 117 on launcher at Peenemünde-West (1944) (WTS).

of up to 10,500 m were allegedly reached. Not a single rocket was, however, tested in conjunction with an airborne target.

In addition to the flight testing of the Hs 117 A1 and A2, investigations were being carried out at Henschel on improved versions of the "Schmetterling." The designation Henschel Project S2 related to a rocket with a forward horizontal stabilizer and a rear wing (Fig. 7.24) [34]. The S2 was especially developed for a "blind attack" against high-speed, jet-propelled aircraft flying at low altitudes. The rocket was to be guided by a radio beam into the vicinity of the target and thereafter a "target seeker" was to guide the rocket automatically to the target (bombers that could cruise at high altitudes and speeds of up to 720 km/h). Henschel considered the development of a high-frequency target detector and

Fig. 7.23   Plan view of antiaircraft rocket Henschel Hs 117 A1 "Schmetterling" (summer 1944) [7].

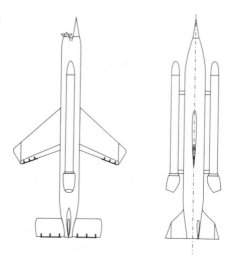

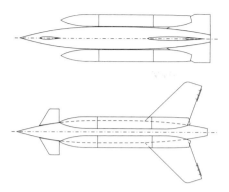

Fig. 7.24 Project Henschel S2b (1945) [8].

suitable radar devices with a high resolution as especially difficult in its accomplishment. Both technologies existed only in a very rudimentary way in Germany at the end of the war. Henschel believed that a pure subsonic rocket would suffice to effectively fight the mentioned targets.

Project S2 was a horizontal-wing configuration to be controlled by aileron and elevator. The forward horizontal stabilizer ("canard") indicates that an improved maneuverability had been considered. The wing was the same as in the case of the Hs 117. The same trailing-edge spoilers were also employed for the lateral control. The complete horizontal stabilizer was adjustable. There were no wind-tunnel measurements that could provide information concerning the stability of the canard configuration. The rocket was to guide itself along the guide beam until the target seeker had located the target. Instead of the two boosters, now four boosters were employed. The S2 was recommended to the RLM for development in January 1945; however, this was not achieved anymore because the technical prerequisites were simply not given. The project S2 comes, nevertheless, already very close to the type of antiaircraft rockets employed today.

In 1944, *H. Voepel* of Henschel undertook studies concerning a version of an antiaircraft rocket with supersonic capability. This rocket was designed as tailless configuration, which was to reach *Mach* numbers of up to 1.5. True to the *Wagner* philosophy, this rocket was to have a design that largely avoided *Mach* number associated problems. Designated "Zitterrochen (electric ray)," *Voepel* suggested a tailless configuration with a cylindrical fuselage and a wing with a rhombus-like planform (see Fig. 7.25). The control of this rocket was to be solely accomplished by spoiler control surfaces at the wing trailing edge. Three wind-tunnel models with different wings were built for this configuration. The wings had the same area but different aspect ratios (2.0, 1.0, 0.5). All wings possessed biconvex circular-arc airfoils with 50% maximum-thickness locations. The thickness ratio was different for the three wings and changed with the aspect ratio from 6.33, 3.7, to 2%. The wind-tunnel measurements were performed at the AVA at subsonic and supersonic speeds

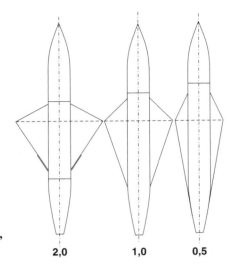

Fig. 7.25   Project Henschel "Zitterrochen" with three different aspect ratios [9].

2,0   1,0   0,5

[35]. In the case of all three wings, the measurements showed hardly any change in the center-of-pressure location with angle of attack and increasing *Mach* number. The missile was, however, statically unstable (positive $dC_m/dC_A = dC_m/dC_L$) with a tendency towards a neutral behavior with increasing *Mach* number. This instability could have been remedied with a more suitable center-of-gravity location or trimming.

The first wing with an aspect ratio of 2.0 was probably of sufficient interest to Henschel to also investigate it without the disturbing influence of a fuselage [36]. The wing was exposed to the airflow from two directions, once in a normal way with the blunt edge (leading edge) forward and once with the pointed edge (trailing edge) forward. The latter resembles the measurements on a delta wing with an aspect ratio of 2.0. The measurements showed in the case of the wing with the airflow from the rear at speeds of up to $M = 0.85$ no increase in the zero drag, while in the case of the "normal" airflow direction a pronounced increase in the zero drag occurred (Figs. 7.26 and 7.27). This effect was certainly caused by the larger sweep angle of the "delta wing." Also with regard to the lift coefficient, the wing with the airflow from the rear (delta wing) was superior. The delta wing provided a higher lift coefficient, at the same angle of attack and a nearly unchanged drag coefficient (higher $C_A/C_W = C_L/C_D$). Both configurations showed a similar behavior in the center-of-pressure location. In both speed regimes the center-of-pressure location remained unchanged, but was different at subsonic and supersonic speeds. The difference was, however, in the case of the "delta wing" considerably smaller than for a standard wing. This led to the conclusion that the delta wing caused less trim problems during the transition from subsonic to supersonic speeds. It is not known whether these

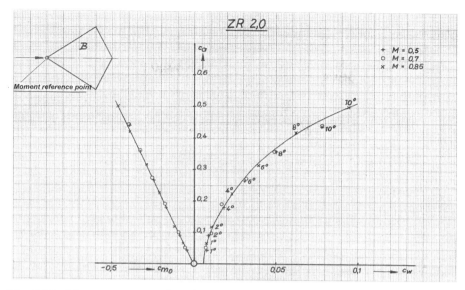

Fig. 7.26 Measurements on the wing of the "Zitterrochen" at subsonic speeds [10].
Axes: $c_a = c_L$, $c_w = c_D$, c_{m0}.

important findings about the qualities of the delta wing were also provided
to other aircraft companies.

As a medium-range antiaircraft rocket, the "Schmetterling" was also of
interest to the Soviet Union. As in the case of the "Wasserfall (Waterfall)" the

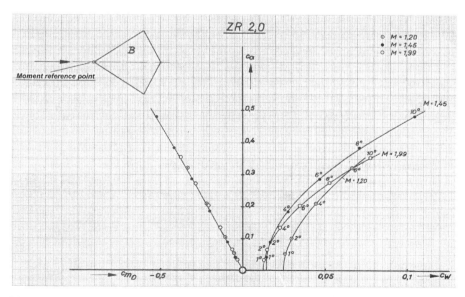

Fig. 7.27 Measurements on the wing of the "Zitterrochen" at supersonic speeds. Axes:
$c_a = c_L$, $c_w = c_D$, c_{m0}.

difficulty was to obtain the complete design documents for the reconstruction of the rocket. It was still possible to build three examples of the "Schmetterling" at the institute "Berlin" though without the command module for the control of the rocket from the ground.

The "Schmetterling," rebuilt in the Soviet Union, received the designation R-102. Altogether 20 rockets of the type R-102 were built at the research institute of the Armament Ministry (NII-88). Soviet experts criticized the complexity of the rocket that impeded the production. Especially the fuel tanks were in need of improvement. The German actuators were of poor quality and worked only at room temperatures. Because of the lack of important components, the assembly of the first examples took until the middle of 1949. The flight testing commenced at the proving ground Kapustin Jar in the autumn of the same year. Of the twelve rockets launched, five crashed immediately after takeoff. After installing an improved Soviet control system, further launches were carried out achieving the demanded performance. The weak point was the lack of an automatic control on a radio guide beam. The manual control by the line-of-sight procedure was too inaccurate and did not allow a military application. The "Schmetterling" was, in addition, a pure subsonic rocket, which could no longer be deployed against high-speed jet aircraft [29]. All work was, therefore, stopped at the beginning of 1950.

Based on the experience with the R-102, the development of an improved rocket designated R-112 commenced at the NII-88. Three variants were pursued, the R-112A with target seeker, the R-112B with warhead, and the R-112C with ramjet propulsion. The R-112 was a supersonic rocket with a maximum speed of 700 m/s. The rocket had wings located forward and delta-shaped stabilizers with control surfaces (Fig. 7.28). As in the case of the "Schmetterling," the launch was to take place from a gun mount. The fuel tanks had been altered, and the fuel switched to kerosene instead of Tonka. For the fuel supply the gas pressure generator was used, which had already been used on the R-101. Two control methods were to be employed: one optical using the line-of-sight procedure, and the other automatic by means of a target seeker.

A development parallel to the R-112 was the R-117 that differed in the arrangement of the tanks and the equipment. The tanks of the R-117 were now part of the load-bearing structure. The tip of the rocket carried a proximity fuse. As a guidance method the line-of-sight procedure as well as guidance on a beam to a collision course was envisaged. Both projects were to be combined to a new development in 1951. During the same year, work on both projects was stopped [37].

The experience gained with the rockets R-102 and the further developments R-112 and R-117 formed the basis for the development of the SA-2 "Guideline," which was launched from a mobile launcher. German experts

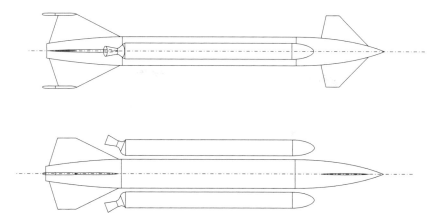

Fig. 7.28   Project of a Soviet antiaircraft rocket (R-112B) based on the "Schmetterling" [11].

still had a large share in the development of this rocket after 1951. The German experts were especially responsible for the control of the rocket by radio beam and for the flight controller.

Jürgen Michels writes in his book about the postwar development of German secret weapons abroad that some of the rockets of the type "Schmetterling (Butterfly)" reached France where they were further developed with the aid of German experts [38]. The direct successor of the "Schmetterling" was, therefore, the experimental rocket "SE 4100." The "SE 4100" was, contrary to the "Schmetterling," a cross-wing configuration with four swept wings. The stabilizers were also arranged crosswise. The rocket was launched from a gun mount by means of two boosters arranged laterally. Whether the "Schmetterling" was also an example in the development of the antiaircraft rocket "PARCA" can no longer be safely ascertained. It is, however, sure that German experts were also involved in the development of this rocket.

7.3.3 ANTIAIRCRAFT ROCKET MESSERSCHMITT "ENZIAN (GENTIAN)"

The antiaircraft rocket "Enzian (Gentian)" of the Messerschmitt Company goes back to the design of glide bombs that *Alexander Lippisch* had already drafted during his time as head of the "Department L" of the Messerschmitt A.G. in Augsburg [39]. Figure 7.29 shows the design of the long-range glider FG 10 of December 1941 whose similarity to the rocket-propelled aircraft Messerschmitt Me 163 is unmistakable. The FG 10 was dropped by a carrier aircraft and was remotely controlled to hit targets at larger distances. As propulsion, a liquid-fuel rocket was envisaged. The further development of

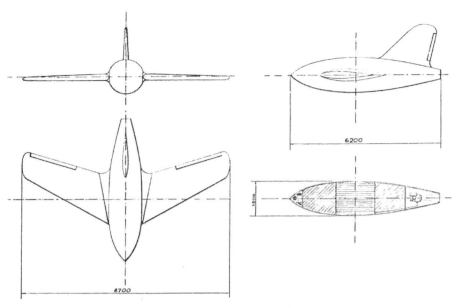

Fig. 7.29   Long-range glide bomb Lippisch FG 10 [12].

this design was assigned to *Hermann Wurster*, a former test pilot for the Messerschmitt Company.

On the basis of this configuration, Messerschmitt started in 1942 its own development study concerning a Fliegerabwehrrakete (FR) (antiaircraft rocket). In mid-1943 a first design designated FR 1 evolved from these studies under the supervision of *Wurster* (see [9], p. 12). The FR 5 resulted from the FR 1 in several design steps for which Messerschmitt finally received a development contract from the RLM at the end of 1943. After the transfer of the project department of the Messerschmitt Company to Oberammergau, the development was continued there. For secrecy reasons the project department was named "Oberbayerische Forschungsanstalt Oberammergau" (OFO) (Upper-Bavarian Research Establishment Oberammergau). The antiaircraft rocket now called "Enzian" with a projected operational altitude of 15 km and a range of 25 km was positioned between the demands of Projects 3 and 4 of the RLM specifications. The "Enzian" was designed as a subsonic rocket for a *Mach* number range of up to a maximum of $M = 0.9$. A liquid-fuel rocket was scheduled as propulsion. The launch was carried out from a rotatable gun mount by means of four boosters to be discarded. The rocket was to be guided to the target by sight via a control stick.

The design FR-5 formed the basis for the first experimental model, the "Enzian" E-1. The design of the E-1 commenced in January 1944. Figure 7.30 shows side views of the missile. The "Enzian" E-1 was a tailless

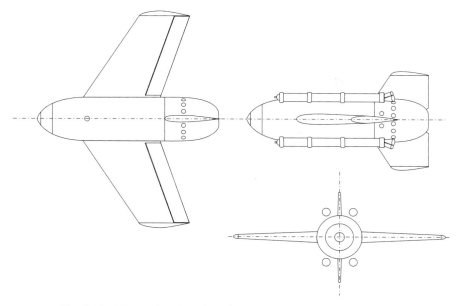

Fig. 7.30   Three-view drawing of the Messerschmitt "Enzian" E-1.

horizontal-wing configuration similar to the Messerschmitt Me 163. Like the antiaircraft rockets "Schmetterling" and "Feuerlilie," the missile was symmetrical in two planes. The wing had a sweep of 30 deg (25% chord line) and was untwisted. Being a tailless configuration, the wing sweep was primarily required for stability reasons, having, of course, also a positive effect on the critical *Mach* number. As wing airfoil section, a symmetrical NACA 0020 airfoil with 20% relative thickness and a maximum-thickness location of 30% chord was selected. The airfoil thickness tapered down, starting at the wing root, linearly up to the wing tip to a NACA 0010 airfoil with a 10% relative thickness. The complete wing trailing edge was taken up by a combined aileron/elevator arrangement with a constant chord of 0.26 m. This trailing-edge flap was aerodynamically balanced and could be adjusted as an elevator with a deflection of ± 10 deg. As aileron, the deflection was restricted to ± 5 deg. The control surfaces were driven by a motor of the Siemens K-12 autopilot [40]. Wind-tunnel measurements showed that the combination aileron/elevator kept an almost constant effectiveness up to a *Mach* number of $M = 0.9$. The two vertical stabilizers, arranged symmetrically, did not possess any rudders. The control of the rocket was carried out exclusively via elevator and ailerons. The fuel tank for the Walther engine Rl 210b was installed such that the center of gravity shifted only within narrow limits when emptying the tanks. The center-of-gravity range was chosen such that there was within the entire *Mach* number range sufficient stability about the pitch and yaw axes. With regard to the lateral stability, the missile exhibited a neutral spiral-dive behavior.

The "Enzian" was designed for a maximum load factor of 8 g. The maximum angle of attack permitted at subsonic speeds was $\alpha = 16$ deg. The "Interrogation Reports" provide different information about the minimum turn radius, which the "Enzian" E-1 could fly at the maximum load factor. According to *Hugh L. Dryden*, turn radii of down to 500 m could be flown at the maximum *Mach* number. In pursuing a maneuvering fighter aircraft, this would not have been sufficient. According to another interrogation, the minimum turn radius close to the ground was 250 m at a load factor of 4 g.

The short length and large diameter fuselage of the rocket did not correspond to the ideas of a slender body that was to fly at high subsonic *Mach* numbers. One of the people questioned rightfully criticized the fact that the very large curvature would lead to shock waves followed by flow separations. This would, in turn, have led to a reduction in the critical *Mach* number. The blunt base together with the exhaust jet equally caused a strong aft-body drag. The fuselage was cylindrical in the area of the wing junction, which *Wurster* considered advantageous since the penalizing wing-fuselage interference would be avoided.

Wurster quoted a critical *Mach* number of $M = 0.86$ for the "Enzian." Unfortunately, there are no high-speed wind-tunnel measurements that could confirm this statement. An original "Enzian" E-1 was still investigated in the large subsonic tunnel A3 of the LFA in March 1945 (Fig. 7.31) [41]. Results of these measurements are, unfortunately, not available either. At the time of the development of the "Enzian," the essential information concerning the effect of the airfoil thickness, the sweep angle, the maximum-thickness location, and the wing aspect ratio on the critical *Mach* number were known. Also the effect of the exhaust jet on the aftbody drag had been thoroughly investigated in conjunction with the A4 development at the HVA in Peenemünde. Similar investigations also took place at the LFA Braunschweig in conjunction with the development of the missile "Hecht (Pike)." During

Fig. 7.31 "Enzian" E-1 mounted in the large wind-tunnel A3 of the LFA Braunschweig, 1945.

the development of the Messerschmitt Me 262 and later of the experimental aircraft Messerschmitt P 1101, use had been made of this knowledge. Obviously these findings of high-speed research had not entered the development of the "Enzian" E-1. For a swept wing with a 30-deg sweep, a thickness of 20%, and a maximum-thickness location of 30%, the critical *Mach* number should not have been higher than $M = 0.75$ [42]. With the effect of the short length and large diameter fuselage, one would estimate the critical *Mach* number to be even lower. According to *Wurster*, the "Enzian" E-1 with the Walter-engine at an altitude of 7000 m reached a speed of 240 m/s corresponding to $M = 0.78$.

The rocket was launched from a modified gun mount with a 7-m long ramp (Fig. 7.32). The boosters accelerated the projectile to a speed of 24 m/s after leaving the ramp, this being sufficient for the aerodynamic control. However, the control was only activated after the drop of the boosters to ensure a safe separation of the boosters from the rocket. The liquid-fuel rocket engine had a burn time of 45 s. Because of the pressure decrease within the pressure vessel during the operation, the thrust was reduced at the end of the combustion period to half of its initial value. A gyroscope provided the stabilization about the roll axis. To induce a roll angle via the command module, the roll gyroscope was deflected by means of an actuator. To improve the roll stabilization, a rate gyroscope was later added to the attitude gyroscope. The installation of pitch stabilization proved to be unnecessary because the "Enzian" possessed sufficient natural stability.

About 60 examples of the "Enzian" E-1 were manufactured and 38 tested at the artillery test range Karlshagen (Usedom). Cine-theodolites were employed for the flight-test evaluation. Some of the rockets were equipped with a telemetry transmitter, which returned the control commands carried out.

The flight testing of the E-1 commenced in May 1944 (see [8], p. 142). During the first 12 flights, the flight performance with the Walter engine was to be investigated. A second subject of the investigations was the longitudinal stability after leaving the launching ramp and dropping the boosters. The

Fig. 7.32 "Enzian" E-1 on launcher at the antiaircraft artillery proving grounds Karlshagen (courtesy of Deutsches Museum München).

trials were performed without remote control and with fixed control surfaces. The first flights lasted only a few seconds and regularly ended with a crash of the rocket. The cause was related to the boosters. The nozzles of the four boosters had to be adjusted such that the thrust axes of all four boosters intersected at a point upstream of the center of gravity to achieve stable conditions. Obviously this adjustment in case of the first flights was too inaccurate. The boosters produced strong moments that could not be compensated by the natural stability. There is also the question whether all of the boosters used, Type Rheinmetall RI 503, had exactly the same thrust. A pressure compensation valve was to guarantee that a defined thrust prevailed. Another problem concerned the separation of the boosters. These were, if possible, to break free simultaneously from the rocket so that asymmetrical conditions did not arise. By attaching small wings to the boosters, a safe separation from the rocket was finally accomplished.

The flights No. 13 to 20 served to test the roll stabilization. The elevator position was blocked. It turned out that the gyroscope employed did not suffice. The gyroscope was replaced by a model from the Horn Company of Leipzig, which finally met the requirements. The next ten flights were carried out with the remote control activated. The roll stabilization was supplemented by a rate gyroscope, which improved the natural-frequency oscillation behavior about the roll axis. A pitch stabilization to balance perturbations about the pitch axis was, in addition, tested. It turned out, however, that the delays due to the actuator negated the benefits. The artificial pitch stabilization was, as a result, forgone because the natural stability was sufficient. Altogether 22 flights were carried out until the end of 1944. Flight durations of up to 82 s were achieved [18]. The last flights, Nos. 31 to 38, served to investigate flight performance, stability, and controllability. According to *Wurster*, these flights proceeded satisfactorily. Speeds of up to 240 m/s at altitudes of 7000 m were reached.

The "Enzian" E-1 was considered as an interim solution of the E-4, which was to be equipped with the more powerful rocket engine Konrad VFK 613. The "Enzian" E-4 was to become the series-production model of this antiaircraft rocket. The development commenced in January 1944. The origin of this development was the missile FR-6. Wind-tunnel measurements on the FR-6 configuration are identified for the time period April 1 until July 31, 1944, and from September 1 until November 30, 1944, in the wind tunnel A1 of the LFA in Braunschweig (Fig. 7.33) [43]. The E-4 differed externally only little from the E-1. The fuselage was, at the same diameter, slimmer with the nose of the fuselage better suited for high-speed flight. The latter would have been beneficial to the critical *Mach* number. The wing was unchanged except for the combined aileron/elevator flaps that no longer extended over the entire span. Figure 7.34 shows side views of the "Enzian" E-4. The tanks for fuel and oxidizer were skillfully arranged within the fuselage so that

Fig. 7.33 Wind-tunnel model of the FR-6 missile in the wind-tunnel A1 of the LFA Braunschweig, 1944.

hardly any center-of-gravity shift occurred during the operation of the rocket. Because there are no results of high-speed wind-tunnel measurements available, only little can be said about the critical *Mach* number. According to *Wurster*, the critical *Mach* number of the E-4 was $M = 0.9$. Because the airfoil-section thickness was not changed and merely the fuselage was somewhat slimmer, one may assume that the critical *Mach* number was not much higher than in the case of the E-1. The "Enzian" E-4 was to be completely produced of wood (Table 7.4). No single example was, however, built by the end of the war.

The design study FR-4 of 1943 envisaged a cross-wing configuration. This concept developed into the design E-5 by February 1945 (see [8], p. 143).

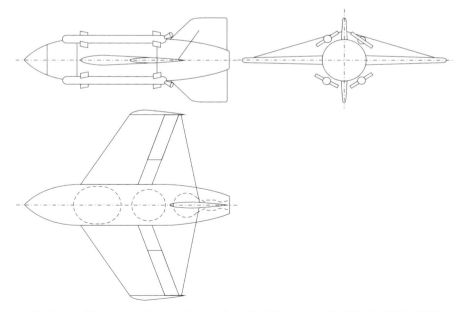

Fig. 7.34 Side views of the antiaircraft rocket Messerschmitt "Enzian" E-4 [13].

TABLE 7.4 "ENZIAN" E-4 TECHNICAL DATA

Length	4.10 m
Span	4.05 m
Fuselage diameter	0.88 m
Wing sweep	30 deg (25% chord line)
Wing airfoil section	NACA 0020 (root), NACA 0010 (tip)
Wing area	4.9 0 m²
Aspect ratio	3.35
Launch mass	1900 kg
Propulsion	Konrad VFK 613 with 20 kN maximum thrust
Booster	4 Rheinmetall RI 503 with 13.5 kN thrust each
Maximum *Mach* number	$M = 0.9$
Range/Altitude	25 km/13,500 m

The "Enzian" E-5 was planned as a supersonic rocket that was to operate in a *Mach* number range of $1.6 < M < 2.0$. The most striking difference compared to the E-4 was the wings, arranged crosswise, with a sweep of 55 deg (25% chord line). All four wings were equipped with trailing-edge flaps. The wings were appreciably thinner than in the case of the E-4. The wing thickness decreased from 7% at the root to 4% at the wing tip. There is no information available about the airfoil selected. One may assume, however, that for a missile cruising at supersonic speeds, a more aft location of the maximum-thickness was selected. The information on the fuselage is contradictory. According to *Wurster*, the fuselage was to have a length of 6 m, and the nose of the fuselage was to be considerably more slender than in the case of the E-4. The copy of an original drawing of the "Enzian" E-5 shows, however, the same fuselage as the E-4 [44]. Figure 7.35 shows side views of the "Enzian" E-5. The engine was the same as in the case of the E-4, however a slightly increased thrust was planned. The "Enzian" E-5 was naturally stable in the pitch and yaw axes. A roll stabilizer ensured stabilization about the roll axis. Control was accomplished by wing flaps that functioned as elevator and rudder (in Cartesian coordinates instead of polar coordinates).

Retrospectively, it is hard to tell whether the "Enzian" E-5 really would have reached flight speeds of up to $M = 2$ because there are no results of wind-tunnel measurements at supersonic speeds available. Wing aspect ratio and wing thickness of the E-5 were small, both having a positive effect on the neutral-point movement within the critical *Mach* number range. The wing sweep was sufficiently high so that one could expect subsonic flow on the wing up to $M = 2$. It may be assumed that the missile did not have any stability problems when crossing the transonic range especially when this occurred quickly. The optimum fuselage design was surely not chosen. Measurements that were carried out by *Theodor Zobel* in 1943 in the LFA high-speed wind-tunnel A2 on bodies of revolution with different maximum-thickness locations

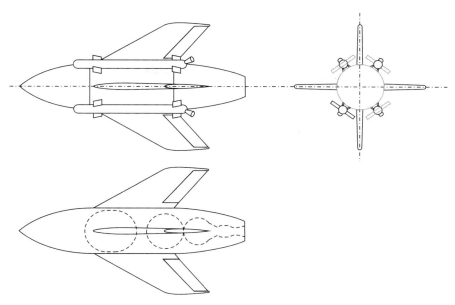

Fig. 7.35 Project Messerschmitt "Enzian" E-5 [14].

(see Sec. 2.3.2) showed that the critical *Mach* number increases with increasing maximum-thickness location [45]. A higher slenderness ratio would also have contributed to this effect. There was, of course, the desire to retain in the development of the E-5 as many components of the E-4 as possible.

By the end of the war, an operational model of the "Enzian" E-4 did not exist. The engine anticipated for the E-4 had not been tried, and the control was not yet sufficiently tested. Whether the "Enzian" could have been successfully deployed against a high-altitude target remains doubtful. One may also doubt that the E-5 has been the model for the later antiaircraft rocket "Hawk" as is claimed in the literature. None of the German antiaircraft rockets have ever been tested against an actual target. All statements that there would have been an effective defense against the great bomber offensives in the summer of 1943 if the development had started in time in 1941 are to be doubted.

7.4 ANTIAIRCRAFT ROCKET "WASSERFALL (WATERFALL)" OF THE HVA PEENEMÜNDE

Of the antiaircraft rocket developments, the "Wasserfall (Waterfall)" rocket, developed by the HVA in Peenemünde, was the technically most ambitious antiaircraft rocket project. The "Wasserfall" corresponded to the guidelines of Project 4, a liquid-fuel rocket with a range of 50 km, to be launched vertically, which was to fight targets at altitudes of up to 20 km. An automatic control by a radio beam pointed at the target was planned.

Shortly after the approval of the antiaircraft rocket development program in September 1942, the project planning for the antiaircraft rocket designated C2 commenced at the HVA. *Ludwig Roth*, a collaborator of *Wernher von Braun*, was responsible for the development of the C2. After the transfer of the entire A-4 development team to the Elektomechanische Werke Karlshagen (Electromechanical Works Karlshagen), newly founded in the autumn of 1944, the antiaircraft rocket development was continued there. The requirements regarding altitude and range resulted in the design of a rocket that was to fly at supersonic speeds. To achieve the objective as quickly as possible, *Roth* let himself be closely guided in his design by the long-range missile A4, for which sufficient documents were available. The C2 was to become a single-stage liquid-fuel rocket like the A4, which was launched vertically and then changed over to a flight trajectory aimed at the target. During the vertical ascent, the C2 was stabilized by jet rudders that were to be ejected after about ten seconds. After this time, the rocket would have reached a speed of 150 m/s, and the aerodynamic control would have become effective (see [8], p. 134).

Besides the development of the engine and the new fuel feed system, the aerodynamics was one of the focal points in the further development of the rocket. Unlike the A4, the C2 was to become a highly maneuverable projectile. That is why the stabilizers were equipped with considerably larger aerodynamic control surfaces. To guarantee the necessary maneuverability, the rocket was, in addition, to be equipped with wings.

In 1940 the A4 development team of Peenemünde had projected a winged A4 that was to reach distances of up to 600 km. During extensive wind-tunnel tests at the Luftschiffbau (Dirigible Construction Company) Zeppelin in Friedrichshafen, various wing forms were investigated for their suitability [46]. The result of these investigations was a 45-deg swept wing with an area of 13.5 m². In October 1942, *W. Dornberger*, responsible for the A4 development program, stopped the further development of this rocket, designated A9, to concentrate all available personnel on the A4. The project A9 was taken up again in mid-June 1944. Also responsible for this project was *Roth*, who now had to spread his scarce staff over two projects. Under strong pressure from the SS, *von Braun* and *Roth* had to arrive quickly at concrete results. There remained no time for wind-tunnel tests especially considering that the Peenemünde supersonic tunnel had just been shipped to Bavaria and set up again at Kochel. According to the documents of 1942, one wing was built and mounted on the A4 (Fig. 7.36). Similar to the "Wasserfall," the stabilizers were fitted with large control surfaces because the rocket had to be guided within the atmosphere on a shallower trajectory (Fig. 7.37). *Von Braun* and *Roth* were aware of the fact that crosswind would represent a serious danger during the launch of the rocket. The wings did not possess any ailerons, and also the jet rudders would not have been able to balance rolling moments arising. Because the rocket was a makeshift model and not a new development,

Fig. 7.36 Long-range rocket A4b (courtesy of the Deutsches Museum München).

it was designated A4b. The first winged A4b took off on December 27, 1944, and crashed shortly after the launch because the ensuing rotation of the rocket about the vertical axis could not be compensated [47].

The considerations concerning the A9 configuration and the wind-tunnel measurements of 1942 have surely played a role in the development of the "Wasserfall (Waterfall)." The only documents the aerodynamicists could fall back on in the development of the C2 were the wind-tunnel measurements on the A4. The first test results had shown that the A4 crossed the sound barrier flawlessly. This suggested a configuration that should not differ too strongly from the A4. The wind-tunnel measurements on the C2 commenced

Fig. 7.37 Aerodynamic control surfaces on A4b (courtesy of the Deutsches Museum München).

in March 1943 under the supervision of *R. Hermann* in the Peenemünde supersonic tunnel. This tunnel was transferred to Kochel, where work continued.

The C2 was equipped with wings arranged crosswise so that with a change in direction no rotation about the longitudinal (roll) axis had to be carried out. The rocket was designed for a load factor of 12. In a tight turn, the wings had to generate a lift of 18 tons [48]. Instead of a cross wing, an annular wing was investigated at first, then, however, rejected because drag at supersonic speeds was too high. The annular wing would have provided constant lift conditions independent of the respective axis of rotation of the rocket during the banked flight. No advantage was seen in swept wings because the sweep would have had little effect due to the large fuselage diameter relative to the span. However, it may also be possible that aeroelastic problems due to the high wing loads were feared. Because it was known that very low-aspect-ratio wings at high angles of attack caused high drag, the angle of attack at subsonic speeds was limited to 15 deg and at supersonic speeds to 8 deg [49]. However, this restriction was also required for stability reasons as will be explained in more detail later. At a roll angle of $\varphi = 45$ deg, the stability parameter dC_m/dC_A (dC_m/dC_L) at angles of attack of $\alpha > 15$ deg changed sign (see Fig. 7.46).

A further requirement was to keep the center-of-pressure location of the rocket throughout the entire operational range up to $M = 2.5$ essentially constant such that there were no large demands on the drives of the control surfaces. It was aimed, therefore, to keep the center of pressure, if possible, close to the center of gravity so that the moments to be generated by the control surfaces remained small. Ideally, this meant that the center of pressure followed the center of gravity when emptying the fuel tanks. This ideal situation was not attainable because the location of the center of pressure also depended on the angle of attack. Because the rocket did not possess an automatic stabilization system, a minimum distance to the center of gravity had to be kept to ensure sufficient stability. Figure 7.38 shows the time-dependent development of dynamic pressure, *Mach* number, center of pressure, and center-of-gravity locations (see [48], p. 24). To be able to meet these demands, numerous models with quite different wings were built and tested (see Fig. 7.39).

The HVA had developed their own procedure to determine the stability [50]. The rocket model was suspended free to rotate about the pitch axis. Because the pitch axis passed through the center of gravity of the model, it did not have any moments wind off. Figure 7.40 shows this model support in the wind tunnel. The model support could be set at an angle of attack of up to $\alpha \pm 45$ deg and the zero-moment condition registered.

Also in the design of the rear fuselage, the A4 configuration was closely followed. In the case of the A4, extensive wind-tunnel measurements had been carried out with a simulated exhaust jet to determine its effects at sub- and supersonic speeds [51]. In case of the "Wasserfall," in addition the

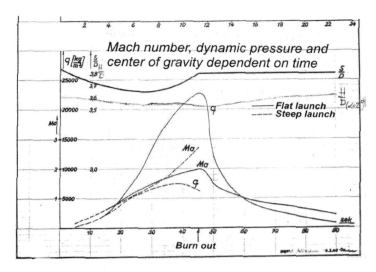

Fig. 7.38 *Mach* **number (*Ma*), dynamic pressure (*q*), center of gravity, and center-of-pressure locations (*S/D* and *H/D*) dependent on time [15]; axis: time (sec).**

influence of the flowfield, induced by the exhaust jet, on the control surfaces was of particular interest. Here, special measurements were carried out at Kochel (Fig. 7.41), and the results from the A4 measurements were consulted. An effect on the control surfaces was not noticed. Oscillation measurements showed that the location of the pressure point was also not changed. As already observed in the case of the A4, the aerodynamic damping about the yaw and pitch axes increased.

The first layout, designated C2/E1, possessed trapezoidal wings and an empennage reduced in size compared to the A4 (Fig. 7.42). The four wings had a biconvex circular-arc airfoil section with a slightly rounded nose. They were not equipped with control surfaces. The entire control occurred solely by the empennage control surfaces. Related to the theoretical center-of-gravity dependence, stability for this configuration at an angle of attack of $\alpha = 0$ was

Fig. 7.39   Various wind-tunnel models of the "Wasserfall."

Fig. 7.40 Wind-tunnel support to determine zero-moment positions [16].

Fig. 7.41 Wind-tunnel measurements on the model of the "Wasserfall" with the simulation of the exhaust jet at Kochel (courtesy of Deutsches Museum München).

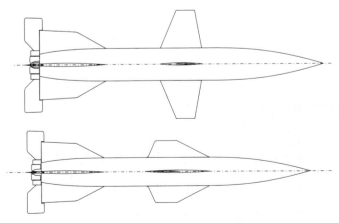

Fig. 7.42 Side view of the "Wasserfall" models C2/E1, version d, (top) and C2/E2 (bottom).

not to be attained. By a rearward shift of the wings, a forward shift of the center of gravity, and an enlargement of stabilizers and jet flaps, the desired stability at $\alpha = 0$ was achieved. Because of the changing center-of-pressure location when crossing the transonic range, this configuration turned out, however, to be unstable within the supersonic regime. The changed center-of-gravity location would also have led to an increased load on the actuators.

To remedy the problem, the HVA tried to eliminate the difference in the center-of-pressure locations at subsonic and supersonic speeds. The approach was the introduction of a wing shape largely similar to the form of the stabilizers. The result was the configuration C2/E2 with a wing of an extremely low aspect ratio ($\Lambda < 1$) (Fig. 7.42, bottom). On this configuration, three-component measurements as well as pressure distribution measurements were performed. Figure 7.43 shows the results of the pressure distribution measurements at subsonic and supersonic speeds (see [48], p. 27). To prevent the center-of-pressure movement within the supersonic regime, the trailing edge of the wing was swept slightly forward. In this way, the almost constant center-of-pressure location depicted in Fig. 7.44 could be obtained (see [48],

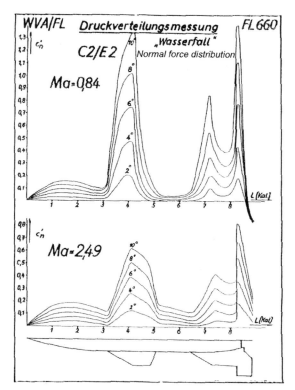

Fig. 7.43 Pressure distribution measurements on the "Wasserfall" C2/E2, distribution of the normal force coefficient, C_n', over the length of the model, L [17].

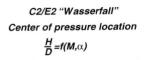

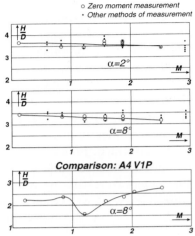

Fig. 7.44   Center-of-pressure location (H/D) of the "Wasserfall" C2/E2 for two different angles of attack (replotted from [18]).

p. 28). These results were gained purely experimentally. There was simply a place for time-consuming analytical investigations.

The aerodynamic theory of low-aspect-ratio wings was in essence available in 1943. Also related results of wind-tunnel measurements at Göttingen and Braunschweig existed, which could have been consulted. How far this knowledge was used is not sure. Anyway, in the report of *H. Kurzweg* concerning the aerodynamic development of the "Wasserfall," the relevant literature is not quoted. The experimental effort that was spent at the HVA to answer this question illuminates the tremendous difficulties faced at that time and the enormous pressure on all involved in this project.

The most favorable arrangement of the wings with respect to the empennage was also considered. It had, in any case, to be a symmetrical arrangement to avoid rolling moments during maneuvers. It was feared that in the case of a parallel arrangement of wings and stabilizers that the downwash of the wings would impinge on them at small angles of attack. To avoid this, the wing was rotated by 45 deg with respect to the stabilizers. However, wind-tunnel tests on this configuration, designated C2/E3, showed that this position led at high angles of attack to intolerable interference effects and, most of all, to too large forces on the control surfaces. For this reason the 45-deg arrangement was finally rejected again.

The "spin-free" condition of the rocket was considered as an essential prerequisite for a perfect flight. A rotation of the rocket about its longitudinal axis had to be avoided such that gyrating oscillations did not occur (see [48],

p. 16). Aerodynamic rolling moments arise due to manufacturing asymmetries or crosswinds affecting wings and empennage. A roll stabilizer was to prevent rotations about the longitudinal (roll) axis. Originally, control surfaces were planned for the wing trailing edges. Because it was too difficult to route the drive mechanisms of the control surfaces through the thin wing, the control surfaces at the stabilizers had to take over their function. To gain the required information about the order of magnitude of the rolling moments about the longitudinal (roll) axis, spin measurements were carried out in the wind tunnel. The testing technique applied was similar to the one used to determine the zero pitch-moment positions. The result was the specification that control-surface deflections of $\eta = \pm 2$ deg should be sufficient to compensate unintentional rolling moments.

Another aerodynamic problem that required intensive investigations was the correct design of the aerodynamic control surfaces. The requirement for high maneuverability at sub- and supersonic speeds resulted in completely different requirements concerning the control-surface actuators compared to the A4. The aerodynamic control surfaces were mounted on a common shaft with the jet rudders. The latter took on the stabilization of the rocket during the first ten seconds of the launch. At a sufficiently high dynamic pressure, the rocket was solely controlled by the aerodynamic control surfaces. The following limits were established for the control surface design that resulted from the demands of controllability and the estimate of structural loads:

- Maximum deflection, $\eta = \pm 25$ deg
- Normal force to be applied per control surface, 10 kN
- Permitted control-surface deflection dependent on the angle of attack, subsonic flow, $0.3 \leq \eta \leq 1$
- supersonic flow, $1.0 \leq \eta \leq 2.0$

A further demand resulted from the limited performance of the available actuators. Flight tests with hydraulic actuators (Autopilot D17), already used in aircraft design, were unsatisfactory [52]. The utilization of suitable electric motors was uncertain. The hinge moments of the control surfaces were, therefore, limited to a maximum of 500 Nm. Under these conditions, a large number of control-surface designs were investigated. Wind-tunnel measurements were mainly carried out in the supersonic tunnel of the Wasserbau Versuchsanstalt (WVA) (Hydraulic Engineering Test Organization) in Kochel [53]. Additional measurements on large models at subsonic speeds were carried out at the Luftschiffbau Zeppelin and at the Luftfahrtforschungsanstalt München (LFM) (Aeronautical Research Establishment). Measurements on an original control surface were carried out in the high-speed wind-tunnel A2 of the LFA up to $M = 0.8$. The measurements in the A2 served to prove the flutter safety. These measurements were scheduled for the end of March

1945. It is not known whether they could be completed [54]. The objective of all investigations was to find a control-surface design that caused only an insignificant movement of the control-surface center of pressure between the subsonic and supersonic flow regime, such that the permitted hinge moment of the control surface was not exceeded. From the requirement of control-surface stability, it followed that the control-surface center of pressure was always to be located downstream of the control-surface hinge line. Variables of the investigation were the control-surface chord, the design of the aerodynamic compensation, control surfaces with geared tabs, and, last but not least, spoiler-type control surfaces as they were successfully used in a similar form on the antiaircraft rocket "Schmetterling." Moment measurements with trailing-edge spoilers showed a dramatic drop in the effectiveness at supersonic speeds, and so this form of control surface was rejected. More promising were tests with tabs, which could, however, not be continued due to the war situation.

As a result of these measurements, two control-surface designs emerged, the control surfaces R12 and R21. Figure 7.45 shows the hinge moments for both control surfaces at subsonic and supersonic speeds. The parameter of the plots is the angle of attack α. The control surface R12 was unstable at small deflections $(dC_R/d\eta < 0)$ and became stable for deflections of $\eta > 10$ deg. The reason for this may be the aerodynamic balance that positions the center of pressure at subsonic speeds upstream of the hinge line. After crossing the transonic range, the center of pressure moved downstream, and stable hinge-moment dependences occurred. The control surface R21 had a smaller chord and, therefore, generated smaller hinge moments. The characteristic dependence of the hinge moments is similar; the gradients are, however, not as steep, which indicates a lesser increase of the hinge moment in case of a control surface deflection.

Figure 7.46 shows the controllability curves that are a measure of the ability of the control surface to produce a certain rocket angle of attack (trim conditions). The configuration C2/E2 was equipped with the control surface R21 and C2/E3 with the control surface R12. If the rocket is tilted about an axis located within the plane of the wing (roll angle $\varphi = 0$), stable conditions for both types of control surface are shown by the curves. At a roll angle of $\varphi = 45$ deg, the curves show at subsonic speeds a pronounced S-shape that indicates an instability within this angle-of-attack range ($\alpha > 15$ deg). Because the angle of attack at subsonic speeds was, anyway, not to exceed 15 deg, this instability was tolerable.

Stability investigations on the complete rocket with different control-surface deflections at subsonic speeds showed a strong dependence of the stability parameter dC_m/dC_A (dC_m/dC_L) on the angle of attack α and the control surface deflection η (see Fig. 7.47). At a roll angle of $\varphi = 45$ deg, the configuration C2/E3 (control surface R12) exhibited an unstable range at angles of

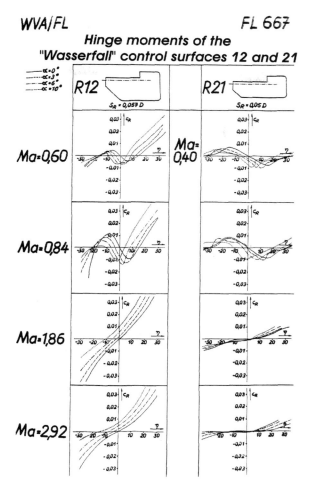

Fig. 7.45   Hinge moments (C_R) for two different control-surface geometries dependent on the control surface position (angle) η and the *Mach* number. Parameter of the plots is the angle of attack α [18].

attack between $\alpha = 15$ deg and $\alpha = 25$ deg. Also here again, however, the limitation of the angle of attack to a maximum of 15 deg holds. The pitching moments are stable for two different center-of-gravity locations at supersonic speeds.

These investigations, which had not yet been completed by the end of the war, clearly show the problems that the engineers were facing at that time. For all airflow directions within the flight envelope, the rocket had to be statically stable. The control effectiveness had to be sufficient to obtain the required maximum load factor of 12. Furthermore the control-surface hinge moments were to stay small enough to be able to use the available actuators.

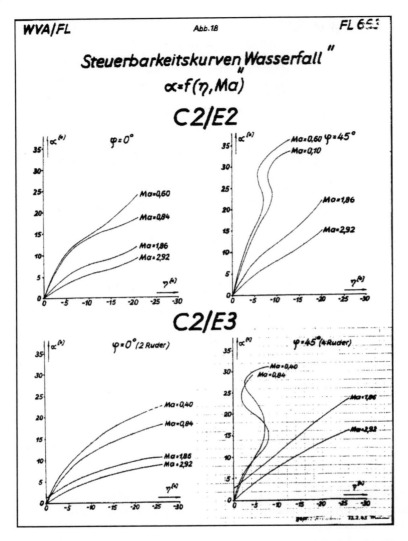

Fig. 7.46 Controllability plots (trim conditions) of the "Wasserfall" models C2/E2 and C2/E3. On the right-hand side, the empennage is rotated with respect to the wing by $\varphi = 45$ deg (η = control-surface deflection, $Ma = M$) [20].

The results of these investigations did not satisfy the engineers. The cause was considered to be the influence of the wing surfaces, located upstream, on stabilizers and control surfaces (see [48], p. 10). It was assumed that with sufficient stability of the missile without control surfaces no more problems would have occurred. A solution of the problem could have been the installation of smaller control surfaces at the nose of the rocket. However, corresponding investigations could not be performed anymore.

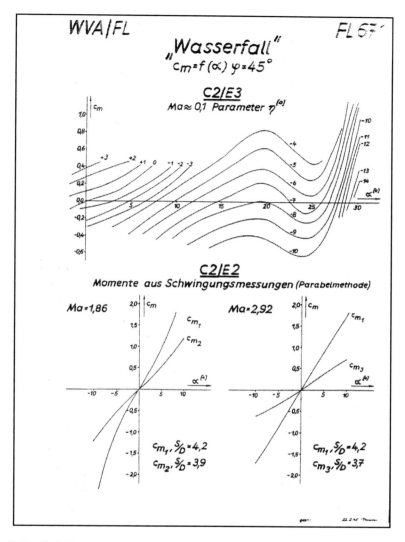

Fig. 7.47 **Stability investigations on "Wasserfall" models: top—pitching-moment coefficient** c_m **vs angle of attack** α **with the control-surface deflection** η **as parameter; bottom—pitching-moment dependence at two different center-of-gravity locations** S/D **[21]; legend: moments determined by oscillation measurements (Parabola method).**

The HVA wanted to know whether the "Wasserfall" rocket, with all its problems, would be able to hit a target at high altitudes. In mid-1943 the Deutsches Forschunginstitut für Segelflug (DFS) (German Research Institute for Gliding) in Ainring received a contract from the Flakversuchsstelle (Artillery Test Center) Karlshagen to build a simulator that would allow the simulation of the target approach of the rocket and that could also be used for training the personnel. This device could be used in the training of operators of an antiaircraft rocket battery (rocket launchers controlled by a simple command unit) [55]. *E. Fischer* had already built training devices for the control of glide bombs by means of the line-of-sight procedure, and he was responsible for designing the simulator. For the "Wasserfall" the automatic guidance by a guide beam was now to be simulated. For solving the equations of motion, electromechanical analog computers could be used, which had already played a role during the development of the A4. The entire equipment was to be set up in a large hangar. Target and rocket were represented by models that moved according to the control commands within a frame (see Fig. 7.48). The entire range of operation (battle space) of the "Wasserfall" could be represented within this setup. At the end of the war, this simulator was almost completed; it was shipped to the United States in the autumn of 1945.

Although the wing problem and the control-surface problem, not to mention the automatic beam guidance problem, had not been satisfactorily solved by the end of 1943, the RLM demanded a flight test of the "Wasserfall" by the end of the year [56]. The trials carried out up to August 1944 suffered primarily from the poor performance of the engine [18]. To test the controllability, the simple stick control had to be fallen back on as the only available method. Up to August, it still did not work properly. These defects could, however, be remedied to some extent by the end of the year. The last examples were equipped with the Autopilot D18 and the control-surface actuators of

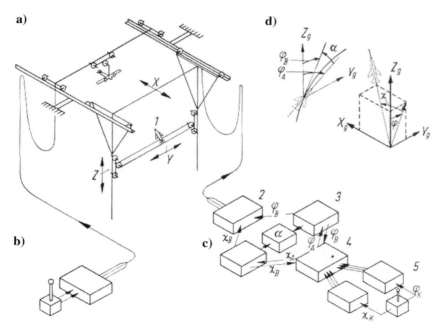

Fig. 7.48   Training simulator for antiaircraft rockets of the DFS Ainring [22].

TABLE 7.5   C2/E2 "WASSERFALL" TECHNICAL DATA

Length with control surfaces	7.85 m
Fuselage diameter	0.88 m
Span	1.89 m
Wing area (one plane)	1.181 m²
Leading-edge sweep	50 deg
Aspect ratio	0.43
Airfoil section	biconvex circular-arc airfoil
Takeoff weight	3500 kg
Thrust	78.4 kN (8 tons)
Maximum angle of attack, subsonic	15 deg
Maximum angle of attack, supersonic	8 deg
Maximum load multiple	12 g
Flight speed at burnout	up to 800 m/s at an altitude of 15 km
Maximum altitude	15 km
Maximum range	24 km

the Siemens K12 autopilot. This used electric motors with gears, which proved quite successful. Up to the end of the year, 22 rockets were launched. Here, speeds of up to 470 m/s and altitudes of up to 14,000 m were reached (Table 7.5). The automatic stabilization finally worked reliably when crossing the critical area around $M = 1$. Because of the poor performance of the actuators, only simple maneuvers could be flown. Up to the termination of the program, several interim versions of the "Wasserfall" rocket were built and tested (Figs. 7.49 and 7.50). Altogether 35 rockets were launched; however, most of them failed. Some of these rockets carried the telemetry

Fig. 7.49 "Wasserfall" C2/E1d shortly prior to the launch at Peenemünde (courtesy of Deutsches Museum München).

Fig. 7.50 Final version of the "Wasserfall" C2/E2 ready for launch off the Greifswald Oie (courtesy of Deutsches Museum München).

transmitter "Messina," which was to provide information about the systems and the flight progress [49]. According to another source, 28 "Wasserfall" rockets were launched with 25 failures [57]. These activities, the many still unresolved problems concerning the engine and the remote control, as well as the number of failures, clearly show how far away it still was from operational readiness by the end of the war.

The suggestion for a smaller version of the "Wasserfall" was made by the design department of the Elektromechanische Werke (Electromechanical Works) at the end of 1944 [58]. This suggestion corresponded to the desire of the RLM to save high-grade steel and rocket fuel. The small "Wasserfall" was a scaled-down version of the full-scale configuration that needed only 30% of the amount of Salbei (nitric acid fuel) of the full-scale rocket but showed, nevertheless, rewarding possibilities of application. Since the beginning of 1945, there was still no chance to guide a rocket on a guide beam; it was from the start designed for control by sight. The small "Wasserfall" was not built anymore. The technical data are, however, given for comparison in Table 7.6.

7.4.1 TRIALS AND FURTHER DEVELOPMENT IN THE SOVIET UNION

After the occupation of Central Germany by the Soviet troops, besides the long-range missile A4, the German antiaircraft rockets also aroused the interest of the Soviet experts. To be able to use this technology in the Soviet

TABLE 7.6 SMALL "WASSERFALL" TECHNICAL DATA

Length with control surfaces	5.68 m
Fuselage diameter	0.64 m
Span	1.36 m
Leading-edge sweep	50 deg
Aspect ratio	0.43
Airfoil section	biconvex circular-arc airfoil
Takeoff mass	1450 kg
Thrust	28.4 kN (2.9 tons)
Maximum load multiple	2 g
Maximum range	20 km
Maximum altitude	12 km

Union, a research institute, designated "Berlin," was founded in Berlin. This institute was directly subordinated to the special committee of the Counsel of Ministers of the Soviet Union in Germany. Primary task of the institute "Berlin" was the restoration of the technical documentation and the reconstruction of the antiaircraft rockets "Wasserfall," "Schmetterling," and "Rheintochter."

Matthias Uhl [29] reports in detail about the activities of the institute "Berlin," on the political and economic environment wherein the employees of the institute worked, as well as on the transfer of all activities to the Soviet Union. Results from the retesting in the Soviet Union and subsequent improvements suggested by Soviet experts are considered here, allowing conclusions concerning the problems that had already occurred during German testing.

The German and Soviet engineers faced great problems. After the evacuation of Peenemünde and other development and production sites in Central Germany prior to the arrival of the Soviet troops, all documents and components were brought to safety. The main share of the scientific and technical documentation as well as the rockets already completed fell into the hands of the Americans. In the part of Germany occupied by the Soviets, only fragments of technical drawings, single components, demolished test facilities, and destroyed production sites remained. Under the supervision of the Soviet experts, the German employees still remaining in the Soviet occupational zone were to carry out a complete reconstruction of the antiaircraft rockets up to and including flight testing until the end of 1948.

The remaining components of the antiaircraft rockets were the starting point for an unprecedented engineering effort. Not only the rocket, including propulsion, fuel tanks, pumps, etc., had to be reconstructed, but also the complete rocket control from the stabilization gyroscopes to the control-surface actuators. There was as well a lack of instrumentation, command equipment for the remote control, laboratories, and many other things. Not even the rocket fuels were available. Only under the greatest of difficulties was the institute "Berlin" successful in completing the technical documentation and two examples of the "Wasserfall" and one of the "Rheintochter" by the end of October 1946. A little later, three examples of the "Schmetterling" followed. Control and guidance systems, needed to carry out flight tests, were not available.

The German experts and all facilities were transferred to the Soviet Union on October 22, 1946. The German rocket experts were integrated into the newly founded research institute of the Armament Ministry "NII 88." There, the "Wasserfall" was to be further developed into the Soviet configuration R-101 and the "Schmetterling" into the model R-102. They wanted to test the "Rheintochter" but did not consider a further development. The "Wasserfall" was considered by the Soviet experts as the most promising candidate regarding a basis for their own development.

The rockets, still built in Germany, obviously only served as visual objects because they were not complete. The Soviet experts were generally of the opinion that the rocket was in many aspects in need of improvements. For flight testing the "Wasserfall," therefore, the improved Soviet version R-101 was envisaged from the beginning. During 1947, the German experts had performed the following tasks [37]:

1) Reconstruction of a command module (transmitter and receiver) for a remote line-of-sight procedure.

2) Construction and testing of such a device for the "Wasserfall."

3) Development of a system for the automatic target search during the final phase of the target approach.

4) Concept of a control system for the automatic guidance of the "Wasserfall" on a radio beam.

5) Investigations concerning the aerodynamics of the "Wasserfall" and development of a new aerodynamic concept.

Up to the end of 1948, 30 test examples of the R-101 were built at the NII 88 under extensive utilization of German components. Twelve examples of the R-101 were launched in January 1949 at the test field Kapustin Jar. The expected performance was not achieved. The reason was that the thrust was too low (7.6 tons instead of 8 tons) and the thrust burn time was shorter, 35 to 40 s instead of 46 s The rocket did, therefore, not achieve the demanded terminal velocity and altitude. It rotated, furthermore, about its roll axis and experienced motions similar to the pendulum of a clock around a center of rotation outside the body. The problems were related to the play in the control-surface actuators being too large. In addition, the deployment of all four control surfaces for the roll stabilization proved to be problematic; only two control surfaces would have been better. The design of the jet rudders was also in need of a modification [59].

The propulsion problem, still existing, led to an improved model of the R-101 designated R-109. Instead of the fuel combination nitric acid/Tonka, nitric acid/kerosene was used. The nitrogen pressure vessel was replaced by a powder-based gas generator. In this way more space for fuel was gained, and the weight could be reduced. Of the model R-109, six rockets were built and successfully launched.

After improvements of the fuel feed system and the flight control, the flight tests with the R-101 resumed in November 1949. The flight performance was now satisfactory; however, control at supersonic speeds was almost impossible. After further improvements of the flight control, another 25 rockets were launched during the winter of 1950–1951. Controls and controllability were now satisfactory; however, there was still a lack of a suitable system for the target acquisition and the control of the terminal flight phase. As was already recognized in Germany, the control by sight was not a solution. A radio-beam control with a target identification by radar was being developed, but was not available in the foreseeable future. Also a passive search head, called "Strahl (Beam)," was built, but the development was stopped for lack of suitable ground-based equipment and test instrumentation [60].

Of interest here is the deployment of the German target seeker "Max." "Max A" was an active target detector operating on a wavelength of 3 cm. "Max" was envisaged for the final approach control of the rockets and the provision of the necessary corrections to the actuators of the control surfaces. An important component of "Max" was a proximity fuse that was urgently needed for the antiaircraft rockets [61]. Designated R-108, several R-101 rockets were equipped and tested with this target seeker head. It is not known whether the R-108 was deployed against airborne targets. This would have needed the control by guide beam, which was not yet available at that time.

This was also ultimately the reason why the work on the R-101 was stopped on August 17, 1951. On the other hand the objective of the Soviet experts was accomplished, namely, to achieve and to surpass the German development status of 1945. From that date they could continue to work without the support of the German experts.

Because of the analysis and the flight tests of the rocket "Wasserfall," the following improvements were introduced in cooperation with the Soviet experts:

1) Aerodynamics: Relocation of the control surfaces to the tip of the rocket.

2) Flight control: Improved flight controller with regard to stabilization and control, droppable jet rudders, control via radio beam, and target seeker head (active and passive).

3) Propulsion: Change to kerosene and gas generator, improvement of the combustion chamber.

4) Ground: Development and construction of a radio-beam transmitter to control the rocket.

Fig. 7.51 Soviet antiaircraft rocket Lawotschkin R-113/S-25 "Berkut" (SA-1 "Guild," 1953). Starting point for this rocket was the German "Wasserfall."

All experience and improvements concerning the R-101 entered into the design of the R-113/S25 "Berkut" (SA-1 "Guild"). The S25 (Fig. 7.51) was a rocket, taking off vertically, with jet rudders for the stabilization during the launch phase. The jet rudders were dropped after turning the rocket at the target level onto the radio guide beam. Both the rocket and the target were tracked by radar. The positions of the rocket and the target were constantly compared and the rocket kept on a collision course. *Kurt Kracheel* writes in his memoirs that he still worked together with a group of specialists on the flight control system and the radio-beam control of this rocket until after 1951 [62]. The German experts accompanied the development of this rocket up to the operational readiness. The R-113/S25 was successfully deployed for the first time against an airborne target in May 1953.

7.4.2   FURTHER DEVELOPMENT IN THE UNITED STATES

In spring 1945, when the Americans first received information about the antiaircraft rocket "Wasserfall," antiaircraft rockets of their own were already under development ("Lark," especially developed for fighting kamikaze attackers, and the "Nike-Ajax," an antiaircraft rocket to fight high-altitude jet bombers). Both developments have their origin in requirements of the U.S. Navy (for the "Lark") and the U.S. Army (for the "Nike-Ajax") in 1944. How much German technology has been used in the development of these rockets is unknown. The Americans had a considerable lead in the areas of flight control, automatic navigation (control), active and passive target detectors, as well as radar technology over similar developments in Germany. Also with regard to the development of smaller liquid-fuel rockets, they possessed sufficient experience. The Americans were, therefore, not necessarily dependent on German know-how in the development of their antiaircraft rockets.

The antiaircraft rocket Fairchild "Lark" was guided by means of a radio guide beam into the proximity of the target. The final approach was accomplished by a passive radar seeker, which reacted to the radiation reflected by the target after having been sent out by tracking radar. Such a system had already been installed on the glide bomb "Bat" and in spring 1945 successfully deployed against Japanese ships. The configuration had been defined in January 1945 (Fig. 7.52). The "Lark" possessed a liquid-fuel rocket engine and was launched from a mobile launcher by means of boosters. The maximum speed was limited to $M = 0.85$ in order not to enter the compressible flow regime [63].

Fig. 7.52 First American antiaircraft rocket Fairchild XSAM-N-2 "Lark" (1948).

The deployment of the first German aircraft with a jet engine in summer 1944 quickly made it obvious that these aircraft could not be combated with conventional antiaircraft weapons. In view of their own jet bomber program, the U.S. Army Air Corps resolved in January 1945 to create an antiaircraft rocket for the deployment against high-altitude, high-speed jet bombers. The Western Electric Company and the Bell Telephone Laboratories came to the conclusion in a three-month study, which lasted until May 1945, that the planned antiaircraft rocket should combat aircraft at speeds of up to 1000 km/h and at altitudes of up to 18,000 m. Only a highly maneuverable supersonic rocket, which had to be guided on a radio beam into the vicinity of the target, was possible. A computer was to constantly compare the positions of rocket and target and to guide the rocket to a predetermined collision point. When the collision point had been reached, the computer automatically generated the ignition command for the explosive charge. The liquid-fuel rocket was to take off vertically with the aid of a solid-fuel booster, accelerate to supersonic speed, and then turn onto the guide beam. The Douglas Company was contracted with the aerodynamic layout. As a suitable shape for a supersonic missile, Douglas considered a slender pointed fuselage. At the rear fuselage, four triangular stabilizers with control surfaces for the roll stabilization were installed. At the forward fuselage, four low-aspect-ratio control-surface fins, adjustable as a whole, were installed for the control about the pitch and yaw axes (Fig. 7.53). Wind-tunnel measurements up to $M = 1.72$ were carried out in the only supersonic tunnel available at the time in the United States at the Aberdeen Proving Grounds of the Ballistic Research Laboratories. Up to the end of this study, no

Fig. 7.53 World's first operational antiaircraft rocket Douglas SAM-A-7 "Nike-1" (1951).

information about German antiaircraft rockets had been available [64]. The development philosophy tended to consider the rocket as lost equipment and to keep it, correspondingly, as simply as possible and to integrate expensive technological devices only into the ground-based equipment.

After six years of development in November of 1951 at the test range "White Sands," the first downing of a target aircraft succeeded. The "Nike-Ajax" was the first operational antiaircraft system worldwide and secured the United States a technological lead over all other countries.

The Joint Chiefs of Staff of the Allied Forces in Europe announced in July 1945 the transfer of about 350 German scientists and engineers to the United States, where they were to help the Americans with their knowledge in the war against Japan. Among them were 100 experts related to the long-range missile A4 and the antiaircraft rocket "Wasserfall." The development of the "Wasserfall" was to be completed and the rocket eventually deployed against Japan [65]. The first experts, among them *Wernher von Braun*, arrived, however, in the United States only after the Japanese surrendered in September 1945. Others followed until the end of the year.

Probably thought of as an alternative to the "Nike-Ajax," the development of an antiaircraft rocket based on the "Wasserfall" commenced in 1946 under contract of the U.S. Army Ordnance Corps. It may be assumed that German experts were involved in this development. The latter was part of the so-called "Hermes Program" that also included the reconstruction and further develop-ment of the A4 rocket. Test data transmitter and flight control were tested on an A4. During 1950, several "Hermes A-1" rockets were tested at the White Sands Proving Grounds. The tests were completely satisfactory. From the launch up into the supersonic regime, the rocket was stable and controllable within the entire operational range. Probably under the impression of the success of the "Nike-Ajax" program, the development of the antiaircraft rocket was stopped in October 1951, but the program for a medium-range missile continued. The "Hermes A-1" rocket was similar in its dimensions to the "Wasserfall" but lighter than this configuration and, therefore, required less thrust. Whether the Americans benefited from the "Hermes Program," running parallel to the "Nike-Ajax" program, in the development of the "Nike-Ajax" or whether German experts still cooperated in the "Nike-Ajax" development program can no longer be determined. It is, however, interesting that the Russians in the development of the R-113/S25 followed a similar approach as the Americans in the case of the "Nike-Ajax."

REFERENCES

[1] F. *Trenkle*, "Die deutschen Funklenkverfahren bis 1945," AEG-Telefunken Aktiengesellschaft, Ulm 1982, pp. 72–92.

[2] O. W. *von Renz*, "Deutsche Flug-Abwehr," 20 Jahrhundert, Verlag Mittler & Sohn, Berlin, 1960, p. 175.

[3]   W. v. Axthelm, "Übersicht über den Entwicklungsstand und die Entwicklungsabsichten der Flakartillerie," Az. 67, No. 1373/42 g.Kdos., Sept. 18, 1942, DM.

[4]   W. v. Axthelm, "Taktisch-technische Forderungen für die Entwicklung von Flakraketen," Az. 67, No. 1730/42 g.Kdos.," Sept. 25, 1942, DM.

[5]   "Schreiben Chef der Heeresrüstung," Az. 72 n-p, Wa Prüf 11/HAP No. 977/42 g.Kdos., Nov. 22. 1942, DM.

[6]   F. Trenkle, "Die deutschen Funkmessverfahren bis 1945," Motorbuch Verlag Stuttgart, 1979, p. 30.

[7]   H. H. Kurzweg, "The Aerodynamic Development of the V-2," ed. By Th. Bennecke, A. W. Quick, "History of German Guided Missiles Development," AGARDOGRAPH 20, Verlag E. Appelhans & Co, Braunschweig, 1957, pp. 50–68.

[8]   Th. Benecke, K.-H. Hedwig, J. Hermann, "Flugkörper und Lenkraketen," Die deutsche Luftfahrt, Band 10, Bernard & Graefe Verlag, Bonn, 1987.

[9]   M. Griehl, "Deutsche Flakraketen bis 1945," Waffen-Arsenal, Sonderband 67, Podzun-Pallas-Verlag GmbH, Wölfersheim-Berstadt, 2002.

[10] G. Braun, P. Retert, "Entwurf eines Versuchsmodells F 22 zur Entwicklung einer Flakrakete, LFA-Bericht," May 21, 1941, ZWB 12884 (Deutsches Museum München); this report is a fragment of text without figures.

[11] W. Wernitz, "Research and Development of the Guided Missile 'Feuerlilie'," ed. by Th. Bennecke, A. W. Quick, "History of German Guided Missiles Development," AGARDOGRAPH 20, Verlag Appelhans & Co, Braunschweig, 1957, p. 419.

[12] "Schreiben Forschungsführung, Zusammenfassender Bericht über den Stand der Forschungsaufgaben für ferngelenkte, aufsteigende Flugkörper im Bereich der Forschungsführung (Feuerlilie)," Berlin, Feb. 2, 1943, BA-MA RH/8/I 1239.

[13] B. Krag, "Die Luftfahrtforschungsanstalt in Braunschweig-Völkenrode 1936–1945, ed. by R. Ahlers, G.–Sauerbeck, "Geschichte des Forschungsstandortes Braunschweig-Völkenrode," Verlag E. Appelhans & Co, Braunschweig, 2003, pp. 32–35.

[14] "Tätigkeitsbericht der Abteilung AF des Instituts für Aerodynamik für die Zeit vom 1.4.1942 bis March 31, 1943," DLR Archiv Köln.

[15] "Notes on Interrogation of Dr. Gerhard Braun," July 18, 1945, Doc. No. SR/237/CFW.

[16] "ETO Ordnance Technical Intelligence Report No. 325," June 23, 1945.

[17] R. Smelt, "A Critical Review of German Research on High-Speed Airflow," Journal of the Royal Aeronautical Society, Vol. L, No. 432, 1946, p. 925.

[18] "Vorläufige Flugerprobung von Flakraketen," Erprobungsbericht Br. B. Nr. 21/45, Flakversuchsstelle der Luftwaffe, Karlshagen, Jan. 5, 1945.

[19] R. H. Norris, "(CIOS): Interrogation Report on Gerhard Braun, regarding Flak Rockets," May 23, 1945.

[20] A. Hafer, "Flugmechanische Forschung im Zweiten Weltkrieg: das Beispiel der Deutschen Forschungsanstalt für Luftfahrt in Braunschweig," ed. by H. Schubert, "Luftfahrtforschung, Luftfahrtindustrie, und Luftfahrtwirtschaft in Braunschweig," Texte einer Vortragsveranstaltung der DGLR-Fachgruppe 12 'Geschichte der Luft- und Raumfahrt' am, May 18, 1990, p. 2.9.

[21] H. Strobel, "Leistungsrechnung für die Pulver-Flak-R," Mitteilung BSM/StF, Peenemünde Oct. 15, 1942, Archiv der Wehrtechnischen Studiensammlung Koblenz, Rheinmetall-Borsig Akte H 0238.

[22] "Schreiben Chef H Rüst," Az. r 0055 Wa Prüf 11/II, Bb.No. 1104/42 g.Kdos., Dec. 7, 1942, DM.

[23] A. Fricke, "Beschreibung des 1. Entwurfes einer nach Sicht gesteuerten Flakrakete ('Rheintochter' 1. Entwicklungsstufe)," Rheinmetall-Borsig AG, Bericht LB 19, April 1943, Archiv der Wehrtechnischen Studiensammlung, Koblenz, Akte H 0238, G 1340/3/012.

[24] H. Ludwieg, "Windkanalmessungen im Unter- und Überschallbereich am Modell der Flakrakete 'Rheintochter'," AVA Bericht 43/H/20, July 3, 1943.

[25] "Flugbahnauswertung vom Dezember 1943," Rheinmetall-Borsig Dokument, Archiv der Wehrtechnischen Studiensammlung, Koblenz, Rheinmetall-Borsig, Akte H 0238, G 1340/3/008.

[26] G. Brennecke, G. Keydell, "Messungen an einem rollwinkelunempfindlichen Fluggerät (Gerät R1)," AVA Bericht 44/W/59, Dec. 12, 1944.

[27] A. Fricke, Müller, "Beschreibung des 2. Entwurfes einer nach Sicht oder elektrischer Ortung gesteuerten Flakrakete mit Flüssigkeits- oder Pulverantrieb ('Rheintochter' R 3f und R 3p) Rheinmetall-Borsig Bericht LB 28," Sept. 1944, Archiv der Wehrtechnischen Studiensammlung Koblenz, Akte H 0238, G 1340/3/011.

[28] G. Keydell, G. Brennecke, "Messungen an einem rollwinkelunempfindlichen Fluggerät," AVA Bericht 44/W/27 vom 31.7.1944.

[29] M. Uhl, "Stalins V-2, Der Technologietransfer der deutschen Fernlenkwaffentechnik in die UdSSR und der Aufbau der sowjetischen Raketenindustrie 1945 bis 1959," Wehrtechnik und wissenschaftliche Waffenkunde Band 14, Bernard & Graefe Verlag, Bonn, 2001, p. 134.

[30] H. A. Wagner, "Guidance and Control of the Henschel Missiles," ed. by Th. Benecke, A. W. Quick, "History of German Guided Missiles Development," AGARDOGRAPH No. 20, Verlag E. Appelhans & Co., Braunschweig, 1957.

[31] H. A. Wagner, "Ferngelenkte Gleitbomben," in "Sonderprobleme der Fernlenkung," Schriften der Akademie der Luftfahrtforschung, Berlin 1942.

[32] J. Henrici, D. Mandel, "Aerodynamik des 'Schmetterling'," in "Ferngesteuerte Flugkörper nach Arbeiten von Professor Wagner," Archiv der Wehrtechnischen Studiensammlung, Koblenz, G 1500 V107.

[33] O. Walchner, "Messungen am Gerät Hs 117 bei Unterschallgeschwindigkeiten," AVA Bericht: 44/H/11, April 29, 1944.

[34] "Projekte S II, Projektbeschreibung der Firma Henschel," Jan. 14, 1945, Archiv der Wehrtechnischen Studiensammlung, Koblenz, G 1340/3/005.

[35] O. Walchner, "Windkanalmessungen am Henschel-Gerät 'Zitterrochen' bei Unter- und Überschallgeschwindigkeiten," AVA Bericht: 44/H/13, July 10, 1944.

[36] O. Walchner, "Windkanalmessungen am Flügel des Henschelgerätes 'Zitterrochen' bei Unter- und Überschallgeschwindigkeiten," AVA Bericht: 44/H/22, Oct. 24, 1944.

[37] M. D. Jewtifjew, "Aus der Geschichte des Aufbaus eines Flugabwehrsystems in Russland," Verlag Wusowskaje Kniga, Moskau, 2000.

[38] J. Michels, "Peenemünde und seine Erben in Ost und West," Bernard & Graefe Verlag, 1997, p. 262.

[39] A. Lippisch, "Ein Dreieck fliegt—Die Entwicklung der Delta-Flugzeuge bis 1945," Motorbuch Verlag Stuttgart, 1976, p. 85.

[40] L. Lawrence, "Great Enzian Flak-Rocket Development," F-IR-6-RE 1946 (DM); contains 6 "Interrogation Reports" of Dr. Wurster. (After the occupation of the Messerschmitt works, Dr. Wurster was interrogated regarding the development of the 'Enzian' by experts engineers of Bell Aircraft, General Electric, Curtiss Wright, and de Havilland Aircraft, as well as by members of the CIOS-Team 367. The most prominent interrogator was Hugh L. Dryden of NACA. Unclear if 'Enzian' E-1 or E-4 is mentioned.)

[41] S. Eicke, "Terminplan ab Januar 1945 für den Windkanal A3," Bundesarchiv-Militärarchiv Freiburg, RL39/350.

[42] H. Schlichting, E. Truckenbrodt, "Aerodynamik des Flugzeugs," zweiter Band, Springer-Verlag, Berlin-Göttingen-Heidelberg, 1960, p. 174.

[43] Scholkemeier, Lindemann, Kopfermann, "Tätigkeitsberichte des Instituts für Aerodynamik der LFA für den A1," DLR-Archiv Köln.

[44] H. J. Ebert, J. B. Kaiser, K. Peters, "Willy—Messerschmitt—Pionier der Luftfahrt und des Leichtbaus," Die deutsche Luftfahrt, Band 17, Bernard & Graefe Verlag Bonn, 1992, p. 285.

[45] Th. Zobel, "Der Hochgeschwindigkeitskanal der Luftfahrtforschungsanstalt Hermann Göring," Vortrag, June 4, 1943, Jahrbuch 1943/1944 der deutschen Akademie der Luftfahrtforschung, 1944, p. 140.

[46] J. Engelmann, "Geheime Waffenschmiede Peenemünde," Podzun-Pallas-Verlag, Friedberg, pp. 40–47.

[47] M. J. Neufeld, "The Rocket and the Reich," Harvard Univ. Press, Cambridge, MA, U.S., 1999, pp. 248–251.

[48] H. Kurzweg, "Die aerodynamische Entwicklung der Flakrakete 'Wasserfall'," Wasserbau-Versuchsanstalt GmbH, München, WVA-FL Archiv No. 171, March 15, 1945.

[49] R. H. Reichel, "Die ferngesteuerte Flabrakete C2 'Wasserfall'," INTERAVIA, 6 Jahrgang, No. 10, 1951.

[50] Kretschmer, "Ermittlung momentenfreier Lagen von Modellen im Unterschallwindkanal," Wasserbau-Versuchsanstalt Kochelsee, Kochel, WVA FL/Ae Archiv No. 66/137, June 30, 1945.

[51] R. Lehnert, "Widerstandsbeiwerte für das A4V1P mit Berücksichtigung des Strahl- und Reibungseinflusses für Unter- und Überschallgeschwindigkeit—Untersuchung der Strahlexpansion," HVA Aerodynamisches Institut, Archiv No. 66/105, Peenemünde, March 24, 1943.

[52] M. J. Neufeld, "The Rocket and the Reich," Harvard University Press, Cambridge, MA, U.S., 1999, pp. 234.

[53] P. P. Wegener, H. U. Eckert, "Ruderentwicklung im Windkanal zur Raketensteuerung im Unter- und Überschall," Wasserbau-Versuchsanstalt München, April 25, 1945.

[54] W. Eicke, "Terminpläne für die Windkanäle des Instituts für Aerodynamik und des Instituts für Gasdynamik ab March 1, 1945," Bundesarchiv-Militärarchiv Freiburg, RL 39/350.

[55] E. Fischel, "Contributions to the Guidance of Missiles," ed. by Th. Benecke, A. W. Quick, "History of German Guided Missiles Development," Verlag Appelhans & Co., Braunschweig, 1957.

[56] M. J. Neufeld, "The Rocket and the Reich," Harvard University Press, Cambridge, MA, U.S., 1999, pp. 236–238.

[57] R. W. Kluge, C. L. Fay, "U.S. Technical Industrial Investigation Committee," CIOS Target No., Aeronautical Subcommittee, Aug. 1945, "Kap. 2: Self Propelled Missiles," p. 27.

[58] L. Roth, "Entwurf einer kleinen Flakrakete vom Typ 'Wasserfall'," Elektromechanische Werke GmbH, Bb.No. 21/72/45 g.Kdos., Jan. 1945 (DM).

[59] A. N. Olenik, "Bericht über eine Dienstreise an den Leiter der Planungsbehörde für die Militärindustrie," M.I. Malachow, Dec. 13, 1949.

[60] N. Kirpitschnikow, "Auskunft an den Vorsitzenden des Komitees 2 im Regierungsrat der Sowjetunion," Genosse Bulganin, Jan. 5, 1949.

[61] G. Güllner, "Summary of the Development of High-Frequency Homing Devices," ed. by Benecke, Quick, "History of German Guided Missiles Development," AGARDOGraph 20, Verlag Appelhans & Co, Braunschweig, 1957.

[62] K. Kracheel, "Flugreglerentwicklungen deutscher Spezialisten in der UdSSR (1946 bis 1958)," ed. by H. Schubert, "Die Tätigkeit deutscher Luftfahrtingenieure und –wissenschaftler im Ausland nach 1945," DGLR Blätter zur Geschichte der Deutschen Luft- und Raumfahrt V, Bonn-Bad Godesberg, 1992.

[63] F. I. Ordway, R. C. Wakeford, "International Missile and Spacecraft Guide," McGraw Hill, New York, U.S., 1960, p. 100.

[64] *M. T. Cagle*, "Historical Monograph Development, Production and Deployment of the Nike Ajax Guided Missile System 1945–1959," U.S. Army Ordnance Missile Command, Redstone Arsenal, AL, U.S., June 1959.

[65] *K. J. Weitze*, "Guided Missiles at Holloman Air Force Base, Test Programs of the United States Air Force in Southern New Mexico, 1947–1970," in "Cultural Resources Publication No. 5," Air Command Holloman Air Force Base, NM, U.S., Nov. 1997.

Chapter 8

TRANSFER OF GERMAN HIGH-SPEED AERODYNAMICS AFTER 1945

BURGHARD CIESLA AND BERND KRAG

8.1 INTRODUCTION

A confidential study of the development of German high-speed aerodynamics during WWII existed at the National Advisory Committee for Aeronautics (NACA) in the United States in October 1946 [1]. It questioned why the Germans were so far ahead of the Allies at the end of the war. Named as a decisive reason was the enormous effort that went into rocket development that, in turn, confronted German aeronautical research with problems, whose solution became only possible by treading radically new paths. The NACA study also contains the argument that it was less the resourcefulness of the Germans, but rather the war that had forced Germany, in comparison to other aeronautically-oriented nations, to rapidly adopt a new research dimension.

Indeed, due to the allied superiority in aeronautical equipment, the aerial warfare against Germany caused the Germans to search intensively for technological escapes. Shortly after the war, the American Senator *Albert D. Thomas*, chairman of a congressional aeronautical committee, brought it to a point: "We won the aerial warfare against the Germans with muscle and not with intellect. We suffocated them with the pure weight of our aircraft" (see [2], p. 359). Be that as it may, when this "muscle" released its forces over German towns and industrial centers, all forces were mobilized in Germany in search of effective antidotes. Jet and rocket-propelled fighters, antiaircraft rockets, flying bombs, and long-range missiles represented without doubt such effective antidotes, which were only possible due to the "visionary encouragement of radically new ways." Aeronautical research of the Allies did not have to meet this challenge to arms technology during the war. In the end and unique up to that time, the Germans possessed an experience, systematically compiled research data and a battery of supersonic wind tunnels which amazed the allied victors."

"If we are not too proud to make use of German-born information, much benefit can be derived from it and we can advance from where Germany left off" [3]. So answered the head of the American Air Technical Intelligence, *Donald Putt*, in May 1946 during a presentation in response to the question why America should certainly use the knowledge and experience of the Germans. History has shown that America had not been too proud and it has also shown that the British, French, Russians, and other nations acted the same way. The knowledge and technology transfer from Germany, occurring after 1945, took place according to a principle that is still valid today: The benefit legitimizes the action.

8.2 SCIENTIST AND TECHNOLOGY TRANSFER AFTER 1945

Transfer of German and Austrian scientists to the victors after WWII and the comprehensive utilization of the German scientific-technical "know-how" is a historical novelty. In no other previous or subsequent wars has technology transfer to the victor played such a prominent role. The victors acquired, more or less systematically, scientific, technical, commercial, and economic intellectual property, and they took scientists and technicians—voluntarily or by force—into their services. Firstly the scientific-technical competence was questioned. Valid was the principle: Legitimation by usefulness. The people responsible for and active in the business of knowledge and technology transfer acted worldwide according to this pragmatic basic pattern. Nowhere in the world wanted and no power could forgo the technical expertise of Germany [4, 5].

The deployment of new weapon systems by Germany during the war had made it abruptly clear to the Allies that, on their side, there were considerable technological gaps. Especially German long-range missiles (A4/V2), deployed since September 1944, caused for some time an atmosphere of helplessness and concern at the allied commands. Here, it was seriously considered, among other things, using long-range missiles as a potential carrier system for chemical or bacteriological weapons. It mattered at the same time that the missiles approached without advance warning and that there were no means of defense. With the dark humor typical of the British, it was explained later: "If you heard the detonation at impact, you had already survived; if not, there were no further personal problems" [6]. In a similar way the deployment of the German jet and rocket fighters was also of concern. The novel jet fighters of the type Me 262 succeeded during a single air battle in February 1945 to down 25 American B-17 long-range bombers. The American 8th Air Fleet again lost 24 of her bombers on March 18, 1945, to an Me 262 attack, and on March 18, 1945, the new German jet fighters downed 17 British Lancaster bombers during an air battle. In view of this loss, the Allied Air Command considered to at least stop the daytime attacks on Germany [7, 8].

The Allies generally considered the deployment of the new German weapons as serious danger. To avoid further bad surprises, the military advance on the continent was accelerated in 1944–1945, and reconnaissance missions to assess the German technological progress were established. From December 1943, a reconnaissance mission under the codename ALSOS to assess the German nuclear weapons development assumed a central position [9, 10]. In addition the British and Americans were informed "in real time," so to speak, about the use of the new German weapons technology in Europe and, moreover, about the technology transfer from Germany to Japan by listening to the radio traffic [11].

The British decoding center was located at Bletchley Park, about 50 kilometers away from London, where since the beginning of the war the encoded German radio traffic ("Enigma") was decoded with growing success. The deciphering of the German radio traffic, but also breaking the code used by the Japanese embassy in Berlin, provided the Allies with detailed information about deployment plans, troop movements, supply problems, and most of all, also about the new German weapons developments [12–14]. The "Code breakers" of Bletchley Park discovered the German "wonder weapons" very early. The decoded radio messages advised the British and Americans about production sites, test facilities, deployment plans, and also about problems. The decoding experts were, for instance, able to extract from the German radio traffic in February 1945 that the deployment of the Me 262 had to be strictly limited due to the fuel shortages [7, 15].

The existence of Bletchley Park was, however, known to only a small and select group of people because the Germans were not supposed to know that the British and Americans knew what they actually were not supposed to know. Therefore, the origin of the information was also veiled from their own military command. Thus the certain knowledge concerning, among other things, the reduced deployment of the Me 262 in February 1945 could only indirectly enter the discussion concerning the possible suspension of the bomb raids mentioned previously. In other words, the truth had to be told "differently" to their own ranks. Nevertheless, the opinion that the German code had not been broken was held by the former German staff officers for more than twenty years after the end of WWII. Only the British publications of the seventies revealed the whole extent of the "battle for the code" during WWII [16].

In this way, the Allies learned, moreover, of the German-Japanese technology contacts via submarines and also heard, among other things, what the enemy knew about their own intentions. The Japanese Military Attaché in Berlin had, for instance, in the summer of 1944 sent a radio message to Tokyo stating that the Americans after their victory over Germany intended to bring at least 20,000 German engineers and scientists to the United States to use their knowledge and abilities in the fight against Japan [14, 15, 17]. In view of the uncertainty concerning the duration of the war in the Pacific, the American and British military authorities had the threatening visions of Japanese flying bombs, rocket fighters, or rocket attacks on towns on the west coast of the United States launched off Japanese submarines. It must be considered here that in the spring of 1945, a successful completion of the nuclear bomb development in the United States could not be foreseen at all. It was assumed that the war against Japan would last another one and a half to two years [18]. The question over and over again was: What if the Japanese military deployed German "wonder weapons"? The story of the German submarine "U 234" shows how real these fears had been. The U 234 sailed for Japan with some German weapons experts, two Japanese, and was jam-packed with technical documents (Me 163, Me 262, rocket documents) and material such as uranium oxide on a secret mission in March 1945. However, the "last boat" did not reach Japan but was captured by the escort destroyer USS Sutton and towed to America [17, 20].

As long as the war against Japan lasted, nobody in Washington or London had needed to be convinced that they had to be pragmatic about the scientist and technology transfer from Germany. For this purpose, the U.S. Army Air Force founded a special reconnaissance operation with the cover name "Lusty," which cooperated with the British Royal Air Force. The central objective was to shorten the war with Japan by using German "know-how" [21]. During the occupation of parts of the later Soviet occupational zone by the Americans and British between April and June of 1945, 1500 to 1600 scientists and technicians together with their families were, in addition to tons of documents and equipment, removed. The more German "know-how" reached the United States and Great Britain, the faster and more certain, so the motto, was a victory over Japan possible. The instruction was that the evacuated German experts were to be secured and kept at the ready "as a subservient reservoir of technical intelligence" to be used in an effort essential to the war. It certainly also mattered greatly that one did not want to leave everything to the Russians, but at that time it was not yet the guiding principle of the technology transfer of the Americans and British. The deployment of the American nuclear bombs at the beginning of August 1945 led, however, unexpectedly to the quick Japanese surrender. The argument regarding Japan, prevailing up to that time, had become redundant overnight [19, 22–25].

The Americans and British now had to consider a more subtle approach. The American Trade Secretary, *Henry A. Wallace*, finally described taking along German weapons experts in a memorandum to the American president dated December 4, 1945, as "intellectual reparations." He emphasized here that the engagement of leading German experts was wise and logical, because such a transfer would constitute a large advantage for American science and industry. The trade secretary urged that the United States could not forgo the technical expertise of the Germans due to the leading technological role of the United States in the world and the profits to be expected and, most of all, considering her military superiority. He concluded his memorandum with the remark that one should equate the utilization of German know-how and her bearers of knowledge with "intellectual reparations" of the defeated Germany to the United States. However, the concept of "intellectual reparations" is still controversial, both with regard to the question in how far an allocation to the reparations is adequate and also considering the aspect of the financial assessment of this kind of service. There is no practicable method that allows such services to be assessed, as the value and the economic yield can only be determined afterwards—ex post facto—and then extremely inaccurately. Generally, any quantification is highly speculative, and there is, up to the present, no clear definition within existing international law.

The different postwar transfer programs of the Allied victors such as, for instance, "Lusty," "Overcast" (United States/Great Britain), "Darwin Panel Scheme" (Great Britain), "Project Paper Clip," which bundled up all American activities, or the Soviet transfer action (Verbringungsaktion), later designated "Ossawakim," were in strong contradiction to the Allied occupational policy. According to the political and economic principles of the Allied victors, Germany was to be completely disarmed and demilitarized. Regarding militarily relevant research, these basic principles were more precisely specified in the control-board law No. 25 of April 29, 1946. According to this, any research related to science and for military purposes and its practical application was to be prevented. Research in fields with military potential had to be supervised and led into a peaceful direction. The law became effective on May 7, 1946, and ordered the dissolution and elimination of all technical military organizations in Germany and prohibited applied research in nine scientific fields, among them nuclear physics, aeronautics, and rocket and radar development. Thousands of German scientists and technicians were thus no longer able to work in their previous occupational fields. But the principle "Legitimation by Usefulness" rendered it possible that many Germans could continue to work on military research and development projects on the order of the respective occupational power [5, 26].

The German, British, and American "Laboratory Warriors" had, moreover, demonstrated the potential of the new scientific "weapons armories" with the rocket and aircraft development, the deciphering complex at Bletchley Park, and especially with the nuclear bombs [27]. Until the end of the war, the military-industrial complex within the United States, Great Britain, and Germany had become a military-industrial-academic complex. The Soviet Union also possessed in 1945 a powerful armament industry with scientific research and development facilities, but in comparison to the defeated Germany or the United States she was far behind. The American nuclear bombs and the cold war caused the Soviet Union to put every effort into attaching the same value to the academic component of the armament business as the Americans had already done. But not only the Soviet Union, also countries like France, Canada, Australia, South Africa, Spain, Sweden, the

Netherlands, Switzerland, Argentina, Brazil, and later countries like China, India, and Egypt did all they could to set up or expand, more or less successfully, a military-industrial-academic complex. One thing was clear worldwide: with WWII useful military science and technology had become a prerequisite for having a voice in world politics [19, 28–30].

At the end of the war, the head of the American Army Air Force, Gen. *Henry H. Arnold* [31], explained in plain language:

> We have won this war and it is no longer of interest to me. I do not think that we should waste our time by discussing whether we have gained the victory by pure superior strength or by a qualitative superiority. Only one thing shall interest us. What does the future of air power and the aerial warfare look like? Where do the new inventions like jet propulsion, rockets, radar and nuclear energy lead to?

After the war, this thinking caused above all for German inventors, scientists, or technicians what could become for any other German a trap: "... his involvement in the German past. There was no power on Earth, which believed it could do without geniuses—be they real or imaginary. Neither the United States nor France, England let alone the Soviet Union could turn up their noses at this"([2], pp 588–589).

All victorious powers competed, therefore, in securing German experts and their know-how. The hunt for German scientists and technicians, already starting during the final stages of the war, was actually not a hunt for the Germans, who frequently approached the victors voluntarily, but it was a hunt between the hunters. After the victory over Japan, a main argument of the American and British occupational policy regarding the utilization of experts was, for example, that, in the interest of national security, the German specialists were denied the other victorious powers, that is, "Policy of Denial." However, during the first two postwar years, this denial aimed mainly at the Soviet Union was also applied to Latin America and Spain. In the course of the unfolding of the cold war, the policy of denial changed. As of about 1947, the migration to South America was, for example, silently tolerated. Moreover, members of the British Commonwealth were offered German scientists or these countries were allowed to look for suitable persons within West Germany. The policy of denial of people with armament-related knowledge focused at the end of the 1940s on China, North Korea, the Soviet Union, and the other East European countries [27].

Today, it is hard to say how many German and Austrian scientists and technicians left or had to leave Germany after the war. Approximately from 6000 to 7000 scientists, engineers, technicians, and foreman left Germany. Of these, about 3000 experts were taken to the Soviet Union, about 1000 German scientists and technicians up to the mid-1950s went to the United

States, and it is assumed between 2000 and 3000 experts emigrated to France. The number of German specialists in Great Britain is assumed to be 800 to 900. In the order of 500 to 600 specialists emigrated to South America, Australia, New Zealand, South Africa, Canada, Spain, and other countries. But all these are rough estimates. The data are uncertain, but most certain are the numbers for the Soviet Union and the United States [19, 26, 32, 33]. Generally, it can be stated that the planning and operations concerning the appropriation of the armament, defense, and personnel capacities of the Third Reich constituted one of the first steps into the Cold War. After all, the arms race during the post-war years was decisively determined by the military-related technical innovations of the Third Reich [30] (see Fig. 8.1).

The complexity of the topic and the important role of the aeronautical knowledge after 1945 is explained, among other things, by the speech of an American secret service officer at the Country Club of Dayton, Ohio, on the evening of May 6, 1946. The Air Force Technical Center Wright Field, located not far from the town, had invited the local high society to an evening event. The reason for this was heightened public interest in what was happening at the Air Force test center in recent months. More and more German aeronautical experts had been arriving whose presence was no longer a secret to the public. There was criticism by the residents, but also by members of the Air Force who were outraged that the government now cooperated with the former enemy. Therefore, the Air Force took the offensive and tried to justify why these Germans and their knowledge were of great importance to America. At a country club in Dayton, Ohio, the late Gen.

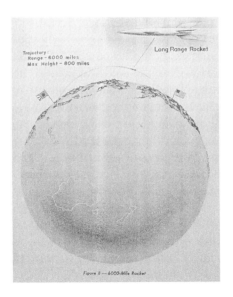

Fig. 8.1 How a long-range missile launched in the United States approaches Japan; the attack on America was demonstrated to *Hitler* by a similar presentation. The approach of a Soviet long-range missile on New York is again presented in a Soviet propaganda movie about the importance of rockets at the end of the 1940s. ("Where we stand" by the head of the Army Air Force Scientific Advisory Group, *Theodore von Kármán*, August 1945.)

Donald L. Putt, head of the Air Technical Intelligence of Wright Field, presented the following reasons [3]:

> I think it will be clear … that the many tons of material, documents and the scientists which have been brought to Wright Field represent the end product or results of the expenditure of millions of dollars for research facilities, research work, and millions of man hours expended in the construction of these facilities and their operation. The material, documents and scientists were screened and selected from those fields in which the Germans were ahead of our own state of research and development. The developments in these fields are now of first order importance to us because they provide the simplest means of obtaining previously unheard-of speeds in air transportation with the possibility of flight high in the stratosphere and some day, perhaps, interplanetary transportation. So, I might ask, with this information in our possession, should we or the American tax payer spend additional time and money to supplant facilities and work which has already been accomplished by German Science and is now available for our use with the expenditure of a relatively small amount of money and effort. If we are not too proud to make use of this German-born information, much benefit can be derived from it and we can advance from were the Germans left off.

However, the benefit of the employment of German scientists, engineers, and technicians for the United States, pointed out by *Putt*, caused again and again criticism and discussions. The excitement about the fact that the American government had, by "utilizing" former hostile scientists and engineers, set "usefulness" above being true to principles, showed up again and again in different press reports and letters of protest. The "Society for the Prevention of World War III" or the "Federation of American Scientists" protested, among other things, against the U.S. government knowingly violating American laws and valid norms, despite being responsible for the upholding of these same laws and norms. However, the cold war caused the excitement to quickly dissipate. The pragmatic line fast gained acceptance in America when it was pointed out that the new enemy—the Soviet Union— did not hesitate to use the Germans by the thousands (Fig. 8.2). During the cold war, the Germans in the United States were no longer asked what they did during the Nazi era, but it was now only of interest whether they had had any contact with communism [5, 19, 34, 35].

The topic of the emigration of scientists and technicians from Germany after WWII has caused confusion and surprise even up to the present day. The international press reported in August 1999, for example, that the Australian government had secretly brought German scientists and technicians to Australia after the end of WWII. For almost half a century, this transfer remained mysterious. The occupation of the Germans was concealed, so the

IVAN is watching you

Fig. 8.2 A propaganda poster that reads "IVAN is watching you ... because you and all of us have set out to build more and more better weapons—to do it faster all the time. We must use every bit of know-how and intensive skill we have to improve our methods—to turn out more and more for every hour we work. Only this way we become military strong" (*Aviation Week*, August 27, 1951).

official statement of the Australian government, to prevent the expected public protests and to advance that country's own weapons development. At the time, it had never been the intention to bring Germans with a Nazi past into the country: The Australian government emphasized in the summer of 1999 that Australia had not been a refuge for Nazi scientists. Nevertheless, seven war criminals, searched for by the United Nations, were also among the 127 German scientists and technicians, known by now, who had worked in rocket research on the Fifth Continent. Moreover, the *Sydney Morning Herald* reported in August 1999 that the secret immigration to Australia between 1946 and 1951 was expressly desired by the Americans and the British, because it was feared that the German experts would otherwise fall into the hands of the Soviets [36, 37].

In this regard, it seems quite an irony of history that due to the disastrous working and living conditions, Russian researchers started after the collapse of the Soviet Union to migrate to foreign countries. A creeping "academic-technical migration" occurred primarily to the United States, to Israel, and to Western Europe [38]. But countries such as Iraq, Iran, Libya, and North Korea tried hard to get elite Russian scientists and technicians. The United States, as winners of the Cold War, was as before confronted with the question as to how the scientific-technical capacities of the former opponent should be treated. The answer of the U.S. government in January 1992 was similar to the one given after 1945: "We can get real advantages. And if we don't, someone else will" [39].

8.3   HIGH-SPEED AERODYNAMICS IN THE UNITED STATES

8.3.1   AMERICAN AERONAUTICAL RESEARCH PRIOR TO 1945

At the time of the Volta Congress in 1935, the topic of high-speed flight within, at that time, the futuristic sonic and supersonic speed range was of exotic character. After the presentation of *Adolf Busemann* concerning the "swept-wing effect," the organizer of the Volta Congress, Gen. *Arturo Crocco* at an evening reception, brought the general opinion of the "Community" concerning the value of such research with his famous sketch—a swept-wing aircraft driven by a propeller—aptly to the point. The considerations of the young German aerodynamicist *Busemann* were, without doubt, of academic interest and fascinating but seemed in no way of practical relevance. At the research establishments of the leading aeronautical nations—United States, France, Great Britain, Italy, Soviet Union, Germany, and Japan—other tasks were on the agenda in the mid-thirties.

The level of the research equipment of these leading aeronautical nations was roughly similar to that prior to the outbreak of the war. The information exchange within the community was largely unrestrained up to 1939, and visits and return visits between the leading research establishments were carried out despite the worldwide political polarizations. For the construction of a new wind tunnel, the experience of the colleagues in Switzerland, France, Italy, Great Britain, the United States, and in Germany was generally invaluable. Managers and collaborators of these research establishments were invited or received visits when the construction of a new wind tunnel was planned anywhere in the world. For this reason *Theodore von Kármán*, who had worked since 1930 in the United States, followed in 1937 an invitation to the Soviet Union to advise the TsAGI Central Aero- and Hydrodynamics Institute, Zhukovsky, Moscow on the construction of a high-speed wind tunnel [40]. Moreover, delegations from Germany, the Soviet Union, and Japan repeatedly visited American research installations to be inspired with regard to the development of their own research setups [41]. In turn, the Americans received on their return visits one or another indication concerning activities that were actually top secret. By the autumn of 1936 the NACA knew that the Germans were quietly building an ultramodern aeronautical research center at Völkenrode near Braunschweig. *George W. Lewis* of the NACA wrote after his visit to Europe in his trip report: "At Braunschweig, it is planned to build four new tunnels." *Lewis* judged with regard to the equipment standard at Langley Field [42]:

> It is my opinion at the present time that the equipment at Langley Field is equal to or better than the equipment in the German research laboratories for the pursuance of aeronautical research, but the personnel of the German research laboratories is larger in number, and the engineers have had an opportunity of having special training, which has not been afforded to many of our own engineers.

At the beginning of WWII, specific plans existed in the United States concerning the construction of new wind tunnels in the field of high-speed aerodynamics. At the NACA, an "Aerodynamics Subcommittee" was founded for that purpose, which was to deal with the tight situation in the area of high-speed wind tunnels [43–45]. In an assessment of the degree of utilization of the NACA facilities, it was explained [46]:

> In 1939, the increased volume of the development work required for the investigation of the numerous types of airplanes resulted in an inadequate supply of equipment solely intended for development work. As a result, the wind tunnels of the NACA were placed on 24-hour day operation in order to handle the increased volume of development work and at the same time maintain a proper balance of fundamental research so that the advancement of basic aeronautical science would not be hampered by the increased demands of development.

Besides the considerations at the NACA and the military, there were also plans within the American industry. Especially here, it had become obvious that the companies needed a greater independence and more flexible access with regard to the aerodynamic investigations during the prototype development. The Americans followed a path that the German aircraft industry had already pursued since the 1920s. Here, the necessity that aerodynamic measurements should also be carried out on location, that is, at the companies, had already been recognized much earlier. The dilemma that all American aircraft companies encountered in the 1930s showed itself, for example, at Boeing in Seattle, Washington. For the development of the Boeing "Stratoliner 307" and the long-range bombers B-17 and B-29, the measurement possibilities in the wind tunnel of the University of Washington, at the California Institute of Technology (Caltech), and the NACA were totally insufficient. Because of the crash of the prototype of a "Stratoliner 307" in March 1939, Boeing also lost their chief aerodynamicist *Ralph Cram*. The tragedy could possibly have been avoided if sufficient wind-tunnel testing prior to the flight trials had been done. Boeing drew their conclusions and planned the construction of their own wind tunnel. Among others, Boeing won as an adviser *Theodore von Kármán* of the Caltech who pleaded for designing the facility as a high-speed tunnel. Decisive here was that in 1941 the new "Chief of Flight and Research," *Edmund T. Allen*, took up his job at Boeing. In 1941 he had obtained knowledge of the new jet engine of the British and anticipated—encouraged by *von Kármán*—earlier than anyone else in his field the beginning of a new epoch. The decision to build a high-speed wind tunnel at Boeing was made in August 1941. However, *Allen* did not experience the confirmation of his idea anymore; he died in a crash of a B-29 prototype in 1943. The first test runs of the new tunnel took place in February 1944, and the high-speed wind tunnel was fully operational at Boeing at the end of the war [47–50].

As well as the high-speed wind tunnel at Boeing, two further "supersonic tunnels" went into operation during 1944 and 1945: A wind tunnel at the Army

Aberdeen Proving Ground in Maryland and one at the NACA Ames Aeronautical Research Laboratory in California [51]. The commissioning of the facilities only at the end of the war, however, indicates that nothing really changed during the war regarding the weakness of American aeronautical research already showing in 1939. There was a lack of possibilities for a flexible and timely access to high-speed aerodynamic research facilities. When the Army Air Force in 1943 requested proposals for the development of a jet bomber, the aircraft companies Boeing, Convair, Martin, and North American, charged with the task, literally queued up at the high-speed wind tunnels available at NACA and Caltech in 1944 due to the limited utilization times [47].

Altogether up to 1945, the effort concerning high-speed aerodynamics within American aeronautical research was determined more by enthusiasm than by a strict program. At the end of the war, no systematic investigations concerning the effectiveness of the swept wing existed in the United States. The American aerodynamicist *Robert T. Jones* called attention to the application of the swept wing to high-speed aircraft at the end of 1944, and a corresponding NACA report was published at the beginning of 1945 (Fig. 8.3); however, his considerations were dismissed by the NACA as "wild ideas." In the literature it is, moreover, pointed out that *Jones* had possibly known the theory of *Busemann*, because the Volta Congress of 1935 had not been a secret conference and the results were published worldwide. It remains a fact that the initial ideas for many research tasks related to high-speed aerodynamics and commencing after 1945 came from Germany. American aeronautical research

Fig. 8.3   In the book published in 1990 by R. T. Jones concerning wing theory, he also reports on his early considerations at the end of 1944 regarding the "swept-wing effect" (R. T. Jones, "Wing Planforms for High Speed Flight," NACA TR 863, 1945, 1947).

adopted the fundamentals developed in Germany and continued the research [52, 53].

At the end of the war, this was seen at the NACA, in principle, similarly. However, with regard to the research of *Jones* concerning the "swept wing," it was emphasized that he had come to his conclusions "without prior knowledge of the German work." In October 1946 the NACA drew the following first conclusion with reference to the publications of *Busemann*, the systematic research in Germany, and the American considerations concerning the swept wing [1]:

> The most advanced theoretical work in Germany pertaining to the lift and drag of wings at supersonic speeds is that of *Busemann*, 1942–43. This paper has provided a starting point for several important theories for wings in this country. The drag of triangular wings had been treated by *Jones* and *Puckett* and the lift has been treated by *Brown* and *Stewart*. In contrast to the supersonic test data on specific models in Germany, systematic tests were made in this country on triangular wings at supersonic speeds.
>
> Whereas the Germans conceived the idea of sweepback in 1940, it is significant that not one combat airplane appeared in the war that fully utilized the advantage of sweep. It remained for *Jones* at the NACA to first theoretically solve the case of drag of swept wings at supersonic speeds. The sweep theory was published without prior knowledge of the German work by *Jones* of the NACA in 1945. Although the Germans compiled a large amount of test data on sweepback, present efforts to apply these results to missile design are failing.

With regard to high-speed wind tunnels, the NACA clarified: "There is no question that the Germans have been much more aggressive than any other country in the development and operation of supersonic test equipment such as supersonic tunnels."

But it was also clearly seen that, although, for example, the wind tunnels at the Bavarian town of Kochel were extraordinary, Germans made, overall, "no particularly outstanding contributions to supersonic wind-tunnel design." The Italian high-speed wind-tunnel of Guidonia was considered as an innovative solution. But *Busemann* was, without a doubt, right when he explained in 1945 to the Americans: "Good tunnels were not needed for wartime research and development" [1].

8.3.2 *INFORMATIVE MISSIONS, ARMS SHOW, AND PROJECT "PAPERCLIP"*

The German attack on London with flying bombs (Fi 103/V1) and long-range missiles (A4/V2) as well as secret information by the secret services concerning novel German arms developments caused Army Air Force Gen. *Henry H. Arnold* in late summer of 1944 to take the opportunity, after a victory, to press the United States to systematically adopt the German technological advancements. *Arnold* contacted *Theodore von Kármán* in September 1944 and convinced him to take on the directorship of a special advisory group at the Air Force. The group was to examine the German "know-how" and make suggestions for an efficient technology transfer. *Von Kármán* immediately recognized the importance of this request. He agreed and started to draw up a list of persons who were considered suitable for this task. Towards the end of 1944, the list was finally completed, and the Army Air Force Scientific Advisory Group (AAF-SAG) was founded (Fig. 8.4). The group consisted of 31 outstanding scientists and experts from American aeronautical research and the aircraft industry [4, 21, 31, 54–56]. Shortly before the surrender of Germany and in the context of the reconnaissance operation "Lusty," the group moved to Europe and examined German aeronautical research. The American visitors had quickly found a special interest in the results concerning high-speed aerodynamics.

The American wind-tunnel expert *Frank L. Wattendorf*, who on his visit to the LFA Völkenrode reacted at first glance with astonishment at the facilities, immediately sent a memorandum to his native country (Fig. 8.5). In this he suggested setting up a similar research center in the United States, because only on this basis could the superiority of America as aeronautical power and the national safety in the future be guaranteed [5].

The reports written after the visit to Europe clearly show the significance of the German research and development status to the consolidation of the

Fig. 8.4   Members of the AAF-SAG.

AAF SCIENTIFIC ADVISORY GROUP

Dr. Th. von Karman
Director

Colonel F. E. Glantzberg   Dr. H. L. Dryden
Deputy Director, Military   Deputy Director, Scientific

Lt Col G. T. McHugh, Executive
Capt C. H. Jackson, Jr., Secretary

CONSULTANTS

Dr. C. W. Bray	Dr. A. J. Stosick
Dr. L. A. DuBridge	Dr. W. J. Sweeney
Dr. Pol Duwez	Dr. H. S. Tsien
Dr. G. Gamow	Dr. G. E. Valley
Dr. I. A. Getting	Dr. F. L. Wattendorf
Dr. L. P. Hammett	Dr. F. Zwicky
Dr. W. S. Hunter	Dr. V. K. Zworykin
Dr. I. P. Krick	Colonel D. N. Yates
Dr. D. P. MacDougall	Colonel W. R. Lovelace II
Dr. G. A. Morton	Lt Col A. P. Gagge
Dr. N. M. Newmark	Lt Col F. W. Williams
Dr. W. H. Pickering	Major T. F. Walkowicz
Dr. E. M. Purcell	Capt C. N. Hasert
Dr. G. B. Schubauer	Mr. M. Alperin
Dr. W. R. Sears	Mr. I. L. Ashkenas
	Mr. G. S. Schairer

Fig. 8.5   Notebook entry of *Hugh L. Dryden* **concerning the AAF-SAG visit at the LFA Völkenrode and at Göttingen in May 1945.**

research structures in the United States performed somewhat later. Of special importance was the interim report of *von Kármán* of August 1945 concerning the status in the development of the German aeronautical and rocket-related research (Fig. 8.6). He gave his report the title "Where we stand," and in view of the obvious arrears, *von Kármán* asked: "Where we shall go"? In the case of "Where we stand," *von Kármán* addressed first high-speed aerodynamics and recommended for the next years an enormous buildup and expansion program related to the American research and development structures. In his assessment he explained that one had to learn the following lesson from the Germans ([57], p. 16):

> Leadership in the development of these new weapons of the future can be assured only by uniting experts in aerodynamics, structural design,

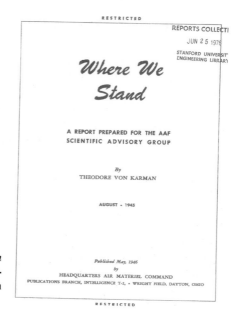

Fig. 8.6 Report of *Theodore von Kármán* concerning the status of German aeronautical research and rocket development in August 1945.

electronics, servomechanisms, gyros, control of devices, propulsion, and warheads under one leadership, and providing them with facilities for laboratory and model shop production in their specialties and with facilities for field tests. Such centers must be adequately supported by the highest ranking military and civilian leaders and must be adequately financed, including the support of related work on special aspects of various problems at other laboratories and the support of special industrial developments.

The Americans and British also founded the Air Document Research Center (ADRC) in London, England, and the Air Document Division, Intelligence T-2, at Wright Field in Ohio, United States. The ADRC was responsible for the transfer of research reports and equipment in the area of German aeronautical research. The establishment of the London documentation center also goes back to the initiative of the American, Gen. *Arnold*, who pressed for a restructuring of the document evaluation due to the chaotic material accumulation at the end of the war. The task of the ADRC was the systematic collection and processing of the German research documents for the later use by the military, industry, and science in Great Britain and the United States. The material, sifted and processed in London, was then sent to Wright Field, to the Air Document Division [3]. Until May 1946, the Americans collected about 1200 tons of reports on aeronautics- and rocket-related topics in Germany. For aeronautical equipment it was about 2500 tons. The first German aeronautical experts also arrived in the United States in September 1945. The number increased through 1948 to about 200 persons, who mostly worked at the Air Force Test Center Wright Field near Dayton, Ohio [4].

With regard to the equipment and the captured weapons, a special shipment and a related weapons show is worth mentioning here. In July 1945, the British aircraft carrier "HMS Reaper" reached the United States with 30 German aircraft and tons of equipment onboard. The freight was first brought to the Air Force Base Freeman Field, Indiana, for storage and testing [21]. Also the captured Japanese aircraft were sent there. The Air Force shipped some of the captured aircraft and missiles in October 1945 to Wright Field, Ohio. They were to be presented to the American public at a large weapons show (Fig. 8.7). Between October 13 and 21, 1945, everyone could observe what the Germans had created during the war in "military hardware" related to aircraft and rocket development. On the opening day of the weapons show, 200,000 people came, and on the following day the number of visitors was 300,000. The *Dayton Journal* described the bustle at the "Fairground" in its issue of October 15, 1945, as follows [58]:

> Mile-long rows of tent absorbed the large crowd, believed to be the largest ever attracted to a southwestern Ohio event. Not until the mid-afternoon air show brought spectators out of the tents to gaze skyward at the nation's newest planes was the size of the crowd realized. Aerial weapons which previously were merely words repeated often in newspapers or by radio became real objects of warfare over the weekend. The enemy's most dreaded planes and missiles were seen and touched. Adults who never looked inside a plane before climbed into cockpits, children were allowed to operate controls of all kinds and untold thousands of photographs were taken.

Among the visitors of the "Army Air Force Fair" was the aeronautical pioneer *Orville Wright*, who prophetically mentioned to an accompanying

Fig. 8.7 Army Air Force Fair at Wright Field in October 1945.

a)

HEADQUARTERS, ARMY AIR FORCES
WASHINGTON

9 JUL 1945   9 JUL 1945

MEMORANDUM FOR ASSISTANT CHIEF OF STAFF, G-2, WDGS

Subject: Return of German Scientific Personnel for Temporary
Technical Exploitation

1. Pursuant to existing War Department policies and contingent upon provisions thereof, it is requested that arrangements be made for the return of
the following German scientific personnel for interrogation and temporary exploitation by the Air Technical Service Command, Wright Field:

Name	Priority	Location	Specialization
Dr. Otto Lutz	1	LFA Braunschweig	Aerodynamics (Propulsive)
Dr. R. Dirksen	4	LFA Braunschweig	Structures & Materials
Professor Blenck	4	LFA Braunschweig	Supervision
Dr. Kerris	2	LFA Braunschweig	Instruments
Mr. Knackstedt	1	LFA Braunschweig	Supersonic Flow
Mr. Neuber	4	LFA Braunschweig	Theory of Elasticity
Mr. Herzog	4	LFA Braunschweig	Materials
Dr. E. Eckert	1	LFA Braunschweig	Aerodynamics of Engines
Dr. Kilpert	3	LFA Braunschweig	Combustion
Dr. Seilschopp	3	LFA Braunschweig	Fuel Chemistry
Professor Rossmann	4	LFA Braunschweig	Weapons
Mr. Schussler	4	LFA Braunschweig	Ballistics
Mr. Hackemann	4	LFA Braunschweig	Weapons (Function)
Mr. Doetsch	4	LFA Braunschweig	Theory of Ballistics
Dr. A. Betz	1	Gottingen	Applied Aerodynamics
Dr. W. Encke	2	Gottingen	Axial Fans & Compressors
Dr. W. Ritz	3	Gottingen	Ceramic Turbine Components
Dr. O. Walchner	1	Gottingen	Supersonic Wind Tunnels
Dr. A. Kussner	4	Gottingen	Flutter & Vibration
Dr. W. Tollmien	1	Gottingen	Aerodynamics
Dr. W. Herrmann	1	Kochel	Supersonic Wind Tunnels
Dr. H. Kurzweg	1	Kochel	Supersonic Wind Tunnels
Dr. G. Eber	1	Kochel	Mach 7-10 Wind Tunnel
Dr. H. Ramm	1	Kochel	Supersonic Measures
Dr. G. Arnold	1	Kochel	Supersonic Measures
Dr. W. Schwaiger	1	Otztal	Large Wind Tunnels
Gen. Dornberger	2	Peenemunde (G.FK.)	Rockets - V-2
Dr. Von Braun	2	Peenemunde (G.FK.)	Rockets - V-2
Dr. F. A. F. Schmidt	2	Garmisch-Partenkirchen	Aerodynamics (Propulsive)
Dr. F. Bollenrath	4	Sonthofen	Structures & Materials
Dr. Voigt	1	Messerschmidt (Ober.)	Aircraft Design
Dr. F. Weinig	2	Stuttgart	Compressors & Engines
Dr. W. Kamm	2	Stuttgart	Power Plants

Fig. 8.8   a) List of names (July 1945), and b) selection priorities (February 1946).

officer on sight of the captured German weapons: "now they fly without propellers, soon they will fly without wings" [58]. Looking at the Me 262, the Me 163, or the long-range missile A4/V2 literally the same question that had already been asked by *von Kármán* with his interim report a few weeks before could be asked "Where shall we go?".

But not only tons of research reports, equipment, and weapons came to the United States (Fig. 8.8). Shortly after the end of the war, the scientist-related program "Overcast" was initiated and adopted in July 1945 to—as already mentioned—first shorten the war in the Pacific with the aid of German scientists and technicians. The German scientists involved in rocket research and, most of all, the aeronautical specialists, such as, for instance, aerodynamicists, belonged to the chosen ones of the first hour.

The Americans replaced "Overcast" in March 1946 with project "Paperclip," better fitted to the new conditions after the war. Between September 1945 and January 31, 1946, 129 German scientists and technicians were brought overseas via "Overcast." Of these, about 80% were part of the rocket people around *Wernher von Braun* who were attended to by the Army. Via "Paperclip," about another 400 German experts and their families came into

b)

WAR DEPARTMENT
OFFICE OF THE ASSISTANT SECRETARY
WASHINGTON, D. C.

Noted~Asst. Secy. of War

1 February 1946

MEMORANDUM FOR MR. PETERSEN

SUBJECT: Military exploitation of German scientists.

At about the time of VE Day, the War Department, largely at the instigation of Judge Patterson and Dr. Bowles, determined to proceed with a vigorous program for military exploitation of German scientists. The operation was designated "Overcast" and was originally thought of principally as a means of getting information for use against the Japanese. The arrangement was for a joint Army-Navy operation. G-2 was designated as the administering agency, AAF and Ordnance being the principal users and exploiters.

After VJ Day the project was re-examined and our requirements were re-assessed. Dr. Bowles took the position that, in view of the political considerations involved, our requirements should be kept at an absolute minimum. For this reason, four criteria were laid down, as follows:

(1) Only outstanding scientists in fields in which the Germans were notably superior to us could be chosen;

(2) Only scientists who could not be efficiently exploited in the field would be brought to this country;

(3) Scientists would not be brought to this country who were politically objectionable;

(4) Scientists would be brought to this country only on a voluntary basis.

Germans brought to this country for exploitation were to remain in military custody and were to be exploited and returned as promptly as possible. They were not to have access to classified military information except where that was required for their exploitation.

These rules having been laid down, Mr. McCloy obtained informal clearance from the State Department, which indicated that it had no objection as long as the scientists were maintained in military custody under purely temporary arrangements.

Contemporaneously with the making of these arrangements, agreements were entered into with the British in the CCS 870 series. The agreement is very general and provides for substantial equality in allocation of scientists between the British and the U.S., with provision for exchange of results. The agreement relates only to

Declassified
NND 140067
By ___ S M ___ NARS, Date ___

Fig. 8.8   (Continued)

the United States (Figs. 8.9 and 8.10). The employment structure had changed considerably in 1948 due to the individual branches of the Services. Of the 500 specialists who worked at this time under project "Paperclip" in the United States, now about 40% were with the Air Force, 34% with the Army, 16% with the Navy, and 10% were under the auspices of the Department of Commerce [59].

At the same time, there were discussions at the departments responsible whether the Germans should be sent back home again, because their knowledge about new war-related weapons technologies had largely been "absorbed." The Germans stayed, moreover, merely as "guests" of the American military under avoidance of the immigration laws in the United States. Opposing their return was, however, the fact that the Germans had obtained sensitive insights into American conditions. There was a fear that they could report upon their return their American experience to the "other

RESTRICTED

NAME	DEPEN-DENTS IN US	SPON. DEPT.	SPON. AGENCY	EMPLOYING AGENCY	CUSTODIAN	RESIDENTIAL ADDRESS
BAEUMKER, Adolf G. H.	AIR		AMC	25 October 1948	3	28 December 1949
BALJE, Otto-Erich	AIR		AMC	17 January 1949	2	3 February 1950
BALL, Erich K. A.	ARMY		ORD	9 September 1949	4	13 July 1950
BARTH, Hans	COMM			12 July 1950	0	
BARTHALOMAEUS, Hans B. A.	AIR		TAU	20 April 1949	3	3 July 1950
BAUER, Alfred F.	COMM			21 February 1949	4	23 February 1950
BAUSCHINGER, Oskar H.	ARMY		ORD	11 August 1949	3	13 July 1950
BEDUERFTIG, Hermann F.	ARMY		ORD	11 April 1949	3	13 October 1949
BEEK, Gerd W. de	ARMY		ORD	15 March 1949	3	3 November 1949
BEER, Heinz P. A.	AIR		AMC	10 January 1949	1	6 October 1949
BEICHEL, Rudi	ARMY		ORD	11 August 1949	2	14 July 1950
BEIER, Anton	ARMY		ORD	14 September 1949	2	14 July 1950
BEINERT, Hans H.	AIR		TAU	14 October 1949	1	17 August 1949
BEISCHER, Dietrich E.	NAVY		BuAer	9 May 1949	4	6 January 1950
BERGELER, Herbert R.	ARMY		ORD	19 April 1949	4	12 September 1949
BERKNER, Hans O.	AIR		AMC	21 March 1949	5	13 July 1950
BERNDT, Rudi J.	AIR		AMC	10 January 1949	3	16 January 1950
BIELITZ, Friedrich K.	AIR		AMC	25 October 1948	3	29 March 1950
BIELSTEIN, Hans O.	AIR		AMC	1 May 1950	0	
BOEHM, Josef M.	ARMY		ORD	2 June 1949	4	13 February 1950
BOCCIUS, Walther G. C.	AIR		AMC	4 October 1948	3	24 October 1949
BOTH, Eberhard	ARMY		SIG	28 March 1949	4	18 October 1949
BOTTENHORN, Hermann A.	AIR		AMC	6 December 1948	2	12 September 1950
BRAUN, Magnus H.A.M. von	ARMY		ORD	28 June 1949	0	
von BRAUN, Werner	ARMY		ORD	2 November 1949	3	24 August 1950
BREDTSCHNEIDER, Kurt E. B.	COMM			26 May 1949	1	24 April 1950
BRILL, Rudolph F.	COMM			17 June 1949	2	5 July 1950
BRUCKMANN, Bruno W.	AIR		AMC	21 February 1949	5	6 July 1950
BUCHHOLD, Theodor A.	ARMY		ORD	14 October 1948	2	26 September 1949
BUECHEL, Erwin H.	AIR		TAU	23 March 1949	0	
BUETTNER, Konrad J. K.	AIR		TAU	6 May 1949	2	14 September 1950
BUROSE, Walter W. B.	ARMY		ORD	15 March 1949	1	12 January 1950
BUSEMANN, Adolf	NAVY		BuAer	25 July 1949	4	20 February 1950
DAHM, Werner K. M.	ARMY		ORD	9 December 1949	1	22 September 1950
DANNENBERG, Konrad K.	ARMY		ORD	22 April 1949	2	Wife: 14 Sep. 1949 Son: 11 Oct. 1949
DHOM, Friedrich (Fritz)	ARMY		ORD	2 May 1949	3	4 January 1950
DIEHL, Karl L.	COMM			22 September 1949	2	8 May 1950
DOELHOFF, Friedrich List von	AIR		AMC	26 May 1948	2 / 1	12 August 1949 / 20 February 1950
DOERICK, Herbert O.	ARMY		ORD	2 May 1949	3	24 April 1950
DOEPP, Philipp von	AIR		AMC	27 September 1948	2	10 August 1949
DONATH, Ernst E.	ARMY		QMC	25 July 1949	3	17 February 1950
DRABGER, Walter W.	NAVY		BuSaD	5 December 1949	0	
DRAWE, Gerhard P.	ARMY		ORD	19 July 1949	1	20 February 1950
DUELL, Bernhard	ARMY		QMC	24 January 1949	0	
DUELL, Gertraud (Mrs. Bernhard)	ARMY		QMC	24 January 1949	0	
DUERR, Friedrich	ARMY		ORD	14 April 1949	3	2 November 1949
EBER, Gerhard R. E.	NAVY		BuOrd	20 December 1948	3 / 1	24 February 1950 / 22 March 1950
EISENHARDT, Otto K.	ARMY		ORD	15 March 1949	3	9 December 1949

Fig. 8.9   Partial "Paperclip" personnel list of November 20, 1950.

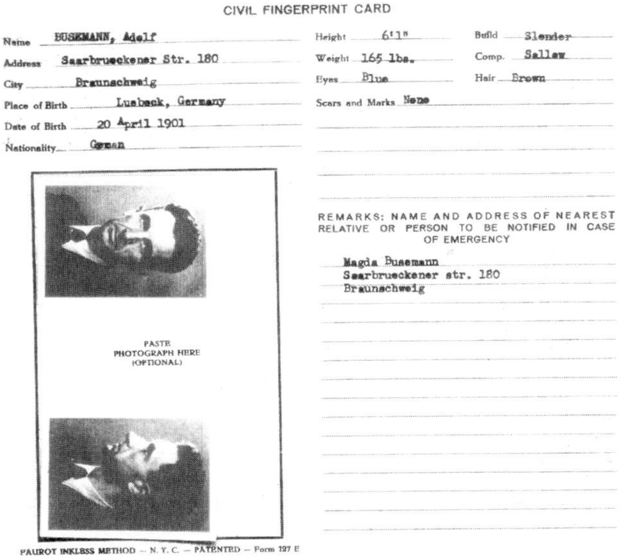

CIVIL FINGERPRINT CARD

Name   BUSEMANN, Adolf
Address   Saarbrueckener Str. 180
City   Braunschweig
Place of Birth   Lusbeck, Germany
Date of Birth   20 April 1901
Nationality   German

Height   6'1"   Build   Slender
Weight   165 lbs.   Comp.   Sallow
Eyes   Blue   Hair   Brown
Scars and Marks   None

REMARKS: NAME AND ADDRESS OF NEAREST
RELATIVE OR PERSON TO BE NOTIFIED IN CASE
OF EMERGENCY

Magda Busemann
Saarbrueckener str. 180
Braunschweig

PASTE
PHOTOGRAPH HERE
(OPTIONAL)

FAUROT INKLESS METHOD — N. Y. C. — PATENTED — Form 127 E

Fig. 8.10 "Paperclip" personnel file of the aerodynamicist *Adolf Busemann*. The personnel file was composed of 20 pages and contained, besides brief evaluations, a list of publications and contracts of employment, also two affidavits concerning the conduct of *Busemann* during the Third Reich by *William Threlfall* and *Herbert Seiferth*. *Busemann* stayed until 1947 in Great Britain and then came to the United States.

side." It was, therefore, decided in 1948 to naturalize the roughly 500 German specialists and their 1200 relatives. Up to 1950, about 90% of the 500 "Paperclip" experts had finally received the official "Resident Alien" status, which allowed them to become American citizens after five years.

8.3.3 CONSOLIDATION OF RESEARCH STRUCTURES

The Aeronautical Research Establishment Völkenrode near Braunschweig had, as already indicated, deeply impressed the members of the AAF-SAG during their visit in May 1945. Moreover, the information about the Heeresversuchsanstalt (HVA) (Test Establishment of the Army) Peenemünde-Ost confirmed that the Germans had succeeded in the area of aircraft and rocket development to establish a unique combination of basic, experimental, and applied research. To the AAF this seemed again as "the lesson to be learned from the activities" in Germany. The utilization of the German organizational principle in American aeronautical research was considered as urgently needed. In this regard, *von Kármán* formulated in his interim report of August 1945 the basic objectives. In view of the activities in high-speed aerodynamics, he summed up: "We cannot hope to secure air superiority in any future conflict without entering the supersonic speed range" ([57], p. 7).

A committee was, therefore, created at the Army Air Force in October 1945 to draft up plans for an "Air Engineering Development Center" (AEDC). In early December 1945, it presented "an exhaustive study of existing research and development facilities, both foreign and domestic, as well as projected plans for new aircraft and missiles." Besides the Army Air Force, the Navy Bureau of Aeronautics, the Civil Aeronautics Administration of the Ministry of Trade, the NACA, and the aircraft industry developed their respective ideas concerning future aeronautical research. For the coordination of the activities, the NACA formulated on March 21, 1946, an "Aeronautical Research Policy." The latter was to avoid an unnecessary duplication of research facilities and parallel research efforts. As a result, the "Special Committee on Supersonic Facilities" was created at the NACA, which consisted of members of the Army Air Force (after 1947 USAF), Navy, industry, and the NACA. In October 1946, the committee finally introduced a 10-year program with costs of about 2 billion U.S. dollars—a financial order of magnitude that corresponded to that of the nuclear-bomb project during the war. Within the following years, the program was repeatedly modified and once more presented to the Congress in 1949 as "Unitary Wind Tunnel Plan." The plan envisaged now the construction of high-speed wind tunnels at universities, industrial enterprises, as well as at civil and military research establishments at costs of more than 1 billion U.S. dollars. Within the consolidation and new construction program, finally passed by Congress in 1950, the AEDC of the Air Force, since 1947 an independent branch of the services, had become the focal point [60].

The AEDC was ceremoniously inaugurated at Tullahoma, Tennessee, on June 25, 1951 (Fig. 8.11). In honor of Gen. *Arnold*, who died in 1950, the center was named Arnold Engineering Development Center. During the 1950s, the AEDC developed into the "largest complex of flight simulation test facilities" in the western world. The Americans used German experience

Fig. 8.11   AEDC in the 1990s.

The United States Air Force's
Arnold Engineering Development Center
An Air Force Systems Command Test Facility

in the construction; strictly speaking the AEDC was a mixture of Peenemünde, Völkenrode, Kochel, and Ötztal [60, 61].

8.4 EXAMPLES OF AMERICAN AIRCRAFT

8.4.1 BOEING'S XB-47

Among the members of the AAF-SAG was also the chief development engineer *George S. Schairer* of Boeing. Together with *Theodore von Kármán* he met, at the Aeronautical Research Establishment in Völkenrode near Braunschweig, *Adolf Busemann* and heard, first hand so to speak, of the German research concerning the swept-wing effect.

During the conversation between *von Kármán* and *Busemann* concerning the swept-wing effect, which *Schairer* attended, the importance of this aerodynamic idea very quickly became clear to the Boeing engineer. At that time, Boeing was working on a first jet-bomber project. Difficulties had repeatedly arisen, especially with regard to the aerodynamics. Gen. *Donald Putt* reported about the first jet-bomber program of the AAF in the context of an "Oral History Program" of the U.S. Air Force in April 1974 in retrospective [62]:

> Just before I left Wright Field to go overseas in January 1945, the last program I went on the design competition for was what was to be our first jet bomber. There were four competitors. There was North American, Convair, Boeing, and Martin. ... All of them were straight-wing aircraft, very conventional looking except for hanging some engines on the wings that had no propeller on them. They were perhaps more streamlined and refined aerodynamically than prior aircraft, but basically they were pretty conventional designs. Boeing had presented a couple of configurations varying the placement of the engine, one where they were in the fuselage, tucked in close to the fuselage, another conventionally hung out on the wing, but still a straight wing.

This project *Schairer* had on his mind when he followed the discussions between *Busemann* and *von Kármán*. Among other things, *Busemann* was asked how and when he had had the idea of the swept wing, and *Busemann* reminded *von Kármán* of his lecture at the Volta Congress of 1935. On the reaction of *von Kármán*, *Putt* remarked later: "I remember *von Kármán* kind of slapping his head, 'oh yes,' he remembered" [62]. *Schairer* immediately reacted on the information of *Busemann* and wrote a detailed letter on May 10, 1945 (Fig. 8.12), to his company in Seattle [63]:

> We are seeing much of German aerodynamics. They are ahead of us in a few items which I will mention. The Germans have been doing extensive work on high-speed aerodynamics. This has led to one very important discovery. Sweepback or sweepforward has a very large effect on (the)

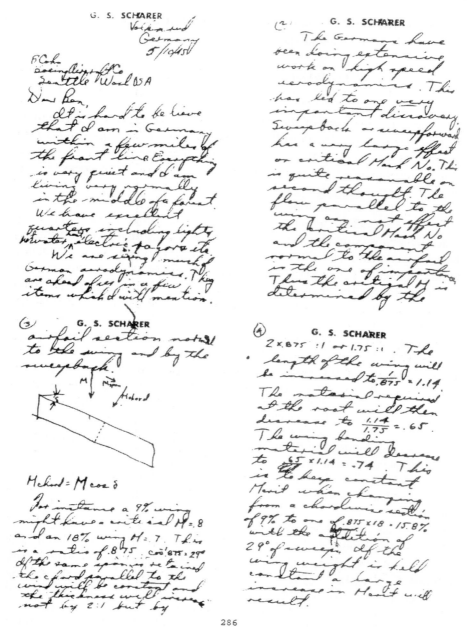

286

Fig. 8.12 Letter of the Boeing chief development engineer *George Schairer* out of Völkenrode on May 10, 1945 (courtesy of Boeing Historical Archive, The Boeing Company).

critical *Mach* number. This is quite reasonable on second thought. The flow parallel to the wing cannot affect the critical *Mach* number and the component normal to the airfoil is the one of importance. Thus the critical

M is determined by the airfoil section normal to the wing and by the sweepback.

But it still took some time before the information given in *Schairer's* letter achieved practical results. Only when the Boeing engineer returned again from Europe and also detailed German test reports concerning swept-wing investigations were available was the swept-wing concept taken into consideration in late summer of 1945 in the current project at Boeing. However, the jet engines were at that time still integrated into the fuselage. The known configuration of the later B-47, where the jet engines were located underneath the wings, was finally systematically tested at Boeing in the autumn of 1945 [64].

However, at this point we return to the history, the development, and significance of the project to Boeing. The bomber B-47 "Stratojet" of Boeing was the first large jet aircraft with a high-aspect-ratio swept wing, which was manufactured in large numbers. With the development of the highly efficient wing for the prototype XB-47, Boeing secured for itself for decades predominance in civil and military aircraft design. Competing with other aircraft companies, the change to an unproven new technology meant a great risk. The courageous step towards the "swept wing," however, paid off for Boeing. The B-47 became the starting point for a series of successful bomber and commercial aircraft. Starting with the eight-engine large jet bomber B-52 of transcontinental range (the first truly successful commercial aircraft with jet propulsion), the Boeing 707, the jumbo-jet Boeing 747, and up to the "fly-by-wire" Boeing 777 provided a market-dominating position for this great aerospace company. Without the company-owned high-speed wind tunnel, completed at the end of the war, this spectacular development would, however, hardly been conceivable.

The origins of the B-47 "Stratojet" can be traced back to the year 1943. In June 1943, the USAAF asked leading U.S. aircraft manufacturers to commence a study for a high-speed, jet-propelled reconnaissance aircraft or a medium-size bomber [65]. In November 1944, the USAAF announced the requirements for such a medium jet-propelled bomber. The latter envisaged a range of 5600 km (3500 miles), an operational ceiling of 13,500 m (45,000 ft), and a maximum speed of 885 km/h (550 mph). Boeing, Martin, Convair, and North American Aviation responded to this request for proposals.

The first design of Boeing, the Model 424, was a reduced version of the long-range bomber B-29 with four jet engines (Fig. 8.13). The engines were attached in twin-nacelles under the wing. The main landing gear was retracted into the engine nacelles. The wing had an aspect ratio of 12 and a thickness of only 10% along the entire span [66]. The airfoil corresponded to a symmetrical NACA airfoil of the 65-series with a maximum-thickness location of 40%. The Model 424 conformed to the state of the art in the United States at

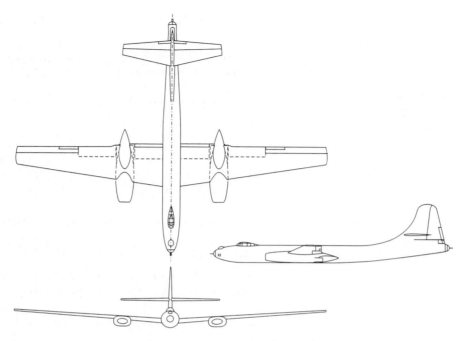

Fig. 8.13   Boeing Model 424 (Spring 1944).

that time and the recommendations of the NACA regarding the design of long-range, high-speed aircraft, that is, aerodynamically clean fuselage, thin wing with an aspect ratio as high as possible, and a laminar airfoil with a maximum-thickness location as far rearward as possible. Wind-tunnel measurements on the Model 424 soon showed that this design would not be better than the competition. The project XB-46 of Convair conveys a good impression of what the completed Model 424 of Boeing would have looked like (Fig. 8.14).

Nevertheless the XB-46 reached a speed of 877 km/h at an altitude of 4575 m during trials, not far from the requirements of the USAAF [67]. A wing with 10% thickness was most likely the utmost that could be imagined at the time, considering the acceptable weight and wing volume needed for additional fuel, for an aircraft with the required range.

Fig. 8.14   Jet bomber Convair XB-46 (1947).

With the Model 424 and the jet engines of the time, the demanded speed of 885 km/h was not attainable. The installation of the jet engines caused the Boeing engineers the greatest headaches. The only possible way to reduce the interference drag between wing and engine, and hence reach higher critical *Mach* numbers, was the installation of the jet engines within the fuselage. In December 1944, Boeing brought out the Model 432 (Fig. 8.15). The four jet engines were now buried in the fuselage above the wings and were provided with air by two lateral air intakes.

The landing gear was also completely accommodated within the fuselage. The USAAF was interested in this design and charged Boeing with a more detailed study under the designation XB-47. Similar orders were issued to North American for the XB-45, to Convair for the XB-46, and to Martin for the XB-48. The Model 432 was employed for a variety of wind-tunnel investigation, especially in the high-lift range (Fig. 8.16). The attachment of the necessary lift augmentation devices to the thin wing was problematic.

In April 1945, Boeing received information from the NACA concerning the advantage of wing sweep at high flight speeds. For the same engine performance, sweeping the wings promised about a 10% higher speed and longer range. The theory, developed by *Robert T. Jones* [68], was, however, not yet

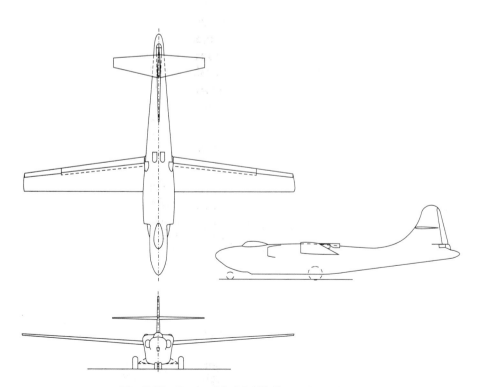

Fig. 8.15   Boeing Model 432 (December 1944).

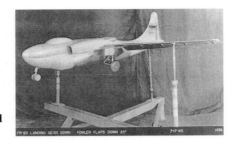

Fig. 8.16   Boeing Model 432 wind-tunnel model.

confirmed by wind-tunnel measurements and, therefore, aroused skepticism in the industry.

While Boeing in Seattle continued to work on the conventional straight wings, as previously described, the chief engineer of the company, *George S. Schairer*, traveled with the AAF-SAG to Europe and sent firsthand information concerning the swept-wing effect to Seattle. With a swept wing, the aerodynamic problems of the Model 432 could be remedied, and a clear advantage over the competitors could be achieved. A visit of the SAG team headed by *von Kármán* to the Junkers aircraft plants in Dessau is not documented; however, the members of the SAG team were able to inform themselves in detail from the wind-tunnel data about the jet bomber Junkers Ju 287 [47]. Because of its poor aeroelastic qualities, the forward-swept wing was out of the question for the Boeing project [69] (see Sec. 4.3.1 and Fig. 4.17). Of interest to the members of the SAG was also the fact that within the German aeronautical industry, independent of aeronautical research establishments, wind-tunnel measurements were being performed on a large scale to find, for instance, favorable locations for the jet-engine nacelles, at the time this also being an important design parameter for Boeing. Somewhat later, work commenced at Boeing on a version of the Model 432 with a 35-deg swept wing (Fig. 8.17). The empennage remained initially unswept.

A document shown in Fig. 8.18 by *Siegfried Günter*, prepared on the order of the Allies during summer 1945, summarized the state of knowledge at that time concerning the effect of wing sweep on the speed and range of jet aircraft [70]. It had a large influence on the decision for a 35-deg swept wing. In this report, a tailless aircraft, a conventional aircraft with fuselage, and an aircraft

Fig. 8.17   Boeing Model 432 wind-tunnel model with swept wings.

Influence of Fuselage and Sweepback Angle on

Range and Speed of Turbo-Jet Airplanes.

(Part II)

Siegfried Günter

(Project Design Office of Ernst Heinkel A.G.)

Landsberg, 31 August 1945

Fig. 8.18   Report of *Siegfried Günter* of August 31, 1945 (cover sheet).

with swept wings were compared based on the *Breguet* range equation. The 35-deg swept wing with a thickness ratio of 12% represented an optimum with regard to the drag minimization at transonic speeds, weight, wing volume, and flight characteristics. Larger wing sweep would have led to an increase in the rolling moment due to side-slip and hence to flying qualities problems at high angles of attack. In addition, the effectiveness of the ailerons and the landing flaps would have been dramatically decreased, which would have led to unreasonably high landing speeds (see Sec. 4.3.2). This report is based on a multitude of wind-tunnel measurements, carried out at the Aerodynamische Versuchsanstalt Göttingen (AVA) (Aerodynamic Test Establishment) and Deutsche Versuchsanstalt für Luftfahrt (DVL) (German Aeronautical Test Establishment), and theoretical considerations. *Günter* pointed out that his report could only provide indications and that for the design of a specific aircraft and the solution of specific problems dedicated wind-tunnel investigations were still required.

The aeronautical research establishments carried out wind-tunnel measurements on wings where parameters like aspect ratio, airfoil thickness, sweep, etc. were varied, and provided the industry with the necessary information for the design of a wing. However, except for the Ju 287 of Junkers, no further aircraft with a large-aspect-ratio swept wing had been designed for a specific mission by the German industry up to the end of the war. So after the end of the war, Boeing was instrumental in developing for the first time a large-aspect-ratio swept wing for an aircraft with a clearly defined mission.

Supported by documents from Germany and NACA and building on their own experience in the design of long-range aircraft, Boeing started an extensive test program not only on the Model 432 but also on individual wings. The objective was the selection of the right airfoil section for the wing, the attachment of the landing flaps, and the design of slats, spoilers, and ailerons. More than 5000 wind-tunnel hours were spent on the design of the wing. Without the company-owned high-speed wind tunnel this demanding task could not have been accomplished. The design was changed in September 1945 due to a requirement of the USAAF to install six engines (Fig. 8.19). The version of the XB-47 now designated Model 448 possessed a central air intake at the nose of the aircraft; the additional jet engines were accommodated within the rear fuselage.

Because of the concentration of the jet engines within the fuselage and their close proximity to the fuel tank and the hydraulic lines, the USAAF feared problems regarding the fire hazard and maintenance of the engines. The Air Force recommended, therefore, attaching the engines outside the fuselage on the wings. The Boeing engineers thus faced the same problem that had already worried them on the Model 424. Again German reports concerning the optimum arrangement of the engine nacelles on the wings were consulted. High-speed wind-tunnel investigations at Heinkel and

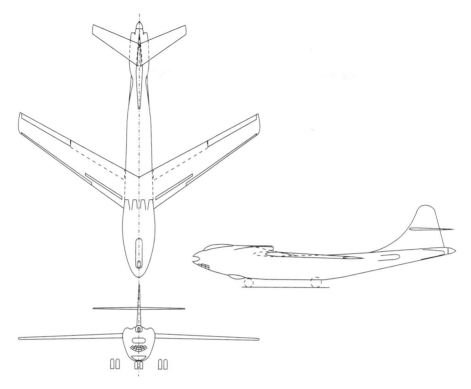

Fig. 8.19   Boeing Model 448 (September 1945).

Junkers had shown that large forward and aft positions of the nacelles with respect to the wings appreciably reduced the interference drag at transonic speeds (compare Chapter 6.3). Also larger vertical distances between the engine and the wing were advantageous as long as the pylons did not get too long [71]. To find the optimum position for the engine nacelles, Boeing built a wind-tunnel model with model jet engines. The position of the engines in front of the wings with a vertical distance such that the exhaust jet did not impinge on the deflected landing flaps proved to be favorable. The German measurements were too pessimistic with regard to the low nacelle position. The influence of the pylons was far less than predicted.

The design 448 was completely changed in October 1945. The new configuration 450-1-1 had a more slender fuselage, and the engines were now housed in nacelles underneath the wings (Fig. 8.20). Two engines were housed in each twin-nacelle favorably positioned with respect to the flow on pylons below the wings. The two other engines were attached at the wing tips.

With the pylon-mounting, Boeing had found the aerodynamically most favorable solution that also offered, at the same time, the largest possible safety in case of an engine fire. The landing gear was, as in the case of the

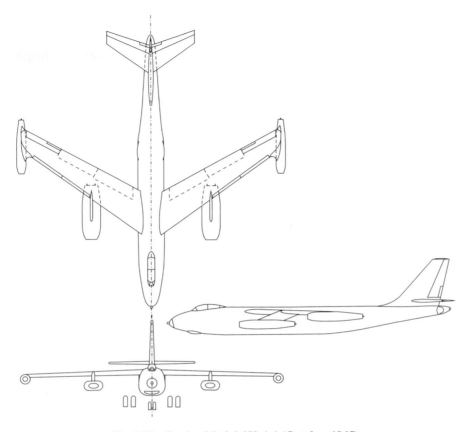

Fig. 8.20   Boeing Model 450-1-1 (October 1945).

Model 432, retracted into the fuselage. The wing was again the subject of intensive investigations. Computations showed that a 10%-thick wing (measured in the freestream direction) had a higher critical *Mach* number than a 12%-thick wing but resulted in a reduction of range by 384 km (200 miles) due to the higher structural weight. The problem of the aileron reversal, already a potential danger with the 12%-thick wing, would have intensified in case of a 10%-thick wing. Boeing, therefore, decided in favor of a 12%-thick wing for the Model 450.

Considerable effort was required in the selection of the right airfoil sections for the wing. *Reynolds* number and wing sections determine the boundary-layer behavior and thus, in the end, also the flight characteristics of the aircraft especially in the high-lift domain. The *Günter* report recommended a NACA 0012-0,55-40 airfoil section as optimum for a slender 35-deg swept wing (12% thickness, relative nose radius 0.55, 40% maximum-thickness location). However, *Günter* also said at the same time that with regard to the airfoil sections still further measurements, especially with the high-speed airfoils recently developed in Germany, were required. The choice of a good high-speed airfoil still did not allow any conclusion concerning its suitability at high lift. The German aeronautical industry had already thought about it, though primarily in conjunction with fighter aircraft. The opinion prevailed that the nose radius was of lesser importance for the critical *Mach* number than for the high-lift range. Larger nose radii were advantageous especially with regard to the "tip-stall" problem occurring on swept wings. *Hans Gropler* employed the previously mentioned airfoil for the Ju 287, however, with a larger nose radius (0.88) to improve the high-lift characteristics. To approach the pressure distribution of the airfoil with the smaller nose radius at high speeds, the airfoil was slightly cambered (1.5% at 23% chord). This represented in his opinion a good compromise between the high-lift behavior and the critical *Mach* number. The mean camber line was to be shaped in a way that, at zero angle of attack, $\alpha = 0$ deg, the pitching moment disappeared ($C_m = 0$) and the pressure distribution was like the one of a non-cambered airfoil at high speeds. Measurements in the context of the Ju 287 development were said to have shown that this airfoil had an even somewhat higher critical *Mach* number than a symmetrical airfoil [72]. With regard to the "tip-stall" problem, experts unanimously shared the opinion within the aircraft industry that extendable slats were the best solution. Wing twist would have only little effect.

The first wind-tunnel measurements within the high-lift range with a symmetrical airfoil showed a premature flow separation along the entire span. Boeing engineers finally achieved an improvement in the stall behavior with an airfoil of the NACA 65-series with a slightly cambered forward section and with a mean camber line Type 130. The measurements showed that this airfoil camber generated less drag at higher lift coefficients than airfoils with a larger maximum-camber location [73]. In addition, the inflection point in the airfoil contour (cusp) in the downstream area, typical of the NACA laminar airfoils, was removed [74] following the findings gained in Germany during the development of high-speed airfoils [72]. At high speeds this airfoil was not worse than the original symmetrical airfoil with a small nose radius. The design philosophy for this airfoil was so similar to that of *Hans Gropler* for the Ju 287 that one could almost think Boeing had simply adopted the Ju 287 airfoil. The extensive airfoil investigation showed, however, the effort Boeing had undertaken in the selection of a suitable airfoil. The comparison with Junkers clearly shows that engineers arrive at quite similar solutions for the same design objectives. Drag measurements along the span showed that the minimum critical *Mach* number of $M = 0.875$ was reached in midspan of the wing and was also not lower close to the fuselage.

As also stated in German reports, the NACA wind-tunnel test program had shown that the lateral stability decreases with increasing sweep. This effect becomes greater at low speeds and high angles of attack (during takeoff

Fig. 8.21   Boeing Model 450-1 wind-tunnel model with drooped wing tips.

and landing). Tests with drooped wing tips yielded a pronounced improvement in the stability characteristics [75]. Consequently, Boeing also investigated a version of the Model 450 with "drooped wing tips" (Fig. 8.21). Because of the resulting strength and weight problems, the drooped wing tips were discarded again.

The Model 450-1-1 had, however, a pronounced "low *Mach* (number) pitch-up" problem. Repositioning of the outer engines to a more inward position led to an improvement. The vortex pair coming off the engine nacelles produced a positive interference with the wing flow thus preventing a premature separation of the flow on the outer wing [76]. The engine arrangement led to a reduction of the wing loads. This advantage was used to enlarge the span with the result of a lower drag and an increased range. The new engine arrangement and the enlargement of the span resulted in the Model 450-2-2. Fuselage and landing-gear arrangement remained unchanged compared to the predecessor model. The fuel for the aircraft was completely accommodated within the fuselage because the thin wing did not provide any space for fuel. To provide more space for fuel, Boeing changed the design once more, and the Model 450-3-3 emerged (Fig. 8.22).

The greatest difference compared to the Model 450-2-2 was the tandem landing gear. The resulting weight saving and the increased available internal volume allowed higher loading capacity. The aircraft could, however, not rotate any further due to the landing gear being in a far rearward position. The front leg of the landing gear was, therefore, lengthened thus generating a larger angle of attack during takeoff. Because of the higher drag, the flight trials later showed that the takeoff distance was also increased. This was the reason for the installation of 18 booster rockets on the early B-47 models. The Model 450-3-3 corresponded largely to the prototype XB-47. For all models, three-part slats were planned on the outer wing. The first prototype of the XB-47 was, however, built without slats. The flight testing of the configurations of the B-47 with slats showed that, with retracted slats, the aircraft also exhibited a docile behavior in the high-lift range; therefore, Boeing discontinued the slats on later configurations, thus saving weight [77]. Flight

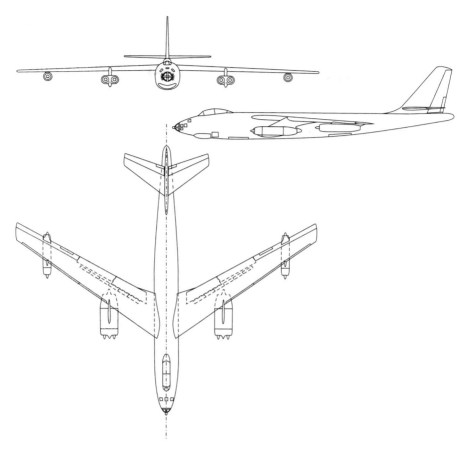

Fig. 8.22   Boeing Model 450-3-3, the final version of the XB-47 (November 1945).

tests at high speeds had, in addition, shown that slats not completely retracted caused a considerable drag [78].

Considerable importance was attached to the dimensioning of the roll and yaw control. It had to ensure that the aircraft remained controllable with asymmetric engine failure under unfavorable conditions. A large Fowler-flap, whose outer element could be deployed as aileron, provided sufficient controllability about the longitudinal axis also in case of a strong crosswind. The aeroelastic characteristics of the wing were intensively investigated on an elastically similar wind-tunnel model. Because of the elasticity of the wing, the aileron effectiveness decreased with increasing dynamic pressure. The ailerons became ineffective at a speed of 450 KIAS (Knots Indicated Air Speed) [76]. Therefore, in addition a spoiler control was provided. Later, the flight tests showed, however, that the aileron effectiveness at high dynamic pressures was still sufficient so that the spoiler control could be omitted [79].

Because at high dynamic pressures large control-surface forces are to be expected, an irreversible hydraulic control was envisaged from the start. In case of a failure of the hydraulics, the pilot could revert to a manual control with the aid of auxiliary control surfaces (tabs).

The elasticity of the wing raised another problem that had almost led to large design changes. The center of pressure shifted inboard and upstream simultaneously due the bending of the wing at large dynamic pressures or load multiples and the bending-torsion coupling of the swept wing (see Fig. 4.10). The outer wings acted here as an additional elevator. This led to an unstable neutral-point location that could no longer be compensated by the elevator. For the pilot, this meant that his aircraft could become unstable at higher load multiples, which was not acceptable according to the guidelines for the handling qualities. An analysis of the elastic behavior finally showed that the rear fuselage with the horizontal stabilizer was also deformed under load. Fortunately for the aircraft, the changed flow direction at the horizontal stabilizer produced a counter-rotating moment, which just compensated the effect of the shift in the neutral-point location.

As the wing of the XB-47 was already being manufactured, further wind-tunnel measurements were carried out on the model. Measurements with a wake rake downstream of the wing showed that flow separation first occurred on the outer wing. This conflicted with the opinion prevailing at that time, and also held by the German experts, namely, that due to the wing-fuselage interference, premature flow separation would occur close to the fuselage. For this reason, airfoil sections near the fuselage should be as thin as possible. The XB-47 possessed constant airfoil sections along the span, which greatly simplified the production process. However, fuel could not be accommodated inside the thin wing. This new knowledge was, however, only incorporated on the Boeing B-52 that had a thicker wing in the proximity of the fuselage and could carry about 60% of the fuel.

In April 1946 Boeing received an order from the USAAF for the construction of two prototypes of the Model 450-3-3. Two and a half years later, on September 12, 1946, the XB-47 was rolled out in Seattle, and on December 17, 1947, the first flight took place [64]. The flight tests indicated insufficient damping of the Dutch-roll motion (gyrating oscillation). The wing tips bent upwards due to elasticity and removed the negative dihedral of the wing. This problem could, however, be solved by the installation of a yaw damper. The elasticity of the wing had, however, also an advantage. Sweep and elasticity reduced the lift increase $dC_A/d\alpha$ ($dC_L/d\alpha$) and thus reduced the gust sensitivity of the aircraft. A "high *Mach* (number) pitch-up" problem was revealed at high-speed flight maneuvers with load multiples of $n > 1$. The cause was flow separation downstream of the shock on the outer wing at *Mach* numbers of $M > 0.83$. However, the aircraft was to be flown at load multiples of 1.15 g and roll angles of up to 30 deg without problems. The attachment of vortex generators on the outer wing improved the flight characteristics also within this range.

The climb performance of the XB-47 was excellent. Thanks to the elasticity and the yaw damper the stability about all axes was very good in gusty conditions. The landing with the tandem landing gear was not simple. Because of the slow response of the jet engines of that time, the approach was carried out at higher engine rpm and with a brake parachute deployed in order to be able to bring the engines quickly up to full thrust in case it was necessary to abort the landing [77].

The first prototype did not quite meet the expectations of Boeing with regard to the flight performance. The test pilots were, however, enthusiastic because the XB-47 could be flown like a fighter aircraft and not like a portly bomber. Only the second prototype with the more powerful General Electric J47-GE-3 engines reached a speed of more than 965 km/h. The demands of the USAAF were thus more than met and the competition clearly outclassed. The XB-47 became the basis for the further success story of the company (Fig. 8.23). This was aptly expressed by the aerodynamicist *William H. Cook* at the beginning of the 1990s when he gave his book about Boeing the title *The Road to the 707.*

8.4.2 NORTH AMERICAN AVIATION XP-86

In the fifties the North American F-86 "Sabre" with its swept wing and empennage became the epitome of the modern high-speed fighter aircraft, the absolute "jet fighter." The aircraft impressed by its elegance and with regard to flight performance and flight characteristics was also an excellent design. By its deployment in the Korean War, the F-86 achieved fame and the aura of an all-out superior fighter aircraft. With a downing ratio of 14:1 against the Russian MiG-15, the F-86 advanced to one of the most successful fighter aircraft. The success of this aircraft is also reflected in the fact that the F-86 belonged to the first choice in many countries when equipping their air forces. Also the Air Force of the Federal Republic of Germany received its first F-86 in 1952 from Canada. The aircraft was extremely popular with the pilots and served until 1966 in four fighter squadrons until it was replaced by the Lockheed F-104 "Star Fighter" [80]. An extensive literature exists concerning the development of the F-86 and her numerous variants. The question that

Fig. 8.23 Boeing B-47 "Stratojet" during flight.

concerns us is what use was made of German documents in the development
of the F-86.

In 1941 the American aeronautical industry received information from
Britain concerning their jet engine development. During flight testing of the
first American jet aircraft, the Bell XP-59, at the end of 1942, a British jet
engine was installed and its usefulness demonstrated. After that, further
design studies of jet-propelled fighter aircraft were presented by the American
aeronautical industry. In November 1944 such a project came from the North
American Aviation (NAA) Company. On request from the U.S. Navy, the
company started with the design of a jet-propelled fighter aircraft that was to
operate from aircraft carriers. An axial-flow jet engine of American design
was envisaged as propulsion. The starting point of the development was the
P-51H, the last version of the successful "Mustang," whose technology was
to a large part adopted for the new design [81]. Under the in-house designation NA-134, a new design evolved that could not deny its descent from the
P-51 (Fig. 8.24).

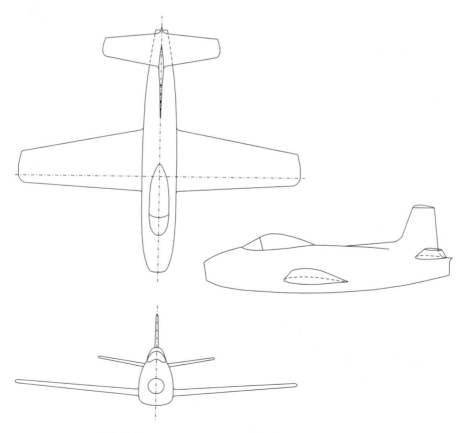

Fig. 8.24   North American Model NA-134 (1944).

The NACA laminar airfoil had already proven a success as high-speed airfoil on the P-51. Because speeds of up to 885 km/h were envisaged, the chief aerodynamicist of NAA, *Ed Horkey*, decided on a thin laminar airfoil of the NACA 64-series to avoid, if possible, compressibility effects. The airfoil (NACA 641-112, $a = 0.6$) with a relative thickness of 12% and a maximum-thickness location of 40% remained unchanged along the span. The slightly cambered mean camber line was adopted from the P-51H airfoil. The Model NA-134 was, interestingly enough, also investigated with slats in the area of the ailerons [82]. In January 1945, the Navy ordered three prototypes designated XFJ-1. The first XFJ-1 took off on November 27, 1946 (Fig. 8.25). During the flight trials, a maximum speed of 872 km/h at an altitude of 4900 m was reached. In the spring of 1947, an XFJ-1 achieved a *Mach* number of $M = 0.87$ in a dive, that is, the highest *Mach* number reached up to that date, by an American fighter aircraft [83].

In May 1945, the USAAF started to become interested in the project NA-134. The Air Force demanded a fighter aircraft of medium range with a maximum speed of more than 960 km/h (600 mph). The NAA responded to the inquiry of the USAAF with the configuration NA-140 (Fig. 8.26), a design derived from the configuration NA-134. The NAA received an order for the construction of three prototypes, designated XP-86 (Fig. 8.27), on May 18, 1945. The NA-140 had a more slender fuselage than the Model NA-134, and the empennage was placed further back.

The wing had the same airfoil section as the NA-134, however, with a smaller thickness of only 11% at the wing root and 10% at the wing tip. The chief aerodynamicist *Ed Horkey* saw the only possibility to noticeably increase the speed in comparison to the NA-134 in the choice of a thinner airfoil. The nose of the fuselage was slightly changed to adopt the AN/APG-5 radar range finder. As propulsion, the axial jet engine General Electric GE TG-180 was envisaged. The maximum speed was to be 940 km/h at an altitude of 3000 m ($M = 0.8$), still far away from the demanded 965 km/h. The NAA knew that the Model NA-140 would not be better than the competing Republic XP-84. If the maximum speed could not be considerably increased, then that could have been the end of the XP-86 [93].

Fig. 8.25   North American XFJ-1 (N. Avery, "North American Aircraft 1934–1998," Narkewicz/Thompson Publisher, 1998).

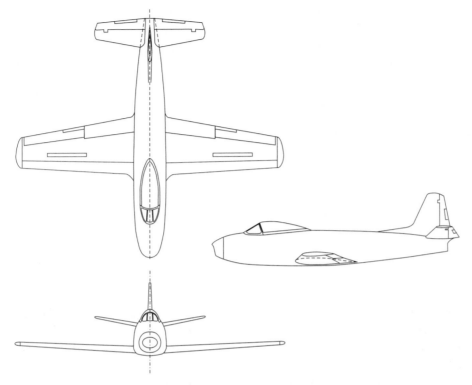

Fig. 8.26   North American Model NA-140 (XP-86, May 1945).

When *George S. Schairer* of Boeing sent his well-known letter concerning the advantages of wing sweep from Germany to his boss, *Benedict Cohn*, on May 10, 1945, he also asked him to inform other interested parties of the U.S. aeronautical industry about his "discovery." *Ed Horkey* of the NAA also belonged to the ones *Cohn* informed. The information and documents concerning the swept wing, arriving little by little from Germany, literally inspired many an engineer within the U.S. aeronautical industry. Thus a colleague of *Edgar Schmued*, the

Fig. 8.27 North American XP-86 with straight wings [85].

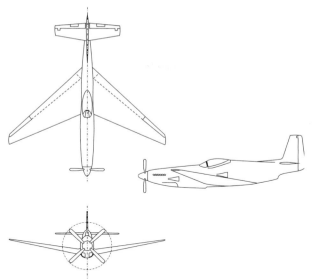

Fig. 8.28   Study of the North American P-51H with swept wings.

creator of the famous P-51 "Mustang," designed a version of the P-51H (Fig. 8.28) with swept wings and hybrid propulsion (piston and jet engine). This curious design was, however, not a precursor of the XP-86 and was also not connected with the development of this aircraft [84].

The head of the project aerodynamics office at the NAA, *Larry Green*, understood enough "technical German" to be able to evaluate the research reports from Germany. He had, in addition, participated in the interrogation of German engineers and scientists [85], and the statements of his colleagues who were part of the expert teams that had traveled in the wake of the American troops through Europe were also available [86]. Primarily the swept-wing topic was of great interest to *Green*. Of special interest was here the Me 262 HG II and its 35-deg swept-wing configuration.

A report of *G. Koch* of the LFA Braunschweig-Völkenrode [87] and the known report of *H. Ludwieg* of the AVA in Göttingen [88] concerning the first measurements on swept wings belonged to the German documents of primary interest to *Green*. While both reports provided a confirmation of the theory of the yawed wing established shortly before by *R. T. Jones*, a further German report by *B. Göthert* of the DVL Berlin was of great practical importance because the design of a swept wing for a specific application was concerned [89]. These reports were complemented by reports on wind-tunnel measurements on the Messerschmitt Me 262 HG II configuration.

The swept wing was, moreover, not unknown to *Green*. Three years before he had carried out measurements on a swept wing for the experimental aircraft Curtiss XP-55 in the wind tunnel of Caltech and observed an unpleasant characteristic of the swept wing, the "tip stall." At that time, this problem, important to the controllability of an aircraft, could not be solved. *Green* was, therefore, very skeptical, whether the utilization of the swept wing on the XP-86 would not cause the same problems.

A practical solution had already been recommended a year before by the NACA. The NACA had carried out wind-tunnel measurements on swept wings in the context of the development of tailless aircraft at Northrop and discovered that a slat was best suited to prevent flow separation at the wing tip at high-lift coefficients [90]. Whether *Green* was informed about this report is not known, however, it is quite possible that he knew from interrogation of German engineers that due to the automatic slats, the Messerschmitt Me 262 did not have any "tip-stall" problems. Moreover, *Green* knew the experimental aircraft P 1101 of Messerschmitt, which, as a captured aircraft, was at Wright Field for trials. The aircraft also possessed such a slat to prevent "tip-stall" problems.

From mid-1942 in Germany, after the high-speed characteristics of the swept wing had been dealt with, the high-lift behavior moved increasingly into the center of the interest (see Secs. 2.3.2 and 2.3.3). The German aerodynamicists had recognized that the effectiveness of standard landing flaps and ailerons decreased with increasing sweep. However, high-lift coefficients and good lateral controllability were especially important during the landing

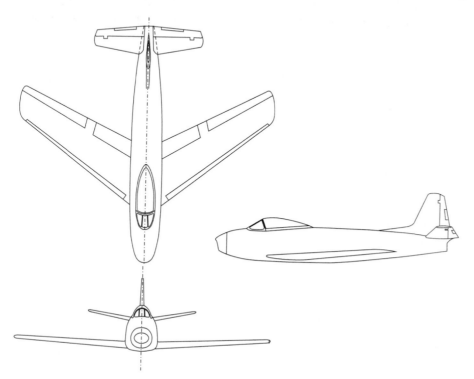

Fig. 8.29   North American NA-140 with swept wings (August 1945).

approach. Especially the Messerschmitt Company worked in Germany on the solution of the high-lift problem. On behalf of the many investigations carried out with regard to this topic at German research establishments, we shall quote the report of *G. Brennecke*, who investigated the effectiveness of different high-lift devices on a 35-deg swept wing [91]. An important message of this report concerned the "tip-stall" problem. *Brennecke* showed that with the aid of a slat in the area of the ailerons the aircraft could be well controlled. However, according to statements of former employees of the NAA, among the German reports the NAA had received none were about investigations that provided any information concerning the high-lift behavior thus confirming the statements of the German experts [84].

The studies of the German documents by the Americans showed that it is possible to increase the critical *Mach* number by $\Delta M = 0.1$ using a 35-deg swept wing. Therefore, the NAA started in June 1945 to project a version with swept wings, parallel to the straight-wing NA-140. The wind-tunnel measurements on the Me 262 HG II configuration finally convinced *Green* of the advantages of this wing. Therefore, the 35-deg swept wing of the Me 262 HG II configuration with an aspect ratio of 5 and a taper ratio of 0.5 became, after all, the model for the XP-86 swept wing [85]. It was realized simultaneously that slats were needed to be able to successfully meet the "tip-stall" problem although there were reservations at the NAA concerning this

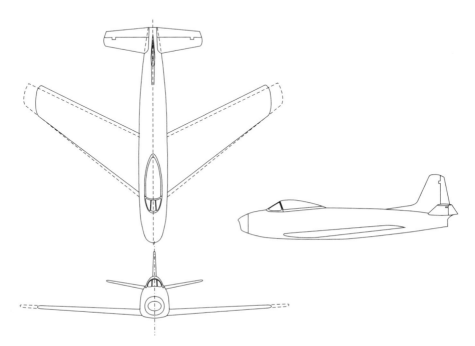

Fig. 8.30   North American Model 140 with swept wings of aspect ratios of 5 and 6.

"German" solution. The NAA had a complete Me 262 wing delivered from Wright Field to study a suitable slat system. However, in the end, wind-tunnel tests were not carried out with this wing.

With the choice of this wing, the NAA also abandoned the slightly cambered NACA laminar airfoil of the NA-140 (Fig. 8.29) and selected the symmetrical high-speed airfoil of the Me 262 HG II, which Messerschmitt had at the end of the war envisaged for the P 1101. Seen in the freestream direction, the airfoil had a relative thickness of 9.8% at the wing root (Airfoil 0009,8) and at the tip a thickness of 9%. In the forward part, the airfoil consisted of an ellipse with the maximum thickness at 40% of the wing chord and ended without an inflection point in the contour in a point. The nose radius was substantially larger than in case of the laminar airfoils of the so-called NACA 64-series, that is, advantageous for the attachment of slats and, at the same time, also advantageous with regard to the high-lift behavior. The renunciation of a well-tried American airfoil can only be explained by the tight schedule of the complete project. With the wing and airfoil choice, the NAA had decided in favor of a quickly available overall solution, which could be assumed to have already been adequately investigated by the DVL in Germany. A model of the XP-86 with a swept wing and slats along the entire leading edge was completed in September 1945 and tested in the NAA low-speed wind tunnel.

The wind-tunnel measurements showed that, although the "tip-stall" problems were not yet completely solved, the slats provided an acceptable stall behavior. With this result, the aerodynamicists of the NAA went to Wright Field to discuss the new concept with the head of the research and development department of the Air Force. The NAA received on November 1, 1945, permission to go ahead with the development of the XP-86 swept wings. At that time, the employees of the loads department raised doubts because of the high structural weight. As a result, the swept-wing investigation of B. Göthert was consulted, which had shown that with an aspect ratio of 6, a better lift-to-drag ratio could be achieved. As a result, the design was changed. Compared to the Göthert wing ($\lambda = 0.5$), the taper ratio was slightly increased to save structural weight. At the same time, the laminar airfoil of the NA-140 with straight wing was reinvestigated (Fig. 8.30). Until the beginning of November 1945, extensive wind-tunnel measurements were carried out on both configurations in the NAA low-speed tunnel. Over all, there were measurements related to about 150 different slat configurations, but the intense "pitch-up" in the case of the wing with the larger aspect ratio could not be overcome.

The structural experts had in the meantime developed a new structural design that promised a reduction of the weight. This was the reason for the NAA to return to the old wing with an aspect ratio of 5. The decision for this wing came officially in March 1946. In this way, the experts of the NAA tried to reach an optimum between flight characteristics and maximum lift. The experts decided in favor of an automatically adjustable slat as had also been deployed on the Me 262. The Me 262 slat mechanism was adopted with some improvements. Instead of the two-part slat of the Me 262, the XP-86 possessed a slat consisting of four segments (Fig. 8.31). The slat had, in addition, a constant chord over the entire span that had already been recognized by *W. Voigt* as advantageous regarding the stall behavior. The reason for the choice of a segmented slat may well have been the fear that aeroelastic deformations of the wing may have led to problems when activating the slats.

Towards the end of 1946, measurements were performed in the high-speed wind tunnel of the NACA Ames Research Center. Here, flow separation showed up along the entire wing trailing edge. The experts recognized the excessive trailing-edge angle of the airfoil as the cause. By an extension of the airfoil chord by 10.76 cm, the problem could be solved and the wing of the XP-86 obtained its final shape.

All control surfaces were aerodynamically compensated and the control-surface gap sealed on the inside. The control-surface operation was carried out via hydraulic boosters with force feedback. The entire horizontal stabilizer

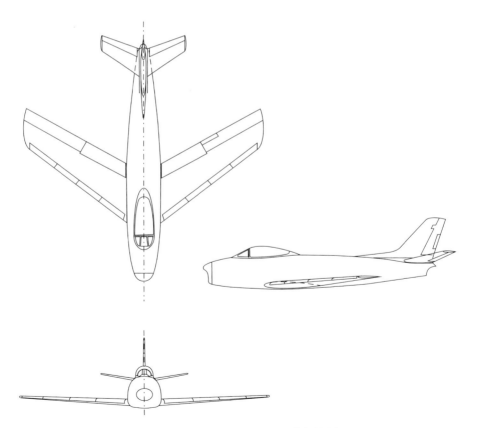

Fig. 8.31   North American XP-86 (1946).

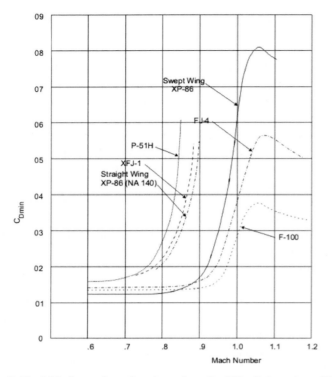

Fig. 8.32 Critical *Mach* number of various aircraft of North American ([85], p. 15).

could be adjusted for trimming purposes by an electromotor. Dive brakes were mounted at the sides of the rear fuselage at the level of the horizontal stabilizer. A comparison of the critical *Mach* numbers for the drag rise of the P-51H to the XP-86 shows the progress that was achieved in flight performance with the aid of the swept wing (Fig. 8.32).

The XP-86 (Fig. 8.33) took off on its maiden flight on October 1, 1947, with test pilot *George Welch* at the controls. During the same year, *Chuck*

Fig. 8.33 Prototype North American XP-86 during flight (courtesy of USAF Museum).

Yeager carried out the first supersonic flight with the XP-86 at a *Mach* number of $M = 1.06$. The flight trials showed that the complicated slat system had to be improved. The slats already extended during climb, which was not permissible. An asymmetric extension of the slat segments during certain flight maneuvers or during spinning tests was even far more problematic. The slat mechanism was, therefore, repeatedly changed until a satisfactory operation had been achieved.

The test pilots were impressed by the good-natured stall behavior of the XP-86 although flow separation (stall) commenced with a hardly noticeable buffeting. Gyrating oscillations (Dutch-roll) only showed up at lift coefficients of $C_L > 1$, that is, they were not relevant within the normal operational range. The flight tests showed that the slats also opened without any problems at higher *Mach* numbers without causing adverse effects. The NAA engineers were surprised by the swift decrease in maximum lift with increasing *Mach* number. Maximum lift at high *Mach* numbers was reached at maneuvers with high load factors (level turns and pull outs). Interestingly enough only slight buffeting showed up at these maneuvers. The lateral stability was judged to be very good. The longitudinal stability was investigated at NAA up to $M = 0.92$. It was criticized that the stick-force gradient (stick force per g) at rear center-of-gravity locations became almost indifferent. Overall more serious was, however, that the aircraft showed the tendency to turn into a dive (tuck under) above $M = 0.88$ and at $M > 0.9$ was inclined to stall abruptly. This flight behavior could not have been predicted from the wind-tunnel measurements at Ames (NACA) [92].

However, all of these faults could be remedied during flight trials making the F-86 "Sabre" into what it still is today in the memory of numerous pilots—an ideal fighter aircraft (Table 8.1). In September 1948, Maj. *Dick Johnson* increased the world speed record to 1080 km/h with an F-86A. The development and flight testing of the F-86 provided a tremendous technology jump for the NAA and secured the company a lead over her competitors in the development of supersonic fighter aircraft.

TABLE 8.1   XP-86 TECHNICAL DATA

Span	37 ft (11.3 m), 1/7/16 in
Aspect ratio	4.79; in March 1946 definitely fixed
Sweep	35 deg (25% chord line)
Airfoil normal to the 25% chord line	0011,6 (root), 0010,4 (tip) [93][a]
Engine	General Electric TG-180 with 4000 lb.s.t
Maximum speed (horizontal)	>650 mph (1045.8 km/h) flown
Maximum speed	$M = 1.06$ in a dive on October 19, 1947
Critical *Mach* number	0.875 (wind-tunnel measurements)

[a]By the later enlargement of the airfoil chord, the relative thickness decreased somewhat.

8.5   HIGH-SPEED AERODYNAMICS IN THE SOVIET UNION

8.5.1   AERONAUTICAL RESEARCH PRIOR TO 1945

Compared to the initial conditions, Soviet aeronautical research experienced a considerable upswing between the two World Wars. The main institution was TsAGI, founded in Moscow in 1918. At the beginning of the 1930s, TsAGI was moved to Zhukovsky near Moscow where a large modern scientific center arose. Further outstanding research establishments were the Moscow Aeronautical Research Institute, the Central Institute of Air Engines (TsIAM), as well as further test centers of the Air Force and the aircraft industry. In the 1930s, the number of employees went into the thousands. Just at TsAGI, approximately 3500 employees worked there in the autumn of 1936 [42].

The equipment standard of the Soviet research institutes stayed, however, below that of other leading aeronautical nations. Among others, a large Soviet committee of experts, who visited Germany in the context of agreements within the German-Soviet nonaggression pact (August 1939) at the end of 1939, drew attention to the existing deficits. During its visit, the delegation got to know different German aeronautical research establishments and aircraft companies. After their return, some members of the delegation reported about their impressions at a meeting of the People's Commissioner's Office of the Aeronautical Industry. By stressing the merits of German aeronautical research and of the aircraft industry, the committee at the same time, severely criticized their own incomplete and poor equipment, the missing advanced research, and the unsatisfactory cooperation and organization [94].

Moreover, many positive development approaches literally fizzled out due to the political "purging" under Stalin. A number of capable scientists, designers, and engineers lost their lives or disappeared over night in one of the detention camps. A typical example is, for instance, the execution of the director of TsAGI, *N. M Charlamow*, in 1938 [40]. In this connection, a system of scientific-technical special design offices, colloquially called "Scharaschkas," were established where scientists, engineers, and technicians were forced to work as prisoners on military research and development [95].

After the Volta Congress (Sec. 1.3), there were attempts at TsAGI to catch up in the field of high-speed aerodynamics (wind-tunnel facilities, research topics) and, most of all, to build up the research facilities required. After all, the conference proceedings of the Volta Congress, and hence also the lecture of *Busemann*, were available in Russian in 1939, that is, the research results and theoretical approaches presented at the Volta Congress could be studied by a larger Soviet expert audience (Fig. 8.34).

However, from 1941 the war generally restricted fundamental research to a minimum and stopped innovative development projects, such as a

А. Буземан.

ПОДЪЕМНАЯ СИЛА ПРИ СВЕРХЗВУКОВЫХ СКОРОСТЯХ.

Введение.

Хорошее соотношение подъемной силы и сопротивления крыла при малых скоростях полета становится хуже, когда скорость приближается к значению скорости звука. Возникает поэтому вопрос, можно ли при сверхзвуковых скоростях получить опять большое значение подъемной силы при малом сопротивлении и как можно этого добиться? Этому вопросу и будет посвящено мое сообщение. Другие доклады будут касаться вопросов о технике экспериментального исследования в этом направлении и современных результатов произведенных испытаний, поэтому я могу ограничиться изложением тех возможных решений вопроса, которые даются теорией.

Область явлений, к которым будут относиться излагаемые теоретические соображения и выводы, будет существенно ограничена тем, что я буду рассматривать явления при исключительно сверхзвуковых скоростях, когда ни в одной точке поля нет скорости, меньшей скорости звука. Обусловливается это ограничение тем, что для таких состояний движения, где скорости частично выше скорости звука, а частично ниже, мы имеем только частные решения некоторых отдельных задач. С другой стороны, при обтекании твердых тел в потоке со сверхзвуковой скоростью скорости могут опускаться частично в некоторых областях движения ниже скорости звука только в том случае, когда твердое тело имеет тупой передний конец или поставлено в потоке под большим углом атаки. В таком случае в окрестности застойной точки в передней части тела мы получаем скорости, ниже скорости звука, и очень высокие давления, в то время как в задней части тела мы не наблюдаем области со столь же большими давлениями, действующими в обратном направлении. В силу последнего в подобного рода состояниях движения, с частным появлением дозвуковых скоростей, едва ли возможно искать какие-либо практические преимущества. Нужно, однако, признать это утверждение недоказанным с полной строгостью, так как соображения и методы, относящиеся к чисто сверхзвуковому движению, такого доказательства не дают.

Fig. 8.34 Russian translation of *Busemann's* lecture at the Volta Congress.

rocket-propelled aircraft development. A consolidation of the research base was largely neglected, and the work of the aeronautical research establishments concentrated on the requirements of the war. Only the deployment of the new German weapons developments during summer/fall of 1944 led the Moscow government to specific measures to copy, reconstruct, or adopt these technology advances [96].

8.5.2 TECHNOLOGY TRANSFER (1945–1946)

After WWII, just as the other allied victors, the Soviet Union used to a large degree the technical and scientific know-how of the defeated German Reich. The German specialists were regarded by the Soviet Union as an important means to be able to keep up in the future East–West military conflict. However, regarding the medium- and long-term perspectives, it showed very quickly that the Soviet Union, despite her great interest in the German weapons technologies, was not interested in integrating German experts. In adopting German aircraft and rocket technology, the Soviet Union followed a transfer concept that was "aimed at the delayed competence acquisition." The intention was that the Soviet designers and technicians should catch up with the technological development and then advance this development on their own [30, 97].

A break in the Soviet technology transfer from Germany occurred on February 25, 1945. On this day, with the Instruction No. 7590 of the State Committee of Defense (GKO), Stalin ordered the formation of a "special committee." This committee coordinated all matters related to the dismantling of industrial facilities, the transfer of technology, as well as the confiscation of cultural goods and other German assets. Soviet institutions, research establishments, universities, and the Academy of Sciences urged employees and leading executives to visit Germany. These visitors obtained from the "special committee" officers' ranks, uniforms, and extraordinary authority to assess and transfer German know-how to the Soviet Union. The top priority orders to take over high-technology items, like long-range missiles or jet engines, very quickly forced the evaluation teams of the "special committee" to encourage a careful dismantling practice. The formal dismantling of these research institutes and production plants, that is, the packaging into boxes and the fast removal turned out to be counterproductive and quickly highlighted the failure of the intended technology transfer [24, 98, 99]. Within a short time, it was, therefore, decided that the "scientific disarmament" of Germany should not be allowed to take place as a formal dismantling action. Within the Soviet occupational zone, special design offices were, therefore, established in 1945–1946 where, on the order of the "special committee," intensive studies of the German technologies and scientific progress took place, in order "to come in the interest of our country to a systematic and fuller exploitation of the scientific forces of Germany," as a representative of the Soviet Academy of Sciences remarked ([24], p. 44).

Parallel to the setup of the system of research and development centers ("special design offices") in the Sowjetische Besatzungszone (SBZ) (Soviet Occupation Zone), various peoples commissar's offices in the Soviet Union were instructed to set up corresponding industries, such as, for instance, a rocket and nuclear industry. The background was the fact that in the Soviet Union the necessary industrial structures were not available or the ones available had to be newly aligned. The reconstruction and study of the German technologies attained extraordinary significance in the resolution of these tasks. In the SBZ, the Soviet aeronautical industry itself maintained in May 1946 four large special design offices with a total staff of more than 5000 engineers, technicians, and laborers. The number had increased to more than 8000 employees in October 1946. The number of persons employed at the central facilities and its branches for the rocket development near Nordhausen was of a similar order of magnitude (about 7000). The detailed Soviet specialists are also contained in these figures. However, their number can only be guessed and was probably never more than 10%.

During April and May 1946, it was decided to complete the reconstruction and development tasks at the special design offices in the SBZ and to prepare the transfer of important German experts to the Soviet Union. During the last days of October 1946, the special design offices of the aircraft industry and other arms-related industries in the SBZ were dissolved. More than 80% of all German specialists working later in the USSR were moved to the USSR during the large transfer action of October 22, 1946. A further dismantling action, this time concerning very innovative industrial structures, spread through the SBZ. About half of the 3000 German specialists sent between 1945 and 1947 to the USSR came from the aircraft and engine industry. Of those alone, 64% came from Junkers [24, 30, 34, 96, 100].

With the transfer to the USSR, the chapter "aircraft industry" ended in the SBZ/GDR for eight years. The factories were completely dismantled. In the USSR, the German aeronautical and engine specialists were, in essence, settled at two locations: Podberesje and Kuibyschew (today Samara). Podberesje is located about 120 kilometers north of Moscow. Here, the center of the German aircraft construction in the USSR, the so-called OKB 1 and 2 (Opytnoye Konstruktorskoye Byuro) (Experimental Projects Design Office) was established. There, about 50% of all German aircraft specialists in the USSR were assembled. The Soviet Ministry of Aeronautics concentrated primarily on the German engine specialists at Kuibyschew, a city located about 700 kilometers southeast of Moscow. Only a few German specialists were employed as individual researchers [19, 24, 94].

8.5.3 EFFECTS OF THE TECHNOLOGY TRANSFER

While the fighting in Berlin was still raging, the first group of Soviet experts arrived at the DVL in Berlin-Adlershof on April 29, 1945. There had not been any heavy battles around the DVL, and so the buildings had remained largely undamaged. The Germans themselves had not dismantled these facilities. Only documents and research reports had been hidden by members of the institutes on the premises of the DVL, for example, the safes with the documents were integrated into the walls of the bomb shelters [101].

The DVL was of great importance to Soviet aeronautical research because the evaluation commissions here received a comprehensive insight into the progress of German aeronautics research. Besides a large collection of German research reports (Fig. 8.35), tons of equipment were shipped to TsAGI and TsIAM in the Soviet Union. Moreover a documentation center was established (Office of New Technology) at TsAGI, where—as in the case of the Americans and British—German reports were translated and further distributed to the relevant institutions. The ideas and results available with these German research reports were of great importance because they, most of all, allowed a reorientation towards fields largely ignored up to then [19, 94, 101].

The test reports concerning the investigation of aircraft and their components in the wind tunnel of the DVL between 1939 and 1944 probably belonged to the most valuable discoveries. The later head of the Moscow TsAGI, _G. S. Büschgens_, remarked, "the reports of _Göthert_, which included documents on tests on airfoils and wings of different sweep at various _Mach_ numbers, were the most interesting. Here, an evaluation of the effect of sweep on M_{cr} and the beginning of the increase in wave drag was presented" ([94], p. 132). Indeed, with the availability of the Russian translation of the German

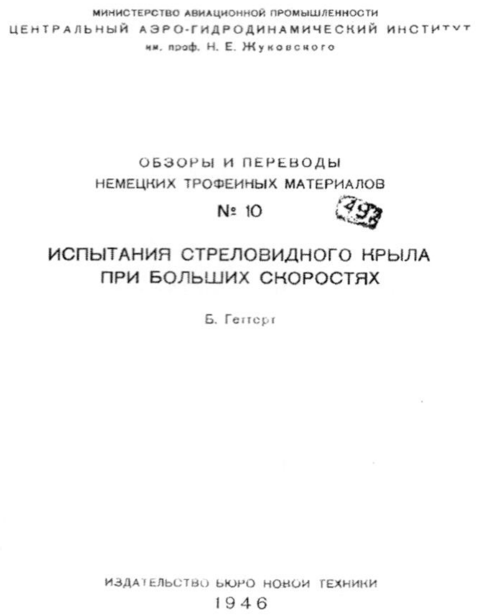

МИНИСТЕРСТВО АВИАЦИОННОЙ ПРОМЫШЛЕННОСТИ
ЦЕНТРАЛЬНЫЙ АЭРО-ГИДРОДИНАМИЧЕСКИЙ ИНСТИТУТ
им. проф. Н. Е. Жуковского

ОБЗОРЫ И ПЕРЕВОДЫ
НЕМЕЦКИХ ТРОФЕЙНЫХ МАТЕРИАЛОВ

№ 10

ИСПЫТАНИЯ СТРЕЛОВИДНОГО КРЫЛА
ПРИ БОЛЬШИХ СКОРОСТЯХ

Б. Геттерт

Fig. 8.35 TsAGI Report No. 10/1946 by
Göthert **about high-speed measurements on**
a swept wing, in Russian.

ИЗДАТЕЛЬСТВО БЮРО НОВОЙ ТЕХНИКИ
1946

test reports, intensive research quickly commenced at TsAGI concerning the swept-wing effect and high-speed aerodynamics (Figs. 8.36–8.38). The very large research groups of TsAGI were headed by *S. A. Christianowitsch* and *W. W. Struminskij*. The first systematic laboratory reports of TsAGI date from 1947 [40, 94, 102].

The first Soviet team of experts at the DVL in April 1945 consisted of seven members. The head of this group was Maj. Gen. *N. I. Petrow*, the

Fig. 8.36 TsAGI Laboratory Report No.
4/1947 concerning the aircraft Number
"152" with swept wings.

Центральный Аэро-Гидродинамический Институт
имени проф. Н.Е. Жуковского

ЛАБОРАТОРИЯ № 4.

УТВЕРЖДАЮ:
Зам. Начальника ЦАГИ
п/академик
⋅⋅⋅ 1947г.　　　　　　　　　　/С.А. Христианович/

О Т Ч Е Т

ПО ИССЛЕДОВАНИЮ МОДЕЛИ САМОЛЕТА № 152 С РЕАК-
ТИВНЫМ ДВИГАТЕЛЕМ И СТРЕЛОВИДНЫМИ КРЫЛЬЯМИ
в аэродинамических трубах Т-1-2 и Т-5 ЦАГИ
по заданию завода № 301.

Работа № 4307

Нач. отдела № 2
канд. техн. наук　　　　　- З.П.Горский

Ведущие инженеры:
Старший инженер　　　　　- В.Г. Табачников
Нач.расчетной гр.　　　　- К.С. Петрова　　*1947*

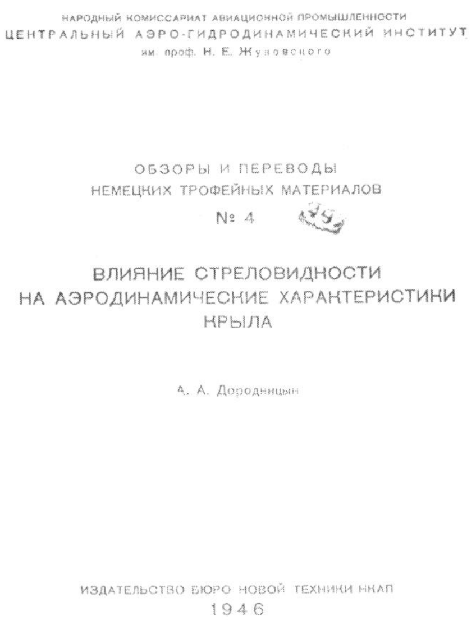

НАРОДНЫЙ КОМИССАРИАТ АВИАЦИОННОЙ ПРОМЫШЛЕННОСТИ
ЦЕНТРАЛЬНЫЙ АЭРО-ГИДРОДИНАМИЧЕСКИЙ ИНСТИТУТ
им. проф. Н. Е. Жуковского

ОБЗОРЫ И ПЕРЕВОДЫ
НЕМЕЦКИХ ТРОФЕЙНЫХ МАТЕРИАЛОВ
№ 4

ВЛИЯНИЕ СТРЕЛОВИДНОСТИ
НА АЭРОДИНАМИЧЕСКИЕ ХАРАКТЕРИСТИКИ
КРЫЛА

А. А. Дородницын

ИЗДАТЕЛЬСТВО БЮРО НОВОЙ ТЕХНИКИ НКАП
1946

Fig. 8.37 Summarizing TsAGI report concerning swept-wing research (1947).

Director of the Institute for Aircraft Equipment. Other representatives of central Soviet research establishments included *W. W. Wladimirow* (Deputy Director of the Institute for Aircraft Engines), *K. N. Surschin* (Deputy Director of TsAGI), *P. S. Ambarzumjan* (Deputy Director of the Military Institute for Aircraft Engines), *Sosim* (Deputy Head of the Flight Testing Center), *G. N. Abramowitsch* (Deputy Director of the Research Institute 1), and *B. E. Tschertok* (scientific employee at the Research Institute 1) belonged to the team. [94]. The specific instructions from Moscow were,

Fig. 8.38 TsAGI design drawings of different swept wings (around 1946).

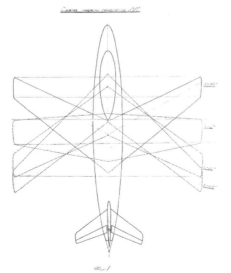

the task of the commission consists in the dismantling, value conserva-
tion and shipping to Moscow of all German experimental aircraft and
engines, the aircraft equipment, and the aggregates and material for their
production, scientific-technical examples of work, laboratory equipment,
wind tunnels, instrumentation, libraries, and scientific archives. The
commission has to take up work immediately after the occupation of the
respective locations, scientific centers, and industrial areas of Germany
by the Soviet troops. ([94], p. 131)

As already explained, the approach of the expert commission was at first
determined by the principle of the formal dismantling, that is, the Soviet officers
selected "on the scene" and ordered the immediate shipment of the facilities,
equipment, and documents of the DVL. However, the sense of this approach was
quickly questioned by the Soviet participants. Doubts essentially existed whether
the equipment and documents, packed into boxes, would really reach their
Soviet addressees and whether one would appreciate at home the value of the
captured goods—without closer explanations and the knowledge of the relevant
context. The procedure, therefore, changed within a short time period and the
Soviet expert teams remained more than a year at the DVL [24, 96, 101].

Although the majority of the employees had already left weeks before,
mostly in the western direction, about 50 employees stayed at the DVL.
Among them was the head of the DVL, *Günther Bock*. The employees, and
primarily *Bock*, were intensively questioned and charged with the preparation
of reports. The interrogation documents and reports have been maintained at
the archives of TsAGI. *Bock* still remained at the DVL in Berlin until the
summer of 1946. In July 1946, he was finally brought to TsAGI at Zhukovsky
near Moscow where he worked as one of the few "individual researchers" up
to 1953 as an advisor. According to information of the Soviet State Security
Service, *Bock* "was only in a very limited way" exploited. Altogether under
instruction of the experts, the Soviet dismantling teams took 436 pieces of
machinery, two presses, and 6245 other instruments, equipment, and facility
components from the DVL. The dismantled property went to TsAGI and the
TsIAM at or near Moscow [94, 103].

With this new knowledge, Moscow defined during the summer of 1945 the
task of the Soviet expert commissions and the institutions within the Soviet
Union. The head of TsAGI received the corresponding task to ensure "the study
of all scientific research and experimental documents" that were taken from
Germany and were related to the aerodynamics of jet engines and novel type
of engines [94, 96]. Primarily the modernization of their own research facilities
was urgently needed. The high-speed wind tunnel "T 106," operating at TsAGI
since 1943, was simply not up to the new tasks. Equipment of the DVL wind
tunnel was presumably used at TsAGI, and German experience was employed
in modernizing the tunnel. After the completion of the modernization, the wind
tunnel received in 1947 the designation "T 106 M" and could be deployed up

to *Mach* numbers of $M \approx 1.1$. It is very likely that *Bock* advised TsAGI in the modernization and improvement of the high-speed wind tunnel [40].

8.5.4 GERMAN AIRCRAFT AND ENGINE DEVELOPMENT*

8.5.4.1 BOMBER OF JUNKERS. Two German design offices (in Russian OKB) were established at Podberesje after the transfer action of October 1946. Employees of Junkers worked at the OKB 1 on the development of bombers with jet propulsion. The Ju 287, already modified in the SBZ as EF-131, was tested for the first time in the USSR during the winter of 1946–1947 (also see Secs. 5.3 and 6.3). Parallel to this, there were tests with an experimental single-seater (EF-126 with an Argus-pulse engine), which had been developed by Junkers at the end of the war. Also a mock-up of the large jet bomber EF-132, still projected at Dessau, was completed. All three aircraft were, however, in the end not successful. They were taken off the development program of the OKB 1 in 1947. Insufficient "usefulness" was given as a reason. The EF-131 was thereafter fundamentally modified and tested as EF-140. The EF-140 was to become a high-speed jet bomber operating at high altitudes. The prominent feature of this version, as in the case of the Ju-287 and the EF-131, was again the swept-forward wings. The testing of this aircraft commenced during 1948–1949, but, in the end, extensive modifications to the design caused a deterioration of the flight characteristics of the aircraft "140." Also work on this aircraft was finally stopped (compare here Chapter 6.3). The Soviet director of the two German OKBs at Podberesje changed in the autumn of 1948. The management was now taken over by the aircraft designer *Semjon Michailowitsch Alexejew*, who had already worked under the well-known aircraft designer *Lawotschkin*. The bomber project "150" was introduced to him by the OKB 1 at the end of January 1949. Unlike the EF-131 and EF-140, this aircraft had a 35-deg swept-back wing.

Long time delays occurred frequently due to different technical standards. When plans of the wing and fuselage of the "150" were forwarded to production, the factory tasked with the manufacturing reacted strongly against the German standards in the design plans. However, the Russian standards lists made little sense, as most of the Germans could not speak any Russian. This meant the translation of the lists during 1949–1950 led to delays of several months. In the meantime, the interest in the bomber "150" declined further because the competing designs Il-28 and TU-16 of Soviet design offices of Iljuschin and Tupolew, which were larger and better organized, had become ready for series production.

With a tandem landing gear, the arrangement of the jet engines on forward-swept pylons under the wings, a T-tail empennage, and the swept high-wing

---

*References [19, 24, 30, 94, 104–107] were used in preparing this section.

Fig. 8.39   Aircraft "152."

design, the bomber "150" combined at the time some unusual innovations and was technically advanced. Between 1951 and May 1952, flight trials took place. After an accident of the experimental aircraft during landing on May 9, 1952, the Soviet OKB management decided not to rebuild the aircraft. In 1953, work on the "150" was stopped, and the design of a civilian version of the bomber, the passenger aircraft "152" started, which after 1954 was further developed in the GDR (Fig. 8.39).

8.5.4.2   HIGH-SPEED AIRCRAFT OF SIEBEL/DFS.   Besides the OKB 1, the OKB 2 still existed at Podberesje where employees of different German aircraft companies were active. These were gathered in Halle, at the Siebel Works, by the Soviet occupational power within a special design department. The high-speed aircraft "DFS 346," shown in Fig. 8.40, was also transferred from there to the Soviet Union in 1946. After intensive wind-tunnel tests, the problem of the premature flow separation on swept wings was solved by equipping the

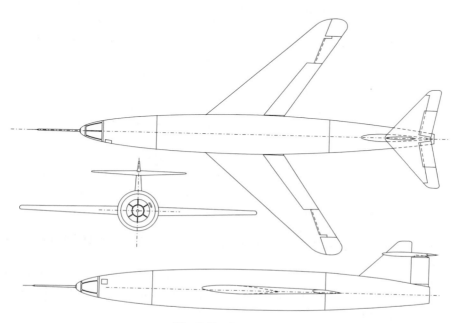

Fig. 8.40   DFS 346.

upper sides of the wings with boundary-layer fences (also see Sec. 6.5). But here one could also not speak of a speedy continuation of the work. At the beginning, there was a lack of a suitable carrier aircraft. In addition, there were problems with the special fuel, and there was a lack of important material and equipment. As in the case of the OKB 1, in addition, the language problems led to delays because the documents had to be translated and adapted to the Soviet standards. Soviet design offices "borrowed" the designs and test results thus impeding the work further.

In 1948 the "346" was first tested as a powerless glider. The experimental version "346 V-1" was deployed one year later, however, only equipped with an engine dummy due to the lack of an operational rocket engine. The "346 V-2," for which documents no longer exist, was presumably used for static strength tests. In June 1950, the "346 V-3," equipped with a "Soviet-modified" *Walter* engine was ready. But initially this version was tested with the engine inoperative. The tests with engine only commenced in August 1951 and ended on September 14 with the crash of the test specimen. The German test pilot *Wolfgang Ziese*, who with interruptions had tested all operational versions of the "346" in the Soviet Union, was able to save himself by ejecting with his pressure-proof cockpit and a parachute jump in denser air (see Sec. 5.4). The end for the "346" program followed thereafter, and also the design of the delta-fighter project "486" was dropped at the OKB 2.

There were several reasons for this. Already in 1947, the opinion was held in the Soviet Ministry of Aeronautics that the Germans should be sent back home again, because their knowledge had been "skimmed off" scientifically, they were too expensive, and their activity had reached a point where they had to become more familiarized with the Soviet development projects should their work continue to be profitable. But the latter was out of the question, at least in the area of aircraft design, because at most supporting German development work for the purpose of a Soviet "skim off" was considered acceptable. Moreover, the competition among the Soviet design offices was already intense enough without the Germans. On the other hand, there were in the meantime sufficient answers to most questions that had to be settled with regard to the DFS 346. More powerful engines and experimental facilities also existed, which made such tests with carrier aircraft superfluous. The race for higher *Mach* numbers and international recognition was, in addition, to remain a "purely" Soviet matter.

It was, therefore, necessary to slowly "cool down" the Germans with regard to their knowledge and to prepare their departure. Between 1950 and 1954, the aeronautical specialists returned in several "waves" from the Soviet Union. Many of the experts involved in the testing of the "346" after 1945, found further employment in the aircraft industry of the GDR, which existed only for a short time (1954 to 1961) but where the passenger aircraft "152," developed at the OKB 1, was to be put into production.

8.5.4.3 ENGINES OF JUNKERS AND BMW. German engine design and part of the guided-missile and aircraft-equipment development were concentrated in Kuibyschew/Samara. Just as in Podberesje, the individual company teams—except for the combination of the Siebel people and aircraft specialists of other companies at the OKB 2—worked largely separated from each other. In Kuibyschew, Junkers people from Dessau (about 350 persons), the BMW specialists from Stassfurt (about 250 persons), and the experts from Askania, Berlin, (about 50 persons) worked as teams. All three groups were administratively assigned to the Test Works No. 2 of the Ministry for Aircraft Industry of the USSR.

The Junkers people worked—according to the Soviet stipulations—at first on the more powerful jet engine Jumo 012 and the BMW people on different versions of the BMW jet engines. However, at first the BMW colleagues failed due to a number of material and supply problems because they had largely relied on the Soviet side. On the other hand, the Junkers people primarily used material brought from Germany. Shortly before completing the 100-hour test run, the Junkers engine also failed. The development of these projects was stopped, and the two teams combined. From 1948, they continued to work together on the turboprop engine Jumo 022, which was envisaged for long-range transport aircraft. However, the development and testing of the jet-engine project turned out to be very difficult because suitable test beds were not available. Thus, a three-stage water brake as well as corresponding assemblies and test beds for equipment had to first be set up. The Soviets accepted this engine in 1950. Thereafter, a more powerful version was still demanded. The engines of the Soviet aircraft of the type Tu-95/114 or An 22, still flying today, go back to these German developments in the Soviet Union.

8.6 INFLUENCE OF GERMAN AERONAUTICAL RESEARCH IN OTHER COUNTRIES (1945–1948)

8.6.1 EARLIEST POSTWAR YEARS

The obvious success of the technology transfer from Germany to the United States and the Soviet Union as well as the role of both nations during the Cold War has frequently pushed the developments in Great Britain, France, and other countries into the background.

In the case of Great Britain, it must be taken into account that the British technology transfer with regard to the organization and to personnel was coordinated with the United States. Washington, D.C., and London exchanged information, equipment, and German experts with each other during and after the war, all within the framework of the activities of different secret service organizations. With regard to the estimated 800 to 900 German experts in Great Britain, one must generally consider the aspect of the short stay as well

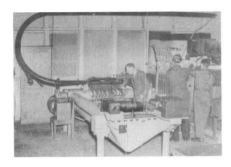

Fig. 8.41 Inauguration of the first Dutch supersonic wind tunnel in 1948; it was designed by German aerodynamicist *Siegfried Erdmann* who worked from 1946 at the NLL.

as the exchange with or passing on to the United States or third countries [108–112].

Furthermore, the countries connected to the British Empire, such as Australia, Canada, New Zealand, and South Africa (Commonwealth), or such traditional allies of England as the Netherlands, also profited. In 1946, for example, the German aerodynamicist *Siegfried Erdmann* started to work at the Dutch Nationaal Luchtvaartlaboratorium (NLL) pictured in Fig. 8.41, bringing in his experience gained within the German long-range missile project in the field of the high-speed aerodynamics [113].

The Australian government recruited at least 127 scientists between 1946 and 1951, supported by Great Britain, to work, among other things, on the Australian-British rocket program. London had already asked the Australian government in September 1945 to do so. Later the intention was to counter Soviet or French headhunters having a chance [36]. Canadian evaluation teams dealt within the context of the British technology assessment, British Intelligence Objective Subcommittee (BIOS) (see Fig. 8.42), and extensively with German aeronautical research and rocket development [114]. The Canadian National Research Council (NRC) obtained German jet-engine technology for testing as well as further scientific instrumentation and equipment from German aeronautical research installations [115].

France used the personnel and material capacities of Germany to a degree comparable to that of the Soviet Union. The French action was, however, watched by the American and British allies with great distrust and there were many conflicts after 1945 in securing German "know-how." In July 1946, the War Department in Washington even suspected, "unconfirmed but repeated rumors indicate that France and the USSR may be cooperating on the exploitation of the services of German scientists" [116]. The Americans and British were particularly annoyed that the French had managed with the help of German engineers to move the 8-m high-speed wind tunnel from the Ötztal to Modane. At that time, besides the German experts, France had also the most modern high-speed wind tunnel at her disposal [117].

Within the Soviet sphere of influence, Czechoslovakia tried after 1945 to go her own way. The basis for this was essentially the consolidation of the

WARTIME AERONAUTICAL RESEARCH & DEVELOPMENT IN GERMANY

by J. J. Green, M.E.I.C., F.R.Ae.S., *Air Transport Board, Ottawa*

R. D. Hiscocks, M.E.I.C., *The de Havilland Aircraft of Canada Limited, Toronto*

J. L. Orr, A.F.R.Ae.S., *National Research Council, Ottawa*

Part II

●   The Aerodynamic Research Establishment at Göttingen

The Aerodynamische Versuchsanstalt zu Göttingen, known for short as the A.V.A. was the premier centre for basic aerodynamic research in Germany. It was here that Prandtl, Betz and many others were able to make those important contributions to aerodynamic theory which won for Göttingen its international reputation. The Kaiser Wilhelm Institute for Flow Mechanics, described later, was actually situated within the grounds of A.V.A. During the War, additional wind tunnel facilities were constructed in an abandoned salt mine, a few miles north of Göttingen, where the old mine workings provided excellent camouflage.

The A.V.A. was founded as an incorporated body, the industry being assessed for test work done on its behalf. The accounts of the A.V.A. were examined annually by the Government and grants from the Air Ministry were adjusted to allow for fluctuation in the support received from the industry.

There was a total staff of about 700 at A.V.A. of whom some 30-50 were scientists, 100 were technical assistants and 100 were girl computers who were later used in the laboratories as technicians to re-

place the men who had entered military service. The most important work of the A.V.A., apart from routine tests for the aircraft manufacturers, was in the field of fundamental aerodynamics, and this was not restricted to aeronautics alone. In general, the equipment of the A.V.A. was smaller and less impressive than that of the newer laboratories such as L.F.A. and L.F.M. and much of it was quite old. On the other hand the calibre and ability of the staff appeared to be generally superior, judging by the results obtained.

Institute for Low-Speed Wind Tunnels

This institute was equipped with six tunnels at Göttingen and two low turbulence tunnels at Reyershausen. Of the tunnels at Göttingen, with the exception of a variable density tunnel with an elliptic jet, all were small and of low speed. Most of the tunnels were old and have already been described in the literature, none of them incorporating any unconventional features. *Behaviour of High Speed Wings at Low Speeds*

One of the major projects in the Institute for low speed tun-

nels was the investigation of the behaviour of high speed aircraft at the low speeds encountered during take-off and landing. Models of several high speed jet aircraft used for this purpose were seen in the laboratories. The primary concern was the achievement of low speed control and stability for the swept-back wing which was coming into extensive use in Germany for high speed aircraft. The considerable advantage of the swept-back wing in delaying the onset of compressibility effects had been demonstrated in the high speed tunnels at A.V.A. and had become widely accepted in Germany. Unfortunately, sweepback has an adverse effect on lateral stability, maximum lift coefficient and on the effectiveness of controls and high lift devices, due principally to "out-flow" effects.

It was found that a swept-forward wing gave the same improvement in delaying compressibility effects, together with improved lateral stability and maximum lift coefficient, but had the disadvantage of an unfavourable rolling moment with yaw. In order to improve the maximum lift coefficient, a nose flap (Fig. 15) was developed. It appeared to be the only means of effectively increasing the maximum lift of swept-back wings. In a typical case, with 35° of sweepback, the nose flap gave a maximum lift coefficient of 2.0, in conjunction with conventional 20 per cent trailing edge flaps, representing a 35 per cent improvement over the wing without nose flap. To

12

Fig. 8.42   BIOS report "Wartime Aeronautical Research & Development in Germany" (1945).

Czechoslovakian industry structures within the context of the armament activities of the SS after 1942–1943. Czechoslovakia used the available personnel and material capacities—provided these had not been confiscated or dismantled by the Soviet Union after the end of the war—for the setup of her own armament industry and air force. Czechoslovakia produced between 1946 and 1949 small quantities of a copy of the Me 262 [7].

Switzerland and Sweden, who remained neutral during WWII, also profited in Europe from the German know-how in aeronautics, and that independently of the larger victorious nations and their allies. Among other things, documents concerning the experimental swept-wing jet fighters P 1101 and P 1111 had reached Switzerland via an employee of the Messerschmitt Company shortly before the end of the war. In autumn 1945 the Swedish aircraft engineer *Fried Wanström* of the SAAB Aircraft Company also had been able to look at the test reports and project documents in Switzerland. Similarly to *George Schairer* a short time before at the Aeronautical Research Establishment Völkenrode, the Swede *Wanström* also immediately recognized the great importance of the "swept-wing effect" and ensured the realization of this aerodynamic idea at SAAB. The project planning was completed in the autumn of 1946, and the prototype of the swept-wing jet fighter Saab 29 "Tunnan (Barrel)" took off successfully on September 1, 1948 (Fig. 8.43). The series production commenced around 1950 at the underground manufacturing facilities of SAAB. The SAAB 29 belonged, together with the North American F-86 "Sabre" (1947) and the Mikojan-Gurewitsch MiG 15 (1947), worldwide to the first jet fighters with swept wings ready to go into series production [53].

The de Havilland D.H. 108 "Swallow" with swept wings, which first flew in Great Britain in 1946 was, on the other hand, less successful because all three prototypes crashed (compare Chapter 6.2.3). All the same, the data obtained found its way into the design of the D.H. 110 "Sea Vixen" and the development work on the world's first jet liner, the D.H. 106 "Comet." However, after the experience with the D.H. 108 the aircraft originally planned with only a vertical stabilizer was also fitted with a horizontal stabilizer ([53], p. 96).

Switzerland was also working on a jet fighter with swept wings. Moreover, members of the AAF-SAG (see Sec. 8.3) traveled in early June 1945 to Switzerland to visit the local research facilities and the Brown Boveri Company (BBC), which had, during the war, delivered compressors and components for wind-tunnel facilities in Germany [118, 119].

Fig. 8.43 Jet fighter Saab J-29 with swept wings.

Fig. 8.44   Pulqui II.

Further third-world countries like Argentina and Brazil only received during the Cold War in 1947–1948 official permission of the Americans and the British to look in Germany for suitable people. South America became at the end of the 1940s the fifth emigration center of German aeronautical researchers and aircraft specialists. An estimated 150 German experts followed the offers of Argentina and Brazil from 1947, with more than 70% of the Germans working in Argentina. Of these, approximately half worked in Argentine aircraft development [26]. With regard to the swept wing, the development work of the German aircraft specialists under *Kurt Tank* has to be mentioned. *Tank* had received at the end of 1947 the instruction to develop a jet fighter. On the basis of the project Ta 183, designed by the aerodynamicist *Hans Multhopp* during the last year of war, the group under *Tank* created the fighter aircraft "Pulqui II" (Fig. 8.44) with swept wings (40-deg sweep). The aircraft took off successfully for the first time during June 1950 [26, 120].

The developments in Spain, Egypt, and India, which took place under German influence, commenced in the 1950s and will not be the subject of this presentation, because the swept-wing effect had, at that time, already become common knowledge and was long part of the basic know-how of any future aircraft engineer or aerodynamicist. Moreover, the allied research bans in aeronautics were lifted in the mid-fifties, and in both German states a reconstruction of aeronautical research and the aircraft industry commenced.

8.6.2 DEVELOPMENT IN GREAT BRITAIN

As previously mentioned, British and American secret service organizations cooperated closely in discovering, assessing, securing, and processing the German know-how and in the takeover of scientists. The organizations called themselves Field Information Agency Technical (FIAT), Combined Intelligence Objectives Subcommittee (CIOS), or Joint Intelligence Objectives Agency (JIOA). With regard to aeronautical research and high-speed aerodynamics, the ADRC, already mentioned, and the Air Document Division played a special role. It should be mentioned here that the Anglo-American cooperation was not without its difficulties. On the one hand there was generally a historical prejudice on both sides, and on the other hand both the Americans and the British suspected that their respective partner was not disclosing everything at the exchange of the gained information [3, 121].

Fig. 8.45 Group photograph of the German scientists accommodated early 1947 at Farnborough Court; seated from left to right: *Gerald Klein* (2), *Adolf Busemann* (3), *Walter Tolmien* (4), *Ernst Schmidt* (6), *Hermann Schlichting* (8); center row: *Karl Doetsch* (2), *Hermann Neubert* (3), *Hans Multhopp* (7); upper row: *Hermann Jordan* (5), *Gerhard Sissingh* (6), *Dietrich Küchemann* (7).

The aeronautical research facilities at Völkenrode near Braunschweig and at Göttingen, located in the British occupational zone, were largely dismantled by the British after the war. The largest portion of the German aeronautical experts worked at the Royal Aircraft Establishment (RAE) in Farnborough (Fig. 8.45). With regard to the occupation of the Germans, it seems that, in comparison to the United States, only a small number was employed in Great Britain for a longer period of time. After one or two years, many of the German experts were released again or were sent on request of the Americans to the United States. *Adolf Busemann* thus received in the context of the Project "Paperclip" in 1947 an employment contract from the U.S. Navy and moved with his family to the United States.

One of the aerodynamicists who stayed in Great Britain was *Dietrich Küchemann* who was later involved in the design of the delta wing of the Anglo-French supersonic passenger aircraft "Concorde" [117, 120]. During the first postwar years, he was also decisively involved in the swept-wing investigations at Farnborough. The importance attributed in Great Britain to the swept-wing effect is also seen by the fact that the "Tailless Aircraft Advisory Committee" for the promotion of high-speed flight, founded in 1943, was renamed in 1948 as the "Swept-Wing Committee" directly charged with the coordination of swept-wing investigations in theory, wind-tunnel tests, flight tests, and project work in industry and research [122].

8.6.3 DEVELOPMENT IN FRANCE

In 1947 the French authorities selected from the almost 100,000 German prisoners of war about 6700 men as "spécifiquement recrutés." These were mostly members of technical professions and were to work in France in different industrial areas, but also in research and development projects. German experts in the fields of rocket, missile, torpedo, engine, aircraft, helicopter, and tank development as well as experts in material science, navy-related fields, ballistics, and handguns were at the same time brought to France after the war. But these were by far not yet all the areas [123]. Regarding the employment of German specialists in France, an order of magnitude of 2000 to 3000 persons may be assumed, which approximately corresponded with the transfer of personnel to the Soviet Union. Just as in the case of the other allied victors, the share of aircraft, engine, and rocket experts was probably the largest [111, 112, 117, 120].

Concerning the research in the field of the high-speed flight, for instance, the activity of the flight meteorologist and head of the former Deutsches Forschungsinstitut für Segelfug (DFS) (German Research Institute for Gliding), *Walter Georgii*, at the Arsenal de l'Aéronautique south of Paris was of importance. In 1946 he received an offer from the Americans but decided in favor of France because a larger number of his former colleagues at the DFS were able to follow him there and were given opportunities to work again in their fields of research [124]. The Arsenal de l'Aéronautique was an aeronautical research center with known research facilities, workshops, and possibilities for the construction of prototypes. Here, research on the DFS 346 related to high-speed flight, supersonic aerodynamics, and aeroelasticity was continued under *Eugen Sänger*. In 1946 the French created another research institute, comparable to the DVL in Berlin-Adlershof, that is, the Office National d'Études et de Recherches Aéronautiques (ONERA) (French Aerospace Lab). Here former employees of the DVL, such as *Wilhelm Flügge* and *Kurt Maguerre*, carried out intensive research concerning the swept wing. At ONERA, furthermore, computational methods concerning the flutter behavior of aircraft (*Johannes Dörr*) were developed and strength-related research was carried out. The first French aircraft with swept wings, the "Sud-Quest S.O. 6020/6025," went on flight trials in 1948 [117, 120, 122].

Moreover, many German experts within the western occupational zones, who decided, for instance, against employment in the United States and for a continuation of their professional career in France, let themselves be guided in their decision by the principle, "it is easier to walk home than to swim home" ([111], p. 180).

After 1945 the German aeronautical researchers became wanderers between the political systems. Their personal records (curricula vitae) reflected the collective experience of the "century of catastrophes." The transfer of the

German technology to the countries of the allied victors provided many an aeronautical researcher with the experience that scientific and technical elites are extremely important, legitimizing themselves in essence by their usefulness to their respective political system. Their specific knowledge and their experience after 1945 were absolutely important to both sides of the east–west conflict.

REFERENCES

[1] National Archives, RG 255, NACA, Box 190.

[2] *H. Mönnich,* "BMW: Eine deutsche Geschichte. Piper, München/Zürich, 1991.

[3] Bolling Air Force Base, Project Paperclip, A 2055.

[4] *M. Judt, B. Ciesla,* "Technology Transfer out of Germany after 1945," Harwood Academic Publishers, Amsterdam, the Netherlands, 1996, pp. 11–25.

[5] *M. Walker* (ed.), "Science and Ideology: A Comparative History," Routledge, London, 2003, pp. 156–185.

[6] *B. Johnson,* "Streng Geheim: Wissenschaft und Technik im Zweiten Weltkrieg," Paul Pietsch Verlage, Himberg bei Wien, o. J., p. 187.

[7] *H. Morgan,* "Me 262 'Sturmvogel/Schwalbe'," Motorbuchverlag, Stuttgart, 1996, pp. 76–77, 130–145, 158–159.

[8] *H. R. Borowski* (ed.), "Military Planning in the Twentieth Century," Office of Air Force History, Washington, DC, U.S., 1986, p. 123.

[9] *S. A. Goudsmit,* Alsos. Henry Schumann, New York, 1947.

[10] *D. Hoffmann,* "Operation Epsilon: Die Farm-Hall-Protokolle oder die Angst der Alliierten vor der deutschen Bombe," Rowohlt, Berlin, 1993, pp. 10–13.

[11] *F. H. Hinsley,* "British Intelligence in the Second World War: Its Influence on Strategy and Operations," Cambridge University Press, 5. vols, London, 1979–1990.

[12] *D. Kahn,* "The Codebrakers," Scribner, New York, U.S. 1996, pp. 969–984.

[13] *H. Sebag-Montefiore,* "Enigma: The Battle for The Code," Phoenix, London, 2001.

[14] *M. Smith,* "Enigma entschlüsselt: Die 'Codebreakers' von Bletchley Park," Heyne, München, 2000, pp. 244–276.

[15] *F. H. Hinsley,* "British Intelligence in the Second World War," Abridged Version, Cambridge Univ. Press, London, 1993, pp. 414–443, 567–574, 593.

[16] *F. L. Bauer,* "Entzifferte Geheimnisse: Methoden und Maximen der Kryptologie," Springer Verlag, Berlin, 1997, pp. 204, 205.

[17] *J. M. Scalia,* "In geheimer Mission nach Japan U 234," Motorbuch Verlag, Stuttgart, 2002, pp. 23–29.

[18] *J. Colville,* "Downing Street Tagebücher 1939–1945," Goldmann Verlag, Berlin, 1991, p. 405.

[19] *Chr. Buchheim* (ed.), "Wirtschaftliche Folgelasten des Krieges in der SBZ/DDR," Nomos Verlagsgesellschaft, Baden-Baden, 1995, pp. 79–109.

[20] *H. Boog* (ed.), "Luftkriegführung im Zweiten Weltkrieg," Ein internationaler Vergleich, Mittler Verlag, Herford/Bonn, 1993, pp. 272–275.

[21] *W. E. Samuel,* "American Raiders: The Race to Capture the Luftwaffe's Secrets," Univ. Press of Mississippi, Jackson, MS, U.S., 2004, pp. 95–122, 248–352, 371–399.

[22] *K.-D. Henke,* "Die amerikanische Besetzung Deutschlands," R. Oldenbourg Verlag, München, 1995, pp. 742–776.

[23] *B. Ciesla,* "Ihre Deutschen waren besser als unsere," Sächsische Zeitung, Oct. 6, 1995, p. 2.

[24] *Chr. Mick*, "Forschen für Stalin. Deutsche Fachleute in der sowjetischen Rüstungsindustrie 1945–1958," R. Oldenbourg Verlag, München/Wien, 2000, pp. 34, 42–65, 80–86.

[25] *M. Herrmann*, "Project Paperclip: Deutsche Wissenschaftler im Dienste der U.S. Streitkräfte nach 1945," dissertation Univ. Erlangen-Nürnberg, Nürnberg, 1998, pp. 75–80.

[26] *R. Stanley*, "Transfer von Rüstungstechnologie nach Lateinamerika durch Wissenschaftsmigration: Deutsche Rüstungsfachleute in Argentinien und Brasilien 1947–1963," dissertation FU Berlin, Berlin 1996, pp. 65–80, 94–100, 144.

[27] *T. Shachtman*, "Laboratory Warriors: How Allied Science and Technology Tipped the Balance in World War II," Perenial, New York, 2003, pp. 248–333.

[28] *J. Ph. Baxter III*, "Scientists Against Time," M.I.T. Press, Cambridge, MA, U.S., 1968.

[29] *St. W. Leslie*, "The Cold War and American Science: The Military-Industrial-Academic Complex at MIT and Stanford," Columbia Univ. Press, New York, 1993.

[30] *R. Karlsch*, *J. Laufer*, "Sowjetische Demontagen in Deutschland 1944–1949," Duncker & Humblot, Berlin, 2002, pp. 187–225.

[31] *Th. v. Kármán*, "Die Wirbelstrasse. Mein Leben für die Luftfahrt," Hoffmann und Campe, Hamburg, 1968, pp. 320–325.

[32] *M. Wala*, *U. Lehmkuhl*, "Technologie und Kultur. Europas Blick auf Amerika vom 18. bis zum 20. Jahrhundert," Böhlau Verlag, Köln/Weimar/Wien, 2000, pp. 175–177.

[33] *C. Glatt*, "Reparations and the Transfer of Scientific and Industrial Technology from Germany," dissertation European Univ. Inst., Florence, Italy, 1994.

[34] *B. Ciesla*, "Der Spezialistentransfer in die UdSSR und seine Auswirkungen in der SBZ und DDR," Aus Politik und Zeitgeschichte, Dec. 3, 1993, B49–50/93, pp. 24–31.

[35] *M. J. Neufeld*, "Die Rakete und das Reich. Wernher von Braun, Peenemünde und der Beginn des Raketenzeitalters," Brandenburgisches Verlagshaus, Berlin, 1997, pp. 322–323.

[36] *Frankfurter Rundschau*, Aug. 17, 1999, p. 2.

[37] *Neue Züricher Zeitung*, Aug. 19, 1999, p. 3.

[38] *Die Zeit*, Dec. 27, 1991, p. 62.

[39] *The Washington Post*, Jan. 23, 1992, p. A25.

[40] *G. S. Bjuschgens*, *E. L. Bedrshizkij*, "ZAGI—Zentr Awiazionnoj Nauki," Nauka, Moskwa, 1993, pp. 51–53, 77, 79.

[41] National Archives, RG 255, NACA, Box 114.

[42] *George W. Lewis*, "Report on Trip to Germany and Russia, Sept.–Oct. 1936," National Archives, RG 255, NACA.

[43] National Archives, RG 255, NACA, Box 1998.

[44] National Archives, RG 255, NACA, Box 172.

[45] National Archives, RG 255, NACA, Box 2010.

[46] National Archives, RG 255, NACA, Box 2004.

[47] *W. Cook*, "The Road to the 707," TYC Publishing Company, Bellevue, WA, U.S., 1991, pp. 32–34, 122–123, 139, 146, 162.

[48] National Archives, RG 255, NACA, Box 2001.

[49] National Air and Space Museum, Theodore von Kármán Papers, Box 57.17.

[50] *W. Cook*, "The First Boeing High-Speed Wind Tunnel," Boeing Company, Seattle, WA, U.S., pp. 1–14.

[51] *J. D. Anderson*, "A History of Aerodynamics," Cambridge Univ. Press, Cambridge, UK, 1997, p. 436.

[52] *R. T. Jones*, "Wing Theory," Princeton Univ. Press, Princeton, NJ, U.S. 1990, p. 91.

[53] *W. Radinger*, *W. Schick*, "Messerschmitt Geheimprojekte," Aviatic Verlag, Planegg, 1991, pp. 121–122.

[54] *M. H. Gorn*, "Harnessing the Genie: Science and Technology Forecasting for the Air Force 1944–1986," Office of Air Force History, Washington, DC, U.S., 1988, pp. 11–58.

[55] M. H. Gorn, "The Universal Man: Theodore von Kármán's Life in Aeronautics," Smithsonian Institution Press, Washington, DC, U.S., 1992, pp. 93–110.

[56] D. A. Daso, "Hap Arnold and the Evolution of American Airpower," Smithsonian Institution Press, Washington, DC, U.S., 2000, pp. 187–214.

[57] Th. v. Kármán, "Where We Stand: A Report of the AAF Scientific Advisory Group, August 1945," Headquarters Air Material Command, Wright Field, OH, U.S., 1946.

[58] F. C. Lynch, "The 1945 Army Air Forces Fair—Forerunner of Today's Air Force Orientation Group," Aerospace Historian, Vol. 28, No. 2, 1981, p. 104.

[59] Jürgen Kocka (ed.), "Historische DDR-Forschung," Akademie Verlag, Berlin, 1993, pp. 287–301.

[60] National Air and Space Museum, Theodore von Kármán Papers, Box 88.4.

[61] National Archives, RG 255, NACA, Box 189.

[62] D. L. Putt, Interview No. 724 on April 1–3, 1974, USAAF Oral History Story Program, Maxwell Air Force Base, Signatur 36112, pp. 107–109.

[63] L. Bölkow, "Mit dem Pfeilflügel zum Hochgeschwindigkeitsflug," 50 Jahre Turbostrahlflug, DGLR-Bericht 89-05, Bonn, 1989, pp. 253, 286.

[64] M. W. Geer, "Boeing's Ed Wells," Univ. of Washington Press, Seattle, WA, U.S., 1992, pp. 113–115.

[65] http://www.daveswarbirds.com/usplanes/aircraft/XB-47.htm, retrieved Dec. 2009.

[66] Boeing Archives, Signature 7824.

[67] L. S. Jones, "U.S. Bombers," Aero Publishers, Fallbrook, CA, U.S., 1974, p. 156.

[68] R. T. Jones, "Wing Plan Forms for High-Speed Flight," NACA TN No. 1033, June 1945.

[69] Letter G. S. Schairer to B. Krag of Dec. 1, 1989.

[70] S. Günter, "The Influence of Fuselage and Wing Sweepback on the Range and Speed of Turbo-Jet Airplanes," Landsberg, July 1945, Boeing-Archiv, Signature 9798.

[71] G. Schulz, "Aerodynamische Regeln für den Einbau von Strahltriebwerksgondeln," Zeitschrift für Flugwissenschaften, Heft 5, 1955, p. 123.

[72] R. W. Kluge, C. L. Fay, "CIOS," Aeronautical Subcommittee, August 1945, pp. 29–37.

[73] Boeing Archives, Signature 7824A.

[74] Boeing Archives, Signature 7824, sheet 10.

[75] W. Letko, A. Goodman, "Preliminary Wind-Tunnel Investigation at Low Speed of Stability and Control Characteristics of Swept-Back Wings," NACA TN No. 1046, April 1946.

[76] George S. Schairer, "On the Design of Early Large Swept-Wing Aircraft," DGLR Symposium, 50 Jahre Turbostrahlflug, München, Oct. 26–27, 1989, DGLR-Bericht 89-05, pp. 289–321.

[77] R. M. Robbins, W. H. Cook, "Boeing baut Bomber mit Pfeilflügeln—Entwurfsgrundsätze und Flugeigenschaften der B-47 'Stratojet'," Interavia, 8 Jahrgang, No. 1, 1953, pp. 32–34.

[78] J. E. Steiner, "Evolutionary Aspects of Large Swept-Wing Aircraft," DGLR Symposium, 50 Jahre Turbostrahlflug, München, Oct. 26–27, 1989, DGLR-Bericht 89-05, pp. 323–349.

[79] P. M. Bowers, "The Boeing B-47," Profile Publications, No. 83.

[80] P. Hoeveler, "In erster Reihe: F-86 Sabre bei der Luftwaffe," Klassiker der Luftfahrt, No. 6, 2002.

[81] L. Davis, "North American F-86 Sabre: Wings of Fame," vol. 10, Aerospace Publishing Ltd., 1998.

[82] Boeing Archives, Drawing NA-134.

[83] "First of the Furies," Royal Air Force Flying Review, Vol. XVII, No. 3, Dec. 1961, p. 31.

[84] Corerspondance B. Krag with N. M. Blair, former NAA, May–Oct. 2004.

[85] M. M. Blair, "Evolution of the F-86," Evolution of Aircraft Wing Design Symposium, AIAA, 1980, pp. 3, 7.

[86] T. F. Walkowicz, "Birth of Sweepback," Air Force Magazine, April 1952, p. 72.

[87] G. Koch, "Druckverteilungsmessungen am schiebenden Tragflügel," LGL Bericht 156, 1942.

[88] H. Ludwieg, "Pfeilflügel bei hohen Geschwindigkeiten," LGL Bericht 127, 1940.

[89] B. Göthert, "Hochgeschwindigkeitsmessungen an einem Pfeilflügel (Pfeilwinkel φ = 35°)," DVL FB 1813, 1942.

[90] C. H. Donlan, "An Interim Report on the Stability and Control of Tailless Aircraft," NACA Report 796, 1944.

[91] G. Brennecke, "Auftriebssteigerung beim Pfeilflügel," AVA Bericht 43/W/35, July 17, 1943.

[92] Boeing Archives, Signature 9037.

[93] E. Shacklady, "The North American F-86A Sabre," Profile Publications, No. 20.

[94] D. A. Sobolew, "Deutsche Spuren in der sowjetischen Luftfahrtgeschichte," Mittler Verlag, Hamburg/Berlin/Bonn, 2000, pp. 109, 130–137, 185–224, 238–239, 288.

[95] L. L. Kerber, "Stalin's Aviation Gulag. A Memoir of Andrei Tupolev and the Purge Era," Smithsonian Institution Press, Washington, DC, U.S., 1996.

[96] M. Uhl, "Stalin's V-2: Der Technologietransfer der deutschen Fernlenkwaffentechnik in die UdSSR und der Aufbau der sowjetischen Raketenindustrie 1945 bis 1959," Bernard & Graefe Verlag, Bonn, 2001, pp. 21–74.

[97] U. Albrecht, R. Nikutta, "Die sowjetische Rüstungsindustrie," Westdeutscher Verlag, Opladen, 1989, pp. 77–97.

[98] P. N. Knyschewskij, "Moskaus Beute: Wie Vermögen, Kulturgüter und Intelligenz nach 1945 aus Deutschland geraubt wurden," Olzog Verlag, München/Landsberg am Lech, 1995, pp. 21–34.

[99] K. Akinscha, G. Koslow, "Beutekunst: Auf Schatzsuche in russischen Geheimdepots," Deutscher Taschenbuchverlag, München, 1995, pp. 62–69.

[100] H. Mehringer, M. Schwartz, H. Wentker (eds.), "Erobert oder befreit?," R. Oldenbourg Verlag, München, 1999, pp. 71–92.

[101] B. E. Tschertok, "Raketen und Menschen," Elbe-Dnjepr-Verlag, Klitzschen, 1997, pp. 51–53.

[102] ZAGI-Archiv, "Bericht No. 4/1946," Bericht 22/1947.

[103] Günther Bock, "Die deutsche Luftfahrtforschung im Jahre 1945. Aussagen von Günther Bock vor einer sowjetischen Kommission am 17. September 1945," Eigenverlag Peter Korrell, Reprint, Wolfenbüttel, 1998.

[104] J. Michels, J. Werner, "Luftfahrt Ost 1945–1990," Bernard & Graefe Verlag, Bonn, 1994, pp. 18–165.

[105] D. Hoffmann, K. Macrakis (eds.), "Naturwissenschaft und Technik in der DDR," Akademie Verlag, Berlin, 1997, pp. 193–211.

[106] H. Walther, "Deutsche Spezialisten für die Sowjetunion," Flugzeug Classic, Heft 4-10, 2003.

[107] B. Ciesla, "DFS 346: Überschall-Forschungsprojekt aus Deutschland," Flug Revue, Dezember 1997, pp. 84–87.

[108] National Archives, RG 18, Overcast/Paperclip, Box 12.

[109] C. G. Lasby, "Project Paperclip: German Scientists and the Cold War," Atheneum, New York, 1971.

[110] J. Gimbel, "Science, Technology, and Reparations: Exploitation and Plunder in Postwar Germany," Stanford Univ. Press, Stanford, CA, U.S., 1990.

[111] T. Bower, "Verschwörung Paperclip. NS-Wissenschaftler im Dienste der Siegermächte," List Verlag, München, 1987.

[112] L. Hunt, "Secret Agenda: The United States Government, Nazi Scientists, and Project Paperclip," 1945 to 1990. St. Martin's Press, New York, 1991.

[113] S. F. Erdmann, "Deutsch-Niederländische Odyssee im Anlauf der Raumfahrt," DUP Satellite, Delft, the Netherlands, 2001.

[114] J. J. Green, J. L. Orr, R. D. Hiscocks, "Wartime Aeronautical Research & Development in Germany," Engineering Institute of Canada N.D.C.N. 1919, 1949.

[115] St. T. Koerner, "Technology Transfer from Germany to Canada after 1945," Comparative Technology Transfer and Society, Vol. 2, No. 1, April 2004, pp. 99–123.

[116] Intelligence Review, War Department, No. 24, July 25, 1946, p. 49.

[117] DGLR (ed.), "Die Tätigkeit deutscher Luftfahrtingenieure und –wissenschaftler im Ausland nach 1945," DGLR, Bonn-Bad Godesberg, 1992, pp. 14–15, 165–170, 188–199.

[118] Johns Hopkins Univ., Milton S. Eisenhower Library, Special Collections, MS. 147, Box 2.10.

[119] National Archives, RG 255, NACA, Box 2007.

[120] E. H. Hirschel, H. Prem, G. Madelung, "Luftfahrtforschung in Deutschland," Bernard & Graefe Verlag, Bonn, 2001, pp. 307–321.

[121] J. Gimbel, "Deutsche Wissenschaftler in britischem Gewahrsam," Vierteljahreshefte für Zeitgeschichte, 3/1990, pp. 459–471.

[122] W. Heinzerling, "Die Geschichte des Pfeilflügels (2): Die projektmässige Anwendung des Pfeilflügels," Jahrbuch der DGLR IV/1981, pp. 71–75.

[123] U. Albrecht, "Rüstungsfragen im deutsch-französichen Verhältnis (1945–1960)," Vortragsmanuskript auf dem Kolloquium "Deutsch-französische Wirtschaftsbeziehungen 1945–1960," Paris, France, Dec. 8–10, 1994.

[124] W. Georgii, "Forschen und Fliegen," Verlag H.M. Hauschild, Bremen, 1997, p. 312.

[125] H. Schlichting, E. Truckenbrodt, "Aerodynamik des Flugzeugs Teil 1," Springer Verlag, Berlin, Göttingen, Heidelberg, 1962, pp. 359–361.

[126] B. Etkin, "Dynamics of Flight," John Wiley & Sons, Inc., New York, London, Sydney, 1965, p. 88.

[127] W. Eisenmann, "Das Seitenstabilitätsverhalten eines Flugzeugs mit Pfeilflügeln" Ministry of Supply (Air) Völkenrode, Bericht No. AGD 1039/G, April 22, 1946.

[128] R. P. Hallion, "On the Frontier—Flight Research at Dryden, 1946–1981," NASA History Series, NASA SP; 4303, 1984, pp. 58, 91.

[129] W. H. Cook, "The Road to the 707," TYC Publishing Company, Washington, 1991, p. 173.

[130] H. Ludwieg, "Verbesserung der kritischen Machzahl von Tragflügeln durch Pfeilung," Festschrift zum 60. Geburtstag von A. Betz, AVA Göttingen, 1945.

APPENDIX: FLIGHT MECHANICAL PROBLEMS

Wing sweep has a considerable influence on the flight mechanical behavior of the aircraft. This must already be taken into account by the aircraft engineer during the design so that the guidelines for stability and controllability are being adhered to.

The demand for a static stability about the pitch axis provides that the aircraft returns by itself, after a disturbance about the pitch axis, back into the initial position. Thus, the component of the resulting aerodynamic force $(-)Z$ normal to the wing chord line must be located downstream of the center of

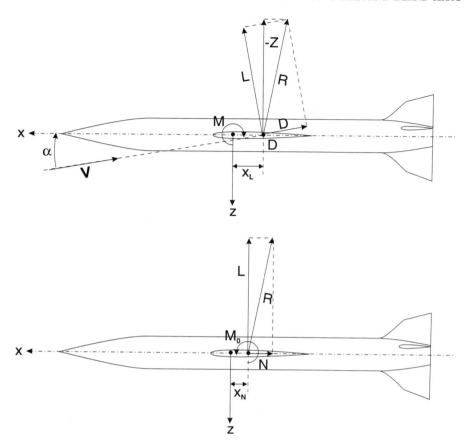

Fig. 8.46 Center of pressure and neutral-point locations (nomenclature: L = lift, D = center of pressure, M = pitching moment referred to the center of gravity, M_0 = pitching moment independent of lift, N = neutral point, R = resulting aerodynamic force, V = flight speed, D = drag, X_L = center-of-pressure location, X_N = neutral-point location, Z = normal force, α = angle of attack).

gravity. The position where the aerodynamic force acts on is called center of pressure D. Figure 8.46 illustrates this relation. The pitching moment M about the center of gravity is generated by this aerodynamic force. At an enlargement of the angle of attack, Z increases and produces in case of static stability an opposing moment. In a nondimensional notation this pitching moment amounts to [125]:

$$C_m = C_{m0} + (dC_m/dC_L)\,(C_L) \tag{8.1}$$

C_L is the lift coefficient and C_{m0} is a moment independent of lift. C_{m0} is dependent on the airfoil geometry and is for symmetrical airfoils equal to zero. In the case of small angles of attack α, one may approximate the lift by $L = -Z$. The moment M about the center of gravity can then be written as

$$M = -X_L L \tag{8.2}$$

X_L is the distance between the center of pressure D and the reference point for the moment (center of gravity). The center-of-pressure location is then in the nondimensional notation:

$$X_L/l = -C_m/C_L \tag{8.3}$$

The length of l is the wing chord. With Eq. (8.1), Eq. (8.3) can be written as

$$X_L/l = -C_{m0}/C_L - dC_m/dC_L \tag{8.4}$$

The expression C_{m0}/C_A describes the pressure-point movement dependent on lift or angle of attack. One can also represent the pitching moment M by the zero-moment, independent of lift, and a force at a distance X_N from the moment reference point:

$$M = M_0 - X_N L \tag{8.5}$$

with X_N being the distance to the neutral point N, that is, the neutral-point location. In a nondimensional notation, the previous equation becomes

$$C_m = C_{m0} - (X_N/l) C_L \tag{8.6}$$

With Eq. (1), one obtains the neutral-point location as

$$X_N/l = -dC_m/dC_L \tag{8.7}$$

The expression dC_m/dC_L is also described as "stability measure (parameter)" of the longitudinal motion; it is less than zero in the case of stability.

The relation of Eq. (8.7) is independent of the swept wing and also holds for an entire configuration. The flight mechanical coefficients C_L, C_{m0}, and C_m are not constant for an aircraft or rocket, but are dependent on the freestream direction, for example, the angle of attack, the *Mach* number, and the configuration. The pitching moment of an aircraft is permanently being trimmed, either manually by the pilot or it happens automatically by an auto-trim system. However, natural stability can only be ensured if the stability parameter is less than zero.

The swept wing offers itself to tailless aircraft due to its longitudinal extent. In this case, static stability can be achieved by wing twist because the outer wings are generally located downstream of the center of gravity. The ailerons, located on the outer wings, can, therefore, simultaneously be employed as elevator. Because of the short lever arm of this combined aileron/elevator arrangement, large center-of-pressure movements due to angle-of-attack changes are hard to trim. One used for such tailless aircraft in the past was the so-called "center-of-pressure fixed airfoils" whose mean line exhibited a slight S-form. The tailless aircraft of the *Horten* brothers and of *Alexander Lippisch*, for example, the Me 163, are given here as examples of this design. Such airfoils are little suited for high-speed aircraft. The delta wing is more advantageous because it possesses, due to its larger longitudinal extent, longer lever arms for the control surfaces located at the wing trailing edge.

Also at lateral freestream conditions during yawed flight, the swept wing behaves differently than a straight wing. Figure 8.47 shows as example a swept wing in yawed flight. The components of the flight speed V normal to the wing leading edge, V_n, are effective in determining lift. This then generates an asymmetrical lift distribution and thus a negative rolling moment. This yaw-induced rolling moment acts in the same sense as the dihedral of the wing. This, in turn, may lead at large lift coefficients to problems [126]. Investigations at the Institute of Gas Dynamics of the LFA have shown that the amplitude ratio of the coupled roll-yaw oscillation (gyrating or Dutch-roll oscillation) strongly increases at higher altitudes and higher lift coefficients [127]. This effect can be reduced by a negative dihedral as can be seen on many aircraft of the first generation with swept wings. The forward-swept wing behaves opposite. The flow conditions at yawed flight generate a positive

Fig. 8.47 Flow conditions on the swept wing (nomenclature: V = flight speed, V_n = component of the flight speed normal to the 25% chord line, β = angle of sideslip, φ = sweep angle).

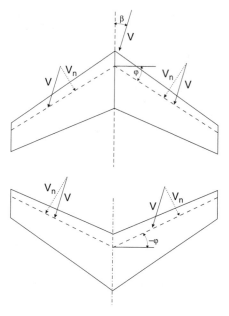

rolling moment, which leads to an increase of the yaw angle. To compensate this effect, one can increase the dihedral of the wing or enlarge the vertical stabilizer correspondingly. Overall the forward-swept wing is less problematic regarding the lateral stability.

Sweep and aspect ratio of the wing have an influence on the dynamic behavior of the aircraft. For instance, roll damping is reduced by decreasing the span and increasing sweep. Also yaw damping is affected this way. Besides the effect of the vertical stabilizer, the additional speed on the wing due to the rotation about the yaw (vertical) axis, an additional induced drag, is generated acting against (opposing) the rotation. At a very high sweep, this damping rotational motion may disappear or reverse its effect (see [126], p. 175). Roll and yaw damping provide a dampened gyrating oscillation. Older high-performance aircraft often have highly swept wings of a low aspect ratio. Such aircraft then have often a weakly damped Dutch-roll oscillation. If in addition a long fuselage with an unfavorable mass distribution is added, an instability threatens, which caused a yaw-roll coupling, or inertia-coupling, in case of the first high-performance jets dangerous flight conditions [128]. Large aileron deflections at high dynamic pressures led quickly to roll rates due to the low roll damping. The inertia forces produced by the rotation led to an excitation of pitch and yaw movements, which quickly built up large amplitudes hard to control by the pilot. This problem could be remedied in the case of the North American YF-100 "Super Sabre" by an enlargement of the vertical stabilizer area (higher yaw damping). An artificial stabilization by pitch and yaw dampers provides in the case of modern aircraft good handling characteristics.

The spanwise lift distribution on the wing has a large effect on the high-lift behavior of the wing. Figure 8.48 shows schematically the lift distribution on three wings at high lift coefficients. The lift maximum is in case of the swept-back wing located in the outer wing area. During stall, the flow will separate here first ("tip stall"), and the lift on the outer wing will break down. This causes an upstream movement of the center of pressure, which is reflected in an abruptly occurring pitching moment. On the swept-back wing, the boundary layer flows with the velocity component parallel to the 25% chord line

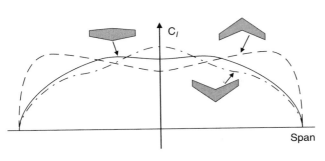

Fig. 8.48 Lift distribution along the span of three wings (nomenclature: C_l = local lift coefficient).

C_l

Span

towards the wing tip. There, the boundary layer thickens and causes a premature separation. This has an effect on the aileron effectiveness, and especially
at high angles of attack aileron deflection may provoke stalling of the flow.
During the landing approach, this may lead to problems. If maneuvers are
being performed at higher speeds with large load multiples, the same problems can occur. Remedies are automatic slats, boundary-layer fences on the
upper wing surface, or especially designed wing leading edges.

The lift maximum is in the case of the forward-swept wing located in the
proximity of the fuselage while the outer wing is relieved. The ailerons are
thus always located in the area of a "healthy" flow keeping their effectiveness
also at high-lift coefficients. This was the reason why the forward-swept wing
was of interest to the aircraft designers despite its problematic aeroelastic
characteristics. The boundary layer flows in case of the forward-swept wing
towards the fuselage, which can lead there to problems at high-lift coefficients. The wing flow was stabilized close to the fuselage of the Ju 287 and
the HFB 320 by a retractable slat.

The effectiveness of landing flaps is also affected by sweep. The forward-
swept wing behaves here less favorably than the swept-back wing because
the trailing edge exhibits a larger sweep than the leading edge. At asymmetrical flight conditions, this can lead to undesirable rolling moments as in the
case of the HFB 320; see Sec. 6.3. During the development of the first aircraft
with swept wings, the engineers saw the main problems within the high-lift
regime. A lot of wind-tunnel time was invested to develop suitable slats and
landing flaps, which provided sufficiently high lift coefficients during the
landing approach to keep the landing (touch-down) speeds within the range
of conventional aircraft.

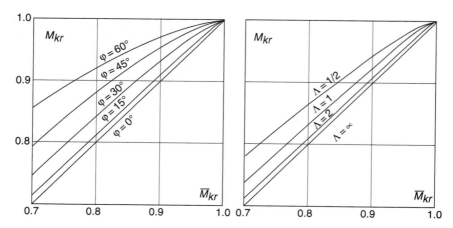

**Fig. 8.49   Effect of wing sweep and aspect ratio on the critical *Mach* number (nomencla-
ture: M_{kr} = critical *Mach* number, $\overline{M}_{kr}$ = critical *Mach* number of the unswept wing)
[130].**

The aeroelastic characteristics of the swept wing are treated in detail in Chapter 4. The elastic deformation of the wing has, however, also an effect on the flight behavior. At higher dynamic pressures, the bending-torsion coupling causes in the case of the swept-back wing a reduction in the effective angle of attack of the outer wing. A redistribution of the lift occurs, and the center of pressure moves towards the inside and upstream. Regarding the load on the wing structure, this is favorable; there is, however, considering flight mechanics the danger that the longitudinal motion (pitch) becomes unstable. This point was missed during the development of the Boeing XB-47 and would have seriously endangered the project had not simultaneously the deformation of the rear fuselage compensated this effect by reducing the angle of attack at the horizontal stabilizer [129]. For the forward-swept wing the danger exists that the effective angle of attack of the outer wing is enlarged by the torsion divergence. The effect is similar to the one in the case of the swept-back wing: center of pressure shifts upstream and causes a destabilization of the longitudinal motion.

As already indicated in Chapter 7 concerning the antiaircraft rockets, decreasing the wing aspect ratio has a similar effect on the critical *Mach* number as has wing sweep. A comparison between the sweep effect and the influence of the wing aspect ratio on the critical *Mach* number is given in Fig. 8.49 [130]. One recognizes that an aspect ratio of $\Lambda = 0.5$ increases the critical *Mach* number by the same amount as a wing with a sweep of $\varphi = 45$ deg. The low-aspect-ratio wing has advantages over a swept wing in the case of antiaircraft rockets, which must withstand maneuver loads of 10 g and more. Such a wing may at lower weight be relatively stiffly constructed and does also not have the aeroelastic problems of the swept wing (coupling of bending and torsion).

Appendix A

ABBREVIATIONS AND ACRONYMS

A&AEE	Aeroplane and Armament Experimental Establishment	United Kingdom
AAF-SAG	Army Air Force Scientific Advisory Group	United States
ADRC	Air Document Research Center	United States
ADS	Air Disarmament Squadron	Western Allies
AECMA	Association Européenne des Constructeurs de Matériel Aérospatiale	France
AEDC	Arnold Engineering Development Center; known as Air Engineering Development Center before 1951	Tullahoma, Tennessee, United States
AEG	Allgemeine Elektrizitäts Gesellschaft (General Electric Co.)	Germany
AG	Aktiengesellschaft (Stock Corporation)	Germany
AGARD	Advisory Group for Aerospace Research and Development (see NATO)	United States
AMC	Air Material Command	United States
AMD	Avions Marcel Dassault	France
Ar	Arado Flugzeugwerke (Arado Aircraft Company)	Germany
As	Argus-Motoren-Gesellschaft (Argus Engine Company)	Germany
AVA	Aerodynamische Versuchsanstalt Göttingen (Aerodynamic Test Establishment, Göttingen)	Germany
BAL-Mtt	Bauaufsicht der Luftwaffe bei Messerschmitt (Construction Supervision of the Air Force at Messerschmitt)	Germany
BBC	Brown Boveri Cie.	Switzerland
Bd	Band (volume)	
BDLI	Bundesverband der Deutschen Luft- und Raumfahrt-Industrie (Federal Association of the German Aerospace Industry)	Germany
BIOS	British Intelligence Objective Subcommittee	United Kingdom
BMBF	Bundesministerium für Bildung und Forschung (Federal Ministry of Education and Research)	Germany
BMW	Bayerische Motorenwerke (Bavarian Motor Company)	Germany
BRAMO	Brandenburgische Motorenwerke GmbH (Brandenburg Motor Company GmbH)	Germany
BRD	Bundesrepublik Deutschland	
BSAR	DLR-Archiv Braunschweig (DLR Archive, Braunschweig)	Germany

BVF	Bauvorschriften für Flugzeuge (design regulations for aircraft)	
CAFT	Combined Advances Field Team	
Caltech	California Institute of Technology	Pasadena, California, United States
CEV	Centre d'Essais en Vol	France
CFD	computational fluid dynamics	
CIOS	Combined Intelligence Objectives Subcommittee	United States
DB	Daimler-Benz AG (Daimler-Benz Corporation)	Germany
DE	Höchste Prioritäten Stufe (highest priority level)	
DFG	Deutsche Forschungsgemeinschaft (German Research Association)	Germany
DFL	Deutsche Forschungsanstalt für Luftfahrt (German Aeronautical Research Establishment; also German Research Establishment of Aeronautics)	Germany
DFS	Deutsches Forschungsinstitut für Segelflug (German Research Institute for Gliding)	Germany
DFVLR	Deutsche Forschungs- und Versuchsanstalt für Luft- und Raumfahrt (German Aerospace Research and Test Establishment; also German Research and Test Establishment of Aeronautics and Space)	Germany
DGF	Deutsche Gesellschaft für Flugwissenschaften (German Society of Flight Sciences)	Germany
DGLR	Deutsche Gesellschaft für Luft- und Raumfahrt Lilienthal-Oberth (German Aerospace Society Lilienthal-Oberth)	Germany
D.H.	De Havilland Aircraft Co. (United Kingdom)	
DLR	Deutsches Zentrum für Luft- und Raumfahrt (German Aerospace Center)	Germany
DM	Deutsches Museum München (German Museum Germany)	Germany
DVG	Drahtlos-Telegraphie und Luftelektrische Versuchsstation Gräfelfing (Radio-telegraphic and Air-Electricity Test Station, Gräfelfing, Germany)	
DVL	Deutsche Versuchsanstalt für Luftfahrt (German Test Establishment of Aeronautics; also German Aeronautical Test Establishment)	Germany
EAFB	Edwards Air Force Base	California, United States
ECM	electronic countermeasures	
EF	Entwicklungsflugzeug (development aircraft)	
EFNER	Ecole du Personel Navigont d'Essais et de Réception	France
E-Stelle	Erprobungsstelle der Luftwaffe (E-Center; Flight-test Center of the German Air Force)	Germany
ETH	Eidgenössische Technische Hochschule Zürich (Swiss Institute of Technology)	Switzerland
ETW	European Transonic Wind Tunnel	
e.V.	Eingetragener Verein (registered society)	
FAA	Federal Aviation Authority	United States

FAI	Fédération Aéronautique Internationale (World Air Sports Federation)	France
FALU	Fachnormenausschuss für Luftfahrt (Special Standards Committee for Aeronautics)	Germany
FB	Forschungsbericht (research report)	
Feudin'54	die feindlich gestimmten 54er; interne Streitereien (Feuding'54; internal disputes)	
FFA	Flygtekniska Försöksanstalten (Aeronautical Research Institute)	Stockholm, Sweden
FFM	Flugwissenschaftliche Forschungsanstalt München (Research Establishment of Flight Sciences, Munich)	Germany
FFO	Flugfunk-Forschungsinstitut Oberpfaffenhofen (Air Radio Research Institute, Oberpfaffenhofen)	Germany
FGZ	Forschungsanstalt Graf Zeppelin (Research Establishment Count Zeppelin)	Germany
FH	Forschungsgemeinschaft Halle (Research Association Halle)	Germany
FIAT	Field Information Agency Technical	United States
FIST	Flugtechnisches Institut Stuttgart (Flight-Technical Institute, Stuttgart)	Germany
FKFS	Forschungsinstitut für Kraftfahrzeugwesen und Fahrzeugmotoren (Research Institute of Motor Vehicles and Motor-Vehicle Engines)	Stuttgart, Germany
Flak	Fliegerabwehrkanone (antiaircraft artillery)	
FMG	Funkmessgerät (radio instrument; also radio meter)	
FPS	Forschungsinstitut Physik der Strahlantriebe (Research Establishment for Jet-Propulsion Physics)	Stuttgart, Germany
FR	Fliegerabwehrrakete (antiaircraft rocket)	
FRG	Federal Republic of Germany (Bundesrepublic Deutschland)	Germany
FVA	Flugwissenschaftliche Vereinigung Aachen (Flight Sciences Association, Aachen)	Germany
FVA-Prague	Flugtechnische Versuchsanstalt Prag (Flight-Technical Test Establishment)	Prague
FVW	Faserverbund-Werkstoff (fiber-reinforced material)	
Fw	Focke-Wulf Company	Germany
GALCIT	Guggenheim Aeronautical Laboratory at the California Institute of Technology	Pasadena, California, United States
GAMM	Gesellschaft für angewandte Mathematik und Mechanik (Society of Applied Mathematics and Mechanics)	Germany
GDR	German Democratic Republic	Germany
GFK	Glasfaserverstärkter Kunststoff (glass-fiber reinforced plastics)	
GHH	Gute Hoffnung Hütte (Metallurgical Plant, Good Hope)	Germany
GKO	State Committee of Defense	Russia
GL	General Luftfahrzeugmeister (Head of Technical Office, RLM)	

GL/Flak E (GL/C-E)	General Luftfahrzeugmeister (Special Division for Development of Antiaircraft Weapons, RLM)	
GM	Zusatzeinspritzung eines Sauerstoffträgers in Motoreu (additional injection of an oxygen carrier; also oxygen enrichment)	
GmbH	Gesellschaft mit beschränkter Haftung (limited liability company)	
GOAR/ GöAR	DLR-Archiv Göttingen (DLR Archive, Göttingen)	Germany
HFB	Hamburger Flugzeugbau (Hamburg Aircraft Construction Company)	Germany
HG	Hochgeschwindigkeitsprogramm (High-speed Program)	Germany
Ho	Horten-Flugzeugbau (Horten Aircraft Company)	Germany
HOG	Hermann-Oberth-Gesellschaft (Hermann-Oberth Society)	
HS	Henschel Flugzeugwerke AG (Henschel Aircraft Company AG)	Kassel, Germany
HST	High-speed tunnel	
HVA	Heeresversuchsanstalt (Test Establishment of the Army)	Germany
HWA	Heereswaffenamt (Weapons Office of the Army; also German Army Weapons Office)	Germany
HWK	Hellmuth Walter Raketentriebwerke, Kiel (Hellmuth Walter Rocket Engines)	Germany
ICAS	International Council of the Aeronautical Sciences	
INA	Internationale Normalatmospäre (international standard atmosphere)	
JES	Journal of Engineering Sciences	
JFM	Junkers Flugzeug- und Motorenwerke (Junkers Aircraft and Motor Works)	Dessau, Germany
JG	Jagdgeschwader (Fighter Squadron)	Germany
JIOA	Joint Intelligence Objectives Agency	Western Allies
KG(J)	Kampfgeschwader (Jagd) [Fighter Squadron (Hunter)]	Germany
KIAS	knots indicated air speed	
Kobü	Konstruktionsbüro (Design Office)	
KPAR	DLR-Archiv Köln-Porz (DLR Archive, Cologne-Porz)	Germany
KWG	Kaiser-Wilhelm-Gesellschaft zur Förderung der Wissenschaften (Emperor-Wilhelm Society for the Promotion of Science)	Germany
KWI	Kaiser-Wilhelm-Institut (Emperor-Wilhelm Institute)	Germany
LFA	Luftfahrtforschungsanstalt Hermann Göring, Braunschweig (Aeronautical Research Establishment, Braunschweig)	Germany
LFM	Luftfahrtforschungsanstalt München (Aeronautical Research Establishment Munich)	Germany
LFW	Luftfahrt Forschungsanstalt Wien (Aeronautical Research Establishment Vienna)	Austria

LGL	Lilienthal-Gesellschaft für Luftfahrtforschung (Lilienthal Society for Aeronautical Research)	Germany
LuFo	Deutsche Luftfahrtforschung (German Aeronautical Research)	Germany
MAC	mean aerodynamics chord	
MAI	Moskauer Luftfahrtinstitut (Moscow Aeronautical Research Institute)	Russia
MAP	Ministry of Aircraft Production	Russia
MC	Macchi-Castoldi	Italy
MiG	Mikojan-Gurewitsch	
M.I.T.	Massachusetts Institute of Technology	Cambridge, Massachusetts, United States
ML	Motor-Luftstrahl-Antrieb (piston-motor/air-jet engine; also turbojet propeller drive)	
MPG	Max-Planck-Gesellschaft zur Förderung der Wissenschaften (Max-Planck-Society for the Promotion of Science)	Germany
MS	Messstrecke (test section)	
MSL	mean sea level	
MTU	Motoren- und Turbinen-Union (Engine and Turbo-machinery Association)	Germany
MVA	Modellversuchsanstalt für Aerodynamik (Model Test Establishment for Aerodynamics)	Germany
MW	Methanol-Wassergemisch (methanol–water mixture)	
MZM	Motorzweigwerk Magdeburg (Engine Factory Branch Magdeburg)	Germany
NAA	North American Aviation Company	United States
NACA	National Advisory Committee for Aeronautics	United States
NACA TM	NACA Technical Memorandum	United States
NAE	National Aeronautical Establishment	Canada
NASM	National Air and Space Museum	United States
NATO	North Atlantic Treaty Organization	
NII-88	Forschungsinstitut des Ministeriums für Bewaffnung (Research Institute of the Armament Ministry)	Germany
NLL	Nationaal Luchtvaartlaboratorium	Amsterdam, The Netherlands
NOL	Naval Ordnance Laboratory	Silver Spring, Maryland, United States
NPL	National Physical Laboratory	Teddington, United Kingdom
NRC	Canadian National Research Council	Canada
OFO	Oberbayrische Forschungsanstalt Oberammergau (Upper-Bavarian Research Establishment Oberammergau, Messerschmitt Project Office)	Germany
OHG	Offene Handelsgesellschaft (Open Business Council)	
OKB	Opytnoye Konstruktorskoye Byuro (Special Design Office)	Russia

OKH	Oberkommando des Heeres (Supreme Command of the Army)	Germany
ONERA	Office National d'Etudes et de Recherches Aéronautiques (known today as Aérospatiales) (French Aerospace Lab)	France
OS	operation surgeon	
PTL	Propeller-Turbinen-Luftstrahl-Antrieb (propeller-turbine-jet-propulsion; also turboprop propulsion)	
RADAR	Radio Detection and Ranging	
RAE	Royal Aeronautical Establishment	United Kingdom
RAeS	Royal Aeronautical Society	United Kingdom
RAF	Royal Air Force	United Kingdom
RLM	Reichsluftfahrtministerium (State Ministry of Aeronautics/German Air Ministry)	Germany
RM	Reichsmark [Reichs (German) mark]	
RRG	Rhön-Rositten-Gesellschaft (Rhön-Rositten Society)	Germany
RTO	Rückstoss-Turbine Null (recoil or thrust turbine zero)	
S	Sondertriebwerk (special engine)	
SBZ	Sowjetische Besatzungszone (Soviet Zone of Occupation)	
SMF	Securité Militaire Française	
SNCASO	Société Nationale de Constructions Aeronautiques de Sud Ouest (National Society for Aeronautical Design, South–West)	France
SNECMA	Société Nationale d'Etude et de Construction de Moteurs d'Aviation (National Society for Research and Design of Aeronautical Engines)	France
SoE	Sonderentwicklung (special development)	
SWC	Swept Wing Committee	
TH	Technische Hochschule (Institute of Technology/ Technical University)	
TL	Turbinen-Luftstrahl-Antrieb (turbo-air-jet propulsion; also turbo-jet propulsion)	
TsAGI	Zentralinstitut für Luftstrahltriebwerke Aero- und Hydromechanik, Zhukovsky, Moskau (Central Aero- and Hydrodynamics Institute, Zhukovsky, Moscow)	Russia
TsIAM	Zentrales Institute für Luftstrahltriebwerke (Central Institute of Air Engines)	Russia
UM	Untersuchung/Messung (investigations/measurements)	
USAAF	U.S. Army Air Force	United States
VDI	Verein Deutscher Ingenieure (Association of German Engineers)	Germany
VKI	Von Kármán Institute for Fluid Dynamics	Brussels, Belgium
Wa A	Heeres Waffenamt (Army Weapons Office)	Germany
WaPrüf II	Amt für Waffenprüfung und Entwicklung II (Office of Weapons Proof and Development II)	
WGL	Wissenschaftliche Gesellschaft für Luftfahrt (Scientific Society for Aeronautics)	Germany

WTS	Wehrtechnische Studiensammlung Koblenz (Collection of Defense-related Studies Koblenz; also Federal Archive, Koblenz, Germany)	
WVA	Wasserbau Versuchsanstalt, Kochel (Hydraulic Engineering Test Organization)	Germany
WWII	Second World War	
ZAMM	Zeitschrift für angewandte Mathematik und Mechanik (Journal of Applied Mathematics and Mechanics) Wiley-VCH Verlag Gmbh (publisher)	Germany
ZFW	Zeitschrift für Flugwissenschaften (Journal of Flight Sciences)	
ZWB	Zentrale für wissenschaftliches Berichtswesen, (Center for Scientific Publications)	Berlin, Germany

ENGLISH SPELLING OF EUROPEAN TOWNS AND COUNTRIES

German Towns	English Spelling
Aachen	Aix-la-Chapelle
Berlin	Berlin
Bonn	Bonn
Bonn-Bad Godesberg	Bonn-Bad Godesberg
Braunschweig	Brunswick
Braunschweig-Völkenrode	Brunswick-Voelkenrode
Freiburg	Freiburg
Friedberg	Friedberg
Göttingen	Goettingen
Hamburg	Hamburg
Hannover	Hanover
Karlshagen	Karlshagen
Kassel	Cassel
Koblenz	Coblenz
Kochel	Kochel
Köln	Cologne
München	Munich
Nürnberg	Nuremberg
Opladen	Opladen
Peenemünde	Peenemuende
Planegg	Planegg
Stuttgart	Stuttgart

European Towns/Countries	
Amsterdam	Amsterdam
Italien	Italy
Moskau	Moscow
Niederlande	Netherlands
Österrreich	Austria
Roma	Rome
Schweiz	Switzerland
Wien	Vienna
Zürich	Zurich

Notions in the References

Band	volume
Bericht	report
FB*	research report
Festschrift	publication celebrating an event
Heft	issue
Mitteilung	note
S. 1-51	page 1-51
Teil	part
Verlag	publishing house

*Further information see Appendix A.

SOURCES OF ORIGINAL REPORTS

Accademia Nazionale dei Lincei
Della Lungare
00165- Roma
Italy
http://www.lincei.it/modules.php?name=Content&pa=showpage&pid=60
e-mail: segreteria@lincei.it

Archiv zur Geschichte der
Max-Planck-Gesellschaft
Boltzmannstr. 14
D-14195 Berlin-Dahlem
Germany
http://www.archiv-berlin.mpg.de/wiki/english.php
e-mail: mpg-archiv@archiv-berlin.mpg.de

Boeing Company
Michael J. Lombardi
Manager/ Historian
Historical Services
Seattle, WA 98124-2207
USA
e-mail: michael.j.lombardi@boeing.com

Bundesarchiv
Postfach 450569
D-12175 Berlin
Germany
e-mail: c.lorenz@barch.bund.de

Deutsches Museum
Museumsinsel 1
D-80558 München
Germany
http://www.deutsches-museum.de/en/archives/
e-mail: archiv@deutsches-museum.de

Deutsches Zentrum für Luft- und Raumfahrt (DLR) e.V.
Zentrales Archiv
Bunsenstr. 10
D-37073 Göttingen
Germany
http://www.dlr.de/100jahre/en/
e-mail: Jessika.Wichner@dlr.de

Sächsische Landesbibliothek-
Staats- und Universitätsbibliothek
Zweigbibliothek Bauingenieurswesen/Verkehrwissenschaften
D-01054 Dresden
Germany
http://www.slub-dresden.de/en/
e-mail: zw30info@slub-dresden.de

Smithsonian Institution Archives
Reference Archivist
MRC 507
PO Box 37012
Washington, D.C. 20013-7012
http://siarchives.si.edu/sia/main_contact.html
e-mail: osiaref@si.edu

Technische Universität Braunschweig
Institute of Fluid Mechnanics-Archiv
Prof. Dr.-Ing. Rolf Radespiel
Bienroder Weg 3
D-38106 Braunschweig
Germany
e-mail: r.radespiel@tu-bs.de

INDEX

CONTRIBUTING AUTHORS

Meier, Hans-Ulrich, Prof. Dr.-Ing. habil. Study of mechanical engineering/aerospace technology at the TH Braunschweig. 1964 Associate Scientist at the Aerodynamics Department of the Aerodynamic Test Establishment (AVA) Göttingen. 1970 Ph.D. from the TU Braunschweig. 1973–1974 Visiting Associate Professor at the University of Maryland and Visiting Consultant at the Naval Ordnance Laboratory (NOL). 1977 Head of the Department Boundary Layers (successor of Dr. J. C. Rotta). 1974–1997 Scientific Officer of the American–German (US-FRG) Data Exchange Agreement "Viscous and Interacting Flow Field Effects." 1983–1988 Head of the DFVLR-Messerschmitt-Bölkow-Blohm (MBB) Working Association "Missile Aerodynamics." 1984 lectureship in fluid-mechanical measurement techniques. 1986 habilitation (right to lecture) and since 1990 Extracurricular Professor at the TU Clausthal. 1988–1998 Director of the German–Dutch Wind Tunnel (DNW). 1991 Honorary Professor at the Nanjing Aeronautical Institute (NAI), PR China. 1994–2003 Board of Directors of the von Kármán Institute of Fluid Dynamics, Rhode Saint Genése, Belgium. 1999–2003 Head of the DGLR Department Fluid- and Gasdynamics. More than 60 scientific publications in journals and books. 2005 "48th Ludwig Prandtl Memorial Lecture," 76th Annual Scientific Conference of the GAMM (Gesellschaft für Angewandte Mathematik und Mechanik), Luxembourg.

Ciesla, Burghard, Priv.-Doz. (Lecturer), Dr. phil. habil. Apprenticeship as radio technician. Study of history at the Humboldt University Berlin (Dipl.-Historian). 1990 Doctorate in economic history (Dr. oec). 1991–1992 Research Fellow at the German Historical Institute in Washington, D.C. Collaborator at the Forschungsschwerpunkt

Zeithistorische Studien (Research Focus Contemporary History) respectively at the Zentrum für Zeithistorische Forschung (Center of Contemporary History Research) (ZZF) in Potsdam. 1997–2001 employee at the Institute of History of the University Potsdam. 2004 habilitation at the University Potsdam (right to lecture contemporary history). Since 2001 freelance historian on various scientific, media, and exhibition projects. Extended teaching and research stays in the United States, Austria, and Japan. Editor and author of books and more than 40 articles on the scientific, technical, economic, and contemporary history of the 20th century.

Försching, Hans, Prof. Dr.-Ing. habil. 1949 Apprenticeship as machinery mechanic. 1951 study of mechanical engineering at the TH Karlsruhe. After graduation in 1955, test engineer at the Daimler-Benz AG (Corporation) Gaggenau. 1958–1963 Associate Scientist at the Department of Aeroelasticity of the AVA Göttingen. 1962 Ph.D. from the TU Braunschweig. 1963–1965 Chief Engineer for structural dynamics and aeroelasticity within the Messerschmitt development team for the HA 300 at Heluan, Egypt. 1968 habilitation at the TU Braunschweig. 1972 Director of the Institute of Aeroelasticity of the DFVLR-AVA Research Center (RC) Göttingen. 1980–1982 Acting Head of the Institute of Structural Mechanics at the DFVLR-RC Braunschweig. 1975 Extracurricular Professor at the TU Braunschweig. Over 100 publications and books. 1974 textbook *Fundamentals of Aeroelasticity* which was translated into Russian and Chinese. Visiting professorships abroad. Member of scientific organizations, such as the AGARD Structures and Materials Panel SMP (Chairman 1988–1990), Senate of the DFVLR and DGLR. Member of the editorial boards of various professional journals, including the *Journal of the Flight Science* and the *Journal of Engineering Sciences*. Scientific awards and prizes, among others the AGARD Scientific Achievement Award (1991), promotional award with research fellowship of the Japan Society for the Promotion of Science (JSPS), Aachen and München Prize (Award) for Technology and Applied Natural Sciences (1999).

Galleithner, Hans, Dipl.-Ing. (TU) Study of mechanical engineering/aeronautical technology at the TH (TU) München. 1967 flight test at Dornier Company, Oberpfaffenhofen (Do 31 program). Since 1972 employee of the DFVLR/DLR, 1972–73 EPNER, Istres, France, education as flight-test engineer. 1975–2004 cooperation military flight testing of the DLR Institute of Flight Mechanics, flight systems technology with the E-Stelle (Flight-Test Center of the Air Force)/WTD 61. 1980–88, Head of the DLR Flight Department, Oberpfaffenhofen. Since January 2005 retired.

Involved in numerous civil and military flight-test programs, among others: DLR Falcon E, Alpha-jet, Tornado, 1983–1984 F-15/F-16 (Edwards AFB), 1990–1992 X-31 (Palmdale/Edwards AFB), Grob Strato 2C, Egrett. 1994–2004 Eurofighter. 1996–2003 NATO RTO Flight Test Technology Team (responsible for the AGARDograph Series 160/300). Since 1996 Lectureship "Aircraft Trials," FHS (Technical College) München.

Heinzerling, Werner, Dipl.-Ing. (TU) Study of mechanical engineering/aeronautics at the TU München. 1964–1967 Scientific Assistant at the Institute of Fluid Mechanics. 1967–1989 Development Engineer at Messerschmitt-Bölkow-Blohm/ Dasa Aircraft Branch. 1973 Head of the Department Experimental Aerodynamics. 1976 Head of the Department Project Aerodynamics; development; and series attendance of the MRCA Tornado, pre-development Eurofighter, planner of aerodynamic test facilities, including the German-Dutch Wind Tunnel (DNW), the European Transonic Wind Tunnel (ETW), and the Wind-Tunnel Committee of the European Aeronautical Industry AECMA. 1973–1989 Advisory Board for Flight Physics of the German Museum München (1984 Oskar-Miller Badge). 1986–1996 Member of the Senate and Head of the Expert Field Aerospace History of the DGLR. 1989–2004 Head of the Section Aerospace at the German Museum München with set up of the Branch "Flight Hangar Schleissheim." Since 1992 Member of the Board of the Royal Aeronautical Society/München Branch. 2000 Lecturer for fluid-technical testing at the FH (Technical College) München and 2005 for aircraft aerodynamics at the FH Augsburg. Numerous publications concerning aircraft aerodynamics and aeronautical history. Glider pilot since 1958.

Krag, Bernd, Dr.-Ing. 1961–1967 Study of mechanical engineering/aerospace technology at the TH Braunschweig. 1975 Ph.D. from the TU Braunschweig (Dr.-Ing.). Associate Scientist at the Chair of Flight Mechanics of the TU Braunschweig. 1972 joined the Institute of Flight Mechanics of the DFVLR Research Center Braunschweig. Worked on control configured vehicles (CCV), active control (gust reduction), modeling and system identification, databases for training simulators, trailing (wake) vortex problems. 1981–1986 Head of the GARTEUR Action Group "Parameter Identification from Flight-Test Data." From 1993–2002 Head of the Department Winged Aircraft of the Institute of Flight-Systems Technology of the DLR Research Center Braunschweig. Member of the DGLR and the "Study Group Braunschweig's Aeronautical History." Over 30 publications in referenced professional journals and as contributions to scientific conferences.

Since retiring from DLR, several publications concerning aeronautical research in Braunschweig (DLR-local Chronicle Braunschweig, "History of the Aeronautical Research Establishment Braunschweig-Völkenrode," DLR special issue commemorating the 100th anniversary of the first powered flight, contribution to "35 Years German Aerospace Center").

Schubert, Helmut, Dipl.-Ing., Dipl.-Wirtschafts-Ing. (Economics-Ing.) Study mechanical engineering at the TU Berlin and TU München, M.B.A. at the TU München. Since 1964 at the MAN Turbomortoren Company as Development Engineer within the Engineering Sciences Department. Participation in the development of the engines RB.153-61, RB.193-12 and RB.199-34R. 1973 Head of the Staff Office Technical Information of the MTU Aero Engines, München.

Since retirement in 2000, adviser to the MTU Aero Engines GmbH (Ltd.). Author of several reference books on aeronautical propulsion and aeronautical history, lectures and journal publications about the history of aeropropulsion technology.

Member of the Aerospace Advisory Board the German Museum München, the AIAA, the ASME, the GBSL, the RAES, and the LPC. Long-time head of the DGLR Section on Aerospace History.

Stanewsky, Egon, Dr.-Ing. (Translator) 1955–1961 Study of mechanical engineering/aeronautical engineering at the TH Aachen. 1961–1965 scientific employee at the Institute of Applied Gas Dynamics of the DVL at Aachen and later at Cologne-Porz-Wahn. Carried out research at subsonic and supersonic speeds. 1965–1971 Scientist Associate at the Lockheed-Georgia Company, Marietta, GA, U.S. Design and construction of a high-Reynolds number transonic wind tunnel and corresponding research. 1971–1980 Scientist at the Institute of Experimental Fluid Mechanics of the DFVLR Göttingen (now DLR) with predominantly experimental research at transonic speeds. 1981–1986 and 1991–1997 Head of the High-Speed Aerodynamics Department of the Institute of Experimental Fluid Mechanics. 1981 Ph.D. from the Technical University Berlin. 1987–1990 Director (acting) of the Institute of Experimental Fluid Mechanics of the DLR. 1990–1991 Adjunct Professor at the Pennsylvania State University. 1993–1995 Project Coordinator for the European Project EUROSHOCK-Drag Reduction by Passive Shock Control and editor of the corresponding book (Volume 56 of *Notes on Numerical Fluid Mechanics*, Vieweg Publishing). 1996–1999 Project Coordinator for the European Project EUROSHOCK II—Drag Reduction by Shock and Boundary-Layer Control and editor of the corresponding book (Volume 80 of *Notes on Numerical Fluid Mechanics and Multidisciplinary Design*, Springer Publishing). 1998–2000 Senior-Scientist of the DLR. 2000 Retired. 2001–2002 translation from German into English of major parts of *Luftfahrtforschung in Deutschland* (*Aeronautical Research in Germany, from Lilienthal until Today*), Vol. 30 of *Deutsche Luftfahrt* published by Bernard & Graefe.

SUPPORTING MATERIALS

Many of the topics introduced in this book are discussed in more detail in other AIAA publications. For a complete listing of titles in the AIAA Library of Flight Series, as well as other AIAA publications, please visit www.aiaa.org.

AIAA is committed to devoting resources to the education of both practicing and future aerospace professionals. In 1996, the AIAA Foundation was founded. Its programs enhance scientific literacy and advance the arts and sciences of aerospace. For more information, please visit www.aiaafoundation.org.